Polarized Light and the Mueller Matrix Approach

Series in Optics and Optoelectronics

Handbook of Optoelectronic Device Modeling and Simulation (Two-Volume Set)
Joachim Piprek

Handbook of Optoelectronics, Second Edition (Three-Volume Set)
John P. Dakin, Robert G. W. Brown

Optical MEMS, Nanophotonics, and Their Applications
Guangya Zhou, Chengkuo Lee

Thin-Film Optical Filters, Fifth Edition
H. Angus Macleod

Laser Spectroscopy and Laser Imaging
An Introduction
Helmut H. Telle, Ángel González Ureña

Fourier Optics in Image Processing
Neil Collings

Holography
Principles and Applications
Raymond K. Kostuk

An Introduction to Quantum Optics, Second Edition
Photon and Biphoton Physics
Yanhua Shih

Polarized Light and the Mueller Matrix Approach, Second Edition
José J. Gil, Razvigor Ossikovski

For more information about this series, please visit: https://www.crcpress.com/Series-in-Optics-and-Optoelectronics/book-series/TFOPTICSOPT

Polarized Light and the Mueller Matrix Approach

Second Edition

José J. Gil
Razvigor Ossikovski

CRC Press
Taylor & Francis Group
Boca Raton London New York

CRC Press is an imprint of the
Taylor & Francis Group, an **informa** business

Second edition published 2022
by CRC Press
6000 Broken Sound Parkway NW, Suite 300, Boca Raton, FL 33487-2742

and by CRC Press
4 Park Square, Milton Park, Abingdon, Oxon, OX14 4RN

CRC Press is an imprint of Taylor & Francis Group, LLC

First edition published by CRC Press 2016

Library of Congress Cataloging-in-Publication Data
Names: Gil Pérez, José Jorge, author. | Ossikovski, Razvigor, author.
Title: Polarized light and the Mueller matrix approach / José J. Gil, Razvigor Ossikovski.
Description: Second edition. | Boca Raton : CRC Press, 2022. |
Series: Series in optics and optoelectronics |
Includes bibliographical references and index.
Identifiers: LCCN 2021050506 (print) | LCCN 2021050507 (ebook) |
ISBN 9780367407469 (hardback) | ISBN 9781032215112 (paperback) |
ISBN 9780367815578 (ebook)
Subjects: LCSH: Electromagnetic waves–Polarization. | Polarization (Light)
Classification: LCC QC441 .G55 2022 (print) |
LCC QC441 (ebook) | DDC 535.5/2–dc23/eng/20211217
LC record available at https://lccn.loc.gov/2021050506
LC ebook record available at https://lccn.loc.gov/2021050507

ISBN: 978-0-367-40746-9 (hbk)
ISBN: 978-1-03-221511-2 (pbk)
ISBN: 978-0-367-81557-8 (ebk)

DOI: 10.1201/9780367815578

Typeset in Palatino LT Std
by Newgen Publishing UK

José J. Gil dedicates this book to Mercedes.
Razvigor Ossikovski dedicates this book to Bojidar, Vania, Marie and Anne,
for their support, encouragement and patience.

Contents

Preface

Polarization is a fundamental property of electromagnetic waves, profound knowledge of which, in terms of both mathematical formulation and physical interpretation, is required in various fields of classical and quantum physics. *Polarimetry*, which most generally refers to the various procedures and methods for the measurement and analysis of physical properties related to polarization and its transformation by the effect of material media, actually covers a large, rapidly increasing, range of applications in science and technology: astronomy and astrophysics; atmospheric and environmental studies; remote sensing; photonics and nanophotonics; fiber optics telecommunications; chemical engineering; medicine and biology; materials science; optics industry; plasma physics; LCD devices; thin films and layered media; microwave devices; quantum teleportation, etc.

The main objective of this book is to integrate, in a comprehensive and consistent manner, the basic concepts of the polarization phenomena from the double point of view of the states of polarization of electromagnetic waves, as well as of the transformations of these states of polarization by the action of material media. Recent decades have been characterized by successive contributions that constitute today the consolidated basis of our knowledge of the field, and that deserve to be put together and described in a detailed monographic treatise. Despite the indubitable interest of a number of nonlinear polarization phenomena, the subjects dealt with here are focused only on linear effects, covering most of the practical situations in polarimetry.

For several years, works on near-field phenomena and nanotechnologies have motivated the study and characterization of *three-dimensional* (3D) states of polarization of electromagnetic waves, beyond the conventional *two-dimensional* (2D) approaches. Thus, fundamental concepts like Stokes parameters, polarization matrix, degree of polarization, spin, nonregularity, etc. must be appropriately extended to three dimensional formulations. *2D polarization states* are characterized by the fact that the electric field of the wave evolves in a fixed plane while, in general, *3D states* require the consideration of the three components of the electric field of the wave regardless of the reference frame considered. Chapters 1 and 2 are respectively devoted to the analysis of the basic concepts related to 2D and 3D polarization states, including their mathematical representation through appropriate structures like the *Stokes vector* and the *polarization matrix*, as well as to the study of some useful in practice quantities derived from these structures. Further, the space–time and space–frequency formulations combining the concepts of polarization and coherence of electromagnetic waves are studied and interpreted in the light of the recent contributions to this fundamental branch of physical optics.

One of the concepts that plays a central role in the content of this book is that of the *coherency matrix*, defined in terms of measurable quantities and associated either with electromagnetic waves or with material media. It has the structure of a two-dimensional (for 2D polarization states), three-dimensional (for 3D polarization states) or four-dimensional (for material media) statistical covariance matrix. Moreover, its scope is not limited to the optical frequencies, but can be rather applied to any kind of electromagnetic radiation, as well as to the linear transformation of the state of polarization of the radiation resulting from its interaction with material media.

The action of material media, linearly transforming the state of polarization of the electromagnetic waves interacting with them, is studied in several consecutive chapters.

The *nondepolarizing* linear transformations of the states of polarization are formulated in Chapter 3 through the Jones and Mueller-Jones approaches, including the physical interpretation of the mathematical operations in those spaces, as well as such fundamental concepts like *passivity* (the physical restriction of not amplifying the intensity of the interacting electromagnetic waves) and *reciprocity* (the behavior of the medium when incident and emerging electromagnetic beams are interchanged).

Chapter 4 is dedicated to the particular forms and properties of the Jones and Mueller matrices associated with different basic types of nondepolarizing media like *retarders* (linear, circular or elliptical), *diattenuators* (linear, circular and elliptical) and *pseudorotators*. The main approaches for the *serial decomposition* of the Jones and Mueller matrices associated with nondepolarizing media are also analyzed and interpreted. These serial decompositions correspond to equivalent systems composed of cascades of simple components that exert consecutively their effects on the polarization state of the interacting electromagnetic wave.

The mathematical representation of the polarimetric action of depolarizing systems requires the consideration of the concept of general *Mueller matrix*; that is, a 4×4 real matrix whose structure corresponds to a *physically realizable linear transformation* of the Stokes parameters of the incident electromagnetic wave into the Stokes parameters of the emerging wave. Thus, Chapter 5 deals with the foundations of the notion of Mueller matrix, applicable to both depolarizing and nondepolarizing interactions, and relying on well-defined statistical mixtures of basic nondepolarizing interactions. Most generally, the Mueller matrix contains the total amount of information one is able to obtain from the interaction of polarized radiation with a linear-response medium. As a result of the statistical picture for physical realizability, a fundamental characteristic property of a Mueller matrix is that it has an associated Hermitian matrix with the structure and properties of a covariance matrix.

Concepts like passivity, reciprocity, *polarimetric purity* (which refers to the measure of how close is a given system to a nondepolarizing one), as well as certain decompositions related to the essential algebraic structure of Mueller matrices, are also considered and generalized throughout respective sections of Chapter 5. In particular, the so-called *normal form* (together with the associated *symmetric decomposition*) of a Mueller matrix leads to the key concepts of *type-I* and *type-II Mueller matrices*, which have distinct physical and mathematical natures and cover, in a complementary way, the entire set of Mueller matrices. In order to be comprehensible to readers coming from different fields, Chapter 5 also reviews certain alternative conventions that are commonly used in remote sensing and *synthetic aperture radar polarimetry*, together with the rules for conversion between alternative definitions.

The content of Chapter 6 goes more deeply into the analysis of the physical parameters associated with Mueller matrices such as *diattenuation, polarizance, indices of polarimetric purity, polarization entropy*, and *anisotropy coefficients*, as well as the physical quantities that are invariant under certain types of transformations. The last section of Chapter 6 is devoted to the synthesis of an arbitrary depolarizing Mueller matrix **M** through the smooth and continuous transformation of a *reference pure Mueller matrix* associated with **M**, thus providing a comprehensive view on the nature of the properties of depolarizing systems.

As a direct consequence of the statistical origin of the concept of Mueller matrix, any given depolarizing system can be conceived as a *parallel combination* of nondepolarizing components, that is, any given Mueller matrix can be expressed, in a variety of ways, as a weighted sum of nondepolarizing Mueller matrices whose respective weights add to one. This *arbitrary decomposition* along with other interesting parallel decompositions are described and interpreted in Chapter 7, which also deals with the closely related concept of *polarimetric subtraction* of a given component from the parallel combination.

With the ready availability of Mueller matrix polarimeters and the continuously increasing complexity of the materials and media under investigation, the interpretation of an experimentally obtained Mueller matrix in terms of physical properties of the medium is of ever growing importance today. Indeed, in numerous practical cases it is not possible to directly relate the polarimetric response of the medium to its elementary properties (dichroism, birefringence, optical activity ...) through rigorous electromagnetic (EM) theory modelling. An alternative approach to this problem is algebraically decomposing the experimental Mueller matrix into simpler components without any explicit reference to an EM model. The various matrix decomposition approaches at hand can be grouped into two classes, *serial* (or *product*) and *parallel* (or *sum*) decompositions.

A powerful tool for the analysis of any, experimental, theoretical or simulated, depolarizing Mueller matrix **M**, complementary to the various parallel decompositions, is provided by the *serial decompositions* of **M** in terms of ordered products of particularly simple Mueller matrices. They represent a generally depolarizing Mueller matrix as a product of the Mueller matrices of basic optical components such as diattenuators, retarders and canonical depolarizers. The potential benefit of applying the algebraic approach to experimentally obtained Mueller matrices e.g., from biological samples, is twofold. First, the algebraic methodology is universal, in contrast to modelling the optical response of the sample. That is, algebraic decompositions are applicable to any experimental Mueller matrix, whether an EM model describing the medium under investigation exists or not. This allows the experimentalist to obtain immediate physical information on the sample – through the representation of the latter as a chain of elementary optical components – even in the absence of any optical model or (most often) when the latter is either too complex or not accurate enough. The second advantage of the algebraic approach stems from the standard representation of every Mueller matrix in a canonical form playing the role of an *optical equivalent system* having the same polarimetric response as the medium represented by the original matrix. The equivalent system approach thus makes it possible to perform a formal comparison, in terms of polarimetric properties, of Mueller matrices of various physical origins. The serial decompositions are described, interpreted and compared in the respective sections of Chapter 8, which also includes a detailed and comprehensive analysis of the important subset of *singular Mueller matrices*.

The early contributions of Jones to the differential formulation of the Jones matrices, which was later translated by Azzam to nondepolarizing Mueller matrices, has been the subject of renewed interest in the last years. A series of works have extended this formal framework, particularly relevant for the description of *continuous media*, to the general case of depolarizing systems. The differential approach, based on the physical picture of continuously distributed polarization and depolarization components, parallels and complements the product decomposition approach whereby depolarization is modeled as a spatially localized "lump" phenomenon. It allows one to characterize a continuous depolarizing medium in terms of six elementary polarization properties and a 3×3 complex covariance matrix describing the depolarization. Within the framework of the statistical interpretation of the differential formalism, the polarization and depolarization components of the differential Mueller matrix are identified physically with the mean values and the variances-covariances of the fluctuating polarization properties. The general formalism, as well as the specific algebraic quantities and notions related to the concept of *differential Mueller matrix*, such as the six elementary polarization properties, are introduced, analyzed and physically interpreted in Chapter 9.

The intricate structure of Mueller matrices has long underpinned the necessity of developing appropriate geometric representations of the polarimetric properties of material media. The most useful approaches providing a geometric viewpoint on the polarization effects exhibited by different types of media are presented and discussed in a systematic way in Chapter 10. In particular, it is shown that the definition of an appropriate set of *characteristic ellipsoids* provides a meaningful, visual, and readily interpretable representation of any depolarizing Mueller matrix. The geometric counterparts of the principle polarization properties of media, like depolarization, polarizance-diattenuation (dichroism), and retardance (birefringence) are also analyzed. Chapter 10 also includes additional useful approaches, such as the *two-vector* and *five-vector representations* of nondepolarizing and depolarizing systems, respectively.

The random or systematic errors together with the inherent noise, as well as the intrinsic averaging nature of the measurement process, unavoidably transform the Mueller matrix responses of media and systems, expected to be nondepolarizing on physical grounds, into slightly depolarizing or even, unphysical ones. A large set of procedures, generally referred to as matrix filtering, consisting in obtaining the optimal (in a certain sense) nondepolarizing estimate to an experimentally determined depolarizing or unphysical Mueller matrix is dealt with in Chapter 11. Besides the widely used covariance filtering method, this chapter likewise presents a special class of parallel decompositions, called integral decompositions, that allows one to obtain not only a nondepolarizing estimate but also the uncertainties of the related polarization parameters. Another class of matrix filtering approaches comprises the virtual experiment and instrumental filtering methods; these provide an optimal nondepolarizing estimate when the modulation scheme of the polarimetric system used to measure the Mueller matrix to be filtered is known. Finally, when the response of the medium or system is assumed nondepolarizing, partial polarimetric systems yielding incomplete (i.e., missing a row or a column) Mueller matrices are sometimes used; Chapter 11 reports the procedure of completing a partial experimental Mueller matrix to a full, 16-element one.

In summary, this monograph on the polarization properties of both electromagnetic waves and material media provides a general and unified view, as well as detailed description, analysis and interpretation, of the fundamental concepts and main approaches underlying this field of physical optics. The presentation of the theoretical and mathematical foundations is combined with appropriate physical interpretations, illustrative figures, as well as with a number of selected practical examples taken from experiments. The authors believe that this book will be useful not only to beginners willing to get acquainted with the fundamentals, but also to confirmed workers in the field looking for a deeper grasp of the topic.

Preface to the second edition

A number of very significant new results on polarization optics have been published in recent years, which solve important pending problems and, for the sake of completeness, self-consistency and rigor, clearly justify their integration in this second edition of the book. Despite the fact that many subjects in polarization theory offer promising lines of research, the basic theoretical core of the structure and physical interpretation of polarization matrices and Mueller matrices is today well founded and consolidated, which allows for a better achievement of objectives already established for the first edition.

Thus, although the general structure of the contents has been preserved, a detailed and comprehensive revision and improvement has been carried out, so that, in addition to the reformulation and correction of certain topics, a significant amount of new content has been incorporated either through new sections (and even through the new Chapter 11), or by modifying existing sections. Moreover, certain nonessential contents of the first edition have been removed – for instance the summaries at the end of the chapters, whose contents are clearly covered by the revised introductions. Also, since the quantum formulation of polarization states deserves a stand-alone treatise, the basic information on this subject in the previous version of Chapters 1 and 2 has been removed.

With the aim of providing a useful guide to the reader who is already familiar with the first edition, and leaving aside a large number of minor improvements, corrections and references, the main new contents are briefly summarized below.

Chapters 1 and 2. Concept of polarization time; definition of the polarization matrix (or coherency matrix) in terms of measurable quantities, applicable even to highly polychromatic light; statistical nature of measurable polarization properties; spectral polarization matrix; concept of spin of a polarization state; formulation of partially coherent composition of mixed states; explicit formulation and geometric representation of sets of orthonormal 3D Jones vectors (which constitutes a new fruitful framework for the meaningful interpretation of the important features of three-dimensional polarization states); new precise formulation of the coefficients of the pure components of the arbitrary decomposition; specific study of discriminating states and the concept of nonregularity; smart decomposition of a polarization state; general geometric representation of polarization states through the polarization object, consisting of the rigid combination of an ellipsoid and a vector.

Chapter 3. Physical interpretation of the trace of a nondepolarizing Mueller matrix; formalisms based on coherency vector, covariance vector, matrix states and quaternion states.

Chapter 4. Symmetric and diagonal retarders.

Chapter 5. Synthesis of a general Mueller matrix through an ensemble average of partially coherent interactions, statistical parameterization of a Mueller matrix and very simple general characterization based on it; concept of Jones generators; structure of the polarimetric information contained in a general Mueller matrix; depolarization generated by partial spectral or spatial coherence

Chapter 6. Interpretation of the degree of spherical purity as an index of statistical dimension (in the light of the statistical integral formulation); physical interpretation of quantities of $\mathbf{M}$ that are invariant under dual-retarder transformations ($\det \mathbf{M}$ and others); general formulation of eigenvalue-based depolarization metric spaces, which provides a well-founded additional support for certain important properties like the indices of polarimetric purity.

Chapter 7. Generalization of the concept of arbitrary decomposition; analysis of the minimal and maximal numbers of retarders and diattenuators as arbitrary parallel components (Table 7.1).

Chapter 8. General revision and update.

Chapter 9. Spatial evolution of depolarization within the differential Mueller matrix formalism.

Chapter 10. Classification and geometric representation of singular Mueller matrices (Tables 10.4 and 10.5); new five-vector representation of Mueller matrices based on the integral statistical formulation.

Chapter 11. This entirely new chapter is devoted to the main theoretical approaches for the filtering of measured Mueller matrices. We refer the reader to the Preface and to Chapter 11 itself for the description of the new contents.

Acknowledgements

The authors are deeply indebted to Drs. Ignacio San José, Oriol Arteaga and Kurt Hingerl for their constant support and contributions on topics dealt with in this book. They would also like to express their gratitude and acknowledge to Drs. Tero Setälä and Alfredo Luis and for their careful review and valuable suggestions and comments on Chapters 1 and 2. This acknowledgement is extended to Celia Martínez Nebra for her graphic designs, and also to the great number of colleagues around the world who have contributed with interesting proposals, observations and corrections.

José J. Gil is also very grateful to Drs. Ari T. Friberg, Tero Setälä and Andreas Norrman for their constant collaboration and for the fruitful discussions on a number of topics that have been collected in this second edition of the book.

Razvigor Ossikovski is very thankful to his colleagues and collaborators for their constant support without which his contribution to this book would have been impossible. In particular, special thanks go to Drs. Enrique Garcia-Caurel and Michel Stchakovsky for the stimulating exchange of ideas and encouragement.

Authors

José Jorge Gil earned his PhD in Physics from the University of Zaragoza in 1983. While he was a research student he developed an original dual rotating absolute Mueller polarimeter and introduced some new concepts, such as the depolarization index. He has been a professor at the University of Zaragoza from 1983, where he has led a large number of R&D projects in Physics as well as in e-Learning technologies and methodologies. He was also the general manager of the R&D Department of the Spanish company BGL (1991–1996), where he led the development of new wireless systems for interactive meetings, which deserved the Tecnova award from the Spanish Industry Ministry in 1993. He is author of a large number of scientific articles and some patents. He was the recipient of the G. G. Stokes Award 2013 from the International Society for Optics and Photonics (SPIE) in recognition of his "collection of rigorous mathematical descriptions of polarization that are used widely to interpret experimental data."

Razvigor Ossikovski is an alumnus (X88 PEI) of the International Programme of the Ecole Polytechnique, France, where he also earned his MSc (1992) and PhD (1995) degrees in Physics. He likewise holds an Engineer's degree (1991) from Rousse Polytechnic, Bulgaria. He held R&D engineeer and team leader positions at the companies HORIBA Jobin Yvon, Corning Inc. and HighWave Optical (1995–2003) before taking his current academic position as assistant professor (2003) and, after his habilitation, associate professor (2010) and full professor (2018) in Physics at the Ecole Polytechnique, LPICM (Laboratory of Physics of Interfaces and Thin Films). His current research interests are the theory of polarimetry (specifically, Mueller matrix algebra and phenomenological models), as well as experimental tip-enhanced Raman spectroscopy (TERS). He has authored or co-authored five patents, two books, four book-chapters and more than 200 publications.

1

Polarized Electromagnetic Waves

1.1 Introduction: Nature of Polarized Electromagnetic Waves

Due to the specific nature of electromagnetic waves, the electric field of the wave evolves in time at any given point **r** in space. The complete description of the electromagnetic wave at point **r** requires the knowledge of four field vectors, namely the electric field strength **E**, the electric displacement density (electric induction) **D**, the magnetic field strength **H**, and the magnetic flux density (magnetic induction) **B**.

In general, the microscopic forces exerted by the electric field of the wave on matter are much larger than the forces produced by the magnetic field, so the temporal evolution of the electric field is chosen as the representative of the property called polarization.

When, in particular, the electromagnetic wave propagates in an isotropic medium, the electric and magnetic strengths are tangent to the wavefront at the considered point **r** (Figure 1.1). Nevertheless, in the case of an anisotropic medium, it is **D**, and not **E**, what is tangent to the wavefront at each given point **r**. Thus, unlike the case of isotropic media, for anisotropic media the temporal evolution of **D** is usually taken as the representative of the polarization of the electromagnetic wave (Cloude 2009, p. 13), even though we shall use the symbol **E** without distinction in the mathematical formulation of polarization.

In spite of the fact that we sometimes use the term *light*, in general and unless otherwise indicated, the contents of this book are applicable to the whole range of frequencies of the electromagnetic spectrum. Whenever appropriate, we include particular indications for some examples that are relative to specific spectral ranges.

Electromagnetic waves are produced by accelerated charges as well as by any kind of nuclear (e.g. gamma radiation), atomic and molecular emissions, so that the electromagnetic wave can be considered as composed of a very high number of more or less independent contributions. An electromagnetic wave is said to be monochromatic in the ideal case where its spectral width is zero, and hence its coherence time τ is infinite. The end point of the electric field vector of a monochromatic wave describes a fixed ellipse (*polarization ellipse*) lying in a stable plane (*polarization plane*) and the wave is said to be totally polarized. In a great variety of real situations, the spectral width $\Delta\nu$ is very narrow compared to the mean frequency $\overline{\nu}$ of the wave, which then is said to be *quasimonochromatic* ($\Delta\nu << \overline{\nu}$). In most cases in practice, the coherence time $\tau = 1/\Delta\nu$ (which determines a lower limit for the *polarization time* τ_p during which, on average, the polarization ellipse is stable) is much larger than the mean natural period $T_0 = 1/\overline{\nu}$ of the wave. The polarization time can be larger than the coherence time, as evidenced, for instance, in the case of fully

DOI: 10.1201/9780367815578-1

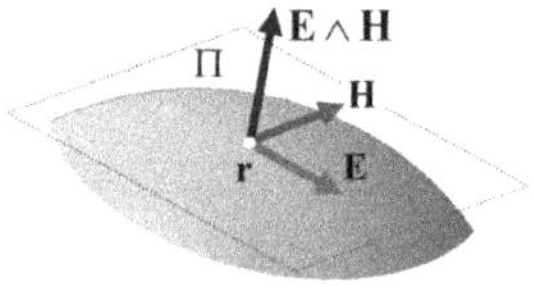

FIGURE 1.1
When the electromagnetic wave propagates in an isotropic medium, both electric and magnetic strength vectors lie in a plane Π which is tangent to the wavefront at the considered point **r**.

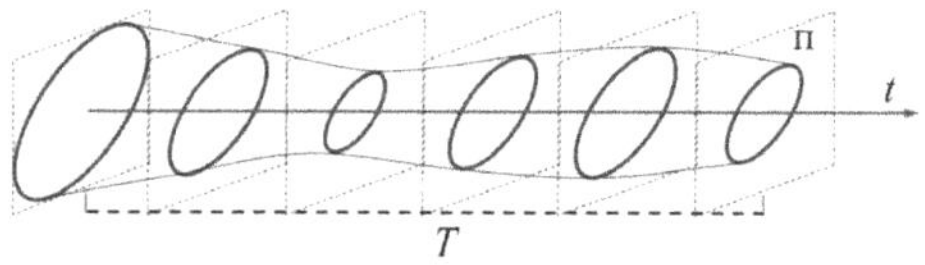

FIGURE 1.2
Despite intensity fluctuations, totally polarized 2D states are characterized by the fact that the shape of the polarization ellipse remains constant during the measurement time T.

polarized light, which is compatible even with relatively small values of τ (see Section 1.8). Another temporal scale to be considered is the measurement time T. In general, polychromatic waves behave as monochromatic for time intervals shorter than the coherence time τ so that, given a measurement time T, and leaving aside temporal intensity fluctuations (Mandel 1963), the following cases can be distinguished:

(a) *Totally polarized states*. A well defined polarization ellipse is realized for time intervals comparable to T_0 and it remains constant during the measurement time T (i.e., $\tau_p > T$), Figure 1.2.
(b) *2D polarization states.* The fluctuating electric field vector evolves in a fixed plane Π during the measurement time. Π is tangent to the wavefront at point **r**, so that the direction **k** perpendicular to Π can be considered a stable and well defined direction of propagation. Then, by taking a coordinate system that includes the direction **k** as the Z axis, the electric field of the wave is completely determined by two components fluctuating along two orthogonal reference axes XY lying in plane Π (Figure 1.3). Hereafter, we shall refer to this kind of polarization state as *2D states* (two-dimensional states), and
  - When the shape of the polarization pattern evolves fully randomly (but lying in plane Π) the state of polarization is said to be 2D unpolarized.
  - When a mean polarization ellipse can be identified during T, the state is said to be 2D partially polarized.
  - Totally polarized states constitute a subclass of 2D polarization states. Obviously, full polarization implies that the polarization plane is constant.
(c) *3D polarization states* refers to general polarization states for which the description of the electric field of the wave requires considering its three nonzero components. Thus 2D states constitute a subcategory of 3D states and, when appropriate, 3D states that are not 2D states are referred to as *genuine 3D states*.

When $\tau_p > T_0$ (as occurs for quasimonochromatic waves), a well defined polarization ellipse is realized whose shape is stable during time intervals larger than the natural period, while it may undergo changes during time T.

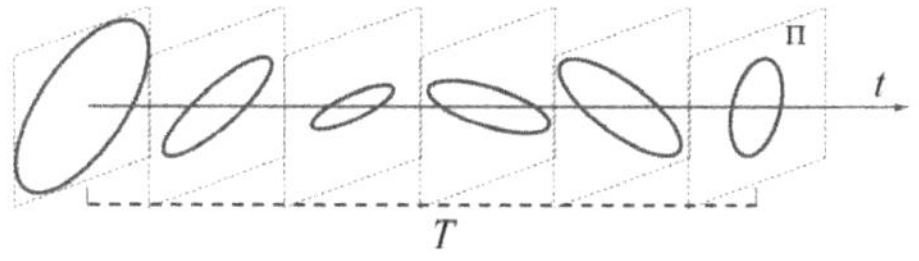

FIGURE 1.3
Partially polarized quasimonochromatic 2D states: the shape of the polarization ellipse changes during the measurement time T, while the plane Π containing it is stable in time.

Polychromatic electromagnetic waves for which the fluctuations of the electric are so fast that no polarization ellipse is formed within a single natural cycle (i.e., during the natural period T_0) are called *sub-cycle fluctuating states*. This may occur for waves with very large spectral bandwidth, as is the case of ultrashort pulses (Porras 2013). Despite the fact that the *instantaneous polarization ellipse* is not well defined, the second order correlations of the electric field components allow for the definition of averaged Stokes parameters, which in turn lead to the definition of an average polarization ellipse. This kind of states can have either 2D or genuine 3D character.

In general, the measurement time T (i.e., the response time of the detector) in an optical experiment is much longer than the coherence time. Typical values of the above indicated time intervals in the optical range are the following: $T_0 \cong 10^{-15}\,\mathrm{s}$, $10^{-9}\,\mathrm{s} \le \tau \le 10^{-4}\,\mathrm{s}$, and $T > 10^{-4}\,\mathrm{s}$ (Loudon 1983; Mandel and Wolf 1995; Brosseau 1998). Moreover, for many practical situations within the microwave range, T is usually shorter than τ, as typically occurs for instance in synthetic-aperture radar polarimetry (*SAR polarimetry*, Cloude 2009; López-Martínez and Pottier 2021).

The assumption of quasimonochromaticity is justified for a broad scope of physical situations, in which case a large number of natural cycles of the wave are realized without changes of both the shape and the polarization plane. Some concepts, like the instantaneous polarization ellipse (Gil 2007), the polarization time (Setälä et al. 2008; Voipio et al. 2010; Shevchenko et al. 2017), and the instantaneous Jones vector (Gil 2007), are well defined for quasimonochromatic waves, while these notions lose their consistency and physical meaning as the spectral profiles move away from the condition of quasimonochromaticity.

However, it is important to keep in mind that many essential notions in polarization theory are still valid even for waves with arbitrary spectral profile. This is the case of the central concept of the analytic signal associated with each Cartesian component of the electric field, which allows for its complex representation (Gabor 1946; Mandel and Wolf 1995) as well as the proper formulation of polarization in the frequency domain (Tervo et al. 2004). The same happens with key concepts like the polarization matrix (or coherency matrix, Wiener 1930; Wolf 1959; Barakat 1963), the Stokes parameters (Stokes 1852) and the degree of polarization (Wolf 1959), which admit definitions based directly on measurable parameters from the very phenomenological point of view (Meemon et al. 2008). The relevance of the pioneering work of Soleillet (1929), who introduced matrix structures that are equivalent to polarization matrices of 2D and 3D states, has been discussed and emphasized by Arteaga and Nichols (2018).

Before other considerations, it should be recalled that polarization plays a fundamental role in the interaction between electromagnetic radiation and matter, not only from the classical point view, but also for the very selection rules in quantum mechanics (Shore 1990).

The present chapter is devoted to 2D states of polarization, i.e., those that characterize the electric field evolving into a plane Π that is constant in time, so that the direction of propagation is fixed along the axis Z orthogonal to Π (Figure 1.4). As indicated above, the

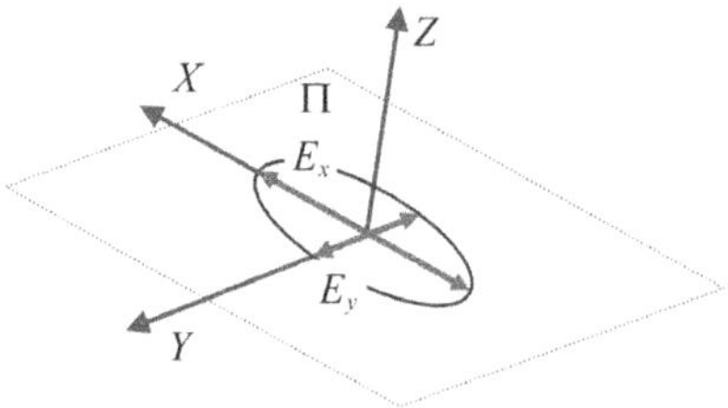

FIGURE 1.4
In the case of waves with arbitrary form, the polarization state is referenced with respect to the local coordinate system constituted by a pair of axes *XY* lying on the local plane Π tangent to the wavefront at the considered point **r**, together with an axis *Z* orthogonal to Π which determines the local direction of propagation.

plane Π is called the *polarization plane* regardless of whether the electromagnetic wave is quasimonochromatic or not. The general case of 3D states of polarization will be dealt with in Chapter 2.

1.2 The Polarization Ellipse

Consider the temporal evolution of the end point of the electric field **E** of a quasimonochromatic electromagnetic wave that, at a given point **r**, is two-dimensional in the sense that there exists a constant plane Π in which **E** lies during the measurement time, and let us take the direction perpendicular to Π as the *Z* axis of the *XYZ* Cartesian coordinate system (Figure 1.4). The components of the electric field can be expressed as

$$\begin{aligned} E_x(z,t) &= A_x(t)\cos\left(\bar{k}z - \bar{\omega}t + \beta_x(t)\right) \\ E_y(z,t) &= A_y(t)\cos\left(\bar{k}z - \bar{\omega}t + \beta_y(t)\right) \end{aligned} \tag{1.1}$$

where $\bar{k}, \bar{\omega}$ are the respective mean values of the wavevector length $k = 2\pi/\lambda_0$ (λ_0 being the wavelength for the vacuum) and of the angular frequency, $\omega = 2\pi\nu$ (ν being the natural frequency); β_x, β_y are the respective phases, and A_x, A_y are the respective amplitudes. Note that, obviously, the cosine functions in Eq. (1.1) can be replaced by sine ones by adding $\pi/2$ to the phases. Moreover, the change of signs of the arguments of the cosine functions preserves the interpretation of $E_x(z,t)$ and $E_y(z,t)$ as the components of the electric field of a wave traveling in the positive *Z* direction, while the choice $\pm(\bar{k}z + \bar{\omega}t)$ would correspond to a wave traveling in the negative *Z* direction.

The polarization of the wave is a concept linked to measurable quantities given by averages over measurement times that involve a very large number of cycles, so that the common harmonic dependence on $\bar{\omega}t + \beta_x(t)$ can be removed from the polarization descriptors. Moreover, the value $z = 0$ can be taken as the reference, in such a manner that the state of polarization, at point **r**, is defined from the variables

$$E_x(t) = A_x(t) \qquad E_y(t) = A_y(t)\cos\left(\delta(t)\right) \tag{1.2}$$

with $\delta(t) \equiv \beta_y(t) - \beta_x(t)$

To analyze the locus traced out by the end point of the electric field **E** of the wave, let us observe that its components always satisfy the following ellipse equation (Goldstein 2011):

$$\left(\frac{E_x(t)}{A_x(t)}\right)^2+\left(\frac{E_y(t)}{A_y(t)}\right)^2-2\left(\frac{E_x(t)E_y(t)}{A_x(t)A_y(t)}\right)\cos\big(\delta(t)\big)=\sin^2\big(\delta(t)\big) \tag{1.3}$$

In the case of quasimonochromatic waves under consideration, variables $E_x(t)$ and $E_y(t)$ fluctuate slowly in comparison to the mean natural period T_0 and therefore, the shape of the ellipse remains constant, at least within time intervals shorter than the coherence time. The so-called *polarization time* τ_p (Figure 1.5) (Setälä et al. 2008; Shevchenko et al. 2009; Voipio et al. 2010; Shevchenko et al. 2017) refers to a measure of the time interval for which, on average, the shape of the polarization ellipse of a given electromagnetic wave is stable. Despite the fact that an observation time equal or larger than T_0 is required for the end point of the electric field vector to draw a complete ellipse, the mentioned slow fluctuation is the reason why the ellipse defined by Eq. (1.3) is usually termed the *instantaneous polarization ellipse*. For waves whose spectral profile is far from the quasimonochromaticity condition (sub-cycle fluctuating states), the concept of instantaneous polarization ellipse loses its applicability due to the fluctuations that the field undergoes during T_0.

Depending on the nature and characteristics of the field fluctuations, the cases of totally polarized, partially polarized and unpolarized 2D states can be distinguished (provided that the polarization plane is constant during the measurement time, as required for the case of 2D states under consideration).

Many common light sources such as the Sun, light bulbs or flames emit unpolarized light. In general, the interaction of light (and of other kinds of electromagnetic waves) with matter via scattering, transmission, refraction and reflection produces or modifies the state of polarization to some extent.

Unpolarized light is also called *natural light*. Certain artificial light sources as lasers or radar antennas typically emit polarized light. Other natural or artificial electromagnetic waves outside the optical range (which covers infrared, visible and ultraviolet) as, for example X-ray, microwaves, and radio waves are commonly emitted with a certain degree of polarization, and the states of polarization change through some kinds of interaction with matter.

In the ideal case of monochromatic waves, the variables A_x, A_y and δ are constant in time and hence, both the shape and size of the polarization ellipse are fixed.

The *instantaneous intensity*

$$I(t)\equiv A_x^2(t)+A_y^2(t) \tag{1.4}$$

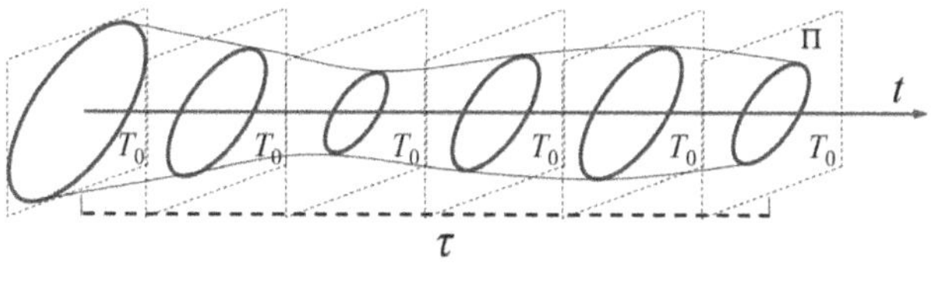

FIGURE 1.5
The field variables of a quasimonochromatic wave fluctuate slowly in comparison to the mean natural period T_0 and therefore the shape of the ellipse remains constant during the *polarization time* τ_p, which involves a large number of natural cycles.

is a measure of the power of the electric field of the electromagnetic wave at the point **r** considered and time t. Therefore, leaving aside the theoretical interest of considering the instantaneous intensity, measurable intensities refer to an average of the fluctuating $I(t)$ during the measurement time.

While the ratio $A_y(t)/A_x(t)$ as well as the phase difference $\delta(t)$ may remain constant (totally polarized states), the instantaneous intensity fluctuates so that, regardless of the constancy or not of the shape of the polarization ellipse, its size varies during a typical measurement time, resulting in an average size that is proportional to the corresponding average intensity (Mandel 1963)

$$I = \langle A_x^2(t) + A_y^2(t) \rangle = a_x^2 + a_y^2 \qquad \left[a_x^2 \equiv \langle A_x^2(t) \rangle \quad a_y^2 \equiv \langle A_y^2(t) \rangle \right] \tag{1.5}$$

where $\langle \; \rangle$ indicates time average over the measurement time.

Obviously, the intensity can also be directly defined as the time averaged sum of the squares of the strengths of the real Cartesian components of the electric field $I = \langle E_x^2(t) + E_y^2(t) \rangle$ and therefore such definition is valid for waves of arbitrary spectral bandwidth.

Thus, a totally polarized state is fully described by the characteristic parameters of the polarization ellipse, namely the intensity I (average size), the ratio $a_y/a_x \equiv \tan\alpha$ $(0 \le \alpha \le \pi/2)$, and the phase shift δ $(0 \le \delta < 2\pi)$. Alternatively, the polarization ellipse can be characterized through I together with the azimuth ϕ $(0 \le \varphi < \pi)$ and the ellipticity angle χ $(-\pi/4 \le \chi \le \pi/4)$ (Figure 1.6), which are related to a_x, a_y, α and δ by the equations (Goldstein 2011)

$$\tan 2\varphi = \frac{2a_x a_y}{a_x^2 - a_y^2}\cos\delta = \tan 2\alpha \cos\delta \qquad \sin 2\chi = \frac{2a_x a_y}{a_x^2 + a_y^2}\sin\delta = \sin 2\alpha \sin\delta \tag{1.6}$$

The polarization ellipse is represented for different values of δ in Figure 1.7.

- Values $\delta = 0, \pi$ correspond to linearly polarized states $(\chi = 0)$, regardless of the value of α.
- Values $\delta = \pi/2, 3\pi/2$ correspond to states for which the semiaxes of the polarization ellipse lie along the reference axes XY, or to right-handed and left-handed circularly polarized states respectively.
- Intermediate values of δ correspond to states with $\chi \neq 0$. Positive and negative values of the ellipticity angle correspond respectively to right-handed and left-handed elliptical polarized states. Right-handed and left-handed circular polarized states correspond respectively to the particular combined values $(\delta = \pi/2, \chi = \pi/4)$ and $(\delta = \pi/2, \chi = -\pi/4)$.

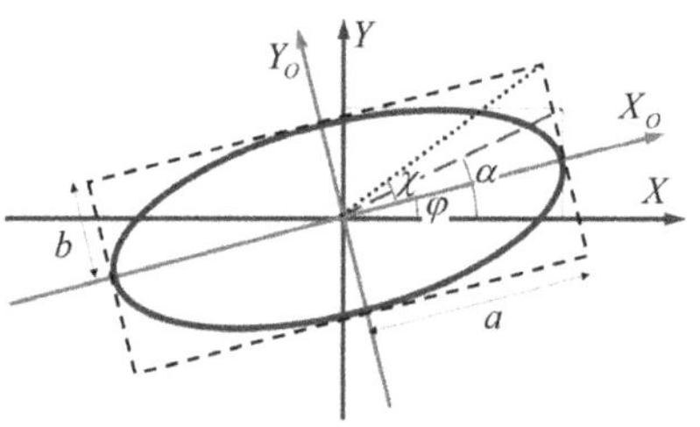

FIGURE 1.6
The polarization ellipse is characterized by $I = a^2 + b^2$ together with the azimuth ϕ $(0 \le \varphi < \pi)$ and the ellipticity angle χ $(-\pi/4 \le \chi \le \pi/4)$; a and b being the semiaxes of the polarization ellipse.

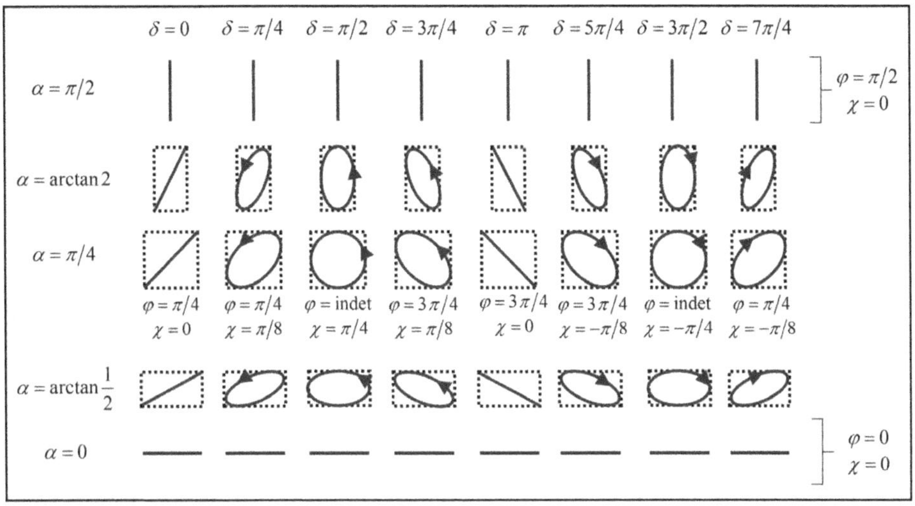

FIGURE 1.7
Values $\delta = 0, \pi$ correspond to linearly polarized states ($\chi = 0$). Values $\delta = \pi/2, 3\pi/2$ correspond to states whose polarization ellipse has its semiaxes aligned along the reference axes X and Y, or to right handed and left handed circularly polarized states respectively. Intermediate values of δ correspond to states with $\chi \neq 0$. Positive and negative values of the ellipticity angle correspond respectively to right handed and left handed elliptical polarized states. Right handed and left handed circular polarized states correspond respectively to the particular combined values ($\delta = \pi/2$, $\alpha = \pi/4$) and ($\delta = 3\pi/2$, $\alpha = \pi/4$). For a fixed value of α, the area of the ellipse is maximal when $\delta = \pi/2$ or $\delta = 3\pi/2$, and it is zero when $\delta = 0$ or $\delta = \pi$ (linearly polarized states).

Note that throughout this book we assume the common convention that right handed states correspond to counterclockwise handedness with respect to an observer towards whom the wave travels (i.e., the wave propagates towards the reader in Figure 1.7).

The *ellipticity* is defined as the ratio $b/a \equiv |\tan\chi|$, where b and a are the respective minor and major axes of the polarization ellipse, and χ (with $-\pi/4 \le \chi \le \pi/4$) is the *ellipticity angle* (positive for right-handed polarization states and negative for left-handed ones). The *eccentricity* of the polarization ellipse is defined as $\sqrt{1-\tan^2\chi}$.

The area A of the polarization ellipse of a totally polarized state is given by

$$A = \pi a b = \pi a_x a_y |\sin\delta| = (\pi/2) I |\sin 2\chi| \tag{1.7}$$

so that, as expected, the normalized area $\hat{A} \equiv A/I$ has its maximum value $\hat{A}_{max} = \pi/2$ for circularly polarized states, while its minimum $\hat{A}_{min} = 0$ corresponds to linearly polarized states.

Even though the concept of polarization applies in principle to the evolution of the electric field of the wave at a given point **r**, it is straightforward to extend this concept to all points distributed on a certain region of a common wavefront and sharing a common state of polarization. In these cases of polarization states relative to waves with arbitrary form, the polarization is referenced with respect to the local coordinate system constituted by a pair of axes XY lying on the local polarization plane Π tangent to the wavefront surface, together with an axis Z orthogonal to Π (Figure 1.4).

1.3 The Analytic Signal Representation

Following the common practice in polarization optics, it is very advantageous to use the analytic signal representation where the components of the electric field are described through respective complex variables. For a detailed description and rigorous foundation of the analytic signal representation of the wavefield we refer the reader to Gabor (1946) and Born and Wolf (2005).

Consider an arbitrary Cartesian reference system to represent the three mutually orthogonal real components $E_x(t)$, $E_y(t)$and $E_z(t)$ of the electric field vector $\mathbf{E}(t)$ of a general polarization state and assume the condition, always satisfied by physical fields in nature, that they are zero-mean square integrable stochastic processes, i.e.,

$$\int_{-\infty}^{\infty} E_k^2(t)\,dt < \infty \quad [k = x, y, z] \tag{1.8}$$

and that, in addition, $E_x(t)$, $E_y(t)$ and $E_z(t)$ are temporally stationary, at least in wide sense. The stochastic analytic signal representations $\varepsilon_x(t)$, $\varepsilon_y(t)$ and $\varepsilon_z(t)$ associated with the real fields $E_x(t)$, $E_y(t)$ and $E_z(t)$ are defined as the components of the complex vector

$$\boldsymbol{\varepsilon}(t) = \int_0^{\infty} \boldsymbol{\varepsilon}(\omega)\, e^{-i\omega}\, d\omega \qquad \left[\boldsymbol{\varepsilon}(t) \equiv \left(\varepsilon_x(t), \varepsilon_y(t), \varepsilon_z(t)\right)^T\right] \tag{1.9}$$

where $\boldsymbol{\varepsilon}(\omega)$ is the frequency-domain vector representation

$$\boldsymbol{\varepsilon}(\omega) = \frac{1}{2\pi}\int_{-\infty}^{\infty} \mathbf{E}(t)\, e^{i\omega}\, dt \tag{1.10}$$

Since $\mathbf{E}(t)$ is real, then $\boldsymbol{\varepsilon}(\omega)$ satisfies the symmetry property $\boldsymbol{\varepsilon}(\omega) = \boldsymbol{\varepsilon}^*(-\omega)$, where * stands for complex conjugate. Consequently, the lower limit of the integral in Eq. (1.9) has been set to zero (instead of -∞) without loss of generality (Born and Wolf 2005).

The above definition the analytic signals constituting the components of the *analytic signal vector* $\boldsymbol{\varepsilon}(t)$ is consistent with their expressions as

$$\varepsilon_x(t) \equiv E_x(t) + i\tilde{E}_x(t) \qquad \varepsilon_y(t) \equiv E_y(t) + i\tilde{E}_y(t) \qquad \varepsilon_z(t) \equiv E_z(t) + i\tilde{E}_z(t) \tag{1.11}$$

where $\tilde{E}_x(t)\tilde{E}_y(t)$ and $\tilde{E}_z(t)$ are the stochastic Hilbert transforms of the real components $E_x(t)$, $E_y(t)$ and $E_z(t)$ of the electric field (Wolf 1959)

$$\tilde{E}_k(t) = \frac{1}{\pi}\int_{-\infty}^{\infty} \frac{E_k(t')}{t' - t}\, dt' \quad [k = x, y, z] \tag{1.12}$$

As indicated above, the real field variables $E_k(t)$ are zero-mean, and we assume the realistic physical conditions that they are also temporally stationary (at least, in the wide sense) stochastic processes obeying the same statistics and having the same power spectrum $\phi(\nu)$. Then, the Hilbert transforms $\tilde{E}_k(t)$ obey the same statistics and also have the same power spectrum $\phi(\nu)$ as the real components, so that the analytic signals $\varepsilon_k(t)$ are also zero-mean, temporally stationary (at least in the wide sense), stochastic processes with a common power spectrum given by $4\phi(\nu)$ for $\nu > 0$ and zero for $\nu < 0$ (see Section 1.5.1 for a brief summary of statistical concepts like stochastic process, stationarity, etc.).

The customary analytic signal representation of a 2D polarization state is retrieved from the above general 3D formulation by taking the Z axis along the direction normal to the polarization plane containing the fluctuating electric field, so that $E_z(t)=0$ (i.e., $\varepsilon_z(t)=0$) and therefore only the two non-vanishing components $\varepsilon_x(t)$, $\varepsilon_y(t)$ of $\boldsymbol{\varepsilon}(t)$ are considered.

1.4 The Jones Vector

Under a wide variety of experimental conditions, the spectral width $\Delta\nu$ can be considered very narrow compared to the central frequency $\bar{\nu}$ of the spectrum of the wave (quasimonochromatic wave). It has been shown that for quasimonochromatic waves (whose real field variables can be expressed as in Eq. (1.1)), the following relations are satisfied with a good approximation in practice (Barakat 1984):

$$\begin{aligned}\tilde{E}_x(t) &\approx A_x(t)\cos\left[\bar{k}z-\bar{\omega}t+\beta_x(t)\right]\\ \tilde{E}_y(t) &\approx A_y(t)\cos\left[\bar{k}z-\bar{\omega}t+\beta_y(t)\right]\end{aligned} \tag{1.13}$$

The variables $\varepsilon_x(t)$ and $\varepsilon_y(t)$ can then be arranged as the components of the following 2×1 complex vector:

$$\boldsymbol{\varepsilon}(t)\equiv\begin{pmatrix}\varepsilon_x\\ \varepsilon_y\end{pmatrix}\equiv e^{i\left[u(t)+\frac{\beta_x(t)+\beta_y(t)}{2}\right]}\begin{pmatrix}A_x(t)e^{-i\delta(t)/2}\\ A_y(t)\,e^{i\delta(t)/2}\end{pmatrix}\qquad\begin{bmatrix}\delta(t)\equiv\beta_y(t)-\beta_x(t)\\ u(t)\equiv\bar{k}z-\bar{\omega}t\end{bmatrix} \tag{1.14}$$

In the case under consideration of quasimonochromatic waves, the polarization time τ_p can be either smaller or larger than the measurement time, while there always hold the inequalities $\tau_p\geq\tau>>T_0$, τ and T_0 being the coherence time and the natural period respectively. Thus, τ_p involves a large number of cycles, while the shape of the polarization ellipse typically remains constant during the polarization time. For time intervals larger than the polarization time, the shape of the polarization ellipse varies to some extent, resulting in partial polarization.

Moreover, measurable quantities as, for instance, the Stokes parameters (Stokes 1852; Fano 1953) which will be considered later, involve necessarily time or ensemble averaging of second-order products, like $\varepsilon_i(t)\varepsilon_j^*(t)$, taken at a fixed point z, so that, as it has been done for the real representation in Eq. (1.1), the global phase factor can be removed in the description of polarization states in terms of observables. Consequently, for fields whose spectral profile is not too far from quasimonochromaticity conditions, the *instantaneous Jones vector* can be defined as

$$\boldsymbol{\varepsilon}(t)=\sqrt{I(t)}\begin{pmatrix}\hat{A}_x(t)\,e^{-i\delta(t)/2}\\ \hat{A}_y(t)\,e^{i\delta(t)/2}\end{pmatrix}\qquad\begin{bmatrix}I(t)\equiv A_x^2(t)+A_y^2(t)\\ \hat{A}_x(t)\equiv A_x(t)/I(t)\quad\hat{A}_y(t)\equiv A_y(t)/I(t)\end{bmatrix} \tag{1.15}$$

Note that $\boldsymbol{\varepsilon}(t)$ is defined up to an unmeasurable global phase factor $e^{i\varphi}$.

The above definition of the instantaneous Jones vector, as a descriptor of the instantaneous polarization state, has been established under the assumption of quasimonochromaticity, which ensures the consistency of the concept of instantaneous polarization ellipse, i.e.,

$A_y(t)/A_x(t)$ and $\delta(t)$ are well defined. Obviously, the analytic signals of the components of the electric field of waves with arbitrary spectral profiles can be arranged as the two components of the corresponding analytic signal vector, but such vector can only be properly called a Jones vector when the field is quasimonochromatic, thus allowing the expression of $\boldsymbol{\varepsilon}(t)$ in Eq. (1.15).

The instantaneous Jones vector, as a function of time, includes all information relative to the temporal evolution of the polarization ellipse and the intensity (*polarization dynamics*, Voipio et al. 2010). $\boldsymbol{\varepsilon}(t)$ is called *instantaneous* in the sense that the possible time dependence of relative amplitudes and relative phase is considered.

Consider now the particular case where the quantities $A_y(t)/A_x(t)$ and δ remain constant in time and consequently the shape of the polarization ellipse remains fixed during the measurement time. The corresponding state of polarization is described by means of the Jones vector (Jones 1941)

$$\boldsymbol{\varepsilon} \equiv \begin{pmatrix} \varepsilon_x \\ \varepsilon_y \end{pmatrix} = \begin{pmatrix} a_x e^{-i\delta/2} \\ a_y e^{i\delta/2} \end{pmatrix} = \sqrt{I} \begin{pmatrix} \hat{a}_x e^{-i\delta/2} \\ \hat{a}_y e^{i\delta/2} \end{pmatrix} \qquad \begin{bmatrix} I \equiv a_x^2 + a_y^2 \\ \hat{a}_x \equiv a_x/I \quad \hat{a}_y \equiv a_y/I \end{bmatrix} \tag{1.16}$$

where a_x and a_y are given by the averages $a_x^2 \equiv \langle A_x^2 \rangle$ and $a_y^2 \equiv \langle A_y^2 \rangle$ of the respective square of the amplitudes $A_x(t)$ and $A_y(t)$ over the measurement time T, while $I \equiv \langle I(t) \rangle$ is the time averaged intensity.

Leaving aside a global phase factor, the Jones vector can also be expressed in terms of the intensity I, the azimuth φ and the ellipticity angle χ as follows:

$$\boldsymbol{\varepsilon} = \sqrt{I} \begin{pmatrix} \cos\chi \cos\varphi - i \sin\chi \sin\varphi \\ \cos\chi \sin\varphi + i \sin\chi \cos\varphi \end{pmatrix} = \sqrt{I} \begin{pmatrix} \cos\varphi & -\sin\varphi \\ \sin\varphi & \cos\varphi \end{pmatrix} \begin{pmatrix} \cos\chi \\ i\sin\chi \end{pmatrix} \tag{1.17}$$

which can be interpreted in the following manner (from right to left): The rightmost vector, a function of χ, represents the Jones vector of an elliptical polarization state whose major and minor semiaxes lie along the reference laboratory axes X and Y respectively; the matrix, a function of φ, is a rotation matrix that rotates the said Jones vector by the clockwise angle φ, and the scalar factor $\sqrt{I}$ is the square root of the averaged intensity of the state.

Thus, a totally polarized state is fully described by its corresponding Jones vector $\boldsymbol{\varepsilon}$, which provides complete information about the polarization ellipse as well as the intensity. The definition (1.16) of the Jones vector is consistent with the fact that total polarization is compatible with the intensity fluctuations that are unavoidable in real physical situations (Mandel 1963). In fact, totally polarized waves maintain the azimuth and ellipticity of the polarization ellipse fixed, whereas the size of the ellipse fluctuates resulting in a *mean intensity* I over the measurement time. Moreover, slow time variations of the Jones vector with respect to the measurement time can be represented by this model (Gil 2007).

Jones vectors have been defined in Eq. (1.16) with respect to a XY coordinate system in the polarization plane Π (Figure 1.3), in such a manner that a generic Jones vector $\boldsymbol{\varepsilon}$, referenced with respect to XY, can be written as

$$\boldsymbol{\varepsilon} = \varepsilon_x \mathbf{e}_x + \varepsilon_y \mathbf{e}_y \qquad \mathbf{e}_x \equiv \begin{pmatrix} 1 \\ 0 \end{pmatrix} \quad \mathbf{e}_y \equiv \begin{pmatrix} 0 \\ 1 \end{pmatrix} \tag{1.18}$$

where the basis vectors $\mathbf{e}_x$ and $\mathbf{e}_y$ represent respective linearly polarized states whose electric fields lie along the axes X and Y.

From a formal point of view, any 2×1 complex vector can be considered a Jones vector, so that the whole set of Jones vectors, together with the standard internal product, the product by a complex scalar, and the Frobenius norm, constitutes a Hilbert space. Thus, generalized orthonormal bases $(\mathbf{e}_1, \mathbf{e}_2)$ other than $(\mathbf{e}_x, \mathbf{e}_y)$ can be defined. A change of coordinate system from XY to $X'Y'$ for the representation of Jones vectors is performed by means of an orthogonal transformation of the form

$$\boldsymbol{\varepsilon}' = \mathbf{Q}(\theta)\,\boldsymbol{\varepsilon} \qquad \mathbf{Q}(\theta) = \begin{pmatrix} \cos\theta & \sin\theta \\ -\sin\theta & \cos\theta \end{pmatrix} \tag{1.19}$$

where the Jones vectors $\boldsymbol{\varepsilon}$ and $\boldsymbol{\varepsilon}'$ represent a common state of polarization with respect to the coordinate systems XY and $X'Y'$ respectively. The orthogonal matrix $\mathbf{Q}(\theta)$ corresponds to a proper counterclockwise rotation, by the angle θ about the axis Z, from the original coordinate system XY to the new axes $X'Y'$ (Figure 1.8). Hereafter, this kind of orthogonal transformations will be called *rotation transformations*.

To avoid confusion, it is important to emphasize that when a real vector $\mathbf{v}$ is rotated by an angle θ about the axis Z, but maintaining fixed the reference system XY, the transformation results in a new vector $\mathbf{w}$, whose expression with respect to the system XY is given by $\mathbf{w} = \mathbf{Q}(-\theta)\mathbf{v} = \mathbf{Q}^T(\theta)\mathbf{v}$. In other words, two Jones vectors linked by a rotation transformation only differ in the azimuth ($\varphi' = \varphi - \theta$) and correspond either to the same state of polarization, but represented with respect to two different reference systems XY and $X'Y'$, or to two different polarization states that only differ on their azimuths. Observe that the ellipticity angle χ is an intrinsic parameter of the polarization state, irrespective of the Cartesian coordinate system XY used to represent it.

Moreover, in general, any pair of complex vectors $(\mathbf{e}_1, \mathbf{e}_2)$ satisfying $\mathbf{e}_1^\dagger \mathbf{e}_2 = 0$ and $\mathbf{e}_1^\dagger \mathbf{e}_1 = \mathbf{e}_2^\dagger \mathbf{e}_2 = 1$, where the superscript † denotes conjugate transpose, constitutes a generalized orthonormal basis. Thus, pairs of mutually orthogonal linear, elliptical or circular states can be used as generalized reference bases by transforming the canonical basis $(\mathbf{e}_x, \mathbf{e}_y)$ through *unitary transformations* like

$$\boldsymbol{\varepsilon}' = \mathbf{U}\boldsymbol{\varepsilon} \quad \left(\mathbf{U}^\dagger = \mathbf{U}^{-1}\right) \tag{1.20}$$

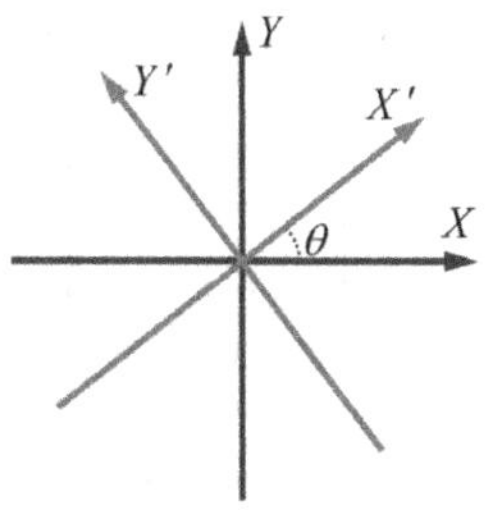

FIGURE 1.8
A change of coordinate system from XY to $X'Y'$ for the representation of Jones vectors is performed through an orthogonal transformation of the form $\varepsilon' = \mathbf{Q}(\theta)\,\varepsilon$ where $\mathbf{Q}$ corresponds to a proper counterclockwise rotation by the angle θ, about the axis Z, from the original coordinate system XY to the new axes $X'Y'$.

Observe that rotation transformations are those where the unitary matrix **U** is real valued (i.e., **U** is orthogonal), and therefore rotation transformations constitute a subclass of unitary transformations.

Particularly interesting alternative bases are the linear +45° and linear –45° ($\mathbf{e}_{+\pi/4}, \mathbf{e}_{-\pi/4}$) defined by the basis vectors

$$\mathbf{e}_{+\pi/4} \equiv \frac{1}{\sqrt{2}}\begin{pmatrix}1\\1\end{pmatrix} \quad \mathbf{e}_{-\pi/4} \equiv \frac{1}{\sqrt{2}}\begin{pmatrix}1\\-1\end{pmatrix} \quad \left[\mathbf{U}=\mathbf{Q}=\frac{1}{\sqrt{2}}\begin{pmatrix}1 & 1\\1 & -1\end{pmatrix}\right] \tag{1.21}$$

and the right and left handed circular ($\mathbf{e}_r, \mathbf{e}_l$), defined by the basis vectors

$$\mathbf{e}_r \equiv \frac{1}{\sqrt{2}}\begin{pmatrix}1\\i\end{pmatrix} \quad \mathbf{e}_l \equiv \frac{1}{\sqrt{2}}\begin{pmatrix}1\\-i\end{pmatrix} \quad \left[\mathbf{U}=\frac{1}{\sqrt{2}}\begin{pmatrix}1 & 1\\i & -i\end{pmatrix}\right] \tag{1.22}$$

It should be noted that in some treatises the signs of the second components of the above vectors appear interchanged because the alternative convention for handedness is taken ($\chi > 0$ for clockwise handedness with respect to an observer towards whom the wave travels).

Despite the fact that, unless otherwise stated, in this book the polarization states will be considered referenced with respect to the basis ($\mathbf{e}_x, \mathbf{e}_y$), the generic notation $\boldsymbol{\varepsilon} = \varepsilon_1 \mathbf{e}_1 + \varepsilon_2 \mathbf{e}_2$ will be used in order to indicate the validity of the mathematical expressions regardless of the particular basis chosen.

Generalized bases containing complex components are very useful for some purposes, for instance, representing a pure state as a coherent superposition of a right handed and a left handed circularly polarized states. However, such types of generalized bases involving imaginary parameters, while being algebraically acceptable, are not physically realizable as laboratory coordinate systems. In fact, only rotation transformations of the form $\mathbf{e}_{i'} = \mathbf{Q}\,\mathbf{e}_i$ $(i = x, y)$, where **Q** is a 2×2 orthogonal matrix, are admissible for generating physically realizable laboratory coordinate systems.

The scalar product of two Jones vectors $\boldsymbol{\mu}$ and $\mathbf{v}$ is defined as

$$\boldsymbol{\mu}^{\dagger}\mathbf{v} = \left(\mu_1^*, \mu_2^*\right)\begin{pmatrix}v_1\\v_2\end{pmatrix} = \mu_1^* v_1 + \mu_2^* v_2 \tag{1.23}$$

and the squared absolute value $|\boldsymbol{\varepsilon}|^2 = \boldsymbol{\varepsilon}^{\dagger}\boldsymbol{\varepsilon} = I$ of a given Jones vector $\boldsymbol{\varepsilon}$ is precisely the intensity of the corresponding state of polarization. Two pure states represented by respective Jones vectors $\boldsymbol{\mu}$ and $\mathbf{v}$ are said to be orthogonal when $\boldsymbol{\mu}^{\dagger}\mathbf{v} = 0$. Moreover, the product of a Jones vector $\boldsymbol{\varepsilon}$ by a complex number $a \equiv |a|e^{i\alpha}$ produces a new Jones vector $\boldsymbol{\varepsilon}' = a\boldsymbol{\varepsilon}$ where $|a|$ scales the amplitude while the phase factor $e^{i\alpha}$ modifies the phase.

Another interesting topic on Jones vectors is the coherent superposition, at a given point **r**, of two totally polarized states whose respective polarization ellipses lie in a common plane Π (Figure 1.9). The composed state at point **r** is totally polarized and its Jones vector is given by the addition of the Jones vectors of the mutually coherent components

$$\boldsymbol{\varepsilon} = \boldsymbol{\varepsilon}_1 + \boldsymbol{\varepsilon}_2 \tag{1.24}$$

The relative phase of the addends plays an important role, and therefore the difference between the respective phases of $\boldsymbol{\varepsilon}_1$ and $\boldsymbol{\varepsilon}_2$ should be taken into account under the summation.

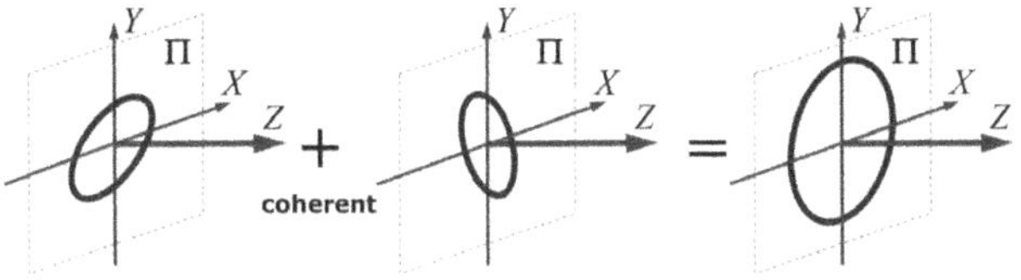

FIGURE 1.9
Coherent superposition, at a given point **r**, of two totally polarized states whose respective polarization ellipses lie in a common plane Π. The composed state at point **r** is totally polarized and its Jones vector is given by the addition of the Jones vectors of the mutually coherent components.

To complete this summary of the properties and physical interpretation of Jones vectors, consider a Jones vector $\boldsymbol{\varepsilon}$ and its relation to $\boldsymbol{\varepsilon}^r$ that corresponds to the same physical situation but reversing the direction of propagation. From the definition of the Jones vector, a change of the sign of the second component is required, while Eq. (1.16) shows that the required time reversal does not produce any additional change in a stable Jones vector, so that $\boldsymbol{\varepsilon}^r = \mathrm{diag}(1,-1)\boldsymbol{\varepsilon}$, where $\mathrm{diag}(1,-1)$ denotes a diagonal matrix with diagonal elements $(1,-1)$.

It should be stressed that, due to the very definition of the Jones vector, it never can represent a partially polarized state, but the use of Jones vectors is restricted to totally polarized (or *pure*) states. In the case of partially polarized states, the azimuth and/or the ellipticity of the polarization ellipse varies during the measurement time and an alternative mathematical description, different from that used in the Jones approach, is necessary in order to take into account all parameters that characterize completely the state of polarization.

1.5 Polarization Matrix and Stokes Vector

The electric field of electromagnetic waves fluctuates randomly, which is caused by the chaotic nature of the light emission processes together with the possible random behavior of the propagation medium; therefore, as indicated in previous sections, the description of polarization states through measurable quantities requires statistical treatments. In fact, due to the rapid temporal fluctuations of the electromagnetic field, the correlations of the components of the field must be considered rather than the amplitude fluctuations themselves, which cannot be detected directly.

The experimental detection of polarization states is performed by means of sets of intensity measurements that are directly linked to the Stokes parameters of the state, which in turn determine the corresponding *Stokes vector* and *polarization matrix*. Such equivalent mathematical structures characterizing completely the polarization states are studied along respective subsections, which are preceded by a summary of the main statistical concepts involved.

1.5.1 Statistical Nature of Measurable Polarization Properties

The analytic signals of the components of the electric field of a given wave, at a given point **r**, are zero mean variables that can be considered stochastic processes whose complete statistical description, equivalent to the bivariate joint probability density function for the two real components of the electric field of the wave, requires in general the knowledge of all their n-order moments. In the particular case of waves with a Gaussian spectral profile, as occurs with thermal light, the second-order moments are sufficient (Brosseau 1998). Nevertheless, there are important cases in practice where higher-order moments play an important role, especially for radiation emitted by certain artificial sources as well as in the quantum domain. In the second-order approach, polarization refers to the second-order moments of the analytic signals $(\varepsilon_1, \varepsilon_2)$ at a given fixed point in space.

Besides the essential statistical nature of physical phenomena producing electromagnetic radiation, the medium in which the field propagates may also exhibit random features. Furthermore, the detection procedure involves finite spatial and temporal intervals that necessarily affect the measured quantities (a very simple example is given by the ideal concept of monochromatic wave, which, despite the physical impossibility of generating it, it cannot be detected in strict sense because it would require infinite measurement time). Therefore, statistics play a key role in the description of polarized light.

For readers who are not familiar with the basic statistical concepts involved in polarization theory, a brief summary of them is presented below. For more comprehensive treatises we redirect the reader to Mandel and Wolf (1995) and Goodman (2015).

Consider a *random experiment* (an experiment whose outcome cannot be predicted) whose possible different outcomes are numbers (these outcomes being taken as mutually exclusive). Such collection of mutually exclusive elementary events can be regarded as the values of a certain variable x, which is known as a *random variable* or *variate*. When the possible values of x can be arranged as a countable set of numbers, $x_1, x_2, x_3 \ldots$, then x is said to be a *discrete random variable*. For each possible outcome x_i of a discrete variable there corresponds a probability p_i, so that the probabilities sum to one, $\sum p_i = 1$.

1.5.1.1 Stochastic (or Random) Process

When the random values of the variable x are no longer countable and form a continuum within a certain interval, the concept of discrete random variable is generalized to that of a random or stochastic process or function. Therefore, to appropriately formulate the mathematical description, it is introduced a continuous parameter t (such as time), that labels the variates. $x(t)$ is called a *random process* or a *random function of t*.

Since the dependence on t of $x(t)$ is not deterministic, its values can only be described in a statistical manner by means of a *probability distribution* or *probability density*, so that for each value of t, $x(t)$ is a random variable within a certain domain, with probability density $p(x,t)$, which satisfies $\int p(x,t)dx = 1$. Note that because of the dependence on t, $p(x,t)$ represents an infinite family of probability distributions (instead of a single one). The set of all variates x at all times t constitutes the random process $x(t)$.

The concept of random process is necessary for the description of variables whenever fluctuations are present, and therefore it is applied to a great variety of physical phenomena, including optical coherence and polarization (Mandel and Wolf 1995).

1.5.1.2 Ensemble Average

The expectation value of x at time t is given by $\langle x(t) \rangle = \int x(t)p(x,t)\,dx$. Moreover, the set of all possible realizations or samples of $x(t)$ can alternatively be considered as the random process. The *ensemble* of $x(t)$ is defined as the collection of all possible realizations.

In practice, a sample function (or realization) describes the outcome of a measurement as a function of time t. Nevertheless, the repetition of the trial will generally produce realizations that are different each other and that can be labeled successively $x^{(1)}(t)$, $x^{(2)}(t)$ … Then, the expectation of x at time t can be obtained by averaging over the ensemble of all realizations, so that the expressions below constitute equivalent definitions of the *ensemble average*

$$\langle x(t)\rangle = \lim_{N\to\infty}\frac{1}{N}\sum_{r=1}^{N} x^{(r)}(t) = \int x(t)p(x,t)dx \tag{1.25}$$

1.5.1.3 Stationarity

In many physical situations the random functions of time have the property that the nature of the fluctuations does not change with time. That is to say, when all possible probability distributions involving the random variable are invariant under arbitrary translations of the origin of time, the process is called *statistically stationary* (also *strictly stationary, strongly stationary* or *strict-sense stationary*).

A weaker form of stationarity, called *stationarity in the wide sense* (or *weak stationarity*), is also considered for stochastic processes for which the mean is independent on time and the time dependence of the second-order correlation function (or *covariance function*) is realized only through the difference $\tau = t_2 - t_1$, that is, both the mean and the correlation function do not change by shifts in time.

1.5.1.4 Ergodicity

There are many practical cases in which every realization of the ensemble holds the same statistical information about the stationary random process as any other realization. In such cases the time averages of the different realizations are all equal and coincide with the ensemble average, and the stochastic process is then said to be *ergodic*. Thus, any single realization contains all relevant statistical information and can be taken as representative. For a ergodic stochastic process all realizations are statistically similar and only differ in detail.

Ergodicity implies stationarity, while the converse is not true. Moreover, provided the second-order moment is finite, strong stationarity implies weak stationarity (Bendat and Piersol 2010). In certain cases ergodicity is defined with subtle differences with respect to the standard one, so that it is common to find the term *stationary* and *ergodic*, which emphasizes the genuine properties associated with the separate definitions of both stationarity and ergodicity.

Anyway, as pointed out by Wolf (1982), most, if not all, stationary random processes that occur in nature are ergodic. In what follows, throughout the entire book unless otherwise indicated, electromagnetic fields will be considered stationary and ergodic processes.

1.5.1.5 Gaussian random processes

Random processes involving two or more normally (Gaussian) distributed random variables are quite frequent in the description of physical situations. A stationary Gaussian process is necessarily ergodic. All statistical properties of a Gaussian process are determined by the mean and the second-order correlation function. Thus, weak stationarity of a Gaussian process implies strong stationarity. Moreover, a linear transformation of a Gaussian leads to a Gaussian process. Furthermore, the Fourier transform of a Gaussian function is again a Gaussian function. All the above concepts are straightforwardly extended to stochastic processes involving n variables, as for instance 2D polarization states ($n = 2$), 3D polarization states ($n = 3$) and Mueller states ($n = 4$, see Chapter 5).

Regarding electromagnetic waves, there are important cases, as for instance light emitted by thermal sources (such as incandescent matter or gas discharges), where the components E_x and E_y of the electric field can be considered temporally stationary Gaussian random processes with equal power spectrum, in which case the corresponding analytic signals (i.e., the components of the instantaneous Jones vector), satisfy the same statistical properties as those of E_x and E_y (Barakat 1985).

1.5.2 2D Polarization Matrix

A proper description of the second-order polarization properties (at a given point **r** in space) relies on the concept of *polarization matrix* (or *coherency matrix*) $\mathbf{\Phi}$ defined as (Wiener 1930; Wolf 1959; Barakat 1963)

$$\mathbf{\Phi} = \langle \boldsymbol{\varepsilon}(t) \otimes \boldsymbol{\varepsilon}^{\dagger}(t) \rangle = \begin{pmatrix} \langle \varepsilon_1(t)\varepsilon_1^*(t)\rangle & \langle \varepsilon_1(t)\varepsilon_2^*(t)\rangle \\ \langle \varepsilon_2(t)\varepsilon_1^*(t)\rangle & \langle \varepsilon_2(t)\varepsilon_2^*(t)\rangle \end{pmatrix} \tag{1.26}$$

where $\boldsymbol{\varepsilon}$ is the 2D analytic signal vector of the wave, $\otimes$ stands for the Kronecker product, and the brackets indicate time averaging over the measurement time

$$\langle x(t)\rangle = \lim_{T\to\infty} \frac{1}{T}\int_0^T x(t)\,dt \tag{1.27}$$

As a result of this definition, $\mathbf{\Phi}$ is a 2×2 covariance matrix (i.e., a positive semidefinite Hermitian matrix) that contains all second-order measurable information on the 2D state of polarization (including intensity). Under the assumption that the stochastic processes $(\varepsilon_1, \varepsilon_2)$ are stationary and ergodic, the brackets can alternatively be considered as ensemble averaging, where $\boldsymbol{\varepsilon}(t)$ in Eq. (1.26) represents a sample realization.

Matrix $\mathbf{\Phi}$ is the equal-point and equal-time correlation matrix of the components of the analytic signal vector $\boldsymbol{\varepsilon}$ (see Section 1.6.1.1). Its diagonal elements are the averaged intensities associated with each of the two components, while its off-diagonal elements are measures of the equal-time correlations between the mutually orthogonal analytic signals ε_1 and ε_2 at point **r**.

It is sometimes advantageous that the stationary fluctuating fields are presented and analyzed in the frequency domain rather than in the time domain. The *spectral polarization matrix* is defined from the Fourier transform of the equal-point mutual coherency matrix (see Section 1.6) and can always be expressed as

$$\mathbf{\Phi}(\omega) = \langle \boldsymbol{\varepsilon}(\omega) \otimes \boldsymbol{\varepsilon}^{\dagger}(\omega) \rangle_{\omega} = \begin{pmatrix} \langle \varepsilon_1(\omega)\varepsilon_1^*(\omega)\rangle_{\omega} & \langle \varepsilon_1(\omega)\varepsilon_2^*(\omega)\rangle_{\omega} \\ \langle \varepsilon_2(\omega)\varepsilon_1^*(\omega)\rangle_{\omega} & \langle \varepsilon_2(\omega)\varepsilon_2^*(\omega)\rangle_{\omega} \end{pmatrix} \tag{1.28}$$

where the field variables are stochastic processes and $\langle\ \rangle_{\omega}$ denotes ensemble averaging over the ensemble of all *equivalent monochromatic realizations* of the field components at frequency ω. It is important to keep in mind that, while this kind of average results in a function of the same variable ω as that of the individual stochastic processes, $\mathbf{\Phi}(\omega)$ should not be confused with the polarization matrix of a monochromatic wave; in words of Wolf (2007): "The distinction between a monochromatic field and an ensemble of monochromatic fields of the same frequency is crucial..."

The corresponding polarization matrix is then obtained as

$$\mathbf{\Phi}=\int_0^{\infty}\mathbf{\Phi}(\omega)\,d\omega \tag{1.29}$$

It is remarkable that the polarization matrix can alternatively be defined as in Eq. (1.38) from the very phenomenological definition of the Stokes parameters (see Section 1.5.3), and therefore the concept of $\mathbf{\Phi}$ is applicable to any kind of 2D polarization state with arbitrary spectral profile. The definition of the polarization matrix will be generalized in Chapter 2 for the general case of 3D polarization states, for which the components of the electric field do not necessarily evolve in a fixed plane.

Regardless of the underlying approach considered (either time or frequency averaging, or phenomenological definition) $\mathbf{\Phi}$ has always the form of a covariance matrix (i.e., $\mathbf{\Phi}$ is a positive semidefinite Hermitian matrix) and therefore its two eigenvalues are non-negative. Such eigenvalue constraints constitute a complete set of necessary and sufficient conditions for a Hermitian matrix $\mathbf{\Phi}$ to be formally a polarization matrix, i.e., to represent a particular 2D state of polarization of an electromagnetic wave at a given point in space.

The elements ϕ_{ij} $(i,j=1,2)$ of $\mathbf{\Phi}$ can be written as follows in terms of the corresponding standard deviations σ_1,σ_2 (the diagonal elements of $\mathbf{\Phi}$, σ_1^2 and σ_2^2, being the corresponding variances) and the complex degree of mutual coherence μ

$$\mathbf{\Phi}=\begin{pmatrix}\sigma_1^2 & \mu\,\sigma_1\sigma_2\\ \mu^*\sigma_1\sigma_2 & \sigma_2^2\end{pmatrix} \tag{1.30}$$

where

$$\sigma_1^2\equiv\phi_{11}=\langle|\varepsilon_1(t)|^2\rangle \qquad \sigma_2^2\equiv\phi_{22}=\langle|\varepsilon_2(t)|^2\rangle \qquad \mu=\phi_{12}\big/\sqrt{\phi_{11}\,\phi_{22}} \tag{1.31}$$

For some purposes it is useful to consider the *polarization density matrix* $\hat{\mathbf{\Phi}}\equiv\mathbf{\Phi}/\mathrm{tr}\mathbf{\Phi}$ (i.e., the intensity-normalized polarization matrix), which in terms of the jargon used for density matrices representing quantum states, contains complete information on the *populations* and *coherences* associated with the polarization state (Fano 1953; 1957).

Polarization matrices inherit, as an underlying reference basis, the generalized reference basis $\mathbf{e}_1,\mathbf{e}_2$ used to represent the analytic signals of the two components of the field variables. Unless otherwise stated, $\mathbf{\Phi}$ will be referenced with respect to the underlying canonical basis constituted by the orthonormal set of column vectors $(1,0)^T,(0,1)^T$ (where the superscript T stands for transposition). Moreover, regardless of the underlying reference basis considered, the polarization matrix $\mathbf{\Phi}$ can be expressed as a linear expansion, with real coefficients, on the following matrix basis constituted by the three Pauli matrices plus the identity matrix:

$$\boldsymbol{\sigma}_0=\begin{pmatrix}1&0\\0&1\end{pmatrix} \quad \boldsymbol{\sigma}_1=\begin{pmatrix}1&0\\0&-1\end{pmatrix} \quad \boldsymbol{\sigma}_2=\begin{pmatrix}0&1\\1&0\end{pmatrix} \quad \boldsymbol{\sigma}_3=\begin{pmatrix}0&-i\\i&0\end{pmatrix} \tag{1.32}$$

Observe that notations σ_1, σ_2 (plain letters) are used for the variances in Eq. (1.30) and $\boldsymbol{\sigma}_i$ (bold letters) are used for the Pauli matrices (in order to preserve the common notations used in related works), this should not lead to confusion because $\boldsymbol{\sigma}_i$ are matrices while the variances σ_1, σ_2 are scalar quantities. Note also that, when dealing with certain formulations involving Pauli matrices in quantum physics and quantum optics, the Pauli matrices $\boldsymbol{\sigma}_2$ and $\boldsymbol{\sigma}_3$ are frequently labeled as $\boldsymbol{\sigma}_3$ and $\boldsymbol{\sigma}_2$ respectively (Sheppard et al. 2016b; 2017).

The set of linearly independent matrices $\boldsymbol{\sigma}_i$ has well known interesting properties as hermiticity, $\boldsymbol{\sigma}_i = \boldsymbol{\sigma}_i^\dagger$; trace-orthogonality, $\mathrm{tr}(\boldsymbol{\sigma}_i \boldsymbol{\sigma}_j) = 2\delta_{ij}$ (δ_{ij} being the Kronecker delta), and $\boldsymbol{\sigma}_i^2 = \boldsymbol{\sigma}_0$. Therefore $\boldsymbol{\sigma}_i$ are also unitary and, except for $\boldsymbol{\sigma}_0$, are traceless.

Note finally that for any polarization matrix $\boldsymbol{\Phi}_p$ associated with a totally polarized state, there is always an *equivalent Jones vector* $\boldsymbol{\varepsilon}$ that satisfies $\boldsymbol{\Phi}_p = \boldsymbol{\varepsilon} \otimes \boldsymbol{\varepsilon}^\dagger$, regardless of the spectral profile of the electromagnetic wave at the point $\mathbf{r}$ considered.

1.5.3 Stokes Vector

As mentioned above, it is straightforward to show that $\boldsymbol{\Phi}$ always admits the following linear expansion (Falkoff and Macdonald 1951; Fano 1953):

$$\boldsymbol{\Phi} = \frac{1}{2} \sum_{i=0}^{3} s_i \boldsymbol{\sigma}_i \tag{1.33}$$

whose real coefficients s_i are given by

$$s_i = \mathrm{tr}(\boldsymbol{\Phi} \boldsymbol{\sigma}_i) \qquad (i = 0,1,2,3) \tag{1.34}$$

or, in the explicit form

$$s_0 = \phi_{11} + \phi_{22} \qquad s_1 = \phi_{11} - \phi_{22} \qquad s_2 = \phi_{12} + \phi_{21} \qquad s_3 = i(\phi_{12} - \phi_{21}) \tag{1.35}$$

The quantities s_0, s_1, s_2, s_3 are the so-called *Stokes parameters* (Stokes 1852) and constitute a complete set of measurable parameters. In fact, the 2D Stokes parameters are observables that admit a direct phenomenological definition from a series of experimental measurements performed through a Stokes polarimeter, namely (Stokes 1852; Wiener 1930; Clarke and Grainger 1974)

- s_0 is the intensity, given by the sum of the intensities associated with the components of the electric field with respect to any orthonormal generalized basis: $s_0 = I_x + I_y = I_{+45^\circ} + I_{-45^\circ} = I_r + I_l = I_{\mathbf{e}_1} + I_{\mathbf{e}_2}$.
- s_1 is the difference between the intensities corresponding to the components of the electric field with respect to the canonical basis $(\mathbf{e}_x, \mathbf{e}_y)$ (see Eq. (1.18)), $s_1 = I_x - I_y$, with $s_1 = 1$ ($\varphi = 0$, $\chi = 0$) for linear-horizontal polarized states and $s_1 = -1$ ($\varphi = \pi/2$, $\chi = 0$) for linear-vertical polarized states. A simple procedure for the measurement of the parameter s_1 of a plane wave consists of two consecutive intensity measurements (I_x and I_y) by placing a linear polarizer (also called analyzer) at 0° and 90° before the detector.
- s_2 is the difference between the intensities corresponding to the components of the electric field with respect to the basis $(\mathbf{e}_{+\pi/4}, \mathbf{e}_{-\pi/4})$ (see Eq. (1.21)), $s_2 = I_{+\pi/4} - I_{-\pi/4}$, with

$s_2 = 1$ ($\varphi = \pi/4$, $\chi = 0$) for +45° linearly polarized states and $s_2 = -1$ ($\varphi = 3\pi/4$, $\chi = 0$) for –45° linearly polarized states. A simple procedure for the measurement of the parameter s_2 of a plane wave consist of two consecutive intensity measurements ($I_{+\pi/4}$ and $I_{-\pi/4}$) by placing a linear polarizer at +45° and –45° before the detector.

- s_3 is the difference between the intensities corresponding to the components of the electric field with respect to the basis $(\mathbf{e}_r, \mathbf{e}_l)$ (see Eq. (1.22)), $s_3 = I_r - I_l$, with $s_3 = 1$ for right handed circular-polarized states and $s_3 = -1$ for left handed circularly-polarized states. The handedness σ is determined by the sign of s_3: $\sigma \equiv s_3/|s_3|$ ($\sigma = +1$ for right-handed states and $\sigma = -1$ for left-handed states). A simple procedure for the measurement of the parameter s_3 of a plane wave consists of two consecutive intensity measurements (I_r and I_l) by placing a right-circular analyzer and a left-circular analyzer before the detector. Note that a circular polarizer can be achieved by the serial combination of a linear polarizer and a quarter-wave plate whose respective eigenaxes make an angle of 45°.

As pointed out by Barakat (1963), the work of Wiener (who was not aware of the previous Stokes findings) includes an operational procedure for the experimental determination of the Stokes parameters (obviously Wiener did not use that name) and predated the well-known work of Fano (1957) showing the connection between the Stokes parameters, polarization matrix, and the Pauli spin matrices.

From the phenomenological definition of the Stokes parameters, it follows that a set of four real parameters s_0, s_1, s_2, s_3 can formally be considered a set of Stokes parameters if and only if they satisfy the inequalities

$$s_0 \geq 0 \qquad s_0^2 - s_1^2 - s_2^2 - s_3^2 \geq 0 \tag{1.36}$$

Even though the indicated notation is commonly used in many related works, it is important to warn the reader that the Stokes parameters are also frequently denoted as I, Q, U, V or I, M, C, S.

The Stokes parameters are usually arranged as a 4 × 1 *Stokes vector* **s**

$$\mathbf{s} \equiv (s_0, s_1, s_2, s_3)^T \tag{1.37}$$

(when appropriate, Stokes vectors and other column vectors are expressed in the horizontal notation used above).

The polarization matrix $\mathbf{\Phi}$ associated with a given set of Stokes parameters s_0, s_1, s_2, s_3 (which are always well defined for any 2D polarization state in terms of measurable intensities) can always be written as follows in terms of phenomenological observable quantities:

$$\mathbf{\Phi} = \frac{1}{2}\begin{pmatrix} s_0 + s_1 & s_2 - i s_3 \\ s_2 + i s_3 & s_0 - s_1 \end{pmatrix} \tag{1.38}$$

The non-negativity of $\mathbf{\Phi}$ (i.e., the positive semidefiniteness of $\mathbf{\Phi}$) is directly derived from the following characteristic constraints of **s**:

$$\mathrm{tr}\mathbf{\Phi} = s_0 \geq 0 \qquad 4\det\mathbf{\Phi} = s_0^2 - s_1^2 - s_2^2 - s_3^2 \geq 0 \tag{1.39}$$

For further calculations it is useful to note that the relation between $\mathbf{\Phi}$ and $\mathbf{s}$ can also be expressed as

$$\mathbf{s} = \sqrt{2}\boldsymbol{\mathcal{L}}\boldsymbol{\varphi} \tag{1.40a}$$

where $\boldsymbol{\mathcal{L}}$ is the matrix

$$\boldsymbol{\mathcal{L}} = \frac{1}{\sqrt{2}}\begin{pmatrix} 1 & 0 & 0 & 1 \\ 1 & 0 & 0 & -1 \\ 0 & 1 & 1 & 0 \\ 0 & i & -i & 0 \end{pmatrix} \tag{1.40b}$$

and the *polarization covariance vector* $\boldsymbol{\varphi}$ is defined as the column vector whose components are the elements of the polarization matrix arranged in the following manner:

$$\boldsymbol{\varphi} \equiv (\varphi_0, \varphi_1, \varphi_2, \varphi_3)^T = (\phi_{11}, \phi_{12}, \phi_{21}, \phi_{22})^T \tag{1.41}$$

Eq. (1.35) shows that, obviously, the information contained in $\mathbf{s}$ (or $\boldsymbol{\varphi}$) is completely equivalent to that provided by $\mathbf{\Phi}$.

It should be noted that the term "vector" is used for the 4-tuple Stokes vectors in a very wide sense. The multiplication of a Stokes vector $\mathbf{s}$ by a real scalar c, produces a Stokes vector $\mathbf{s}' = c\,\mathbf{s} = \mathbf{s}\,c$ if and only if $c \geq 0$. The resultant Stokes vector $\mathbf{s}'$ represents the same state of polarization as $\mathbf{s}$ up to a positive scale factor that only affects the intensity, $I(\mathbf{s}') = c\,I(\mathbf{s})$. Moreover, in general, the addition of two Stokes vectors $\mathbf{s}_1$ and $\mathbf{s}_2$ only has physical meaning in the form $\mathbf{s} = \mathbf{s}_1 + \mathbf{s}_2$, and not as a subtraction $\mathbf{s}_1 - \mathbf{s}_2$. In fact, the addition represents the incoherent superposition of two 2D states whose respective polarization ellipses lie in a common plane (the general case where the polarization planes of the components do not coincide will be studied in Chapter 2). The intensity of the resultant Stokes vector is given by the sum of the intensities of the superimposed states $I(\mathbf{s}) = I(\mathbf{s}_1) + I(\mathbf{s}_2)$. There are certain situations where the subtraction can be physically admissible, as for instance the superposition represented by $\mathbf{s} = \mathbf{s}_1 + \mathbf{s}_2$, where the Stokes vector $\mathbf{s}_1$ can be considered as the result of the *polarimetric subtraction* $\mathbf{s}_1 = \mathbf{s} - \mathbf{s}_2$ in the sense that $\mathbf{s}_2$ is a Stokes vector that added to $\mathbf{s}_1$ gives a Stokes vector $\mathbf{s} = \mathbf{s}_1 + \mathbf{s}_2$, which represents the incoherent superposition of $\mathbf{s}_1$ and $\mathbf{s}_2$. The polarimetric subtraction is a relevant concept in polarimetry which will be dealt with in Chapter 7.

Obviously, since negative intensities have no physical meaning, given a Stokes vector $\mathbf{s} \neq \mathbf{0}$, it does not have an inverse Stokes vector with respect to the "+" operation. Consequently, the set of Stokes vectors $\{(s_0, s_1, s_2, s_3)^T \,|\, s_0 \geq 0,\ s_0^2 \geq s_1^2 + s_2^2 + s_3^2\}$ together with the product by a non-negative scalar and the sum constitutes a semi-ring algebraic structure, and not a vector space (Arnal 1990).

Even though the use of generalized bases for the representation of Jones vectors is not unusual, Stokes vectors are always considered as referenced with respect to the underlying canonical real basis $(\mathbf{e}_x, \mathbf{e}_y)$, i.e., to a laboratory coordinate system XYZ, Z being the direction of propagation.

A change of reference frame from XY to $X'Y'$ for the representation of Stokes vectors is performed by means of a *rotation transformation* of the form

$$\mathbf{s}' = \mathbf{M}_G(\theta)\,\mathbf{s} \qquad \mathbf{M}_G(\theta) \equiv \begin{pmatrix} 1 & 0 & 0 & 0 \\ 0 & \cos 2\theta & \sin 2\theta & 0 \\ 0 & -\sin 2\theta & \cos 2\theta & 0 \\ 0 & 0 & 0 & 1 \end{pmatrix} \tag{1.42}$$

where the orthogonal matrix $\mathbf{M}_G(\theta)$ corresponds to a proper counterclockwise rotation by the angle θ about the axis Z, from the original X reference axis to X'. Note that the Stokes vectors $\mathbf{s}$ and $\mathbf{s}' = \mathbf{M}_G(\theta)\,\mathbf{s}$ in Eq. (1.42) represent the same state of polarization, but are referenced with respect to different coordinate systems (XY and $X'Y'$respectively). Thus, when the coordinate system is fixed and are the components of the electric field which are rotated by a counterclockwise angle θ by the action of a medium (a rotator), such transformation from the incident state, $\mathbf{s}$, to the emerging one, $\mathbf{s}'$, is performed by $\mathbf{s}' = \mathbf{M}_G(-\theta)\,\mathbf{s}$, so that $\mathbf{M}_G(\theta)$ can be interpreted as the Mueller matrix of a rotator performing a clockwise rotation of angle θ of the components of the electric filed of the state.

A Stokes vector $\mathbf{s}_p$ satisfying

$$\mathbf{s}_p^T \mathbf{G}\,\mathbf{s}_p = s_{p0}^2 - s_{p1}^2 - s_{p2}^2 - s_{p3}^2 = 0 \left[\mathbf{G} \equiv \operatorname{diag}(1,-1,-1,-1)\right] \tag{1.43}$$

corresponds to a totally polarized state and is called a *totally polarized* or *pure Stokes vector*. The matrix $\mathbf{G}$ represents the Minkowskian metric.

Two pure Stokes vectors $\mathbf{s}_1$, $\mathbf{s}_2$ are said to be mutually orthogonal when their corresponding Jones vectors $\boldsymbol{\varepsilon}_1$, $\boldsymbol{\varepsilon}_2$ are mutually orthogonal ($\boldsymbol{\varepsilon}_1^\dagger \boldsymbol{\varepsilon}_2 = 0$), so that the mutual orthogonality of $\mathbf{s}_1$ and $\mathbf{s}_2$ is expressed by the fact that the scalar product of $\mathbf{s}_1$ and $\mathbf{s}_2$ is zero, $\mathbf{s}_1^T \mathbf{s}_2 = 0$. In other words, a pure Stokes vector $\mathbf{s}_{p1} = (s_0, s_1, s_2, s_3)^T$ is said to be orthogonal to another pure Stokes vector $\mathbf{s}_{p2}$ when $\mathbf{s}_2^T = \mathbf{G}\,\mathbf{s}_1^T = (s_0, -s_1, -s_2, -s_3)^T$.

Multiplication by a positive scalar and additive compositions translate directly from the space of Stokes vectors to the space of 2D polarization matrices, and consequently both formalisms are completely equivalent with regard to their physical interpretation.

A Stokes vector can always be expressed as

$$\begin{gathered} \mathbf{s} = \mathbf{s}_p + \mathbf{s}_u \\ \mathbf{s}_p \equiv \left(\sqrt{s_1^2 + s_2^2 + s_3^2}, s_1, s_2, s_3\right)^T \qquad \mathbf{s}_u \equiv \left(s_0 - \sqrt{s_1^2 + s_2^2 + s_3^2}, 0, 0, 0\right)^T \end{gathered} \tag{1.44}$$

so that $\mathbf{s}$ can be interpreted as an incoherent superposition of a pure state (first addend, hereafter called the *characteristic component*) and an unpolarized state (hereafter called the *2D unpolarized component*). The characteristic component defines the *average polarization ellipse* (or *characteristic polarization ellipse*) of the whole state $\mathbf{s}$. Furthermore, $\mathbf{s}$ can be parameterized as

$$\mathbf{s} = I \begin{pmatrix} 1 \\ \mathcal{P}\cos 2\varphi \cos 2\chi \\ \mathcal{P}\sin 2\varphi \cos 2\chi \\ \mathcal{P}\sin 2\chi \end{pmatrix} = \mathbf{s}_p + \mathbf{s}_u \qquad \left[\mathbf{s}_p = I\mathcal{P} \begin{pmatrix} 1 \\ \cos 2\varphi \cos 2\chi \\ \sin 2\varphi \cos 2\chi \\ \sin 2\chi \end{pmatrix} \quad \mathbf{s}_u = I(1-\mathcal{P}) \begin{pmatrix} 1 \\ 0 \\ 0 \\ 0 \end{pmatrix} \right] \tag{1.45}$$

in terms of the following measurable quantities:

Intensity. The intensity $I = s_0$, is a measure of the average power of the electric field of the electromagnetic wave at the point **r** considered.

Degree of polarization. $\mathcal{P} \equiv \sqrt{s_1^2 + s_2^2 + s_3^2}/s_0$ is the *degree of polarization* of the 2D state represented by **s**. $\mathcal{P}$ is a dimensionless quantity whose values are restricted to $0 \leq \mathcal{P} \leq 1$. The maximum $\mathcal{P} = 1$ corresponds to totally polarized states, which, therefore, can also be represented by respective Jones vectors. Intermediate values $0 < \mathcal{P} < 1$ correspond to partially polarized states; the higher the value of $\mathcal{P}$, the higher the correlation (or mutual coherence) of the field components. The minimum $\mathcal{P} = 0$ corresponds to *2D unpolarized states*, i.e., to states with a completely random temporal distribution of the polarization ellipse (but with fixed polarization plane), or in other words, to states with zero correlation between the two analytic signals of the field components.

Azimuth. The azimuth φ, with $0 \leq \varphi < \pi$, of the direction of the major semiaxis of the characteristic polarization ellipse with respect to the given reference axis X.

Ellipticity. The ellipticity angle χ, with $-\pi/4 \leq \chi \leq \pi/4$, of the characteristic polarization ellipse.
At this point, it is worth summarizing certain concepts related to the above analysis, some of which will be studied in more detail in further sections.
2D unpolarized states entail the equality $I_1 = I_2$ of the intensities associated with the pair of orthogonal components of the electric field with respect to any generalized orthonormal basis $\mathbf{e}_1$, $\mathbf{e}_2$. As pointed out in the seminal works of Stokes (1852) and Verdet (1869), the indicated invariance is an essential and characteristic property of unpolarized states. There are various random distributions which correspond to unpolarized waves. As Ellis and Dogariu have shown (2004a), the measurement of the correlations of the Stokes parameters allows for distinguishing between the different types of unpolarized states.

Characteristic decomposition. Any 2D polarization state **s** can be considered the result of the incoherent superposition of the characteristic component $\mathbf{s}_p$ and the 2D unpolarized component $\mathbf{s}_u$. That is to say, a 2D state represented by a given Stokes vector $\mathbf{s} \equiv (I, s_1, s_2, s_3)^T$ is polarimetrically indistinguishable from an incoherent combination of two 2D states propagating in the same direction, namely, a pure state $\mathbf{s}_p$, with intensity $I_p = \mathcal{P}I$, and a 2D unpolarized state with intensity $I_u = (1-\mathcal{P})I$. For pure states, which are characterized by the equality $\mathcal{P} = 1$, the total intensity is associated with the characteristic component and the shape of the polarization ellipse is constant for time intervals larger than the measurement time. 2D unpolarized states satisfy the equality $\mathcal{P} = 0$, so that the total intensity is associated with the unpolarized component.

Equivalent characteristic Jones vector. While a partially polarized state **s** cannot have an associated Jones vector, the characteristic component $\mathbf{s}_p$ of **s**, being itself a pure state, has an associated *equivalent characteristic Jones vector* $\boldsymbol{\varepsilon}(\mathbf{s}_p)$ regardless of the spectral profile of the wave to which the Stokes vector **s** is associated at the point **r** in space considered. That is, even for polychromatic fields for which there is no well defined instantaneous Jones vector, $\boldsymbol{\varepsilon}(\mathbf{s}_p)$ is always well defined for their pure characteristic component.

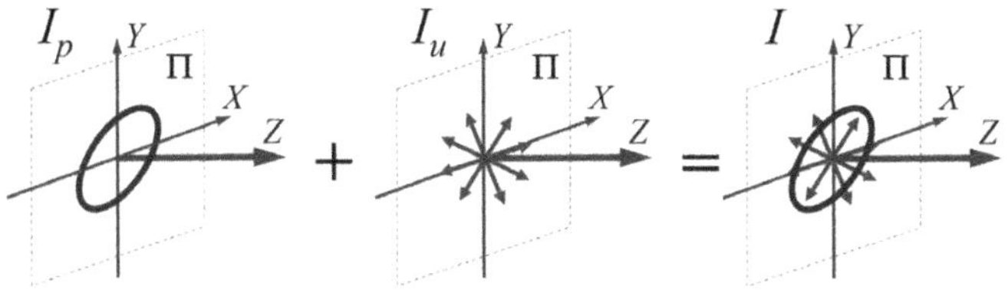

FIGURE 1.10
The degree of polarization $\mathcal{P}$ of a 2D state is defined as the ratio of the intensity I_p of the totally polarized component to the intensity $I = I_p + I_u$ of the entire state.

Characteristic interpretation of the degree of polarization. $\mathcal{P}$ is just the ratio of the intensity $I_p = \mathcal{P}I$ of the characteristic component to the intensity $I = I_p + I_u$ of the entire state (Figure 1.10). Thus, $\mathcal{P}$ is a dimensionless and non-negative quantity limited by the double inequality $0 \le \mathcal{P} \le 1$. Moreover, $\mathcal{P}$ is invariant with respect to any rotation of the underlying coordinate system XY about the direction of propagation Z. Furthermore, from a more general point of view, $\mathcal{P}$ is invariant with respect to any change of the generalized underlying reference basis $(\mathbf{e}_1, \mathbf{e}_2)$, i.e., with respect to any unitary transformation of the basis vectors $(\mathbf{e}_1, \mathbf{e}_2)$, which in turn can be performed by passing the light beam through a retarder (see Chapter 4).

Degree of depolarization. An alternative formulation of the polarimetric purity of a 2D state of polarization is given by the *randomness density*, or *degree of depolarization* defined as $\mathcal{D} \equiv \sqrt{1-\mathcal{P}^2}$, which is a measure of the randomness of the polarization ellipse. Obviously, $\mathcal{D}^2 + \mathcal{P}^2 = 1$ and therefore $0 \le \mathcal{D} \le 1$, with $\mathcal{D} = 0$ for totally polarized states and $\mathcal{D} = 1$ for unpolarized states. Note that the quantity $\mathbf{s}^T\mathbf{G}\,\mathbf{s} = I^2\mathcal{D}^2$ (which is intensity-dependent and has the same dimensions as I^2) is called the *mean randomness* (Barakat 1987a)

Characteristic polarization ellipse. The characteristic component determines the corresponding *characteristic polarization ellipse* with semiaxes a, b given by

$$a^2 = \frac{1}{2}\left(\sqrt{s_1^2 + s_2^2 + s_3^2} + \sqrt{s_1^2 + s_2^2}\right) \qquad b^2 = \frac{1}{2}\left(\sqrt{s_1^2 + s_2^2 + s_3^2} - \sqrt{s_1^2 + s_2^2}\right) \tag{1.46}$$

and whose area is $A = (\pi/2)|s_3| = (\pi/2)I\mathcal{P}|\sin 2\chi|$

Helicity and spin vector of a 2D polarization state. The helicity of a polarization state is given by the Stokes parameter s_3, which is directly linked to the ellipticity angle χ of the characteristic polarization ellipse of the state, $s_3 = I\mathcal{P}\sin 2\chi$, and also coincides with the component of the spin vector in the direction of propagation (Cameron et al. 2012; Crimin et al. 2019) (Gil et al. 2021; Figure 1.11). Since the state of polarization corresponds to a given point $\mathbf{r}$ in space, the information provided by s_3 is directly related to the *intrinsic angular momentum* or *spin* (despite the possibility of considering the helical nature of a complete or partial spatial region of the wavefront and its associated *orbital angular momentum*, Gori et al. 1998; Cameron et al. 2012). The concept of spin of a polarization state is analyzed in more detail in Section 1.10.

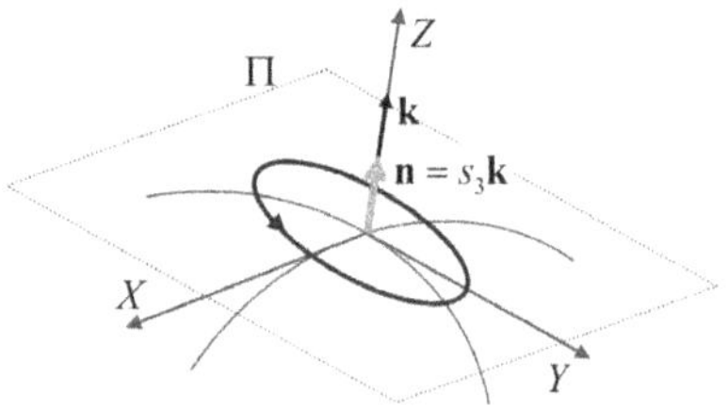

FIGURE 1.11
The absolute value of the Stokes parameter s_3 gives the magnitude of the spin vector **n**, which lies along the axis Z perpendicular to the wavefront.

1.6 2D Space–Time and Space–Frequency Representations of Coherence and Polarization

Electromagnetic waves exhibit randomness due to the random fluctuations associated with the spontaneous or stimulated emission of photons by matter, as well as to random fluctuations in the propagation medium. The degree of correlation of the emission processes caused by myriads of atoms or molecules of the source material located closely to each other leads to a certain correlation of the electric field variables of the emitted electromagnetic wave. The correlation between the analytic signals $\varepsilon_1(\mathbf{r},t)$ and $\varepsilon_2(\mathbf{r},t)$ of the field components of the wave, at a given point **r** and at a given time t, is determined by the degree of mutual coherence μ (see Eqs. (1.30) and (1.31)), which in turn is completely determined by the polarization matrix $\mathbf{\Phi}$ or by the Stokes vector **s**. The correlation between $\varepsilon_1(\mathbf{r}_1,t)$ and $\varepsilon_2(\mathbf{r}_2,t+\tau)$ at different points $\mathbf{r}_1$ and $\mathbf{r}_2$, and at different times t and $t+\tau$, is the origin of interference phenomena that produce fringe patterns in Young's experiments. Therefore, when considering an electromagnetic beam with respect to a certain spatial and temporal domain (and not at a single given point and at equal time as with polarization) the concept of coherence is linked to the above mentioned correlation and fringe visibility. From the assumption of stationarity, the time difference $t_2-t_1\equiv\tau$ is characterized through the single parameter τ, while separated variables t_1 and t_2 have to be considered for non-stationary fields.

1.6.1 2D Representations of Coherence and Polarization

Electromagnetic fields behave generally as ergodic stochastic processes and, as indicated above, the usual formulation of polarization by means of 2D polarization matrices and Stokes vectors corresponds to second-order correlations between the analytic signals $\varepsilon_1(\mathbf{r},t)$ and $\varepsilon_2(\mathbf{r},t)$ of the components of the random electric field vector at a fixed point **r** in space and at a given instant t of time (recall that the said analytic signals are well defined even for polychromatic waves and not exclusively for quasimonochromatic ones, provided the conditions of stationarity and square-integrability indicated in Section 1.3 are satisfied).

Moreover, while polarization refers to the evolution in time of the end point of the electric field of the wave, coherence refers to the correlations between the amplitudes of the electromagnetic modes at different points and times, as well as to their ability to interfere (Glauber 1963).

When the analytic signals $\varepsilon_1(\mathbf{r},t)$ and $\varepsilon_2(\mathbf{r},t)$ have different spatial coherence properties, both the spectral profile and the state of polarization changes upon propagation (James

1994), so that the characterization of the polarization properties of the beam requires the consideration of correlations of the analytic signals $\varepsilon_1(\mathbf{r}_1,t)$ and $\varepsilon_2(\mathbf{r}_2,t+\tau)$ taken at different points (thus, carrying the information on the spatial coherence properties) and at different instants of time, with an arbitrary delay τ between them. That is, while the usual equal-point and equal-time formulation gives the description of polarization states at a given point and without consideration of coherence properties beyond the equal-point and equal-time mutual coherence μ, the complete second-order description of the properties of the electromagnetic beam requires the combined treatment of polarization, spatial coherence, and temporal coherence of the electric field signals. Under this wider scope of representation of the coherence and polarization properties, several physical situations can be studied, as for example the case of completely unpolarized but spatially fully coherent electromagnetic beams (Wolf 2003).

In this section we consider both complementary formulations of the theory of coherence and polarization in the space–time domain (Wolf 1954; Gori 1998) and in the space–frequency domain (Wolf 2007). While the space–time formulation appears as a straightforward extension of the equal-point, equal-time, approach presented in previous sections, the space–frequency domain formulation provides particularly interesting results, as the changes of the state of polarization upon propagation (James 1994).

1.6.1.1 Coherence Matrix

As in the previous sections, let us consider a 2D random, wide-sense stationary, electromagnetic wave represented by the analytic signals of the components of the electric field whose second-order correlation properties are characterized by the *coherence matrix* (Wolf 1954; Mandel and Wolf 1995)

$$\mathbf{J}(\mathbf{r}_1,\mathbf{r}_2,\tau) \equiv \langle \boldsymbol{\varepsilon}(\mathbf{r}_1,t) \otimes \boldsymbol{\varepsilon}^{\dagger}(\mathbf{r}_2,t+\tau)\rangle = \begin{pmatrix} \langle \varepsilon_1(\mathbf{r}_1,t)\, \varepsilon_1^{*}(\mathbf{r}_2,t+\tau)\rangle & \langle \varepsilon_1(\mathbf{r}_1,t)\, \varepsilon_2^{*}(\mathbf{r}_2,t+\tau)\rangle \\ \langle \varepsilon_2(\mathbf{r}_1,t)\, \varepsilon_1^{*}(\mathbf{r}_2,t+\tau)\rangle & \langle \varepsilon_2(\mathbf{r}_1,t)\, \varepsilon_2^{*}(\mathbf{r}_2,t+\tau)\rangle \end{pmatrix} \tag{1.47}$$

where $\mathbf{r}_1$ and $\mathbf{r}_2$ are two arbitrary points in space and τ is an arbitrary time shift. It should be kept in mind that this 2D representation is restricted to quasi-plane waves (paraxial approximation), where the electric field of the wave can be considered as evolving in a fixed plane Π at any point $\mathbf{r}$. In general, the local XY reference axes lie in the local polarization plane Π (which is tangent to the wavefront) so that the underlying coordinate system is position-independent. The general three-dimensional case will be dealt with in Chapter 2.

The 2D, equal-point, coherence matrix is therefore defined as

$$\mathbf{J}(\tau) \equiv \langle \boldsymbol{\varepsilon}(t) \otimes \boldsymbol{\varepsilon}^{\dagger}(t+\tau)\rangle = \begin{pmatrix} \langle \varepsilon_1(t)\, \varepsilon_1^{*}(t+\tau)\rangle & \langle \varepsilon_1(t)\, \varepsilon_2^{*}(t+\tau)\rangle \\ \langle \varepsilon_2(t)\, \varepsilon_1^{*}(t+\tau)\rangle & \langle \varepsilon_2(t)\, \varepsilon_2^{*}(t+\tau)\rangle \end{pmatrix} \tag{1.48}$$

where, according to the usual convention, the spatial dependence on the point $\mathbf{r}$ has been omitted. In particular, the 2D polarization matrix $\mathbf{\Phi}$ is given by the equal-point and equal-time 2D coherence matrix: $\mathbf{\Phi} = \mathbf{J}(0)$.

As pointed out by Ellis and Dogariu (2004b), each element of $\mathbf{J}(\mathbf{r}_1,\mathbf{r}_2,\tau)$ relates directly to a measurable visibility of interference fringes after the radiation from the two pinholes of the Young's interference setup has been passed through specific optical elements. For readers interested in the propagation of the coherence–polarization properties of an electromagnetic beam, we refer them to the *unified theory of coherence and polarization* developed by Wolf (2003), which can extensively be used for the study and analysis of the changes in coherence and polarization of electromagnetic beams on propagation (Korotkova 2017).

1.6.1.2 Space–Time Coherence Stokes Vector

The space–time *coherence Stokes vector* (or space–time *two-point Stokes vector*) is defined as (Ellis and Dogariu 2004b)

$$\mathbf{s}(\mathbf{r}_1,\mathbf{r}_2,\tau) \equiv \begin{pmatrix} j_{11}(\mathbf{r}_1,\mathbf{r}_2,\tau)+j_{22}(\mathbf{r}_1,\mathbf{r}_2,\tau) \\ j_{11}(\mathbf{r}_1,\mathbf{r}_2,\tau)-j_{22}(\mathbf{r}_1,\mathbf{r}_2,\tau) \\ j_{12}(\mathbf{r}_1,\mathbf{r}_2,\tau)+j_{21}(\mathbf{r}_1,\mathbf{r}_2,\tau) \\ ij_{12}(\mathbf{r}_1,\mathbf{r}_2,\tau)-ij_{21}(\mathbf{r}_1,\mathbf{r}_2,\tau) \end{pmatrix} \tag{1.49}$$

The space–time polarization properties at a point $\mathbf{r}$ are determined by the corresponding polarization matrix retrieved as $\mathbf{\Phi} = \mathbf{J}(0)$, or by the corresponding Stokes vector $\mathbf{s} = \mathbf{s}\,(\mathbf{r}_1 = \mathbf{r}_2, \tau = 0)$.

1.6.1.3 Spectral Coherence Matrix

In the same way that partial polarization can be formulated in terms of the correlations between the mutually orthogonal field components at a given point $\mathbf{r}$, the theory of partial polarization may also be formulated in the space–frequency domain in order to study the correlations at each frequency (Wolf 2007). In fact, for certain purposes – for instance the analysis of problems regarding propagation and light-media interaction – the simpler frequency-domain expressions make it advantageous to represent the coherence properties of a field through the spectral point of view.

In what follows, in addition to weak stationarity we assume the condition satisfied by any physical field in nature that correlations $\langle \varepsilon_i(\mathbf{r}_1,t)\, \varepsilon_j^*(\mathbf{r}_2,t+\tau)\rangle$ fall off sufficiently rapidly with $|\tau|$, so that $\mathbf{J}(\mathbf{r}_1,\mathbf{r}_2,\tau)$ is absolutely integrable with respect to τ, which in turn implies that $\mathbf{J}(\mathbf{r}_1,\mathbf{r}_2,\tau)$ is square integrable, that is (Wolf 1981)

$$\int_{-\infty}^{\infty} \left|\mathbf{J}(\mathbf{r}_1,\mathbf{r}_2,\tau)\right| d\tau < \infty \qquad \int_{-\infty}^{\infty} \left|\mathbf{J}(\mathbf{r}_1,\mathbf{r}_2,\tau)\right|^2 d\tau < \infty \tag{1.50}$$

and therefore $\mathbf{J}(\mathbf{r}_1,\mathbf{r}_2,\tau)$ possesses the well defined Fourier transform (Mandel and Wolf 1995; Wolf 2003)

$$\mathbf{W}(\mathbf{r}_1,\mathbf{r}_2,\omega) = \frac{1}{2\pi}\int_{-\infty}^{\infty} \mathbf{J}(\mathbf{r}_1,\mathbf{r}_2,\tau)\, e^{i\omega} d\tau \tag{1.51}$$

which is a continuous function of ω and is called the *spectral coherence matrix* (or *cross spectral density matrix*) of the wave with respect to the given frequency ω and points $\mathbf{r}_1$ and $\mathbf{r}_2$.

Eq. (1.51) forms a Fourier transform pair with its inverse (Mandel and Wolf 1995; Tervo et al. 2004)

$$\mathbf{J}(\mathbf{r}_1,\mathbf{r}_2,\tau) = \int_0^{\infty} \mathbf{W}(\mathbf{r}_1,\mathbf{r}_2,\omega)\, e^{-i\omega\tau}\, d\omega \tag{1.52}$$

where the lower limit is zero, and not $-\infty$ because of the use of the analytic signals $\varepsilon_i(\mathbf{r},t)$ $(i=1,2)$ introduced in Section 1.3 to represent the space–time components of the electric field (Mandel and Wolf 1995). Recall that the above relations have been established under the natural assumption that $|\mathbf{J}(\mathbf{r}_1,\mathbf{r}_2,\tau)|$ falls off rapidly with $|\tau|$, so that $\mathbf{J}(\mathbf{r}_1,\mathbf{r}_2,\tau)$ satisfies Eqs. (1.50) (Wolf 1981).

The subtle, but important, fact should be stressed that while $\mathbf{W}(\mathbf{r}_1,\mathbf{r}_2,\omega)$ is obtained through a Fourier transform of the correlation matrix $\mathbf{J}(\mathbf{r}_1,\mathbf{r}_2,\tau)$ (coherence matrix), this would not necessarily ensure that $\mathbf{W}(\mathbf{r}_1,\mathbf{r}_2,\omega)$ is a correlation matrix by itself (Wolf 2007; Korotkova 2017). Nevertheless, by taking advantage of the *coherent mode representation* (Gamo 1964; Wolf 1982), it has been proven that $\mathbf{W}(\mathbf{r}_1,\mathbf{r}_2,\omega)$ is a valid correlation matrix for which the averaging process is carried out with respect to an ensemble of strictly monochromatic realizations $\boldsymbol{\varepsilon}(\mathbf{r},\omega)$ of the electric field (Tervo et al. 2004). Obviously, such space–frequency monochromatic realizations differ from those space–time ones $\boldsymbol{\varepsilon}(\mathbf{r},t)$ used to define $\mathbf{J}(\mathbf{r}_1,\mathbf{r}_2,\tau)$ in Eq. (1.47), which correspond to the field at the given time instant t, while each realization $\boldsymbol{\varepsilon}(\mathbf{r},\omega)$ plays the role of an *equivalent monochromatic field* at frequency ω (Wolf 1982; Korotkova 2017; Tervo et al. 2004). The coherent modes are fully uncorrelated for the case of weak stationary stochastic processes under consideration, while they are partially correlated for the more general case of non-stationary fields.

The elements of the cross-spectral density matrix can be expressed in the following equivalent forms:

$$w_{ij}(\mathbf{r}_1,\mathbf{r}_2,\omega)=\langle\varepsilon_i(\mathbf{r}_1,\omega)\,\varepsilon_j^*(\mathbf{r}_2,\omega)\rangle_\omega=\frac{1}{2\pi}\int_{-\infty}^{\infty}\left\langle\varepsilon_i(\mathbf{r}_1,t)\,\varepsilon_j^*(\mathbf{r}_2,t+\tau)\right\rangle e^{i\omega\tau}\,d\tau \qquad (1.53)$$

where the average $\langle\ldots\rangle_\omega$ is taken over the ensemble of the monochromatic realizations corresponding to the given single frequency ω (Tervo et al. 2004; Alonso and Wolf 2008). Despite the Fourier transform relation between $\mathbf{J}(\mathbf{r}_1,\mathbf{r}_2,\tau)$ and $\mathbf{W}(\mathbf{r}_1,\mathbf{r}_2,\omega)$, it should be stressed that $\varepsilon_i(\mathbf{r},\omega)$ are not Fourier transforms of the space–time realizations $\varepsilon_i(\mathbf{r},t)$.

From a strict point of view within the framework of the theory of ordinary functions, in the case of stationary fields (as assumed), variables $\varepsilon_i(\mathbf{r}_1,t)$ are neither square-integrable nor absolutely integrable, and therefore the Fourier representation in Eq. (1.52) does not exist for stationary fields. Nevertheless, its use is well justified in practice because, as described by Mandel and Wolf (1995), this problem is overcome by means of the generalization of the Wiener's *generalized harmonic analysis* (Wiener 1930) to stochastic processes.

A detailed and rigorous formulation of the cross-spectral density matrix as a correlation matrix and how it as well as its coherent mode components for the given frequency ω satisfy the Helmholtz and Maxwell divergence equations can be found in Tervo et al. 2004.

Regarding the interesting features of the space–frequency representation, it is also important to recall that it has been proven that, even if a stochastic electromagnetic beam is submitted to a filtering process to be converted into a very narrow-band beam, in general the coherence and polarization properties of the resulting beam do not correspond to those of a monochromatic beam, but are rather conditioned by the spectral coherence properties of the original beam (Lahiri and Wolf 2009; 2010; Partanen et al. 2018). In fact, as shown by Partanen el al. (2018), the normalized spectral polarization and coherence Stokes parameters of the filtered and unfiltered light are identical, the filtered light being strictly cross-spectrally pure. In addition, the same authors demonstrated theoretically and experimentally that the time-domain electromagnetic degree of coherence (see Section 1.6.2.4) of the filtered light does not change when the filter passband is reduced.

The main results of the work of Lahiri and Wolf (2010) are summarized in the following two theorems:

> *Theorem 1. "If a statistically stationary electromagnetic beam of arbitrary bandwidth is filtered to become narrow-band, of mean frequency ϖ, the degree of polarization of the filtered beam in the space–time domain is equal to the spectral degree of polarization at the frequency ϖ of the original beam."*

This theorem implies that, even if a stochastic light beam is filtered to a very narrow frequency range, its degree of polarization may not be equal to unity, but it depend on the spectral polarization properties of the original beam.

> *Theorem 2. "Consider a statistically stationary electromagnetic beam of arbitrary bandwidth. If the beam is filtered to become narrow-band, of mean frequency ϖ, the maximum value, with respect to τ, of the modulus of its space–time degree of coherence $|v_{\varpi}(\mathbf{r}_1,\mathbf{r}_2,\tau)|$ is equal to the modulus $|v(\mathbf{r}_1,\mathbf{r}_2,\varpi)|$ of the spectral degree of coherence of the original beam"* (the definitions of the above mentioned degrees of coherence can be found in Section 1.6.2).

"Theorems 1 and 2 show how to determine the spectral degree of polarization and the spectral degree of coherence of an electromagnetic beam of arbitrary bandwidth, at a particular frequency, from measurements performed in the space–time domain."

The same authors also have been proven that "there are completely polarized light beams with cross-spectral density matrix whose elements do not have a factorized form with respect to its two spatial arguments and hence differ from those representing a monochromatic light beam" (Lahiri and Wolf 2009).

The equal-point coherence–polarization properties of the 2D state of polarization in the space–frequency domain are given by the *spectral polarization matrix*

$$\mathbf{W}(\omega)=\frac{1}{2\pi}\int_{-\infty}^{\infty}\mathbf{J}(\tau)\,e^{i\omega\tau}\,d\tau \tag{1.54}$$

where the dependence on $\mathbf{r}$ has been omitted.

For quasimonochromatic states with very narrow spectral width around a central angular frequency ϖ the spectral polarization matrix $\mathbf{W}(\varpi)$, evaluated at ϖ, provides the same measurable quantities as the usual (equal-point, equal-time) polarization matrix $\mathbf{\Phi}=\mathbf{W}(\varpi)$. In the most general case, $\mathbf{\Phi}$ and $\mathbf{W}(\varpi)$ are not equivalent.

The *intensity distribution* or *spectral density* (at a given point $\mathbf{r}$) is defined as

$$I(\omega)\equiv s_0(\mathbf{r},\mathbf{r},\omega)=\mathrm{tr}\,\mathbf{W}(\mathbf{r},\mathbf{r},\omega) \tag{1.55}$$

1.6.1.4 Spectral Coherence Stokes Parameters

In accordance with the definition of the cross-spectral density matrix, the *spectral coherence Stokes parameters* (or space–frequency *two-point Stokes parameters*) *are defined as* (Korotkova and Wolf 2005; Setälä et al. 2006; Tervo et al. 2009; Friberg and Setälä 2016; Korotkova 2017; Setälä et al. 2021)

$$\mathbf{s}(\mathbf{r}_1,\mathbf{r}_2,\omega)\equiv\begin{pmatrix} w_{11}(\mathbf{r}_1,\mathbf{r}_2,\omega)+w_{22}(\mathbf{r}_1,\mathbf{r}_2,\omega)\\ w_{11}(\mathbf{r}_1,\mathbf{r}_2,\omega)-w_{22}(\mathbf{r}_1,\mathbf{r}_2,\omega)\\ w_{12}(\mathbf{r}_1,\mathbf{r}_2,\omega)+w_{21}(\mathbf{r}_1,\mathbf{r}_2,\omega)\\ iw_{12}(\mathbf{r}_1,\mathbf{r}_2,\omega)-iw_{21}(\mathbf{r}_1,\mathbf{r}_2,\omega)\end{pmatrix} \tag{1.56}$$

Unlike the usual (one-point) Stokes parameters, the spectral coherence Stokes parameters $s_i(\mathbf{r}_1,\mathbf{r}_2,\omega)$ contain information about both coherence and polarization properties and this space–frequency generalized formulation has been demonstrated to obey relatively simple laws (more than for the two-point space–time formulation). As with the cross-spectral density matrix $\mathbf{W}(\mathbf{r}_1,\mathbf{r}_2,\omega)$, one of the interesting applications of the concept of the spectral coherence Stokes vector $\mathbf{s}(\mathbf{r}_1,\mathbf{r}_2,\omega)$ is that it can be used to determine the changes of the one-point Stokes parameters on propagation, including changes in the degree of polarization, as well as changes in the azimuth and ellipticity of the characteristic polarization ellipse. In fact, the use of spectral coherence Stokes parameters has demonstrated to be particularly useful for the formulation of a variety of problems involving propagation, diffraction and interference of random electromagnetic beams (Setälä et al. 2021).

The *spectral Stokes vector* $\mathbf{s}(\omega) \equiv \mathbf{s}(\mathbf{r},\mathbf{r},\omega)$ is determined from $\mathbf{W}(\omega) \equiv \mathbf{W}(\mathbf{r},\mathbf{r},\omega)$ as

$$\mathbf{s}(\omega) = \left[w_{11}(\omega)+w_{22}(\omega), w_{11}(\omega)-w_{22}(\omega), w_{12}(\omega)+w_{21}(\omega), iw_{12}(\omega)-iw_{21}(\omega)\right]^T \quad (1.57)$$

The usual equal-point, equal-time polarization matrix $\mathbf{\Phi}$ and the conventional Stokes parameters are retrieved as

$$\mathbf{\Phi} = \mathbf{J}(0) = \int_0^\infty \mathbf{W}(\omega)\, d\omega \; \mathbf{s} = \mathbf{s}(\mathbf{r}_1 = \mathbf{r}_2, \tau = 0) = \int_0^\infty \mathbf{s}(\omega)\, d\omega \quad (1.58)$$

An efficient procedure to measure the spatial and spectral distributions of the space–frequency coherence Stokes parameters has been designed and experimentally validated by Partanen et al. (2019). The method is based on a digital micromirror device and a diffraction grating, and is demonstrated by using a quartz-wedge depolarizer to prepare a beam with a complicated space–frequency coherence structure. The results obtained show that the polarization and electromagnetic coherence properties may vary with wavelength on a scale less than a nanometer.

1.6.2 Measures of the Degree of Coherence of 2D Electromagnetic Fields

Generally speaking, *coherence* refers to the statistical dependence between the values of an electromagnetic field at two points $\mathbf{r}_1$ and $\mathbf{r}_2$ and two times t and $t+\tau$. This section is devoted to different approaches for the definition of measures of the *degree of coherence*. Interesting summaries of the main concepts involved, as well as the successive approaches, can be found in (Réfrégier and Roueff 2007; Friberg and Dändliker (eds.) 2008; Hassinen 2013; Friberg and Setälä 2016). Further, comprehensive treatises on the characterization of light beams, including coherence and polarization properties (beyond the one-point and fixed-time polarization properties) can be found in (Martínez-Herrero et al. 2009 and Korotkova 2017). A compared analysis of the different definitions of the degree of coherence was performed by Tervo et al. (2012). Among the different approaches described below, special emphasis is put on the space–time and space–frequency versions of the electromagnetic degree of coherence (Tervo et al. 2003; Setälä et al. 2004b; Tervo et al. 2004) which satisfies all the conditions derived from the Glauber criteria (Glauber 1963) for a proper definition of the degree of coherence.

In analogy to the formulation for the mutual coherence of scalar wavefields, let us consider the normalized complex correlation between the analytic signals $\varepsilon_1(\mathbf{r}_1,t)$ and $\varepsilon_2(\mathbf{r}_2,t+\tau)$

$$\mu(\mathbf{r}_1,\mathbf{r}_2,\tau) = j_{12}(\mathbf{r}_1,\mathbf{r}_2,\tau) \Big/ \sqrt{j_{11}(\mathbf{r}_1,\mathbf{r}_1,0)\, j_{22}(\mathbf{r}_2,\mathbf{r}_2,0)} \quad (1.59)$$

The one-point version of $\mu(\mathbf{r}_1,\mathbf{r}_2,\tau)$ is

$$\mu(\tau) = j_{12}(\tau)\Big/\sqrt{j_{11}(0)\, j_{22}(0)} \tag{1.60}$$

whose equal-time version is the complex degree of mutual coherence defined in Eq. (1.31)

$$\mu = j_{12}(0)\Big/\sqrt{j_{11}(0)\, j_{22}(0)} = \phi_{12}\Big/\sqrt{\phi_{11}\,\phi_{22}} \tag{1.61}$$

where ϕ_{ij} are the elements of the polarization matrix $\mathbf{\Phi}$. Observe that $\mu(\mathbf{r}_1,\mathbf{r}_2,\tau)$ is not invariant under rotation transformations (i.e., under rotations of the coordinate system about the direction of propagation) and hence, is not invariant under unitary transformations.

Beyond the above scalar approach, the characterization of the coherence properties of electromagnetic fields involves, in a mixed way, both correlations between the complex amplitudes of the representative analytic signals and the visibility of the interference fringes. In the case of scalar wave representations both attributes converge in such a manner that a single degree of coherence can be defined by jointly taking them into account in a consistent manner. Nevertheless, in the case of vectorial representations (required for a complete characterization of electromagnetic waves, including polarization properties), the scalar degree of coherence does not admit a straightforward generalization, and several parameters have been defined as overall measures of the coherence. Depending on which of the above mentioned complementary properties the interest is focused, different parameters can be defined.

1.6.2.1 Complex Degree of Coherence

A complex parameter that provides a certain overall measure related to both the visibility and the location of the maxima of the interference fringes, is defined from $\mathbf{J}(\mathbf{r}_1,\mathbf{r}_2,\tau)$ as

$$\nu(\mathbf{r}_1,\mathbf{r}_2,\tau) = \frac{\operatorname{tr}\mathbf{J}(\mathbf{r}_1,\mathbf{r}_2,\tau)}{\sqrt{\operatorname{tr}\mathbf{J}(\mathbf{r}_1,\mathbf{r}_1,0)}\,\sqrt{\operatorname{tr}\mathbf{J}(\mathbf{r}_2,\mathbf{r}_2,0)}} = \frac{s_0(\mathbf{r}_1,\mathbf{r}_2,\tau)}{\sqrt{I(\mathbf{r}_1)}\,\sqrt{I(\mathbf{r}_2)}} \tag{1.62}$$

where the quantities $I(\mathbf{r}_i) = s_0(\mathbf{r}_i,\mathbf{r}_i,0)$ $(i = 1,2)$ are the intensities $I(\mathbf{r}_1)$ and $I(\mathbf{r}_2)$ at the respective points. Since $[\operatorname{tr}\mathbf{J}(\mathbf{r}_1,\mathbf{r}_2,\tau)]^* = \operatorname{tr}\mathbf{J}(\mathbf{r}_1,\mathbf{r}_2,-\tau)$ (i.e., $s_0^*(\mathbf{r}_1,\mathbf{r}_2,\tau) = s_0(\mathbf{r}_1,\mathbf{r}_2,-\tau)$) the absolute value of $\nu(\mathbf{r}_1,\mathbf{r}_2,\tau)$ is

$$|\nu(\mathbf{r}_1,\mathbf{r}_2,\tau)| = \frac{\sqrt{\operatorname{tr}\mathbf{J}(\mathbf{r}_1,\mathbf{r}_2,\tau)\,\operatorname{tr}\mathbf{J}(\mathbf{r}_1,\mathbf{r}_2,-\tau)}}{\sqrt{I(\mathbf{r}_1)\, I(\mathbf{r}_2)}} = \frac{\sqrt{s_0(\mathbf{r}_1,\mathbf{r}_2,\tau)\, s_0(\mathbf{r}_1,\mathbf{r}_2,-\tau)}}{\sqrt{I(\mathbf{r}_1)\, I(\mathbf{r}_2)}} \tag{1.63}$$

Note that $|\nu(\mathbf{r}_1,\mathbf{r}_2,\tau)|$ provides a measure of the contrast of the fringe pattern while its phase equals the phase of the fringe pattern.

The space–frequency counterpart of $\nu(\mathbf{r}_1,\mathbf{r}_2,\tau)$ is given by the *spectral degree of coherence* defined by Wolf (2003) as the complex quantity

$$\nu(\mathbf{r}_1,\mathbf{r}_2,\omega) = \frac{\operatorname{tr}\mathbf{W}(\mathbf{r}_1,\mathbf{r}_2,\omega)}{\sqrt{\operatorname{tr}\mathbf{W}(\mathbf{r}_1,\mathbf{r}_1,\omega)}\,\sqrt{\operatorname{tr}\mathbf{W}(\mathbf{r}_2,\mathbf{r}_2,\omega)}} = \frac{s_0(\mathbf{r}_1,\mathbf{r}_2,\omega)}{\sqrt{I(\mathbf{r}_1,\omega)}\,\sqrt{I(\mathbf{r}_2,\omega)}} \tag{1.64}$$

Since $[\mathrm{tr}\,\mathbf{W}(\mathbf{r}_1,\mathbf{r}_2,\omega)]^* = \mathrm{tr}\,\mathbf{W}(\mathbf{r}_1,\mathbf{r}_2,\omega)$ (i.e., $s_0^*(\mathbf{r}_1,\mathbf{r}_2,\omega) = s_0(\mathbf{r}_2,\mathbf{r}_1,\omega)$) the absolute value of $\nu(\mathbf{r}_1,\mathbf{r}_2,\omega)$ is

$$\left|\nu(\mathbf{r}_1,\mathbf{r}_2,w)\right| = \frac{\sqrt{\mathrm{tr}\,\mathbf{W}(\mathbf{r}_1,\mathbf{r}_2,\omega)\;\mathrm{tr}\,\mathbf{W}(\mathbf{r}_2,\mathbf{r}_1,\omega)}}{\sqrt{I(\mathbf{r}_1,\omega)\,I(\mathbf{r}_2,\omega)}} = \frac{\sqrt{s_0(\mathbf{r}_1,\mathbf{r}_2,\omega)\,s_0(\mathbf{r}_2,\mathbf{r}_1,\omega)}}{\sqrt{I(\mathbf{r}_1,\omega)\,I(\mathbf{r}_2,\omega)}} \tag{1.65}$$

In general, for a given pair of points $(\mathbf{r}_1,\mathbf{r}_2)$, the complex quantity $\nu(\mathbf{r}_1,\mathbf{r}_2,\omega)$ varies with frequency and is directly related to measures of fringe visibility in experimental arrangements where the intensities in the pinholes are equal and the polarization properties are not relevant. In fact, the only spectral coherence Stokes parameter involved in $\nu(\mathbf{r}_1,\mathbf{r}_2,\omega)$ is $s_0(\mathbf{r}_1,\mathbf{r}_2,\omega)$.

To inspect the ability of $\nu(\mathbf{r}_1,\mathbf{r}_2,\tau)$ to be a proper measure of the overall degree of coherence corresponding to $\mathbf{J}(\mathbf{r}_1,\mathbf{r}_2,\tau)$, Tervo et al. (2003) observed that $|\nu(\mathbf{r}_1,\mathbf{r}_2,\tau)|$ can take the value zero even for fully coherent electromagnetic beams as, for instance, is the case of a beam whose analytic signals are of the form $\varepsilon_1(\mathbf{r}_1,t) = \varepsilon_2(\mathbf{r}_2,t) = c\exp(-i\omega t)$, c being the complex amplitude and ω the angular frequency. Thus, the information provided by $\nu(\mathbf{r}_1,\mathbf{r}_2,\tau)$ is mainly focused on the interference and visibility, rather than on correlations between the amplitudes of the components of the electromagnetic wave and polarization (Tervo et al. 2003; Luis 2007).

1.6.2.2 Complex Degree of Mutual Polarization

Another coherence parameter is the *complex degree of mutual polarization* between the wave fluctuations at points $\mathbf{r}_1$ and $\mathbf{r}_2$, defined by Ellis and Dogariu (2004b) as

$$V^2(\mathbf{r}_1,\mathbf{r}_2,\tau) = 4\,\frac{s_1^2(\mathbf{r}_1,\mathbf{r}_2,\tau)+s_2^2(\mathbf{r}_1,\mathbf{r}_2,\tau)+s_3^2(\mathbf{r}_1,\mathbf{r}_2,\tau)}{I(\mathbf{r}_1)\,I(\mathbf{r}_2)} \tag{1.66}$$

where $I(\mathbf{r})$ stands for the intensity at point $\mathbf{r}$, so that $V(\mathbf{r}_1,\mathbf{r}_2,\tau)$ contains information on both its phase and magnitude and provides a measure of the similarity between the states of polarization at points $\mathbf{r}_1$ and $\mathbf{r}_2$ in terms of measurable visibilities. In the case of completely correlated fields, $|V(\mathbf{r}_1,\mathbf{r}_2,\tau)|$ gives a measure of the degree of similarity between the states of polarization at the two points $\mathbf{r}_1$, $\mathbf{r}_2$ considered, while the phase of $V^2(\mathbf{r}_1,\mathbf{r}_2,\tau)$ is related to the relative phases of the waves at those points.

A space–frequency version of V can be defined as

$$V^2(\mathbf{r}_1,\mathbf{r}_2,\omega) = 4\,\frac{s_1^2(\mathbf{r}_1,\mathbf{r}_2,\omega)+s_2^2(\mathbf{r}_1,\mathbf{r}_2,\omega)+s_3^2(\mathbf{r}_1,\mathbf{r}_2,\omega)}{I(\mathbf{r}_1,\omega)\,I(\mathbf{r}_2,\omega)} \tag{1.67}$$

1.6.2.3 Intrinsic Degrees of Coherence

The physical requirement of invariance under unitary transformations to be satisfied by a proper definition of a space–time degree of coherence lead Réfrégier and Goudail (2005) to introduce the so-called *intrinsic degrees of coherence*, whose definition relies on the concept of *normalized coherence matrix*

$$\mathbf{M}(\mathbf{r}_1,\mathbf{r}_2,\tau) \equiv \left[\boldsymbol{\Phi}(\mathbf{r}_2,t+\tau)\right]^{-1/2}\mathbf{J}(\mathbf{r}_1,\mathbf{r}_2,\tau)\left[\boldsymbol{\Phi}(\mathbf{r}_1,t)\right]^{-1/2} \tag{1.68}$$

where the effects of the polarization matrices $\boldsymbol{\Phi}(\mathbf{r}_1,t)$ and $\boldsymbol{\Phi}(\mathbf{r}_2,t+\tau)$ have been extracted from the coherence matrix $\mathbf{J}(\mathbf{r}_1,\mathbf{r}_2,\tau)$, reducing it to $\mathbf{M}(\mathbf{r}_1,\mathbf{r}_2,\tau)$.

Observe that Eq. (1.68) is only valid when both $\boldsymbol{\Phi}(\mathbf{r}_1,t)$ and $\boldsymbol{\Phi}(\mathbf{r}_2,t+\tau)$ are invertible (nonsingular) that is, when neither $\boldsymbol{\Phi}(\mathbf{r}_1,t)$ nor $\boldsymbol{\Phi}(\mathbf{r}_2,t+\tau)$ represents a totally polarized state. Nevertheless, Eq. (1.68) can straightforwardly be generalized to the case where one or the two states $\boldsymbol{\Phi}(\mathbf{r}_1,t)$ and $\boldsymbol{\Phi}(\mathbf{r}_2,t+\tau)$ are totally polarized by taking $[\boldsymbol{\Phi}(\mathbf{r}_1,t)]^{-1/2}$ and/or $[\boldsymbol{\Phi}(\mathbf{r}_2,t+\tau)]^{-1/2}$ as the respective pseudoinverses. To do so, consider a singular polarization matrix $\boldsymbol{\Phi}_i(\mathbf{r}_i,t_i)$ whose inverse square root enters in Eq. (1.68), and expand it through the singular value decomposition $\boldsymbol{\Phi}_i(\mathbf{r}_i,t_i)=\mathbf{U}_i\,\mathrm{diag}(a_i,0)\mathbf{U}_i^{\dagger}$, where $\mathbf{U}_i$ is unitary and a_i (with $a_i>0$) is the nonzero singular value. Then $[\boldsymbol{\Phi}_i(\mathbf{r}_i,t_i)]^{1/2}=\mathbf{U}_i\,\mathrm{diag}(\sqrt{a_i},0)\mathbf{U}_i^{\dagger}$ and its pseudoinverse is given by $[\boldsymbol{\Phi}_i(\mathbf{r}_i,t_i)]^{-1/2}=\mathbf{U}_i\,\mathrm{diag}(1/\sqrt{a_i},0)\mathbf{U}_i^{\dagger}$.

Once the *normalized coherence matrix* $\mathbf{M}(\mathbf{r}_1,\mathbf{r}_2,\tau)$ has been properly defined, including the case of totally polarized electromagnetic fields (at points $\mathbf{r}_1$ and $\mathbf{r}_2$ respectively) the intrinsic degrees of coherence are defined as the singular values $\mu_S(\mathbf{r}_1,\mathbf{r}_2,\tau)$ and $\mu_I(\mathbf{r}_1,\mathbf{r}_2,\tau)$ of $\mathbf{M}(\mathbf{r}_1,\mathbf{r}_2,\tau)$, which are given by the square roots of the (non-negative) eigenvalues of the positive semidefinite Hermitian matrix $\mathbf{M}(\mathbf{r}_1,\mathbf{r}_2,\tau)\,\mathbf{M}^{\dagger}(\mathbf{r}_1,\mathbf{r}_2,\tau)$. That is to say, there exists a unitary matrix $\mathbf{N}_2$ such that

$$\mathbf{M}(\mathbf{r}_1,\mathbf{r}_2,\tau)\mathbf{M}^{\dagger}(\mathbf{r}_1,\mathbf{r}_2,\tau)=\mathbf{N}_2(\mathbf{r}_1,\mathbf{r}_2,\tau)\begin{pmatrix}\mu_S^2(\mathbf{r}_1,\mathbf{r}_2,\tau) & 0\\ 0 & \mu_I^2(\mathbf{r}_1,\mathbf{r}_2,\tau)\end{pmatrix}\mathbf{N}_2^{\dagger}(\mathbf{r}_1,\mathbf{r}_2,\tau)\qquad\left[\mu_S^2\geq\mu_I^2\right] \tag{1.69}$$

the columns of $\mathbf{N}_2$ being the *left singular vectors* of $\mathbf{M}(\mathbf{r}_1,\mathbf{r}_2,\tau)$

Similarly,

$$\mathbf{M}^{\dagger}(\mathbf{r}_1,\mathbf{r}_2,\tau)\,\mathbf{M}(\mathbf{r}_1,\mathbf{r}_2,\tau)=\mathbf{N}_1(\mathbf{r}_1,\mathbf{r}_2,\tau)\begin{pmatrix}\mu_S^2(\mathbf{r}_1,\mathbf{r}_2,\tau) & 0\\ 0 & \mu_I^2(\mathbf{r}_1,\mathbf{r}_2,\tau)\end{pmatrix}\mathbf{N}_1^{\dagger}(\mathbf{r}_1,\mathbf{r}_2,\tau) \tag{1.70}$$

where $\mathbf{N}_1$ is the unitary matrix whose columns are the *right singular vectors* of $\mathbf{M}(\mathbf{r}_1,\mathbf{r}_2,\tau)$, so that the singular value decomposition of $\mathbf{M}(\mathbf{r}_1,\mathbf{r}_2,\tau)$ results in

$$\mathbf{M}(\mathbf{r}_1,\mathbf{r}_2,\tau)=\mathbf{N}_2(\mathbf{r}_1,\mathbf{r}_2,\tau)\begin{pmatrix}\mu_S(\mathbf{r}_1,\mathbf{r}_2,\tau) & 0\\ 0 & \mu_I(\mathbf{r}_1,\mathbf{r}_2,\tau)\end{pmatrix}\mathbf{N}_1^{\dagger}(\mathbf{r}_1,\mathbf{r}_2,\tau) \tag{1.71}$$

Observe that the intrinsic degrees of coherence satisfy $0\leq\mu_I^2\leq\mu_S^2\leq 1$ and can be expressed as (Réfrégier and Roueff 2008)

$$\mu_S^2=T(1+Q)\qquad\qquad \mu_I^2=T(1-Q) \tag{1.72a}$$

where

$$T\equiv\frac{1}{2}\mathrm{tr}\left(\mathbf{M}^{\dagger}\mathbf{M}\right)\qquad\qquad Q\equiv\sqrt{1-4\det\left(\mathbf{M}^{\dagger}\mathbf{M}\right)/\left[\mathrm{tr}\left(\mathbf{M}^{\dagger}\mathbf{M}\right)\right]^2} \tag{1.72b}$$

so that

$$Q=\left(\mu_S^2-\mu_I^2\right)/\left(\mu_S^2+\mu_I^2\right)\qquad\qquad T=\left(\mu_S^2+\mu_I^2\right)/2 \tag{1.73}$$

and consequently,

$$T = 0 \Rightarrow \mathbf{J}(\mathbf{r}_1, \mathbf{r}_2, \tau) = \mathbf{0} \qquad T = 1 \Rightarrow \mu_S = \mu_I = 1 \tag{1.74}$$

It has been proven that when the electromagnetic wave is the sum of two statistically independent waves with orthogonal polarizations and respective scalar degrees of coherence μ_1 and μ_2, then $\mu_1 = \mu_S$ and $\mu_2 = \mu_I$ and thus, the choice of μ_S and μ_I as representatives of the intrinsic degrees of coherence corresponds to the common intuition (Réfrégier and Goudail 2005).

The two quantities μ_S and μ_I are akin to the absolute value of the scalar complex degree of coherence, while the unitary matrices $\mathbf{N}_1$ and $\mathbf{N}_2$ provide information akin to the phase of the scalar complex degree of coherence.

Furthermore, it is remarkable that, under arbitrary and independent unitary transformations of the interfering fields (which can be physically realized by interposing birefringent plates into their respective pathways), the maximal value of the fringe contrast is equal to the larger intrinsic degree of coherence μ_S. These transformations can be appropriately formulated in terms of Jones vectors and unitary Jones matrices, which act as retarding (birefringent) modulators (see Chapter 4). Therefore, the difference between the space–time complex degree of coherence $\nu(\mathbf{r}_1, \mathbf{r}_2, \tau)$ defined in Eq. (1.62) and $\mu_S(\mathbf{r}_1, \mathbf{r}_2, \tau)$ is that $\nu(\mathbf{r}_1, \mathbf{r}_2, \tau)$ determines the *actual* fringe contrast under a particular experimental configuration (fixed respective polarization states of the interfering waves), while $\mu_S(\mathbf{r}_1, \mathbf{r}_2, \tau)$ determines the maximum fringe contrast with respect to all possible retarding (birefringent) modulations of the polarization states of the interfering waves. Such deterministic and reversible modulations preserving the statistical dependence of the considered fields can be performed through the appropriate insertion of birefringent plates into the pathways of the waves prior to their interference.

The space–frequency counterpart of the of the intrinsic degrees of coherence is formulated from the *space–frequency normalized coherence matrix*

$$\mathbf{M}(\mathbf{r}_1, \mathbf{r}_2, \omega) \equiv \left[\mathbf{W}(\mathbf{r}_2, \omega)\right]^{-1/2} \mathbf{W}(\mathbf{r}_1, \mathbf{r}_2, \omega) \left[\mathbf{W}(\mathbf{r}_1, \omega)\right]^{-1/2} \tag{1.75}$$

where, as in the space–time formulation, the notation $[\mathbf{W}(\mathbf{r}_i, \omega)]^{-1/2}$ is to be understood in pseudoinverse sense in case of a totally polarized spectral state (i.e., $\det \mathbf{W}(\mathbf{r}_i, \omega) = 0$). Then, the space–frequency intrinsic degrees of coherence $\mu_S(\mathbf{r}_1, \mathbf{r}_2, \omega)$ and $\mu_I(\mathbf{r}_1, \mathbf{r}_2, \omega)$ are obtained as the singular values of $\mathbf{M}(\mathbf{r}_1, \mathbf{r}_2, \omega)$ whose singular value decomposition is expressed as

$$\mathbf{M}(\mathbf{r}_1, \mathbf{r}_2, \omega) = \mathbf{N}_2(\mathbf{r}_1, \mathbf{r}_2, \omega) \begin{pmatrix} \mu_S(\mathbf{r}_1, \mathbf{r}_2, \omega) & 0 \\ 0 & \mu_I(\mathbf{r}_1, \mathbf{r}_2, \omega) \end{pmatrix} \mathbf{N}_1^{\dagger}(\mathbf{r}_1, \mathbf{r}_2, \omega) \tag{1.76}$$

In summary, the concept of coherence (in the order-two Glauber's sense) involves the statistical correlation between the orthogonal components of the field, which is measured through the degree of polarization, and the statistical correlation between the waves at two given points, $\mathbf{r}_1$ and $\mathbf{r}_2$, and at two times, t and $t + \tau$, which is given by the intrinsic degrees of coherence (Réfrégier and Roueff 2007). The coherence matrix can be determined through four independent interferometric arrangements, while the polarization matrices can be measured by using polarimetric techniques (see Section 1.5.3).

1.6.2.4 Electromagnetic Degree of Coherence

In accordance with the coherence criteria stated by Glauber (1963); Tervo et al. (2003) defined the following real parameter as the *electromagnetic degree of coherence* in the space–time domain:

$$\gamma(\mathbf{r}_1,\mathbf{r}_2,\tau)=\frac{\sqrt{\operatorname{tr}\left[\mathbf{J}(\mathbf{r}_1,\mathbf{r}_2,\tau)\,\mathbf{J}(\mathbf{r}_2,\mathbf{r}_1,-\tau)\right]}}{\sqrt{\operatorname{tr}\mathbf{J}(\mathbf{r}_1,\mathbf{r}_1,0)\,\operatorname{tr}\mathbf{J}(\mathbf{r}_2,\mathbf{r}_2,0)}}=\frac{\sqrt{\operatorname{tr}\left[\mathbf{J}(\mathbf{r}_1,\mathbf{r}_2,\tau)\,\mathbf{J}^{\dagger}(\mathbf{r}_1,\mathbf{r}_2,\tau)\right]}}{\sqrt{I(\mathbf{r}_1)\,I(\mathbf{r}_2)}}=\frac{\left\|\mathbf{J}(\mathbf{r}_1,\mathbf{r}_2,\tau)\right\|_2}{\sqrt{I(\mathbf{r}_1)\,I(\mathbf{r}_2)}}\tag{1.77}$$

where

$$\left\|\mathbf{J}(\mathbf{r}_1,\mathbf{r}_2,\tau)\right\|_2\equiv\sqrt{\sum_{k,l=1}^{2}\left|j_{kl}(\mathbf{r}_1,\mathbf{r}_2,\tau)\right|^2}=\sqrt{\sum_{i=0}^{3}\left|s_i(\mathbf{r}_1,\mathbf{r}_2,\tau)\right|^2}\tag{1.78}$$

is the Frobenius norm of $\mathbf{J}(\mathbf{r}_1,\mathbf{r}_2,\tau)$, and $I(\mathbf{r}_1)$ and $I(\mathbf{r}_2)$ are the intensities at the respective points.

The real quantity $\gamma(\mathbf{r}_1,\mathbf{r}_2,\tau)$ provides mixed information on both fringe visibility and polarization and is independent of arbitrary unitary transformations $\mathbf{T}_{R1}\,\boldsymbol{\varepsilon}(\mathbf{r}_1,t)$ and $\mathbf{T}_{R2}\,\boldsymbol{\varepsilon}(\mathbf{r}_2,t+\tau)$ of the states $\boldsymbol{\varepsilon}(\mathbf{r}_1,t)$ and $\boldsymbol{\varepsilon}(\mathbf{r}_2,t+\tau)$ prior to their superposition, where $\mathbf{T}_{R1}$ and $\mathbf{T}_{R2}$ are unitary *Jones matrices* representing the action of birefringent devices. The linear transformations of Jones vectors through the effect of birefringent devices (retarders) will be studied in Chapter 4.

As it will be shown in Section 1.11.1, a peculiarity of $\gamma(\mathbf{r}_1,\mathbf{r}_2,\tau)$ is that its equal-point, equal-time, version reduces to $\gamma(\mathbf{r},\mathbf{r},0)=\sqrt{(1+\mathcal{P}^2)/2}$, a function of the degree of polarization, in contrast to the apparently natural expectation that the degree of coherence has to be equal to one when $\mathbf{r}_1=\mathbf{r}_2$ and $\tau=0$. This specific property can be considered as an essential and fundamental consequence of the difference between the scalar and electromagnetic approaches.

The counterpart of $\gamma(\mathbf{r}_1,\mathbf{r}_2,\tau)$ with respect to the space–frequency domain is given by the following real parameter called the *electromagnetic spectral degree of coherence* (Setälä et al. 2004b; Tervo et al. 2004):

$$\begin{aligned}\gamma(\mathbf{r}_1,\mathbf{r}_2,\omega)&=\frac{\sqrt{\operatorname{tr}\left[\mathbf{W}(\mathbf{r}_1,\mathbf{r}_2,\omega)\,\mathbf{W}(\mathbf{r}_2,\mathbf{r}_1,\omega)\right]}}{\sqrt{\operatorname{tr}\mathbf{W}(\mathbf{r}_1,\mathbf{r}_1,\omega)}\,\sqrt{\operatorname{tr}\mathbf{W}(\mathbf{r}_2,\mathbf{r}_2,\omega)}}\\&=\frac{\sqrt{\operatorname{tr}\left[\mathbf{W}(\mathbf{r}_1,\mathbf{r}_2,\omega)\,\mathbf{W}^{\dagger}(\mathbf{r}_1,\mathbf{r}_2,\omega)\right]}}{\sqrt{I(\mathbf{r}_1,\omega)}\sqrt{I(\mathbf{r}_2,\omega)}}=\frac{\left\|\mathbf{W}(\mathbf{r}_1,\mathbf{r}_2,\omega)\right\|_2}{\sqrt{I(\mathbf{r}_1,\omega)}\sqrt{I(\mathbf{r}_2,\omega)}}\end{aligned}\tag{1.79a}$$

where

$$\left\|\mathbf{W}(\mathbf{r}_1,\mathbf{r}_2,\omega)\right\|_2\equiv\sqrt{\sum_{k,l=1}^{2}\left|w_{kl}(\mathbf{r}_1,\mathbf{r}_2,\omega)\right|^2}=\sqrt{\sum_{i=0}^{3}\left|s_i(\mathbf{r}_1,\mathbf{r}_2,\omega)\right|^2}\tag{1.79.b}$$

It has been proven that the electromagnetic degree of coherence is a measure of both the visibility and the intensity of the fringes, as well as of the modulation contrasts of the three polarization Stokes parameters. Furthermore, $\gamma(\mathbf{r}_1, \mathbf{r}_2, \omega)$ can be determined through four visibility measurements by using suitable waveplates that transform the modulation of any Stokes parameter into an intensity variation (Setälä et al. 2006). This approach has been validated experimentally through an appropriate measurement arrangement described in Partanen et al. (2019).

As required by the coherence conditions stated by Glauber (1963) for total coherence, the field is completely coherent ($\gamma(\mathbf{r}_1, \mathbf{r}_2, \omega) = 1$) at a given frequency ω if and only if the components of the field at that frequency are fully correlated for all pairs of points $\mathbf{r}_1$ and $\mathbf{r}_2$ within the spatial region of observation considered (Setälä et al. 2004b). Note also that, as pointed out by Saastamoinen et al. (2005), an ideal single-mode laser is always spatially fully coherent provided that $\gamma(\mathbf{r}_1, \mathbf{r}_2, \omega)$ is used as the definition of the degree of coherence.

For the common case of stationary electromagnetic fields (at least in the wide sense), Setälä et al. (2004a; 2004b) used the space–time and space frequency versions, $\gamma(\mathbf{r}_1, \mathbf{r}_2, \tau)$ and $\gamma(\mathbf{r}_1, \mathbf{r}_2, \omega)$, of the electromagnetic degree of coherence to get the electromagnetic versions of some interesting well known results of second-order scalar coherence theory (Mehta et al. 1966; Mandel and Wolf 1981; Mandel and Wolf 1995). In particular, in the space–time domain, the following three properties are fully equivalent: "(i) the field is completely coherent in the space–time domain, (ii) the field is monochromatic, and (iii) the electric mutual coherence tensor factors in the two spatial variables and is periodic in the time difference" (Setälä et al. 2004a).

Furthermore, regarding the space–frequency domain, the following two properties are fully equivalent: "(i) the field is spatially completely coherent, and (ii) the electric cross-spectral density tensor factors in the two spatial variables" (Setälä et al. 2004b). The above results are consistent with previous studies where the factorization property of coherence functions is proposed for the definition of complete coherence.

Since both $\gamma(\mathbf{r}_1, \mathbf{r}_2, \tau)$ and $\gamma(\mathbf{r}_1, \mathbf{r}_2, \omega)$ are intensity-normalized measures of coherence, respective versions appropriately weighted by intensities result in the effective degrees of coherence (Tervo et al. 2004; Vahimaa and Tervo 2004; Blomstedt et al. 2007; 2015; Blomstedt 2013).

Friberg and Wolf (1995) and Leppänen et al. (2016), found that no general closed form exists for the relation between the space–time and space–frequency domain versions of the electromagnetic degree of coherence of stationary random fields. However, they described interesting relations that illustrate the similarities and differences between both quantities, which were established from certain considerations covering quasimonochromatic and broadband situations of electromagnetic Gaussian Schell-model beams and blackbody radiation. A very interesting summary of the electromagnetic theory of optical coherence can be found in Friberg and Setälä (2016). These authors emphasized critical and fundamental aspects like:

- $\gamma(\mathbf{r}_1, \mathbf{r}_2, \tau)$ and $\gamma(\mathbf{r}_1, \mathbf{r}_2, \omega)$ satisfy all properties required for a proper definition of the degree of coherence, they are valid for general 3D fields (see Section 2.7.2), and they reduce in a natural manner to the traditional measures for scalar fields (single polarization mode).
- $\gamma(\mathbf{r}_1, \mathbf{r}_2, \tau)$ and $\gamma(\mathbf{r}_1, \mathbf{r}_2, \omega)$ account for the totality of correlations among the electric-field components. Furthermore, they are necessarily nonzero when there are any correlations between the components and vice versa.
- The coherence Stokes parameters constitute a complete representation of the electromagnetic coherence state. Their one-point versions are the conventional Stokes

parameters, which fully characterize the (second-order) polarization state. Such connection becomes explicit in Young's and Michelson's interferometers, where, as a natural extension of the scalar results, the visibilities (modulation contrasts) of all the Stokes parameters on the observation plane give precisely the electromagnetic degrees of coherence.

- When an electromagnetic beam is allowed in to interfere with itself through a Young's two-pinhole setup, the intensity fringes exhibit unit visibility, while the modulation contrasts of the three Stokes parameters yield the degree of polarization.
- In full analogy with the corresponding scalar theory, the intensity correlations associated with fluctuating electromagnetic fields (which can be measured through a Hanbury Brown and Twiss interferometer) are completely characterized by the electromagnetic degrees of coherence (Hassinen et al. 2011a).
- The intimate relation between coherence and polarization in random electromagnetic fields becomes transparent from the link between the coherence Stokes parameters and the electromagnetic degrees of coherence.

An experimental setup for the measurement of the electromagnetic degree of coherence by means of cubical gold (dipolar) nanoscatterers has been designed and validated by Saastamoinen et al. (2020).

Regarding random non-stationary electromagnetic fields, the analysis performed by Al Lakki et al. (2021) shows that they are completely coherent pointwise in the sense of $\mu(\mathbf{r}_1,\mathbf{r}_2,\omega_1,\omega_2)=1$ if and only if the cross-spectral density matrix factors in the space–frequency variables. In addition, complete coherence over a space–frequency domain leads to a coherent mode and implies full temporal coherence and factorization of the coherence matrix.

1.6.2.5 Overall Degree of Coherence

An interesting formulation for the coherence and polarization properties of electromagnetic fields was developed by Luis (2007), who defined an overall measure $D(\mathbf{r}_1,\mathbf{r}_2,\omega)$ of coherence of electromagnetic fields as a distance between correlation matrices, in such a manner that $D(\mathbf{r}_1,\mathbf{r}_2,\omega)$ combines the information on the degree of polarization, the visibility of interference fringes and the correlation between the field variables. The framework that allows this joint treatment for 2D electromagnetic beams relies on the 4×4 *cross-spectral density matrix* defined as

$$\mathbf{M}_L(\mathbf{r}_1,\mathbf{r}_2,\omega)\equiv\begin{pmatrix}\mathbf{W}(\mathbf{r}_1,\mathbf{r}_1,\omega) & \mathbf{W}(\mathbf{r}_1,\mathbf{r}_2,\omega)\\ \mathbf{W}^{+}(\mathbf{r}_1,\mathbf{r}_2,\omega) & \mathbf{W}(\mathbf{r}_2,\mathbf{r}_2,\omega)\end{pmatrix} \tag{1.80}$$

whose normalized form is

$$\hat{\mathbf{M}}_L(\mathbf{r}_1,\mathbf{r}_2,\omega)\equiv\frac{\hat{\mathbf{M}}_L(\mathbf{r}_1,\mathbf{r}_2,\omega)}{\operatorname{tr}\hat{\mathbf{M}}_L(\mathbf{r}_1,\mathbf{r}_2,\omega)}=\frac{\hat{\mathbf{M}}_L(\mathbf{r}_1,\mathbf{r}_2,\omega)}{\operatorname{tr}\mathbf{W}(\mathbf{r}_1,\mathbf{r}_1,\omega)+\operatorname{tr}\mathbf{W}(\mathbf{r}_2,\mathbf{r}_2,\omega)}=\frac{\hat{\mathbf{M}}_L(\mathbf{r}_1,\mathbf{r}_2,\omega)}{I(\mathbf{r}_1,\omega)+I(\mathbf{r}_2,\omega)} \tag{1.81}$$

The quantity $D(\mathbf{r}_1,\mathbf{r}_2,\omega)$ is then defined as the following measure of the distance between a completely uncorrelated and completely unpolarized state, represented by the 4×4 identity matrix $\mathbf{I}_4$, and the normalized 4×4 cross-spectral density matrix $\hat{\mathbf{M}}_L(\mathbf{r}_1,\mathbf{r}_2,\omega)$ (Luis 2007):

$$D^2(\mathbf{r}_1,\mathbf{r}_2,\omega) \equiv \frac{1}{n-1}\operatorname{tr}\left\{\left[\mathbf{I}_4 - n\,\hat{\mathbf{M}}_L(\mathbf{r}_1,\mathbf{r}_2,\omega)\right]^2\right\} = \frac{1}{n-1}\left[n\left\|\hat{\mathbf{M}}_L(\mathbf{r}_1,\mathbf{r}_2,\omega)\right\|_2^2 - 1\right] \quad (1.82)$$

where n is the number of independent scalar waves involved in $\mathbf{M}_L(\mathbf{r}_1,\mathbf{r}_2,\omega)$; so that $n = 2$ applies for one-point measures ($\mathbf{r}_2 = \mathbf{r}_1$) and $n = 4$ applies in general.

The maximum $D(\mathbf{r}_1,\mathbf{r}_2,\omega) = 1$ is achieved when $\hat{\mathbf{M}}_L(\mathbf{r}_1,\mathbf{r}_2,\omega)$ has only one nonzero eigenvalue, which in turn occurs if and only if the field satisfies the Glauber conditions to be fully coherent (i.e., all field variables are fully correlated) and consequently, the wave is totally polarized. The minimum $D(\mathbf{r}_1,\mathbf{r}_2,\omega) = 0$ corresponds to fully uncorrelated states in which case $\hat{\mathbf{M}}_L(\mathbf{r}_1,\mathbf{r}_2,\omega) = \mathbf{I}_4$ for any pair of points in the spatial domain of the observation considered.

As shown by Luis (2007),

$$\begin{aligned}\left\|\mathbf{M}_L(\mathbf{r}_1,\mathbf{r}_2,\omega)\right\|_2^2 &\equiv \operatorname{tr}\left[\mathbf{M}_L^2(\mathbf{r}_1,\mathbf{r}_2,\omega)\right] \\ &= \operatorname{tr}\left[\mathbf{W}^2(\mathbf{r}_1,\mathbf{r}_1,\omega)\right] + \operatorname{tr}\left[\mathbf{W}^2(\mathbf{r}_2,\mathbf{r}_2,\omega)\right] + 2\operatorname{tr}\left[\mathbf{W}^2(\mathbf{r}_1,\mathbf{r}_2,\omega)\right] \\ &= \frac{1}{2}\left(1+\mathcal{P}_1^2\right)I_1^2 + \frac{1}{2}\left(1+\mathcal{P}_2^2\right)I_2^2 + 2I_1I_2\,\gamma^2(\mathbf{r}_1,\mathbf{r}_2,\omega)\end{aligned} \quad (1.83)$$

where $\mathcal{P}_i \equiv \mathcal{P}(\mathbf{r}_i,\omega)$ and $I_i \equiv I(\mathbf{r}_i,\omega)$ are respectively the spectral degree of polarization and the spectral density of the corresponding one-point spectral polarization matrix $\mathbf{W}(\mathbf{r}_i,\mathbf{r}_i,\omega)$, with $i = 1,2$, and $\gamma(\mathbf{r}_1,\mathbf{r}_2,\omega)$ is the electromagnetic degree of coherence.

Thus, $D(\mathbf{r}_1,\mathbf{r}_2,\omega)$ is expressed in terms of the quantities $\mathcal{P}_i$, I_i and $\gamma(\mathbf{r}_1,\mathbf{r}_2,\omega)$, which are all invariant under position-dependent unitary transformations. Therefore, $D(\mathbf{r}_1,\mathbf{r}_2,\omega)$ is independent of the underlying local generalized coordinate system. Note also from Eq. (1.79) that the numerator of $\gamma(\mathbf{r}_1,\mathbf{r}_2,\omega)$, and therefore $D(\mathbf{r}_1,\mathbf{r}_2,\omega)$, is related to a certain extent to the space–frequency degree of mutual polarization.

Consequently, $D(\mathbf{r}_1,\mathbf{r}_2,\omega)$ constitutes an overall formulation of coherence and polarization in the space–frequency domain which satisfies both the invariance to position-dependent unitary transformations and Glauber's criterion for total coherence, while it takes into account the interference fringe visibility, the correlations between the amplitudes of the electromagnetic modes, and the degree of polarization (Luis 2007). As we have seen, $D(\mathbf{r}_1,\mathbf{r}_2,\omega)$ is defined from a four-path interference between four scalar analytic signals, so that its measurement requires a four-branch interferometric experimental setup, whose particular configurations lead to alternative measures that can be focused either on the fringe visibility, on the correlations, or on the polarization properties. For further detailed information on $D(\mathbf{r}_1,\mathbf{r}_2,\omega)$ and its measurement we direct the reader to Luis 2007 and Luis 2010.

1.6.3 Cross-Spectral Purity and Coherence–Polarization Purity

The concept of cross-spectral purity was introduced by Mandel (Mandel 1961; Mandel and Wolf 1976), by means of the scalar approach to optical waves. The extension of this concept to vectorial representations of electromagnetic waves plays an important role in the coherence–polarization theory and has been demonstrated to be very fruitful in addressing certain problems, so that it deserves to be considered in order to go deeper in the understanding of the coherence–polarization phenomena.

Given an interference experiment of two electromagnetic fields with equal normalized spectra, the normalized spectrum of the superposition is, in general, different from the spectral profiles of the interfering waves. When there is a region in the interference plane where the normalized spectral profile remains unaltered, the waves considered are said to be *cross-spectrally pure* within this region. Mandel showed that such waves exhibit the so-called *reduction property*, which is that the complex degree of mutual coherence can be factored into two parts that characterize spatial and temporal coherence separately. When applied to the degree of mutual coherence $\mu(\mathbf{r}_1, \mathbf{r}_2, \tau)$ of two analytic signals representative of the components of the electric field (see Eq. (1.59)), the scalar cross-spectral purity condition is expressed as

$$\mu(\mathbf{r}_1, \mathbf{r}_2, \tau) = \mu(\mathbf{r}_1, \mathbf{r}_2, \tau_0)\, \mu(\mathbf{r}_1, \mathbf{r}_2, \tau - \tau_0) \tag{1.84}$$

where τ_0 stands for the time delay for which the absolute value $|\mu(\mathbf{r}_1, \mathbf{r}_2, \tau)|$ of the degree of mutual coherence is largest. Eq. (1.84) establishes the factorization of $\mu(\mathbf{r}_1, \mathbf{r}_2, \tau)$ into spatial coherence $\mu(\mathbf{r}_1, \mathbf{r}_2, \tau_0)$ and temporal coherence $\mu(\mathbf{r}_1, \mathbf{r}_2, \tau - \tau_0)$, parts as a necessary condition for the cross-spectral purity. Moreover, Mandel and Wolf found the additional condition that the spectral degree of coherence of cross-spectrally pure scalar fields is necessarily of the form

$$\mu(\mathbf{r}_1, \mathbf{r}_2, \omega) = \mu(\mathbf{r}_1, \mathbf{r}_2, \tau_0)\, e^{i\omega\tau_0} \tag{1.85}$$

and thus, the absolute value of the degree of coherence is independent of frequency and $|\mu(\mathbf{r}_1, \mathbf{r}_2, \omega)| = |\mu(\mathbf{r}_1, \mathbf{r}_2, \tau_0)|$ (Friberg and Wolf 1995).

To analyze in more detail the concept of cross-spectral purity in the context of polarized electromagnetic waves, consider the zeroth components $s_0(\mathbf{r}_1, \mathbf{r}_2, \tau)$ and $s_0(\mathbf{r}_1, \mathbf{r}_2, \omega)$ of the space–time $\mathbf{s}(\mathbf{r}_1, \mathbf{r}_2, \tau)$ and space–frequency $\mathbf{s}(\mathbf{r}_1, \mathbf{r}_2, \omega)$ coherence Stokes vectors, whose normalized values are defined as

$$\hat{s}_0(\mathbf{r}_1, \mathbf{r}_2, \tau) = \frac{s_0(\mathbf{r}_1, \mathbf{r}_2, \tau)}{\sqrt{s_0(\mathbf{r}_1, \mathbf{r}_1)\, s_0(\mathbf{r}_2, \mathbf{r}_2)}} = \frac{\mathrm{tr}\,\mathbf{J}(\mathbf{r}_1, \mathbf{r}_2, \tau)}{\sqrt{\mathrm{tr}\,\mathbf{J}(\mathbf{r}_1, \mathbf{r}_1, 0)\ \mathrm{tr}\,\mathbf{J}(\mathbf{r}_2, \mathbf{r}_2, 0)}} \tag{1.86}$$

$$\hat{s}_0(\mathbf{r}_1, \mathbf{r}_2, \omega) = \frac{s_0(\mathbf{r}_1, \mathbf{r}_2, \omega)}{\sqrt{s_0(\mathbf{r}_1, \mathbf{r}_1, \omega)\, s_0(\mathbf{r}_2, \mathbf{r}_2, \omega)}} = \frac{\mathrm{tr}\,\mathbf{W}(\mathbf{r}_1, \mathbf{r}_2, \omega)}{\sqrt{\mathrm{tr}\,\mathbf{W}(\mathbf{r}_1, \mathbf{r}_1, \omega)\ \mathrm{tr}\,\mathbf{W}(\mathbf{r}_2, \mathbf{r}_2, \omega)}} \tag{1.87}$$

where $s_0(\mathbf{r}, \mathbf{r}, \omega) = \mathrm{tr}\,\mathbf{W}(\mathbf{r}, \mathbf{r}, \omega) = I(\mathbf{r}, \omega)$ is the spectral density at point $\mathbf{r}$ considered.

These expressions allow us to characterize the coherence–polarization in a Young's interference experiment (Figure 1.12), arranged in such a way that the normalized spectral densities of the wave at the points $\mathbf{r}_1$ and $\mathbf{r}_2$ of the pinholes are the same at every frequency, i.e., $s_0(\mathbf{r}_2, \omega) = a s_0(\mathbf{r}_1, \omega)$, with $a > 0$.

The state of polarization on the observation plane exhibits, in general, a periodical modulation. In the case of unpolarized states it has been shown that they can be divided into two types. The so-called type-I, or *pure-unpolarized states* (observe that the term *pure* refers here to cross-spectral purity and not to polarimetric purity) are those that remain unpolarized in the Young's interference experiment. The type-II or *impure-unpolarized states* produce partially polarized states in the said interference. It has been proven that the general condition for a state of polarization to be preserved, at a particular frequency ω, under Young's interference can be expressed as (Hassinen et al. 2013)

$$\mathbf{W}(\mathbf{r}_1, \mathbf{r}_2, \omega) = f(\mathbf{r}_1, \mathbf{r}_2, \omega)\, \widehat{\mathbf{W}}(\omega) \tag{1.88}$$

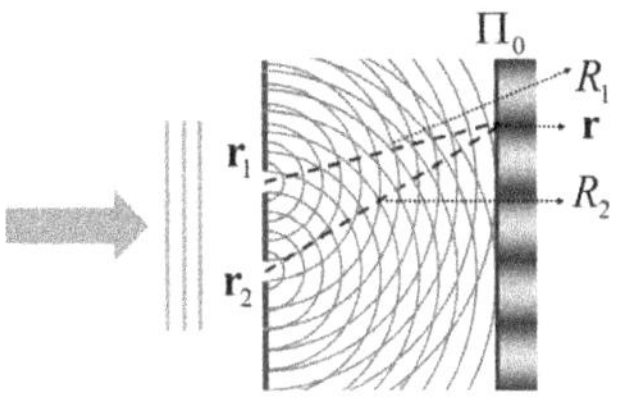

FIGURE 1.12
Young's interference setup.

where $f(\mathbf{r}_1,\mathbf{r}_2,\omega)$ is an arbitrary scalar function and $\hat{\mathbf{W}}(\omega)$ is the normalized cross-spectral density matrix characterizing the states of polarization at the pinholes located at points $\mathbf{r}_1$ and $\mathbf{r}_2$

$$\hat{\mathbf{W}}(\omega)=\frac{\mathbf{W}(\mathbf{r}_1,\mathbf{r}_1,\omega)}{\operatorname{tr}\mathbf{W}(\mathbf{r}_1,\mathbf{r}_1,\omega)}=\frac{\mathbf{W}(\mathbf{r}_2,\mathbf{r}_2,\omega)}{\operatorname{tr}\mathbf{W}(\mathbf{r}_2,\mathbf{r}_2,\omega)} \tag{1.89}$$

Eq. (1.88) is called *condition of polarization purity* (or better, *condition of coherence–polarization purity*) in the space–frequency domain, so that two electromagnetic waves can be *pure* with respect to coherence–polarization if their spectral density matrix separates into polarization and spatial coherence parts and, in addition, the normalized polarization part is equal to the normalized polarization matrix of the original wave (at the pinholes).

It is worth noting that electromagnetic fields whose analytic signals ε_1 and ε_2 of the respective orthogonal components satisfy the condition $\varepsilon_2 = a\,\varepsilon_1$, a being a random variable independent of $\mathbf{r}$, satisfy the separability of polarization and spatial modulation required for coherence–polarization purity (Hassinen et al. 2013). This kind of separability corresponds to the concept of *non-quantum entanglement* (Simon et al. 2010b, Qian and Eberly 2011), with theoretical importance and potential applications in Photonics.

Let us return to the Stokes parameter representation and recall that, as with the scalar approach, an electromagnetic wave is considered cross-spectrally pure if there is a region around a given point $\mathbf{r}_0$ in the observation plane Π_0 in which the normalized spectral density is the same as at the pinholes. In the case of a quasimonochromatic wave, the condition of electromagnetic cross-spectral purity is given by the pair of relations

$$\hat{s}_0(\mathbf{r}_1,\mathbf{r}_2,\tau)=\hat{s}_0(\mathbf{r}_1,\mathbf{r}_2,\tau_0)\,\hat{s}_0(\mathbf{r}_1,\mathbf{r}_2,\tau-\tau_0) \tag{1.90}$$

$$\hat{s}_0(\mathbf{r}_1,\mathbf{r}_2,\omega)=\hat{s}_0(\mathbf{r}_1,\mathbf{r}_2,\tau_0)\,e^{i\omega\tau_0} \tag{1.91}$$

which are analogous to the pair of necessary and sufficient conditions (1.84) and (1.85) for a scalar wave field to be cross-spectrally pure. Condition (1.90) of separability of temporal and spatial coherence is called the *electromagnetic version of the reduction property* (Hassinen et al. 2009).

Moreover, by taking into account the inequality $|\hat{s}_0(\mathbf{r}_1,\mathbf{r}_2,\tau)|\le 1$, the *reduction condition* (1.90) entails that, for any τ, $|\hat{s}_0(\mathbf{r}_1,\mathbf{r}_2,\tau)|\le|\hat{s}_0(\mathbf{r}_1,\mathbf{r}_2,\tau_0)|$, and therefore, for quasimonochromatic waves, τ_0 is the time delay that corresponds to the maximum fringe visibility. In addition, condition (1.91) shows that $|\hat{s}_0(\mathbf{r}_1,\mathbf{r}_2,\omega)|=|\hat{s}_0(\mathbf{r}_1,\mathbf{r}_2,\tau_0)|$ and consequently $|\hat{s}_0(\mathbf{r}_1,\mathbf{r}_2,\omega)|$ is independent of frequency. A measure of the closeness to cross-spectral purity of an electromagnetic wave at two given points $(\mathbf{r}_1, \mathbf{r}_2)$, is given by the *degree of electromagnetic cross-spectral purity* defined as (Hassinen et al. 2009)

$$\mu_{\text{csp}}(\mathbf{r}_1,\mathbf{r}_2) = 1 - \left|\hat{s}_0(\mathbf{r}_1,\mathbf{r}_2,\tau_0)\right|^2 \frac{\int_0^\infty \left|\psi(\mathbf{r}_1,\mathbf{r}_2,\omega) - \hat{s}_0(\mathbf{r}_1,\omega)\right|^2 d\omega}{\int_0^\infty \left[\left|\psi(\mathbf{r}_1,\mathbf{r}_2,\omega)\right|^2 + \left|\hat{s}_0(\mathbf{r}_1,\omega)\right|^2\right] d\omega} \tag{1.92}$$

where

$$\psi(\mathbf{r}_1,\mathbf{r}_2,\omega) = \hat{s}_0(\mathbf{r}_1,\omega)\frac{\hat{s}_0(\mathbf{r}_1,\mathbf{r}_2,\omega)}{\hat{s}_0(\mathbf{r}_1,\mathbf{r}_2,\tau_0)} e^{-i\omega\tau_0} \tag{1.93}$$

and $\hat{s}_0(\mathbf{r}_1,\omega) = \hat{s}_0(\mathbf{r}_2,\omega)$ is the common normalized spectral density at points $\mathbf{r}_1$ and $\mathbf{r}_2$

$$\hat{s}_0(\mathbf{r}_1,\omega) = \frac{s_0(\mathbf{r}_1,\omega)}{\int_0^\infty s_0(\mathbf{r}_1,\omega)\, d\omega} \tag{1.94}$$

The electromagnetic treatment of cross-spectral purity leads to a variety of physical situations beyond the scalar formulation. For example, a superposition of two cross-spectrally impure states of polarization (in general, elliptical) can be cross-spectrally pure. Conversely, a superposition of two mutually orthogonal cross-spectrally pure states (as, for instance, the components of the electric field of a given electromagnetic wave) can be cross-spectrally impure.

Consider now the complete spectral Stokes vectors $\mathbf{s}(\mathbf{r}_1,\omega)$ and $\mathbf{s}(\mathbf{r}_2,\omega)$ characterizing the states of polarization at the two pinholes located at points $\mathbf{r}_1$ and $\mathbf{r}_2$. These, besides the zeroth parameters considered above, also include the remaining three Stokes parameters corresponding to the respective characteristic components of the states of polarization of the wave at points $\mathbf{r}_1$ and $\mathbf{r}_2$. Again, let us assume that the normalized spectral Stokes parameters of the wave at points $\mathbf{r}_1$ and $\mathbf{r}_2$ (pinholes) are the same at every frequency, i.e., $s_k(\mathbf{r}_2,\omega) = a_k s_k(\mathbf{r}_1,\omega)$, with $a_k > 0$ and $k = 0,1,2,3$.

The coherence–polarization purity condition is satisfied when the spectral distributions of the Stokes parameters are given by Hassinen et al. (2011b)

$$\begin{gathered} s_k(\mathbf{r}_1,\mathbf{r}_2,\omega)\, e^{-i\omega\tau_k} = \hat{s}_k(\mathbf{r}_1,\mathbf{r}_2,\tau_k)\sqrt{s_k(\mathbf{r}_1,\omega)\, s_k(\mathbf{r}_2,\omega)} = f_k(\mathbf{r}_1,\mathbf{r}_2,\tau_k)\sqrt{a_k}\,\left|s_k(\mathbf{r}_1,\omega)\right| \\ (k = 0,1,2,3) \end{gathered} \tag{1.95}$$

where τ_k is the time delay corresponding to the respective point $\mathbf{r}_k$ considered in the observation plane Π_O of interference, $f_k(\mathbf{r}_1,\mathbf{r}_2,\tau_k)$ is a frequency-independent function, and $\hat{s}_k(\mathbf{r}_1,\mathbf{r}_2,\tau_k)$ is just the normalized coherence Stokes parameter in the space–time domain

$$\hat{s}_k(\mathbf{r}_1,\mathbf{r}_2,\tau_k) \equiv \frac{s_k(\mathbf{r}_1,\mathbf{r}_2,\tau_k)}{\sqrt{s_k(\mathbf{r}_1,\mathbf{r}_1,0)\, s_k(\mathbf{r}_2,\mathbf{r}_2,0)}} \qquad (k = 0,1,2,3) \tag{1.96}$$

Obviously, the above normalization is only feasible when $s_k(\mathbf{r}_1,\mathbf{r}_1,0) \neq 0$ (or, equivalently, $s_k(\mathbf{r}_2,\mathbf{r}_2,0) \neq 0$), in which case the conditions for the coherence–polarization purity (or cross-spectral purity) of the Stokes parameters can be formulated as (Hassinen et al. 2011b)

$$\hat{s}_k(\mathbf{r}_1, \mathbf{r}_2, \tau) = \hat{s}_k(\mathbf{r}_1, \mathbf{r}_2, \tau_k)\, \hat{s}_k(\mathbf{r}_1, \mathbf{r}_2, \tau - \tau_k) \tag{1.97}$$

$$\hat{s}_k(\mathbf{r}_1, \mathbf{r}_2, \omega) = \hat{s}_k(\mathbf{r}_1, \mathbf{r}_2, \tau_0)\, e^{i\omega\tau_k} \tag{1.98}$$

in complete analogy with the previous results.

Thus, for a cross-spectrally pure normalized spectral Stokes parameter $\hat{s}_k(\mathbf{r}_1, \omega)$, the corresponding spectral coherence Stokes parameter is a product of a spatial coherence term and a temporal coherence term. Moreover, $|\hat{s}_k(\mathbf{r}_1, \mathbf{r}_2, \omega)| = |\hat{s}_k(\mathbf{r}_1, \mathbf{r}_2, \tau_0)|$ so that $|\hat{s}_k(\mathbf{r}_1, \mathbf{r}_2, \omega)|$ is independent of frequency.

Let us note finally that, when all Stokes parameters are cross-spectrally pure, the space–time electromagnetic degree of coherence equals necessarily the space–frequency electromagnetic degree of coherence (Hassinen et al. 2011b).

1.7 Poincaré Sphere

The *Poincaré vector* **u**, associated with a given Stokes vector **s**, defined as

$$\mathbf{u} \equiv \begin{pmatrix} u_1 \\ u_2 \\ u_3 \end{pmatrix} = \begin{pmatrix} \cos 2\varphi \cos 2\chi \\ \sin 2\varphi \cos 2\chi \\ \sin 2\chi \end{pmatrix} \tag{1.99}$$

is immersed in the Stokes vector in such a manner that **s** can be written in the partitioned form

$$\mathbf{s} = I \begin{pmatrix} 1 \\ \mathcal{P}\mathbf{u} \end{pmatrix} = I \begin{pmatrix} 1 \\ \mathbf{p} \end{pmatrix} \tag{1.100}$$

where the vector, $\mathbf{p} \equiv \mathcal{P}\mathbf{u}$, defined as the product of the degree of polarization $\mathcal{P}$ and the Poincaré vector **u**, is called the *polarization vector* (also *Pauli vector* or *Bloch vector*). Therefore, the normalized Stokes vector $\hat{\mathbf{s}}$ defined as

$$\hat{\mathbf{s}} \equiv \mathbf{s}/I \equiv (1, \hat{s}_1, \hat{s}_2, \hat{s}_3)^T \equiv \begin{pmatrix} 1 \\ \mathcal{P}\mathbf{u} \end{pmatrix} \equiv \begin{pmatrix} 1 \\ \mathbf{p} \end{pmatrix} \tag{1.101}$$

constitutes the algebraic counterpart of the geometric representation of the states of polarization by means of the Poincaré sphere (Poincaré 1892) see Figure 1.13.

All possible states of polarization with unit intensity can be represented graphically with respect to an orthogonal coordinate system constituted by the axes $S_1S_2S_3$ corresponding to the respective values of the intensity-normalized elements $\hat{s}_1, \hat{s}_2, \hat{s}_2$

It is important to emphasize the fact that the Poincaré sphere, despite being very useful and meaningful, is defined with respect to an abstract reference system $S_1S_2S_3$ that does not correspond to any physically realizable laboratory coordinate system. While unitary transformations correspond to general rotations of $S_1S_2S_3$ about any axis that passes through the origin of the sphere, rotation transformations correspond to rotations of $S_1S_2S_3$ about the S_3 axis (thus affecting only to the azimuth of the characteristic polarization ellipse).

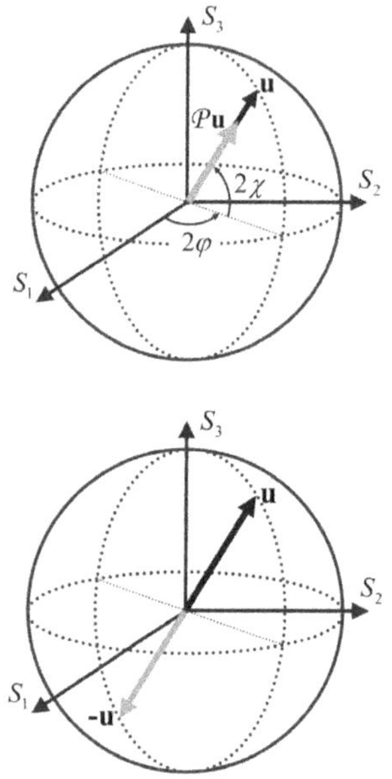

FIGURE 1.13
(a) Representation of normalized states of polarization in the Poincaré sphere. (b) Antipodal points on the surface of the Poincaré sphere correspond to mutually orthogonal pure Stokes vectors.

In the second-order approach, the Poincaré sphere provides a useful geometric representation for all possible 2D states of polarization (Figure 1.13a). The characteristic properties (1.36) of the Stokes parameters show that they define a 4D cone whose generatrix is the axis s_0. The cut of this cone with the plane $s_0 = 1$ gives the *Poincaré sphere*, which is a solid sphere of unit radius where all states have intensity equal to one. Points on the surface ($\mathcal{P} = 1$) represent totally polarized states (pure states) and points inside the sphere ($\mathcal{P} < 1$) represent partially polarized states (mixed states). The origin represents the second-order unpolarized state.

Next, we summarize the main properties of the Poincaré sphere regarding the representation of the different 2D states of polarization:

- The surface ($\mathcal{P} = 1$) corresponds to pure states.
- The origin ($\mathcal{P} = 0$) corresponds to 2D unpolarized states.
- Points inside the Poincaré sphere (i.e., $0 < \mathcal{P} < 1$), correspond to partially polarized states. The smaller the degree of polarization, the closer the state to the origin of the sphere.
- States whose characteristic component is linearly polarized correspond to points contained in the plane $s_3 = 0$. Points $S_{1+}(1,0,0)$, $S_{1-}(-1,0,0)$, $S_{2+}(0,1,0)$ and $S_{2-}(0,-1,0)$ correspond respectively to linearly polarized pure states whose electric fields vibrate along the axes X, Y, +45° and –45°.
- Points on the upper hemisphere correspond to states whose characteristic component has positive ellipticity angle ($0 < \chi \leq \pi/4$). In particular, point $S_{3+}(0,0,1)$ corresponds to right-handed circularly polarized states. It should be noted that in some works the opposite handedness convention is taken, in which case $S_{3+}(0,0,1)$ corresponds to left handed circularly polarized states (Kliger et al. 1990; López-Martínez and Pottier 2021).

- Points in the lower hemisphere correspond to states whose characteristic component has negative ellipticity angle $(-\pi/4 \leq \chi < 0)$. In particular, point $S_{3-}(0,0,-1)$ corresponds to left-handed circularly polarized states.
- Antipodal points on the surface correspond to mutually orthogonal pure states $\mathbf{s} = (1, \mathbf{u}^T)^T$, $\mathbf{s}_\perp = (1, -\mathbf{u}^T)^T$ (Figure 1.13b)

When the electromagnetic wave propagates in a continuous medium as, for instance, light propagating in a birefringent or dichroic medium, the evolution of the state of polarization describes a trajectory in the Poincaré sphere. Depolarizing effects of the medium move the states toward the origin, while *enpolarizing* effects of the medium producing an increase of the degree of polarization move the states toward the surface. General aspects of the polarimetric effects of material media will be dealt with in later chapters.

Due to its simplicity, the Poincaré sphere is, by far, the commonest way for the geometric representation of 2D polarized states. However, it is worth noting that there exist other interesting representations as the stereographic projection plane of the Poincaré sphere (Azzam and Bashara 1987), and the Majorana sphere (which is specific for pure states) (Hannay 1998).

1.8 Polarization Time

As seen in Section 1.2, electromagnetic waves whose spectral profile is not too far from the quasimonochromaticity conditions have a well defined time-dependent instantaneous polarization ellipse and therefore have a well defined instantaneous Jones vector $\boldsymbol{\varepsilon}(t)$ that represents the time-dependent totally polarized states. The temporal evolution of the instantaneous polarization states is referred to as *polarization dynamics* and this section is devoted to summarize the main results of the work of Setälä et al. (2008) dealing with the characterization of the typical time intervals for which a given field maintains constant the shape of the polarization ellipse.

In accordance with Eq. (1.40), the corresponding *instantaneous Stokes vector* is defined as $\mathbf{s}(t) = \mathcal{L}[\boldsymbol{\varepsilon}(t) \otimes \boldsymbol{\varepsilon}^*(t)]$, which obviously satisfies $s_0^2(t) = s_1^2(t) + s_2^2(t) + s_3^2(t)$ (zero randomness). Therefore, the totally polarized $\mathbf{s}(t)$ is fully determined by the three-dimensional vector $\bar{\mathbf{s}}(t) \equiv [s_1(t), s_2(t), s_3(t)]^T$, which can be expressed as $\bar{\mathbf{s}}(t) = s_0(t)\mathbf{u}(t)$, where $s_0(t)$ is the time dependent intensity and $\mathbf{u}(t) \equiv [s_1(t), s_2(t), s_3(t)]^T / s_0(t)$ is the corresponding instantaneous Poincaré vector. As the polarization state evolves in time, $\mathbf{u}(t)$ traces out a path on the Poincaré sphere.

To characterize the temporal evolution of the instantaneous polarization ellipse, Setälä et al. (2008) introduced the following fourth-order correlation function:

$$C(\tau) = \left\langle \bar{\mathbf{s}}^T(t)\, \bar{\mathbf{s}}(t+\tau) \right\rangle = \left\langle \left[\mathbf{u}^T(t)\, \mathbf{u}(t+\tau) \right] s_0(t) s_0(t+\tau) \right\rangle \tag{1.102}$$

where $\langle\rangle$ stands for averaging over time t. Note that the scalar product $\mathbf{u}^T(t)\, \mathbf{u}(t+\tau)$ is closely related to the proximity on the Poincaré sphere of the instantaneous polarization states at times t and $t+\tau$, whose limits are 1, when the states are equal, and -1 when the states are orthogonal. Moreover, factor $s_0(t)s_0(t+\tau)$ plays a non-negligible role in $C(\tau)$ because it constitutes a weight for the scalar product of the Poincaré vectors $\mathbf{u}(t)$ and $\mathbf{u}(t+\tau)$.

Since $C(\tau)$ takes its maximal value $\langle s_0(t)s_0(t+\tau)\rangle$ when the polarization state is constant, an appropriate normalized quantity to characterize the polarization dynamics is given by Setälä et al. (2008)

$$\gamma_p(\tau) = \langle \bar{\mathbf{s}}^T(t)\,\bar{\mathbf{s}}(t+\tau)\rangle / \langle s_0(t)s_0(t+\tau)\rangle \tag{1.103}$$

which provides a measure of the similarity of the polarization states at times t and $t+\tau$, i.e., of the closeness of the states on the Poincaré sphere. $\gamma_p(\tau)$ is constrained to the limits $-1 \le \gamma_p(\tau) \le 1$, while, as expected, $\gamma_p(0) = 1$.

The *polarization time* τ_p is conceived as the time interval over which, on average, the instantaneous polarization state does not significantly change, and can be defined through a specific criterion based on the fact that $\gamma_p(\tau)$ decreases from $\gamma_p(0) = 1$ to a given value, as for instance $\gamma_p(\tau_p) = 1/2$. The corresponding polarization length is defined as $l_p = c\,\tau_p$, c being the speed of light (Setälä et al. 2008).

When the field obeys Gaussian statistics $\gamma_p(\tau)$ adopts the following interesting expression (Setälä et al. 2008):

$$\gamma_p(\tau) = \frac{\mathcal{P}^2 - \gamma^2(\tau) + 2|v(\tau)|^2}{1 + \gamma^2(\tau)} \tag{1.104}$$

where $\mathcal{P}$ is the degree of polarization, $\gamma(\tau)$ is the equal-point electromagnetic degree of coherence, and $v(\tau)$ is the equal-point complex degree of coherence.

The consistency in the definition of $\gamma_p(\tau)$ is also reflected by the fact that $\tau = 0$ corresponds to $\gamma^2(0) = (1+\mathcal{P}^2)/2$ (see Section 1.6.2.4) and $v^2(0) = 1$. Meaningful analyses of the physical significance of the concepts dealt with in this section, together with interesting examples can be found in Setälä et al. 2008, Shevchenko et al. 2009, Voipio et al. 2010, Shevchenko et al. 2012 and Shevchenko et al. 2017.

Even though the concept of instantaneous polarization ellipse has no physical meaning for states with huge bandwidth, as for instance ultrashort light pulses involving a few cycles only, such a concept can be extended by conceiving the evolution of the end point of the electric field through an instantaneous polarization ellipse as for quasimonochromatic pulses (with given instantaneous Stokes parameters) plus an *instantaneous phase* (or *polarization phase*) that specifies the location of the end point of the instantaneous electric field over the ellipse (Porras 2013).

1.9 Intrinsic Polarization Matrix

As seen in Section 1.5, both alternative representations given by the 2D polarization matrix $\mathbf{\Phi}$ and the Stokes vector $\mathbf{s}$ contain complete and equivalent physical information about a 2D state of polarization. The polarimetric purity of a state $\mathbf{\Phi}$ is simply given by the degree of polarization $\mathcal{P}$, which provides a measure of how close to the surface of the Poincaré sphere a state is.

The polarization matrix admits a natural and algebraically immediate generalization to higher dimensional representations and, therefore, it is worth looking deeper into the properties of $\mathbf{\Phi}$.

Let us first note that $\mathbf{\Phi}$ can be expressed as follows in terms of the physical parameters $I,\mathcal{P},\chi,\varphi$ (χ an d ϕ determining the characteristic polarization ellipse and the associated unit Poincaré vector $\mathbf{u}$ in Eq. (1.99)):

$$\begin{aligned}\mathbf{\Phi} &= \frac{I}{2}\begin{pmatrix} 1+\mathcal{P}u_1 & \mathcal{P}(u_2-iu_3) \\ \mathcal{P}(u_2+iu_3) & 1-\mathcal{P}u_1 \end{pmatrix} \\ &= \frac{I}{2}\begin{pmatrix} 1+\mathcal{P}\cos 2\varphi\cos 2\chi & \mathcal{P}(\sin 2\varphi\cos 2\chi - i\sin 2\chi) \\ \mathcal{P}(\sin 2\varphi\cos 2\chi + i\sin 2\chi) & 1-\mathcal{P}\cos 2\varphi\cos 2\chi \end{pmatrix}\end{aligned} \tag{1.105}$$

and, as with Stokes vectors, it can be expressed as the sum of the polarization matrix $\mathbf{\Phi}_p$ of a pure state (*characteristic component*) and the polarization matrix $\mathbf{\Phi}_u$ of a *2D unpolarized state* (both propagating in the same direction),

$$\begin{gathered}\mathbf{\Phi}=\mathbf{\Phi}_p+\mathbf{\Phi}_u \\ \mathbf{\Phi}_p = I\mathcal{P}\hat{\mathbf{\Phi}}_p \qquad \mathbf{\Phi}_u = I(1-\mathcal{P})\hat{\mathbf{\Phi}}_u \\ \hat{\mathbf{\Phi}}_p \equiv \frac{1}{2}\begin{pmatrix} 1+\cos 2\varphi\cos 2\chi & \sin 2\varphi\cos 2\chi - i\sin 2\chi \\ \sin 2\varphi\cos 2\chi + i\sin 2\chi & 1-\cos 2\varphi\cos 2\chi \end{pmatrix} \quad \hat{\mathbf{\Phi}}_u \equiv \frac{1}{2}\begin{pmatrix} 1 & 0 \\ 0 & 1 \end{pmatrix}\end{gathered} \tag{1.106}$$

Eq. (1.106) is the well-known two-dimensional particular expression of the so-called *characteristic*, or *trivial*, decomposition of an n-dimensional polarization matrix (Gil 2007).

As we have seen, both the intensity I and the degree of polarization $\mathcal{P}$ are invariant with respect to changes of the generalized underlying coordinate system $(\mathbf{e}_1,\mathbf{e}_2)$. Coming back to the definition of $\mathbf{\Phi}$, we observe that an arbitrary change of the generalized coordinate system for the representation of the instantaneous Jones vector $\boldsymbol{\varepsilon}$ is given by a 2×2 unitary matrix $\mathbf{U}$, so that $\boldsymbol{\varepsilon}' = \mathbf{U}\boldsymbol{\varepsilon}$ and thus, the polarization matrix that represents the same state of polarization, but referenced with respect to the new basis is given by

$$\mathbf{\Phi}' = \langle \mathbf{U}\boldsymbol{\varepsilon}(t)\otimes\boldsymbol{\varepsilon}^\dagger(t)\mathbf{U}^\dagger\rangle = \mathbf{U}\langle\boldsymbol{\varepsilon}(t)\otimes\boldsymbol{\varepsilon}^\dagger(t)\rangle\mathbf{U}^\dagger = \mathbf{U}\mathbf{\Phi}\mathbf{U}^\dagger \tag{1.107}$$

Unitary transformations preserve the eigenvalues of $\mathbf{\Phi}$, which have the following simple expressions in terms of the intensity I and degree of polarization $\mathcal{P}$:

$$\lambda_1 = \frac{1}{2}I(1+\mathcal{P}) \quad \lambda_2 = \frac{1}{2}I(1-\mathcal{P}) \qquad \left[I = \lambda_1+\lambda_2 = \mathrm{tr}\mathbf{\Phi} \quad \mathcal{P} = \frac{\lambda_1-\lambda_2}{\mathrm{tr}\mathbf{\Phi}}\right] \tag{1.108}$$

Since $\mathbf{\Phi}$ has the mathematical structure of a 2×2covariance matrix, its eigenvalues are non-negative. Consequently, the above expressions allow us to conclude on the non-negativity of I and $\mathcal{P}$, in accordance with their physical meaning.

Thus, we get again the result that I and $\mathcal{P}$ are not only invariant with respect to rotation transformations, but they are furthermore invariant with respect to unitary complex transformations, whose eigenvectors do not correspond to any realizable laboratory coordinate systems in the real domain.

At this point it is important to recall that, as seen in Section 1.6, unitary transformations play a key role in the study of coherence of electromagnetic fields in both space–time and space–frequency domains where the focus is on the correlations of the field variables (determined by the invariant degrees of coherence and by the degree of polarization) rather than in the particular states of polarization of the interfering waves. Nevertheless, when the focus is on polarization, two kinds of unitary transformations should be distinguished, namely orthogonal and nonorthogonal. Orthogonal transformations (hereafter rotation transformations) are represented by real valued unitary matrices (called orthogonal matrices) and can be physically realized by means of rotations of the Cartesian reference axes about the direction of propagation, so that they exclusively affect the azimuth, while the shape of the polarization ellipse is preserved. Nonorthogonal unitary matrices have necessarily off-diagonal elements with nonzero imaginary parts (i.e., cannot be realized through rotations of the laboratory axes) and necessarily produce changes in the ellipticity, so that the shape of the polarization ellipse undergoes the corresponding modification.

Arbitrary rotations of the laboratory coordinate system are formulated mathematically as rotation transformations of the form

$$\mathbf{\Phi}' = \mathbf{Q}\,\mathbf{\Phi}\mathbf{Q}^T \qquad \left(\mathbf{Q}^T = \mathbf{Q}^{-1} \quad \det\mathbf{Q} = +1\right) \tag{1.109}$$

where the orthogonal matrix $\mathbf{Q}$ corresponds to a proper rotation about the axis Z from the original coordinate system XY to the new axes $X'Y'$. Thus, $\mathbf{Q}$ only depends on the rotation angle θ (counterclockwise) and has the form

$$\mathbf{Q}(\theta) = \begin{pmatrix} \cos\theta & \sin\theta \\ -\sin\theta & \cos\theta \end{pmatrix} \tag{1.110}$$

In order to identify the physically invariant quantities involved in the 2D polarization matrix, it is particularly convenient to decompose $\mathbf{\Phi}$ into its real and imaginary parts, $\mathbf{\Phi} = \mathrm{Re}\mathbf{\Phi} + i\,\mathrm{Im}\mathbf{\Phi}$, where the matrix $\mathbf{\Phi}_R \equiv \mathrm{Re}\mathbf{\Phi}$ is symmetric and positive-semidefinite (which thus, in turn, can be considered a polarization matrix representing a certain state of polarization), whereas the imaginary component $\mathbf{\Phi}_I \equiv \mathrm{Im}\mathbf{\Phi}$ is skew-symmetric ($\mathbf{\Phi}_I = -\mathbf{\Phi}_I^T$). Matrix $\mathbf{\Phi}_R$ can always be diagonalized through a particular rotation $\mathbf{Q}_O \equiv \mathbf{Q}(\theta_O)$ of the coordinate system (Dennis 2004)

$$\mathbf{Q}_O\,\mathbf{\Phi}_R\,\mathbf{Q}_O^T = \mathrm{diag}(a_1, a_2) \tag{1.111a}$$

with the choice

$$0 \le a_2 \le a_1 \tag{1.111b}$$

where θ_O is the counterclockwise angle rotated about the propagation axis Z from X to its corresponding transformed axis X_O (recall also that the columns of $\mathbf{Q}_O$ are the eigenvectors of $\mathbf{\Phi}_R$).

The real polarization matrix $\mathbf{\Phi}_R$ defines the so-called *intensity ellipse* (not to be confused with the polarization ellipse whose semiaxes are field amplitudes instead of intensities) whose semiaxes (a_1, a_2) lie along the respective transformed axes $X_O Y_O$ (Figure 1.14). Therefore, $\mathrm{diag}(a_1, a_2)$ can be interpreted as the polarization matrix of a state composed of the incoherent superposition of two linearly polarized pure states

$$\mathrm{diag}(a_1, a_2) = \mathbf{\Phi}_{p1} + \mathbf{\Phi}_{p2} \qquad \left[\mathbf{\Phi}_{p1} \equiv a_1\,\mathrm{diag}(1,0) \quad \mathbf{\Phi}_{p2} \equiv a_2\,\mathrm{diag}(0,1)\right] \tag{1.112}$$

with respective intensities a_1 and a_2 and propagating in the direction $\mathbf{k}$ along the reference axis Z.

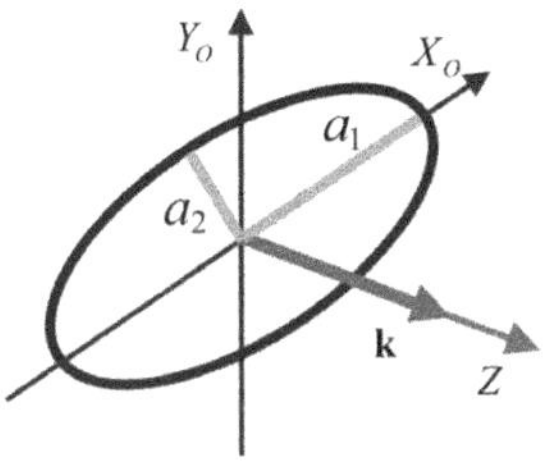

FIGURE 1.14
Intensity ellipse, with semiaxes $a_1 \geq a_2$, representing a mixed state constituted by the incoherent superposition of two mutually orthogonal pure linearly polarized states $\mathbf{\Phi}_{p1}$ and $\mathbf{\Phi}_{p2}$ (with respective intensities a_1, a_2), both propagating in direction **k** along the reference axis Z.

By applying the rotation $\mathbf{Q}_O$ to the whole matrix $\mathbf{\Phi}$, we observe that the real and imaginary parts transform separately, so that we get the transformed polarization matrix (Dennis 2004; Gil 2014a)

$$\mathbf{\Phi}_O \equiv \mathbf{Q}_O \mathbf{\Phi} \mathbf{Q}_O^T = \mathbf{Q}_O \mathbf{\Phi}_R \mathbf{Q}_O^T + i\mathbf{Q}_O \mathbf{\Phi}_I \mathbf{Q}_O^T = \begin{pmatrix} a_1 & 0 \\ 0 & a_2 \end{pmatrix} + i\begin{pmatrix} 0 & -n/2 \\ n/2 & 0 \end{pmatrix} = \begin{pmatrix} a_1 & -in/2 \\ in/2 & a_2 \end{pmatrix}$$
$$\left[0 \leq a_2 \leq a_1 \quad 4a_1 a_2 \geq n^2\right] \tag{1.113}$$

where $n = I\mathcal{P}\sin 2\chi$ is invariant with respect to arbitrary rotation transformations and can be either zero ($\chi = 0$, linearly-polarized characteristic component), positive ($\chi > 0$, right-handed characteristic component), or negative ($\chi < 0$, left-handed characteristic component).

From $\mathbf{\Phi} = \mathbf{Q}_O^T \mathbf{\Phi}_O \mathbf{Q}_O$, we deduce that $\mathbf{\Phi}$ can be parameterized through the following four independent parameters: the orientation angle θ_O, the semiaxes (a_1, a_2) (called the *principal intensities*) of the *intensity ellipse*, and the helicity $n = s_3$ given by the single component (along the Z axis) of the spin vector **n** of the wave at the point **r** considered. Observe that the spin vector of a 2D polarization state is always normal to the *polarization plane* Π containing the polarization ellipse (Gil et al. 2021, see Section 1.10).

At this point, it is important to stress that while $\mathbf{\Phi}_R$ is an auxiliary polarization matrix that should not be confused with $\mathbf{\Phi}$ nor $\mathbf{\Phi}_O$, the *intrinsic polarization matrix* $\mathbf{\Phi}_O$ (Gil 2014a) represents the same state as $\mathbf{\Phi}$, but being referenced with respect to the coordinate system $X_O Y_O$ obtained by a specific counterclockwise rotation transformation (Figure 1.15),

$$\mathbf{\Phi}_O \equiv \mathbf{Q}_O \mathbf{\Phi} \mathbf{Q}_O^T = \begin{pmatrix} \cos\theta_O & \sin\theta_O \\ -\sin\theta_O & \cos\theta_O \end{pmatrix} \frac{1}{2} \begin{pmatrix} s_0 + s_1 & s_2 - is_3 \\ s_2 + is_3 & s_0 - s_1 \end{pmatrix} \begin{pmatrix} \cos\theta_O & -\sin\theta_O \\ \sin\theta_O & \cos\theta_O \end{pmatrix}$$
$$= \frac{1}{2} \begin{pmatrix} s_0 + \sqrt{s_1^2 + s_2^2} & -is_3 \\ is_3 & s_0 - \sqrt{s_1^2 + s_2^2} \end{pmatrix} \tag{1.114}$$

where the rotation angle θ_O is determined by

$$\sin 2\theta_O = s_2 \Big/ \sqrt{s_1^2 + s_2^2} \qquad \cos 2\theta_O = s_1 \Big/ \sqrt{s_1^2 + s_2^2} \tag{1.115}$$

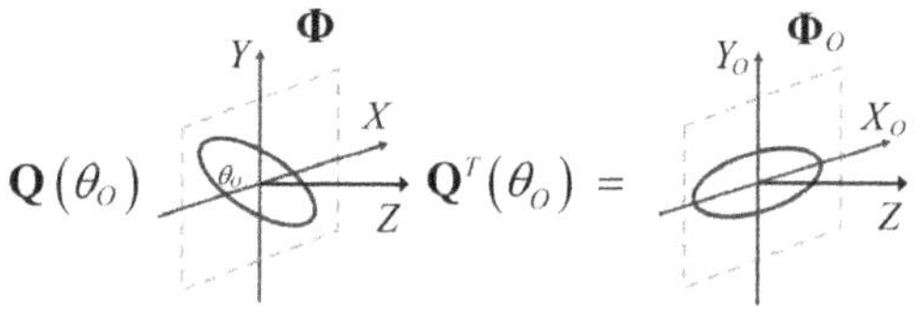

FIGURE 1.15
The 2D polarization matrix $\mathbf{\Phi}$ is transformed to its intrinsic form $\mathbf{\Phi}_O$ through an appropriate rotation about the propagation axis Z.

while the expressions of the principal intensities and the helicity in terms of the original Stokes parameters are the following:

$$a_1 = \frac{1}{2}\left(s_0 + \sqrt{s_1^2 + s_2^2}\right) \qquad a_2 = \frac{1}{2}\left(s_0 - \sqrt{s_1^2 + s_2^2}\right) \qquad n = s_3 \tag{1.116}$$

The fact that the *intrinsic polarization matrix* $\mathbf{\Phi}_O$ is positive semidefinite is consistent with its expression in terms of the intrinsic Stokes parameters, and implies that the quantities (a_1, a_2, n) satisfy the constraining inequalities $a_1 \geq a_2 \geq 0$ and $4a_1a_2 \geq n^2$ indicated in Eq. (1.113), so that, up to the limits set by these restrictive conditions, the three intrinsic quantities (a_1, a_2, n) are independent. Note that the helicity $n = s_3$ is invariant under rotation of the axes XY about the axis Z normal to the plane containing the fluctuating electric field vector.

Furthermore, the intensity I of the wave can expressed in the following alternative forms:

$$I = \mathrm{tr}\mathbf{\Phi} = \lambda_1 + \lambda_2 = \mathrm{tr}\mathbf{\Phi}_O = \mathrm{tr}\mathbf{\Phi}_{p1} + \mathrm{tr}\mathbf{\Phi}_{p2} = a_1 + a_2 \tag{1.117}$$

To inspect the nature and properties of the intrinsic polarization matrix, let us first remark that the *intrinsic Stokes vector* $\mathbf{s}_O$ associated with $\mathbf{\Phi}_O$ is given by

$$\mathbf{s}_O = \left(s_0, \sqrt{s_1^2 + s_2^2}, 0, s_3\right)^T = (a_1 + a_2, a_1 - a_2, 0, n)^T \tag{1.118}$$

where $\mathbf{s} = (s_0, s_1, s_2, s_3)^T$ is the Stokes vector associated with $\mathbf{\Phi}$. The components of $\mathbf{s}_O$ will be denoted as $\mathbf{s}_O \equiv I(1, \mathcal{P}_l, 0, \sigma\mathcal{P}_c)^T$ in terms of the intensity I, the *degree of linear polarization* $\mathcal{P}_l$, the handedness $\sigma \equiv n/|n| = s_3/|s_3|$ and the *degree of circular polarization* $\mathcal{P}_c$ (the intrinsic quantities $\mathcal{P}_l$ and $\mathcal{P}_c$ will be dealt with in more detail in Section 1.11.2). Observe also that from the definition of $\mathbf{\Phi}_O$ it follows that the quantity $\mathcal{P}_l = (a_1 - a_2)/(2I)$ is necessarily nonnegative and thus the negative value $-\sqrt{s_1^2 + s_2^2}$ does not apply to Eq. (1.118). Therefore, the characteristic polarization ellipse with a given azimuth φ with respect to the reference axes XY, has zero azimuth when the state is referenced with respect to the intrinsic coordinate system X_OY_O obtained from XY through a counterclockwise rotation $\theta_O = \varphi$ about the Z axis (Figure 1.16). In the particular case where the polarization state is constituted by any incoherent composition of a circularly polarized state and an unpolarized state, the linear Stokes component vanishes and the intrinsic Stokes vector takes the form $\mathbf{s}_O = I(1, 0, 0, \sigma\mathcal{P}_c)^T$, while the corresponding intensity ellipse degenerates to a circle ($a_1 = a_2$).

Therefore, the rotation produced by the matrix $\mathbf{Q}_O$ that diagonalizes $\mathbf{\Phi}_R$ leads to the cancellation of the third component of the intrinsic Stokes vector $\mathbf{s}_O$ and all *linearly polarized*

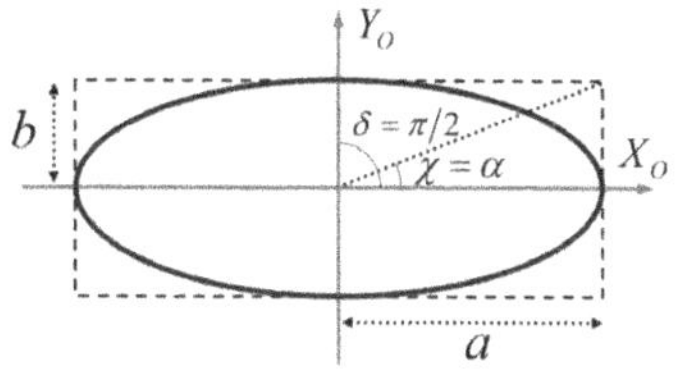

FIGURE 1.16
The characteristic polarization ellipse with a given azimuth ϕ with respect to the coordinate system XY, has zero azimuth when the state is referenced with respect to the intrinsic coordinate system $X_O Y_O$ obtained from XY through a counterclockwise rotation $\theta_O = \varphi$ about the Z axis.

charge is concentrated in the second component of $\mathbf{s}_O$. Consequently, as with the semiaxes of the intensity ellipse, the semiaxes of the intrinsic polarization ellipse associated with $\mathbf{s}_O$ (i.e., with $\mathbf{\Phi}_O$) lie along the *intrinsic reference axes* $X_O Y_O$ (Figure 1.16).

Given a state $\mathbf{s}$ (or $\mathbf{\Phi}$) the characteristic parameters of $\mathbf{s}_O$ (or $\mathbf{\Phi}_O$) can be expressed as (Gil 2015; 2016b)

$$a_1 = \frac{1}{2} I\left(1 + \mathcal{P}_l\right) \qquad a_2 = \frac{1}{2} I\left(1 - \mathcal{P}_l\right) \qquad n = s_3 = I\sigma\mathcal{P}_c \tag{1.119}$$

Moreover, the semiaxes a, b, of the polarization ellipse are given by Eq. (1.46) and can also be expressed as follows in terms of the intrinsic parameters:

$$\begin{aligned} a^2 &= \frac{1}{2}\left(\sqrt{\left(a_1 - a_2\right)^2 + n^2} + a_1 - a_2\right) = \frac{1}{2} I\left(\mathcal{P} + \mathcal{P}_l\right) \\ b^2 &= \frac{1}{2}\left(\sqrt{\left(a_1 - a_2\right)^2 + n^2} - a_1 + a_2\right) = \frac{1}{2} I\left(\mathcal{P} - \mathcal{P}_l\right) \end{aligned} \tag{1.120}$$

For pure states $(\mathcal{P} = 1)$, the principal intensities are given by the squared semiaxes $a_1 = a^2$, and $a_2 = b^2$, while for mixed states $a_1 > a^2$ and $a_2 > b^2$.

It is remarkable that the absolute value $|n| = |s_3| = I\mathcal{P}_c$ of the spin of the state gives a measure of the area $A = \pi ab$ of the polarization ellipse, $|n| = 2A/\pi$.

1.10 Concept of Spin of a Polarization State

As suggested by Poynting (1909) and experimentally observed by Beth (1936), the intrinsic angular momentum of a fully polarized electromagnetic wave at a given point in space is determined by the amount of circular polarization associated with the polarization state. Such an intimate relation between intrinsic angular momentum and circular polarization is also known to hold in the case of random light (for both 2D and 3D approaches), for which the degree of circular polarization is obtained from the imaginary part of the corresponding polarization matrix (Dennis 2004).

Although the concept of polarization refers to the time evolution of the electric field vector at a single point in space, it is worth recalling first the fact that, from a strict theoretical point of view, a perfect plane wave (even if it is circularly polarized) cannot carry angular momentum because both electric and magnetic fields lie on the surface of the entire infinite wavefront (Simmons and Guttman 1970). This apparent paradox is solved by considering real observation conditions (finite wavefront) and from the fact that any measurement or interaction with matter necessarily involves a finite cross section, in such a manner that an axial component of the fields is generated at the contour limiting the wavefront (or the cross section). In fact, when a molecule is interacted by the beam, it can pick up some angular momentum and thus modify the transverse profile of the beam and generate components of the field in the direction normal to the central area of the wavefront (Allen and Padgett 2002; Yurchenko 2002; Stewart 2005b; Barnett et al. 2012).

When dealing with electromagnetic waves, the angular momentum is frequently referred to as *optical angular momentum* and involves two physical components called the *orbital angular momentum* (OAM) and the spin angular momentum (SAM, or simply *spin*). The separability of these well defined physical components involves certain subtle (but important) issues mainly originated by the fact that, while the OAM and the SAM are well defined and constitute respective significant properties of a light beam, in general they are not true angular momenta because their quantized forms do not satisfy the required commutation relations (Van Enk and Nienhuis 1994a; 1994b). Successive relevant approaches and clarifications on this matter have been developed by several authors (Allen et al. 1992; Barnett and Allen 1994; Stewart 2005a; Fernandez-Corbatón et al. 2012; Bialynicki-Birula and Bialynicki-Birula 2011; Bliokh et al. 2013; Bliokh et al. 2014; Barnett et al. 2016; Bliokh et al. 2017; Picardi et al. 2018).

From the point of view of the purposes of this section, focused on the concept of spin of a 2D state, some preliminary important notions are summarized below:

(a) The SAM is associated with the degree of circular polarization of a polarization state, while the orbital angular momentum is associated with helical phase fronts (hence it is not associated with the polarization state, which is realized at a single point, Allen et al. 2003).
(b) In classical optics, the angular momentum of a light beam has two well-defined parts, namely the OAM and the SAM, which are meaningful physical quantities and, for the case of paraxial beams, can properly be separated.
(c) From the point of view of polarization, where the field properties are considered at a fixed single point $\mathbf{r}$, the *spin vector* $\mathbf{n}$ (with dimensions of intensity) is a pseudovector (i.e., it can always be expressed as the vector product of two vectors as shown below), which is well defined from the imaginary part of the polarization matrix. In particular, for 2D states (totally or partially polarized), $\mathbf{n}$ lies along the Z axis orthogonal to the polarization plane, and the sense of $\mathbf{n}$ is determined by the sign of the ellipticity angle χ, which in turn determines the handedness (Barnett et al. 2012).
(d) As seen in Section 1.5.3, the helicity of a 2D state is determined by the Stokes parameter s_3, which is a pseudoscalar quantity, that is, it behaves mathematically like a scalar, but it changes sign under reflection. Such change is consistent with the fact that the spin vector remains unchanged under reflection. Nevertheless, it should be noted that since polarization is defined locally for a given point $\mathbf{r}$ in space, the specific information on the sense of the direction of propagation is contained neither in the polarization matrix nor in the Stokes vector. Inasmuch as the helicity s_3 is the only nonzero component of the spin vector of a 2D state (in the standard representation where the Z axis is normal to the polarization plane), it is usually referred to as the *spin* (while $\mathbf{n}$ is referred to as the *spin vector* of the 2D state considered).

The spin angular momentum vector **S** of a quasimonochromatic wave is composed of two parts, namely the electric and magnetic spin vectors $\mathbf{S}=\mathbf{S}_E+\mathbf{S}_M$ (Bekshaev and Soskin 2007; Berry 2009; Neugebauer, et al. 2015), defined as (with angular momentum dimensions)

$$\mathbf{S}_E=(\varepsilon_o/4\varpi)\langle \mathrm{Im}[\varepsilon^*(t)\wedge\varepsilon(t)]\rangle \quad \mathbf{S}_M=(\mu_o/4\varpi)\langle \mathrm{Im}[\mathbf{h}^*(t)\wedge\mathbf{h}(t)]\rangle \tag{1.121}$$

where the symbol $\wedge$ represents vector product (or cross product), $\langle\ \rangle$ stands for time average over the measurement time (assumed that it involves a number of natural cycles), ϖ is the mean frequency, ε_o is the electric permittivity of the medium (assumed isotropic), μ_o is the magnetic permeability, while vectors $\boldsymbol{\varepsilon}$ and $\mathbf{h}$ are the *electric* and *magnetic* Jones vectors whose respective components are the analytic signals of the components of the electric and magnetic fields of the wave (for which the harmonic dependences are removed, as done for the definition of Jones vector in Eq. (1.15)).

Since the light–matter interaction in the optical range takes place predominantly via the electric field, and in accordance with the polarization description, which is formulated by means of the time evolution of the electric field, hereafter we will consider only the *electric spin*, and we will refer to it simply as the *spin*. Since polarization properties are defined for each single point in space, the quantities in Eq. (1.121) are not volume integrated and are referred to in related works as spin densities, in the sense that they correspond to the point **r** considered. However, the term *density* is also frequently applied to intensity-normalized quantities (as for instance the polarization density matrix), and consequently hereafter we will use the term *spin density vector* to the intensity-normalized version $\hat{\mathbf{n}}=\mathbf{n}/I$ of the spin vector **n**. Indeed, from a polarimetric point of view, we are interested in the magnitude of the spin rather than its dimensions of angular momentum, and the absolute value $|\mathbf{n}|$ of the spin vector of a 2D polarization state is identified with the absolute value $|s_3|$ of the helicity s_3 (with dimensions of intensity), while its intensity-normalized version (hence dimensionless) $\hat{s}_3$, whose absolute value equals the degree of circular polarization $\mathcal{P}_c=|\hat{s}_3|=|\hat{\mathbf{n}}|$, appears as the relevant quantity for many purposes (Berestetskii et al. 1982; Gil 2014a; Gil 2016a; Gil et al. 2021).

For a proper formulation of the electric spin vector of a 2D polarization state, let us consider its fluctuating electric field vector lying in a plane containing the reference axes *XY*, so that the analytic signal vector $\boldsymbol{\varepsilon}$ can be represented as $\boldsymbol{\varepsilon}=(\varepsilon_1,\varepsilon_2,0)^T$ with respect to the Cartesian coordinate system *XYZ*. Then, the dimensionless *spin density vector* $\hat{\mathbf{n}}$ is defined as

$$\begin{gathered}\hat{\mathbf{n}}=\langle \mathrm{Im}[\boldsymbol{\varepsilon}^*(t)\wedge\boldsymbol{\varepsilon}(t)]\rangle/I=\left\langle \mathrm{Im}\left[0,0,\varepsilon_1^*(t)\varepsilon_2(t)-\varepsilon_2^*(t)\varepsilon_1(t)\right]^T\right\rangle/I=(0,0,\hat{s}_3)^T\\ \left[I=\langle\varepsilon_1(t)\varepsilon_1^*(t)\rangle+\langle\varepsilon_2(t)\varepsilon_2^*(t)\rangle \quad \hat{s}_3=\sigma\mathcal{P}_c=\mathcal{P}\sin 2\chi\right]\end{gathered} \tag{1.122}$$

where σ is the handedness ($\sigma=+1$ for right-handed states and $\sigma=-1$ for left-handed states). The above definition shows that the real pseudovector $\hat{\mathbf{n}}$ can be expressed as $\hat{\mathbf{n}}=\hat{s}_3\mathbf{k}$, **k** being the unit vector along the positive direction of the *Z* axis. Therefore, the *spin density* $\hat{n}\equiv\hat{s}_3$ is restricted to $-1\le\hat{n}\le 1$, so that $\hat{n}=+1$ corresponds to right-handed circularly polarized pure states; $\hat{n}=-1$ corresponds to left-handed circularly polarized pure states, and $\hat{n}=0$ is reached when $s_3=0$, i.e., for states that are associated with any incoherent composition of a linearly polarized state and an unpolarized state. Moreover, the absolute value of the spin density vector is precisely the degree of circular polarization $|\hat{\mathbf{n}}|=\mathcal{P}_c$.

Thus, the direction of $\hat{\mathbf{n}}$ is always given by $\mathbf{k}$ for 2D states, and its single component $\hat{s}_3$ is determined univocally by the imaginary part of the polarization matrix $\mathbf{\Phi}$. In fact, $|\hat{s}_3| = \sqrt{2}\|\mathrm{Im}\hat{\mathbf{\Phi}}\|_2$, and therefore, $\mathrm{Im}\mathbf{\Phi} = \mathbf{0}$ if and only if $\hat{\mathbf{n}} = \mathbf{0}$, which means that real valued polarization matrices correspond to states that lack spin.

The spin vector $\mathbf{n}$, as a polarization descriptor (with dimensions of intensity), is given by

$$\mathbf{n} = I\hat{\mathbf{n}} = s_3\,\mathbf{k} = (I\sigma\mathcal{P}_c)\,\mathbf{k} = (I\mathcal{P}\sin 2\chi)\,\mathbf{k} \tag{1.123}$$

Regarding the definition of the spin vector of a polarization state as a vector quantity $\mathbf{S}_E$ with dimensions of angular momentum, the space–time representation allows to do it easily in the case of quasimonochromatic waves, for which the instantaneous Jones vector $\boldsymbol{\varepsilon}_J$ is well defined (and hence, an instantaneous spin angular momentum $\mathbf{S}_E$ is also well defined, at least for a number of natural cycles), so that

$$\mathbf{S}_E = \langle \mathbf{S}_E(t)\rangle = (\varepsilon_o/4\varpi)\,s_3\,\mathbf{k} = (\varepsilon_o/4\varpi)\,\mathbf{n} \tag{1.124}$$

where ϖ is the mean frequency.

In the case of high polychromatic waves, the instantaneous Jones vector may not be well defined and the above definition cannot be applied. Nevertheless, an alternative general definition of the spin angular momentum of a polarization state, applicable for waves of arbitrary spectral profile is achieved as follows through the frequency-domain representation (Gil et al. 2022a):

$$\mathbf{S}_E = \int_0^\infty (\varepsilon_o/4\omega)\left\langle \mathrm{Im}\left[\boldsymbol{\varepsilon}^*(\omega)\wedge\boldsymbol{\varepsilon}(\omega)\right]\right\rangle_\omega d\omega = \varepsilon_o\mathbf{k}\int_0^\infty (1/4\omega)\langle s_3(\omega)\rangle_\omega\, d\omega \tag{1.125}$$

where $\langle\ \rangle_\omega$ stands for ensemble average over the ensemble of equivalent single-frequency monochromatic realizations $s_3(\omega)$, and $\mathbf{k}$ is a unit vector along the Z axis normal to the polarization plane.

The average spin angular momentum can be expressed as follows in terms of $\mathbf{n}$, ε_o and an *equivalent spin frequency* ω_n (Gil et al. 2022a)

$$\mathbf{S}_E = (\varepsilon_o/4\omega_n)\,\mathbf{n} \qquad \left[\omega_n \equiv s_3\Big/\left(\int_0^\infty (1/\omega)\langle s_3(\omega)\rangle_\omega\, d\omega\right) \quad \langle s_3(\omega)/\omega\rangle_{\omega=0} = 0\right] \tag{1.126}$$

In the case of quasimonochromatic fields the equivalent spin frequency coincides with the mean frequency, $\omega_n = \varpi$.

Let us finally consider some interesting aspects of the concept of helicity of a 2D polarization state. From its very definition, vector $\mathbf{n}$ (and $\mathbf{S}_E$) lies necessarily on the Z axis, and obviously, the components of $\mathbf{n}$ in the axes XY are strictly zero, while the component of $\mathbf{n}$ in the direction of propagation (the *polarization helicity*) is none other than s_3. On the other hand, the polarization vector $\mathbf{p} \equiv (\hat{s}_1, \hat{s}_2, \hat{s}_3)^T$ (see Eq. (1.100)) is defined with respect to the abstract Poincaré coordinate system $S_1S_2S_3$, and not with respect to the laboratory reference system XYZ; note that while the characteristic polarization ellipse lies in the plane XY, in general it does not lie in the plane S_1S_2 and, despite the fact that the projection of $\mathbf{p}$ on axis S_3 is precisely the helicity, axis S_3 should not be identified with the Z axis. The above indicated coincidence is just a peculiarity of 2D states, but, as described in Chapter 2, it is no longer applicable for general 3D polarization states.

Furthermore, since a 2D pure polarization state is determined by its spin vector (up to its azimuth and spatial orientation of its polarization plane), in quantum physics polarization of photons is identified with the concept of spin. In fact, the quantum Stokes operators, associated with the corresponding Pauli spin matrices, formally satisfy the commutation relations of an angular momentum of spin 1/2 particles (Luis and Rodil 2014). Although this quantum formulation apparently suggests that **p** may be identified with the spin vector, from a polarimetric point of view, and in accordance to the foundations of this concept (Poynting 1909; Beth 1936; Berestetskii et al. 1982; Dennis 2004), the spin vector is always associated with the degree of circular polarization, and therefore it is identified with **n**, which is exclusively determined by the imaginary part of the polarization matrix. Note, in passing, that the most general representation of quantum spin operators of photons (massless spin 1 particles) requires the use of the eight Gell-Mann matrices instead of the Pauli ones.

As will be shown in Section 2.11, from a classical point of view, the concept of polarization helicity is only well defined for 2D states, and loses its physical meaning for genuine 3D states (i.e., those states that exhibit three nonzero components of the electric field vector with respect to any laboratory coordinate system XYZ).

1.11 Polarimetric Purity

As seen in previous sections, while 2D unpolarized light does not exhibit any physical or geometric anisotropy in the averaged fluctuating field (with respect to the plane Π containing it), the *polarization* of a state always corresponds to certain degree of anisotropy, which is linked to the shape and handedness of the average polarization ellipse (Gil et al. 2019a). It has been proven that such anisotropy can be decomposed into two complementary parts, namely the intensity anisotropy (determined by the degree of linear polarization) and the spin anisotropy (determined by the magnitude of the spin vector) (Gil et al. 2019a).

Polarimetric purity refers to how close is a polarized state to its totally polarized characteristic component, and should be interpreted as a measure of the amount of anisotropy mentioned in the above paragraph. In the case of 2D states of polarization, the polarimetric purity is, therefore, given by the value of the degree of polarization $\mathcal{P}$ which, in turn, is a measure of the distance of a polarization state to the origin of the Poincaré sphere.

Moreover, a 2D polarization state is completely polarized at a point **r** if and only if the analytic signals ε_1, ε_2 of the components of the electric field are statistically similar, i.e., if and only if $\varepsilon_1(t) = a\,\varepsilon_2(t)$, where the complex coefficient a is a deterministic function of **r**, independent of time (Roychowdhury and Wolf 2005).

The concept of *polarimetric purity* should not be confused with that of *cross-spectral purity*, which, as indicated in Section 1.6.3, refers to the separability of the complex degree of coherence into spatial and temporal coherence terms, as well as to the independence with respect to frequency of the absolute value of the spectral degree of coherence.

1.11.1 2D Degree of Polarization

Consider the pair of invariant measures given by the Frobenius norm $\|\mathbf{\Phi}\|_2$ and the *trace norm* $\|\mathbf{\Phi}\|_{tr}$ of $\mathbf{\Phi}$, defined as (Gil 2007)

$$\|\mathbf{\Phi}\|_2 \equiv \sqrt{\mathrm{tr}\mathbf{\Phi}^2} = \sqrt{\mathrm{tr}\mathbf{\Phi}_O^2} \quad \|\mathbf{\Phi}\|_{tr} \equiv \mathrm{tr}\mathbf{\Phi} = \mathrm{tr}\mathbf{\Phi}_O \tag{1.127}$$

so that the intensity and the degree of polarization can be expressed as

$$I = \lambda_1 + \lambda_2 = \|\mathbf{\Phi}\|_{tr} \qquad \mathcal{P} = \frac{\lambda_1 - \lambda_2}{\lambda_1 + \lambda_2} = \sqrt{\frac{2\|\mathbf{\Phi}\|_2^2}{\|\mathbf{\Phi}\|_{tr}^2} - 1} \tag{1.128}$$

Another pair of meaningful norms can be defined for Stokes vectors, namely, the Euclidean norm,

$$\|\mathbf{s}\|_2 \equiv \sqrt{\mathbf{s}^T \mathbf{s}} = \sqrt{s_0^2 + s_1^2 + s_2^2 + s_3^2} = 2\|\mathbf{\Phi}\|_2^2 \tag{1.129}$$

and the *max-norm*,

$$\|\mathbf{s}\|_0 \equiv s_0 = \|\mathbf{\Phi}\|_{tr} \equiv \mathrm{tr}\mathbf{\Phi} \tag{1.130}$$

The intensity and the degree of polarization can also be expressed in terms of these norms of **s** as

$$I = \|\mathbf{s}\|_0 \qquad \mathcal{P} = \sqrt{\|\mathbf{s}\|_2^2 / \|\mathbf{s}\|_0^2 - 1} = \sqrt{\|\hat{\mathbf{s}}\|_2^2 - 1} \qquad \left(\hat{\mathbf{s}} \equiv \mathbf{s}/\|\mathbf{s}\|_0 = \mathbf{s}/s_0\right) \tag{1.131}$$

It is straightforward to show that the above norms satisfy the following inequalities:

$$\frac{1}{2}\|\mathbf{\Phi}\|_{tr}^2 \leq \|\mathbf{\Phi}\|_2^2 \leq \|\mathbf{\Phi}\|_{tr}^2 \qquad \|\mathbf{s}\|_0^2 \leq \|\mathbf{s}\|_2^2 \leq 2\|\mathbf{s}\|_0^2 \tag{1.132}$$

Pure states are characterized by the fact that they satisfy $\|\mathbf{\Phi}\|_2^2 = \|\mathbf{\Phi}\|_{tr}^2$ ($\|\mathbf{s}\|_2 = \sqrt{2}\|\mathbf{s}\|_0$), while the other limit, $\|\mathbf{\Phi}\|_2^2 = \|\mathbf{\Phi}\|_{tr}^2/2$ ($\|\mathbf{s}\|_2 = \|\mathbf{s}\|_0$), occurs in the case of 2D unpolarized states.

Even though the above relations in terms of the indicated norms are relatively trivial, their main interest comes from the fact that they constitute the two-dimensional versions of more general properties that will be retrieved in later sections for 3D polarization matrices associated with 3D polarization states of electromagnetic waves and for 4D covariance matrices representing the polarimetric properties of material media.

Further, the expressions of the 2D degree of polarization in terms of the ordinary and intrinsic Stokes parameters are given by

$$\mathcal{P} = \sqrt{s_1^2 + s_2^2 + s_3^2}/s_0 = \sqrt{\mathcal{P}_l^2 + \mathcal{P}_c^2} \tag{1.133}$$

In the case of 2D states that we are dealing with in the present chapter, the general notion of degree of polarization $\mathcal{P}$ takes the form of the ratio between the intensity of the totally polarized part (i.e., the intensity of the characteristic component) and the total intensity of the state of polarization considered, so that $\mathcal{P}$ can also be expressed in the commonly used form

$$\mathcal{P} = \sqrt{1 - 4\frac{\det\mathbf{\Phi}}{\mathrm{tr}\mathbf{\Phi}^2}} = \sqrt{1 - 4\frac{\det\mathbf{\Phi}}{\|\mathbf{\Phi}\|_2^2}} \tag{1.134}$$

which, unlike the expression of $\mathcal{P}$ in terms of the norms of $\mathbf{\Phi}$ in Eq. (1.128), is not generalizable to higher-order coherency matrices.

In general, the degree of polarization refers to the lack of symmetry of the polarization matrix and it involves two kinds of contributions, namely (1) the lack of symmetry of the diagonal elements of $\mathbf{\Phi}$ (i.e., how different are they: intensity anisotropy), and (2) how different from zero the imaginary part of $\mathbf{\Phi}$ is (spin anisotropy). Such components will be studied in Section 1.11.2.

The expression of $\mathcal{P}$ in Eq. (1.128) can be transformed into (Luis 2005a)

$$\mathcal{P}^2 = 2\,\mathrm{tr}\left[\left(\mathbf{I}_2/2 - \mathbf{\Phi}/\mathrm{tr}\mathbf{\Phi}\right)^2\right] = 2\,\mathrm{tr}\left[\left(\hat{\mathbf{\Phi}}_u - \hat{\mathbf{\Phi}}\right)^2\right] \tag{1.135}$$

where $\mathbf{I}_2$ is the 2×2 identity matrix, which is twice the polarization density matrix $\hat{\mathbf{\Phi}}_u$ of a 2D unpolarized state $\left(\mathbf{I}_2 = 2\hat{\mathbf{\Phi}}_u\right)$ Thus, expression (1.135) shows that $\mathcal{P}$ is a measure of the distance between the polarization density matrix $\hat{\mathbf{\Phi}}_u$ of a 2D unpolarized state and the polarization density matrix $\hat{\mathbf{\Phi}}$ (Luis 2005a). Eq. (1.135) can also be considered a particular case of the polarimetric distance $\mathcal{P}_{a,b}$ between two states $\hat{\mathbf{\Phi}}_a$ and $\hat{\mathbf{\Phi}}_b$ defined as

$$\mathcal{P}_{a,b}^2 = 2\,\mathrm{tr}\left[\left(\hat{\mathbf{\Phi}}_a - \hat{\mathbf{\Phi}}_b\right)^2\right] \tag{1.136}$$

which can easily be converted into the following expression of $\mathcal{P}_{a,b}$ in terms of the difference between the polarization vectors $\mathbf{p}_a$ and $\mathbf{p}_b$:

$$\mathcal{P}_{a,b} = \left|\mathbf{p}_a - \mathbf{p}_b\right| \tag{1.137}$$

The polarimetric distance is thus given by the absolute value of the difference vector between the polarization vectors of the pair of states considered. The maximum distance corresponds to pairs of orthogonal pure states (antipodal in the Poincaré sphere), $\mathcal{P}_{a,a\perp} = \left|\mathbf{p}_a - (-\mathbf{p}_a)\right| = 2$, and the minimum corresponds to the case where the two states are represented by the same point in the Poincaré sphere, $\mathcal{P}_{a,a} = 0$.

To go deeper in the concept of polarimetric purity, let us consider now the coherence–polarization representations and recall that, in general, the concepts of space–time and space–frequency degrees of polarization of a given stochastic electromagnetic beam are not equivalent. Nevertheless, under certain conditions they are directly related; these conditions determine the concept of the so-called *polarization purity* (Lahiri 2013), which is closely related to the concept of cross-spectral purity and should not be confused with the concept of polarimetric purity (which is quantified through the usual degree of polarization). To avoid confusion, as in previous sections, we refer to *polarization purity* as *coherence–polarization purity*.

From the formulation of the one-point cross-spectral density matrix, obtained from the reduction of Eq. (1.51) to the one-point case ($\mathbf{r}_1 = \mathbf{r}_2 \equiv \mathbf{r}$), the spectral degree of polarization is defined as

$$\mathcal{P}(\omega) = \sqrt{2\left\|\mathbf{W}(\omega)\right\|_2^2 / \left\|\mathbf{W}(\omega)\right\|_{tr}^2 - 1} \tag{1.138}$$

so that it is limited by $0 \le \mathcal{P}(\omega) \le 1$. The possibility of spectral partial polarization stems from the fact that a random polychromatic field at a given frequency could be partially coherent and partially polarized. $\mathcal{P}(\omega)$ depends on the temporal coherence properties of the components of the electric field and, naturally, can have different values at different frequencies. Thus, the degree of polarization in the time and in the frequency domains are, in general, not equivalent (Setälä et al. 2009a; Lahiri 2009; Setälä et al. 2010b; Réfrégier et al. 2012).

In the case of a 2D quasimonochromatic wave having a very small bandwidth, this quantity $\mathcal{P}(\omega)$, when computed at its mean frequency, $\mathcal{P}(\varpi)$, equals the usual degree of polarization $\mathcal{P}(\varpi)=\mathcal{P}$ (Lahiri and Wolf 2010). The conditions for the equivalence of the two degrees of polarization have been studied by Setälä et al. (2009a) and formulated in a general form by Lahiri (2013).

To complete this review of the concept of polarimetric purity, let us now consider the *equal-point Luis degree of coherence* $D(\mathbf{r},\mathbf{r},\omega)$ and observe that it is precisely the spectral degree of polarization (Luis 2007),

$$D^2(\mathbf{r},\mathbf{r},\omega)\equiv D^2(\omega)=\frac{1}{n-1}\left[n\frac{4\|\mathbf{\Phi}(\omega)\|_2^2}{4I^2}-1\right]=2\frac{\|\mathbf{\Phi}(\omega)\|_2^2}{I^2}-1=\mathcal{P}^2(\omega)\quad [n=2] \tag{1.139}$$

where the value $n=2$ corresponding to the one-point case $(\mathbf{r}\equiv\mathbf{r}_2=\mathbf{r}_1)$ has been taken.

Moreover, the equal-point $(\mathbf{r}\equiv\mathbf{r}_2=\mathbf{r}_1)$, equal-time $(\tau=0)$, expression of the electromagnetic degree of coherence $\gamma(\mathbf{r}_1,\mathbf{r}_2,\tau)$ defined in Eq. (1.77) is

$$\gamma\equiv\gamma(\mathbf{r},\mathbf{r},0)=\|\mathbf{\Phi}\|_2/\|\mathbf{\Phi}\|_{tr}=\sqrt{\mathrm{tr}\mathbf{\Phi}^2}/\mathrm{tr}\mathbf{\Phi}=\sqrt{\lambda_1^2+\lambda_2^2}/(\lambda_1+\lambda_2) \tag{1.140}$$

where λ_1, λ_2 are the eigenvalues of $\mathbf{\Phi}$, so that $\gamma(\mathbf{r},\mathbf{r},0)$ is an invariant quantity directly related to the degree of polarization $\mathcal{P}$ (Tervo et al. 2003),

$$\mathcal{P}=\sqrt{2\gamma^2-1}\qquad \gamma=\sqrt{(1+\mathcal{P}^2)/2}\qquad [\gamma\equiv\gamma(\mathbf{r},\mathbf{r},0)] \tag{1.141}$$

and therefore, $\mathcal{P}$ and $\gamma(\mathbf{r},\mathbf{r},0)$ provide equivalent information. The value of the equal-point, equal-time, degree of coherence γ is constrained by the limits $1/\sqrt{2}\le\gamma\le 1$. The minimum corresponds to a 2D unpolarized state $\gamma(\mathbf{\Phi}_{u-2D})=1/\sqrt{2}$, while the maximum corresponds to pure states $\gamma(\mathbf{\Phi}_p)=1$.

1.11.2 Components of purity of a 2D polarization state

Despite the fact that the characteristic parameters I, $\mathcal{P}$, ϕ, and χ of a Stokes vector $\mathbf{s}$ provide complete information about the state of polarization represented by $\mathbf{s}$, several different parameters, useful for certain purposes, can be defined from $\mathbf{s}$. We bring up two interesting examples of additional quantities derivable from $\mathbf{s}$, namely, the *degree of linear polarization* $\mathcal{P}_l$ and the *degree of circular polarization* $\mathcal{P}_c$ (Chipman 1995), called the *components of polarimetric purity* of $\mathbf{s}$, and defined as (see Section 1.9)

$$\mathcal{P}_l\equiv\sqrt{s_1^2+s_2^2}/s_0=\mathcal{P}\cos 2\chi\qquad \mathcal{P}_c\equiv|s_3|/s_0=\mathcal{P}|\sin 2\chi| \tag{1.142}$$

$\mathcal{P}_l$ and $\mathcal{P}_c$ are, respectively, measures of the maximum portion of the power of the wave that can be assigned as linearly polarized and circularly polarized. Both $\mathcal{P}_l$ and $\mathcal{P}_c$ are defined as positive parameters $(-\pi/4\le\chi\le\pi/4\Rightarrow 0\le\cos 2\chi)$ satisfying $0\le\mathcal{P}_l\le 1$ and $0\le\mathcal{P}_c\le 1$. The convention taken for the definition of $\mathcal{P}_c$ as a positive parameter is consistent with the definition of $\mathcal{P}_c$ for general 3D states (see Section 2.9.1) and therefore the helicity density $\hat{s}_3$ of the 2D state can be expressed as $\hat{s}_3 = \sigma\mathcal{P}_c$, where $\sigma \equiv s_3/|s_3|$ is the handedness.

Both components of purity $\mathcal{P}_l$ and $\mathcal{P}_c$ are invariant with respect to arbitrary rotation transformations (i.e., rotations of the reference laboratory coordinate system XYZ about the Z axis orthogonal the polarization plane XY (Gil 2016b)). The components of purity satisfy the relation

$$\mathcal{P} = \sqrt{\mathcal{P}_l^2 + \mathcal{P}_c^2} \tag{1.143}$$

so that $\mathcal{P}_l$ and $\mathcal{P}_c$ constitute independent and complementary contributions to the polarimetric purity of **s** (Figure 1.17b). In other words, the set of four quantities $I, \mathcal{P}_l, \sigma\mathcal{P}_c, \varphi$ provides a parameterization of **s** (or $\mathbf{\Phi}$) alternative to that given by the set $I, \mathcal{P}, \varphi, \chi$. When the interest is not focused on the orientation (azimuth φ) of the characteristic polarization ellipse within the polarization plane XY, then it is enough to consider the set $I, \mathcal{P}_l, \sigma\mathcal{P}_c$ of three *rotationally invariant* quantities. In particular, $I, \mathcal{P}_l, \sigma\mathcal{P}_c$ uniquely determine the intrinsic polarization matrix $\mathbf{\Phi}_O$ and the intrinsic Stokes vector $\mathbf{s}_O$ (Gil 2015),

$$\mathbf{\Phi}_O = \frac{1}{2} I \begin{pmatrix} 1+\mathcal{P}_l & -i\sigma\mathcal{P}_c \\ i\sigma\mathcal{P}_c & 1-\mathcal{P}_l \end{pmatrix} \qquad \mathbf{s}_O = I\left(1, \mathcal{P}_l, 0, \sigma\mathcal{P}_c\right)^T \tag{1.144}$$

As seen in Section 1.9, the principal intensities can be expressed as

$$a_1 = I\left(1+\mathcal{P}_l\right)/2 \qquad a_2 = I\left(1-\mathcal{P}_l\right)/2 \tag{1.145}$$

while the lengths a, b of the semiaxes of the characteristic polarization ellipse are given by

$$a = \sqrt{I\left(\mathcal{P}+\mathcal{P}_l\right)/2} \qquad b = \sqrt{I\left(\mathcal{P}-\mathcal{P}_l\right)/2} \tag{1.146}$$

Moreover, the ellipticity $R \equiv b/a = |\tan\chi|$ is another rotationally invariant quantity that can be expressed as a function of the degree of polarization $\mathcal{P}$ and the degree of linear polarization $\mathcal{P}_l$ as $R = \sqrt{\mathcal{P}-\mathcal{P}_l}/\sqrt{\mathcal{P}+\mathcal{P}_l}$.

The normalized area $\hat{A} \equiv A/I$ of the polarization ellipse can be expressed as $\hat{A} = \pi\mathcal{P}_c/2$. This relation, together with Eq. (1.146), reinforces the interpretation of $\mathcal{P}_l$ and $\mathcal{P}_c$ in terms of the shape of the characteristic polarization ellipse.

It should be noted again that, while the polarization matrix $\mathbf{\Phi}$ has two parameters that are *unitarily invariant* (i.e., invariant with respect to unitary transformations) namely, the eigenvalues (λ_1, λ_2) or alternatively, any pair of quantities derived from them, for instance $(I, \mathcal{P})$, $\mathbf{\Phi}$ has three rotationally invariant parameters, namely the set $(I, \mathcal{P}, \chi)$, or any alternative equivalent set of three parameters derived from them, like (a_1, a_2, n) or $(I, \mathcal{P}_l, \sigma\mathcal{P}_c)$. Thus, the pair $(\mathcal{P}_l, \sigma\mathcal{P}_c)$ provides detailed information on the allocation of the purity beyond the overall information provided by the degree of polarization $\mathcal{P} = \sqrt{\mathcal{P}_l^2 + \mathcal{P}_c^2}$.

Figure 1.17a shows a geometric representation of Eq. (1.143) in the Poincaré sphere, where the polarization vector $\mathcal{P}\mathbf{u}$ (**u** being the Poincaré vector of the corresponding Stokes vector **s**) is decomposed into two orthogonal components, namely, a component of magnitude $\mathcal{P}_c$ (that accounts for the amount of spin anisotropy) along the axis S_3 (*circular polarization axis*, also denoted as S_c) and a component of magnitude $\mathcal{P}_l$ (that accounts for the amount of intensity anisotropy) along the *linear polarization axis* S_l in the direction of the projection of **u** on the plane S_1S_2. Thus, in general $\mathcal{P}_l = \mathcal{P}\cos 2\chi$ and $\mathcal{P}_c = \mathcal{P}|\sin 2\chi|$,

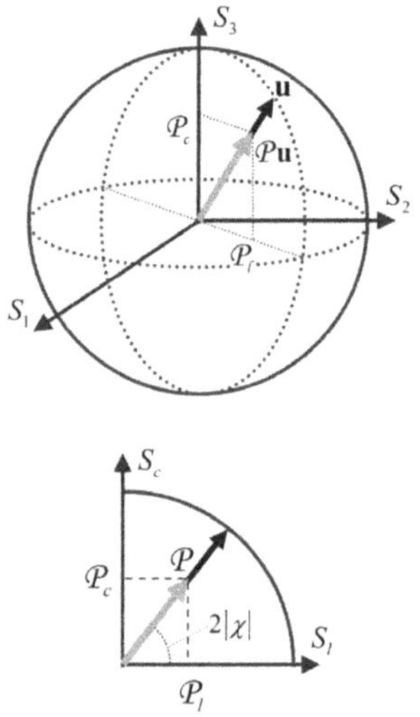

FIGURE 1.17
(a) Geometric representation of the fundamental relation $\mathcal{P} = \sqrt{\mathcal{P}_l^2 + \mathcal{P}_c^2}$: the polarization vector $\mathcal{P}\,\mathbf{u}$ is decomposed into two orthogonal components, namely a circular component of magnitude $\mathcal{P}_c$ along the axis S_3, and a linear component of magnitude $\mathcal{P}_l$ along an axis S_l in the direction of the projection on the plane orthogonal to S_c (S_3). (b) Purity quadrant: the 2D degree of polarization $\mathcal{P}$ is determined by the square average of the degrees of linear and circular polarization, which are represented as the components of purity $\mathcal{P}_l = \mathcal{P}\cos 2\chi$ and $\mathcal{P}_c = \mathcal{P}|\sin 2\chi|$ along the respective linear and circular positive axes S_l and S_c.

with $\mathcal{P}_l = \mathcal{P}$ ($\chi = 0$) for states whose characteristic component is linearly polarized, and $\mathcal{P}_c = \mathcal{P}$ ($\chi = \pm\pi/4$) for states whose characteristic component is circularly polarized. The components of purity summarize, in a meaningful manner, the information of polarization state regarding the way in which the polarimetric purity is distributed between the linear and circular components, which in turn admit a simple geometric representation by means of the *purity quadrant* (Figure 1.17b).

1.11.3 Degree of Mutual Coherence

The degree of mutual coherence μ between the components of the electric field of a given state of polarization is a complex quantity whose absolute value depends on the coordinate system considered. Let us first observe that there always exists a unitary transformation $\mathbf{U}\boldsymbol{\Phi}\mathbf{U}^{\dagger} = \mathrm{diag}(\lambda_1, \lambda_2)$ diagonalizing the polarization matrix $\boldsymbol{\Phi}$, so that the transformed state of polarization represented by the polarization matrix $\mathrm{diag}(\lambda_1, \lambda_2)$ preserves the intensity and the degree of polarization (but does not preserve, in general, the azimuth and the ellipticity) and its degree of mutual coherence is zero. Furthermore, as shown below (Wolf 1959), there always exists a rotation transformation $\boldsymbol{\Phi}_{\psi} = \mathbf{Q}\boldsymbol{\Phi}\mathbf{Q}^T$ (i.e., a rotation of the laboratory Cartesian coordinate system about the axis Z determined by the direction of propagation) such that the transformed polarization matrix $\boldsymbol{\Phi}_{\psi}$ has equal diagonal elements. Therefore, with regard of all possible unitary transformations of $\boldsymbol{\Phi}$, the maximum achievable value of $|\mu|$ corresponds to $\boldsymbol{\Phi}_{\psi}$ and coincides with the degree of polarization (Wolf 1959).

Once the extremal values of $|\mu|$ with respect to unitary transformations of $\boldsymbol{\Phi}$ have been determined, let us recall that only the unitary transformations that are orthogonal preserve the ellipticity of the polarization state, i.e., the transformed state can be considered as the same as that represented by $\mathbf{\Phi}$ but referenced with respect to a rotated Cartesian coordinate system. Therefore, as seen in Section 1.11.2, if we consider the rotationally invariant properties of a given state of polarization, such intrinsic polarization properties are determined by the intensity I, the degree of polarization $\mathcal{P}$ and the ellipticity angle χ. Below, the limits of the absolute value of the degree of mutual coherence of a given polarization state $(I, \mathcal{P}_l, \sigma\mathcal{P}_c)$ are obtained.

From the very definition of μ, it follows that $|\mu|$ can be expressed as follows

$$|\mu| = \sqrt{\frac{\phi_{12}\,\phi_{21}}{\phi_{11}\,\phi_{22}}} = \sqrt{\frac{\mathcal{P}^2 - \mathcal{P}^2\cos^2 2\varphi\cos^2 2\chi}{1 - \mathcal{P}^2\cos^2 2\varphi\cos^2 2\chi}} = \sqrt{\frac{P_c^2 + P_l^2\sin^2 2\varphi}{1 - P_l^2\cos^2 2\varphi}} \tag{1.147}$$

so that, as expected, $0 \le |\mu| \le \mathcal{P}$ (Born and Wolf 2005, p. 629; Wolf 1959). For unpolarized states, the degree of mutual coherence is zero ($\mu = 0$), while for pure states the equality $|\mu| = \mathcal{P} = 1$ holds necessarily. For mixed states, the maximum $|\mu|_{max} = \mathcal{P}$ corresponds to (a) states whose characteristic component is circularly polarized ($\chi = \pm\pi/4$, $\mathcal{P}_l = 0$, $\mathcal{P} = \mathcal{P}_c$), or (b) states whose characteristic polarization ellipse has azimuth $\varphi = \pi/4, 3\pi/4$; that is to say, $|\mu| = \mathcal{P}$ for the Cartesian coordinate systems in which the strengths of the two orthogonal components of the electric field are equal, i.e., in which the Stokes parameter s_1 is zero (Figure 1.18) . In other words, it is always possible to find a coordinate system $X_\psi Y_\psi$ in which $s_1 = 0$, so that $|\mu|$ reaches its maximal value $|\mu|_{max} = \mathcal{P}$. This interesting result, which was obtained by Wolf in a classic paper (Wolf 1959), provides an additional view of the concept of $\mathcal{P}$ as the maximum value of $|\mu|$ with respect to the possible coordinate systems in the polarization plane Π.

Given a polarization state represented by a Stokes vector $\mathbf{s}$ referenced with respect to an arbitrary coordinate system XY, the same state satisfies $|\mu| = \mathcal{P}$ for a system $X_\psi Y_\psi$ rotated with respect to XY by one of the two alternative angles, ψ_1 or ψ_2, determined by

$$\psi_1 \begin{Bmatrix} \sin 2\psi_1 = -s_1\big/\sqrt{s_1^2 + s_2^2} = -\hat{s}_1/\mathcal{P}_l \\ \cos 2\psi_1 = s_2\big/\sqrt{s_1^2 + s_2^2} = \hat{s}_2/\mathcal{P}_l \end{Bmatrix} \quad \psi_2 = \psi_1 + \pi/2 \tag{1.148}$$

By comparing Eqs. (1.148) and (1.115) we see that $\psi_1 = \theta_O + \pi/4$ and consequently $\boldsymbol{\psi}$ and θ_O correspond to coordinate systems forming an angle $\pi/4$ (Figure 1.18), i.e., $|\mu|$ has its maximum value $|\mu|_{max} = \mathcal{P}$ when the state of polarization is referenced with respect to a coordinate system $X_\psi Y_\psi$ which is oriented at 45° or at 135° with respect to the intrinsic coordinate system $X_O Y_O$.

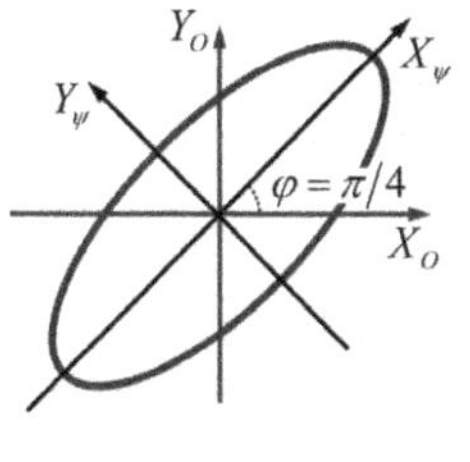

FIGURE 1.18
$|\mu| = \mathcal{P}$ for the laboratory coordinate system in which the Stokes parameter s_1 is zero.

The polarization matrix adopts the following forms when it is referenced with respect to $X_\psi Y_\psi$:

$$\boldsymbol{\Phi}_{\psi_1} = \frac{1}{2}\begin{pmatrix} a_1 + a_2 & a_1 - a_2 - in \\ a_1 - a_2 + in & a_1 + a_2 \end{pmatrix} \qquad \boldsymbol{\Phi}_{\psi 2} = \frac{1}{2}\begin{pmatrix} a_1 + a_2 & -a_1 + a_2 - in \\ -a_1 + a_2 + in & a_1 + a_2 \end{pmatrix} \tag{1.149}$$

Moreover, for a given state of polarization (determined by $(I, \mathcal{P}_l, \sigma\mathcal{P}_c)$) the minimal value of $|\mu|$ is given by Gil (2016d)

$$|\mu|_{\min} = \sqrt{\frac{\mathcal{P}^2 \sin^2 2\chi}{1-\mathcal{P}^2 \cos^2 2\chi}} = \sqrt{\frac{\mathcal{P}_c^2}{1-\mathcal{P}_l^2}} \tag{1.150}$$

which corresponds to states whose characteristic polarization ellipse has azimuth $\varphi = 0, \pi/2$; that is to say, $|\mu|$ takes its minimal value $|\mu|_{\min}$ for the laboratory coordinate systems in which the difference between the strengths of the two orthogonal components is maximal. This condition is satisfied only when $\boldsymbol{\Phi}$ is referenced with respect to the intrinsic coordinate system $X_O Y_O$ or to a coordinate system orthogonal to it. In other words, the intrinsic polarization matrix $\boldsymbol{\Phi}_O$ has the property that $|\mu|$ attains its minimal physically achievable value $|\mu|_{\min}$ given by Eq. (1.150). Therefore, given a polarization state represented by a Stokes vector **s** referenced with respect to an arbitrary coordinate system XY, the same state has the minimum value of $|\mu|$ when referenced with respect to a coordinate system $X_\xi Y_\xi$ rotated with respect to XY by an angle ξ given by either of the following two alternative values (Gil 2016d):

$$\xi_1 = \theta_O = \psi_1 - \pi/4 \qquad \xi_2 = \theta + \pi/2 = \psi_1 + \pi/4 = \psi_2 - \pi/4 \tag{1.151}$$

Unlike $|\mu|_{\max}$, $|\mu|_{\min}$ depends on the ellipticity angle χ, and this dependence is illustrated by means of the following particular cases:

- $|\mu|_{\min} = 0$ for 2D states with zero spin, i.e., 2D states that either lacks characteristic component (unpolarized states), or whose characteristic component is linearly polarized. This result is consistent with the fact that, when $\chi = 0$, the minor semiaxis b of the characteristic polarization ellipse is zero.
- $|\mu|$ has the fixed value $|\mu| = |\mu|_{\min} = |\mu|_{\max} = \mathcal{P}$ for mixed states whose characteristic component has circular polarization (i.e., $\chi = \pm\pi/4$).
- For the intermediate case of a mixed state whose characteristic component has elliptical polarization with $\chi = \pi/8$, then $|\mu|_{\min} = \mathcal{P}/\sqrt{2-\mathcal{P}^2}$.

1.11.4 Polarization Entropy

The polarimetric purity, i.e., the closeness of a mixed state to a pure state, has been characterized by the degree of polarization $\mathcal{P}$ and has been analyzed through several formulations in terms of appropriate norms defined for coherency matrices and Stokes vectors. Nevertheless, for some purposes it is of high interest to consider an alternative view of polarimetric purity by means of the concept of the von Neumann entropy S associated with a 2D state of polarization (Fano 1957; O'Neill 1963; Barakat 1983; Brosseau 1998). Given a 2D polarization density matrix $\hat{\boldsymbol{\Phi}} \equiv \boldsymbol{\Phi}/\mathrm{tr}\boldsymbol{\Phi}$, S is defined as the dimensionless quantity

$$S = -\mathrm{tr}\left(\hat{\mathbf{\Phi}} \log_2 \hat{\mathbf{\Phi}}\right) = -\sum_{i=1}^{2}\left(\hat{\lambda}_i \log_2 \hat{\lambda}_i\right) \tag{1.152}$$

where $\hat{\lambda}_i = \lambda_i / \mathrm{tr}\mathbf{\Phi}$ are the normalized eigenvalues of $\mathbf{\Phi}$.

Even though the Napierian logarithm (ln) is commonly used for the definition of the von Neumann entropy (Von Neumann 1996), the use of the base n logarithm for the definition of the entropy of nD density matrices has the advantage of restricting the value of S to the range $0 \le S \le 1$.

S is interpreted as a measure of the difference in the amount of information between a pure state and the mixed state considered (both with the same intensity), and therefore it has a close relation to $\mathcal{P}$ (Barakat and Brosseau 1996),

$$S = -\mathcal{P}_+ \log_2 \mathcal{P}_+ - \mathcal{P}_- \log_2 \mathcal{P}_- \qquad \left[\mathcal{P}_+ \equiv (1+\mathcal{P})/2 \quad \mathcal{P}_- \equiv (1-\mathcal{P})/2\right] \tag{1.153}$$

so that S is characterized uniquely by $\mathcal{P}$ and decreases monotonically as $\mathcal{P}$ increases. The maximum $S = \log_2 2 = 1$ corresponds to $\mathcal{P} = 0$ (unpolarized state), whereas the minimum $S = 0$ is reached for $\mathcal{P} = 1$ (i.e., when the electromagnetic wave is totally polarized, regardless of its spectral profile) (Brosseau 1998).

The relation of S with the concept of "specific radiation entropy" was introduced by Plank and formulated explicitly by von Laue. Moreover, Barakat (1983) extended the concept of polarization entropy to n pencils of radiation, which was also studied by Barakat and Brosseau (1992), who expressed the normalized specific radiation entropy as the difference of two von Neumann entropies.

Furthermore, Brosseau and Bicout (1994) considered a *thermodynamic view* of the polarization matrix and defined the *temperature* of *polarization*

$$\tau \equiv \frac{2}{\ln(1+\mathcal{P}) - \ln(1-\mathcal{P})} \tag{1.154}$$

which is closely related to the concepts of degree of polarization and polarization entropy. Polarization temperature is zero for pure states and increases monotonically as $\mathcal{P}$ decreases.

As we will see in later chapters, the concept of *polarization entropy*, when related to the depolarizing effects of a material medium, is commonly used to analyze measurements and images obtained in remote sensing through lidar detection and through synthetic aperture radar polarimetry (SAR Polarimetry) in order to detect spatial heterogeneity in a great variety of terrestrial and oceanic surface targets.

1.12 Composition and Decomposition of Two-Dimensional States of Polarization

1.12.1 Coherent Composition and Decomposition of 2D Pure States

As mentioned in Section 1.4, the coherent composition, at a given point **r**, of two pure states whose polarization ellipses lie in the same plane is given by the sum (Figure 1.9)

$$\boldsymbol{\varepsilon} = \boldsymbol{\varepsilon}_1 + \boldsymbol{\varepsilon}_2 \tag{1.155}$$

where $\boldsymbol{\varepsilon}_1$ and $\boldsymbol{\varepsilon}_2$ are the respective Jones vectors of the individual pure states and $\boldsymbol{\varepsilon}$ is the Jones vector of the combined state. Since the combined state is pure, the Jones vector can be considered constant (with intensity equal to the average intensity), so that its associated pure polarization matrix can be calculated as

$$\boldsymbol{\Phi} = \boldsymbol{\varepsilon} \otimes \boldsymbol{\varepsilon}^\dagger = (\boldsymbol{\varepsilon}_1 + \boldsymbol{\varepsilon}_2) \otimes (\boldsymbol{\varepsilon}_1 + \boldsymbol{\varepsilon}_2)^\dagger = \boldsymbol{\Phi}_1 + \boldsymbol{\Phi}_2 + \boldsymbol{\varepsilon}_1 \otimes \boldsymbol{\varepsilon}_2^\dagger + \boldsymbol{\varepsilon}_2 \otimes \boldsymbol{\varepsilon}_1^\dagger \quad (1.156)$$

where $\boldsymbol{\Phi}_1 = \boldsymbol{\varepsilon}_1 \otimes \boldsymbol{\varepsilon}_1^\dagger$, and $\boldsymbol{\Phi}_2 = \boldsymbol{\varepsilon}_2 \otimes \boldsymbol{\varepsilon}_2^\dagger$. Thus, except for the trivial case where $\boldsymbol{\varepsilon}_1 = \boldsymbol{\varepsilon}_2$, the inequality $\boldsymbol{\Phi} \neq \boldsymbol{\Phi}_1 + \boldsymbol{\Phi}_2$ is satisfied (recall that fully coherent composition is assumed here).

Unlike Jones vectors, which are defined in terms of amplitudes and phases, Stokes vectors are defined in terms of intensities and therefore do not carry information on the relative phases of the composed states, so that the sum of Stokes vectors is only applicable for incoherent superpositions. As with any kind of interference phenomenon, the coherent and incoherent superpositions lead to different results that constitute limiting situations of the general case of partially coherent compositions, where the components are partially correlated among them (see Section 1.12.3).

From a polarimetric point of view, totally polarized polychromatic states are undistinguishable from totally coherent states with the same state of polarization. Thus, a peculiarity of the sum of Jones vectors is that, given a pure state $\boldsymbol{\varepsilon}$, it can always be considered a coherent superposition $\boldsymbol{\varepsilon} = \boldsymbol{\varepsilon}_1 + \boldsymbol{\varepsilon}_2 + \ldots + \boldsymbol{\varepsilon}_i \ldots + \boldsymbol{\varepsilon}_n$ of an arbitrary number n of pure states $\boldsymbol{\varepsilon}_i$, so that all such possible compositions are polarimetrically equivalent, because they produce the same polarization matrix $\boldsymbol{\Phi}$ of the combined state.

Recall that the analysis of the composition and decomposition of states of polarization is considered for a given point $\mathbf{r}$. In the case of a number of uniform plane waves propagating along a common direction, the indicated compositions and decompositions can be generalized for the whole wavefront. A similar generalization can be applied to the superposition of pure states with arbitrary wavefronts that overlap in a certain surface region.

1.12.2 Incoherent Composition and Decomposition of 2D Mixed States

Consider now the incoherent superposition, at a given point $\mathbf{r}$, of two states of polarization whose electric fields lie in a common plane and characterized by respective matrices $\boldsymbol{\Phi}_1$ and $\boldsymbol{\Phi}_2$. The polarization matrix $\boldsymbol{\Phi}$ associated with the combined state is given by $\boldsymbol{\Phi} = \boldsymbol{\Phi}_1 + \boldsymbol{\Phi}_2$. The total intensity is $I = \mathrm{tr}\boldsymbol{\Phi} = \mathrm{tr}\boldsymbol{\Phi}_1 + \mathrm{tr}\boldsymbol{\Phi}_2$. Except for the trivial case where $\boldsymbol{\Phi}_1 = \boldsymbol{\Phi}_2$ the resulting state is partially polarized with a degree of polarization $\mathcal{P}$ that decreases as the polarimetric distance between the components increases. Obviously, a given pure state $\boldsymbol{\Phi}_p$ never can be described as the incoherent composition of two different polarization states. Conversely, a given mixed state $\boldsymbol{\Phi}$ can be obtained as the superposition of a number of states $\boldsymbol{\Phi}_i$ satisfying $\boldsymbol{\Phi} = \sum \boldsymbol{\Phi}_i$. All these possible decompositions of $\boldsymbol{\Phi}$ are polarimetrically indistinguishable, i.e., the only physical information obtainable by polarimetric measurements is that provided by $\boldsymbol{\Phi}$.

Thus, a given polarization matrix $\boldsymbol{\Phi}$ representing a 2D mixed state can be submitted to several incoherent decompositions. Since the geometric representation of polarization states in the Poincaré sphere has a direct link with the structure of Stokes vectors, we formulate next the incoherent compositions and decompositions of states in terms of Stokes vectors. All results shown in this section can be translated to the polarization matrix formalism in a straightforward manner.

Let us first consider the so-called *spectral decomposition*,

$$\mathbf{s} = I\begin{pmatrix} 1 \\ \mathcal{P}\mathbf{u} \end{pmatrix} = \frac{1+\mathcal{P}}{2} I\begin{pmatrix} 1 \\ \mathbf{u} \end{pmatrix} + \frac{1-\mathcal{P}}{2} I\begin{pmatrix} 1 \\ -\mathbf{u} \end{pmatrix} \tag{1.157}$$

where the mixed state **s** is represented as the convex sum of two orthogonal pure states (i.e., two pure states that are represented by antipodal points on the surface of the Poincaré sphere, Figure 1.13b). The name *spectral* comes from the fact that the coefficients in the expansion (1.157) are precisely the eigenvalues of the polarization matrix $\mathbf{\Phi}(\mathbf{s})$, so that the relative weight of the pure components in the spectral decomposition is regulated by the degree of polarization $\mathcal{P}$.

Despite the fact that the spectral decomposition has the genuine property that its components are mutually orthogonal, it is a particular example of the infinite number of possible decompositions of a Stokes vector **s** of a mixed state into a sum of two pure Stokes vectors $\mathbf{s}_1$, $\mathbf{s}_2$ (i.e., an incoherent superposition) of the form

$$\mathbf{s}(I,\mathcal{P},\mathbf{u}) \equiv I\begin{pmatrix} 1 \\ \mathcal{P}\mathbf{u} \end{pmatrix} = pI\hat{\mathbf{s}}_p(\mathbf{v}) + (1-p)I\hat{\mathbf{s}}_p(\mathbf{w}) \qquad \left[\hat{\mathbf{s}}_p(\mathbf{v}) \equiv \begin{pmatrix} 1 \\ \mathbf{v} \end{pmatrix} \quad \hat{\mathbf{s}}_p(\mathbf{w}) \equiv \begin{pmatrix} 1 \\ \mathbf{w} \end{pmatrix}\right] \tag{1.158}$$

where the Poincaré vectors **v** and **w** of the pure components are linearly independent unit vectors, and $0 < p < 1$.

Given a mixed state $\mathbf{s}(I,\mathcal{P},\mathbf{u})$, any pure state $I\hat{\mathbf{s}}_p(\mathbf{v})$ is susceptible to be considered a component of **s**. Once a particular (arbitrary) pure component $\hat{\mathbf{s}}_p(\mathbf{v})$ has been chosen, the coefficient p as well as the remainder pure component $\hat{\mathbf{s}}_p(\mathbf{w})$ are uniquely determined by means of (Gil 2007)

$$p = \frac{1-\mathcal{P}^2}{2\left(1-\mathcal{P}\mathbf{u}^T\mathbf{v}\right)} \qquad \mathbf{w} = \frac{\mathcal{P}\mathbf{u} - p\mathbf{v}}{1-p} \qquad (0 < p < 1) \tag{1.159}$$

Provided that **s** is a mixed state, it is straightforward to show that condition $0 < p < 1$ is always satisfied. This *2D arbitrary decomposition* can be geometrically represented in the Poincaré sphere through the decomposition of $\mathbf{s}(I,\mathcal{P},\mathbf{u})$ as the convex sum (i.e., as a linear combination whose positive coefficients sum to one) of two states whose Poincaré vectors **v**, **w** lie in the same plane as **u** (Figure 1.19). Thus, given a mixed state, its incoherent decomposition requires at least two pure components.

Any given mixed state **s** can also be represented by an incoherent composition where some of the addends are mixed states. A particularly interesting example is given by the characteristic or trivial decomposition considered previously in Eqs. (1.44) and (1.106) that can be expressed as

$$\mathbf{s}(I,\mathcal{P},\mathbf{u}) = \mathcal{P}I\hat{\mathbf{s}}_p(\mathbf{u}) + (1-\mathcal{P})I\hat{\mathbf{s}}_u \tag{1.160}$$

where $\hat{\mathbf{s}}_p(\mathbf{u})$ is a normalized pure state with the same Poincaré vector **u** as the mixed state $\mathbf{s}(I,\mathcal{P},\mathbf{u})$, and $\hat{\mathbf{s}}_u = (1,0,0,0)^T$ is a 2D unpolarized state.

Moreover, there exist infinite forms of decomposing a mixed state into three or more pure components. As an example, we can consider the decomposition obtained by applying the spectral decomposition to $\hat{\mathbf{s}}_u$, which can be formulated in terms of any pair of orthogonal

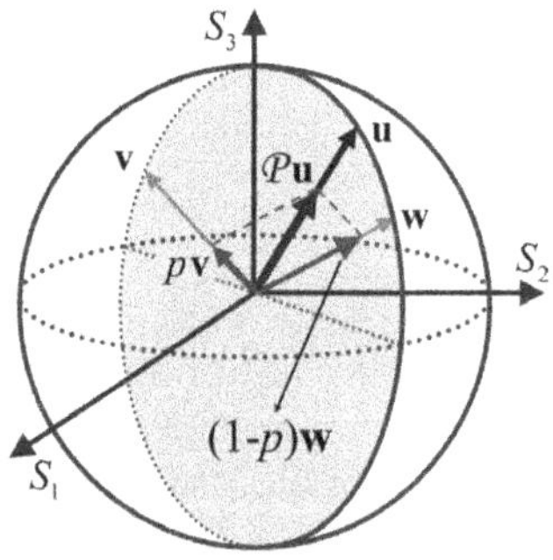

FIGURE 1.19
Arbitrary decomposition of a mixed state into a convex sum of two pure states, the polarization vector of a given state is decomposed as $\mathcal{P}\mathbf{u} = p\mathbf{v}+(1-p)\mathbf{w}$, where **v** can be chosen freely, while such choice determines both p and **w**.

states $\hat{\mathbf{s}}_p(\mathbf{v}), \hat{\mathbf{s}}_p(-\mathbf{v})$ with **v** arbitrary, so that the following type of decomposition into three pure states is obtained:

$$\mathbf{s}(I,\mathcal{P},\mathbf{u}) = \mathcal{P}I\,\hat{\mathbf{s}}_p(\mathbf{u}) + \frac{1-\mathcal{P}^2}{2} I\,\hat{\mathbf{s}}_p(\mathbf{v}) + \frac{(1-\mathcal{P})^2}{2} I\,\hat{\mathbf{s}}_p(-\mathbf{v}) \tag{1.161}$$

1.12.3 Partially Coherent Composition of 2D Pure States

Any partially polarized state can always be considered as a partially coherent composition of a number n of pure states and described as follows:

$$\begin{aligned}
&\boldsymbol{\Phi} = \langle \boldsymbol{\varepsilon}\otimes\boldsymbol{\varepsilon}^\dagger \rangle = \left\langle \left(\sum_{i=1}^{n}\boldsymbol{\varepsilon}_i\right)\otimes\left(\sum_{j=1}^{n}\boldsymbol{\varepsilon}_j\right)^\dagger \right\rangle = \left\langle \left(\sum_{i=1}^{n}\boldsymbol{\varepsilon}_i\right)\otimes\left(\sum_{j=1}^{n}\boldsymbol{\varepsilon}_j^\dagger\right) \right\rangle = \mathbf{X}+\mathbf{Y} \\
&\mathbf{X} \equiv \left\langle \sum_{i=1}^{n}\boldsymbol{\varepsilon}_i\otimes\boldsymbol{\varepsilon}_i^\dagger \right\rangle = \sum_{i=1}^{n}\langle \boldsymbol{\varepsilon}_i\otimes\boldsymbol{\varepsilon}_i^\dagger \rangle = \sum_{i=1}^{n}\boldsymbol{\Phi}_i \\
&\mathbf{Y} \equiv \left\langle \sum_{i,j=1,i\neq j}^{n}\left[(\boldsymbol{\varepsilon}_i\otimes\boldsymbol{\varepsilon}_j^\dagger)+(\boldsymbol{\varepsilon}_j\otimes\boldsymbol{\varepsilon}_i^\dagger)\right] \right\rangle = \sum_{i,j=1,i\neq j}^{n}\left[\langle \boldsymbol{\varepsilon}_i\otimes\boldsymbol{\varepsilon}_j^\dagger \rangle + \langle \boldsymbol{\varepsilon}_j\otimes\boldsymbol{\varepsilon}_i^\dagger \rangle\right]
\end{aligned} \tag{1.162}$$

where $\langle\ \rangle$ stands for time averaging over the measurement time. Matrix **X** has the form of a polarization matrix corresponding to the incoherent composition of the polarization states $\langle \boldsymbol{\varepsilon}_i\otimes\boldsymbol{\varepsilon}_i^\dagger \rangle$ generated by the fluctuating Jones vectors $\boldsymbol{\varepsilon}_i$. Moreover, the cross matrix term **Y** is a Hermitian matrix that is not positive semidefinite and therefore cannot be formally associated with any polarization state.

In the case of fully uncorrelated components $\boldsymbol{\varepsilon}_i$, the average in the expression of **Y** leads to $\mathbf{Y}=\mathbf{0}$ and $\boldsymbol{\Phi}=\mathbf{X}$ (incoherent composition). In the general case of partially correlated components, $\mathbf{Y}\neq\mathbf{0}$, so that **Y** accounts for the necessary correction to be added to **X** in order to provide the resulting polarization matrix $\boldsymbol{\Phi}=\mathbf{X}+\mathbf{Y}$. In the limiting case where the Jones vectors of all components are stable in time and fully correlated among them (fully coherent composition), the composed state has a well defined Jones vector $\boldsymbol{\varepsilon}=\boldsymbol{\varepsilon}_1+\boldsymbol{\varepsilon}_2\ldots+\boldsymbol{\varepsilon}_n$ and the corresponding pure polarization matrix is obtained as $\boldsymbol{\Phi}=\boldsymbol{\varepsilon}\otimes\boldsymbol{\varepsilon}^{\dagger}$.

For realistic natural light sources, the emitted light is commonly composed of a myriad of partially correlated elementary components resulting in unpolarized or partially polarized beams. Nevertheless, many physical phenomena as, for instance, the reflection on a smooth surface or the passing through enpolarizing (dichroic) media, increase the degree of polarization. Furthermore, many artificial sources (like lasers, for instance) produce partially of totally polarized beams because of the partial or total correlation among the waves that make up them.

2

Three-Dimensional States of Polarization

2.1 Introduction

The two-dimensional polarization formalism described in Chapter 1 is applicable to electromagnetic waves whose electric field, at the point **r** considered, evolves in a fixed plane. In the most general case, the three components of the electric field vector **E** of the electromagnetic wave should be considered in order to describe the time evolution of the end point of **E** (which determines the polarization state).

The study and characterization of three-dimensional states of polarization is subject of increasing interest from both theoretical and experimental points of view and consequently, a number of relevant contributions have been published concerning aspects such as:

- The interpretation of physical quantities derived from the eigenvalues and eigenvectors of the 3×3 polarization matrix **R** (Barakat 1977; 1983; Setälä et al. 2002a; 2002b; 2009a; Gil et al. 2004; Ellis and Dogariu 2005a; Luis 2005a; 2005b; Brosseau and Dogariu 2006; Hihoe 2006; Dennis 2007; Gil 2007; Petruccelli et al. 2010; San José and Gil 2011; Gamel and James 2012; 2014; Auñón and Nieto-Vesperinas 2013; Gil et al. 2019b; Sheppard et al. 2020b).
- Geometric interpretation of 3D states (Hannay 1998; Carozzi et al. 2000; Dennis 2004; Saastamoinen and Tervo 2004; Boya and Dixit 2008; Gil and San José 2010; Sheppard 2011; 2012; Gil 2014a; Gil 2021).
- 3D polarization matrix for plane waves and coherent composition of pure (or totally polarized) states (Ellis et al. 2004; Azzam 2011, Sheppard 2014).
- Incoherent composition and decomposition of pure and mixed states (Ellis and Dogariu 2005a; Brosseau and Dogariu 2006; Gil 2007; Gil and San José 2013; Gil 2014a; Gil et al. 2017; Migliaccio et al. 2019).
- 3D polarimetry of light and microwaves (Ellis and Dogariu 2005b; Migliaccio et al. 2015a; 2015b; 2016; 2020; Sorrentino et al. 2017a; 2017b).
- Statistical and coherence properties of 3D states (Tervo et al. 2003; 2004; 2009; Luis 2005b; Ellis et al. 2005; Réfrégier et al. 2006; 2014; Macías–Romero et al. 2011; Friberg and Setälä 2016).

DOI: 10.1201/9780367815578-2

- Generalized and three dimensional Stokes parameters (Roman 1959; Samson 1973; Barakat 1977; Sheppard et al. 2016a; Carozzi et al. 2000; Setälä et al. 2002a; Luis 2005c; 2005d; Korotkova and Wolf 2005; Petrov 2008; Gil 2015).
- Nonregularity (Gil et al. 2017; 2018b; 2021a; Norrman et al. 2019).

As pointed out by Arteaga and Nichols (2018), it should be emphasized again the pioneering contribution of Soleillet (1929), who for the first time introduced a matrix structure equivalent to polarization matrices for 3D states

This chapter deals with the general description of the polarization properties, applicable even when the electric field does not fluctuate in a fixed plane. From the extension of the concept of the analytic signal vector, the Jones vector, the 3D polarization matrix, and the 3D Stokes parameters, as well as some other mathematical structures and physical quantities characterizing the 3D states of polarization are defined and analyzed. Through successive sections, the contents of this chapter also provide answers to questions such as: how to represent 3D states of polarization as combinations of states with a simple physical interpretation; how to interpret in a simple and meaningful manner the 3D Stokes parameters; how to interpret 3D polarization states in terms of meaningful physical quantities; how many different physical situations can be distinguished, and how to classify them.

2.2 3D Analytic Signal and Jones Vector

The definition of the analytic signal vector for general (3D) polarization states has been introduced in Section 1.3. Such representation is applicable for any physical field (with arbitrary spectral profile).

Consider a quasimonochromatic wave of arbitrary form, at a given point $\mathbf{r}$ in space, and let $(\mathbf{e}_1, \mathbf{e}_2, \mathbf{e}_3)$ be a local reference basis of orthonormal vectors along the respective axes XYZ. In this case, in analogy to Eq. (1.14), the time dependence of the components of the analytic signal vector $\boldsymbol{\varepsilon}(t)$ takes the simple form (similar to that of a monochromatic wave, but with a slow time dependence of the respective amplitudes and relative phases of the three components),

$$\boldsymbol{\varepsilon}(t) = \begin{pmatrix} \varepsilon_1(t) \\ \varepsilon_2(t) \\ \varepsilon_3(t) \end{pmatrix} = e^{i(u(t)+\beta_x(t))} \begin{pmatrix} A_x(t) \\ A_y(t)\, e^{i(\beta_y(t)-\beta_x(t))} \\ A_z(t)\, e^{i(\beta_z(t)-\beta_x(t))} \end{pmatrix} \tag{2.1}$$

By removing the time-dependent global phase factor, the 3D *instantaneous Jones vector* is defined as

$$\boldsymbol{\varepsilon}(t) = \begin{pmatrix} \varepsilon_1(t) \\ \varepsilon_2(t) \\ \varepsilon_3(t) \end{pmatrix} = e^{i\gamma} \begin{pmatrix} A_x(t) \\ A_y(t)\, e^{i\delta_y(t)} \\ A_z(t)\, e^{i\delta_z(t)} \end{pmatrix} \tag{2.2}$$

Note that, as with 2D Jones vectors, $\boldsymbol{\varepsilon}(t)$ is defined up to a constant phase factor $\exp(i\gamma)$, which is non-measurable but, as seen in Section 1.12, phase differences among mutually coherent states should be taken into account for the formulation of their coherent composition.

In the above expression, the time dependence of the amplitudes and the phases of the three components of the wave are explicitly indicated. $\boldsymbol{\varepsilon}(t)$ has slow time dependence with respect to the polarization time, and therefore, for time intervals shorter than the polarization time, the polarimetric behavior is equivalent to that of a 2D monochromatic state, and the shape of the polarization ellipse can be considered constant. For time intervals larger than the polarization time the instantaneous Jones vector vary, resulting in 2D or 3D partially polarized states. That is, as with 2D states, the quasimonochromaticity condition ensures that the coherence time τ involves a large number of natural cycles, so that the *instantaneous polarization ellipse* is well defined for time intervals not longer than the polarization time τ_p (with $\tau_p \geq \tau$, see Section 1.8).

Since only the phase differences are measurable, $\delta_x(t)$ has been taken as reference and subtracted from the phases of the remaining two components of $\boldsymbol{\varepsilon}(t)$.

For totally polarized states (i.e., for a fixed polarization ellipse evolving in a fixed plane), the following conditions are satisfied:

$$A_y(t)/A_x(t) = \text{constant} \quad A_z(t)/A_x(t) = \text{constant} \quad \delta_y(t) = \text{constant} \quad \delta_z(t) = \text{constant} \tag{2.3}$$

and the *3D Jones vector* is defined as

$$\boldsymbol{\varepsilon} = e^{i\gamma}\left(a_x, a_y\, e^{i\delta_y}, a_z\, e^{i\delta_z}\right)^T \tag{2.4}$$

where a_x, a_y and a_z are given by the respective averages $a_x^2 \equiv \langle A_x^2 \rangle$, $a_y^2 \equiv \langle A_y^2 \rangle$ and $a_z^2 \equiv \langle A_z^2 \rangle$ over the measurement time T. In this case of totally polarized states, the direction of propagation remains fixed and the shape of the polarization ellipse is constant in a plane Π. The 2D model is then easily reproduced by taking the direction of propagation as the Z reference axis, so that the third component of the 3D Jones vector vanishes.

Let us denote as $(\mathbf{e}_x, \mathbf{e}_y, \mathbf{e}_z)$ and $(\mathbf{e}_{x_O}, \mathbf{e}_{y_O}, \mathbf{e}_{z_O})$ the respective sets of orthonormal unit vectors characterizing the arbitrary coordinate system XYZ and the *intrinsic reference frame* $X_O Y_O Z_O$ defined as that whose axes X_O and Y_O lie precisely along the major and minor semiaxes of the polarization ellipse of the state, while Z_O is normal to the plane $\Pi\,(X_O Y_O)$ containing the polarization ellipse. The *intrinsic Jones vector*, i.e., the Jones vector of a given pure polarization state referenced with respect to its intrinsic reference frame, adopts the form $\boldsymbol{\varepsilon}_O = \sqrt{I}\, e^{i\gamma}(\cos\chi, i\sin\chi, 0)^T$, which depends only on a global phase γ, the intensity I, and the ellipticity angle χ.

The transformation from $\boldsymbol{\varepsilon}_O$ to the Jones vector $\boldsymbol{\varepsilon}$ of the same state but referenced with respect to an arbitrary coordinate system XYZ is obtained by means of the corresponding rotation from $X_O Y_O Z_O$ to XYZ, which is performed through the following steps: (1) a clockwise rotation of angle φ (φ being the *azimuth* of the polarization ellipse) about Z_O, (2) a clockwise rotation of angle θ (*elevation* of the new Z axis) about the transformed Y axis, and (3) a clockwise rotation of angle ϕ (*overall azimuth* of the new X axis) about the transformed

Z axis. The corresponding rotation matrices are the following (with the usual convention that positive angles correspond to counterclockwise rotations):

$$\mathbf{Q}_{-\varphi} \equiv \begin{pmatrix} \cos\varphi & -\sin\varphi & 0 \\ \sin\varphi & \cos\varphi & 0 \\ 0 & 0 & 1 \end{pmatrix} \quad \mathbf{Q}_{-\theta} \equiv \begin{pmatrix} \cos\theta & 0 & -\sin\theta \\ 0 & 1 & 0 \\ \sin\theta & 0 & \cos\theta \end{pmatrix} \quad \mathbf{Q}_{-\phi} \equiv \begin{pmatrix} \cos\phi & -\sin\phi & 0 \\ \sin\phi & \cos\phi & 0 \\ 0 & 0 & 1 \end{pmatrix} \tag{2.5}$$

so that the orthogonal matrix of the composed rotation from $X_OY_OZ_O$ to XYZ is given by (Gil 2018)

$$\mathbf{Q} \equiv \mathbf{Q}_{-\phi}\mathbf{Q}_{-\theta}\mathbf{Q}_{-\varphi} = \begin{pmatrix} c_\theta c_\phi c_\varphi + s_\phi s_\varphi & -c_\theta c_\phi s_\varphi + s_\phi c_\varphi & c_\phi s_\theta \\ -c_\theta s_\phi c_\varphi + c_\phi s_\varphi & c_\theta s_\phi s_\varphi + c_\phi c_\varphi & -s_\phi s_\theta \\ -s_\theta c_\varphi & s_\theta s_\varphi & c_\theta \end{pmatrix} \tag{2.6}$$

where, as in many expressions throughout this book, the trigonometric functions are represented by the abbreviated notations $s_x \equiv \sin x$, $c_x \equiv \cos x$.

A generic pure state with elliptical polarization can be described by means of a polarization ellipse that lies in the plane X_OY_O normal to the direction Z_O defined by the unit vector $\mathbf{e}_{z_O}(\theta,\phi)$, whose major axis is oriented at a clockwise angle φ measured from the direction X_O about Z_O, and whose ellipticity angle is χ. The conventions for the ranges of the angular values are the following: $-\pi \le \phi \le \pi$, $-\pi/2 \le \theta \le \pi/2$; $0 \le \varphi < \pi$ and $-\pi/4 \le \chi < \pi/4$. While χ is intrinsic of the pure polarization state (rotationally invariant), the three remaining angles are relative to the particular Cartesian reference frame used.

When the axis Z_O is maintained under the rotation transformation of the reference axes (i.e., $\theta = \phi = 0$), the expression in Eq. (1.17) of the Jones vector of such a generic pure 2D state embedded into a 3D representation is retrieved

$$\boldsymbol{\varepsilon} = \sqrt{I}e^{i\gamma} \begin{pmatrix} \cos\chi\cos\varphi - i\sin\chi\sin\varphi \\ \cos\chi\sin\varphi + i\sin\chi\cos\varphi \\ 0 \end{pmatrix} \tag{2.7}$$

The same state, when referenced with respect to an arbitrary coordinate system XYZ is given by the Jones vector (Azzam 2011, Azzam and Bashara 1987)

$$\boldsymbol{\varepsilon} = \sqrt{I}e^{i\gamma} \begin{pmatrix} \cos\chi(\sin\phi\sin\varphi + \cos\theta\cos\phi\cos\varphi) + i\sin\chi(\sin\phi\cos\varphi - \cos\theta\cos\phi\sin\varphi) \\ \cos\chi(\cos\phi\sin\varphi - \cos\theta\sin\phi\cos\varphi) + i\sin\chi(\cos\phi\cos\varphi + \cos\theta\sin\phi\sin\varphi) \\ (-\cos\chi\cos\varphi + i\sin\chi\sin\varphi)\sin\theta \end{pmatrix} \tag{2.8}$$

Note that the sign conventions in Eqs. (1.17) and (2.7) have been taken as in (Azzam and Bashara 1987), and are slightly different from the conventions in (Azzam 2011). The procedure for the determination of the angular parameters (ϕ, θ, φ) of a given 3D Jones vector $\boldsymbol{\varepsilon}$ is described in (Gil et al. 2019b)

In the particular case of linearly polarized states ($\chi = 0$) the Jones vector adopts the form

$$\boldsymbol{\varepsilon} = \sqrt{I}\, e^{i\gamma} \begin{pmatrix} \sin\phi \sin\varphi + \cos\theta \cos\phi \cos\varphi \\ \cos\phi \sin\varphi - \cos\theta \sin\phi \cos\varphi \\ -\sin\theta \cos\varphi \end{pmatrix} \tag{2.9}$$

While for circularly polarized states, $\chi = \pm\pi/4$ and $\varphi = 0$, so that

$$\boldsymbol{\varepsilon} = \frac{\sqrt{I}}{\sqrt{2}} e^{i\gamma} \begin{pmatrix} \cos\theta \cos\phi \pm i \sin\phi \\ -\cos\theta \sin\phi \pm i \cos\phi \\ -\sin\theta \end{pmatrix} \tag{2.10}$$

2.3 Sets of Orthonormal 3D Jones Vectors

Despite the fact that any measurable polarization property should be defined in terms the second-order moments (or even higher moments) of the components of the analytic signal vector, it will be shown in further sections that the mathematical structure that underlies the 3D polarization matrix (denoted as **R**, see Section 2.4) and the 3D Stokes parameters can be formulated in terms of sets of three independent 3D complex vectors that, under appropriate physical interpretation, play the role of *equivalent 3D Jones vectors* associated with certain pure states whose incoherent composition produces the given mixed state **R**. In other words, all polarimetric information on 3D polarization states can be described by means of associated sets of mutually orthogonal 3D Jones vectors together with three real and non-negative coefficients that sum to unity.

Therefore, the characterization of 3D polarization states rely strongly on respective associated sets of orthonormal 3D Jones eigenstates $(\hat{\mathbf{u}}_1, \hat{\mathbf{u}}_2, \hat{\mathbf{u}}_3)$, and consequently the geometric and physical interpretation of such sets is an important objective in polarization theory. Unlike the simple geometric interpretation of sets of three-dimensional real orthonormal vectors (which involve a total of three independent parameters), those of complex vectors involve up to six independent parameters (plus three arbitrary phase factors that do not affect the states of polarization) and require a more involved analysis, such as the one that has been performed by Gil (2018a) and Gil et al. (2019b), which is summarized below.

2.3.1 Canonical Set of Orthonormal 3D Jones Vectors

This section deals with the identification of the mathematical formulation, and physical interpretation, of the simplest set of orthonormal 3D Jones vectors, which will serve as the basis from which any other orthonormal set can be constructed in Section 2.3.2.

Consider the Jones vector $\hat{\boldsymbol{\eta}}_1$ of a generic pure state with intensity $I_1 = 1$ that is represented with respect to its own intrinsic reference frame $X_1Y_1Z_1$ (i.e., $\varphi_1 = \phi_1 = \theta_1 = 0$). As seen in Section 2.2, $\hat{\boldsymbol{\eta}}_1$ has the simple form $\hat{\boldsymbol{\eta}}_1 = e^{i\gamma_1}(\cos\chi_1, i\sin\chi_1, 0)^T$. Recall now that the unit Jones vector $\hat{\boldsymbol{\eta}}_2$ whose polarization ellipse lies in the same plane X_1Y_1 (i.e., $\theta_2 = \phi_2 = 0$), with

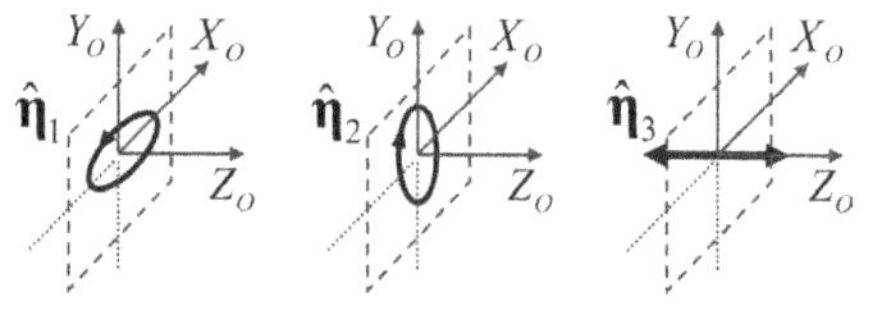

FIGURE 2.1
Canonical set of orthonormal polarization states, represented with respect to the intrinsic reference frame $X_1Y_1Z_1$ of $\hat{\eta}_1$ $(\theta_1 = \phi_1 = \varphi_1 = 0)$. State $\hat{\eta}_2$ corresponds to the configuration $(\theta_2 = \phi_2 = 0,\ \varphi_2 = \pi/2,\ \chi_2 = -\chi_1)$, which is equivalent to $(\theta_2 = 0,\ \phi_2 = \pi/2\ \ \varphi_2 = 0,\ \chi_2 = -\chi_1)$. State $\hat{\eta}_3$ corresponds to the configuration $(\theta_3 = \pm\pi/2,\ \varphi_3 = 0,\ \chi_3 = 0)$ with ϕ_3 arbitrary (Gil 2018; Gil et al. 2019b).

azimuth $\varphi_2 = \pi/2$ and with ellipticity angle $\chi_2 = -\chi_1$, is orthogonal to $\hat{\eta}_1$ and has the form $\hat{\eta}_2 = e^{i\gamma_2}(i\sin\chi_1, \cos\chi_1, 0)^T$. The so-called *canonical set* $(\hat{\eta}_1, \hat{\eta}_2, \hat{\eta}_3)$ is completed with the unit 3D Jones vector $\hat{\eta}_3 = e^{i\gamma_3}(0,0,1)^T$ that corresponds to a linearly polarized state whose electric field vibrates along the Z_1 direction (Figure 2.1).

Note that the canonical set is defined with the convention that the linearly polarized component is $\hat{\eta}_3$, while $\hat{\eta}_1$ and $\hat{\eta}_2$ have nonzero ellipticities, except for the particular case that the three components are linearly polarized $(0 = \chi_1 = \chi_2)$. That is to say, we take the convention that $\hat{\eta}_3$ is linearly polarized in any case, and therefore when $\hat{\eta}_1$ is linearly polarized $(\chi_1 = 0)$, then $\hat{\eta}_2$ and $\hat{\eta}_3$ are necessarily linearly polarized $(\chi_2 = \chi_3 = 0)$.

To avoid possible confusion with the interpretation of the concept of orthogonality of polarization states we emphasize that, as occurs with the common 2D representation of Jones vectors dealt with in Chapter 1, orthogonality of states does not necessarily imply orthogonality between the planes containing the respective polarization ellipses.

2.3.2 General Form of Sets of Orthonormal 3D Jones Vectors

A convenient procedure for the determination of the general form of sets of orthonormal 3D Jones vectors, is to use their representation with respect to the intrinsic reference frame $X_1Y_1Z_1$ of the first component and build generic sets $(\hat{\eta}_1, \hat{\mathbf{v}}_2, \hat{\mathbf{v}}_3)$. This simplifies the interpretation of the relative configuration of the angular parameters of the pure states that constitute this set, while their representation with respect to other coordinate system is easily obtained through a rotation transformation (Gil et al. 2019b).

Any generic Jones vector $\hat{\mathbf{v}}_i$, orthonormal to $\hat{\eta}_1$, can be expressed as a linear combination of the two remaining canonical vectors $(\hat{\eta}_2, \hat{\eta}_3)$ as $\hat{\mathbf{v}}_i = \cos\mu\, e^{i\alpha_i}\,\hat{\eta}_2 + \sin\mu\, e^{i\beta_i}\,\hat{\eta}_3$, with $0 \le \mu \le \pi/2$ and $0 \le \alpha_i, \beta_i \le \pi$. By using this expression of $\hat{\mathbf{v}}_i$, it has been proven that $(\hat{\eta}_1, \hat{\mathbf{v}}_2, \hat{\mathbf{v}}_3)$ constitute the columns of the unitary matrix (Gil 2018).

$$\mathbf{V}_1 = \begin{pmatrix} e^{i\gamma_1}\cos\chi_1 & i e^{i\alpha_2}\cos\mu\sin\chi_1 & i e^{i\alpha_3}\sin\mu\,\sin\chi_1 \\ i e^{i\gamma_1}\sin\chi_1 & e^{i\alpha_2}\cos\mu\,\cos\chi_1 & e^{i\alpha_3}\sin\mu\,\cos\chi_1 \\ 0 & e^{i\beta_2}\sin\mu & -e^{i(\beta_2-\alpha_2+\alpha_3)}\cos\mu \end{pmatrix} \tag{2.11}$$

which depends on the six independent parameters $\chi, \mu, \gamma_1, \alpha_2, \alpha_3, \beta_2$ (two angles and four phases). Here, μ plays the role of an auxiliary parameter that is useful to determine the angular parameters of the shape and orientation of the polarization ellipses of $\hat{\mathbf{v}}_2$ and $\hat{\mathbf{v}}_3$.

To do so, each of the columns 2 and 3 of $\mathbf{V}_1$ should be compared with the corresponding expressions for $\hat{\mathbf{v}}_2$ and $\hat{\mathbf{v}}_3$ in terms of $(\phi_i, \theta_i, \varphi_i, \chi_i)$ $(i = 1,2)$ (see Eq. (2.8)), which leads of a set of coupled trigonometric equations, whose solutions are described and analyzed in detail in Gil et al. 2019b. Such solutions are obtained through those of several complementary cases in such a manner that, in accordance with Eqs. (2.6) and (2.8), the expressions of the components $(\hat{\mathbf{u}}_1, \hat{\mathbf{u}}_2, \hat{\mathbf{u}}_3)$ are given by the columns of the matrix

$$(\hat{\mathbf{u}}_1, \hat{\mathbf{u}}_2, \hat{\mathbf{u}}_3) = \mathbf{Q}_{-\phi_1}\mathbf{Q}_{-\theta_1}\mathbf{Q}_{-\varphi_1}(\hat{\eta}_1, \hat{v}_2, \hat{v}_3) \quad \begin{bmatrix} \hat{\eta}_1 = e^{i\gamma_1}(\cos\chi_1, i\sin\chi_1, 0)^T \\ \hat{v}_2(\gamma_2, \phi_2, \theta_2, \varphi_2, \chi_2) \\ \hat{v}_3(\gamma_3, \phi_3, \theta_3, \varphi_3, \chi_3) \end{bmatrix} \tag{2.12}$$

Moreover, the procedure for the calculation of the characteristic angular parameters $(\phi_i, \theta_i, \varphi_i, \chi_i)$ of a given 3D Jones vector is $\hat{\mathbf{v}}_i$ can be found in (Gil et al. 2019b, Appendix A).

The main features of the sets of orthonormal 3D Jones vectors are summarized below.

- Given a pure polarization state $\hat{\mathbf{\eta}}_1$ and a plane Π_2 with arbitrary spatial orientation, there always exists a pure state $\hat{\mathbf{v}}_2$, orthogonal to $\hat{\mathbf{\eta}}_1$, whose polarization ellipse lies in Π_2.
  The azimuth φ_2 and ellipticity angle χ_2 of $\hat{\mathbf{v}}_2$ depend on the ellipticity angle χ_1 of $\hat{\mathbf{\eta}}_1$ and on the angles ϕ_2 and θ_2 determining the plane Π_2.
  Given $\hat{\mathbf{\eta}}_1$, and once Π_2 is specified, the plane containing the polarization ellipse of the third orthonormal state $\hat{\mathbf{v}}_3$ is fully determined, together with its azimuth φ_3 and ellipticity angle χ_3.
- Given a pure polarization state $\hat{\mathbf{\eta}}_1$, any pure states $\hat{\mathbf{v}}_2$ and $\hat{\mathbf{v}}_3$ that together with $\hat{\mathbf{\eta}}_1$ constitute an orthonormal set, exhibit necessarily a common value for the overall azimuth, i.e., $\phi_3 = \phi_2$, while $\hat{\mathbf{v}}_2$ and $\hat{\mathbf{v}}_3$ have respective elevation angles (θ_2, θ_3) with opposite signs. For the particular case of the canonical set, $\theta_2 = 0$ and $\theta_3 = \pm\pi/2$.
- Once the state $\hat{\mathbf{\eta}}_1$ with its polarization ellipse in the X_1Y_1 plane is taken as reference, the electric fields of the complementary pair $(\hat{\mathbf{v}}_2, \hat{\mathbf{v}}_3)$ vibrate in planes that are not orthogonal to X_1Y_1, except for the case of sets of linearly polarized states.
  The direction containing the fluctuating electric field of the canonical linear component $\hat{\mathbf{\eta}}_3$ is orthogonal to the plane Π_1 of the polarization ellipses of the canonical pair $(\hat{\mathbf{\eta}}_1, \hat{\mathbf{\eta}}_2)$ (Figure 2.1).
- All families of three orthonormal linearly polarized states necessarily satisfy $\phi_2 = \phi_3 = \pm\pi/2$, while their elevation angles form a right angle, $|\theta_2 - \theta_3| = \pi/2$.
- The equality $\phi_2 = \phi_3$, together with the fact that the spin vector of a given pure state is normal to the plane containing its polarization ellipse, implies that the respective spin vectors of any set of orthogonal polarization states lie in a common plane that is orthogonal to the planes containing the polarization ellipses of such states.
- Orthonormal sets where all the components are circularly polarized ($|\chi_1| = |\chi_2| = |\chi_3| = \pi/4$) are not mathematically achievable and therefore are not physically realizable. Orthonormal sets where two components are circularly polarized, necessarily correspond to canonical sets with $\chi_1 = \pm\pi/4$, $\chi_2 = \mp\pi/4$, and $\chi_3 = 0$.

Figure 2.2 illustrates the orientation and ellipticity of the polarization ellipses of states $\hat{\mathbf{v}}(\chi, \varphi, \phi, \theta)$ that are orthogonal to $\hat{\mathbf{\eta}}_1$. Concentric circles represent states with equal elevation angle θ, while straight lines from the origin correspond to states with a fixed overall azimuth ϕ. In all scenarios the ellipticity angle χ reaches zero at $|\theta| = \pi/2$, corresponding to

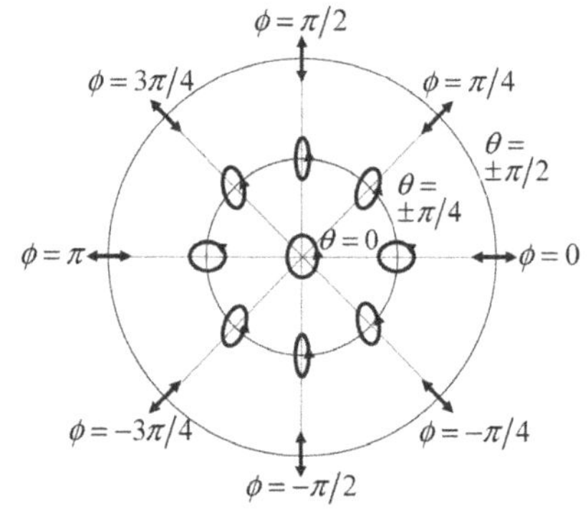

FIGURE 2.2
Representation of the polarization ellipses of states $\hat{\mathbf{v}}(\chi, \varphi, \phi, \theta)$ that are orthogonal to $\hat{\boldsymbol{\eta}}_1$ for different values of ϕ and θ. States in a given circle have equal elevation angle q, while states along straight lines from the origin have equal overall azimuth ϕ. Each ellipse lies in a respective plane whose normal is given by its direction vector $\hat{\mathbf{e}}_z(\theta, \phi)$, so that such planes are tangent to a common sphere and, from a topological point of view, cover all points of this sphere (Gil et al. 2019b).

linear polarization. For states with $\phi = 0, \pi$ (horizontal line), two branches with respective azimuths $\varphi = \pi/2$ and $\varphi = 0$ are separated by an intermediate circular state, such that the ellipticity angle first increases with $|\theta|$ up to $|\chi| = \pi/4$ before decreasing down to $\chi = 0$ at $|\theta| = \pi/2$. States with $|\phi| = \pi/2$ (vertical line), satisfy $\varphi = 0$ and the ellipticity angle reduces monotonically from its maximum at $\theta = 0$ toward $\chi = 0$ at $|\theta| = \pi/2$. For states corresponding to the oblique lines, both φ and χ vary continuously as a function of $|\theta|$ (Gil et al. 2019b).

As illustrative interesting examples, the configurations of the families of sets of orthonormal polarization states satisfying $\phi_2 = \phi_3 = 0$ and $\phi_2 = \phi_3 = \pi/2$ are represented in Figures 2.3 and 2.4 respectively

2.4 3D Polarization Matrix

The 3D description of states of polarization has attracted the interest of a number of researchers for a long time. It was Roman (1959) who first defined a set of 3D Stokes parameters from a particular basis of 3×3 matrices. These matrices are not trace-orthogonal and, consequently, do not constitute the best choice to build such a basis. As Fano pointed out in a seminal paper on the description of states in quantum mechanics by means of density matrices (Fano 1957), an appropriate basis for $n \times n$ density matrices is a set of Hermitian trace-orthogonal operators, as is that constituted by the 3×3 identity matrix together with the eight Gell-Mann matrices. As pointed out in some works, this choice, based on the generators of the SU(3) group, is consistent with a generalization of the polarization algebra to $n \times n$ density matrices (Gil et al. 2004; Gil 2007; San José and Gil 2011; Gil, 2020).

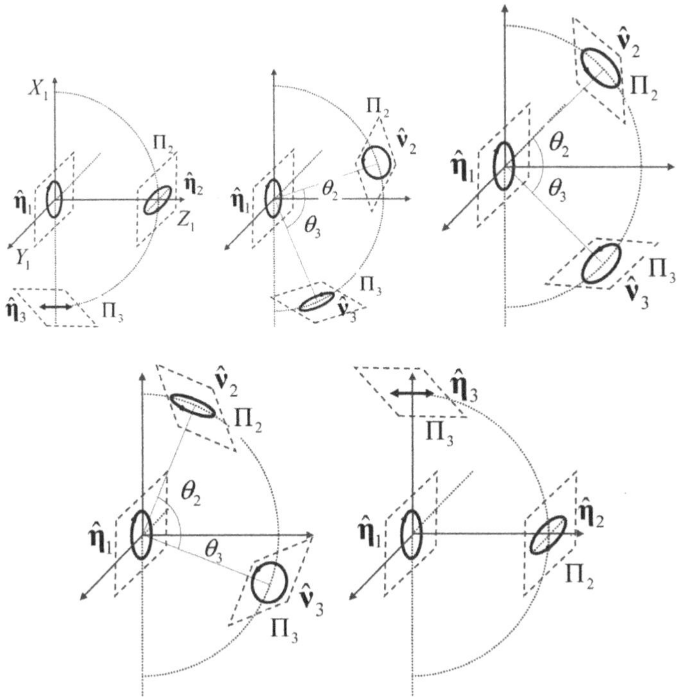

FIGURE 2.3
Physical interpretation of the family of sets of mutually orthogonal states corresponding to $\phi_2 = \phi_3 = 0$. Although all the states correspond to a common fixed point in space, for the sake of clarity they are represented separated. The planes containing the polarization ellipses of $\hat{\mathbf{v}}_2$ and $\hat{\mathbf{v}}_3$ are tangent to the semicircle determined by ϕ_2, and perpendicular to the plane X_1Z_1, while θ_2 and θ_3 are the respective elevation angles about the axis Y_1 (Gil et al. 2019b).

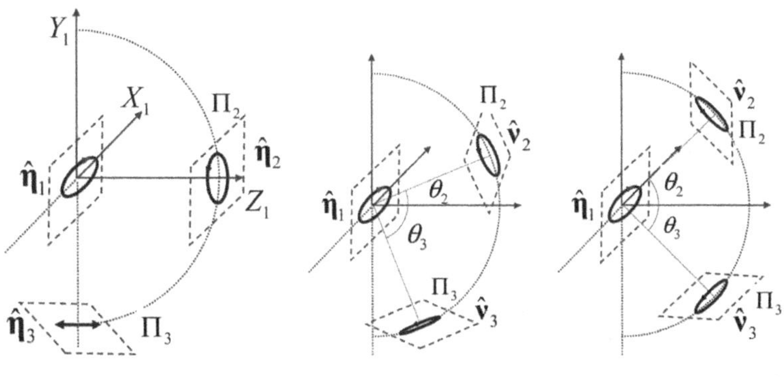

FIGURE 2.4
Physical interpretation of the family of sets of mutually orthogonal states corresponding to $\phi_2 = \phi_3 = \pi/2$. Although all the states correspond to the same point in space, for the sake of clarity they are represented separated. The planes containing the polarization ellipses of $\hat{\mathbf{v}}_2$ and $\hat{\mathbf{v}}_3$ are tangent to the semicircle determined by $\phi_2 = \phi_3 = \pi/2$, and perpendicular to the plane Y_1Z_1, while θ_2 and θ_3 are the respective elevation angles about the axis X_1 (Gil et al. 2019b).

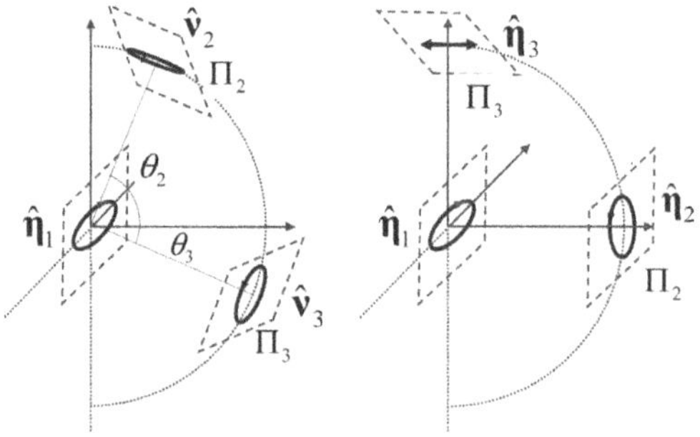

FIGURE 2.4 (Continued)

The *3D polarization matrix* (or *3D coherency matrix*), which contains all the second-order measurable information about the state of polarization (including intensity) of an electromagnetic wave, is defined as the following 3×3 Hermitian matrix:

$$\mathbf{R} = \langle \boldsymbol{\varepsilon}(t)\otimes \boldsymbol{\varepsilon}^{+}(t)\rangle = \begin{pmatrix} \langle \varepsilon_1(t)\,\varepsilon_1^{*}(t)\rangle & \langle \varepsilon_1(t)\,\varepsilon_2^{*}(t)\rangle & \langle \varepsilon_1(t)\,\varepsilon_3^{*}(t)\rangle \\ \langle \varepsilon_2(t)\,\varepsilon_1^{*}(t)\rangle & \langle \varepsilon_2(t)\,\varepsilon_2^{*}(t)\rangle & \langle \varepsilon_2(t)\,\varepsilon_3^{*}(t)\rangle \\ \langle \varepsilon_3(t)\,\varepsilon_1^{*}(t)\rangle & \langle \varepsilon_3(t)\,\varepsilon_2^{*}(t)\rangle & \langle \varepsilon_3(t)\,\varepsilon_3^{*}(t)\rangle \end{pmatrix} \tag{2.13}$$

whose elements are the second-order moments of the zero-mean analytic signals $\varepsilon_i(t)\ (i = 1,2,3)$ (complex random variables) associated with the three (real) Cartesian components of the electric field vector at point $\mathbf{r}$ in space. The brackets $\langle\ \rangle$ indicate time averaging over the measurement time, and in the case of stationary and ergodic fields, $\langle\ \rangle$ can also be interpreted as ensemble averaging over the ensemble of sample realizations.

In full analogy to the coherence–polarization properties of 2D states studied in Chapter 1, matrix $\mathbf{R}$ is the equal-point and equal-time correlation matrix of the components of the analytic signal vector $\boldsymbol{\varepsilon}$. Its diagonal elements are the averaged intensities associated with each of the three components, while its off-diagonal elements are measures of the equal-time correlations between the mutually orthogonal analytic signals ε_1, ε_2 and ε_3 at point $\mathbf{r}$.

Thus, the polarization matrix is defined in terms of nine independent, second-order, time averaged, measurable quantities and also admits a direct phenomenological definition that will be described in Section 2.5 in terms of the associated 3D Stokes parameters.

The *spectral polarization matrix* is defined from the Fourier transform of the equal-point coherence matrix $\mathbf{R}\equiv\mathbf{R}(\mathbf{r})\equiv\mathbf{J}(\mathbf{r},\mathbf{r},\tau=0)$ (see Section 2.7.1) and can always be expressed as $\mathbf{R}(\omega)=\langle\boldsymbol{\varepsilon}(\omega)\otimes\boldsymbol{\varepsilon}^{+}(\omega)\rangle_{\omega}$, where the field variables are stochastic processes and $\langle\ \rangle_{\omega}$ denotes ensemble averaging over the ensemble of all *equivalent monochromatic realizations* of the field components at frequency ω. Recall that, even though this kind of average is itself a function of the same frequency ω as that of the individual realizations, $\mathbf{R}(\omega)$ should not be confused with the polarization matrix of a monochromatic wave (Wolf 2007).

The polarization matrix is obtained by integrating $\mathbf{R}(\omega)$ over all frequencies

$$\mathbf{R} = \int_0^{\infty} \mathbf{R}(\omega)\,d\omega \tag{2.14}$$

From its very definition, **R** is a covariance matrix, and therefore **R** is completely characterized by the fact that it is a positive semidefinite Hermitian matrix. Thus, **R** has three non-negative eigenvalues, and these three constraints constitute a complete set of necessary and sufficient conditions for a Hermitian matrix **R** to be formally a *3D polarization matrix*, i.e. to represent a particular 3D state of polarization.

The diagonal elements of **R** can be interpreted as the intensities associated with the respective *XYZ* components of the electric field, so that the total intensity (a measure of the time-averaged power of the electric field (at point **r**) is given by $I = \operatorname{tr}\mathbf{R} = \langle|\varepsilon_1^2(t)|\rangle + \langle|\varepsilon_2^2(t)|\rangle + \langle|\varepsilon_3^2(t)|\rangle$.

In the particular case of a quasimonochromatic field, the expression of **R** in terms of the amplitudes and relative phases of the field components is

$$\mathbf{R} = \begin{pmatrix} \langle A_x^2(t)\rangle & \langle A_x(t)A_y(t)\,e^{-i\delta_y(t)}\rangle & \langle A_x(t)A_z(t)\,e^{-i\delta_z(t)}\rangle \\ \langle A_x(t)A_y(t)\,e^{i\delta_y(t)}\rangle & \langle A_y^2(t)\rangle & \langle A_y(t)A_z(t)\,e^{i(\delta_y(t)-\delta_z(t))}\rangle \\ \langle A_x(t)A_z(t)\,e^{i\delta_z(t)}\rangle & \langle A_y(t)A_z(t)\,e^{-i(\delta_y(t)-\delta_z(t))}\rangle & \langle A_z^2(t)\rangle \end{pmatrix} \tag{2.15}$$

and $I = \operatorname{tr}\mathbf{R} = \langle A_x^2(t)\rangle + \langle A_y^2(t)\rangle + \langle A_z^2(t)\rangle$.

Going back to the general structure of **R** (and not only to that of quasimonochromatic fields), its statistical properties appear clearly when its elements r_{kl} $(k,l = 1,2,3)$ are written in terms of the corresponding standard deviations σ_i and the complex degrees of mutual coherence μ_{kl} (correlation coefficients), $r_{kl} \equiv \mu_{kl}\,\sigma_k\,\sigma_l$, with $\sigma_k^2 = \langle \varepsilon_k\,\varepsilon_k^*\rangle = r_{kk}$ (variances), and $\mu_{kl} = r_{kl}/\sigma_k\,\sigma_l$, i.e.

$$\mathbf{R} = \begin{pmatrix} \sigma_1^2 & \mu_{12}\,\sigma_1\,\sigma_2 & \mu_{13}\,\sigma_1\,\sigma_3 \\ \mu_{12}^*\,\sigma_1\,\sigma_2 & \sigma_2^2 & \mu_{23}\,\sigma_2\,\sigma_3 \\ \mu_{13}^*\,\sigma_1\,\sigma_3 & \mu_{23}^*\,\sigma_2\,\sigma_3 & \sigma_3^2 \end{pmatrix} \tag{2.16}$$

From the non-negativity of three nested principal minors, we get the following set of explicit and simple necessary and sufficient conditions for **R** to be a covariance matrix and, hence, to be a polarization matrix representing a 3D state of polarization (Gil 2007):

$$r_{11} \geq 0 \qquad 1 \geq \rho_{12} \qquad \det\mathbf{R} = 1 + 2\rho_{12}\,\rho_{23}\,\rho_{13}\cos(\beta_{12} + \beta_{23} - \beta_{13}) - \rho_{01}^2 - \rho_{12}^2 - \rho_{02}^2 \geq 0 \tag{2.17}$$

where $\rho_{kl} \equiv |\mu_{kl}|$ and $\mu_{kl} \equiv \rho e^{\beta_{kl}}$.

Let us finally observe that for any polarization matrix $\mathbf{R}_p$ associated with a totally polarized state, there always exists an *equivalent Jones vector* $\boldsymbol{\varepsilon}$ that satisfies $\mathbf{R} = \boldsymbol{\varepsilon} \otimes \boldsymbol{\varepsilon}^\dagger$, regardless of the spectral profile of the electromagnetic wave at the point **r** considered.

2.5 3D Stokes Parameters

As an appropriate basis for the expansion of **R**, let us now consider the following set of Hermitian and trace-orthogonal matrices constituted by the eight Gell-Mann matrices plus the identity matrix:

$$\boldsymbol{\omega}_{00} \equiv \boldsymbol{\omega}_0 \equiv \begin{pmatrix} 1 & 0 & 0 \\ 0 & 1 & 0 \\ 0 & 0 & 1 \end{pmatrix} \quad \boldsymbol{\omega}_{01} \equiv \boldsymbol{\omega}_1 \equiv \begin{pmatrix} 0 & 1 & 0 \\ 1 & 0 & 0 \\ 0 & 0 & 0 \end{pmatrix} \quad \boldsymbol{\omega}_{02} \equiv \boldsymbol{\omega}_4 \equiv \begin{pmatrix} 0 & 0 & 1 \\ 0 & 0 & 0 \\ 1 & 0 & 0 \end{pmatrix}$$

$$\boldsymbol{\omega}_{10} \equiv \boldsymbol{\omega}_2 \equiv \begin{pmatrix} 0 & -i & 0 \\ i & 0 & 0 \\ 0 & 0 & 0 \end{pmatrix} \quad \boldsymbol{\omega}_{11} \equiv \boldsymbol{\omega}_3 \equiv \begin{pmatrix} 1 & 0 & 0 \\ 0 & -1 & 0 \\ 0 & 0 & 0 \end{pmatrix} \quad \boldsymbol{\omega}_{12} \equiv \boldsymbol{\omega}_6 \equiv \begin{pmatrix} 0 & 0 & 0 \\ 0 & 0 & 1 \\ 0 & 1 & 0 \end{pmatrix} \tag{2.18}$$

$$\boldsymbol{\omega}_{20} \equiv \boldsymbol{\omega}_5 \equiv \begin{pmatrix} 0 & 0 & -i \\ 0 & 0 & 0 \\ i & 0 & 0 \end{pmatrix} \quad \boldsymbol{\omega}_{21} \equiv \boldsymbol{\omega}_7 \equiv \begin{pmatrix} 0 & 0 & 0 \\ 0 & 0 & -i \\ 0 & i & 0 \end{pmatrix} \quad \boldsymbol{\omega}_{22} \equiv \boldsymbol{\omega}_8 \equiv \frac{1}{\sqrt{3}} \begin{pmatrix} 1 & 0 & 0 \\ 0 & 1 & 0 \\ 0 & 0 & -2 \end{pmatrix}$$

The notation with double subscript is justified to emphasize the symmetry in some mathematical expressions, while the alternative notation with single subscript is commonly used in related works and simplifies certain equations. Note that the Pauli matrices are embedded into $\boldsymbol{\omega}_k$.

The basis constituted by the nine matrices $\boldsymbol{\omega}_k$ allows for expanding $\mathbf{R}$ as

$$\mathbf{R} = \frac{1}{3} s_0 \boldsymbol{\omega}_0 + \frac{1}{2} \sum_{k=1}^{8} s_k \boldsymbol{\omega}_k \tag{2.19}$$

where the real coefficients are given by

$$s_k = \operatorname{tr}(\mathbf{R}\boldsymbol{\omega}_k) \;\; (k = 0,\ldots,8) \text{or}\, s_{ij} = \operatorname{tr}(\mathbf{R}\boldsymbol{\omega}_{ij}) \;\; (i,j = 0,1,2,3) \tag{2.20}$$

so that

$$\mathbf{R} = \frac{1}{2} \begin{pmatrix} 2s_{00}/3 + s_{11} + s_{22}/\sqrt{3} & s_{01} - is_{10} & s_{02} - is_{20} \\ s_{01} + is_{10} & 2s_{00}/3 - s_{11} + s_{22}/\sqrt{3} & s_{12} - is_{21} \\ s_{02} + is_{20} & s_{12} + is_{21} & 2s_{00}/3 - 2s_{22}/\sqrt{3} \end{pmatrix} \tag{2.21}$$

The nine real coefficients s_{ij} are called the *3D Stokes parameters* since they are a natural extension of the usual Stokes parameters for 2D states. s_{ij} can be arranged as the elements of the 3×3 *Stokes parameters matrix* $\mathbf{S}$ and their expressions in terms of the elements r_{ij} of the polarization matrix are

$$\begin{array}{lll} s_{00} \equiv r_{11} + r_{22} + r_{33} & s_{01} \equiv r_{12} + r_{21} & s_{02} \equiv r_{13} + r_{31} \\ s_{10} \equiv i(r_{12} - r_{21}) & s_{11} \equiv r_{11} - r_{22} & s_{12} \equiv r_{23} + r_{32} \\ s_{20} \equiv i(r_{13} - r_{31}) & s_{21} \equiv i(r_{23} - r_{32}) & s_{22} \equiv (r_{11} + r_{22} - 2r_{33})/\sqrt{3} \end{array} \tag{2.22}$$

It is obvious that the Poincaré sphere representation is not applicable to the nine 3D Stokes parameters of 3D states of polarization. The constraining inequalities between s_{ij} (derived from the non-negativity of the eigenvalues of $\mathbf{R}$) are more complicated than those of the 2D

Stokes parameters. In fact, these restrictions have the form of inequalities that involve not only $\mathrm{tr}\,\mathbf{R}$ and $\mathrm{tr}\,\mathbf{R}^2$ (as with 2D states), but also $\mathrm{tr}\,\mathbf{R}^3$ (Brosseau 1998).

The 3D Stokes parameters constitute a set of nine independent measurable parameters defined from second-order, time averaged quantities derived from the fluctuating electric field vector, at the point **r** in space considered, and, as with the conventional Stokes parameters of 2D states (see Section 1.5.3), admit a direct phenomenological interpretation. s_{00} is the intensity of the full state, while the remaining eight Stokes parameters can be interpreted as respective differences between the intensities associated with components of the field on certain generalized coordinates (Carozzi 2000; Setälä et al. 2002a). Nevertheless, we refer the reader to the clear and meaningful physical interpretation of the 3D Stokes parameters provided by the intrinsic representation described in Section 2.9 where three of the nine Stokes parameters become zero, while the six *intrinsic Stokes parameters* admit a direct interpretation in terms of the intensity, the degree of linear polarization, the degree of directionality (a measure of the closeness of the 3D state to a 2D one), and the three intrinsic components of the spin density vector of the state.

In analogy to the 2D description of polarization states, a 3D polarization vector **p** is defined as

$$\mathbf{p} \equiv \frac{\sqrt{3}}{2}\left(\hat{s}_1, \hat{s}_2, \hat{s}_3, \hat{s}_4, \hat{s}_5, \hat{s}_6, \hat{s}_7, \hat{s}_8\right)^T \tag{2.23}$$

where $\hat{s}_i \equiv s_i/s_0$ are the intensity-normalized 3D Stokes parameters, in such a manner that its absolute value

$$P_{3D} \equiv |\mathbf{p}| = \frac{\sqrt{3}}{2}\sqrt{\sum_{k=1}^{8}\hat{s}_k^2} \tag{2.24}$$

is constrained by $0 \le P_{3D} \le 1$, where $P_{3D} = 0$ corresponds to completely random 3D states and $P_{3D} = 1$ holds uniquely for pure states, so that P_{3D} gives a proper measure of the *3D degree of polarimetric purity* of the state represented by **s** (Setälä et al. 2002a; Setälä et al. 2002b; Gil et al. 2004; Gil et al. 2017; Gil et al. 2018a; Gil et al. 2019a).

2.6 Composition and Decomposition of 3D States of Polarization

A convenient way for the study and interpretation of the main features and types of three-dimensional states of polarization is provided by the decomposition of the polarization matrix into polarization matrices with simple forms. In the following subsections we analyze procedures for: (1) *coherent composition* of mutually fully coherent states; (2) *partially coherent composition* of pure states; (3) *arbitrary decomposition* of a mixed state in terms of the incoherent superposition of pure states; (4) as a particular case of the arbitrary decomposition, the *spectral decomposition* of a mixed state into three mutually orthogonal pure components whose expansion coefficients are given by the eigenvalues of the polarization matrix **R**; (5) the *characteristic decomposition* of a mixed state into a pure state and up to two mixed states with scaled decreasing degree of polarimetric purity; (6) the *smart* *decomposition* of a mixed state into the two first eigenstates and a 3D unpolarized state, and (7) the *polarimetric subtraction* of a pure state from a mixed state.

2.6.1 Coherent Composition of 3D Pure States

As a preliminary step to the study of partially coherent and incoherent superpositions of polarization states, it is worth considering the coherent composition of pure states, which must be realized through the additive composition of the respective 3D Jones vectors. The difference of the treatment dealt with in this section with respect to that of Section 1.12.1 devoted to 2D states is that here we are considering the coherent superposition of pure states whose polarization ellipses may lie in different planes.

Consider a point **r** in a linear, homogeneous, and isotropic medium where two mutually coherent pure states, characterized by respective 3D Jones vectors $\boldsymbol{\varepsilon}_1$ and $\boldsymbol{\varepsilon}_2$, are superimposed. The resultant state of polarization (at point **r**) is a pure state given by the Jones vector $\boldsymbol{\varepsilon}=\boldsymbol{\varepsilon}_1+\boldsymbol{\varepsilon}_2$. The electric field of the combined state describes a well-defined ellipse which, at point **r**, lies in the plane tangent to the wavefront of the composed wave. The Jones vectors $\boldsymbol{\varepsilon}_1$, $\boldsymbol{\varepsilon}_2$ and $\boldsymbol{\varepsilon}=\boldsymbol{\varepsilon}_1+\boldsymbol{\varepsilon}_2$ have their corresponding associated pure polarization matrices $\mathbf{R}_1$, $\mathbf{R}_2$, and $\mathbf{R}$, satisfying $\mathbf{R} \neq \mathbf{R}_1 + \mathbf{R}_2$. This inequality holds except for the trivial case where $\boldsymbol{\varepsilon}_1=\boldsymbol{\varepsilon}_2$. Thus, the pure states $\boldsymbol{\varepsilon}_1$, $\boldsymbol{\varepsilon}_2$ and $\boldsymbol{\varepsilon}=\boldsymbol{\varepsilon}_1+\boldsymbol{\varepsilon}_2$, have respective polarization planes (viz., the planes containing the respective polarization ellipses) which, in general, are different.

The polarization ellipse of a pure state $\boldsymbol{\varepsilon}$ with nonzero ellipticity lies in a well-defined plane Π, and thus, the direction of propagation of the state (at point **r**) is well defined because it is necessarily perpendicular to Π. On the contrary, a 3D Jones vector representing a linearly polarized state is compatible with any direction of propagation perpendicular to the axis along which the electric field of the electromagnetic wave lies. Consequently, a 3D Jones vector $\boldsymbol{\varepsilon}$ (and hence the corresponding polarization matrix) does not contain intrinsic information about the direction of propagation, but that information can be deduced from $\boldsymbol{\varepsilon}$, either as a fixed direction for states with nonzero ellipticity, or as an arbitrary direction perpendicular to the polarization axis, for linearly polarized states (zero ellipticity). It will be worth keeping in mind this result when we deal later with the incoherent composition and decomposition of linearly polarized states with coincident polarization axes.

Moreover, since the sum of Jones vectors is a Jones vector, given a 3D Jones vector $\boldsymbol{\varepsilon}$, it can always be expressed, in an infinite number of ways, as the sum of a number of 3D Jones vectors.

2.6.2 Partially Coherent Composition of 3D Pure States

The formulation and analysis developed in Section 1.12.3 is entirely applicable to the partially coherent composition of 3D Jones vectors. Observe that to distinguish 3D polarization matrices from 2D ones they are denoted by **R** and $\boldsymbol{\Phi}$ respectively, so that, when applied to 3D polarization states, $\boldsymbol{\Phi}$ in Section 1.12.3 should be interpreted as **R**. The coherent and incoherent compositions of polarization states appear as opposite limiting cases of the general partially coherent superposition.

2.6.3 Arbitrary Decomposition of 3D States

The superposition of mutually incoherent electromagnetic waves at a given point **r** produces a resultant polarized state corresponding to the combined wave. The present section is devoted to analyzing and mathematically formalizing all the physically achievable incoherent compositions and decompositions of 3D states of polarization in terms of pure ones.

Consider the diagonalization $\mathbf{R} = \mathbf{U}\,\mathrm{diag}(\lambda_1, \lambda_2, \lambda_3)\mathbf{U}^{+}$, where $\mathbf{U}$ is the unitary matrix whose columns are the eigenvectors $(\hat{\mathbf{u}}_1, \hat{\mathbf{u}}_2, \hat{\mathbf{u}}_3)$ of $\mathbf{R}$, and $\mathrm{diag}(\lambda_1, \lambda_2, \lambda_3)$ represents the diagonal matrix composed of the non-negative eigenvalues (taken in decreasing order, $\lambda_1 \geq \lambda_2 \geq \lambda_3 \geq 0$). $\mathbf{R}$ can always be expressed as a convex sum of a set of r (with $r \equiv \mathrm{rank}\,\mathbf{R}$) independent 3D pure polarization matrices defined from respective independent complex unit vectors $\mathbf{w}_i$ belonging to the subspace generated by the eigenvectors of $\mathbf{R}$ with nonzero eigenvalues (Gil 2014a; Gil and San José 2013). This *image* space or *range* of $\mathbf{R}$ is denoted as $\mathrm{im}\,\mathbf{R}$ (not to be confused with the imaginary part of $\mathbf{R}$, $\mathrm{Im}\,\mathbf{R}$). The arbitrary decomposition is formulated as (Gil et al. 2017)

$$\mathbf{R} = \sum_{i=1}^{r} p_i \mathbf{R}_{pi} \qquad \left[\mathbf{R}_{pi} \equiv (\mathrm{tr}\,\mathbf{R})\left(\hat{\mathbf{w}}_i \otimes \hat{\mathbf{w}}_i^{\dagger}\right) \qquad p_i = \frac{1}{\sum_{j=1}^{r} \frac{1}{\hat{\lambda}_j}\left|\left(\mathbf{U}^{+}\hat{\mathbf{w}}_i\right)_j\right|^2} \right] \tag{2.25}$$

where $\hat{\lambda}_i \equiv \lambda_i/\mathrm{tr}\,\mathbf{R}$ are the intensity-normalized eigenvalues and $|\ |$ indicates the absolute value (magnitude or Euclidean norm).

It should be stressed that the term *arbitrary* is used in the sense that different sets of r independent unit vectors $\hat{\mathbf{w}}_i$ belonging to $\mathrm{im}\,\mathbf{R}$ allows for generating the r mutually incoherent pure components $\mathbf{R}_{pi} \equiv (\mathrm{tr}\,\mathbf{R})(\hat{\mathbf{w}}_i \otimes \hat{\mathbf{w}}_i^{\dagger})$ of $\mathbf{R}$, with respective weights $p_i > 0$ satisfying $p_1 + p_2 + \ldots + p_r = 1$.

The coefficients p_i can also be calculated trough the simple expression $p_i = 1/(\hat{\mathbf{w}}_i^{\dagger}\mathbf{R}^{-}\hat{\mathbf{w}}_i)$, where $\mathbf{R}^{-}$ is the pseudoinverse of $\hat{\mathbf{R}} \equiv \mathbf{R}/(\mathrm{tr}\,\mathbf{R})$ defined as $\mathbf{R}^{-} = \mathbf{U}\mathbf{D}^{-}\mathbf{U}^{+}$, $\mathbf{D}^{-}$ being the diagonal matrix whose r first diagonal elements are $1/\hat{\lambda}_1, \ldots, 1/\hat{\lambda}_r$ and the last $3-r$ elements are zero (Gil and San José 2019).

Expansion (2.25) provides the method for calculating the coefficients p_i of the polarization matrices $\mathbf{R}_{pi}$ of the pure components. Note that the number of pure components of the arbitrary decomposition is equal to r. When $r = 3$, different sets $\hat{\mathbf{w}}_i$ $(i = 1,2,3)$ of three independent unit vectors can be chosen to generate the respective pure components. When $r = 2$, different sets $(\hat{\mathbf{w}}_1, \hat{\mathbf{w}}_2)$ of two independent unit vectors belonging to im $\mathbf{R}$ can be chosen. When $r = 1$, $\mathbf{R}$ represents a pure state and consequently the arbitrary decomposition becomes a tautology. Hereafter, when appropriate to stress that a given state $\mathbf{R}$ is pure, its coherence matrix will be denoted as $\mathbf{R}_p$ (with the subscript p).

2.6.4 Spectral Decomposition of 3D States

By taking as $\hat{\mathbf{w}}_i$ the eigenvectors $\hat{\mathbf{u}}_i$ $(i = 1,2,3)$ of $\mathbf{R}$, the arbitrary decomposition takes the particular form of the following *spectral decomposition*:

$$\mathbf{R} = \hat{\lambda}_1\left[I\,\mathbf{U}\,\mathrm{diag}(1,0,0)\mathbf{U}^{\dagger}\right] + \hat{\lambda}_2\left[I\,\mathbf{U}\,\mathrm{diag}(0,1,0)\mathbf{U}^{\dagger}\right] + \hat{\lambda}_3\left[I\,\mathbf{U}\,\mathrm{diag}(0,0,1)\mathbf{U}^{\dagger}\right]$$
$$\left[\sum_{i=1}^{3} \hat{\lambda}_i = 1,\ I = \mathrm{tr}\,\mathbf{R}\right] \tag{2.26}$$

where each term in the sum is synthesized from the corresponding eigenvector $\hat{\mathbf{u}}_i$ of $\mathbf{R}$ and its weight in the convex sum is equal to the corresponding normalized eigenvalue $\hat{\lambda}_i$

$$\mathbf{R} = \sum_{i=1}^{3} \hat{\lambda}_i \left[I\left(\hat{\mathbf{u}}_i \otimes \hat{\mathbf{u}}_i^{\dagger}\right)\right] \qquad \left[\sum_{i=1}^{3} \hat{\lambda}_i = 1\right] \tag{2.27}$$

Thus, besides the intensity, the 3D state of polarization $\mathbf{R}$ is determined by its corresponding set $(\hat{\mathbf{u}}_1, \hat{\mathbf{u}}_2, \hat{\mathbf{u}}_3)$ of associated *3D Jones eigenstates* (whose peculiar relative configurations in terms of their polarization ellipses has been studied in Section 2.3) together with their respective relative weights $(\hat{\lambda}_1, \hat{\lambda}_2, \hat{\lambda}_3)$ in the spectral decomposition. That is to say, from a polarimetric point of view, $\mathbf{R}$ is indistinguishable from the incoherent composition of the three pure states with Jones vectors $(\hat{\mathbf{u}}_1, \hat{\mathbf{u}}_2, \hat{\mathbf{u}}_3)$ with respective relative weights $(\hat{\lambda}_1, \hat{\lambda}_2, \hat{\lambda}_3)$.

When referenced with respect to the intrinsic reference frame $X_1Y_1Z_1$ of $\hat{\mathbf{u}}_1$, the set of Jones eigenvectors adopts the form $(\hat{\boldsymbol{\eta}}_1, \hat{\nu}_3, \hat{\nu}_3)$ studied in Section 2.3. In accordance with the convention taken to define $(\hat{\boldsymbol{\eta}}_1, \hat{\nu}_3, \hat{\nu}_3)$, in the particular case that $\hat{\mathbf{u}}_1$ corresponds to a linearly polarized state while $\hat{\mathbf{u}}_2$ and $\hat{\mathbf{u}}_3$ correspond to elliptically polarized ones, the set should be relabeled so as that $\hat{\boldsymbol{\eta}}_1$ is the intrinsic representation of either $\hat{\mathbf{u}}_2$ or $\hat{\mathbf{u}}_3$ instead of that of $\hat{\mathbf{u}}_1$.

Observe that the term *spectral* is used here with reference to the eigenvalue spectrum of $\mathbf{R}$ and without any link to the frequency spectrum of the state represented by $\mathbf{R}$. It should also be noted that, when one of the eigenvalues has a multiplicity higher than one, then the eigenvectors of the corresponding invariant subspace are not unique, and consequently the spectral decomposition is not unique.

It is remarkable that the maximal valued coefficient $p_{\max}$ with respect to any possible arbitrary components of $\mathbf{R}$ is precisely $p_{\max} = \hat{\lambda}_1$ ($\hat{\lambda}_1$ being the largest normalized eigenvalue of $\mathbf{R}$), which is realized for the first spectral component $(\mathrm{tr}\,\mathbf{R})(\hat{\mathbf{u}}_1 \otimes \hat{\mathbf{u}}_1^{\dagger})$, where $\hat{\mathbf{u}}_1$ is the eigenvector of $\mathbf{R}$ associated with $\hat{\lambda}_1$. To prove this, it is enough to observe that the minimal value of the quotient of the expression of p_i in Eq. (2.25) is achieved for $\hat{\mathbf{w}}_i = \hat{\mathbf{u}}_1$, in which case

$$\sum_{j=1}^{r} \frac{1}{\lambda_j} \left|\left(\mathbf{U}^{\dagger} \hat{\mathbf{u}}_1\right)_j\right|^2 = \frac{1}{\lambda_1} \tag{2.28}$$

The detailed procedure to apply the arbitrary decomposition of a given R is described in Section 2.6.7.

2.6.5 Characteristic Decomposition of 3D States

While all components of the arbitrary decomposition of $\mathbf{R}$ are pure states, it is also possible to decompose $\mathbf{R}$ by means of the following *characteristic* (or *trivial*) decomposition (Samson 1973; Gil 2007):

$$\mathbf{R} = \mathbf{U}\,\mathrm{diag}\left(\lambda_1, \lambda_2, \lambda_3\right)\mathbf{U}^{\dagger} = p_1 I \hat{\mathbf{R}}_p + p_2 I \hat{\mathbf{R}}_m + p_3 I \hat{\mathbf{R}}_{u\text{-}3D}$$

$$\begin{bmatrix} \hat{\mathbf{R}}_p \equiv \mathbf{U}\,\mathrm{diag}(1,0,0)\mathbf{U}^{\dagger} \quad \hat{\mathbf{R}}_m \equiv \frac{1}{2}\mathbf{U}\,\mathrm{diag}(1,1,0)\mathbf{U}^{\dagger} \quad \hat{\mathbf{R}}_{u\text{-}3D} \equiv \frac{1}{3}\mathbf{U}\,\mathrm{diag}(1,1,1)\mathbf{U}^{\dagger} = \frac{1}{3}\mathbf{I}_3 \\ p_1 \equiv \hat{\lambda}_1 - \hat{\lambda}_2 \quad p_2 \equiv 2\left(\hat{\lambda}_2 - \hat{\lambda}_3\right) \quad p_3 \equiv 3\hat{\lambda}_3 \end{bmatrix} \tag{2.29}$$

where $\mathbf{I}_3$ is the 3×3 identity matrix and all components have been chosen so as to have the same intensity $I = \mathrm{tr}\,\mathbf{R}$, while the respective coefficients of the components are expressed in terms of the eigenvalues of $\mathbf{R}$, in such a manner that $p_1 + p_2 + p_3 = 1$. Note that $\mathrm{rank}\,\hat{\mathbf{R}}_p = 1$, $\mathrm{rank}\,\hat{\mathbf{R}}_m = 2$ and $\mathrm{rank}\,\hat{\mathbf{R}}_{u\text{-}3D} = 3$, so that the consecutive components have decreasing polarimetric purity.

For a better physical interpretation of important aspects of the characteristic decomposition, the coefficients p_i can be expressed as $p_1 = P_1$, $p_1 = P_2 - P_1$, and $p_3 = 1 - P_2$, where P_1, P_2 are the so-called indices of polarimetric purity (IPP) of the state, defined as $P_1 = \hat{\lambda}_1 - \hat{\lambda}_2$ and $P_2 = 1 - 3\hat{\lambda}_3$ (see Section 2.10.4).

It is remarkable that the pure state $\mathbf{R}_p$, hereafter called the *characteristic component* of $\mathbf{R}$, is precisely the first component of the spectral decomposition, which is determined by the Jones eigenvector $\hat{\mathbf{u}}_1$ of $\mathbf{R}$ with largest eigenvalue.

The *discriminating state* $\mathbf{R}_m$ is given by an equiprobable mixture of the Jones eigenstates $\hat{\mathbf{u}}_1$ and $\hat{\mathbf{u}}_2$ with largest eigenvalues. In general, the polarization ellipses of $\hat{\mathbf{u}}_1$ and $\hat{\mathbf{u}}_2$ lie in different planes, in which case $\mathbf{R}_m$ represents a *genuine 3D state*, in the sense that the fluctuating electric field vector of the mixed state $\mathbf{R}_m$ has three nonzero components, regardless of the laboratory Cartesian coordinate system XYZ taken to represent it. In the limiting case that the polarization ellipses of $\hat{\mathbf{u}}_1$ and $\hat{\mathbf{u}}_2$ lie in a common plane, the set $(\hat{\mathbf{u}}_1, \hat{\mathbf{u}}_2, \hat{\mathbf{u}}_3)$, when referenced with respect to the intrinsic reference frame of $\hat{\mathbf{u}}_1$, takes the form of the canonical set $(\hat{\boldsymbol{\eta}}_1, \hat{\boldsymbol{\eta}}_2, \hat{\boldsymbol{\eta}}_3)$ (see Section 2.3.1) and $\mathbf{R}_m$ has the form $\mathbf{R}_m = \mathbf{R}_{u\text{-}2D} \equiv I\,\mathrm{diag}(1,1,0)$ of a 2D unpolarized state. States $\mathbf{R}$ for which $\mathbf{R}_m = \mathbf{R}_{u\text{-}2D}$ are called *regular*, and represent a limiting case of the general *nonregular states*. States with maximal nonregularity are called *perfect nonregular states*. Due to the key role played by the concepts of nonregularity and discriminating states in the understanding and physical interpretation of 3D polarization states, they will be analyzed in detail in Section 2.12.

The *3D unpolarized state* $\mathbf{R}_{u\text{-}3D}$ represents an equiprobable mixture of any set of orthogonal pure states, i.e., a fully random polarization state.

As with the spectral decomposition, a meaningful physical interpretation of the characteristic decomposition is achieved by taking the first eigenstate $\hat{\mathbf{u}}_1$ (i.e., the eigenvector of $\mathbf{R}$ associated with the largest eigenvalue λ_1) as the reference vector $\hat{\boldsymbol{\eta}}_1$, so that the unit eigenstates take the forms $(\hat{\boldsymbol{\eta}}_1, \hat{\mathbf{v}}_2, \hat{\mathbf{v}}_3)$ studied in Section 2.3. Then the polarization matrix $\mathbf{R}_1$, describing the state $\mathbf{R}$ but referenced with respect to the intrinsic reference frame $X_1Y_1Z_1$ of $\hat{\mathbf{u}}_1$, can be considered an incoherent superposition of the states $(\hat{\boldsymbol{\eta}}_1, \hat{\mathbf{v}}_2, \hat{\mathbf{v}}_3)$, so that the pure and discriminating components of $\mathbf{R}_1$ are

$$\mathbf{R}_p \equiv I\left(\hat{\eta}_1 \otimes \hat{\eta}_1^\dagger\right) \qquad \mathbf{R}_m \equiv I\frac{1}{2}\left[\left(\hat{\eta}_1 \otimes \hat{\eta}_1^\dagger\right) + \left(\hat{v}_2 \otimes \hat{v}_2^\dagger\right)\right] \tag{2.30}$$

Note that the pair $(\hat{\boldsymbol{\eta}}_1, \hat{\mathbf{v}}_2)$ is sufficient for a simple and meaningful physical interpretation of $\mathbf{R}$ in terms of the characteristic decomposition.

Unlike the 2D representation, and except for the particular case of states satisfying $\lambda_2 = \lambda_3$, the characteristic decomposition of $\mathbf{R}$ cannot be performed as a sum of a pure state and a 3D unpolarized state. This is because, in general, pure $n \times n$ density matrices contain $2n - 1$ independent parameters, whereas n D-unpolarized $n \times n$ density matrices are proportional to the $n \times n$ identity matrix and thus contain only one independent parameter, so that the equality between the number n^2 of independent parameters of a generic $n \times n$ density matrix and the sum $2n\left[= (2n-1)+1\right]$ of the independent parameters of a pure $n \times n$ density matrix and a nD-unpolarized $n \times n$ density matrix only holds when $n = 2$ (Gil 2020).

In the case of a 2D polarization state, the well-known decomposition into a pure state and a 2D unpolarized state is recovered.

2.6.6 Smart Decomposition

The pure and discriminating components $\mathbf{R}_p$ and $\mathbf{R}_m$ of the characteristic decomposition of a polarization matrix $\mathbf{R}$ can always be rearranged in such a manner that $\mathbf{R}$ is expressed through the *smart decomposition*

$$\begin{gathered}\mathbf{R}=\frac{P_1+P_2}{2}I\,\hat{\mathbf{R}}_{p1}+\frac{P_2-P_1}{2}I\,\hat{\mathbf{R}}_{p2}+(1-P_2)I\,\hat{\mathbf{R}}_{u-3D}\\ \left[\hat{\mathbf{R}}_{p1}\equiv\hat{\mathbf{u}}_1\otimes\hat{\mathbf{u}}_1^{\dagger}\qquad \hat{\mathbf{R}}_{p2}\equiv\hat{\mathbf{u}}_2\otimes\hat{\mathbf{u}}_2^{\dagger}\qquad \hat{\mathbf{R}}_{u-3D}\equiv\mathbf{I}_3/3\right]\end{gathered}\tag{2.31}$$

where the two first components, $\hat{\mathbf{R}}_{p1}$ and $\hat{\mathbf{R}}_{p2}$, correspond to the respective, mutually orthogonal, Jones eigenstates, $\hat{\mathbf{u}}_1$ and $\hat{\mathbf{u}}_2$, associated to the two largest eigenvalues, $\hat{\lambda}_1$ and $\hat{\lambda}_2$ (with $\hat{\lambda}_1\geq\hat{\lambda}_2\geq\hat{\lambda}_3$), while $\hat{\mathbf{R}}_{u-3D}$ is a 3D unpolarized state (Figure 2.5). The difference between the smart and the spectral decompositions is due to that portions of $\hat{\mathbf{R}}_{p1}$ and $\hat{\mathbf{R}}_{p2}$, with weights $\hat{\lambda}_3$ equal to that of $\hat{\mathbf{R}}_{p3}$ in the spectral decomposition are combined to build $\hat{\mathbf{R}}_{u-3D}$.

When $\mathbf{R}$ is referenced with respect to the intrinsic reference frame $X_1Y_1Z_1$ of $\hat{\mathbf{u}}_1$, the pure components of the smart decomposition take the form $\hat{\mathbf{R}}_{p1}=(\hat{\boldsymbol{\eta}}_1\otimes\hat{\boldsymbol{\eta}}_1^{\dagger})$ and $\hat{\mathbf{R}}_{p2}=(\hat{\mathbf{v}}_2\otimes\hat{\mathbf{v}}_2^{\dagger})$. Therefore, as seen in Section 2.3, the polarization planes Π_1 and Π_2 of $\hat{\mathbf{R}}_{p1}$ and $\hat{\mathbf{R}}_{p2}$ are different in general (Figures 2.2, 2.3 and 2.4), so that $\mathbf{R}$ represents a regular state if and only if either $P_1=P_2$, or Π_1 and Π_2 coincide. Consequently, the directions of the spin density vectors $\hat{\mathbf{n}}_1$ and $\hat{\mathbf{n}}_2$ of the pure components $\hat{\mathbf{R}}_{p1}$ and $\hat{\mathbf{R}}_{p2}$, whose composition

$$\hat{\mathbf{n}}=\frac{P_1+P_2}{2}\hat{\mathbf{n}}_1+\frac{P_1-P_2}{2}\hat{\mathbf{n}}_2\tag{2.32}$$

determines the spin density vector $\hat{\mathbf{n}}$ of $\mathbf{R}$, are in general different and consequently the direction of $\hat{\mathbf{n}}$ is different from that of $\hat{\mathbf{n}}_1$ except for the limiting cases of regular states ($\hat{\mathbf{n}}_2=-\hat{\mathbf{n}}_1$) and perfect nonregular states ($\hat{\mathbf{n}}_2=\mathbf{0}$, $|\hat{\mathbf{n}}|=1/2$, see Section 2.12).

The smart decomposition shows that $\mathbf{R}$ can always be interpreted by means of the following nine independent parameters: the intensity I of the state $\mathbf{R}$; the two orientation angles (θ_1,ϕ_1) that determine the polarization plane of the first pure component $\mathbf{R}_{p1}\equiv I\,\hat{\mathbf{R}}_{p1}$; the azimuth φ_1 and ellipticity angle χ_1 of $\mathbf{R}_{p1}$; the azimuth φ_2 and ellipticity angle χ_2 of the

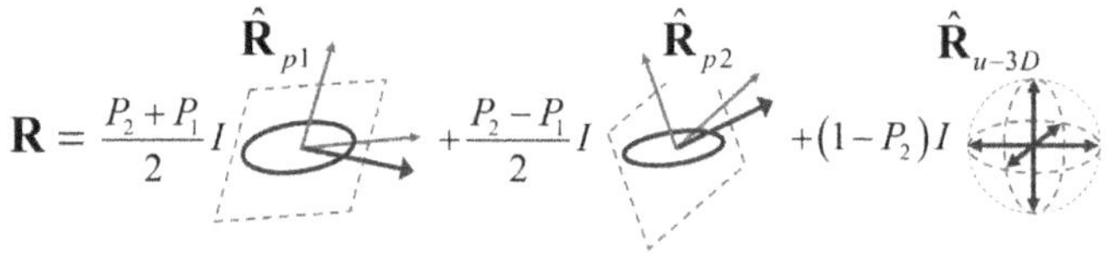

FIGURE 2.5
Smart decomposition of a mixed state $\mathbf{R}$ as the incoherent superposition of its two first eigenstates and a 3D unpolarized state (Gil 2014a).

second pure component $\mathbf{R}_{p2} \equiv I\hat{\mathbf{R}}_{p2}$, and the two indices of polarimetric purity (P_1, P_2) of $\mathbf{R}$ (recall that $\phi_2 = \phi_1$, while the pair φ_2, χ_2 determines θ_2).

In the case of regular states, the number of independent parameters is reduced to seven: I, θ_1, ϕ_1, φ_1, χ_1, P_1 and P_2 and, as expected, 2D states are fully determined by six independent parameters: I, θ_1, ϕ_1, φ_1, χ_1 and P_1, where P_1 is then the 2D degree of polarization $\mathcal{P}$.

2.6.7 Polarimetric Subtraction

Given two polarization matrices $\mathbf{R}_a$ and $\mathbf{R}_b$, their sum $\mathbf{R}_a + \mathbf{R}_b$ represents the incoherent superposition of the states represented by $\mathbf{R}_a$ and $\mathbf{R}_b$. Nevertheless, it is obvious that in general the subtraction $\mathbf{R} - \mathbf{R}_1$, $\mathbf{R}$ and $\mathbf{R}_1$ being arbitrary, does not necessarily produce a positive semidefinite Hermitian matrix, and thus $\mathbf{R} - \mathbf{R}_1$ has not, in general, physical meaning. In the context of 3D polarization matrices, polarimetric subtractions are physically conceivable when $\mathbf{R} - \mathbf{R}_1$ results in a positive semidefinite Hermitian matrix $\mathbf{R}_2 = \mathbf{R} - \mathbf{R}_1$, i.e., the state $\mathbf{R} = \mathbf{R}_1 + \mathbf{R}_2$ can be considered the incoherent composition of the states (pure or not pure) $\mathbf{R}_1$ and $\mathbf{R}_2$. A necessary and sufficient condition for the subtractability of $\mathbf{R}_1$ from $\mathbf{R}$ is given by $\text{rank}(\mathbf{R} + \mathbf{R}_1) = \text{rank}\,\mathbf{R}$ (Gil and San José 2013).

Since the arbitrary decomposition of a given polarization matrix $\mathbf{R}$ provides all physically realizable decompositions of $\mathbf{R}$ in terms of a minimum number of pure components, the polarimetric subtractions with pure subtrahends can be formulated as a comprehensive view of the arbitrary decomposition in the following manner (Gil 2007; Gil and San José 2013):

(1) Given a polarization matrix $\mathbf{R}$ with $2 \le \text{rank}\,\mathbf{R} \le 3$, any three-dimensional complex unit vector $\hat{\mathbf{w}}_1$ belonging to $\text{im}\,\mathbf{R}$ (image subspace of $\mathbf{R}$) can be subtracted from $\mathbf{R}$ producing a remainder polarization matrix. In the case where $\text{rank}\,\mathbf{R} = 3$, any vector $\hat{\mathbf{w}}_1$ necessarily belongs to $\text{im}\,\mathbf{R}$ and thus, it is polarimetrically subtractable.
(2) Synthesize the polarization density matrix $\hat{\mathbf{R}}_{p1} \equiv \hat{\mathbf{w}}_1 \otimes \hat{\mathbf{w}}_1^{\dagger}$.
(3) Calculate the coefficient p_1 of $\hat{\mathbf{R}}_{p1}$ by means of $p_1 = 1/(\hat{\mathbf{w}}_1^{\dagger}\,\mathbf{R}^{-}\hat{\mathbf{w}}_1)$ (see Section 2.6.3).
(4) Calculate the remainder normalized polarization matrix as $\hat{\mathbf{R}}_r = [\hat{\mathbf{R}} - p_1 (\hat{\mathbf{w}}_1 \times \hat{\mathbf{w}}_1^{\dagger})]/(1 - p_1)$.
When $\text{rank}\,\mathbf{R} = 3$ this procedure can be iterated once again as follows:
(5) Take an arbitrary 3D complex unit vector $\hat{\mathbf{w}}_2$ belonging to $\text{im}\,\hat{\mathbf{R}}_r$ and synthesize the normalized polarization matrix $\hat{\mathbf{R}}_{p2} \equiv \hat{\mathbf{w}}_2 \otimes \hat{\mathbf{w}}_2^{\dagger}$ of the second component.
(6) Calculate the coefficient p_2 of $\hat{\mathbf{R}}_{p2}$ in the convex sum as $p_2 = 1/(\hat{\mathbf{w}}_2^{\dagger}\,\mathbf{R}^{-}\hat{\mathbf{w}}_2)$.
(7) Calculate the third pure component and its coefficient through the expressions:

$$\hat{\mathbf{R}}_{p3} = [\hat{\mathbf{R}} - p_1(\hat{\mathbf{w}}_1 \times \hat{\mathbf{w}}_1^{\dagger}) - p_2(\hat{\mathbf{w}}_2 \times \hat{\mathbf{w}}_2^{\dagger})]/(1 - p_1 - p_2), p_3 = 1 - p_1 - p_2.$$

The above polarimetric subtraction provides the procedure to apply the arbitrary decomposition and, beyond its theoretical significance in polarization algebra, constitutes a powerful tool for the analysis of polarimetric measurements. Consider an instrument that provides a set o f measured 2D or 3D Stokes parameters which results from the incoherent composition of the Stokes parameters of the sample under study and some Stokes parameters generated by the instrument itself or by a known source that is not separable; the desired Stokes parameters can be obtained provided the artifact, or extra Stokes parameters, can be obtained or postulated through an independent measurement or criterion. This kind of Stokes subtraction can be applied, for instance, in astronomic and atmospheric polarimetry, where the signal frequently appears combined with a background component (Snick et al. 2014).

2.7 3D Space–Time and Space–Frequency Representations of Coherence and Polarization

2.7.1 3D Representations of Coherence and Polarization

As indicated in Section 1.6, there are interesting cases where the attention is focused on the coherence–polarization properties in a spatial region of the wavefield rather than only at a specific point. The formulation of the space–time and space–frequency representations of 2D states can be generalized to 3D states in the way shown in the next paragraphs.

The space–time coherence–polarization properties of a given 3D electromagnetic wave are appropriately represented by the following *3D coherence matrix* (or *coherence–polarization matrix*):

$$\mathbf{J}(\mathbf{r}_1,\mathbf{r}_2,\tau)=\langle\boldsymbol{\varepsilon}(\mathbf{r}_1,t)\otimes\boldsymbol{\varepsilon}^{\dagger}(\mathbf{r}_2,t+\tau)\rangle$$
$$=\begin{pmatrix}\langle\varepsilon_1(\mathbf{r}_1,t)\,\varepsilon_1^*(\mathbf{r}_2,t+\tau)\rangle & \langle\varepsilon_1(\mathbf{r}_1,t)\,\varepsilon_2^*(\mathbf{r}_2,t+\tau)\rangle & \langle\varepsilon_1(\mathbf{r}_1,t)\,\varepsilon_3^*(\mathbf{r}_2,t+\tau)\rangle\\ \langle\varepsilon_2(\mathbf{r}_1,t)\,\varepsilon_1^*(\mathbf{r}_2,t+\tau)\rangle & \langle\varepsilon_2(\mathbf{r}_1,t)\,\varepsilon_2^*(\mathbf{r}_2,t+\tau)\rangle & \langle\varepsilon_2(\mathbf{r}_1,t)\,\varepsilon_3^*(\mathbf{r}_2,t+\tau)\rangle\\ \langle\varepsilon_3(\mathbf{r}_1,t)\,\varepsilon_1^*(\mathbf{r}_2,t+\tau)\rangle & \langle\varepsilon_3(\mathbf{r}_1,t)\,\varepsilon_2^*(\mathbf{r}_2,t+\tau)\rangle & \langle\varepsilon_3(\mathbf{r}_1,t)\,\varepsilon_3^*(\mathbf{r}_2,t+\tau)\rangle\end{pmatrix} \tag{2.33}$$

where ε_i are the analytic signals of the three components of the analytic signal vector at points $\mathbf{r}_1$ and $\mathbf{r}_2$ and time instants $t_1 \equiv t$ and $t_2 = t+\tau$. As in Eq. (2.13), $\langle\ \rangle$ indicates time average over the measurement time, and in the case of stationary and ergodic fields, the brackets can also be interpreted as ensemble averaging over the ensemble of sample realizations.

The 3D *equal-point coherence matrix* is given by

$$\mathbf{J}(\tau)=\langle\boldsymbol{\varepsilon}(t)\otimes\boldsymbol{\varepsilon}^{\dagger}(t+\tau)\rangle$$
$$=\begin{pmatrix}\langle\varepsilon_1(t)\,\varepsilon_1^*(t+\tau)\rangle & \langle\varepsilon_1(t)\,\varepsilon_2^*(t+\tau)\rangle & \langle\varepsilon_1(t)\,\varepsilon_3^*(t+\tau)\rangle\\ \langle\varepsilon_2(t)\,\varepsilon_1^*(t+\tau)\rangle & \langle\varepsilon_2(t)\,\varepsilon_2^*(t+\tau)\rangle & \langle\varepsilon_2(t)\,\varepsilon_3^*(t+\tau)\rangle\\ \langle\varepsilon_3(t)\,\varepsilon_1^*(t+\tau)\rangle & \langle\varepsilon_3(t)\,\varepsilon_2^*(t+\tau)\rangle & \langle\varepsilon_3(t)\,\varepsilon_3^*(t+\tau)\rangle\end{pmatrix} \tag{2.34}$$

where the spatial dependence has been omitted.

The *space–time 3D coherence Stokes parameters* (also called *3D two-point Stokes parameters*), are defined as (Ellis and Dogariu 2004b)

$$s_{ij} = \mathrm{tr}\left[\mathbf{J}(\mathbf{r}_1,\mathbf{r}_2,\tau)\,\boldsymbol{\omega}_{ij}\right] \tag{2.35}$$

or, in explicit form,

$$
\begin{aligned}
s_{00}(\mathbf{r}_1,\mathbf{r}_2,\tau) &= j_{00}(\mathbf{r}_1,\mathbf{r}_2,\tau)+j_{11}(\mathbf{r}_1,\mathbf{r}_2,\tau)+j_{22}(\mathbf{r}_1,\mathbf{r}_2,\tau)\\
s_{01}(\mathbf{r}_1,\mathbf{r}_2,\tau) &= j_{01}(\mathbf{r}_1,\mathbf{r}_2,\tau)+j_{10}(\mathbf{r}_1,\mathbf{r}_2,\tau)\\
s_{02}(\mathbf{r}_1,\mathbf{r}_2,\tau) &= j_{02}(\mathbf{r}_1,\mathbf{r}_2,\tau)+j_{20}(\mathbf{r}_1,\mathbf{r}_2,\tau)\\
s_{10}(\mathbf{r}_1,\mathbf{r}_2,\tau) &= i\left[j_{01}(\mathbf{r}_1,\mathbf{r}_2,\tau)-j_{10}(\mathbf{r}_1,\mathbf{r}_2,\tau)\right]\\
s_{11}(\mathbf{r}_1,\mathbf{r}_2,\tau) &= j_{00}(\mathbf{r}_1,\mathbf{r}_2,\tau)-j_{11}(\mathbf{r}_1,\mathbf{r}_2,\tau)\\
s_{12}(\mathbf{r}_1,\mathbf{r}_2,\tau) &= j_{12}(\mathbf{r}_1,\mathbf{r}_2,\tau)+j_{21}(\mathbf{r}_1,\mathbf{r}_2,\tau)\\
s_{20}(\mathbf{r}_1,\mathbf{r}_2,\tau) &= i\left[j_{02}(\mathbf{r}_1,\mathbf{r}_2,\tau)-j_{20}(\mathbf{r}_1,\mathbf{r}_2,\tau)\right]\\
s_{21}(\mathbf{r}_1,\mathbf{r}_2,\tau) &= i\left[j_{12}(\mathbf{r}_1,\mathbf{r}_2,\tau)-j_{21}(\mathbf{r}_1,\mathbf{r}_2,\tau)\right]\\
s_{22}(\mathbf{r}_1,\mathbf{r}_2,\tau) &= \left[j_{00}(\mathbf{r}_1,\mathbf{r}_2,\tau)+j_{11}(\mathbf{r}_1,\mathbf{r}_2,\tau)-2j_{22}(\mathbf{r}_1,\mathbf{r}_2,\tau)\right]/\sqrt{3}
\end{aligned}
\tag{2.36}
$$

The 3D polarization properties at point **r** are determined by the corresponding polarization matrix, which is retrieved as $\mathbf{R}=\mathbf{J}(\mathbf{r},\mathbf{r},\tau=0)$, or by the corresponding Stokes parameters matrix $\mathbf{S}(\mathbf{r},\mathbf{r},\tau=0)$.

The formulation of space–frequency representation described in Section 1.6.1, is generalized below for the case of 3D fields. In addition to weak stationarity, it is assumed the condition (satisfied by any physical field in nature) that correlations $\langle \varepsilon_i(\mathbf{r}_1,t)\,\varepsilon_j^*(\mathbf{r}_2,t+\tau)\rangle$ fall off sufficiently rapidly with $|\tau|$, so that $\mathbf{J}(\mathbf{r}_1,\mathbf{r}_2,\tau)$ is absolutely integrable with respect to τ, which in turn implies that $\mathbf{J}(\mathbf{r}_1,\mathbf{r}_2,\tau)$ is square integrable (Wolf 1981), and therefore $\mathbf{J}(\mathbf{r}_1,\mathbf{r}_2,\tau)$ possesses the well defined Fourier transform Wolf (2003)

$$\mathbf{W}(\mathbf{r}_1,\mathbf{r}_2,\omega)=\frac{1}{2\pi}\int_{-\infty}^{\infty}\mathbf{J}(\mathbf{r}_1,\mathbf{r}_2,\tau)e^{i\omega\tau}d\tau \tag{2.37}$$

which is a continuous function of ω and is called the *spectral coherence matrix* (or *cross spectral density matrix*) of the wave with respect to the given frequency ω and to points $\mathbf{r}_1$ and $\mathbf{r}_2$.

Eq. (2.37) forms a Fourier transform pair with its inverse (Tervo et al. 2004)

$$\mathbf{J}(\mathbf{r}_1,\mathbf{r}_2,\tau)=\int_0^{\infty}\mathbf{W}(\mathbf{r}_1,\mathbf{r}_2,\omega)\,e^{-i\omega\tau}\,d\omega \tag{2.38}$$

where, in analogy to the 2D case considered in Section 1.6, the lower limit is zero, and not $-\infty$, because of the use of the analytic signals to represent the space–time components of the electric field (Mandel and Wolf 1995).

Recall that, in spite of $\mathbf{W}(\mathbf{r}_1,\mathbf{r}_2,\omega)$ is obtained through a Fourier transform of $\mathbf{J}(\mathbf{r}_1,\mathbf{r}_2,\tau)$, this, in principle, would not necessarily ensure that $\mathbf{W}(\mathbf{r}_1,\mathbf{r}_2,\omega)$ is a correlation matrix by itself (Wolf 201, Korotkova 2017). Nevertheless, it has been proven that $\mathbf{W}(\mathbf{r}_1,\mathbf{r}_2,\omega)$ is a valid correlation matrix for which the averaging process is carried out with respect to an ensemble of strictly monochromatic realizations $\boldsymbol{\varepsilon}(\mathbf{r},\omega)$ of the electric field (Tervo et al. 2004) (for more details see Section 1.6.1.3). The elements of the cross-spectral density matrix can be expressed in the following equivalent forms:

$$w_{ij}(\mathbf{r}_1,\mathbf{r}_2,\omega)=\langle\varepsilon_i(\mathbf{r}_1,\omega)\,\varepsilon_j^*(\mathbf{r}_2,\omega)\rangle_\omega=\frac{1}{2\pi}\int_{-\infty}^{\infty}\left\langle\varepsilon_i(\mathbf{r}_1,t)\,\varepsilon_j^*(\mathbf{r}_2,t+\tau)\right\rangle e^{i\omega\tau}\,d\tau \tag{2.39}$$

where the average $\langle\ \rangle_\omega$ is taken over the ensemble of the monochromatic realizations corresponding to the given single frequency ω (Tervo et al. 2004; Alonso and Wolf 2008).

The equal-point coherence–polarization properties of the 3D state of polarization in the space–frequency domain are given by the 3D *spectral polarization matrix*

$$\mathbf{W}(\omega)=\frac{1}{2\pi}\int_{-\infty}^{\infty}\mathbf{J}(\tau)\,e^{i\omega\tau}\,d\tau \tag{2.40}$$

which depends on the single spatial argument $\mathbf{r}$ (omitted) and differs from the cross-spectral density matrix $\mathbf{W}(\mathbf{r}_1,\mathbf{r}_2,\omega)$ which depends on two spatial arguments. Therefore $\mathbf{W}(\mathbf{r}_1,\mathbf{r}_2,\omega)$ can be used for the study of the coherence and polarization properties of the corresponding electromagnetic beam on propagation.

In the case of a quasimonochromatic state with very narrow spectral width around a central angular frequency ϖ, the spec tral polarization matrix $\mathbf{W}(\varpi)$, evaluated at ϖ, provides the same measurable quantities as those of the space–time polarization matrix $\mathbf{R}$. In the most general case, $\mathbf{R}$ and $\mathbf{W}(\varpi)$ are not equivalent (Setälä et al. 2009a).

The 3D *intensity distribution* or 3D *spectral density* at a given point $\mathbf{r}$ is given by

$$I(\omega)\equiv\operatorname{tr}\mathbf{W}(\omega) \tag{2.41}$$

Moreover, as an extension to 3D of the 2D spectral coherence Stokes parameters (Korotkova and Wolf 2005; Tervo et al. 2009; Korotkova 2017), the *3D spectral coherence Stokes parameters* (or *space–frequency two-point Stokes parameters*) *are defined as*

$$\begin{aligned}
s_{00}(\mathbf{r}_1,\mathbf{r}_2,\omega)&=w_{00}(\mathbf{r}_1,\mathbf{r}_2,\omega)+w_{11}(\mathbf{r}_1,\mathbf{r}_2,\omega)+w_{22}(\mathbf{r}_1,\mathbf{r}_2,\omega)\\
s_{01}(\mathbf{r}_1,\mathbf{r}_2,\omega)&=w_{01}(\mathbf{r}_1,\mathbf{r}_2,\omega)+w_{10}(\mathbf{r}_1,\mathbf{r}_2,\omega)\\
s_{02}(\mathbf{r}_1,\mathbf{r}_2,\omega)&=w_{02}(\mathbf{r}_1,\mathbf{r}_2,\omega)+w_{20}(\mathbf{r}_1,\mathbf{r}_2,\omega)\\
s_{10}(\mathbf{r}_1,\mathbf{r}_2,\omega)&=i\left[w_{01}(\mathbf{r}_1,\mathbf{r}_2,\omega)-w_{10}(\mathbf{r}_1,\mathbf{r}_2,\omega)\right]\\
s_{11}(\mathbf{r}_1,\mathbf{r}_2,\omega)&=w_{00}(\mathbf{r}_1,\mathbf{r}_2,\omega)-w_{11}(\mathbf{r}_1,\mathbf{r}_2,\omega)\\
s_{12}(\mathbf{r}_1,\mathbf{r}_2,\omega)&=w_{12}(\mathbf{r}_1,\mathbf{r}_2,\omega)+w_{21}(\mathbf{r}_1,\mathbf{r}_2,\omega)\\
s_{20}(\mathbf{r}_1,\mathbf{r}_2,\omega)&=i\left[w_{02}(\mathbf{r}_1,\mathbf{r}_2,\omega)-w_{20}(\mathbf{r}_1,\mathbf{r}_2,\omega)\right]\\
s_{21}(\mathbf{r}_1,\mathbf{r}_2,\omega)&=i\left[w_{12}(\mathbf{r}_1,\mathbf{r}_2,\omega)-w_{21}(\mathbf{r}_1,\mathbf{r}_2,\omega)\right]\\
s_{22}(\mathbf{r}_1,\mathbf{r}_2,\omega)&=\left[w_{00}(\mathbf{r}_1,\mathbf{r}_2,\omega)+w_{11}(\mathbf{r}_1,\mathbf{r}_2,\omega)-2w_{22}(\mathbf{r}_1,\mathbf{r}_2,\omega)\right]/\sqrt{3}
\end{aligned} \tag{2.42}$$

Unlike the one-point 3D spectral Stokes parameters, the generalized 3D Stokes parameters $s_{ij}(\mathbf{r}_1,\mathbf{r}_2,\omega)$ contain information about both coherence and polarization properties.

The notation used for the subscripts of $s_{ij}(\mathbf{r}_1,\mathbf{r}_2,\omega)$ allows for arranging $s_{ij}(\mathbf{r}_1,\mathbf{r}_2,\omega)$ as the elements of the *one-point spectral Stokes parameters matrix*, determined from $\mathbf{W}(\omega)\equiv\mathbf{W}(\mathbf{r},\mathbf{r},\omega)$,

$$\mathbf{S}(\omega)=\begin{pmatrix} w_{00}(\omega)+w_{11}(\omega)+w_{22}(\omega) & w_{01}(\omega)+w_{10}(\omega) & w_{02}(\omega)+w_{20}(\omega)\\ i\left[w_{01}(\omega)-w_{10}(\omega)\right] & w_{00}(\omega)-w_{11}(\omega) & w_{12}(\omega)+w_{21}(\omega)\\ i\left[w_{02}(\omega)-w_{20}(\omega)\right] & i\left[w_{12}(\omega)-w_{21}(\omega)\right] & \left[w_{00}(\omega)+w_{11}(\omega)-2w_{22}(\omega)\right]/\sqrt{3}\end{pmatrix} \tag{2.43}$$

The usual 3D equal-point and equal-time polarization matrix $\mathbf{R}$ is retrieved as

$$\mathbf{R} = \mathbf{J}(0) = \int_0^{\infty} \mathbf{W}(\omega)\, d\omega \tag{2.44}$$

Analogously, the 3D equal-point and equal-time Stokes parameters matrix is given by

$$\mathbf{S} = \mathbf{S}(\tau = 0) = \int_0^{\infty} \mathbf{S}(\omega)\, d\omega \tag{2.45}$$

2.7.2 Measures of the 3D Degree of Coherence of Electromagnetic Fields

A set of measures of the two-point degrees of mutual coherence between the analytic signals $\varepsilon_k(\mathbf{r}_1,t)$ and $\varepsilon_l(\mathbf{r}_2,t+\tau)$ (with $k,l = 1,2,3$), is given by the complex correlation functions

$$\mu_{kl}(\mathbf{r}_1,\mathbf{r}_2,\tau) = \frac{j_{kl}(\mathbf{r}_1,\mathbf{r}_2,\tau)}{\sqrt{j_{kk}(\mathbf{r}_1,\mathbf{r}_1,0)\; j_{ll}(\mathbf{r}_2,\mathbf{r}_2,0)}} \tag{2.46}$$

whose one-point versions are given by

$$\mu_{kl}(\tau) = j_{kl}(\tau)\Big/\sqrt{j_{kk}(0)\; j_{ll}(0)} \tag{2.47}$$

and whose one-point and equal-time, versions are precisely the complex degrees of mutual coherence μ_{kl} defined in Eq. (2.16)

$$\mu_{kl} = j_{kl}(0)\Big/\sqrt{j_{kk}(0)\; j_{ll}(0)} = r_{kl}\Big/\sqrt{r_{kk}\, r_{ll}} \tag{2.48}$$

where r_{kl} are the elements of the polarization matrix $\mathbf{R}$. Let us note that $\mu_{kl}(\mathbf{r}_1,\mathbf{r}_2,\tau)$ are not invariant under orthogonal transformations and, in particular, under rotations of the reference frame.

2.7.2.1 Intrinsic Degrees of Coherence

In analogy to the 2D formulation, the 3D *normalized coherence matrix* is defined as (Réfrégier and Goudail 2005; Réfrégier et al. 2014)

$$\mathbf{M}(\mathbf{r}_1,\mathbf{r}_2,\tau) \equiv \left[\mathbf{R}(\mathbf{r}_2,t+\tau)\right]^{-1/2} \mathbf{J}(\mathbf{r}_1,\mathbf{r}_2,\tau)\left[\mathbf{R}(\mathbf{r}_1,t)\right]^{-1/2} \tag{2.49}$$

where the effects of the 3D polarization matrices $\mathbf{R}(\mathbf{r}_1,t)$ and $\mathbf{R}(\mathbf{r}_2,t+\tau)$ have been extracted from 3D the coherence matrix $\mathbf{J}(\mathbf{r}_1,\mathbf{r}_2,\tau)$, reducing it to $\mathbf{M}(\mathbf{r}_1,\mathbf{r}_2,\tau)$.

As with the definition of the 2D normalized coherence matrix, in the case of a totally polarized state (i.e., $\det\mathbf{R}(\mathbf{r}_i,t) = 0$), the notation $[\mathbf{R}(\mathbf{r}_i,t)]^{-1/2}$ is to be understood in the pseudoinverse sense (see Section 1.6.2.3).

The 3D *intrinsic degrees of coherence* are defined as the singular values $\mu_{Ei}(\mathbf{r}_1,\mathbf{r}_2,\tau)$ (with $i = 1,2,3$) of $\mathbf{M}(\mathbf{r}_1,\mathbf{r}_2,\tau)$ whose singular value decomposition is expressed as

$$\mathbf{M}(\mathbf{r}_1,\mathbf{r}_2,\tau) = \mathbf{N}_2(\mathbf{r}_1,\mathbf{r}_2,\tau)\, \mathrm{diag}(\mu_{E1},\mu_{E2},\mu_{E3})\, \mathbf{N}_1^{\dagger}(\mathbf{r}_1,\mathbf{r}_2,\tau) \qquad (0 \le \mu_{E3} \le \mu_{E2} \le \mu_{E1} \le 1) \tag{2.50}$$

where the space–time dependence of $\mu_{Ei}(\mathbf{r}_1, \mathbf{r}_2, \tau)$ has been omitted, and $\mathbf{N}_1(\mathbf{r}_1, \mathbf{r}_2, \tau)$ and $\mathbf{N}_2(\mathbf{r}_1, \mathbf{r}_2, \tau)$ are unitary matrices.

The space–frequency counterpart of the above formulation is based on the concept of 3D *space–frequency normalized coherence matrix*

$$\mathbf{M}(\mathbf{r}_1, \mathbf{r}_2, \omega) \equiv \left[\mathbf{W}(\mathbf{r}_2, \omega)\right]^{-1/2} \mathbf{W}(\mathbf{r}_1, \mathbf{r}_2, \omega)\left[\mathbf{W}(\mathbf{r}_1, \omega)\right]^{-1/2} \tag{2.51}$$

so that the 3D *space–frequency intrinsic degrees of coherence* $\mu_{Ei}(\mathbf{r}_1, \mathbf{r}_2, \omega)$ $(i = 1, 2, 3)$ are defined from the singular value decomposition of $\mathbf{M}(\mathbf{r}_1, \mathbf{r}_2, \omega)$,

$$\mathbf{M}(\mathbf{r}_1, \mathbf{r}_2, \omega) = \mathbf{N}_2(\mathbf{r}_1, \mathbf{r}_2, \omega)\, \mathrm{diag}(\mu_{E1}, \mu_{E2}, \mu_{E3})\, \mathbf{N}_1^{\dagger}(\mathbf{r}_1, \mathbf{r}_2, \omega) \tag{2.52}$$

An interesting summary of the properties of the 3D intrinsic degrees of coherence, including several important notions like the concept of mean-square coherence and the description of irreversible phenomena, can be found in (Réfrégier et al. 2014).

2.7.2.2 Electromagnetic Degree of Coherence

The 3D version of the electromagnetic degree of coherence (defined for 2D states in Eq. (1.79)) is given by (Tervo et al. 2003)

$$\gamma(\mathbf{r}_1, \mathbf{r}_2, \tau) = \frac{\sqrt{\mathrm{tr}\left[\mathbf{J}(\mathbf{r}_1, \mathbf{r}_2, \tau)\, \mathbf{J}(\mathbf{r}_2, \mathbf{r}_1, -\tau)\right]}}{\sqrt{\mathrm{tr}\,\mathbf{J}(\mathbf{r}_1, \mathbf{r}_1, 0)\, \mathrm{tr}\,\mathbf{J}(\mathbf{r}_2, \mathbf{r}_2, 0)}} = \frac{\left\|\mathbf{J}(\mathbf{r}_1, \mathbf{r}_2, \tau)\right\|_2}{\sqrt{I(\mathbf{r}_1)\, I(\mathbf{r}_2)}} \tag{2.53}$$

where

$$\left\|\mathbf{J}(\mathbf{r}_1, \mathbf{r}_2, \tau)\right\|_2 \equiv \sqrt{\sum_{k,l=1}^{3} \left|j_{kl}(\mathbf{r}_1, \mathbf{r}_2, \tau)\right|^2} \tag{2.54}$$

is the Frobenius norm of $\mathbf{J}(\mathbf{r}_1, \mathbf{r}_2, \tau)$, while $I(\mathbf{r}_1)$ and $I(\mathbf{r}_2)$ are the intensities at the respective points.

The *3D electromagnetic degree of coherence* $\gamma(\mathbf{r}_1, \mathbf{r}_2, \tau)$ of the beam represented by $\mathbf{J}(\mathbf{r}_1, \mathbf{r}_2, \tau)$ is a real quantity, invariant with respect to unitary, position-dependent, transformations of the reference frame. An essential virtue of $\gamma(\mathbf{r}_1, \mathbf{r}_2, \tau)$ is that it satisfies in a natural manner the Glauber's spatial factorization requisite for totally coherent beams.

The *3D space–frequency electromagnetic degree of coherence* is defined as the real parameter (Setälä et al. 2004b)

$$\gamma(\mathbf{r}_1, \mathbf{r}_2, \omega) = \frac{\sqrt{\mathrm{tr}\left[\mathbf{W}(\mathbf{r}_1, \mathbf{r}_2, \omega)\, \mathbf{W}(\mathbf{r}_2, \mathbf{r}_1, \omega)\right]}}{\sqrt{\mathrm{tr}\,\mathbf{W}(\mathbf{r}_1, \mathbf{r}_1, \omega)}\, \sqrt{\mathrm{tr}\,\mathbf{W}(\mathbf{r}_2, \mathbf{r}_2, \omega)}} = \frac{\left\|\mathbf{W}(\mathbf{r}_1, \mathbf{r}_2, \omega)\right\|_2}{\sqrt{I(\mathbf{r}_1, \omega)}\, \sqrt{I(\mathbf{r}_2, \omega)}} \tag{2.55}$$

At a given frequency ω, the field is completely coherent if, and only if, the components of the field at that frequency are fully correlated for all pairs of points $\mathbf{r}_1$ and $\mathbf{r}_2$ within the considered spatial region of observation (Setälä et al. 2004b).

2.7.2.3 Overall Space–Frequency Degree of Coherence

In analogy to the 2D formulation, let us now focus our attention to the 3D version of the overall measure $D(\mathbf{r}_1,\mathbf{r}_2,\omega)$ of coherence of electromagnetic fields, defined in terms of a distance between correlation matrices, in such a manner that $D(\mathbf{r}_1,\mathbf{r}_2,\omega)$ is invariant under unitary transformations. Leaving aside the experimental interferometric arrangement required for its measurement, this quantity provides, at least from a theoretical and formal point of view, an overall measure of the coherence, visibility and polarization properties. The framework that allows this global treatment relies on the 6×6 cross spectral density matrix defined as (Luis 2007)

$$\mathbf{M}_L(\mathbf{r}_1,\mathbf{r}_2,\omega) \equiv \begin{pmatrix} \mathbf{W}(\mathbf{r}_1,\mathbf{r}_1,\omega) & \mathbf{W}(\mathbf{r}_1,\mathbf{r}_2,\omega) \\ \mathbf{W}^{+}(\mathbf{r}_1,\mathbf{r}_2,\omega) & \mathbf{W}(\mathbf{r}_2,\mathbf{r}_2,\omega) \end{pmatrix} \tag{2.56}$$

whose normalized form is

$$\hat{\mathbf{M}}_L(\mathbf{r}_1,\mathbf{r}_2,\omega) \equiv \frac{\mathbf{M}_L(\mathbf{r}_1,\mathbf{r}_2,\omega)}{\operatorname{tr}\mathbf{M}_L(\mathbf{r}_1,\mathbf{r}_2,\omega)} = \frac{\mathbf{M}_L(\mathbf{r}_1,\mathbf{r}_2,\omega)}{\operatorname{tr}\mathbf{W}(\mathbf{r}_1,\mathbf{r}_1,\omega)+\operatorname{tr}\mathbf{W}(\mathbf{r}_2,\mathbf{r}_2,\omega)} = \frac{\mathbf{M}_L(\mathbf{r}_1,\mathbf{r}_2,\omega)}{I(\mathbf{r}_1,\omega)+I(\mathbf{r}_2,\omega)} \tag{2.57}$$

The quantity $D(\mathbf{r}_1,\mathbf{r}_2,\omega)$ is then defined for 3D electromagnetic beams as the following measure of the distance between a completely uncorrelated and completely unpolarized state, represented by the 6×6 identity matrix $\mathbf{I}_{6,}$ and the normalized 6×6 cross spectral density matrix $\hat{\mathbf{M}}_L(\mathbf{r}_1,\mathbf{r}_2,\omega)$:

$$D^2(\mathbf{r}_1,\mathbf{r}_2,\omega) \equiv \frac{1}{n-1}\operatorname{tr}\left\{\left[\mathbf{I}_6 - n\,\hat{\mathbf{M}}_L(\mathbf{r}_1,\mathbf{r}_2,\omega)\right]^2\right\} = \frac{1}{n-1}\left[n\left\|\hat{\mathbf{M}}_L(\mathbf{r}_1,\mathbf{r}_2,\omega)\right\|_2^2 - 1\right] \tag{2.58}$$

where n is the number of independent scalar waves involved in $\mathbf{M}_L(\mathbf{r}_1,\mathbf{r}_2,\omega)$. The value of n to be used is $n=3$ for one-point measures ($\mathbf{r}_2=\mathbf{r}_1$) and $n=6$ in general.

In analogy to the 2D formulation, let us now expand $\| \boldsymbol{M}_L(\boldsymbol{r}_1,\boldsymbol{r}_2,\omega)\|_2^2$ as follows:

$$\begin{aligned}\left\|\mathbf{M}_L(\mathbf{r}_1,\mathbf{r}_2,\omega)\right\|_2^2 &\equiv \operatorname{tr}\left[\mathbf{M}_L^2(\mathbf{r}_1,\mathbf{r}_2,\omega)\right] \\ &= \operatorname{tr}\left[\mathbf{W}^2(\mathbf{r}_1,\mathbf{r}_1,\omega)\right]+\operatorname{tr}\left[\mathbf{W}^2(\mathbf{r}_2,\mathbf{r}_2,\omega)\right]+2\operatorname{tr}\left[\mathbf{W}^2(\mathbf{r}_1,\mathbf{r}_2,\omega)\right] \\ &= \frac{1}{3}I_1^2\left[1+2P_{3D}^2(\mathbf{r}_1,\omega)\right]+\frac{1}{3}I_2^2\left[1+2P_{3D}^2(\mathbf{r}_2,\omega)\right]+2I(\mathbf{r}_1,\omega)I(\mathbf{r}_2,\omega)\,\gamma^2(\mathbf{r}_1,\mathbf{r}_2,\omega)\end{aligned} \tag{2.59}$$

where $P_{3D}(\mathbf{r}_i,\omega)$ is the *spectral degree of polarimetric purity* (see Eq. (2.24)) which will be studied in Section 2.10.2; $I_i \equiv I(\mathbf{r}_i,\omega)$ is the spectral density of the respective one-point spectral polarization matrix $\mathbf{W}(\mathbf{r}_i,\mathbf{r}_i,\omega)$, with $i=1,2$, and $\gamma(\mathbf{r}_1,\mathbf{r}_2,\omega)$ is the 3D electromagnetic degree of coherence.

Since $D(\mathbf{r}_1,\mathbf{r}_2,\omega)$ is expressed in terms of the quantities $P_{3D}(\mathbf{r}_i,\omega)$, $I(\mathbf{r}_i,\omega)$ and $\gamma(\mathbf{r}_1,\mathbf{r}_2,\omega)$, which are all invariant under unitary transformations, $D(\mathbf{r}_1,\mathbf{r}_2,\omega)$ is also invariant under such transformations (which include the particular case of rotation transformations of the local laboratory reference frame).

2.8 Intrinsic Polarization Matrix

Prior to introducing the concept of *intrinsic polarization matrix* it is worth stressing again the difference between the roles played by unitary transformations in coherence theory and in polarization theory. While unitary transformations preserve the correlations between (1) the components of a given wavefield (i.e., correlations measured in terms of the degree of polarization) and (2) the field variables at points $\mathbf{r}_1$ and $\mathbf{r}_2$, and time instants t and $t+\tau$ (i.e., correlations measured in terms of the electromagnetic degree of coherence), nonorthogonal unitary transformations have necessarily to be associated to changes of the polarization state. Nevertheless, orthogonal transformations (i.e., unitary transformations whose matrix is real valued) admit their interpretation as changes of the reference frame (rotation in the 3D space), which maintain the polarization states. Thus, we will distinguish between *complex unitary transformations* (whose associated unitary matrix $\mathbf{U}$ satisfies $\mathrm{Im}\,\mathbf{U} \neq \mathbf{0}$) and *rotation transformations* (whose associated unitary matrix $\mathbf{U}$ is an orthogonal matrix, viz., $\mathrm{Im}\,\mathbf{U} = \mathbf{0}$). Hereafter, orthogonal matrices will be denoted with symbol $\mathbf{Q}$, instead of the symbol $\mathbf{U}$ used for general unitary matrices.

The physically realizable rotations of the laboratory reference frame XYZ are represented by 3×3 orthogonal matrices $\mathbf{Q}$ (with $\det\mathbf{Q}=1$), so that the transformed polarization matrix $\mathbf{R}'$ representing the same state as $\mathbf{R}$ but referenced with respect to the new coordinate system, is given by $\mathbf{R}' = \mathbf{Q}\mathbf{R}\mathbf{Q}^T$. Moreover, it is very useful to consider the decomposition of $\mathbf{R}$ into its real and imaginary parts $\mathbf{R} = \mathrm{Re}\,\mathbf{R} + i\,\mathrm{Im}\,\mathbf{R}$, where $\mathrm{Re}\,\mathbf{R}$ is symmetric and positive-semidefinite, whereas $\mathrm{Im}\,\mathbf{R}$ is skew-symmetric $[\mathrm{Im}\,\mathbf{R} = -(\mathrm{Im}\,\mathbf{R})^T]$. $\mathrm{Re}\,\mathbf{R}$ can always be diagonalized through a particular rotation $\mathbf{Q}_O$ of the reference frame (Dennis 2004)

$$\mathbf{Q}_O\,\mathbf{R}_R\,\mathbf{Q}_O^T = \mathrm{diag}(a_1, a_2, a_3) \qquad (0 \le a_3 \le a_2 \le a_1) \tag{2.60}$$

Note that, since $\mathrm{Re}\,\mathbf{R}$ is a symmetric matrix, it defines an ellipsoid whose semiaxes (a_1, a_2, a_3) are aligned along the respective transformed axes $X_O Y_O Z_O$ (Figure 2.6a). It is important to point out that $\mathrm{diag}(a_1, a_2, a_3)$ can formally be interpreted as the polarization matrix of a state composed of the incoherent superposition of three linearly polarized pure states (Gil 2014a),

$$\begin{gathered}\mathrm{diag}(a_1, a_2, a_3) = \mathbf{R}_{p1} + \mathbf{R}_{p2} + \mathbf{R}_{p3} \\ \mathbf{R}_{p1} \equiv a_1\,\mathrm{diag}(1,0,0) \qquad \mathbf{R}_{p2} \equiv a_2\,\mathrm{diag}(0,1,0) \qquad \mathbf{R}_{p3} \equiv a_3\,\mathrm{diag}(0,0,1)\end{gathered} \tag{2.61}$$

with respective intensities a_1, a_2 and a_3.

The decomposition (2.61) is compatible with a variety of directions of propagation for each component; the only condition is that the direction of propagation for each component is orthogonal to the respective polarization ellipse, degenerating into a segment along the respective axis X_O, Y_O, Z_O (Figure 2.6a). Since, in particular, Z_O is orthogonal to the polarization axes of the linearly polarized states represented by $\mathbf{R}_{p1}$ and $\mathbf{R}_{p2}$, it is always possible, without loss of generality, to choose the axis Z_O (i.e., the axis corresponding to the smaller semiaxis a_3) as the common direction of propagation $\mathbf{k}$ for the pure states $\mathbf{R}_{p1}$ and $\mathbf{R}_{p2}$. Moreover, any axis orthogonal to $\mathbf{k}$ can be considered as the direction of propagation of the third pure component $\mathbf{R}_{p3}$ (Figure 2.6a). Despite the fact that this interpretation is not unique, we stress that it is physically consistent and that, without any additional information, it is polarimetrically indistinguishable from any other possible choice.

By applying the rotation $\mathbf{Q}_O$ to the whole matrix $\mathbf{R}$, we get the transformed polarization matrix (Dennis 2004)

$$\mathbf{R}_O \equiv \mathbf{Q}_O \mathbf{R} \mathbf{Q}_O^T = \mathbf{Q}_O (\mathrm{Re}\,\mathbf{R}) \mathbf{Q}_O^T + i \mathbf{Q}_O (\mathrm{Im}\,\mathbf{R}) \mathbf{Q}_O^T = \mathrm{Re}\,\mathbf{R}_O + i \,\mathrm{Im}\,\mathbf{R}_O$$
$$\mathbf{R}_O \equiv \begin{pmatrix} a_1 & -i n_{O3}/2 & i n_{O2}/2 \\ i n_{O3}/2 & a_2 & -i n_{O1}/2 \\ -i n_{O2}/2 & i n_{O1}/2 & a_3 \end{pmatrix} \tag{2.62}$$
$$\mathrm{Re}\,\mathbf{R}_O \equiv \begin{pmatrix} a_1 & 0 & 0 \\ 0 & a_2 & 0 \\ 0 & 0 & a_3 \end{pmatrix} \quad \mathrm{Im}\,\mathbf{R}_O \equiv \begin{pmatrix} 0 & -i n_{O3}/2 & i n_{O2}/2 \\ i n_{O3}/2 & 0 & -i n_{O1}/2 \\ -i n_{O2}/2 & i n_{O1}/2 & 0 \end{pmatrix}$$

where the real and imaginary parts transform separately and, as occurs with $\mathrm{Im}\,\mathbf{R}$, $\mathrm{Im}\,\mathbf{R}_O$ is skew-symmetric. In accordance with the formulation of the rotation transforming the representation of the 3D Jones vector with respect to its intrinsic reference frame $X_O Y_O Z_O$ to another arbitrary one XYZ described in Section 2.2, the rotation matrix from XYZ to $X_O Y_O Z_O$ considered here can be expressed as follows:

$$\mathbf{Q}_O = \mathbf{Q}_\phi \mathbf{Q}_\theta \mathbf{Q}_\varphi$$
$$= \begin{pmatrix} c_\phi & s_\phi & 0 \\ -s_\phi & c_\phi & 0 \\ 0 & 0 & 1 \end{pmatrix} \begin{pmatrix} c_\theta & 0 & s_\theta \\ 0 & 1 & 0 \\ -s_\theta & 0 & c_\theta \end{pmatrix} \begin{pmatrix} c_\varphi & s_\varphi & 0 \\ -s_\varphi & c_\varphi & 0 \\ 0 & 0 & 1 \end{pmatrix} = \begin{pmatrix} c_\theta c_\phi c_\varphi - s_\phi s_\varphi & c_\theta c_\phi s_\varphi + s_\phi c_\varphi & c_\phi s_\theta \\ -c_\theta s_\phi c_\varphi - c_\phi s_\varphi & -c_\theta s_\phi s_\varphi + c_\phi c_\varphi & -s_\phi s_\theta \\ -s_\theta c_\varphi & -s_\theta s_\varphi & c_\theta \end{pmatrix} \tag{2.63}$$

Obviously, other alternative sequences of three rotations are also physically admissible and realizable.

Thus, since $\mathbf{R} = \mathbf{Q}_O^T \mathbf{R}_O \mathbf{Q}_O$, $\mathbf{R}$ can be parameterized through the following nine independent parameters: the three orientation angles (ϕ, θ, φ); the semiaxes of the *intensity ellipsoid* (denoted by E_I wherever appropriate), called *inertia ellipsoid* by Dennis (2004), given by the *principal intensities* (a_1, a_2, a_3), and the three components (n_{O1}, n_{O2}, n_{O3}) of the *spin vector* $\mathbf{n}_O$ of the state along the respective axes $X_O Y_O Z_O$ (see Section 2.11). It is remarkable that, in general, the orientation of $\mathbf{n}_O$ differs from $\mathbf{k}$. The intensity I of the state $\mathbf{R}_O$ can be calculated through the following alternative ways:

$$I = \mathrm{tr}\,\mathbf{R} = \mathrm{tr}\,\mathbf{R}_O = \lambda_1 + \lambda_2 + \lambda_3 = a_1 + a_2 + a_3 = \mathrm{tr}\,\mathbf{R}_{p1} + \mathrm{tr}\,\mathbf{R}_{p2} + \mathrm{tr}\,\mathbf{R}_{p3} \tag{2.64}$$

The *intrinsic polarization matrix* $\mathbf{R}_O$ (Gil 2014a) is nothing but the polarization matrix of the same state as that represented by $\mathbf{R}$ but referenced with respect to the *intrinsic reference frame* $X_O Y_O Z_O$. Thus, $\mathbf{R}_O$ is positive semidefinite, so that the quantities $(a_1, a_2, a_3;\, n_{O1}, n_{O2}, n_{O3})$ must satisfy the following set of constraining inequalities derived from the non-negativity of the leading principal minors of $\mathbf{R}_O$ (Dennis 2004):

$$a_1 \ge a_2 \ge a_3 \ge 0 \quad 4a_1 a_2 \ge n_{O3}^2 \quad 4a_1 a_3 \ge n_{O2}^2 \quad 4a_2 a_3 \ge n_{O1}^2 \quad 4a_1 a_2 a_3 \ge a_1 n_{O1}^2 + a_2 n_{O2}^2 + a_3 n_{O3}^2 \tag{2.65}$$

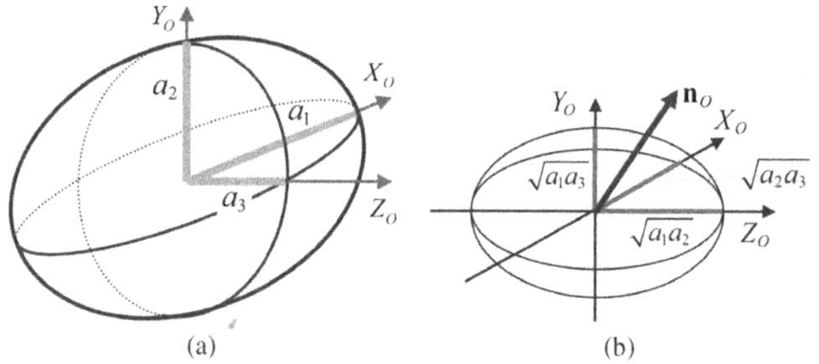

FIGURE 2.6
(a) Without loss of generality, the *intensity ellipsoid* of a state **R**, with semiaxes $a_1 \geq a_2 \geq a_3$, is equivalent to that of a mixed state constituted by the incoherent superposition of three pure linearly polarized states $\mathbf{R}_{p1}$, $\mathbf{R}_{p2}$ and $\mathbf{R}_{p3}$. (b) The *spin ellipsoid* determines the limit for the feasible region for the spin vector; the smaller a_3, the closer to the intrinsic axis Z_O the direction of **n** is (Gil 2021).

Therefore, provided $a_3 > 0$, the feasible values for the components of the spin vector **n** are those that restrict **n** to lie within the (solid) *spin ellipsoid* (or *dual ellipsoid*) defined as (Dennis 2004) (Figure 2.6b)

$$1 \geq n_{O1}^2/(4a_2a_3) + n_{O2}^2/(4a_1a_3) + n_{O3}^2/(4a_1a_2) \tag{2.66}$$

The smaller third principal intensity a_3, the smaller solid angle about the axis Z_O that limits the range of compatible orientations of $\mathbf{n}_O$. When $a_3 = 0, \mathbf{n}_O$ is forced to lie along the axis Z_O, which in this case is precisely the well-defined direction of propagation of the 2D state $\mathbf{R}_O$.

Within the limits set by the restrictive inequalities (2.65), the quantities $(a_1, a_2, a_3; n_{O1}, n_{O2}, n_{O3})$ are independent and are intrinsic to a given polarization matrix **R**. It should be stressed again the fact that the only unitary transformations of the reference frame that are physically realizable as rotations (in the real domain) of the Cartesian coordinate system are those that are orthogonal, and consequently, $\mathbf{R}_O$ (and hence, **R**) possesses six physical invariant parameters instead of the only three physical invariants derivable from the eigenvalues of **R**. Thus, a proper interpretation of the invariant quantities involved in **R** should be performed through rotation transformation.

Let us note finally that transformations of the form $\mathbf{R}' = \mathbf{V}^{+}\mathbf{R}\mathbf{V}$, **V** being a nonorthogonal unitary matrix, can be interpreted as the transformation of **R** due to the effect on **R** of the linear interaction with certain kinds of birefringent devices whose associated 3D Jones matrix has the form of **V**.

2.9 Intrinsic Stokes Parameters

The intrinsic representation of the 3D Stokes parameters provides a simplified and meaningful physical interpretation of the peculiar features of 3D polarization states and leads to the definition of a set of objective descriptors that are described in the present section

2.9.1 Intrinsic Stokes Parameters of a Polarization State

The intrinsic reference frame provides a simplified view of the physically invariant quantities involved in **R**. In particular, the *intrinsic Stokes parameters* can be arranged as the elements of the *intrinsic Stokes parameters matrix* (Gil 2015)

$$\mathbf{S}_O = \begin{pmatrix} a_1+a_2+a_3 & 0 & 0 \\ n_{O3} & a_1-a_2 & 0 \\ -n_{O2} & n_{O1} & (a_1+a_2-2a_3)/\sqrt{3} \end{pmatrix} \equiv I \begin{pmatrix} 1 & 0 & 0 \\ \hat{n}_{O3} & P_l & 0 \\ -\hat{n}_{O2} & \hat{n}_{O1} & P_d/\sqrt{3} \end{pmatrix} \tag{2.67}$$

where, as in other expressions, the subscript "O" indicates that the Stokes parameters are referenced with respect to the intrinsic reference frame. The physical significance of the intrinsic Stokes parameters becomes readily apparent in the above expression in terms of meaningful measurable quantities that are linked directly to the intrinsic phenomenological properties determining a 3D polarization state: the *intensity* I, the *degree of linear polarization* P_l, the *degree of directionality* P_d, and the three intrinsic components $\hat{n}_{Oi}$ of the *spin density vector* $\hat{\mathbf{n}}_O \equiv \mathbf{n}_O/I$ (Gil 2014a; Gil 2015).

The corresponding expression of $\mathbf{R}_O$ in terms of the elements of $\mathbf{S}_O$ is

$$\mathbf{R}_O = \frac{1}{2} I \begin{pmatrix} 2/3+P_l+P_d/3 & -i\hat{n}_{O3} & i\hat{n}_{O2} \\ i\hat{n}_{O3} & 2/3-P_l+P_d/3 & -i\hat{n}_{O1} \\ -i\hat{n}_{O2} & i\hat{n}_{O1} & 2/3(1-P_d) \end{pmatrix} \tag{2.68}$$

In light of the above expressions, we can establish an interpretation of the intrinsic Stokes parameters by means of the following considerations (Gil 2015):

Intensity. As expected, the intensity I is one of the intrinsic Stokes parameters. Obviously, the only restriction on the value of I is that it is non-negative.

Degree of linear polarization P_l. $IP_l = a_1 - a_2$ is the difference between the major and the intermediate semiaxes of the intensity ellipsoid. The intensity-normalized intrinsic Stokes parameter P_l gives a measure of the relative portion of the power of the electromagnetic wave that can be allocated as linearly polarized, and can properly be called the *degree of linear polarization*. The maximal value $P_l = 1$ ($a_2 = a_3 = 0$) is achieved for linearly polarized pure states. The minimal $P_l = 0$ ($a_2 = a_1$) corresponds either to states with fully random polarization ellipse ($a_3 = a_2 = a_1$), or to circularly polarized pure states ($a_2 = a_1$, $a_3 = 0$), or to any combination of them ($0 \le a_3 \le a_2 = a_1$). In the case of 2D states ($P_d = 1$), then $P_l = \sqrt{s_1^2 + s_2^2}/s_0$, thus corresponding to the natural and commonly used definition of the degree of linear polarization.

Degree of directionality, P_d. $P_d = (a_1 + a_2 - 2a_3)/\sqrt{3}I$ is $1/\sqrt{3}$ times the difference between the sum of the normalized intensities of the linear components lying on the plane $X_O Y_O$ and twice the normalized intensity of the component lying in the axis Z_O. The quantity P_d is particularly important for the physical interpretation of the polarization matrix **R** because it provides an appropriate measure of the stability of the polarization plane of the state represented by **R**, which will be studied in detail in later paragraphs. The maximal value $P_d = 1$ ($a_3 = 0$) corresponds uniquely to 2D polarization states (defined as those whose electric field fluctuates in a fixed plane). The maximal value $P_d = 0$ ($a_3 = a_2 = a_1$) corresponds uniquely to 3D unpolarized states.

Spin vector. The three intrinsic Stokes parameters n_{Oi} are the components of the spin vector $\mathbf{n}_O$ of the state along the respective intrinsic reference axes $X_O Y_O Z_O$. The absolute value $|\mathbf{n}_O|$ of $\mathbf{n}_O$ is the magnitude of the spin of the polarization state, which will be referred to as the *spin*. Although the concept of spin vector of a polarization state will be studied in detail in Section 2.11, it is important to emphasize here the fact that, as with the polarization matrix (tensor), the spin vector of a polarization state has a tensor character in the sense that it is invariant with respect to arbitrary rotations of the Cartesian coordinate system, in such a manner that its components vary correspondingly. Thus, the subscript O indicates that the spin vector is referenced with respect to $X_O Y_O Z_O$, while the same spin vector exhibits components n_i, different from n_{Oi}, depending on the reference system considered.

Degree of circular polarization. Obviously, the magnitude $|\mathbf{n}| = |\mathbf{n}_O|$ of the spin vector is independent of the reference frame considered, so that the use of the subscript O in $|\mathbf{n}_O|$ is not necessary and hereafter we will denote it simply as $|\mathbf{n}|$. Observe also that there always exists a pair of coordinate systems $X_{n\pm} Y_{n\pm} Z_{n\pm}$ for which the two first components of $\mathbf{n}$ vanish, so that $\mathbf{n}_n = (0, 0, \pm n)^T$. Except for certain limiting cases, the direction of $\mathbf{n}$ does not coincide with that any of the axes of the intensity ellipsoid. Recall that we are considering the polarization properties at a fixed point $\mathbf{r}$ in the space and not an extensive portion of the wavefront, where the concept of orbital angular momentum may be considered in addition to the spin.

The intensity-normalized versions $\hat{\mathbf{n}} \equiv \mathbf{n}/I$ and $|\hat{\mathbf{n}}| \equiv |\mathbf{n}|/I$ of the spin vector and the spin are termed *spin density vector* and *spin density* respectively (Gil et al. 2021). As it will be shown in Section 2.11, and in accordance with the basic notion of spin of the electric field vector of an electromagnetic plane wave (Poynting 1909; Beth 1936; Berestetskii et al. 1982), $|\hat{\mathbf{n}}|$ is a proper measure of the *degree of circular polarization* $P_c = |\hat{\mathbf{n}}|$ of the polarization state.

The maximal value $P_c = 1$ is achieved for pure circularly polarized states, while the minimal $P_c = 0$ corresponds to mixed states given by an incoherent mixture of three components linearly polarized along the respective axes X_O, Y_O, Z_O, with respective intensities a_1, a_2, a_3.

Intrinsic polarization vector. When referenced with respect to the intrinsic reference frame, the 3D polarization vector $\mathbf{p}$ is expressed as follows in terms of the intensity-normalized intrinsic Stokes parameters:

$$\mathbf{p} = \frac{\sqrt{3}}{2}\left(0, \hat{n}_{03}, P_l, 0, -\hat{n}_{02}, 0, \hat{n}_{01}, P_d/\sqrt{3}\right) \tag{2.69}$$

and thus, the degree of polarimetric purity P_{3D} is given by

$$P_{3D} \equiv |\mathbf{p}| = \sqrt{\frac{3\left(P_l^2 + P_c^2\right) + P_d^2}{4}} \qquad \left[P_c = \sqrt{\hat{n}_1^2 + \hat{n}_2^2 + \hat{n}_3^2}\right] \tag{2.70}$$

Characteristic polarization ellipse. Regardless of the fact that only certain particular polarization states admit a well-defined average polarization ellipse, for many purposes it is interesting to look at the polarization ellipse of the pure component of the characteristic decomposition of **R** (see Eq. (2.29)). The semiaxes a, b of such a *characteristic polarization ellipse* are determined by

$$a^2 = IP_1(\mathbf{R})\left[1+P_l\left(\mathbf{R}_p\right)\right]/2 \qquad b^2 = IP_1(\mathbf{R})\left[1-P_l\left(\mathbf{R}_p\right)\right]/2 \qquad \left[P_1 \equiv (\lambda_1-\lambda_2)/I\right] \quad (2.71)$$

where P_1 is the coefficient of the characteristic pure component in the characteristic decomposition (as will be shown in Section 2.10.4, P_1 is precisely the so-called *first index of polarimetric purity* of **R**). Since $1 = P_l^2(\mathbf{R}_p)+P_c^2(\mathbf{R}_p)$, the *normalized area* $\hat{A} \equiv A/I$ of the characteristic polarization ellipse is $\hat{A} = (\pi/2)P_1(\mathbf{R})P_c(\mathbf{R}_p)$.

Polarization object. A simple geometric view of a generic 3D polarization state is provided by the *polarization object* (Figure 2.7; Gil 2021) composed of the intensity ellipsoid E_I (three parameters: a_1, a_2, a_3) together with the intrinsic representation of the spin vector, which is given by the three components n_{O1}, n_{O2}, n_{O3} or alternatively by its magnitude $|\mathbf{n}|$ and its orientation angles (ϕ_n, θ_n) with respect to the symmetry axes $X_O Y_O Z_O$ of the intensity ellipsoid (Figure 2.7), so that $n_{O1} = |\mathbf{n}|\sin\theta_n \cos\phi_n$, $n_{O2} = |\mathbf{n}|\sin\theta_n \sin\phi_n$ and $n_{O3} = |\mathbf{n}|\cos\theta_n$. Obviously, the orientation of $\mathbf{n}_O$ with respect to E_I is fixed (independent of the reference frame considered), so that the (rigid) combination of E_I and $\mathbf{n}_O$, which define the polarization object, describes the intrinsic polarization properties of the state represented by **R**. When referenced with respect to an arbitrary coordinate system, the three orientation angles (ϕ, θ, φ) should be considered in addition.

Since the intensity, I, plays the geometric role of a scale factor determining the size of the polarization object, the polarization object associated with the polarization density matrix $\hat{\mathbf{R}} \equiv \mathbf{R}/I$ is called the *polarization density object* of the state, composed of the spin density vector $\hat{\mathbf{n}}_O$ and the *polarization density ellipsoid* E, whose semiaxes $(\hat{a}_1, \hat{a}_2, \hat{a}_3)$, given by the *principal variances* $\hat{a}_i \equiv a_i/I$ $(i = 1,2,3)$, are the variances of the Cartesian components $\hat{\varepsilon}_{Oi} \equiv \varepsilon_{Oi}/\sqrt{I}$ of the intrinsic representation of the normalized analytic signal vector $\hat{\boldsymbol{\varepsilon}}_O \equiv \boldsymbol{\varepsilon}_O/\sqrt{I}$.

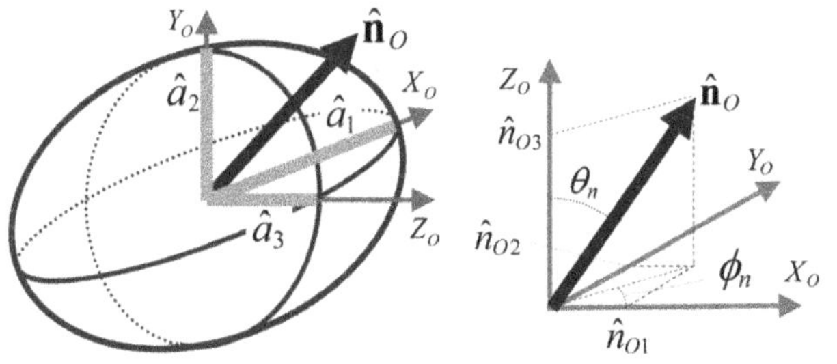

FIGURE 2.7
Any polarization state **R** is fully characterized by its intensity, $I = \mathrm{tr}\,\mathbf{R}$ and its *polarization density object* (a), composed of the polarization density ellipsoid (with semiaxes $\hat{a}_1 \geq \hat{a}_2 \geq \hat{a}_3$), together with its spin density vector $\hat{\mathbf{n}}_O$ (b) (determined by its magnitude $|\hat{\mathbf{n}}_O|$ and its orientation angles ϕ_n, θ_n with respect to the symmetry axes $X_O Y_O Z_O$ of the polarization density ellipsoid). The polarization density object has tensor character in the sense that it is invariant with respect to changes of the reference frame, while their coordinates vary correspondingly. When referenced with respect to an arbitrary coordinate system, the orientation angles (ϕ, θ, φ) of the intensity ellipsoid should be considered in addition (Gil 2021).

The intimate relation between the polarization object and the six intrinsic Stokes parameters $(I, P_l, P_d/\sqrt{3}, \hat{n}_{O1}, \hat{n}_{O2}, \hat{n}_{O3})$ is readily found through the expressions of the principal intensities a_i in terms of the degree of linear polarization, $P_l = \hat{a}_2 - \hat{a}_1$, and the degree of directionality, $P_d = 1 - 3\hat{a}_3$,

$$a_1 = I\hat{a}_1 = I\frac{2+P_d+3P_l}{6} \qquad a_1 = I\hat{a}_2 = I\frac{2+P_d-3P_l}{6} \qquad a_3 = I\hat{a}_3 = I\frac{2(1-P_d)}{6} \tag{2.72}$$

2.9.2 Intrinsic Stokes Parameters for 2D States Embedded into the 3D Representation

In the case of 2D states, characterized by the condition $P_d = 1$, which means that its electric field remains in a fixed plane, this *polarization plane* is precisely the one containing the axes $X_O Y_O$ of the intrinsic reference frame and, therefore, the elements of the third row and column of the intrinsic polarization matrix $\mathbf{R}_O$ become zero ($n_{O1} = n_{O2} = 0$, $P_d = 1$). The 2D intrinsic representation is retrieved through the 2D polarization matrix $\mathbf{\Phi}_O$ embedded into the upper-left 2×2 submatrix of $\mathbf{R}_O$

$$\mathbf{R}_O = \frac{1}{2}I\begin{pmatrix} 1+P_l & -i\hat{n}_{O3} & 0 \\ i\hat{n}_{O3} & 1-P_l & 0 \\ 0 & 0 & 0 \end{pmatrix} \quad \begin{bmatrix} \hat{n}_{O3} = \sigma P_c \\ |\hat{n}_{O3}| = P_c \end{bmatrix} \tag{2.73}$$

where $\hat{n}_{O3} = \hat{s}_3$ is the helicity (or spin) and σ is the handedness ($\sigma = +1$ for right-handed states and $\sigma = -1$ for left-handed states). The degree of polarization for 2D states is retrieved as $\mathcal{P} = \sqrt{P_l^2 + P_c^2}$. It is remarkable that, leaving aside the intensity, the intrinsic Stokes parameters of a totally polarized state are fully determined by the spin: $P_d = 1$, $P_l = \sqrt{1 - P_c^2}$.

The 2D Stokes parameters, referenced with respect to an arbitrary reference frame, XYZ_O are obtained through an arbitrary rotation about the axis Z_O (orthogonal to the polarization plane). The general 3D representation of a 2D state is obtained by applying an arbitrary rotation transforming the intrinsic axes $X_O Y_O Z_O$ to another set of coordinate axes, XYZ, $\mathbf{R} = \mathbf{Q}^T \mathbf{R}_O \mathbf{Q}$ where $\mathbf{Q}$ is the orthogonal matrix expressed in Eq. (2.63) in terms of the set of arbitrary orientation angles (ϕ, θ, φ). Thus, in a generic coordinate system XYZ the spin vector of a 2D state has, in general, three nonzero components $\mathbf{n} \equiv [-\mathrm{Im}(r_{23}), \mathrm{Im}(r_{13}), -\mathrm{Im}(r_{12})]^T$. Obviously, such rotation transformations of a 2D state preserve the orthogonality of $\mathbf{n}$ with respect to the polarization plane $\Pi\,(X_O Y_O)$ (Figure 2.8).

The equation of the plane Π is $s_{21}\,x - s_{20}\,y + s_{10}\,z = 0$, while the direction of $\hat{\mathbf{n}}$, orthogonal to Π, is given by $x/s_{21} = -y/s_{20} = z/s_{10}$ (Sheppard 2014).

The 2D intrinsic Stokes parameters, as well as the components of purity of 2D states have been analyzed in detail in the respective sections of Chapter 1.

There are certain physical situations where the spin of the 2D state is zero (i.e., the polarization matrix is real) and the above procedure for determining the plane Π cannot be applied. These cases correspond to states whose characteristic pure component is linearly polarized and whose discriminating component is a 2D unpolarized state. The 3D intrinsic polarization matrix of such kinds of zero-spin 2D states has the form $\mathbf{R}_O = \mathrm{diag}(a_1, a_2, 0)$ so that, given the polarization matrix $\mathbf{R}$ representing the same state as $\mathbf{R}_O$ but referenced with respect to an arbitrary coordinate frame XYZ, the angles ϕ, θ

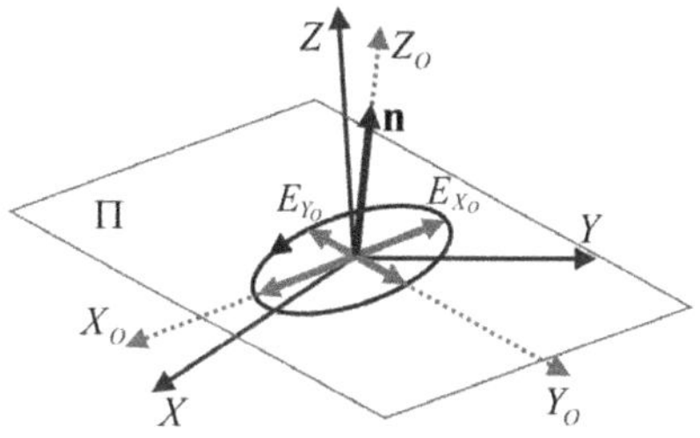

FIGURE 2.8
For 2D polarization states, the spin vector **n** determines the direction along which the electric field has zero projection, i.e., the direction orthogonal to the polarization plane Π containing the fluctuating electric field and the characteristic polarization ellipse.

determining the direction normal to the plane Π, can be obtained from the orthogonal matrix that diagonalizes **R** (Gil et al. 2019b). In the particular case of a pure linearly polarized state, the intrinsic polarization matrix is of the form $\mathbf{R}_O = \mathrm{diag}(a_1, 0, 0)$, and the plane Π is defined up to a rotation about the axis X_O along which the electric field vibrates. The direction cosines (c_1, c_2, c_3) of the polarization axis X_O with respect to the arbitrary reference frame XYZ are expressed as follows in terms of the corresponding 3D Stokes parameters (Sheppard 2014):

$$c_1^2 = \frac{s_{01}\, s_{20}}{2\, s_{00}\, s_{12}}, \quad c_2^2 = \frac{s_{01}\, s_{12}}{2\, s_{00}\, s_{20}}, \quad c_3^2 = \frac{s_{20}\, s_{12}}{2\, s_{00}\, s_{01}} \tag{2.74}$$

Note that the relation $c_1^2 + c_2^2 + c_3^2 = 1$ holds since the 3D Stokes parameters of a pure state are not independent, but are related through some restrictive constraints derived from the condition $\mathrm{rank}\,\mathbf{R}_O = 1$.

2.10 Polarimetric Purity

2.10.1 Norms of 3D Polarization Matrices and Stokes Parameters Matrices

To express certain relations in terms of invariant metric parameters, let us now consider the Frobenius norms of **R** and **S**

$$\|\mathbf{R}\|_2 \equiv \sqrt{\sum_{k,l=1}^{3} |r_{kl}|^2} = \sqrt{\mathrm{tr}\left(\mathbf{R}^{+}\mathbf{R}\right)} = \sqrt{\mathrm{tr}\,\mathbf{R}^2} \qquad \|\mathbf{S}\|_2 \equiv \sqrt{\sum_{i,j=0}^{2} s_{ij}^2} = \sqrt{\mathrm{tr}\left(\mathbf{S}^T\mathbf{S}\right)} \tag{2.75}$$

and define the *trace norm* for **R** and the *max-norm* of **S** as (Gil 2007)

$$\|\mathbf{R}\|_{tr} \equiv \mathrm{tr}\,\mathbf{R} = \left\|\sqrt{\mathbf{R}}\right\|_2^2 = I \qquad \|\mathbf{S}\|_0 \equiv I = \|\mathbf{R}\|_{tr} \tag{2.76}$$

Since $\mathbf{R}$ is a positive semidefinite Hermitian matrix, it is easy to show that $\|\mathbf{R}\|_{tr}$ satisfies all conditions to be a norm. Furthermore, these norms satisfy the following relations:

$$\|\mathbf{R}\|_{tr} = \|\mathbf{s}\|_0 = I \qquad \frac{1}{3}\|\mathbf{R}\|_{tr}^2 \le \|\mathbf{R}\|_2^2 \le \|\mathbf{R}\|_{tr}^2 \qquad \|\mathbf{S}\|_0^2 \le \|\mathbf{S}\|_2^2 \le \frac{7}{3}\|\mathbf{S}\|_0^2 \tag{2.77}$$

A necessary and sufficient condition for a state represented by $\mathbf{R}$ (or $\mathbf{S}$) to be polarimetrically pure is given by the equivalent equalities $\|\mathbf{R}\|_2^2 = \|\mathbf{R}\|_{tr}^2$ and $\|\mathbf{S}\|_2 = \sqrt{7/3}\,\|\mathbf{S}\|_0$, which constitute an objective purity criterion formulated in a symmetric manner with respect to that holding in the 2D case. The lower limit $\|\mathbf{R}\|_2 = \|\mathbf{R}\|_{tr}/\sqrt{3}$, $\|\mathbf{S}\|_2 = \|\mathbf{S}\|_0$, is reached exclusively for unpolarized states.

2.10.2 Degree of Polarimetric Purity

A first definition of the 3D degree of polarization was advanced by Samson (1973) within the scope of geophysical studies of ultra-low frequency magnetic fields. This result was also obtained by Barakat (1983) by formulating the degree of polarization in terms of scalar invariants of the polarization matrix and was studied and interpreted by several authors as the interest in nonparaxial approaches increased (Setälä et al. 2002a; Gil et al. 2004; Luis 2005a).

An overall measure of the polarimetric purity of a 3D state is given by the 3D *degree of polarimetric purity*, or *3D degree of polarization*, defined as (Samson 1973; Setälä et al. 2002a; Gil et al. 2004; 2018a)

$$P_{3D} = \sqrt{\frac{1}{2}\left(3\,\mathrm{tr}\,\hat{\mathbf{R}}^2 - 1\right)} = \sqrt{\frac{1}{2}\left(3\sum_{i=1}^{3}\hat{\lambda}_i^2 - 1\right)} \tag{2.78}$$

where $\hat{\lambda}_i$ are the eigenvalues of the polarization density matrix $\hat{\mathbf{R}} \equiv \mathbf{R}/I$.

This invariant and dimensionless parameter is limited to the interval $0 \le P_{3D} \le 1$, in such a manner that $P_{3D} = 1$ corresponds to the case where $\mathbf{R}$ has a single nonzero eigenvalue (pure state), whereas $P_{3D} = 0$ is reached when the three eigenvalues of $\mathbf{R}$ are equal (fully unpolarized 3D state).

The expressions of P_{3D} in terms of the norms of $\mathbf{R}$ and $\mathbf{S}$ are the following:

$$P_{3D} = \sqrt{\frac{1}{2}\left(\frac{3\|\mathbf{R}\|_2^2}{\|\mathbf{R}\|_{tr}^2} - 1\right)} = \sqrt{\frac{1}{2}\left(3\|\hat{\mathbf{R}}\|_2^2 - 1\right)} \qquad P_{3D} = \frac{\sqrt{3\left(\|\mathbf{S}\|_2^2 - \|\mathbf{S}\|_0^2\right)}}{2\|\mathbf{S}\|_0} = \sqrt{\frac{3}{4}\left(\|\hat{\mathbf{S}}\|_2^2 - 1\right)} \tag{2.79}$$

The above expression of P_{3D} in terms of the polarization density matrix $\hat{\mathbf{R}}$ can be transformed into (Luis 2005a)

$$P_{3D}^2 = |\mathbf{p}(\mathbf{R})|^2 = \frac{3}{2}\mathrm{tr}\left[\left(\frac{1}{3}\mathbf{I}_3 - \hat{\mathbf{R}}\right)^2\right] = \frac{3}{2}\mathrm{tr}\left[\left(\hat{\mathbf{R}}_{u\text{-}3D} - \hat{\mathbf{R}}\right)^2\right] \tag{2.80}$$

where $\mathbf{I}_3$ is the 3×3 identity matrix, which is precisely three times the polarization density matrix $\hat{\mathbf{R}}_{u-3D}$ of a 3D unpolarized state, $\mathbf{I}_3 = 3\hat{\mathbf{R}}_{u-3D}$. Thus, expression (2.80) shows that the degree of polarimetric purity P_{3D} of a state $\mathbf{R}$ can be considered a measure of the distance between the polarization density matrix of the $\hat{\mathbf{R}}$ state and that of a 3D unpolarized state (Luis 2005a).

Eq. (2.80) is a particular case of the polarimetric distance $P_{3D;a,b}$ between two states $\hat{\mathbf{R}}_a$ and $\hat{\mathbf{R}}_b$, defined as (Luis 2005a)

$$P^2_{3D;a,b} = \frac{3}{2}\operatorname{tr}\left[\left(\hat{\mathbf{R}}_a - \hat{\mathbf{R}}_b\right)^2\right] = \left(\mathbf{p}_a - \mathbf{p}_b\right)^2 \tag{2.81}$$

where $\mathbf{p}_a$ and $\mathbf{p}_b$ are the polarization vectors associated with $\hat{\mathbf{R}}_a$ and $\hat{\mathbf{R}}_b$ respectively (Luis 2005a).

The maximum distance $P_{3D;a,b} = 2$ corresponds to pure states satisfying $\mathbf{p}_a + \mathbf{p}_b = \mathbf{0}$, while, as expected, the minimum $P_{3D;a,a} = 0$ corresponds to $\mathbf{p}_a = \mathbf{p}_b$.

Regarding the physical significance of the degree of polarization, it is worth considering the interesting case of thermal light that obeys Gaussian statistics, whose normalized intensity fluctuations are related to the degree of polarization through the expression (Setälä, T. et al. 2004c)

$$\left\langle\left[\Delta I(\mathbf{r},t)\right]^2\right\rangle / \left\langle I(\mathbf{r},t)\right\rangle^2 = \left[1 + 2P^2_{3D}(\mathbf{r})\right]/3 \tag{2.82}$$

where the brackets denote time (or ensemble) averaging, $I(\mathbf{r},t)$ is the instantaneous intensity of the electric field at a space–time point $(\mathbf{r},t)$, and $\Delta I(\mathbf{r},t) = I(\mathbf{r},t) - \langle I(\mathbf{r},t)\rangle$ describes the intensity fluctuations. The physical implications of Eq. (2.82) and its relation to the relation obtained previously by Wolf (1960) for 2D states, have been analyzed by Setälä, T. et al. (2004c) and can also be properly interpreted by replacing P_{3D} by its expression in terms of the indices of polarimetric purity, which will be introduced in Section 2.10.4.

Consider now the space–frequency domain and recall that, from the formulation of the 3D cross-spectral density matrix reduced to the one-point case, $\mathbf{r}_1 = \mathbf{r}_2 \equiv \mathbf{r}$, the 3D *spectral degree of polarimetric purity* is defined as

$$P_{3D}(\omega) = \sqrt{\frac{1}{2}\left(\frac{3\left\|\mathbf{W}(\omega)\right\|_2^2}{\left\|\mathbf{W}(\omega)\right\|_{tr}^2} - 1\right)} \tag{2.83}$$

This expression is equivalent to that presented by Wolf (2007), so that $P_{3D}(\omega)$ is limited by $0 \le P_{3D}(\omega) \le 1$. As in the 2D case, $P_{3D}(\omega)$ may have different values at different frequencies and therefore P_{3D} and $P_{3D}(\omega)$ are, in general, not equivalent.

As seen in Section 2.7.1, in the case of a 3D quasimonochromatic wave with a very small bandwidth, this quantity, when computed at its mean frequency, $P_{3D}(\varpi)$, equals the 3D degree of polarimetric purity, $P_{3D}(\varpi) = P_{3D}$ (Setälä et al. 2009a). Nevertheless it should be kept in mind that, in general, the coherence properties of quasimonochromatic electromagnetic beams differ from those of a monochromatic beam (Lahiri and Wolf 2009; 2010). It is also remarkable that the equal-point 3D version of the *Luis degree of coherence* $D(\mathbf{r},\mathbf{r},\omega)$ is precisely the spectral degree of polarization (Luis 2007)

$$D^2(\mathbf{r},\mathbf{r},\omega) \equiv D^2(\omega) = \frac{1}{n-1}\left(n\left\|\hat{\mathbf{R}}(\omega)\right\|_2^2 - 1\right) = \frac{1}{2}\left(3\left\|\hat{\mathbf{R}}(\omega)\right\|_2^2 - 1\right) = P_{3D}^2(\omega) \tag{2.84}$$

where the value $n = 3$ corresponding to the one-point case ($\mathbf{r} \equiv \mathbf{r}_2 = \mathbf{r}_1$) has been taken.

Moreover, the equal-point ($\mathbf{r} \equiv \mathbf{r}_2 = \mathbf{r}_1$), equal-time ($\tau = 0$), expression for the electromagnetic degree of coherence $\gamma(\mathbf{r}_1, \mathbf{r}_2, \tau)$ defined in Eq. (2.55) is

$$\gamma = \left\|\mathbf{J}(\mathbf{r},\mathbf{r},0)\right\|_2 / I(\mathbf{r}) = \left\|\hat{\mathbf{R}}\right\|_2 = \sqrt{\hat{\lambda}_1^2 + \hat{\lambda}_2^2 + \hat{\lambda}_3^2} \tag{2.85}$$

where $\hat{\lambda}_1$, $\hat{\lambda}_2$, $\hat{\lambda}_3$ are the eigenvalues of $\hat{\mathbf{R}}$, so that γ is an invariant quantity directly related to the 3D degree of polarimetric purity P_{3D} (Tervo et al. 2003):

$$P_{3D} = \sqrt{\left(3\gamma^2 - 1\right)/2} \qquad \gamma = \sqrt{\left(2P_{3D}^2 + 1\right)/3} \tag{2.86}$$

and therefore P_{3D} and $\gamma(\mathbf{r},\mathbf{r},0)$ provide equivalent information. The value of the equal-point, equal-time, degree of coherence γ is constrained by the limits $1/\sqrt{3} \le \gamma \le 1$. The minimum corresponds to a 3D unpolarized state $\gamma(\mathbf{R}_{u-3D}) = 1/\sqrt{3}$, whereas the maximum corresponds to pure states $\gamma(\mathbf{R}_p) = 1$.

2.10.3 Sources of Polarimetric Purity

The concept of polarization is essentially liked to those of anisotropy and asymmetry. A polarization state that does not exhibit any asymmetry in its properties is necessarily totally unpolarized and is described by the polarization matrix $\mathbf{R}_{u-3D} = I\mathbf{I}_3/3$. As the state presents some asymmetry among its principal intensities or some spin, it is *polarized* to some extent. Maximum asymmetry corresponds to totally polarized states $P_{3D} = 1$ (Gil et al. 2019a). Thus, the degree of polarimetric purity provides a measure of the polarimetric asymmetry, which can be caused by different properties that are analyzed in the present section.

By inspecting the expression (2.78) of the degree of polarimetric purity, and taking into account that P_{3D} is invariant with respect to unitary similarity transformations (i.e., $P_{3D}(\mathbf{R}) = P_{3D}(\mathbf{R}_O)$), we find that P_{3D} can be expressed as the following weighted quadratic average of the *degree of linear polarization* P_l, the *degree of circular polarization* P_c and the *degree of directionality* P_d (Gil 2015):

$$P_{3D} = \sqrt{\frac{3}{4}\left(P_l^2 + P_c^2\right) + \frac{1}{4}P_d^2} \tag{2.87}$$

where, as seen in Section 2.9,1, P_l, P_c and P_d are proper measures of complementary sources of purity, which are directly linked to the intrinsic Stokes parameters.

For certain purposes, it is useful to use the *degree of elliptical purity* P_e (Gil et al. 2017), defined as $P_e \equiv \sqrt{P_l^2 + P_c^2}$, with $0 \le P_e \le 1$, which accounts for the combined contributions of the degrees of linear and circular polarization P_l and P_c. The maximum $P_e = 1$ corresponds to a pure state, and the minimum $P_e = 0$ ($P_l = P_c = 0$) corresponds to any combination of a 2D unpolarized state and a 3D unpolarized state. Thus $P_{3D} = \sqrt{3P_e^2/4 + P_d^2/4}$, showing that the polarimetric purity can be considered as composed of two separate contributions, namely the degree of elliptical purity and the degree of directionality.

Let us now recall that nonorthogonal unitary transformations of **R** lead to transformed states of polarization that are different from that represented by **R**, while orthogonal transformations of **R** lead to states of polarization that can be considered the same as **R** but referenced with respect to rotated laboratory reference frames. Therefore, while **R** has three independent physical parameters that are invariant with respect to unitary transformations (i.e., with respect to transformations of generalized reference bases), namely the set of eigenvalues $(\lambda_1, \lambda_2, \lambda_3)$, **R** has six independent parameters that are physically invariant with respect to arbitrary rotations of the laboratory reference frame, namely the set composed of the principal intensities (a_1, a_2, a_3), or equivalently (I, P_l, P_d), together with the intrinsic components $(\hat{n}_{O1}, \hat{n}_{O2}, \hat{n}_{O3})$ of the spin density vector $\hat{\mathbf{n}}$, or equivalently $(|\hat{\mathbf{n}}|, \phi_n, \theta_n)$, where ϕ_n and θ_n are the azimuth and elevation angles of $\hat{\mathbf{n}}$ with respect of the symmetry axes $X_O Y_O Z_O$ of the intensity ellipsoid (Figure 2.6a). Therefore, the sources of asymmetry can be classified into two types, namely *the intensity anisotropy* (i.e., the anisotropy linked to the intensity ellipsoid) and the *spin anisotropy* (linked to the magnitude of the spin density vector).

The intensity anisotropy is quantified by the *dimensionality index* (Norrman et al. 2017; Gil et al. 2019a) defined as

$$d \equiv \sqrt{\frac{(\hat{a}_1 - \hat{a}_2)^2 + (\hat{a}_1 - \hat{a}_3)^2 + (\hat{a}_2 - \hat{a}_3)^2}{2}} = \sqrt{\frac{3P_l^2 + P_d^2}{2}} \tag{2.88}$$

whose values are restricted to the interval $0 \le d \le 1$, with $d = 0$ for states whose intensity ellipsoid is a sphere, and therefore are fully intensity-symmetric, and $d = 1$, for which the intensity ellipsoid takes the most asymmetric form, given by a linear segment corresponding to a pure linearly polarized state. Observe that d is given by a weighted square average of the two components of purity P_l and P_d that have been defined in terms of the principal intensities, while d is irrespective of the value of the remaining component of purity, P_c.

Since the polarization matrix **R** is Hermitian, it exhibits certain asymmetry as long as its imaginary part (skew-symmetric matrix $\mathrm{Im}\,\mathbf{R}$) is nonzero. In fact, in the intrinsic reference frame the correlations between the analytic signals of the electric field components are given by the (imaginary) off-diagonal elements (cross-correlations). Moreover, from Eq. (2.62) it follows that

$$\| \mathrm{Im}\,\hat{\mathbf{R}} \|_2 = |\hat{\mathbf{n}}|/\sqrt{2} = P_c/\sqrt{2} \tag{2.89}$$

which, as seen previously, is invariant under rotation transformations of the reference coordinate system, so that the associated *spin anisotropy* (or *correlation asymmetry*, Gil 2020) increases with $|\hat{\mathbf{n}}|$. Therefore, the intensity and spin anisotropies, d and P_c, constitute complementary sources of asymmetry (polarimetric purity) that allow for expressing P_{3D} as the following weighted square average of them: $P_{3D} = \sqrt{d^2 + 3P_c^2/2}$.

The name *dimensionality index* used to refer to d derives from the fact that it determines the *polarimetric dimension* $D_l \equiv 3 - 2d$ (Norrman et al. 2017), which is a measure of the effective polarimetric dimensions involved in the polarization state: $D_l = 1$ is exclusive of 1D light ($P_l = 1$, which corresponds to pure linearly polarized states); values $1 < D_l \leq 2$ are characteristic of 2D light ($P_l < 1$, $P_d = 1$), and $2 < D_l \leq 3$ are solely achieved by genuine 3D states ($P_d < 1$).

2.10.4 Indices of Polarimetric Purity

P_{3D} takes into account not only the stability of the shape of the polarization ellipse, but also the stability of the polarization plane. Thus, for a 2D unpolarized electromagnetic wave ($a_1 = a_2 > a_3 = 0$, $P_d = 1$), the 2D degree of polarization is zero while $P_{3D} = 1/2$, which reflects the lack of symmetry derived from the fact that the intensity ellipsoid is a circle and not a sphere. An analysis of the structure of the polarimetric purity of a 3D state, complementary to that established in terms of the components of purity (P_l, P_c, P_d), can be performed by inspecting the relative differences between the eigenvalues $(\hat{\lambda}_1 \geq \hat{\lambda}_2 \geq \hat{\lambda}_3)$ of the polarization density matrix $\hat{\mathbf{R}}$. An appropriate pair of *indices of polarimetric purity* is defined as (San José and Gil 2011)

$$P_1 = \hat{\lambda}_1 - \hat{\lambda}_2 \qquad P_2 = \hat{\lambda}_1 + \hat{\lambda}_2 - 2\hat{\lambda}_3 \tag{2.90}$$

These dimensionless quantities are restricted by $0 \leq P_1 \leq P_2 \leq 1$. Thus, while the pair (P_1, P_2) (hereafter called the IPP when appropriate) provides detailed information on the structure of the polarimetric purity of $\mathbf{R}$ (Gil et al. 2017), P_{3D} represents an overall measure of polarimetric purity. It can be calculated from P_1 and P_2 by the weighted quadratic average (San José and Gil 2011)

$$P_{3D} = \sqrt{3P_1^2 + P_2^2}/2 \tag{2.91}$$

Eqs. (2.78) and (2.91) can be rearranged to obtain another interesting expression of P_{3D} as the following homogeneous quadratic measure of all relative differences between the eigenvalues (Samson 1973; Barakat 1977; Gil 2007):

$$P_{3D}^2 = \frac{1}{2}\sum_{\substack{i,j=1 \\ i<j}}^{3} p_{ij}^2 \qquad \left[p_{ij} \equiv \frac{\lambda_i - \lambda_j}{\mathrm{tr}\mathbf{R}}\right] \tag{2.92}$$

To inspect the physical meaning of the IPP, let us recall the characteristic decomposition of $\mathbf{R}$, given by Eq. (2.29) which can be reformulated in the following manner:

$$\begin{gathered}\mathbf{R} = P_1 I\hat{\mathbf{R}}_p + (P_2 - P_1) I\hat{\mathbf{R}}_m + (1 - P_2) I\hat{\mathbf{R}}_{u-3D} \\ \hat{\mathbf{R}}_p \equiv \mathbf{U}\,\mathrm{diag}(1,0,0)\mathbf{U}^\dagger \quad \hat{\mathbf{R}}_m \equiv \frac{1}{2}\mathbf{U}\,\mathrm{diag}(1,1,0)\mathbf{U}^\dagger \quad \hat{\mathbf{R}}_{u-3D} \equiv \frac{1}{3}\mathrm{diag}(1,1,1)\end{gathered} \tag{2.93}$$

where $\mathbf{R}_p = I\hat{\mathbf{R}}_p$ is the pure characteristic component, $\mathbf{R}_m = I\hat{\mathbf{R}}_m$ is the discriminating component (Gil et al. 2021) and $\mathbf{R}_{u-3D} = I\hat{\mathbf{R}}_{u-3D}$ is the 3D unpolarized component.

Thus, P_1 provides a measure of the portion $I_p \equiv P_1 I$ of intensity corresponding to the characteristic pure component with respect to the total intensity I. On the other hand, $(1 - P_2)I$ is the portion of intensity corresponding to the 3D unpolarized component and, therefore,

P_2 gives a measure of the portion $P_2 I$ of intensity corresponding to the component that is not 3D unpolarized

$$P_1 I \hat{\mathbf{R}}_1 + (P_2 - P_1) I \hat{\mathbf{R}}_m = \mathbf{R} - (1 - P_2) I \hat{\mathbf{R}}_{u-3D} \tag{2.94}$$

i.e., $P_2 = (I - I_{3D})/I$, where $I_{3D} \equiv (1 - P_2) I$ is the intensity of the 3D unpolarized part of the state.

P_1 can be interpreted as the relative portion of power corresponding to the totally polarized component of the state and, in the particular case of 2D states, P_1 becomes the usual 2D degree of polarization. Further, P_2 is the relative portion which is not totally depolarized (viz., is not 3D-depolarized), so that P_2 measures the ratio between the power remaining once the 3D-unpolarized part has been subtracted and the total power of the whole state **R**.

The difference between P_1 and P_2 arises clearly from the characteristic decomposition of **R** in Eq. (2.93) because the equality $P_1 = P_2$ holds if and only if $\lambda_2 = \lambda_3$, i.e., the discriminating component $\mathbf{R}_m$ vanishes, in which case **R** can be considered as composed of a totally polarized component $\mathbf{R}_p$ and a 3D unpolarized component $\mathbf{R}_{u-3D}$.

Let us now observe that Eq. (2.91) suggests the possibility to define the angle ψ as

$$\sin\psi = \sqrt{3}\, P_1/(2P_{3D}) \qquad \cos\psi = P_2/(2P_{3D}) \qquad \left[\tan\psi = \sqrt{3}\, P_1/P_2\right] \tag{2.95}$$

so that the indices of polarimetric purity can be written as

$$P_1 = \left(2/\sqrt{3}\right) P_{3D} \sin\psi \qquad P_2 = 2P_{3D} \cos\psi \tag{2.96}$$

Note that ψ is limited to values within the interval $0 \le \psi \le \pi/3$. Thus, for a given value of P_{3D}, ψ provides a measure of the balance between the relative weights of the pure component and the sum of the pure and the discriminating components. The minimum $\psi = 0$ corresponds to states that lack pure component. Moreover, when $\psi = \pi/3$, P_1 reaches its maximal achievable value, $P_1 = P_2 = P_{3D}$.

By taking advantage of the analytic expressions for $\lambda_1, \lambda_2, \lambda_3$ obtained by Sheppard (2011) and Sheppard el al. (2020a) from the *del Ferro-Cardano-Tartaglia-Vieta* solution for the cubic equation with real roots, it can be proven that the angle ψ satisfies the following relation:

$$\cos 3\psi = \cos\psi\left(4\cos^2\psi - 3\right) = \frac{3P_{3D}^2 + 27\det\hat{\mathbf{R}} - 1}{2P_{3D}^3} \qquad \left[P_{3D} > 0\right] \tag{2.97}$$

(excluded the case $P_{3D} = 0$, where $P_1 = P_2 = 0$ and ψ is then an undetermined angle without physical meaning).

An additional way of interpreting the physical significance of the IPP is obtained by inspecting the role they play in the spectral and arbitrary decompositions. By considering the possible values of the indices of purity, we observe that they are directly related to the purity structure of **R** and provide more detailed information than the value of $r \equiv \operatorname{rank}\mathbf{R}$ alone. As a general overview, the following cases can be distinguished (Gil and San José 2010):

(1) When $0 \le P_1 \le P_2 < 1$ $(r = 3)$, **R** can be considered as composed of three pure states $\mathbf{R} = p_1 \mathbf{R}_{p1} + p_2 \mathbf{R}_{p2} + p_3 \mathbf{R}_{p3}$ (recall the arbitrary decomposition in Eq. (2.25) where

$\mathbf{R}_{pi} \equiv (\mathrm{tr}\,\mathbf{R})\,\hat{\mathbf{w}}_i \otimes \hat{\mathbf{w}}_i^{\dagger}$ are arbitrary provided that their generating unit complex vectors $\hat{\mathbf{w}}_i$ are linearly independent; the coefficients p_i have to be calculated for each particular set of three pure components). If, in particular, $P_2 = 0$ (and hence, $P_1 = 0$), then $\mathbf{R} = \mathbf{R}_{u-3D}$, so that it represents a 3D unpolarized state.

(2) When $0 \leq P_1 < P_2 = 1$ ($r = 2$), $\mathbf{R}$ can be considered as composed of two pure states $\mathbf{R} = p_1\,\mathbf{R}_{p1} + p_2\,\mathbf{R}_{p2}$ (note that when $P_2 = 1$ the 3D-unpolarized component $\mathbf{R}_{u-3D}$ of the characteristic decomposition of $\mathbf{R}$ vanishes, but this does not ensure that $\mathbf{R}$ represents a 2D state).

(3) Finally, when $P_1 = 1$ (and hence, $P_2 = 1$ and $r = 1$), $\mathbf{R}$ corresponds to a pure state.

2.10.5 3D Purity Space

Regarding the polarimetric purity of a 3D polarization state, the CP provide information on the sources of purity (P_l, P_c, P_d), while the IPP provide complementary information in terms of the relative weights of the components in Eq. (2.93), regardless of the sources of purity. Such differences between the physical meanings of the CP and the IPP, become clear when considering the two following different polarization states: (a) an equiprobable incoherent mixture of two linearly polarized states whose electric fields vibrate in mutually orthogonal directions ($P_1 = 0,\ P_2 = 1,\ P_l = P_c = 0,\ P_d = 1$) and (b) an equiprobable incoherent mixture of a linearly polarized state whose electric field vibrates along a given direction Z and a circularly polarized one whose polarization plane is orthogonal to Z ($P_1 = 0,\ P_2 = 1,\ P_l = 1/4,\ P_c = 1/2,\ P_d = 1/4$) (Gil et al. 2018b). Both combinations share equal values for their respective IPP but very different CP values.

Since the quantitative statistical structure of purity of 3D states involves two parameters, namely the indices of polarimetric purity P_1, P_2 (IPP) (instead of the single parameter $\mathcal{P}$, the degree of polarization, for 2D states), different physical situations can be identified from the analysis of the feasible region for the values of P_1 and P_2 (Figure 2.9a) (Gil 2007).

- **Edge AB** (vertices A and B excluded). The IPP take the values $P_1 = 0,\ 0 < P_2 < 1$, corresponding to $0 < P_{3D} < 1/2$ and $\lambda_1 = \lambda_2 > \lambda_3 > 0$. The system is equivalent to an incoherent combination of three pure contributions. The two more significant components of the spectral decomposition have equal intensities $\lambda_1 = \lambda_2 > \lambda_3$.
- **Edge BC**. The polarization state is equivalent to an incoherent composition of two pure states. Vertex C corresponds to pure states ($P_1 = P_2 = 1$), while vertex B corresponds to a discriminating state.
- **Edge AC** (vertex C excluded). The IPP take the values $0 < P_2 = P_1 < 1$, corresponding to $P_{3D} = P_2 = P_1$, and $\lambda_1 > \lambda_2 = \lambda_3 > 0$. This region represents states where P_1 has its maximum value compatible with P_2. The two less significant components of the spectral decomposition have equal intensities. Vertex A ($P_1 = P_2 = 0$, $P_{3D} = 0$) corresponds to 3D unpolarized states.
- Points within the **triangle ABC** (edges AB, BC and AC excluded). The IPP take values restricted by $0 < P_1 < P_2 < 1$, corresponding to $0 < P_{3D} < 1$ and $\lambda_1 > \lambda_2 > \lambda_3 > 0$. The 3D mixed states represented in this area are equivalent to incoherent compositions of three pure states with different intensities.

Given a fixed value of P_{3D}, it is compatible with different pairs of values for P_1 and P_2. The values of these three parameters are restricted by the following inequalities (Gil and San José 2010):

$$0 \le P_1 \le P_{3D} \le P_2 \le \min(2P_{3D}, 1) \tag{2.98}$$

As noted above, the equality $P_{3D} = P_1 = P_2$ is satisfied along the segment AC, where P_1 has its maximum feasible value for a given value of P_2. Next, we analyze a set of representative iso-purity lines (elliptical segments in the purity space) where the values of P_1, P_2 are compatible with $P_{3D} = 0, 1/4, 1/2, 3/4$, and 1 (Figure 2.9b).

- $P_{3D} = 0$. The feasible segment is restricted to point A, where $P_1 = P_2 = 0$, and corresponds to a 3D unpolarized state.
- $P_{3D} = 1/4$. It corresponds to the elliptical segment DE. Different pairs of values of P_1, P_2 are allowed from $D(1/4, 1/4)$ to $E(0, 1/2)$. The maximum value for P_2 is $P_2 = 1/2$.
- $P_{3D} = 1/2$. It corresponds to the elliptical segment FB. Different pairs of values of P_1, P_2 are allowed from $F(1/2, 1/2)$ to $B(0, 1)$. The maximum value of P_2 is $P_2 = 1$, which occurs when $P_1 = 0$.
- $P_{3D} = 3/4$. It corresponds to the elliptical segment GH. Different pairs of values of P_1, P_2 are allowed from $G(3/4, 3/4)$ to $H(\sqrt{5/12}, 1)$. The maximum value of P_2 is $P_2 = 1$, which corresponds to $P_1 = \sqrt{5/12}$.
- $P_{3D} = 1$. This maximal polarimetric purity corresponds to point C, where $P_1 = P_2 = 1$ (pure state).

As described in Section 2.12, in general $P_d \le P_2$ and $P_e \ge P_1$, with the limiting equalities $P_d = P_2$ and $P_e = P_1$ being characteristic of regular states. Such relations indicate that the discriminating component of nonregular states can exhibit certain amounts of linear and circular degrees of polarization. The peculiar features and differences between the IPP and the CP can be observed by means of the combined representation of P_2 versus P_1 and P_d versus P_e in the *purity figure* shown in Figure 2.9c. The feasible region for $P_2(P_1)$ is given by the triangle ABC, while the feasible region for $P_d(P_e)$ is constituted by such triangle together with the shadowed area ACIA (which, except for the straight segment AC is exclusive for nonregular states) (Migliaccio et al. 2020).

For both representations $P_2(P_1)$ and $P_d(P_e)$ points A ($P_{3D} = 0$) and C ($P_{3D} = 1$) represent respectively a fully unpolarized state $\mathbf{R}_{u-3D}$ and a pure state $\mathbf{R}_p$. For a given value of P_{3D} points $P_2(P_1)$ and $P_d(P_e)$ are located in a common elliptical branch but both points coincide

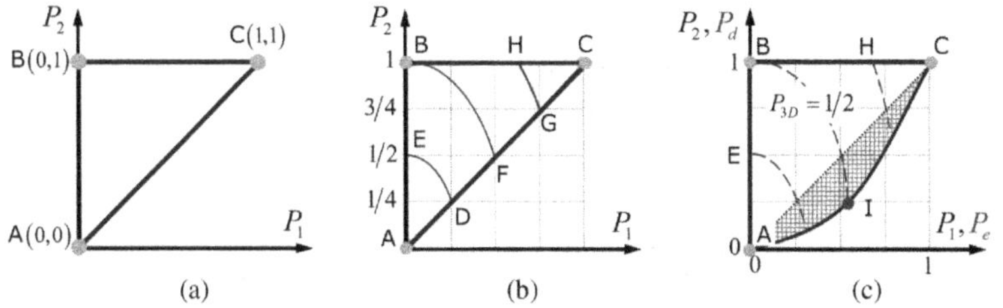

FIGURE 2.9
Purity Space. (a) Feasible region for P_1, P_2 in the 2D purity space. (b) Iso-purity curves in the purity space for different values of P_{3D} (Gil and San José 2010). (c). The feasible region for $P_2(P_1)$ is determined by the triangle ABC, while the feasible region for $P_d(P_e)$ includes also the area ACIA; for both representations, points A and C correspond to respective states $\mathbf{R}_{u-3D}$ and $\mathbf{R}_p$; except for the straight segment AC, the area ACIA is exclusive for nonregular states while points inside the triangle ABC can be regular or not, in general.

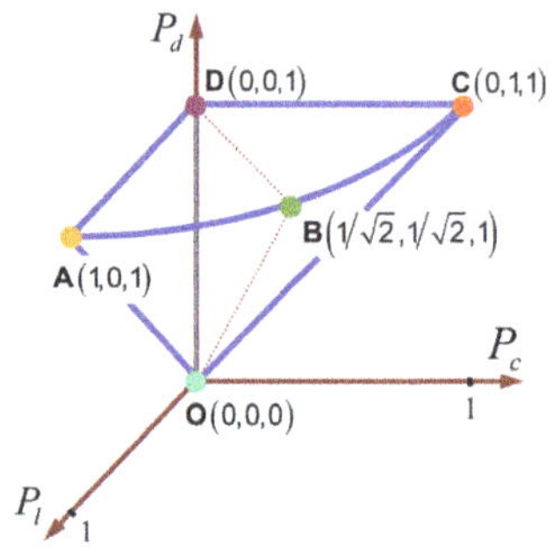

FIGURE 2.10
Feasible region for the components of purity P_l, P_c, and P_d for regular states (Gil 2015).

if and only they correspond to a regular state. As an illustrative example, the elliptical branch $P_{3D} = 1/2$ containing points B and I is represented in Figure 2.9c. Depending on whether both points, $P_2(P_1)$ and $P_d(P_e)$, coincide, the represented states are regular or not. In particular, the branch $P_{3D} = 1/2$ contains all discriminating states (i.e., of the form $\mathbf{R}_m$, with $P_1 = 0$ and $P_2 = 1$). Point I represents a state with maximum nonregularity, which is given by a discriminating state $\mathbf{R}_m$ featuring $P_c(\mathbf{R}_m) = 1/2$ and $P_l(\mathbf{R}_m) = 1/4$ (called a *perfect nonregular state*, Gil et al. 2018b; see Section 2.12).

The feasible region for the CP of 3D states is determined by relation $P_{3D}^2 = 3(P_l^2 + P_c^2)/4 + P_d^2/4$ (Eq. (2.87)) between the components of purity P_l, P_c, P_d and the overall degree of polarimetric purity P_{3D}, which in the particular case of regular states should be combined with $P_l^2 + P_c^2 \leq P_d^2$ (Figure 2.10). The feasible region of the CP of regular states is given by points in the truncated cone $P_l^2 + P_c^2 = P_d^2 \leq 1$ whose generatrix is the axis P_d, and restricted to the positive octant $P_l \geq 0$, $P_c \geq 0$, $P_d \geq 0$, limited by planes $P_l = 0$, $P_c = 0$, $P_d = 1$ (Gil 2015).

The origin O corresponds to 3D unpolarized states; points in the circular positive quadrant AC correspond to totally polarized states ($P_1 = P_{3D} = 1$); points in the plane triangle OAD correspond to states whose characteristic polarization ellipse has no ellipticity (linearly polarized characteristic component); points in the plane triangle OCD correspond to states whose characteristic component is circularly polarized, and points in the region ACD correspond to 2D states ($P_d = 1$).

2.10.6 Degrees of Mutual Coherence of a 3D Polarization State

The degrees of mutual coherence $|\mu_{ij}| = \sqrt{r_{ij}\, r_{ji} / r_{ii} r_{jj}}$ between the analytic signals of the field components ε_1, ε_2 and ε_3 of a 3D state of polarization are relative quantities which depend on the reference frame considered. As we did in Section 1.11.3, devoted to the 2D degree of mutual coherence between the analytic signals ε_1 and ε_2 of the orthogonal components of a given 2D state of polarization, we consider below the extremal values of $|\mu_{ij}|$.

Regarding unitary transformations, let us recall that there always exists a unitary transformation $\mathbf{URU}^{\dagger} = \mathrm{diag}(\lambda_1, \lambda_2, \lambda_3)$ that diagonalizes $\mathbf{R}$, so that the transformed 3D state of polarization represented by the polarization matrix $\mathrm{diag}(\lambda_1, \lambda_2, \lambda_3)$ preserves the intensity and the IPP (but not, in general, other characteristic properties of the polarization state $\mathbf{R}$) and all $|\mu_{ij}|$ vanish. Furthermore, there always exists an orthogonal transformation $\mathbf{R}_E = \mathbf{QRQ}^T$ such that the transformed polarization matrix $\mathbf{R}_E$ has equal diagonal elements, and it can be proved that, regarding unitary transformations (i.e., including orthogonal transformations), the maximum achievable values of $|\mu_{ij}|$ correspond to $\mathbf{R}_E$ (Gil 2016d).

Once the extremal values of $|\mu_{ij}|$ with respect to unitary transformations of **R** have been considered, let us stress that only the unitary transformations that are orthogonal (rotation transformations) preserve the intrinsic properties of the polarization state, i.e., the transformed state can be considered as the same as that represented by **R** but referenced with respect to a rotated Cartesian reference frame.

Having in mind the idea of analyzing properties related to a given state of polarization (characterized by its intrinsic Stokes parameters), let us now consider the limits for $|\mu_{ij}|$ with respect to arbitrary rotations of the Cartesian reference frame.

It is straightforward to prove that the minimum values of $|\mu_{ij}|$ with respect to arbitrary rotations in the real space are achieved when the reference frame is precisely the intrinsic one $X_O Y_O Z_O$, where the differences between the strengths of the components of the electric field are maximal, so that (Gil 2016d),

$$|\mu_{12}|^2_{\min} = \frac{n_{O3}^2}{4a_1 a_2} \qquad |\mu_{13}|^2_{\min} = \frac{n_{O2}^2}{4a_1 a_3} \qquad |\mu_{23}|^2_{\min} = \frac{n_{O1}^2}{4a_2 a_3} \qquad [a_3 > 0] \tag{2.99}$$

and therefore, the intrinsic polarization matrix $\mathbf{R}_O$ has the property that $|\mu_{ij}|$ reach their minimal achievable values $|\mu_{ij}|_{\min}$ with respect to rotations of the reference frame. From Eq. (2.99), it follows that all the three $|\mu_{ij}|_{\min}$ are zero-valued for states with zero spin. Moreover, for circularly polarized states, $|\mu_{12}|_{\min} = |\mu_{12}|_{\max} = 1$ and $|\mu_{13}|_{\min} = |\mu_{23}|_{\min} = 0$.

Regarding the maximal values of $|\mu_{ij}|$ for nonpure states, as indicated at the beginning of this section, given a Hermitian matrix **H**, it is always possible to find an orthogonal matrix **Q** such that the transformed Hermitian matrix $\mathbf{QHQ}^T$ has equal diagonal elements (Setälä et al. 2002a). This operation can be carried out, for instance, through the Bendel–Mickey algorithm (Bendel and Mickey 1978). Thus, there exists a reference frame $X_E Y_E Z_E$ in which the strengths of the three orthogonal components of the electric field are equal (i.e., the diagonal elements of the transformed polarization matrix $\mathbf{R}_E$ are equal), in which case it can be demonstrated that $|\mu_{ij}|$ reach their maximum values with respect to arbitrary rotations. This condition is achieved through a transformation $\mathbf{R}_E = \mathbf{Q}^T(\phi_E, \theta_E, \varphi_E)\,\mathbf{R}_O\,\mathbf{Q}(\phi_E, \theta_E, \varphi_E)$ of the intrinsic polarization matrix (see Eqs. (2.62) and (2.63)), where the required values for the angles $(\phi_E, \theta_E, \varphi_E)$ depend on the values of the intrinsic Stokes parameters $(I, P_l, P_c, n_{O1}, n_{O2}, n_{O3})$. The transformed polarization matrix can be expressed as follows:

$$\mathbf{R}_E = \frac{1}{2} I \begin{pmatrix} 2/3 & \hat{s}_{E01} - i\hat{n}_{E3} & \hat{s}_{E02} + i\hat{n}_{E2} \\ \hat{s}_{E01} + i\hat{n}_{E3} & 2/3 & \hat{s}_{E12} - i\hat{n}_{E1} \\ \hat{s}_{E02} - i\hat{n}_{E2} & \hat{s}_{E12} + i\hat{n}_{E1} & 2/3 \end{pmatrix} \tag{2.100}$$

where $I = a_1 + a_2 + a_3 = s_{E00}$, and $\hat{s}_{Eij}$ $(i, j = 0,1,2)$ are the intensity-normalized Stokes parameters corresponding to $\mathbf{R}_E$ (with $s_{E11} = s_{E22} = 0$). The maximal values for $|\mu_{ij}|$ with respect to orthogonal transformations can be expressed in terms of the Stokes parameters as (Gil 2016d)

$$|\mu_{12}|_{\max} = 3\sqrt{\hat{s}_{E01}^2 + \hat{n}_{E03}^2}\Big/2 \qquad |\mu_{13}|_{\max} = 3\sqrt{\hat{s}_{E2}^2 + \hat{n}_{E02}^2}\Big/2 \qquad |\mu_{23}|_{\max} = 3\sqrt{\hat{s}_{E12}^2 + \hat{n}_{E1}^2}\Big/2 \tag{2.101}$$

2.10.7 3D Polarization Entropy

From the description of the von Neumann entropy for n-dimensional density matrices (Von Neumann 1996) studied by Barakat (1983; 1996b), Brosseau (1998) and other authors, the 3D polarization entropy can be defined as

$$S_{3D} = -\mathrm{tr}\left(\hat{\mathbf{R}}\log_3 \hat{\mathbf{R}}\right) = -\sum_{i=1}^{3}\left(\hat{\lambda}_i \log_3 \hat{\lambda}_i\right) \tag{2.102}$$

where $\hat{\lambda}_1, \hat{\lambda}_2, \hat{\lambda}_3$ are the eigenvalues of the polarization density matrix $\hat{\mathbf{R}}$. S_{3D} is expressed as follows in terms of the indices of polarimetric purity of $\hat{\mathbf{R}}$(Gil 2007):

$$\begin{gathered} S_{3D} = \frac{Q_+}{3}(1-\log_3 Q_+) + \frac{Q_-}{3}(1-\log_3 Q_-) + \frac{Q}{3}[1-\log_3 Q] \\ \left[Q_+ \equiv 1 + P_2/2 + 3P_1/2 \qquad Q_- \equiv 1 + P_2/2 - 3P_1/2 \qquad Q \equiv (1-P_2)\right] \end{gathered} \tag{2.103}$$

As stated in Section 1.11.4, although the Napierian logarithm (ln) is commonly used for the definition of the von Neumann entropy, the use of the base n logarithm for the definition of the entropy of nD density matrices has the advantage of restricting the values of S_{nD} to the range $0 \le S \le 1$.

S_{3D} is characterized uniquely by P_1 and P_2. The maximum $S_{3D} = \log_3 3 = 1$ corresponds to a 3D unpolarized state ($P_1 = P_1 = P_{3D} = 0$), whereas the minimum $S_{3D} = 0$ is reached for a pure state ($P_1 = P_1 = P_{3D} = 1$).

Since the quantitative structure of polarimetric purity is determined by the pair of parameters P_1, P_2, it is possible to define two respective partial entropies $S(P_1)$ and $S(P_2)$ through the expressions

$$S(P_i) \equiv -\left\{\frac{1}{2}(1+P_i)\log_2\left[\frac{1}{2}(1+P_i)\right] + \frac{1}{2}(1-P_i)\log_2\left[\frac{1}{2}(1-P_i)\right]\right\}, (i=1,2) \tag{2.104}$$

These invariant quantities contain objective information on polarimetric randomness equivalent to that provided by the IPP. The condition $0 \le P_1 \le P_2 \le 1$ on the indices of purity has its respective counterpart $0 \le S(P_2) \le S(P_1) \le 1$ for the partial entropies.

The thermodynamic view of the 2D polarization matrix (Brosseau and Bicout 1994) can be generalized to 3D states by means of the concept of 3D *temperature* of *polarization*, defined as

$$\tau_{3D} \equiv \frac{2}{\ln(1+P_{3D}) - \ln(1-P_{3D})} \tag{2.105}$$

The temperature of polarization is zero for pure states and increases monotonically as P_{3D} decreases.

2.11 The Concept of Spin of a 3D Polarization State

The concept of spin of a 2D polarization state has been dealt with in Section 1.10. When generalized to 3D polarization states, interesting features arise that are peculiar of 3D states and deserve a specific analysis to which the present section is devoted.

To introduce the definition of the spin vector of a 3D polarization state, let us consider the spectral polarization matrix $\mathbf{W}(\boldsymbol{\omega}) = \langle \boldsymbol{\varepsilon}(\omega) \otimes \boldsymbol{\varepsilon}^{\dagger}(\omega) \rangle_{\omega}$ (see Section 2.7.1) where the ensemble average is realized over the monochromatic realizations $\boldsymbol{\varepsilon}(\omega)$ of the analytic signal vector of the field. Note that, even though each realization $\boldsymbol{\varepsilon}(\omega)$ corresponds to a pure state, the polarization ellipses of the different (pure) realizations lie, in general, in different planes, so that the use of a three-dimensional Cartesian coordinate system, common for all realizations, is required, so that

$$\boldsymbol{\varepsilon}(\omega) \otimes \boldsymbol{\varepsilon}^{\dagger}(\omega) = \mathrm{Re}\left[\boldsymbol{\varepsilon}(\omega) \otimes \boldsymbol{\varepsilon}^{\dagger}(\omega)\right] + \mathrm{Im}\left[\boldsymbol{\varepsilon}(\omega) \otimes \boldsymbol{\varepsilon}^{\dagger}(\omega)\right]$$
$$\mathrm{Im}\left[\boldsymbol{\varepsilon}(\omega) \otimes \boldsymbol{\varepsilon}^{\dagger}(\omega)\right] = \begin{pmatrix} 0 & -i\,\eta_3(\omega)/2 & i\,\eta_2(\omega)/2 \\ i\,\eta_3(\omega)/2 & 0 & -i\,\eta_1(\omega)/2 \\ -i\,\eta_2(\omega)/2 & i\,\eta_1(\omega)/2 & 0 \end{pmatrix} \quad (2.106)$$

where $\boldsymbol{\eta}(\omega) \equiv [\eta_1(\omega), \eta_2(\omega), \eta_3(\omega)]^T$ corresponds to the conventional definition (but with dimensions of intensity) of the electric spin vector of the pure state represented by $\boldsymbol{\varepsilon}(\omega)$

$$\boldsymbol{\eta}(\omega) = \mathrm{Im}[\boldsymbol{\varepsilon}^{*}(\omega) \wedge \boldsymbol{\varepsilon}(\omega)] = \mathrm{Im}\begin{pmatrix} \varepsilon_2^{*}(\omega)\,\varepsilon_3(\omega) - \varepsilon_3^{*}(\omega)\,\varepsilon_2(\omega) \\ -\varepsilon_1^{*}(\omega)\,\varepsilon_3(\omega) + \varepsilon_3^{*}(\omega)\,\varepsilon_1(\omega) \\ \varepsilon_1^{*}(\omega)\,\varepsilon_2(\omega) - \varepsilon_2^{*}(\omega)\,\varepsilon_1(\omega) \end{pmatrix} \quad (2.107)$$

The ensemble average, at a given frequency ω, of the spin vector realizations $\boldsymbol{\eta}(\omega)$ is therefore determined by the imaginary parts of the elements $w_{ij}(\omega)$ $(i, j = 1,2,3)$ of the spectral polarization matrix $\mathbf{W}(\omega)$

$$\mathbf{n}(\omega) = \langle \boldsymbol{\eta}(\omega) \rangle_{\omega} = 2\left(-\mathrm{Im}\,w_{23}(\omega), \mathrm{Im}\,w_{13}(\omega), -\mathrm{Im}\,w_{12}(\omega)\right)^T \quad (2.108)$$

Consequently, in accordance with Eq. (2.44) the average spin vector $\mathbf{n}$ associated with a 3D state $\mathbf{R}$ is given by

$$\mathbf{n} = \int_0^{\infty} \mathbf{n}(\omega)\, d\omega = 2\left(-\mathrm{Im}\,r_{23}, \mathrm{Im}\,r_{13}, -\mathrm{Im}\,r_{12}\right)^T \quad (2.109)$$

where r_{ij} $(i, j = 1,2,3)$ are the elements of $\mathbf{R}$. The above definition holds for fields with arbitrary spectral profile and can readily be formulated with angular momentum dimensions by replacing $\boldsymbol{\eta}(\omega)$ by the spin angular momentum $(\varepsilon_o/4\omega)\boldsymbol{\eta}(\omega)$, ε_o being the electric permittivity of the medium (assumed isotropic). The integrated value of $\mathbf{S}_E(\omega)$ over all frequencies provides the spin angular momentum $\mathbf{S}_E$, which can be expressed as $\mathbf{S}_E = (\varepsilon_o/4\omega_n)\mathbf{n}$ in terms of $\mathbf{n}$, ε_o and an *equivalent spin frequency* $\omega_n \equiv \varepsilon_o \mathbf{n}/4\mathbf{S}_E$ (Gil et al. 2022a). For the case of quasimonochromatic fields ω_n coincides with the central frequency ϖ.

The term *spin density vector* $\hat{\mathbf{n}}$ of a polarization state refers to the intensity-normalized version $\hat{\mathbf{n}} \equiv \mathbf{n}/I$.

2D states are defined as those whose electric field fluctuates in a fixed intrinsic plane Π $X_O Y_O$ and the spin vector $\mathbf{n}$ lies along the direction of propagation Z_O orthogonal to Π. The (single) z-component s_3 of $\mathbf{n}$ is termed the helicity of the polarization state. In the general case of 3D states, there is no well-defined direction of propagation, so that there is no longer a well-defined projection of the spin vector $\mathbf{n}$ on such direction. In fact, as seen in Section 2.8, $\mathbf{n}$ has in general three components along the respective intrinsic reference axes $X_O Y_O Z_O$ that constitute the symmetry axes of the intensity ellipsoid. Therefore, the chiral character of the polarization state is determined by the three intrinsic components (n_{O1}, n_{O2}, n_{O3}) of $\mathbf{n}$ and not by a single parameter.

The concept of spin vector is formally applicable for 2D and 3D density matrices. Nevertheless, such notion is not directly generalizable for $n\times n$ density matrices with $n>3$ because the number $n(n-1)/2$ of *intrinsic coherences* (the generalization of the intrinsic components of the spin vector), which characterize the chirality properties of the state, exceeds the number n of *intrinsic populations* (the generalization of the principal intensities) (Gil 2020).

2.12 Discriminating States and the Concept of Nonregularity

The component $\mathbf{R}_m$ of the characteristic decomposition $\mathbf{R} = P_1\mathbf{R}_p + (P_2 - P_1)\mathbf{R}_m + (1-P_2)\mathbf{R}_{u-3D}$ of a polarization matrix $\mathbf{R}$ (see Section 2.6.5) is called the *discriminating component* because its properties critically determine those of the whole polarization state $\mathbf{R}$. The general form of a discriminating component $\mathbf{R}_m$ is $\mathbf{R}_m = (1/2)\,\mathbf{U}\,\mathrm{diag}(1,1,0)\,\mathbf{U}^\dagger$, where $\mathbf{U}$ is the unitary matrix that diagonalizes the full polarization matrix $\mathbf{R}$. States $\mathbf{R}_m$ are called *discriminating states* regardless of whether they are the discriminating components of respective full matrices, or represent independent polarization states.

2.12.1 Canonical Representation of a Discriminating State

When the state $\mathbf{R}_m$ is referenced with respect to the coordinate system $X_m Y_m Z_m \equiv Z_3(-Y_3)X_3$, where $X_3Y_3Z_3$ is the intrinsic reference frame of the (pure) eigenstate $\hat{\mathbf{u}}_3$ associated with the smaller eigenvalue λ_3 of $\mathbf{R}$, $\mathbf{R}_m$ takes the simple form (Gil et al. 2018b)

$$\mathbf{R}_{m3} = \frac{1}{2} I \begin{pmatrix} 1 & 0 & 0 \\ 0 & \cos^2\chi_3 & -i\cos\chi_3 \sin\chi_3 \\ 0 & i\cos\chi_3 \sin\chi_3 & \sin^2\chi_3 \end{pmatrix} \tag{2.110}$$

where χ_3 is the ellipticity angle of the state $\hat{\mathbf{u}}_3 = e^{i\gamma}(0, i\sin\chi_3, \cos\chi_3)^T$ (referenced with respect to $X_m Y_m Z_m$). Thus, because of the double degeneracy of the nonzero eigenvalue of $\mathbf{R}_{m3}$, it can be expressed in infinite ways as equiprobable incoherent mixtures of pairs $(\hat{\mathbf{v}}_1, \hat{\mathbf{v}}_2)$ of mutually orthogonal states that, in addition, are orthogonal to $\hat{\mathbf{u}}_3$ (note that, although $\hat{\mathbf{v}}_1$ and $\hat{\mathbf{v}}_2$ are eigenstates of $\mathbf{R}_{m3}$, in general they are not eigenstates of the full state $\mathbf{R}$). The particular choice $\hat{\mathbf{v}}_1 = e^{i\alpha_1}(1,0,0)^T$, $\hat{\mathbf{v}}_2 = e^{i\alpha_2}(0, -i\cos\chi_3, \sin\chi_3)^T$ corresponds to the following simple decomposition of $\mathbf{R}_{m3}$ (Gil et al. 2018b):

$$\mathbf{R}_{m3} = \frac{1}{2} I \hat{\mathbf{R}}_{l-x} + \frac{1}{2} I \hat{\mathbf{R}}_{e-x}$$
$$\hat{\mathbf{R}}_{l-x} \equiv \hat{\mathbf{v}}_1 \otimes \hat{\mathbf{v}}_1^\dagger = \begin{pmatrix} 1 & 0 & 0 \\ 0 & 0 & 0 \\ 0 & 0 & 0 \end{pmatrix} \qquad \hat{\mathbf{R}}_{e-x} \equiv \hat{\mathbf{v}}_2 \otimes \hat{\mathbf{v}}_2^\dagger = \begin{pmatrix} 0 & 0 & 0 \\ 0 & \cos^2\chi_3 & -i\cos\chi_3 \sin\chi_3 \\ 0 & i\cos\chi_3 \sin\chi_3 & \sin^2\chi_3 \end{pmatrix} \tag{2.111}$$

TABLE 2.1
Summary of the Properties of a Discriminating State $\mathbf{R}_m$ in Terms of its Descriptors P_1, P_2, P_e, P_d

Regular	$P_d(\mathbf{R}_m)=P_2(\mathbf{R}_m)=1$	$P_c(\mathbf{R}_m)=0$	$P_e(\mathbf{R}_m)=P_1(\mathbf{R}_m)=0$
Nonregular	$P_d(\mathbf{R}_m)<P_2(\mathbf{R}_m)=1$	$P_c(\mathbf{R}_m)>0$	$P_e(\mathbf{R}_m)>P_1(\mathbf{R}_m)=0$
Perfect NR	$P_d(\mathbf{R}_m)=1/4,\ P_2(\mathbf{R}_m)=1$	$P_c(\mathbf{R}_m)=1/2$	$P_e(\mathbf{R}_m)=\sqrt{5}/4,\ P_1(\mathbf{R}_m)=0$

TABLE 2.2
Summary of the Properties of a General State $\mathbf{R}$ in Terms of its Descriptors P_1, P_2, P_e, P_d

Regular	$P_d(\mathbf{R})=P_2(\mathbf{R})$	$P_c(\mathbf{R})\geq 0$	$P_e(\mathbf{R})=P_1(\mathbf{R})$
Nonregular	$P_d(\mathbf{R})<P_2(\mathbf{R})$	$P_c(\mathbf{R})>0$	$P_e(\mathbf{R})>P_1(\mathbf{R})$

which is advantageous for the study of the peculiar features of discriminating states because it shows that they can always be interpreted as equiprobable incoherent combinations of a linearly polarized pure state $\hat{\mathbf{v}}_1$ and an elliptically polarized pure state $\hat{\mathbf{v}}_2$ whose polarization plane is orthogonal to the direction of vibration of the electric field of $\hat{\mathbf{v}}_1$ (Figure 2.11a).

In the particular case that $\chi_3 = 0$ (i.e., $\hat{\mathbf{u}}_3$ is a linearly polarized state), then $\mathbf{R}_{m3}$ is a real-valued matrix that has the form $\mathbf{R}_{u-2D} = (I/2)\,\mathrm{diag}(1,1,0)$ of the polarization matrix of a 2D unpolarized state, so that the state $\mathbf{R}$ is said to be *regular*, and is characterized by the fact that its characteristic decomposition has the form $\mathbf{R} = P_1\mathbf{R}_p + (P_2 - P_1)\mathbf{R}_{u-2D} + (1-P_2)\mathbf{R}_{u-3D}$. Thus, a polarization state $\mathbf{R}$ is regular if and only if either $P_2 = P_1$ (in which case $\mathbf{R} = P_1\mathbf{R}_p + (1-P_1)\mathbf{R}_{u-3D}$) or its discriminating state is regular. From Eq. (2.111), and taking into account that the eigenvalues of $\mathbf{R}_m$ are invariant under rotation transformations, it follows that, provided $P_2 \neq P_1$, the following statements are totally equivalent: (1) $\mathbf{R}$ is regular; (2) $\mathbf{R}_m$ is regular; (3) $\mathbf{R}_m = \mathbf{R}_{u-2D}$; (4) $\mathrm{Im}\,\mathbf{R}_m = \mathbf{0}$; (5) $P_l(\mathbf{R}_m) = 0$; (6) $P_c(\mathbf{R}_m) = 0$; (7) $\mathbf{n}(\mathbf{R}_m) = \mathbf{0}$; (8) $\chi_3 = 0$; (9) $P_d(\mathbf{R}_m) = 1$; (10) $P_e(\mathbf{R}) = P_1(\mathbf{R})$; (11) $P_d(\mathbf{R}) = P_2(\mathbf{R})$, and (12) $\mathbf{R}_m$ is a 2D state (Gil et al. 2018b).

Among the above characteristic properties of regular states, it is remarkable that, regardless of the values $P_1(\mathbf{R})$ and $P_2(\mathbf{R})$ of the IPP of $\mathbf{R}$, the equality $P_d(\mathbf{R}) = P_2(\mathbf{R})$, which is equivalent to $P_e(\mathbf{R}) = P_1(\mathbf{R})$, is satisfied if and only if $\mathbf{R}$ represents a regular state. Conversely, nonregular states are characterized by either of the equivalent inequalities $P_d(\mathbf{R}) < P_2(\mathbf{R})$ or $P_e(\mathbf{R}) > P_1(\mathbf{R})$ (Tables 2.1 and 2.2).

2.12.2 Degree of Nonregularity

Nonregularity is a property of polarization states that reflects certain critical and peculiar features of them and therefore deserves to be properly quantified. Interesting examples of nonregular states are found for evanescent waves excited in total internal reflection of partially polarized plane-waves (Norrman et al. 2019) and tightly focused light beams (Chen et al. 2020). To define a measure of the degree of nonregularity P_N of $\mathbf{R}$ as a distance from $\mathbf{R}$ to a regular state, let us first observe that the eigenvalues (in decreasing order) of $\hat{\mathbf{R}}_m$ are given by (Gil et al. 2018b)

$$\hat{m}_1 = 1/2 \qquad \hat{m}_2 = \left(\cos^2\chi_3\right)/2 \qquad \hat{m}_3 = \left(\sin^2\chi_3\right)/2 \tag{2.112}$$

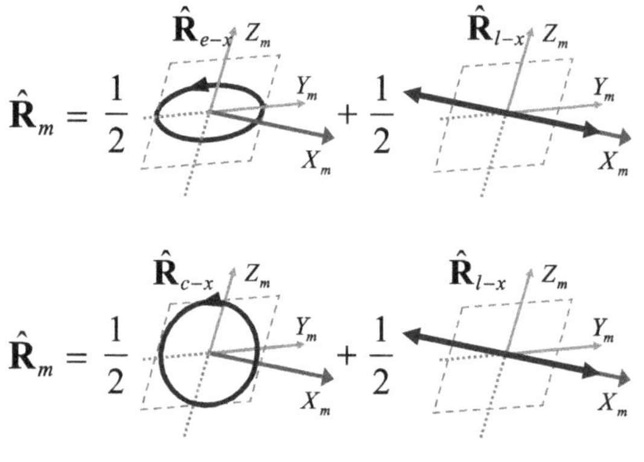

FIGURE 2.11
(a) A nonregular discriminating state $\mathbf{R}_m$ can always be interpreted as an equiprobable mixture of an elliptically polarized state $\mathbf{R}_{e-z}$ and a linearly polarized state $\mathbf{R}_{l-z}$ whose electric field vibrates along a direction Z_m that is orthogonal to the polarization plane X_mY_m of $\mathbf{R}_{c-z}$ (b) In the case of a perfect nonregular state, the component $\mathbf{R}_{e-z}$ is circularly polarized (denoted as $\mathbf{R}_{c-z}$) (Gil et al. 2018b).

so that the degree of nonregularity of a discriminating state $\mathbf{R}_m$ is defined as $P_N(\mathbf{R}_m) = 4\hat{m}_3$, $\hat{m}_3$ being the smaller eigenvalue of $\hat{\mathbf{R}}_m$, with $0 \le P_N(\mathbf{R}_m) \le 1$, so that the minimum $P_N(\mathbf{R}_m) = 0$ corresponds to the limiting case of regular discriminating states, while nonregular discriminating states are characterized by $P_N(\mathbf{R}_m) > 0$ with the maximum $P_N(\mathbf{R}_m) = 1$ corresponding to *perfect nonregular states*.

By considering the weight $(P_2 - P_1)$ of the discriminating state $\mathbf{R}_m$ in the characteristic decomposition of $\mathbf{R}$, the degree of nonregularity of $\mathbf{R}$, is defined as $P_N(\mathbf{R}) = 4(P_2 - P_1)\hat{m}_3$.

Note that perfect nonregularity implies $P_2 = 1$, $P_1 = 0$ together with $\hat{m}_3 = 1/4$ (i.e., $\chi_3 = \pm\pi/4$), so that a perfect nonregular state can always be interpreted (among others possible representations) as an equiprobable incoherent mixture of a linearly polarized state $\mathbf{v}_1 \equiv \mathbf{v}_{l-x} = e^{i\delta_1}\sqrt{I}(1,0,0)^T$ and a circularly polarized state $\mathbf{v}_{c-x} = e^{i\delta_2}\sqrt{I/2}(0,1,i)^T$ whose polarization circle lies in a plane Y_mZ_m orthogonal to the axis X_m along which the electric field of $\mathbf{v}_1$ fluctuates (Figure 2.11b).

The very notion of nonregularity implies that the degree of nonregularity $P_{Nm} \equiv P_N(\mathbf{R}_m)$ of a discriminating state $\mathbf{R}_m$ it is closely related to the polarimetric dimension $D_I(\mathbf{R}_m)$. In fact (Gil et al. 2018b)

$$D_I(\mathbf{R}_m) = 3 - \sqrt{1 - (3/4)P_{Nm}\left[2 - P_{Nm}\right]} \tag{2.113}$$

so that $2 \le D_I(\mathbf{R}_m) \le 5/2$, with $D_I(\mathbf{R}_m) = 2$ for regular discriminating states and $D_I(\mathbf{R}_m) = 5/2$ for perfect nonregular states.

The CP of a discriminating state can be expressed as follows in terms of P_{Nm}:

$$P_l(\mathbf{R}_m) = \frac{1}{4}P_{Nm} \qquad P_c(\mathbf{R}_m) = \frac{1}{2}\sqrt{P_{Nm}(2 - P_{Nm})} \qquad P_d(\mathbf{R}_m) = 1 - \frac{3}{4}P_{Nm} \tag{2.114}$$

and, accordingly, are restricted as $0 \le P_l(\mathbf{R}_m) \le 1/4$, $0 \le P_c(\mathbf{R}_m) \le 1/2$, and $1/4 \le P_d(\mathbf{R}_m) \le 1$.

Consequently, the degree of nonregularity of a full state $\mathbf{R}$ that contains $\mathbf{R}_m$ as its discriminating component can be expressed as follows in terms of the CP of $\mathbf{R}_m$:

$$P_N = 4(P_2 - P_1)P_l(\mathbf{R}_m) = (P_2 - P_1)\left[1 - \sqrt{4P_c^2(\mathbf{R}_m)}\right] = (4/3)\left[1 - P_d(\mathbf{R}_m)\right] \tag{2.115}$$

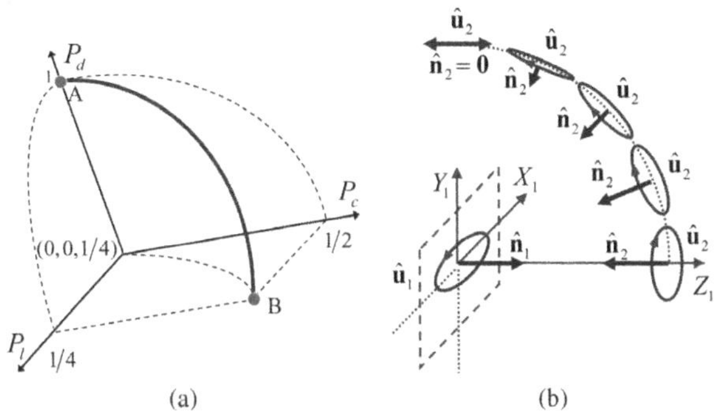

FIGURE 2.12
(a) Feasible region for the components of purity of a discriminating state $\mathbf{R}_m$, given by a curve inscribed in an elliptical cylinder whose basis has semiaxes $1/4$ and $1/2$. Perfect nonregular states correspond to point B $(1/4, 1/2, 1/4)$, and, as P_d increases, the degree on nonregularity decreases up to point A $(0,0,1)$, which corresponds to a regular discriminating state $\mathbf{R}_m = \mathbf{R}_{u\text{-}2D}$. (b) Representation of a family of pairs of orthogonal eigenstates $(\hat{\mathbf{u}}_1, \hat{\mathbf{u}}_2)$ of a polarization state, with respective spin density vectors $\hat{\mathbf{n}}_1$ and $\hat{\mathbf{n}}_2$. All eigenstates are realized in a common point in space but have been separated for the sake of clarity. The absolute value of $\hat{\mathbf{n}}_2$ decreases from $|\hat{\mathbf{n}}_2| = |\hat{\mathbf{n}}_1|$ (canonical pair) down to $|\hat{\mathbf{n}}_2| = 0$ as the elevation angle θ_2 increases from $\theta_2 = 0$ up to $\theta_2 = \pi/2$ (Gil et al. 2019b; 2021).

Thus, regular states are characterized by $P_l(\mathbf{R}_m) = P_c(\mathbf{R}_m) = 0$ and $P_d(\mathbf{R}_m) = 1$, while perfect nonregular states correspond to the values $P_1(\mathbf{R}) = 0$ and $P_2(\mathbf{R}) = 1$, together with $P_l(\mathbf{R}_m) = P_d(\mathbf{R}_m) = 1/4$ and $P_c(\mathbf{R}_m) = 1/2$.

The feasible region for the CP of a discriminating state is represented in Figure 2.12a and is determined by the curve AB lying in the surface of an elliptical cylinder whose basis has semiaxes $1/4$, along the positive branch of axis $P_l(\mathbf{R}_m)$, and $1/2$, along the positive branch of axis $P_c(\mathbf{R}_m)$.

As expected, $P_d(\mathbf{R}_m) = 1$ (point A) is characteristic of regular states ($\mathbf{R}_m = \mathbf{R}_{u\text{-}2D}$). As $P_d(\mathbf{R}_m)$ decreases, the degree of nonregularity of $\mathbf{R}_m$ increases up to the limiting situation of a perfect nonregular state represented by point B, where $P_d(\mathbf{R}_m)$ takes its minimal achievable value $P_d(\mathbf{R}_m) = 1/4$, while the degrees of linear and circular polarization reach their maximal values $P_l(\mathbf{R}_m) = 1/4$ and $P_c(\mathbf{R}_m) = 1/2$.

2.12.3 Dependence of Spin Vector on Nonregularity

Regular states have been defined as those for which either (a) $P_2 = P_1$, so that the characteristic decomposition of $\mathbf{R}$ has the form $\mathbf{R} = P_1\mathbf{R}_p + (1 - P_1)\mathbf{R}_{u\text{-}3D}$, where the discriminating component does not take place, or (b) the discriminating component $\mathbf{R}_m$ of $\mathbf{R}$ corresponds to a 2D-unpolarized state ($\mathbf{R}_m = \mathbf{R}_{u\text{-}2D} = (1/2)\,\mathrm{diag}(1,1,0)$). By comparing the respective expressions $\mathbf{R}_p = \hat{\mathbf{u}}_1 \otimes \hat{\mathbf{u}}_1^\dagger$ and $\mathbf{R}_{u\text{-}2D} = (1/2)(\hat{\mathbf{u}}_1 \otimes \hat{\mathbf{u}}_1^\dagger + \hat{\mathbf{u}}_2 \otimes \hat{\mathbf{u}}_2^\dagger)$ of the pure and discriminating components of a regular state in terms of the eigenstates $\hat{\mathbf{u}}_1$ of $\hat{\mathbf{u}}_2$ of $\mathbf{R}$, it follows that necessarily $\hat{\mathbf{n}}_1 = -\hat{\mathbf{n}}_2$ and, since $\mathbf{R}_{u\text{-}2D}$ and $\mathbf{R}_{u\text{-}3D}$ lack spin, the spin density vector $\hat{\mathbf{n}}$ of a regular state is proportional to the spin density vector $\hat{\mathbf{n}}_p$ of the pure component $\mathbf{R}_p$, $\hat{\mathbf{n}} = P_1\hat{\mathbf{n}}_p$.

Nevertheless, in the general case of nonregular states, the discriminating component $\mathbf{R}_m$ exhibits nonzero amounts of linear polarization and spin (see Section 2.12.2) and the spin density vector $\hat{\mathbf{n}}$ of the state $\mathbf{R}$ is given by the composition $\hat{\mathbf{n}} = P_1\hat{\mathbf{n}}_p + (P_2 - P_1)\hat{\mathbf{n}}_m$ of the spin density vectors $\hat{\mathbf{n}}_p$ and $\hat{\mathbf{n}}_m$ of $\mathbf{R}_p$ and $\mathbf{R}_m$ respectively. Further, by considering the relations $\hat{\mathbf{n}}_p = \hat{\mathbf{n}}_1$ and $\hat{\mathbf{n}}_m = (\hat{\mathbf{n}}_1 + \hat{\mathbf{n}}_2)/2$ together with the possible relative configurations of pairs of orthogonal polarization states studied in Section 2.3, the analysis and interpretation of the possible magnitude and direction of the total spin in terms of those of the pure and discriminating components can be performed.

Figure 2.12b illustrates the relative configurations of pairs of orthonormal eigenstates $\hat{\mathbf{u}}_1$ of $\hat{\mathbf{u}}_2$ and their respective spin density vectors $\hat{\mathbf{n}}_1$ and $\hat{\mathbf{n}}_2$. Since $\hat{\mathbf{n}}_1$ and $\hat{\mathbf{n}}_2$ share a common origin, they determine a well defined plane orthogonal to the polarization planes Π_1 and Π_2 of $\hat{\mathbf{u}}_1$ of $\hat{\mathbf{u}}_2$. Π_1 and Π_2 are in general different, and only coincide in the case of regular states. The magnitude of $\hat{\mathbf{n}}_2$ varies monotonically from $|\hat{\mathbf{n}}_2| = |\hat{\mathbf{n}}_1|$ $(\hat{\mathbf{n}}_2 = -\hat{\mathbf{n}}_1)$ down to $|\hat{\mathbf{n}}_2| = 0$, while the degree of nonregularity P_N increases for increasing values of the angle θ between $\hat{\mathbf{n}}_1$ and $\hat{\mathbf{n}}_2$ ($\theta = \pi - \arccos[\hat{\mathbf{n}}_2^T \hat{\mathbf{n}}_1 / (|\hat{\mathbf{n}}_1||\hat{\mathbf{n}}_2|)]$) (Gil et al. 2021).

2.13 Invariant Quantities of a 3D Polarization State

As seen Section 2.8, the polarization matrix $\mathbf{R}$, referenced with respect to an arbitrary Cartesian reference frame XYZ, can be transformed to the intrinsic polarization matrix $\mathbf{R}_O$ through the orthogonal similarity transformation in Eq. (2.62), determined by the three angles (ϕ, θ, φ), so that $\mathbf{R}_O$ depends on up to six independent parameters that are physically invariant under rotation transformations. Different sets of six independent intrinsic parameters can be considered, as for instance the principal intensities together with the intrinsic components of the spin vector. This section deals with a comparative summary of the physically invariant quantities derivable from a given 3D polarization matrix $\mathbf{R}$.

Let us first recall that $\mathbf{R}$ and $\mathbf{R}_O$ share the same eigenvalue spectrum $(\lambda_1, \lambda_2, \lambda_3)$ and therefore quantities derivable from $\lambda_1, \lambda_2, \lambda_3$ as the intensity I, the indices of polarimetric purity P_1, P_2, and the degree of polarimetric purity $P_{3D} = \sqrt{3P_1^2 + P_2^2}/2$ are physically invariant quantities.

Moreover, the components of purity of $\mathbf{R}$, namely, the degree of linear polarization P_l, the degree of circular polarization P_c and the degree of directionality P_d are independent of the laboratory reference frame considered. The same happens with the intrinsic components and the absolute value of the spin vector $\mathbf{n}_O$. Other additional invariant parameters are the principal intensities a_1, a_2, a_3, the degree of elliptical purity, and a number of other intrinsic quantities that can be derived from the above indicated like, for instance, the semiaxes a, b and the area of the characteristic polarization ellipse. Anyway, as said above, the number of independent invariant quantities is six, which is precisely the number of parameters involved in the intrinsic polarization matrix and in the polarization object (Figure 2.7). A summary of the main invariant quantities of a 3D polarization sate is shown in Table 2.3

TABLE 2.3
Invariant Quantities of a 3D Polarization State

Principal intensities	a_1, a_2, a_3
Eigenvalues of **R** and $\mathbf{R}_O$	$\lambda_1, \lambda_2, \lambda_3$
Intensity	$I = a_1 + a_2 + a_3 = \lambda_1 + \lambda_2 + \lambda_3$
Intrinsic components of the spin vector	$\mathbf{n}_O \equiv (n_{O1}, n_{O2}, n_{O3})^T$
Absolute value of the spin vector	$\lvert\mathbf{n}\rvert = \sqrt{n_{O1}^2 + n_{O2}^2 + n_{O3}^2}$
Orientation angles of $\mathbf{n}_O$ with respect to the intrinsic reference frame	$\tan\phi_n = n_{O2}/n_{O1}$ $\tan\theta_n = \sqrt{n_{O1}^2 + n_{O2}^2}/n_{O3}$
Degree of linear polarization	$P_l = \hat{a}_1 - \hat{a}_2$
Degree of circular polarization	$P_c = \lvert\mathbf{n}\rvert/I$
Degree of elliptical purity	$P_e = \sqrt{P_l^2 + P_c^2}$
Degree of directionality	$P_d = 1 - 3\,\hat{a}_3$
IPP	$P_1 = \lambda_1 - \lambda_2$. $P_2 = 1 - 3\hat{\lambda}_3$
Degree of polarimetric purity	$P_{3D}^2 = \dfrac{3(P_l^2 + P_c^2) + P_d^2}{4}$ $P_{3D}^2 = \dfrac{3P_1^2 + P_2^2}{4}$
Polarimetric dimension	$D_I \equiv 3 - \sqrt{2\left[(\hat{a}_1 - \hat{a}_2)^2 + (\hat{a}_1 - \hat{a}_3)^2 + (\hat{a}_2 - \hat{a}_3)^2\right]}$
Degree of nonregularity	$P_N(\mathbf{R}) = 4(P_2 - P_1)\hat{m}_3$
Semiaxes of the characteristic polarization ellipse	$a = \sqrt{IP_1(\mathbf{R})\left[1 + P_l(\mathbf{R}_p)\right]/2}$ $b = \sqrt{IP_1(\mathbf{R})\left[1 - P_l(\mathbf{R}_p)\right]/2}$
Ellipticity of the characteristic polarization ellipse	$R \equiv b/a = \lvert\tan\chi\rvert = \sqrt{1 - P_l(\mathbf{R}_p)}\Big/\sqrt{1 + P_l(\mathbf{R}_p)}$
Polarization entropy	$S_{3D} = -\mathrm{tr}\left(\hat{\mathbf{R}}\log_3\hat{\mathbf{R}}\right)$ (See Eq. (2.103))
Extremal absolute values for the degrees of mutual coherence with respect to orthogonal transformations (rotations of the reference frame)	$\lvert\mu_{12}\rvert_{\min} = \sqrt{n_{O3}^2/4a_1a_2}$ $\lvert\mu_{13}\rvert_{\min} = \sqrt{n_{O2}^2/4a_1a_3}$ $\lvert\mu_{23}\rvert_{\min} = \sqrt{n_{O1}^2/4a_2a_3}$ $\lvert\mu_{12}\rvert_{\max} = 3\sqrt{\hat{s}_{E01}^2 + \hat{n}_{E3}^2}/2$ $\lvert\mu_{13}\rvert_{\max} = 3\sqrt{\hat{s}_{E02}^2 + \hat{n}_{E2}^2}/2$ $\lvert\mu_{23}\rvert_{\max} = 3\sqrt{\hat{s}_{E12}^2 + \hat{n}_{E1}^2}/2$
Temperature of polarization	$\tau_{3D} \equiv \dfrac{2}{\ln(1 + P_{3D}) - \ln(1 - P_{3D})}$

2.14 Interpretation of the Polarization Matrix

The physical interpretation of 3D polarization states constitutes a primary objective in polarization theory that has attracted the interest of several authors and that can easily be performed by using the results dealt with in the previous sections of this chapter. In fact, the peculiar features of 3D polarization states, whose analysis in terms of the 3D Stokes parameters does not provide an intuitive and simple view of them, become clear

TABLE 2.4
Classification of 2D States $(P_d = 1)$

$P_d = 1$ (2D states) $\Rightarrow P_2 = 1, P_1 = P_e, P_N = 0, P_{3D} \ge 1/2$				
$P_e = 1$ (pure states)			$P_e = 1$ (2D mixed states)	
$\mathbf{R} = \mathbf{R}_p$			$\mathbf{R} = P_1\mathbf{R}_p + (1 - P_1)\mathbf{R}_{u-2D}$	
$P_l = 1$ $\Updownarrow$ $P_c = 0$	$0 < P_l < 1$ $\Updownarrow$ $0 < P_c < 1$	$P_l = 0$ $\Updownarrow$ $P_c = 1$	$0 \le P_e < 1$	$P_1 = 0$ $\mathbf{R} = \mathbf{R}_{u-2D}$
$D_I = 1$	$1 < D_I < 2$	$D_I = 2$	$1 < D_I < 2$	$D_I = 2$
Linear	Elliptical	Circular	Partially polarized	Unpolarized
Arbit. decomp.: Single pure component	Arbit. decomp.: Single pure component	Arbit. decomp.: Single pure component	Arbit. decomp.: Two pure states sharing a common polarization plane Π	Arbit. decomp.: Arbitrary pair of pure orthogonal states with equal intensities and sharing a common polarization plane Π
Indep. parameters: I, ϕ, θ, φ $(\chi = 0)$	Indep. parameters: $I, P_c, \phi, \theta, \varphi$ $(0 < \lvert\chi\rvert < \pi/4)$	Indep. parameters: I, ϕ, θ $\lvert\chi\rvert = \pi/4$ $\sigma = \chi/\lvert\chi\rvert = \pm 1$	Indep. parameters: $I, P_c, P_l, \phi, \theta, \varphi$ $(0 < \lvert\chi\rvert < \pi/4)$	Indep. parameters: I, ϕ, θ
Principal variances: $0 = \hat{a}_3 = \hat{a}_2,\ \hat{a}_1 = 1$	Principal variances: $0 = \hat{a}_3,\ 4\hat{a}_1\hat{a}_2 < P_c^2$	Principal variances: $0 = \hat{a}_3, \hat{a}_2 = \hat{a}_1 = 1/2$	Principal variances: $0 = \hat{a}_3 < \hat{a}_2 < \hat{a}_1$	Principal variances: $0 = \hat{a}_3, \hat{a}_2 = \hat{a}_1 = 1/2$
Spin density vector: $\hat{\mathbf{n}}_O = \mathbf{0}\ (P_c = 0)$	Spin density vector: $\mathbf{0} \ne \hat{\mathbf{n}}_O,\ \hat{\mathbf{n}}_O \perp \Pi$	Spin density vector: $P_c = 1,\ \hat{\mathbf{n}}_O \perp \Pi$	Spin density vector: $P_c < 1,\ \hat{\mathbf{n}}_O \perp \Pi$	Spin density vector: $P_c = 0,\ \hat{\mathbf{n}}_O \perp \Pi$
Polarization object: Figure 2.13.a	Polarization object: Figure 2.13.b	Polarization object: Figure 2.13.c	Polarization object: Figure 2.14 (left)	Polarization object: Figure 2.14.(right)
Characteristic decomposition: $\mathbf{R} = \mathbf{R}_p$			Characteristic decomp.: $\mathbf{R} = P_1\mathbf{R}_p +$ $(1 - P_1)\mathbf{R}_{u-2D}$ $(P_1 > 0)$ Figure 2.15 (left)	Characteristic decomp.: $\mathbf{R} = \mathbf{R}_{u-2D}$ $(P_1 = 0)$ Figure 2.15 (right)

Source: Gil 2021.

TABLE 2.5
Classification of Genuine 3D States $(P_d < 1)$

$P_d < 1$(genuine 3D states)	
Regular 3D mixed state	Nonregular mixed state
$P_N = 0$ $\Updownarrow$ $P_1 = P_e \Leftrightarrow P_2 = P_d$	$0 < P_N \le 1$ $\Updownarrow$ $P_1 < P_e \Leftrightarrow P_2 > P_d$
$2 < D_I \le 3$	$2 < D_I \le 3$
Arbitrary decomposition: Three pure states	Arbitrary decomposition: $(P_d < 1)$: three pure states $P_2 = 1$: two pure states with different polarization planes

TABLE 2.5 (Continued)
Classification of Genuine 3D States ($P_d = 1,\ P_1 < 1$)

Independent parameters: $I, P_l, P_d, P_c, \phi, \theta, \varphi$ $0 < \hat{a}_3 \le \hat{a}_2 \le \hat{a}_1$ $0 \le P_c \quad \left(\hat{\mathbf{n}} \parallel Z_O\right)$ Polarization density object: 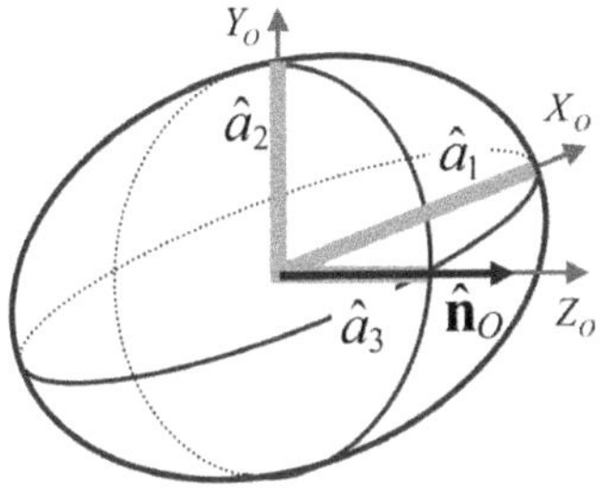 Characteristic and smart decompositions: Figure 2.16	Independent parameters: $I, P_l, P_d, P_c, \phi_n, \theta_n, \phi, \theta, \varphi$ $0 < \hat{a}_3 \le \hat{a}_2 \le \hat{a}_1$ $0 < P_c,\ 0 < P_c(\mathbf{R}_m) \le 1/2,\ \left\lvert \hat{\mathbf{n}}_{O3} \right\rvert \ne P_c$ Polarization density object: 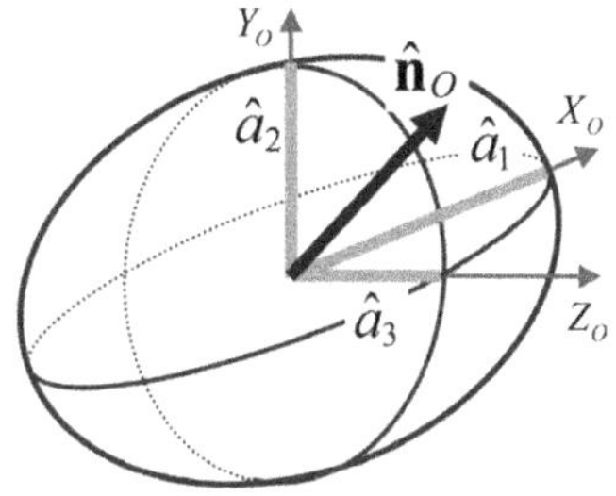 Characteristic and smart decompositions: Figure 2.17

TABLE 2.6
Classification of Discriminating States ($P_2 = 1,\ P_1 = 0$)

$P_2 = 1, P_1 = 0$ (discriminating states, $\mathbf{R} = \mathbf{R}_m$)

$P_N = 0$ ⇕ $P_d = 1$	$0 < P_N < 1$ ⇕ $P_1 < P_e \Leftrightarrow P_2 > P_d$ $(P_c < 1/2, P_l < 1/4, P_d < 1/4)$	$P_N = 1$ ⇕ $P_1 = 0, P_2 = 1$ $P_l = P_d = 1/4, P_c = 1/2$
2D unpolarized state Independent parameters: I, ϕ, θ	Nonregular discriminating state Independent parameters: $I, P_c, \phi, \theta, \varphi$	Perfect nonregular state Independent parameters: I, ϕ, θ
$\mathbf{R} = \mathbf{R}_{u\text{-}2D}$ $= \frac{I}{2}\begin{pmatrix} 1 & 0 & 0 \\ 0 & 1 & 0 \\ 0 & 0 & 0 \end{pmatrix}$	$\mathbf{R} = \frac{I}{2}\mathbf{Q}^T \begin{pmatrix} 1 & 0 & 0 \\ 0 & c_{\chi_3}^2 & -is_{\chi_3}c_{\chi_3} \\ 0 & is_{\chi_3}c_{\chi_3} & s_{\chi_3}^2 \end{pmatrix} \mathbf{Q}$ $[\mathbf{Q} = \mathbf{Q}(\phi, \theta, \varphi)]$	$\mathbf{R} = \frac{I}{4}\mathbf{Q}^T \begin{pmatrix} 2 & 0 & 0 \\ 0 & 1 & -i \\ 0 & i & 1 \end{pmatrix} \mathbf{Q}$ $[\mathbf{Q} = \mathbf{Q}(\phi, \theta)]$
$\hat{a}_3 = 0,\ \hat{a}_2 = \hat{a}_1 = 1/2$ Arbitrary/smart decomposition: Arbitrary pair of pure orthogonal states sharing a common polarization plane Π	$0 < \hat{a}_3 < \hat{a}_2 < \hat{a}_1 = 1/2,\ \hat{a}_2 + \hat{a}_3 = 1/2$ Arbitrary/smart decomposition: Pairs of pure orthogonal states with different polarization planes, so that they satisfy $(\hat{\mathbf{n}}_1 + \hat{\mathbf{n}}_2)/2 = \hat{\mathbf{n}} \quad \left[\lvert\hat{\mathbf{n}}\rvert < 1/2\right]$	$\hat{a}_3 = \hat{a}_2 = 1/4,\ \hat{a}_1 = 1/2$ Equiprobable mixture of a linearly polarized state $\hat{\mathbf{u}}_1$ and a circularly polarized state $\hat{\mathbf{u}}_2$ whose polarization planes are orthogonal $\hat{\mathbf{n}}_2/2 = \hat{\mathbf{n}} \quad \left[\lvert\hat{\mathbf{n}}\rvert = \lvert\hat{\mathbf{n}}_2\rvert/2 = 1/2\right]$
Independent parameters: I, ϕ, θ	Independent parameters: $I, P_c, \phi, \theta, \varphi$	Independent parameters: I, ϕ, θ

(continued)

TABLE 2.6 (Continued)
Classification of Discriminating States R_{p1}

$\hat{\mathbf{n}} = 0$ ($P_c = 0$)	$0 < P_c < 1/2$	$P_c = 1/2$ ($\hat{\mathbf{n}}_O = \Vert X_O$)
Polarization density object:	*Polarization density object:*	Polarization density object:
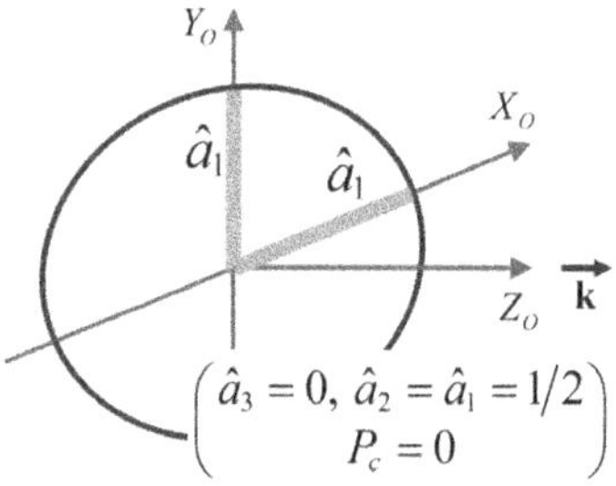	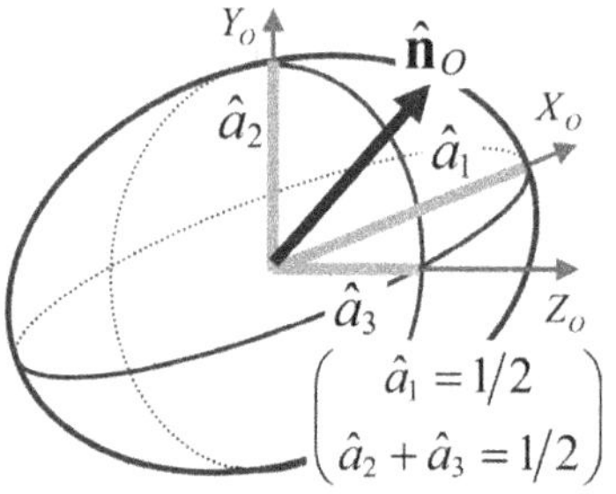	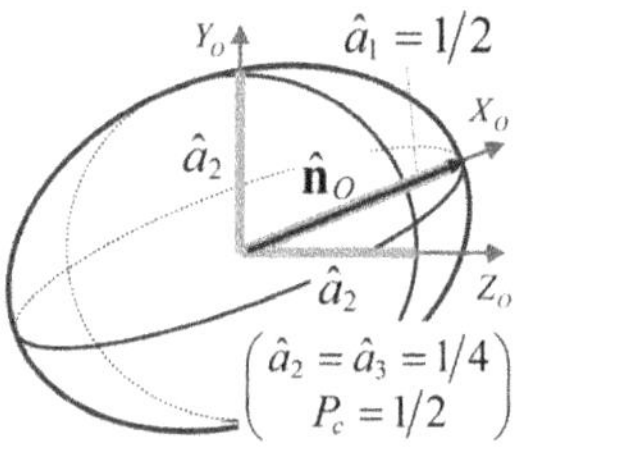
Canonical decomposition:	Canonical decomposition:	Canonical decomposition:
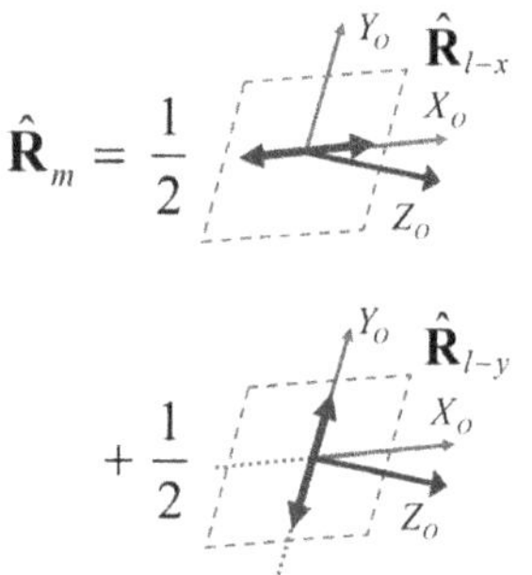	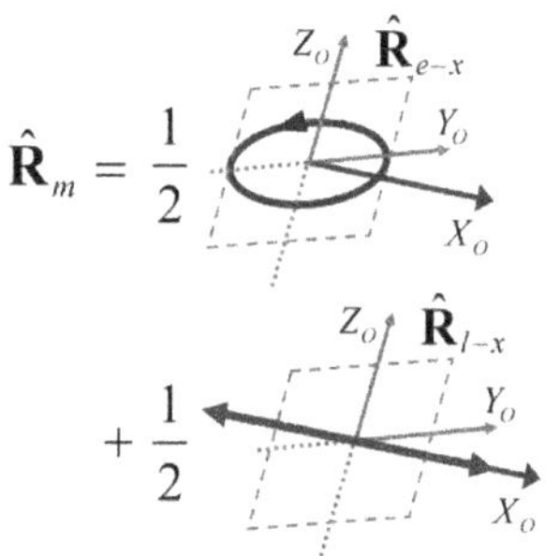	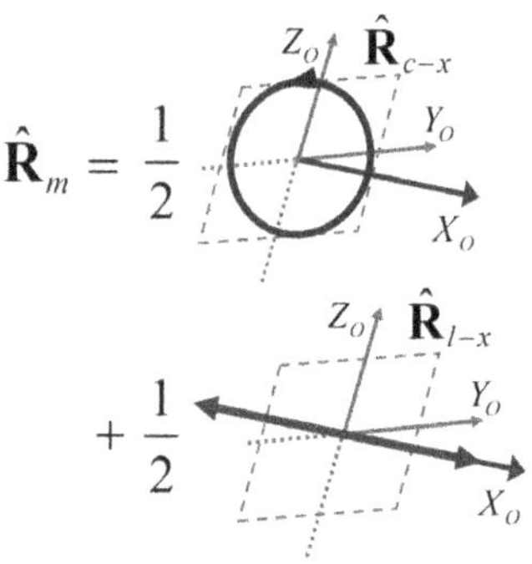

when the polarization states are described through the arbitrary, characteristic and smart decompositions, and interpreted with the help of the polarization object and descriptors like the IPP (P_1, P_2), the CP (P_l, P_c, P_d), the degree of elliptical purity (P_e), the degree of polarimetric purity (P_{3D}), the degree of nonregularity (P_N), and the polarimetric dimension (D_I). In this section, among other contents, we recover and expand the results presented in (Gil 2014a). To make easier the case analyses and discussions, they are performed with the help of Tables 2.4, 2.5 and 2.6.

Despite the obvious fact that any physical polarization state is realized in the 3D real space and therefore all polarization states are 3D states, for the sake of clarity we distinguish between *2D states*, characterized by $P_d = 1$ (i.e., the electric field evolves in a fixed plane) and *genuine 3D states*, characterized by $P_d < 1$ (i.e., the electric field vector has three nonzero components regardless of the reference frame considered).

Even though 2D polarization states have already been studied in detail in Chapter 1, they are revisited and classified in Table 2.4 in light of the polarization descriptors introduced in the previous sections. Since, in general, $P_d \le P_2 \le 1$, the equality $P_d = 1$ implies $P_2 = P_d = 1$ and $P_N = 1$ (2D states are a subcategory of regular states).

Pure states exhibit necessarily a fixed polarization plane, and therefore they constitute a subclass of 2D states. Pure states are characterized by $1 = P_d = P_2 = P_1 = P_e$, or equivalently $P_{3D} = 1$, with an intensity ellipsoid degenerated into an ellipse. Pure states exhibit in general elliptical polarization (Figure 2.13b) and include the limiting cases of linearly polarized states ($P_l = 1$), which correspond to the minimal polarimetric

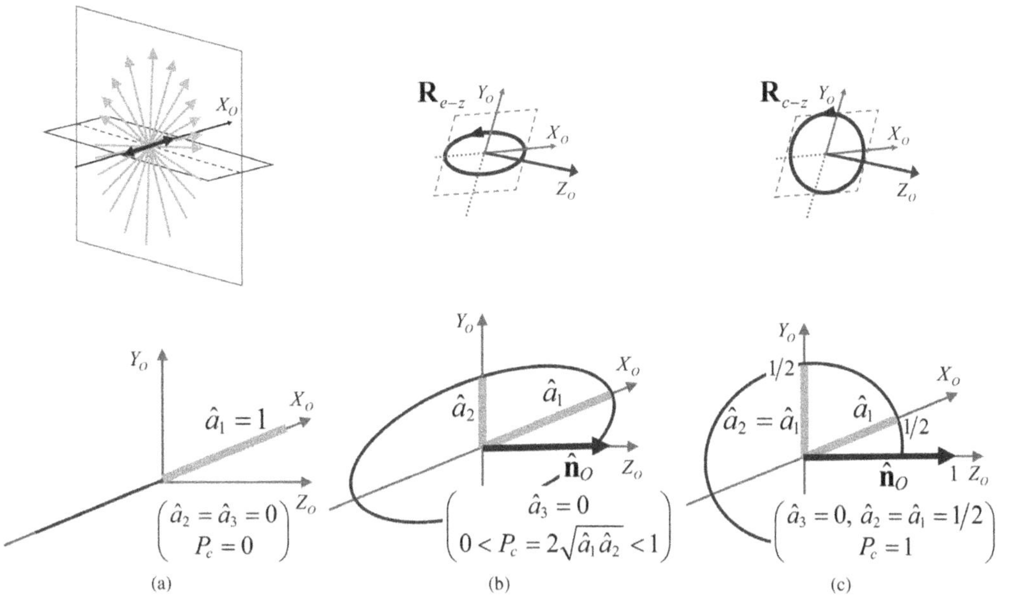

FIGURE 2.13
(a) Representation of a linearly polarized pure state as the incoherent superposition of an arbitrary number of linearly polarized states with the same polarization axis along X_O, but with different directions of propagation perpendicular to X_O; a linearly polarized state has a single nonzero principal variance and lacks spin. (b) An elliptically polarized state is characterized by a well defined polarization ellipse lying in fixed polarization plane; it has two nonzero principal variances and one zero principal variance, while the spin density vector has nonzero magnitude and lies along the direction of propagation Z_O. (c) A circularly polarized state is a pure state whose spin density has its maximum value $P_c = 1$ (Gil 2021).

dimension $D_I = 1$ and to an intensity ellipsoid degenerated into a straight segment (Figure 2.13a) and circularly polarized states ($P_c = 1$), whose intensity ellipsoid is given by a circle and exhibit the maximal achievable absolute value of the spin density vector $|\hat{\mathbf{n}}| = 1$ (Figure 2.13c).

2D mixed states (or 2D partially polarized states, Table 2.4) are characterized by $1 = P_d = P_2$ and $P_1 = P_e < 1$. The characteristic decomposition of a 2D mixed state consists of a combination of a pure state $\mathbf{R}_p$ and a 2D partially polarized state $\mathbf{R}_{u-2D}$, both components sharing a common polarization plane (Figure 2.14a). (Note, in passing, that the composition of a pure state $\mathbf{R}_p$ and a state $\mathbf{R}_{u-2D}$ whose polarization planes do not coincide produces a genuine 3D state). The case of a regular discriminating state ($P_2 = P_d = 1, P_1 = 0$) corresponds precisely to $\mathbf{R}_{u-2D}$ (Figure 2.14b). Observe also that the smart decomposition of a 2D mixed state is given directly by its spectral decomposition.

The types of genuine 3D states ($P_d < 1$)are summarized in Table 2.5, whose columns, from left to right, are devoted to the cases of (a) regular genuine 3D states ($P_d < 1, P_N = 0$), whose discriminating component is a 2D unpolarized state, and (b) nonregular states. The polarization planes of the eigenstates $\hat{\mathbf{u}}_1$ and $\hat{\mathbf{u}}_2$ associated with the respective eigenvalues λ_1 and λ_2 of $\mathbf{R}$ (with the customary convention $\lambda_1 \geq \lambda_2 \geq \lambda_3$) are denoted by Π_1 and Π_2. The achievable values for the polarization descriptors, respective sets of independent parameters, as well as the polarization objects are shown in Table 2.5, while the corresponding characteristic and smart decompositions are shown in Figures 2.15 and 2.16.

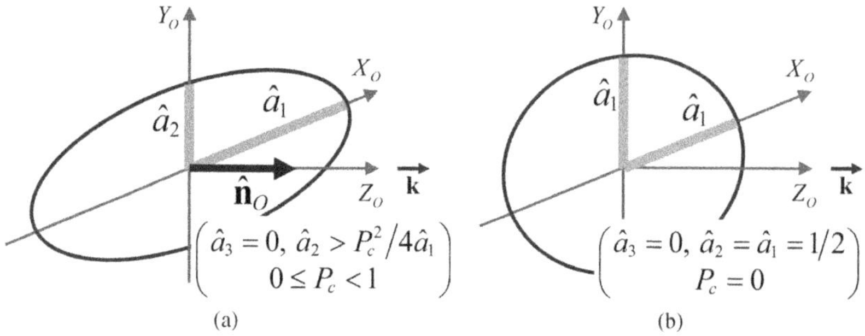

FIGURE 2.14
(a) Polarization object of a 2D mixed state ($P_d = 1, P_e = P_1 < 1$)· (b) When ($P_e = 0$), the state is itself a regular discriminating state $\mathbf{R}_{u-2D} = I\,\hat{\mathbf{R}}_{u-2D}$ (Gil 2021).

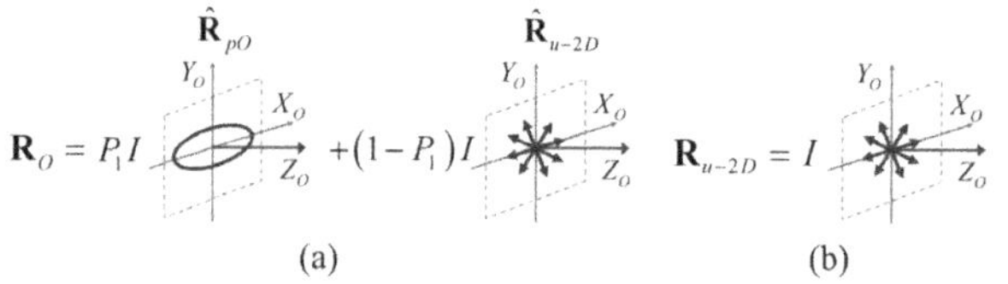

FIGURE 2.15
(a) Intrinsic characteristic representation of a 2D mixed state (i.e., $P_d = 1$, $P_1 < 1$) as the incoherent superposition of a pure state and a 2D unpolarized whose polarization planes coincide. (b) 2D unpolarized state: the electric field fluctuates fully randomly in a fixed plane (Gil 2014a; 2021).

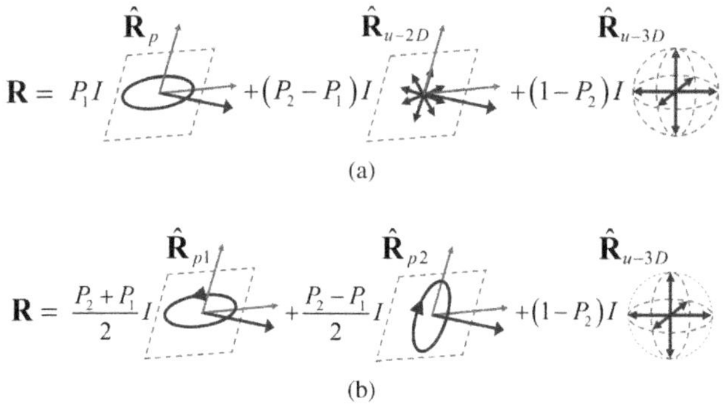

FIGURE 2.16
Regular genuine 3D state. (a) *Characteristic decomposition*: the discriminating component is a 2D unpolarized state R_{u-2d} whose polarization plane coincides with that of the pure component R_p (Gil 2014a). (b) *Smart decomposition*: both pure components share a common polarization plane and correspond to a canonical pair of mutually orthogonal states (see Section 2.3).

Due to the critical role played by the discriminating states for the interpretation of polarization states, Table 2.6 shows the main features of (from left to right) (a) regular discriminating states, (b) partially nonregular discriminating states, and (c) perfect nonregular states. Since any discriminating state has a doubly degenerate nonzero eigenvalue, its smart decomposition can be performed in infinite ways.

$$\mathbf{R} = P_1 I \;\; \hat{\mathbf{R}}_p \;\; + (P_2 - P_1) I \;\; \hat{\mathbf{R}}_m \;\; + (1 - P_2) I \;\; \hat{\mathbf{R}}_{u-3D}$$

(a)

$$\mathbf{R} = \frac{P_2 + P_1}{2} I \;\; \hat{\mathbf{R}}_{p1} \;\; + \frac{P_2 - P_1}{2} I \;\; \hat{\mathbf{R}}_{p2} \;\; + (1 - P_2) I \;\; \hat{\mathbf{R}}_{u-3D}$$

(b)

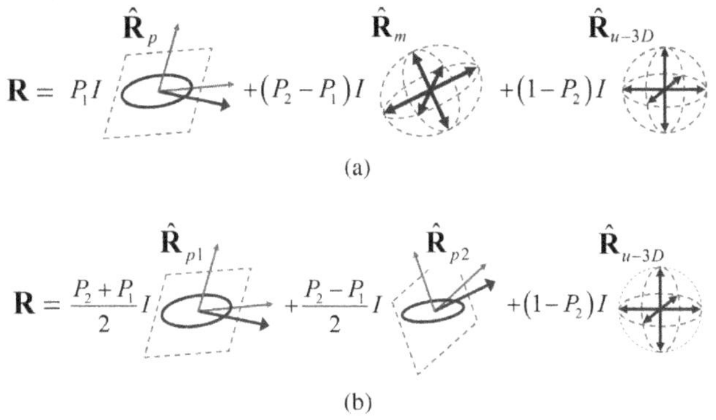

FIGURE 2.17

Nonregular polarization state. (a) *Characteristic decomposition*: the discriminating component $\mathbf{R}_m$ exhibits nonzero spin, while all its three principal intensities are nonzero, $0 < m_3 \leq m_2 \leq m_1$ (Gil 2014a). (b) *Smart decomposition*: the pure components $\mathbf{R}_{p1}$ and $\mathbf{R}_{p2}$ have different polarization planes (see Section 2.3).

Regular discriminating states have the simple form $\mathbf{R}_{u-2D}$ and are built by equiprobable incoherent compositions of arbitrary pairs of mutually orthogonal states with a common polarization plane, as for instance two linearly polarized states whose electric fields vibrate along two orthogonal directions embedded in the polarization plane. Partially nonregular discriminating states are built by equiprobable incoherent compositions of certain pairs of mutually orthogonal states with different polarization planes, including the combination of a linearly polarized state and an elliptically polarized state whose polarization planes are orthogonal. The case of perfect nonregular states corresponds to the limiting case equivalent to an equiprobable incoherent mixture of a linearly polarized state and a circularly polarized state whose polarization planes are orthogonal.

3

Nondepolarizing Media

3.1 Introduction

The interaction of electromagnetic waves with matter is based fundamentally on the atomic or molecular absorption-emission of photons. Leaving aside nonlinear effects, which cover a very interesting variety of physical situations with important applications in science and industry, there exists a wide range of phenomena where the interaction is substantially linear, as is usually the case in reflection, refraction (in general, dispersive) and scattering. The present chapter is devoted to the effects on the polarization of the electromagnetic waves upon their linear interaction with media that exhibit deterministic polarimetric behavior, thus preserving the degree of polarization of totally polarized incident fields.

Consider the linear interaction of a totally polarized electromagnetic wave with a medium (material or structure) interposed in the pathway of the wave. In general, different portions of the medium interact with the electromagnetic radiation in different ways, so that the emerging (transmitted or reflected) beam is a coherent, partially coherent, or an incoherent superposition of the waves resulting from the particular interactions with the various portions. For a sufficiently small portion of the medium, the emerging pencil can be considered totally polarized, i.e., having a 2D pure state of polarization (in the sense of the terminology introduced in Chapters 1 and 2). In fact, the emerging beam is a mixture of photons emerging from each elementary molecular interaction that are necessarily totally polarized.

When a totally polarized electromagnetic plane wave interacts with a homogeneous and nondispersive linear medium, provided the interfaces between the surrounding media of propagation and the medium in question are smooth (with respect to the wavelength), the emerging beam remains totally polarized. As a first example, consider the reflection and refraction of an electromagnetic plane wave on a plane interface between two media with different refraction indices n_1 and n_2, at a given incidence angle θ, and recall the Fresnel relations that provide the magnitudes and phases of the parallel and perpendicular components of the field with respect to the incidence plane (Figure 3.1). The totally polarized incident state can be characterized by a 2D Jones vector $\boldsymbol{\varepsilon}$ (Jones 1941) referenced with respect to a coordinate system constituted by the axes Z (along the direction of propagation), Y (parallel to the incidence plane Π_I), and X (orthogonal the incidence plane). The reflected and refracted states are likewise totally polarized and, consequently, can be characterized by respective 2D Jones vectors $\boldsymbol{\varepsilon}_r$ and $\boldsymbol{\varepsilon}_t$, where $\boldsymbol{\varepsilon}_r$ is referenced with respect to axes X (orthogonal to the incidence plane Π_I), Y_r (lying in the incidence plane) and Z_r (along the direction of the reflected beam), while $\boldsymbol{\varepsilon}_t$ is referenced with respect to axes X, Y_t (lying in the incidence plane Π_I) and Z_t (along the direction of the refracted beam)

DOI: 10.1201/9780367815578-3

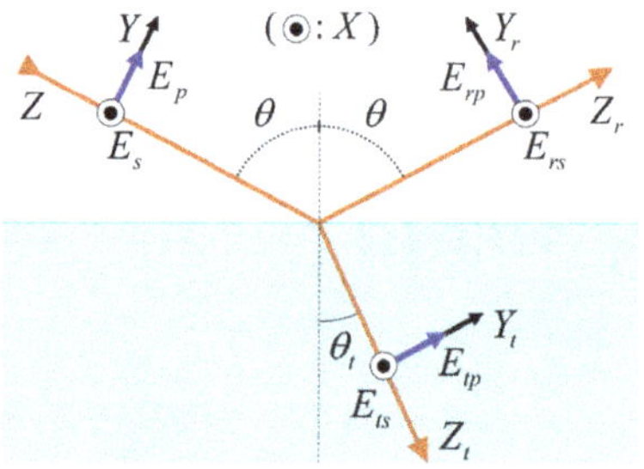

FIGURE 3.1
Reflection and refraction of the parallel (p) and perpendicular (s) components of the electric field of a plane wave at the interface between two media that exhibit different refraction indices n_1 (upper medium) and n_2 (lower medium). The refraction angle is determined from $\sin\theta_t = \sin\theta/n$.

(Figure 3.1). Since the 2D electromagnetic wave is *transverse*, i.e., the vibration of the electric field takes place in the plane normal to the propagation direction (specified by the Z axis), only the components of the electric field parallel and perpendicular to the incidence plane are nonzero. These are commonly referred to as p- and s-components respectively; see Figure 3.1.

The Fresnel relations constitute a linear transformation which, when referenced with respect to the indicated Cartesian coordinate systems, can be expressed as

$$\boldsymbol{\varepsilon}_r = \mathbf{T}_r \boldsymbol{\varepsilon} \qquad \boldsymbol{\varepsilon}_t = \mathbf{T}_t \boldsymbol{\varepsilon} \tag{3.1}$$

where $\mathbf{T}_r$ and $\mathbf{T}_t$ are the 2×2 diagonal real matrices $\mathbf{T}_r \equiv \mathrm{diag}(r_p, r_s)$ and $\mathbf{T}_t \equiv \mathrm{diag}(t_p, t_s)$, which constitute particular examples of *Jones matrices* (Jones 1941). The explicit expressions of transformations (3.1) in terms of the elements of the Jones vectors $\boldsymbol{\varepsilon}_r$, $\boldsymbol{\varepsilon}_t$ and the elements of $\mathbf{T}_r$ and $\mathbf{T}_t$ are the following:

$$\begin{pmatrix} \varepsilon_{rp} \\ \varepsilon_{rs} \end{pmatrix} = \begin{pmatrix} r_p \varepsilon_p \\ r_s \varepsilon_s \end{pmatrix} \qquad \begin{pmatrix} \varepsilon_{tp} \\ \varepsilon_{ts} \end{pmatrix} = \begin{pmatrix} t_p \varepsilon_p \\ t_s \varepsilon_s \end{pmatrix} \tag{3.2}$$

where

$$\begin{aligned} r_p &= \frac{n^2\cos\theta - \sqrt{n^2 - \sin^2\theta}}{n^2\cos\theta + \sqrt{n^2 - \sin^2\theta}} \qquad r_s = \frac{\cos\theta - \sqrt{n^2 - \sin^2\theta}}{\cos\theta + \sqrt{n^2 - \sin^2\theta}} \\ t_p &= \frac{2n\cos\theta}{n^2\cos\theta + \sqrt{n^2 - \sin^2\theta}} \qquad t_s = \frac{2\cos\theta}{\cos\theta + \sqrt{n^2 - \sin^2\theta}} \end{aligned} \qquad (n \equiv n_2/n_1) \tag{3.3}$$

are the well-known Fresnel relations. A Mueller matrix representation of the Fresnel relations, equivalent to the above Jones formulation, can be found in (Collett 1971).

Another representative example is the propagation of a totally polarized quasi-monochromatic plane wave through a birefringent plate with linear retardance Δ, assumed nondispersive within the spectrum of the incident wave (Figure 3.2). By taking a laboratory coordinate system whose axes X and Y are aligned with the respective fast and slow birefringence axes of the waveplate, the Z axis lying along the direction of propagation of the wave, the Jones vector $\boldsymbol{\varepsilon}$ of the incoming state is transformed as

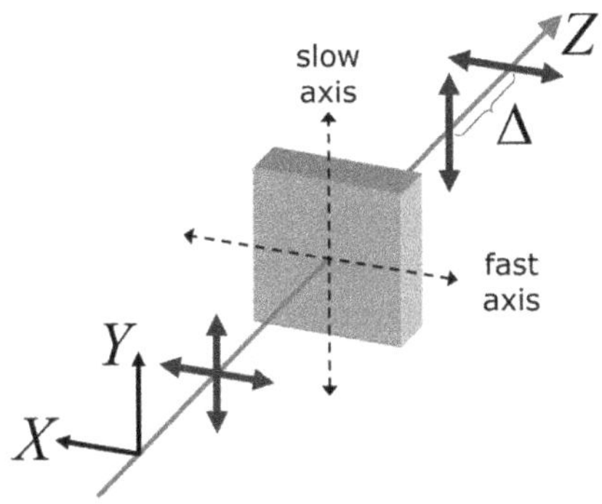

FIGURE 3.2
Transmission of a plane wave through a linear retarder.

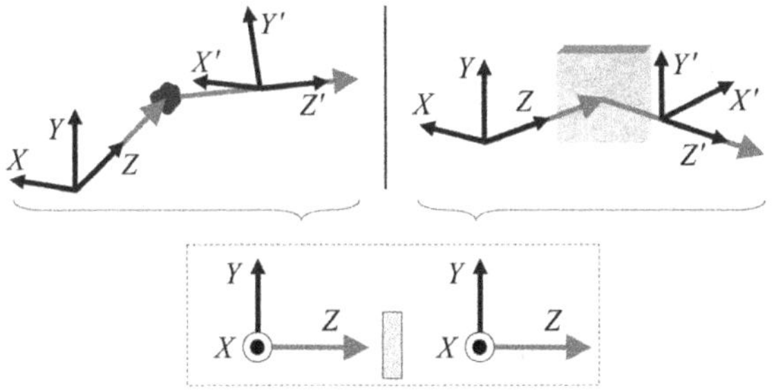

FIGURE 3.3
Local coordinate systems for incident and emerging polarization states. Examples for scattering and reflection.

$$\boldsymbol{\varepsilon}' = \mathbf{T}_{RL0}\,\boldsymbol{\varepsilon} \tag{3.4}$$

where the Jones matrix $\mathbf{T}_{RL0}$ is a diagonal unitary matrix of the form

$$\mathbf{T}_{RL0} \equiv \begin{pmatrix} e^{i\Delta/2} & 0 \\ 0 & e^{-i\Delta/2} \end{pmatrix} \tag{3.5}$$

Most generally, linear interactions where both the incident, $\boldsymbol{\varepsilon}$, and emerging, $\boldsymbol{\varepsilon}'$, states of the electromagnetic beams are *totally polarized* can be represented by a transformation $\boldsymbol{\varepsilon}' = \mathbf{T}\boldsymbol{\varepsilon}$ where the 2×2 complex matrix $\mathbf{T}$ is the so-called Jones matrix (Jones 1941). In that case, the medium is said to be *nondepolarizing* (or *pure*). In this chapter, we shall encounter several interesting examples of Jones matrices which illustrate a variety of possible physical situations.

Usually, as in the above examples, the Jones formalism disregards the change of direction of propagation produced by the interaction, provided both the incident and emerging polarization states have fixed propagation directions (or, in the terminology introduced in Chapter 2, both incident and emerging states are 2D states, but not necessarily with the same direction of propagation). Thus, in general, different local coordinate systems including the respective directions of propagation as the reference axes Z and Z' (Figure 3.3) are used for incident and emerging states. That is, from a polarimetric point of view, the nondepolarizing linear interaction given by a Jones matrix can be represented as in Figure 3.3 where the incident and emerging directions of propagation have been disposed along a common Z axis.

A material medium behaves, in general, in different manners with respect to different kinds of interaction. For example, it can be linear and nondepolarizing with respect to reflection, as well as with respect to incident waves satisfying particular spectral or coherence features. For each angle of incidence and for each kind of incident wave, the medium is represented by a specific Jones matrix. In fact, the Jones matrix corresponding to reflection on a dielectric (or metallic) medium can be expressed as function of the incidence angle. Moreover, Jones matrices can be formulated as functions of the central frequency or of other properties of the interacting wave.

Therefore, in general, a Jones matrix depends on a number of particular circumstances such as the medium considered, the central frequency of the incident electromagnetic quasi-monochromatic wave, the relative orientation of the direction of propagation of the incident wave with respect to the medium, the relative orientation of the emerging direction of observation (with respect to the medium), the location of the reference coordinate planes relative to which the polarization ellipses of the incident and emerging polarization states are considered, etc. (Azzam and Bashara 1987). Thus, provided the interactions are linear and nondepolarizing, a material medium can be represented by different Jones matrices depending on the interaction conditions and on the particular properties of the incident waves.

Due to the essentially nondepolarizing behavior of the elementary interactions whose coherent or incoherent combination provides the response of a material medium with respect to a electromagnetic beam, any mathematical formulation of the linear transformations of the states of polarization must be consistent with the fact that these transformations can be considered as statistical mixtures of nondepolarizing interactions (Parke 1948). This is the reason why any appropriate model for the representation of the polarization transformations produced by material media on electromagnetic waves must be constructed from the basic formulation of nondepolarizing transformations which, in turn, can be represented without loss of generality by the Jones calculus (i.e., Jones vectors for polarization states and Jones matrices for material media).

In summary, in the case of nondepolarizing (or *pure*) linear material systems, pure states are transformed into pure states through basic nondepolarizing interactions. As we shall see, these basic interactions involve the birefringence and dichroic properties of the system. In the most general case of *depolarizing* media, the *incoherent* superposition of the emerging light pencils results in depolarization phenomena.

3.2 Basic Polarimetric Interaction: Jones Calculus

3.2.1 The Jones Matrix

The polarimetric transformations of pure states into pure states by the interaction with a nondepolarizing medium (or *pure* medium, from a polarimetric point of view) can be represented mathematically in terms of the Jones formalism as (Jones 1941, Swindell 1975)

$$\boldsymbol{\varepsilon}' = \mathbf{T}\,\boldsymbol{\varepsilon} \tag{3.6}$$

where $\mathbf{T}$ is the Jones matrix of the medium (specific for the given interaction), while $\boldsymbol{\varepsilon}'$ and $\boldsymbol{\varepsilon}$ are the emerging and incident Jones vectors. That is to say, any nondepolarizing system can be represented by its corresponding Jones matrix and vice-versa.

Despite the fact that a given medium behaves as nondepolarizing, in general both the incident and emerging electromagnetic beams can be partially polarized. Such states cannot be described by Jones vectors, but can be rather represented by 2D polarization matrices $\mathbf{\Phi}$, whose transformation by the medium, represented by $\mathbf{T}$, is given by

$$\mathbf{\Phi}' = \langle \boldsymbol{\varepsilon}' \otimes \boldsymbol{\varepsilon}'^{\dagger} \rangle = \langle \mathbf{T}\boldsymbol{\varepsilon} \otimes (\mathbf{T}\boldsymbol{\varepsilon})^{\dagger} \rangle = \langle \mathbf{T}\boldsymbol{\varepsilon} \otimes \boldsymbol{\varepsilon}^{\dagger}\mathbf{T}^{\dagger} \rangle = \mathbf{T}\langle \boldsymbol{\varepsilon} \otimes \boldsymbol{\varepsilon}^{\dagger} \rangle \mathbf{T}^{\dagger} = \mathbf{T}\mathbf{\Phi}\mathbf{T}^{\dagger} \tag{3.7}$$

where, as in the previous chapters, $\langle\ \rangle$ denotes time average over the measurement time (or alternatively, ensemble average, when ergodicity of the analytic signals ε_1 and ε_2 constituting the components of the Jones vector $\boldsymbol{\varepsilon}$ can be assumed, as is usually the case).

To any Jones matrix can be applied the singular value decomposition

$$\mathbf{T} = \mathbf{T}_{R2}\,\mathrm{diag}(p_1, p_2)\,\mathbf{T}_{R1} \tag{3.8}$$

where $\mathbf{T}_{R1}$ and $\mathbf{T}_{R2}$ are unitary matrices and $\mathbf{T}_{DL0} \equiv \mathrm{diag}(p_1, p_2)$ is a diagonal matrix whose nonzero elements are the real non-negative singular values p_1 and p_2. The *principal amplitude coefficients* p_1, p_2 are the positive square roots of the non-negative eigenvalues of matrix $\mathbf{T}\mathbf{T}^{\dagger}$; $\mathbf{T}_{R2}$ is the unitary matrix that diagonalizes $\mathbf{T}\mathbf{T}^{\dagger}$, $\mathbf{T}\mathbf{T}^{\dagger} = \mathbf{T}_{R2}\,\mathrm{diag}(p_1^2, p_2^2)\,\mathbf{T}_{R2}^{\dagger}$; and $\mathbf{T}_{R1}$ is the matrix that diagonalizes $\mathbf{T}^{\dagger}\mathbf{T}$, $\mathbf{T}^{\dagger}\mathbf{T} = \mathbf{T}_{R1}^{\dagger}\,\mathrm{diag}(p_1^2, p_2^2)\,\mathbf{T}_{R1}$. The origin of the notation $\mathbf{T}_R$ used for unitary Jones matrices is due to the fact that, as we shall see, they represent birefringent media (retarders).

The squares of the singular values of $\mathbf{T}$, which play a central role in Jones algebra, are given by the expressions (Barakat 1987a)

$$p_1^2 = \frac{1}{2}\mathrm{tr}(\mathbf{T}^{\dagger}\mathbf{T})\left\{1 + \sqrt{1 - \frac{4\det(\mathbf{T}^{\dagger}\mathbf{T})}{\mathrm{tr}^2(\mathbf{T}^{\dagger}\mathbf{T})}}\right\} \qquad p_2^2 = \frac{1}{2}\mathrm{tr}(\mathbf{T}^{\dagger}\mathbf{T})\left\{1 - \sqrt{1 - \frac{4\det(\mathbf{T}^{\dagger}\mathbf{T})}{\mathrm{tr}^2(\mathbf{T}^{\dagger}\mathbf{T})}}\right\} \tag{3.9}$$

Moreover, if we consider the isolated effect of the matrix $\mathrm{diag}(p_1, p_2)$, we observe that, as in the example of the reflection considered above, its action on an incident Jones vector $\boldsymbol{\varepsilon}$ is a selective scaling of the amplitudes of the analytic signals representative of the components of the electric field of the wave,

$$\begin{pmatrix} p_1 & 0 \\ 0 & p_2 \end{pmatrix}\begin{pmatrix} \varepsilon_1 \\ \varepsilon_2 \end{pmatrix} = \begin{pmatrix} p_1\,\varepsilon_1 \\ p_2\,\varepsilon_2 \end{pmatrix} \tag{3.10}$$

As a result, the intensity of the pure emerging state is modified as follows with respect to the intensity of the incident pure state:

$$I' = \boldsymbol{\varepsilon}'^{\dagger}\boldsymbol{\varepsilon}' = p_1^2|\varepsilon_1|^2 + p_2^2|\varepsilon_2|^2 \neq \boldsymbol{\varepsilon}^{\dagger}\boldsymbol{\varepsilon} = I \tag{3.11}$$

The quantities p_1^2 and p_2^2 are called the *principal intensity coefficients* of $\mathbf{T}$ (also called *principal intensity transmittances* or *maximum and minimum gain*).

Leaving aside certain artificial experimental arrangements where the medium is a system containing intensity amplifiers (Opatrný and Perina 1993), both natural and manmade objects are passive, i.e., they do not amplify the light intensity, but generally reduce it to some extent. As limiting situations, transparent systems correspond to the ideal case of media that preserve the intensity, while opaque systems produce zero emerging intensity (so that the polarimetric analysis in transmission does not apply). Polarimetric techniques usually deal with the characterization and measurement of the polarimetric properties of a great variety of material targets in science, industry, medicine, remote sensing, etc., where the material samples are inherently passive. Thus, passivity is a physical condition that must be taken into account in the mathematical characterization of the polarimetric properties of material media.

The condition for a Jones matrix to represent a passive (or *passive-realizable*) pure medium arises from the physical restriction that the ratio between the intensities of the emerging and incident beams, must be less than the unit for any incident state of polarization. It is straightforward to show that this condition, called *passivity condition* (also *transmittance condition* or *gain condition*), can be formulated in terms of the largest principal intensity coefficient p_1^2 of $\mathbf{T}$ as follows (Gil 1983, Barakat 1987a):

$$p_1^2 \leq 1 \tag{3.12}$$

Regarding all passive Jones matrices proportional to $\mathbf{T}$, the one with minimal attenuation of intensity is given by the *passive representative* Jones matrix $\tilde{\mathbf{T}} \equiv \mathbf{T}/p_1$.

Hereafter, we will distinguish between passive Jones matrices and wide-sense Jones matrices (passive or not). Unless otherwise indicated, Jones matrices will be assumed passive, i.e., in the strict sense, a 2×2 complex matrix $\mathbf{T}$ is a Jones matrix if it satisfies the passivity condition (3.12).

Furthermore, like Jones vectors, Jones matrices are usually defined within the not measurable arbitrary global phase factor $\exp(i\varphi)$, so that matrices $\mathbf{T}$ and $e^{i\varphi}\mathbf{T}$ are equivalent and indistinguishable through intensity measurements determining the overall transformation $\mathbf{T}$ of the incident Jones vector $\boldsymbol{\varepsilon}$ into the emerging one, $\boldsymbol{\varepsilon}'$. Nevertheless, the phase factor $\exp(i\varphi)$ has to be considered when $\boldsymbol{\varepsilon}'$ is coherently superimposed with other Jones vectors as occurring in interference experiments or, more generally, in multipath propagating coherent states which are recombined prior the measurement. Thus, it is possible to establish a criterion to specify a representative matrix of the equivalence class $e^{i\varphi}\mathbf{T}$ with arbitrary φ. Unless otherwise stated, we shall assume the choice of a representative of $\mathbf{T}$ satisfying $\det\mathbf{T} \in \mathbb{R}$. Whatever the specific choice made, the number of measurable independent parameters of such a representative Jones matrix is seven.

3.2.2 Jones Algebra and Its Physical Interpretation

The set of Jones matrices, together with the internal matrix product, the multiplication by a scalar, and the additive compositions, constitutes the Jones algebra. Next, we summarize the mathematical operations that can be realized on Jones matrices and their physical interpretation.

3.2.2.1 Product of Jones Matrices

Given a pair of Jones matrices $\mathbf{T}_1$ and $\mathbf{T}_2$, the product $\mathbf{T}_2\mathbf{T}_1$ represents the successive serial actions of the media represented by them (Figure 3.4) (Jones 1941):

$$\mathbf{T}_2\mathbf{T}_1\,\boldsymbol{\varepsilon} = \mathbf{T}_2\boldsymbol{\varepsilon}' = \boldsymbol{\varepsilon}'' \tag{3.13}$$

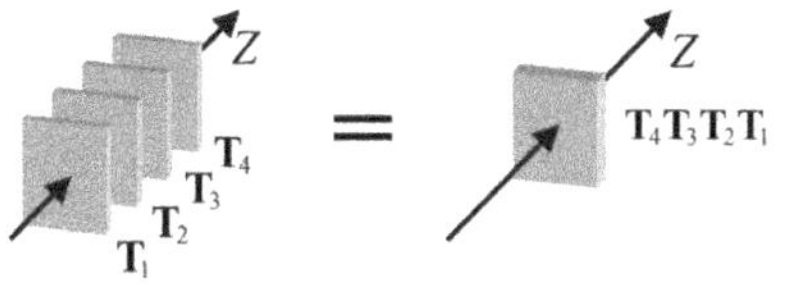

FIGURE 3.4
The polarimetric effect of a cascade of nondepolarizing elements is represented by the ordered product of the corresponding Jones matrices.

The product of Jones matrices is not commutative, $\mathbf{T}_2\mathbf{T}_1 \neq \mathbf{T}_1\mathbf{T}_2$, which reflects the physical fact that the overall polarimetric effect of the serial actions of the media represented by $\mathbf{T}_1$ and $\mathbf{T}_2$ depends on their order in the serial arrangement. As an example, let us consider the serial combination of a horizontal diattenuator and a $\pi/2$ rotator (the main types of Jones matrices will be dealt with in later sections), represented respectively by

$$\mathbf{T}_{DL0} \equiv \begin{pmatrix} p_1 & 0 \\ 0 & p_2 \end{pmatrix} \qquad \mathbf{T}_{R0,\pi/2} \equiv \begin{pmatrix} 0 & 1 \\ -1 & 0 \end{pmatrix} \tag{3.14}$$

so that

$$\mathbf{T}_{DL0}\,\mathbf{T}_{R0,\pi/2} = \begin{pmatrix} 0 & p_1 \\ -p_2 & 0 \end{pmatrix} \qquad \mathbf{T}_{R0,\pi/2}\,\mathbf{T}_{DL0} = \begin{pmatrix} 0 & p_2 \\ -p_1 & 0 \end{pmatrix} \tag{3.15}$$

By recalling the standard properties of matrices, let us note that two Jones matrices commute if and only if they can be diagonalized simultaneously. A comprehensive summary of properties of Jones matrices can be found in the original set of papers of Jones (Swindell 1975).

3.2.2.2 Product of a Jones Matrix and a Scalar

In accordance with the singular value decomposition in Eq. (3.8), given a Jones matrix $\mathbf{T}$ and an arbitrary complex number $c \equiv |c|e^{i\varphi}$, the product $c\mathbf{T}$ can be expressed as

$$c\mathbf{T} = e^{i\varphi}\begin{pmatrix} |c| & 0 \\ 0 & |c| \end{pmatrix}\mathbf{T}_{R2}\begin{pmatrix} p_1 & 0 \\ 0 & p_2 \end{pmatrix}\mathbf{T}_{R1} = e^{i\varphi}\,\mathbf{T}_{R2}\begin{pmatrix} |c|\,p_1 & 0 \\ 0 & |c|\,p_2 \end{pmatrix}\mathbf{T}_{R1} \tag{3.16}$$

so that, disregarding the overall phase factor $\exp(i\varphi)$, the multiplication is equivalent to the effect of an isotropic attenuator represented by $\mathbf{T}_c \equiv |c|\,\mathrm{diag}(1,1)$. Thus, provided $|c| \leq 1$, $\mathbf{T}_c$ and hence $c\mathbf{T}$, are passive Jones matrices, so that $c\mathbf{T}$ differs from $\mathbf{T}$ only by the overall attenuation and phase factors. If $|c| > 1$, then $\mathbf{T}_c$ is not passive-realizable and, regardless of whether $c\mathbf{T}$ is passive or not, $c\mathbf{T}$ cannot be interpreted as the serial combination $\mathbf{T}_c\mathbf{T}$.

3.2.2.3 Determinant and Norms of a Jones Matrix

The singular value decomposition of $\mathbf{T}$ in Eq. (3.8) leads to the well-known result in matrix algebra that the absolute value of the determinant of a given Jones matrix $\mathbf{T}$ is given by the product of its singular values (i.e., the square root of the product of the two principal intensity coefficients)

$$|\det \mathbf{T}| = p_1\, p_2 \tag{3.17}$$

The *Frobenius norm* of **T** is defined as

$$\|\mathbf{T}\|_2 = \sqrt{\operatorname{tr}\left(\mathbf{T}^{+}\mathbf{T}\right)} = \sqrt{\sum_{i,j=1}^{2}\left|t_{ij}\right|^2} = \sqrt{p_1^2 + p_2^2} = \sqrt{2\, m_{00}} \tag{3.18}$$

where m_{00} stands for the *mean intensity coefficient* of **T**, and therefore, $\|\mathbf{T}\|_2^2$ is the sum of the principal intensity coefficients of **T**, which in turn is equal to twice m_{00}. The notation m_{00} and its meaning as the intensity coefficient for incident unpolarized light will be justified later within the framework of the Mueller matrix formalism.

The *nuclear norm* of **T** is defined as the sum of its singular values

$$\|\mathbf{T}\|_{*} = \operatorname{tr}\sqrt{\mathbf{T}^{+}\mathbf{T}} = p_1 + p_2 \tag{3.19}$$

so that the following relation holds between the determinant and norms of **T**:

$$\|\mathbf{T}\|_{*}^2 = \|\mathbf{T}\|_2^2 + 2|\det \mathbf{T}| \tag{3.20}$$

3.2.2.4 The Inverse of a Jones Matrix

From the singular value decomposition $\mathbf{T} = \mathbf{T}_{R2}\, \operatorname{diag}(p_1, p_2)\, \mathbf{T}_{R1}$ of a nonsingular passive Jones matrix **T** ($0 < p_2 \le p_1 < 1$) we get immediately that the inverse matrix of **T** is given by

$$\mathbf{T}^{-1} = \mathbf{T}_{R1}^{+}\, \operatorname{diag}\left(1/p_1, 1/p_2\right) \mathbf{T}_{R2}^{+} \tag{3.21}$$

which is not a passive Jones matrix since producing certain amplification for some or all incident states. Nevertheless, given a real positive number c such that $c \le p_1$, $c\,\mathbf{T}^{-1}$ is a passive Jones matrix.

In the case of **T** being singular ($0 = p_2 < p_1 \le 1$) a pseudoinverse matrix can be defined as

$$\mathbf{T}^{-} \equiv \mathbf{T}_{R1}^{+}\, \operatorname{diag}\left(1/p_1, 0\right) \mathbf{T}_{R2}^{+} \tag{3.22}$$

so that

$$\mathbf{T}^{-}\mathbf{T} = \mathbf{T}\mathbf{T}^{-} = \operatorname{diag}(1,0) \tag{3.23}$$

As for the inverse of a nonsingular Jones matrix, the pseudoinverse of a singular Jones matrix is not passive, but passivity can be achieved through the multiplication of $\mathbf{T}^{-}$ by a real positive number c such that $c \le p_1$.

3.2.2.5 Additive Composition of Jones Matrices

From a strictly mathematical point of view, leaving aside the passivity criterion, it is straightforward to observe that (1) a given Jones matrix **T** can always be expressed, in an

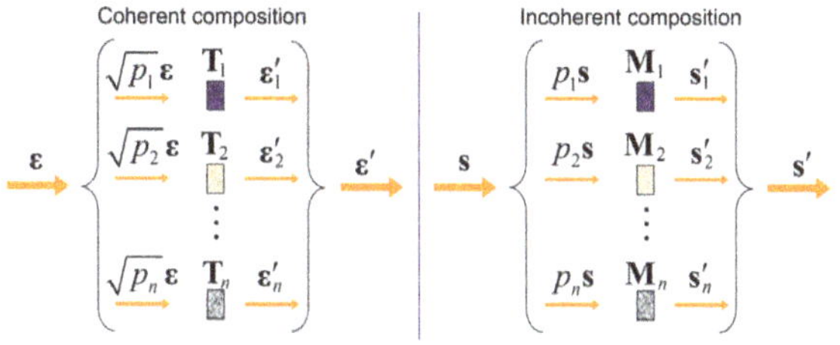

FIGURE 3.5
The coherent composition of polarimetric interactions should be represented through the Jones formalism, while the incoherent composition should be represented through the Stokes–Mueller formalism.

infinite number of ways, as a sum $\mathbf{T} = \sum_{k=1}^{n} \mathbf{T}_k$ of an arbitrary number n of Jones matrices $\mathbf{T}_k$ $(k = 1,\dots,n)$, and (2) any sum of Jones matrices is a Jones matrix. Nevertheless, from a physical point of view, additive compositions (also called *parallel compositions*) of Jones matrices require a careful treatment in order to ensure the physical realizability with respect to passivity. To do so, consider a spatially inhomogeneous material sample composed of a number n of elements (or tiles) characterized by respective Jones matrices $\mathbf{T}_k$ $(k = 1,\dots,n)$ with respective size ratios (relative cross-sections) p_k $\left(\sum_{k=1}^{n} p_k = 1\right)$ (i.e., p_k is the ratio of the area A_k of the kth element to the total area A covered by the light beam). Next, consider a totally polarized spatially homogeneous light beam with intensity I and Jones vector $\boldsymbol{\varepsilon} = \sqrt{I}\,\hat{\boldsymbol{\varepsilon}}$ $\left(\hat{\boldsymbol{\varepsilon}}^{+}\hat{\boldsymbol{\varepsilon}} = 1\right)$ impinging upon the area A of the sample. The beam is shared into n pencils falling on the respective elements (Figure 3.5). Then, provided the emerging beams (with respective Jones vectors $\boldsymbol{\varepsilon}'_k = \mathbf{T}_k \sqrt{p_k}\,\boldsymbol{\varepsilon}$) are coherently superimposed, the Jones vector of the emerging light beam is given by

$$\boldsymbol{\varepsilon}' = \sum_{k=1}^{n} \boldsymbol{\varepsilon}'_k = \sum_{k=1}^{n} \mathbf{T}_k \sqrt{p_k}\,\boldsymbol{\varepsilon} = \left(\sum_{k=1}^{n} \sqrt{p_k}\,\mathbf{T}_k\right)\boldsymbol{\varepsilon} \tag{3.24}$$

and therefore, the composed Jones matrix of the whole sample is

$$\mathbf{T} = \sum_{k=1}^{n} \sqrt{p_k}\,\mathbf{T}_k \quad \left(\sum_{k=1}^{n} p_k = 1\right) \tag{3.25}$$

As indicated above, the rule used in Eqs. (3.24) and (3.25) for the additive composition of Jones matrices is only valid in physical situations where the emerging light pencils are coherently combined and is consistent with the fact that the intensity I of the incident beam is shared into the proportion $I_k = p_k I$ for each distinct element k of the sample. Note that the fulfillment of the passivity criterion by all $\mathbf{T}_k$ implies the passivity of $\mathbf{T}$.

Note also that additive compositions or decompositions of Jones matrices are sensitive to the phase shifts caused by the addends. Therefore, the addends have to be represented by Jones matrices including their own independent phases. In other words, the choice $\det \mathbf{T} \in \mathbb{R}$ has to be applied for the representative of the overall Jones matrix resulting from the sum, but not for the Jones matrices of the addends. When the temporal, spatial or spectral average involves fluctuations of the relative phases of the pure components, an appropriate (but not unique) criterion to represent the phases of such components is to write each Jones matrix as $\mathbf{T}_k = e^{i\phi_k}\,\widehat{\mathbf{T}}_k$ with $\det \widehat{\mathbf{T}}_k \in \mathbb{R}$. This criterion has been used in

FIGURE 3.6
When the incident and emerging directions of the polarization states are interchanged, the Jones matrix **T** must be replaced by its associated reciprocal Jones matrix $\mathbf{T}^r$.

Eqs. (3.24) and (3.25), where all phase shifts among the emerging Jones vectors are taken into account through the differences among the absolute phases of the respective Jones matrices, so that the coefficients have been taken as the square rots of the real relative cross sections p_k. Obviously, complex coefficients c_k can also formally be considered instead of p_k, provided $\sum |c_k|^2 = 1$. The indicated criterion will also be used in Chapter 5 for the synthesis of Mueller matrices from partially coherent or incoherent parallel compositions of Jones matrices.

Obviously, the practical usefulness of the additive decomposition of a given Jones matrix into a particular set of parallel components relies on the appropriateness of the underlying physical model for the multipath or interference phenomenon under consideration.

3.2.3 Reciprocity in Jones Matrices

Let us consider a Jones matrix **T** representing the polarimetric action of a pure system on incident electromagnetic beams which reach the system from a certain direction. In general, when the beams pass through the system (or are reflected, or scattered, by it) in the *reverse* direction (i.e., the incident and emerging directions are interchanged, Figure 3.6), the *reciprocal* (or *reverse*) Jones matrix $\mathbf{T}^r$ that characterizes the system is $\mathbf{T}^r = \mathrm{diag}(1,-1)\,\mathbf{T}^T\,\mathrm{diag}(1,-1)$ (Sekera 1966; Bhandari 2008). There are certain exceptions to this rule for systems involving magnetooptic phenomena (where $\mathbf{T}^r = \mathbf{T}$). In order to avoid any possible confusion, let us note that in the seminal works of Jones (1941) the effect of diag(1,−1) due to the change $Y \leftrightarrow -Y$ in the inversion of the direction of propagation was omitted, and this is the reason why $\mathbf{T}^r = \mathbf{T}^T$ has been reported erroneously in some works as the rule for reciprocity.

3.2.4 Changes of Coordinate System and Rotated Jones Matrices

The transformations of Jones matrices when represented with respect to different coordinate systems are derived from the rule (1.19) that holds for the transformation of Jones vectors. Consider the action of a pure medium **T** on a pure state $\boldsymbol{\varepsilon}$ observed with respect to a coordinate system *XY*: $\mathbf{T}\boldsymbol{\varepsilon} = \boldsymbol{\varepsilon}'$, and write the analogous equation $\boldsymbol{\varepsilon}'_{X'Y'} = \mathbf{T}_{X'Y'}\,\boldsymbol{\varepsilon}_{X'Y'}$ for a coordinate system *X'Y'* rotated through a counterclockwise angle θ with respect to *XY*. Then, from the rule for the rotation transformations of Jones vectors in Eq. (1.19), it follows that

$$\boldsymbol{\varepsilon}'_{X'Y'} = \mathbf{Q}(\theta)\,\boldsymbol{\varepsilon}' = \mathbf{Q}(\theta)\,\mathbf{T}\boldsymbol{\varepsilon} = \mathbf{Q}(\theta)\,\mathbf{T}\,\mathbf{Q}(-\theta)\,\boldsymbol{\varepsilon}_{X'Y'} = \mathbf{T}_{X'Y'}\,\boldsymbol{\varepsilon}_{X'Y'}$$

$$\mathbf{T}_{X'Y'} = \mathbf{Q}(\theta)\mathbf{T}\,\mathbf{Q}(-\theta) \qquad \mathbf{Q}(\theta) = \begin{pmatrix} \cos\theta & \sin\theta \\ -\sin\theta & \cos\theta \end{pmatrix} \qquad \left[\mathbf{Q}(-\theta) = \mathbf{Q}^T(\theta)\right] \tag{3.26}$$

It is important to keep in mind the tensor-like nature of Jones matrices, that is, $\mathbf{T}_{X'Y'} = \mathbf{Q}(\theta)\,\mathbf{T}\,\mathbf{Q}(-\theta)$ is the Jones matrix of the same medium as that represented by $\mathbf{T}$, but is referenced with respect to the rotated coordinate system $X'Y'$. Moreover, when the medium is rotated through a clockwise angle θ about the Z axis orthogonal to the plane XY, the Jones matrix of the rotated medium with respect to the same coordinate system XY is given by

$$\mathbf{T}(\theta) = \mathbf{Q}(-\theta)\,\mathbf{T}\,\mathbf{Q}(\theta) \tag{3.27}$$

Obviously, the change of coordinate system preserves the physically invariant quantities as the singular values, the trace, the determinant and the norms of $\mathbf{T}$.

It is worth noting that $\mathbf{Q}(\theta)$ is a particular form of an orthogonal Jones matrix called a *rotator*. As we shall see in later sections, a rotator can physically be realized by means of a circular retarder.

3.3 Pure Mueller Matrices

3.3.1 The Concept of Pure Mueller Matrix

In the previous section we have considered the mathematical formulation, by means of Jones matrices, of the nondepolarizing transformation of polarization states represented either by Jones vectors (only applicable to totally polarized or pure states) or by polarization matrices (applicable to both pure and mixed states). Nevertheless, a state of polarization can also be represented by the corresponding Stokes vector $\mathbf{s}$, whose transformations by nondepolarizing media are the subject of this section. As a first step, let us arrange the elements ϕ_{ij} $(i,j = 1,2)$ of the polarization matrix $\mathbf{\Phi}$ as components of the *polarization covariance vector*

$$\begin{gathered}\boldsymbol{\varphi} \equiv (\phi_0, \phi_1, \phi_2, \phi_3)^T \equiv \langle \boldsymbol{\varepsilon} \otimes \boldsymbol{\varepsilon}^* \rangle = \left(\langle \varepsilon_1 \varepsilon_1^* \rangle, \langle \varepsilon_1 \varepsilon_2^* \rangle, \langle \varepsilon_2 \varepsilon_1^* \rangle, \langle \varepsilon_2 \varepsilon_2^* \rangle\right)^T \\ [\phi_0 \equiv \phi_{11},\ \phi_1 \equiv \phi_{12},\ \phi_2 \equiv \phi_{21},\ \phi_3 \equiv \phi_{22}]\end{gathered} \tag{3.28}$$

(Note that a *covariance vector* associated with a pure Mueller matrix will be introduced in Sec. 3.4.2, and we term $\boldsymbol{\phi}$ the *polarization covariance vector* in order to make clear the distinction between the two vector structures.)

The transformation (3.7) of the state $\mathbf{\Phi}$ by a nondepolarizing medium represented by the Jones matrix $\mathbf{T}$ can be expressed in terms of the polarization covariance vector as follows:

$$\boldsymbol{\varphi}' = \langle \boldsymbol{\varepsilon}' \otimes \boldsymbol{\varepsilon}'^* \rangle = \langle (\mathbf{T}\boldsymbol{\varepsilon}) \otimes (\mathbf{T}\boldsymbol{\varepsilon})^* \rangle = \langle (\mathbf{T} \otimes \mathbf{T}^*)(\boldsymbol{\varepsilon} \otimes \boldsymbol{\varepsilon}^*) \rangle = (\mathbf{T} \otimes \mathbf{T}^*)\boldsymbol{\varphi} \tag{3.29}$$

The polarization covariance vector $\boldsymbol{\varphi}$ and the corresponding Stokes vector $\mathbf{s}$ are related through (Parke 1948)

$$\mathbf{s} = \sqrt{2}\,\mathcal{L}\,\boldsymbol{\varphi} \tag{3.30a}$$

with

$$\mathcal{L} = \frac{1}{\sqrt{2}} \begin{pmatrix} 1 & 0 & 0 & 1 \\ 1 & 0 & 0 & -1 \\ 0 & 1 & 1 & 0 \\ 0 & i & -i & 0 \end{pmatrix} \tag{3.30b}$$

The matrix $\mathcal{L}$ connecting the two vector representations is unitary $(\mathcal{L}^{-1} = \mathcal{L}^{\dagger})$ and will further appear in a number of algebraic expressions. By combining Eqs. (3.29) and (3.30) we get

$$\mathbf{s}' = \sqrt{2}\,\mathcal{L}\,\boldsymbol{\varphi}' = \sqrt{2}\,\mathcal{L}\left(\mathbf{T} \otimes \mathbf{T}^{*}\right)\boldsymbol{\varphi} = \mathcal{L}\left(\mathbf{T} \otimes \mathbf{T}^{*}\right)\mathcal{L}^{-1}\mathbf{s} \tag{3.31}$$

where the Stokes vectors are transformed by the 4×4 matrix

$$\mathbf{M}(\mathbf{T}) \equiv \mathcal{L}\left(\mathbf{T} \otimes \mathbf{T}^{*}\right)\mathcal{L}^{-1} \tag{3.32}$$

which is defined from the corresponding Jones matrix $\mathbf{T}$ of the medium. $\mathbf{M}(\mathbf{T})$ is called *pure Mueller matrix* (also *nondepolarizing* or *Mueller-Jones matrix*). Therefore, a pure Mueller matrix provides complete information on how a given nondepolarizing medium transforms an arbitrary polarization state (within the fundamental linear interaction assumption).

Pure Mueller matrices constitute a subset of the larger category of matrices representing physical linear transformations of Stokes vectors which will be dealt with in Chapter 5. Jones (1947), being aware of the matrix formulation developed by H. Mueller in a series of lectures during 1945–46 at the Massachusetts Institute of Technology, termed *Mueller matrices* the complete set of these matrices. Nevertheless, as pointed out by Jones himself, as well as by Barakat (1987a), it was Soleillet (1929) who first formulated the linear polarimetric transformations. These were also considered, in matrix formulation, by Perrin (1942) in describing the action of various scattering media. A description of Mueller's approach can be found in the PhD thesis of N. G. Parke (1948).

Hereafter, when appropriate, pure Mueller matrices will be denoted as $\mathbf{M}(\mathbf{T})$ or $\mathbf{M}_{J}$, in order to distinguish them from general depolarizing Mueller matrices which will be studied later.

The elements of $\mathbf{M}(\mathbf{T})$, denoted as m_{ij}, are real and can be expressed in terms of $\mathbf{T}$ as

$$m_{ij} = \frac{1}{2}\,\mathrm{tr}\left(\boldsymbol{\sigma}_{i}\,\mathbf{T}\boldsymbol{\sigma}_{j}\,\mathbf{T}^{\dagger}\right) \tag{3.33}$$

where $\boldsymbol{\sigma}_{i}$ is the set composed of the 2×2 identity matrix and the Pauli matrices (taken in the order that is commonly used in polarization optics which differs from certain quantum formulations where subscripts 2 and 3 are interchanged),

$$\boldsymbol{\sigma}_{0} = \begin{pmatrix} 1 & 0 \\ 0 & 1 \end{pmatrix}, \quad \boldsymbol{\sigma}_{1} = \begin{pmatrix} 1 & 0 \\ 0 & -1 \end{pmatrix}, \quad \boldsymbol{\sigma}_{2} = \begin{pmatrix} 0 & 1 \\ 1 & 0 \end{pmatrix}, \quad \boldsymbol{\sigma}_{3} = \begin{pmatrix} 0 & -i \\ i & 0 \end{pmatrix} \tag{3.34}$$

The explicit expressions of m_{ij} in terms of the elements t_{kl} $(k,l=1,2)$ of $\mathbf{T}$ are

$$\begin{array}{ll}
2m_{00}=t_{11}^{*}t_{11}+t_{12}^{*}t_{12}+t_{21}^{*}t_{21}+t_{22}^{*}t_{22} & 2m_{01}=t_{11}^{*}t_{11}+t_{21}^{*}t_{21}-t_{12}^{*}t_{12}-t_{22}^{*}t_{22}\\
2m_{02}=t_{11}^{*}t_{12}+t_{21}^{*}t_{22}+t_{12}^{*}t_{11}+t_{22}^{*}t_{21} & 2m_{03}=i\left(t_{11}^{*}t_{12}+t_{21}^{*}t_{22}-t_{12}^{*}t_{11}-t_{22}^{*}t_{21}\right)\\
2m_{10}=t_{11}^{*}t_{11}+t_{12}^{*}t_{12}-t_{21}^{*}t_{21}-t_{22}^{*}t_{22} & 2m_{11}=t_{11}^{*}t_{11}+t_{22}^{*}t_{22}-t_{21}^{*}t_{21}-t_{12}^{*}t_{12}\\
2m_{12}=t_{12}^{*}t_{11}+t_{11}^{*}t_{12}-t_{22}^{*}t_{21}-t_{21}^{*}t_{22} & 2m_{13}=i\left(t_{11}^{*}t_{12}+t_{22}^{*}t_{21}-t_{21}^{*}t_{22}-t_{12}^{*}t_{11}\right)\\
2m_{20}=t_{11}^{*}t_{21}+t_{21}^{*}t_{11}+t_{12}^{*}t_{22}+t_{22}^{*}t_{12} & 2m_{21}=t_{11}^{*}t_{21}+t_{21}^{*}t_{11}-t_{12}^{*}t_{22}-t_{22}^{*}t_{12}\\
2m_{22}=t_{11}^{*}t_{22}+t_{21}^{*}t_{12}+t_{12}^{*}t_{21}+t_{22}^{*}t_{11} & 2m_{23}=i\left(t_{11}^{*}t_{22}+t_{21}^{*}t_{12}-t_{12}^{*}t_{21}-t_{22}^{*}t_{11}\right)\\
2m_{30}=i\left(t_{21}^{*}t_{11}+t_{22}^{*}t_{12}-t_{11}^{*}t_{21}-t_{12}^{*}t_{22}\right) & 2m_{31}=i\left(t_{21}^{*}t_{11}+t_{12}^{*}t_{22}-t_{11}^{*}t_{21}-t_{22}^{*}t_{12}\right)\\
2m_{32}=i\left(t_{21}^{*}t_{12}+t_{22}^{*}t_{11}-t_{11}^{*}t_{22}-t_{12}^{*}t_{21}\right) & 2m_{33}=t_{22}^{*}t_{11}+t_{11}^{*}t_{22}-t_{12}^{*}t_{21}-t_{21}^{*}t_{12}
\end{array}\tag{3.35}$$

Conversely, by expressing the elements t_{kl} $(k,l=1,2)$ of $\mathbf{T}$ in polar form, $t_{kl}=|t_{kl}|e^{i\theta_{kl}}$, t_{kl} can be calculated from m_{ij} in the following manner

$$\begin{array}{ll}
2|t_{11}|^{2}=m_{00}+m_{01}+m_{10}+m_{11} & 2|t_{12}|^{2}=m_{00}-m_{01}+m_{10}-m_{11}\\
2|t_{21}|^{2}=m_{00}+m_{01}-m_{10}-m_{11} & 2|t_{22}|^{2}=m_{00}-m_{01}-m_{10}+m_{11}
\end{array}$$

$$\begin{array}{ll}
\cos(\theta_{12}-\theta_{11})=\dfrac{m_{02}+m_{12}}{\sqrt{(m_{00}+m_{10})^{2}-(m_{01}+m_{11})^{2}}} & \sin(\theta_{12}-\theta_{11})=\dfrac{-(m_{03}+m_{13})}{\sqrt{(m_{00}+m_{10})^{2}-(m_{01}+m_{11})^{2}}}\\
\cos(\theta_{21}-\theta_{11})=\dfrac{m_{20}+m_{21}}{\sqrt{(m_{00}+m_{01})^{2}-(m_{10}+m_{11})^{2}}} & \sin(\theta_{21}-\theta_{11})=\dfrac{m_{30}+m_{31}}{\sqrt{(m_{00}+m_{01})^{2}-(m_{10}+m_{11})^{2}}}\\
\cos(\theta_{22}-\theta_{11})=\dfrac{m_{22}+m_{33}}{\sqrt{(m_{00}+m_{11})^{2}-(m_{10}+m_{01})^{2}}} & \sin(\theta_{22}-\theta_{11})=\dfrac{m_{32}-m_{23}}{\sqrt{(m_{00}+m_{11})^{2}-(m_{10}+m_{01})^{2}}}
\end{array}\tag{3.36}$$

A real 4×4 matrix is a pure Mueller matrix if and only if it can be expressed as in Eq. (3.32) in terms of a Jones matrix. Moreover, any pure Mueller matrix $\mathbf{M}_J\equiv\mathbf{M}(\mathbf{T})$ satisfies the characteristic property (Xing 1992)

$$\begin{array}{c}
\mathbf{G}\mathbf{M}_J^T\,\mathbf{G}\mathbf{M}_J=\sqrt{\det\mathbf{M}_J}\,\mathbf{I}_4\\
\left[\mathbf{G}\equiv\operatorname{diag}(1,-1,-1,-1)\qquad \mathbf{I}_4\equiv\operatorname{diag}(1,1,1,1)\qquad \sqrt{\det\mathbf{M}_J}=p_1^2p_2^2\right]
\end{array}\tag{3.37}$$

where the auxiliary pure "N-matrix" of $\mathbf{M}_J$, defined as $\mathbf{N}_J\equiv\mathbf{G}\mathbf{M}_J^T\,\mathbf{G}\mathbf{M}_J=\sqrt{\det\mathbf{M}_J}\,\mathbf{I}_4$, satisfies

$$\mathbf{N}_J\equiv\mathbf{G}\mathbf{M}_J^T\,\mathbf{G}\mathbf{M}_J=\mathbf{G}\mathbf{M}_J\mathbf{G}\mathbf{M}_J^T=\mathbf{M}_J^T\,\mathbf{G}\mathbf{M}_J\mathbf{G}=\mathbf{M}_J\mathbf{G}\mathbf{M}_J^T\,\mathbf{G}\tag{3.38}$$

Relation (3.37) provides the following set of ten equalities (Xing 1992):

$$\begin{array}{ll} m_{00}^2 - m_{10}^2 - m_{20}^2 - m_{30}^2 = \sqrt{\det \mathbf{M}_J} & m_{01}^2 - m_{11}^2 - m_{21}^2 - m_{31}^2 = -\sqrt{\det \mathbf{M}_J} \\ m_{02}^2 - m_{12}^2 - m_{22}^2 - m_{32}^2 = -\sqrt{\det \mathbf{M}_J} & m_{03}^2 - m_{13}^2 - m_{23}^2 - m_{33}^2 = -\sqrt{\det \mathbf{M}_J} \\ m_{00}m_{01} = m_{10}m_{11} + m_{20}m_{21} + m_{30}m_{31} & m_{00}m_{02} = m_{10}m_{12} + m_{20}m_{22} + m_{30}m_{32} \\ m_{00}m_{03} = m_{10}m_{13} + m_{20}m_{23} + m_{30}m_{33} & m_{01}m_{02} = m_{11}m_{12} + m_{21}m_{22} + m_{31}m_{32} \\ m_{01}m_{03} = m_{11}m_{13} + m_{21}m_{23} + m_{31}m_{33}, & m_{02}m_{03} = m_{12}m_{13} + m_{22}m_{23} + m_{32}m_{33} \end{array} \tag{3.39}$$

Through a subsequent elimination of $\sqrt{\det \mathbf{M}_J}$, Eqs. (3.39) give a set of nine equalities that completely characterize a pure Mueller matrix. In the case of a singular pure Mueller matrix (whose associated Jones matrix $\mathbf{T}$ is also singular, i.e., $p_2 = 0$) $\sqrt{\det \mathbf{M}_J} = 0$ should be substituted in the first four of the above equations.

It should be noted that, as follows from the very definition of a pure Mueller matrix from its corresponding Jones matrix, several equivalent sets of nine characteristic equalities, other than Eqs. (3.39), can be obtained. As an example, let us observe that, given a Jones matrix $\mathbf{T}$, its conjugate transpose $\mathbf{T}^\dagger$ is also a Jones matrix and that furthermore, $\mathbf{M}(\mathbf{T}^\dagger) = [\mathbf{M}(\mathbf{T})]^T$ so that, given a pure Mueller matrix $\mathbf{M}_J$, its transposed matrix $\mathbf{M}_J^T$ is also a pure Mueller matrix and thus, any relation between the elements of $\mathbf{M}_J$ must also be satisfied by the elements of $\mathbf{M}_J^T$. Consequently, Eqs. (3.39) are also satisfied if the subscripts i and j are interchanged

$$\begin{array}{ll} m_{00}^2 - m_{01}^2 - m_{02}^2 - m_{03}^2 = \sqrt{\det \mathbf{M}_J} & m_{10}^2 - m_{11}^2 - m_{12}^2 - m_{13}^2 = -\sqrt{\det \mathbf{M}_J} \\ m_{20}^2 - m_{21}^2 - m_{22}^2 - m_{23}^2 = -\sqrt{\det \mathbf{M}_J} & m_{30}^2 - m_{31}^2 - m_{32}^2 - m_{33}^2 = -\sqrt{\det \mathbf{M}_J} \\ m_{00}m_{10} = m_{01}m_{11} + m_{02}m_{12} + m_{03}m_{13} & m_{00}m_{20} = m_{01}m_{21} + m_{02}m_{22} + m_{03}m_{23} \\ m_{00}m_{30} = m_{01}m_{31} + m_{02}m_{32} + m_{03}m_{33} & m_{10}m_{20} = m_{11}m_{21} + m_{12}m_{22} + m_{13}m_{23} \\ m_{10}m_{30} = m_{11}m_{31} + m_{12}m_{32} + m_{13}m_{33} & m_{20}m_{30} = m_{21}m_{31} + m_{22}m_{32} + m_{23}m_{33} \end{array} \tag{3.40}$$

It is also worth noting that the set of pure Mueller matrices is not completely characterized by the fact that they transform pure Stokes vectors into pure Stokes vectors. Additional constraints are embedded into the definition of pure Mueller matrix, or in Eqs. (3.39), so that, for instance, matrices like $\mathbf{G}$ and other *improper* orthogonal matrices cannot be derived from a Jones matrix.

Obviously, due to the biunivocal relation between the representative Jones matrix $\mathbf{T}$ ($\det \mathbf{T} \in \mathbb{R}$) and the corresponding pure Mueller matrix $\mathbf{M}(\mathbf{T})$, $\mathbf{M}(\mathbf{T})$ depends on up to seven independent measurable quantities.

The polarimetric transformation of the Stokes vector $\mathbf{s}$ of an incident 2D polarization state into the Stokes vector $\mathbf{s}'$ of an emerging polarization state by the effect of a linear passive medium can be represented, in general, as $\mathbf{s}' = \mathbf{M}\mathbf{s}$, where the 4×4 real matrix $\mathbf{M}$ is the so-called Mueller matrix. $\mathbf{M}$ is depolarizing in general. While the present chapter is mainly focused on the properties of nondepolarizing Mueller matrices derivable from corresponding Jones matrices, the characterization and properties of general Mueller matrices will be dealt with in future chapters. Nevertheless, it is worth pointing out in advance that a 4×4 real matrix $\mathbf{M}$ is a (passive) Mueller matrix if and only if it can be expressed as a convex sum $\sum p_i \mathbf{M}_{Ji}$ of (passive) pure Mueller matrices $\mathbf{M}_{Ji}$, with the weights p_i satisfying $\sum p_i = 1$ (Gil 2007). This means that not all 4×4 real matrices are Mueller matrices and, in particular, not all 4×4 real matrices transforming Stokes vectors into Stokes vectors are Mueller matrices. The restrictions come from the fact that any transformation $\mathbf{s}' = \mathbf{Ms}$ must be physically realizable, in at least one way, as a superposition of elementary nondepolarizing interactions. The physical and mathematical treatment of these considerations will be dealt with in detail in Chapter 5.

3.3.2 Partitioned Form of a Mueller Matrix

Any (pure or nonpure) Mueller matrix can be expressed in a convenient partitioned form that simplifies the treatment of interesting properties (Robson 1974; Xing 1992),

$$\mathbf{M} = m_{00}\hat{\mathbf{M}} \qquad \hat{\mathbf{M}} \equiv \begin{pmatrix} 1 & \mathbf{D}^T \\ \mathbf{P} & \mathbf{m} \end{pmatrix}$$

$$\left[\mathbf{D} \equiv \frac{1}{m_{00}}(m_{01}, m_{02}, m_{03})^T \qquad \mathbf{P} \equiv \frac{1}{m_{00}}(m_{10}, m_{20}, m_{30})^T \qquad \mathbf{m} \equiv \frac{1}{m_{00}} \begin{pmatrix} m_{11} & m_{12} & m_{13} \\ m_{21} & m_{22} & m_{23} \\ m_{31} & m_{32} & m_{33} \end{pmatrix} \right] \tag{3.41}$$

where the three-component vectors $\mathbf{D}$ and $\mathbf{P}$ are respectively called the *diattenuation vector* and the *polarizance vector* of $\mathbf{M}$ (Lu and Chipman 1996). The magnitudes of these vectors are called *diattenuation*, $D \equiv |\mathbf{D}|$ and *polarizance*, $P \equiv |\mathbf{P}|$ (Bird and Shurcliff 1959; Shurcliff 1962). From the general expression (3.32) of a pure Mueller matrix as a function of its corresponding Jones matrix it is straightforward to show that pure Mueller matrices satisfy the equality $P = D$ (Gil 1983). Nevertheless, this equality is not a necessary condition for general (depolarizing) Mueller matrices. Further, the mean intensity coefficient of $\mathbf{M}$ (i.e., transmittance or reflectance for unpolarized incident states) is given by m_{00}.

The *degree of polarimetric purity* of $\mathbf{M}$ is given by the *depolarization index* (Gil 1986)

$$P_\Delta = \sqrt{D^2 + P^2 + \|\mathbf{m}\|_2^2}\Big/\sqrt{3} \tag{3.42}$$

where $\|\mathbf{m}\|_2$ stands for the Frobenius norm of the normalized submatrix $\mathbf{m}$ of $\hat{\mathbf{M}}$.

As indicated above, when dealing with Mueller matrices it is useful to distinguish the cases of (a) pure Mueller matrices (also called Mueller-Jones or nondepolarizing matrices, studied throughout the present chapter) which are associated with systems whose depolarization index satisfies $P_\Delta = 1$, so that they do not depolarize any totally polarized incident state, and (b) nonpure or depolarizing Mueller matrices which depolarize some or all totally polarized incident states ($P_\Delta < 1$).

3.3.3 Reciprocity Properties of Mueller Matrices

Leaving aside systems involving magnetooptic effects whose reciprocity property ($\mathbf{T}^r = \mathbf{T}$) differs from the rule that usually holds for Jones matrices ($\mathbf{T}^r = \mathrm{diag}(1,-1)\,\mathbf{T}^T\,\mathrm{diag}(1,-1)$), the pure Mueller matrix $\mathbf{M}^r$ representing the same pure medium as $\mathbf{M}$, but with reversed directions of propagation, is obtained from (Sekera 1966; Schönhofer and Kuball 1987)

$$\mathbf{M}^r = \mathbf{X}\mathbf{M}^T\mathbf{X} \qquad \mathbf{X} \equiv \mathrm{diag}(1,1,-1,1) \tag{3.43}$$

It is worth observing that, since the conjugate transpose $\mathbf{T}^\dagger$ of the Jones matrix $\mathbf{T}$ is also a Jones matrix, then the transposed $\mathbf{M}^T = \mathbf{M}(\mathbf{T}^\dagger)$ of the pure Mueller matrix $\mathbf{M} = \mathbf{M}(\mathbf{T})$ is in turn a pure Mueller matrix. In addition, notice that the matrix **X**, pre- and post-multiplying $\mathbf{M}^T$, is not a Mueller matrix since it is not expressible in the form of Eq. (3.32) although transforming totally polarized incident states into totally polarized emerging ones. Like **G**, **X** is an example of an improper orthogonal matrix, i.e., an orthogonal matrix with a negative determinant.

Finally, note that in some works the rule for obtaining the reverse Mueller matrix has been formulated in the incorrect form $\mathbf{M}^r = \mathrm{diag}(1,1,1,-1)\mathbf{M}^T\,\mathrm{diag}(1,1,1,-1)$, whose origin is the incorrect identification $\mathbf{T}^r = \mathbf{T}^T$ derived from certain results due to Jones (1941).

3.3.4 Passivity Condition for Pure Mueller Matrices

The decomposition (3.8) of a Jones matrix **T** can be translated to $\mathbf{M}(\mathbf{T}) \equiv \mathbf{M}_J$, resulting in

$$\mathbf{M}_J = \mathbf{M}_{R2}\,\mathbf{M}_{DL0}\,\mathbf{M}_{R1}, \tag{3.44a}$$

with

$$\mathbf{M}_{R1} = \mathbf{M}(\mathbf{T}_{R1}) = \begin{pmatrix} 1 & \mathbf{0}^T \\ \mathbf{0} & \mathbf{m}_{R1} \end{pmatrix} \qquad \mathbf{M}_{R2} = \mathbf{M}(\mathbf{T}_{R2}) = \begin{pmatrix} 1 & \mathbf{0}^T \\ \mathbf{0} & \mathbf{m}_{R2} \end{pmatrix} \tag{3.44b}$$

where $\mathbf{0} \equiv (0,0,0)^T$, $\mathbf{m}_{R1}$ and $\mathbf{m}_{R2}$ are 3×3 proper orthogonal matrices ($\mathbf{m}_{Ri}^{-1} = \mathbf{m}_{Ri}^T$, $\det \mathbf{m}_{Ri} = +1$; $i = 1,2$) and

$$\mathbf{M}_{DL0} = \mathbf{M}(\mathbf{T}_{DL0}) = \frac{1}{2}\begin{pmatrix} p_1^2+p_2^2 & p_1^2-p_2^2 & 0 & 0 \\ p_1^2-p_2^2 & p_1^2+p_2^2 & 0 & 0 \\ 0 & 0 & 2p_1p_2 & 0 \\ 0 & 0 & 0 & 2p_1p_2 \end{pmatrix} \tag{3.44c}$$

where p_1, p_2 are the singular values of $\mathbf{T}_{DL0}$. $\mathbf{M}_{DL0}$ can also be expressed in terms of the mean intensity coefficient m_{00} (i.e., the intensity coefficient for unpolarized light) and the diattenuation-polarizance angle κ defined from $D \equiv \cos\kappa$, $0 \le \kappa \le \pi/2$,

$$\mathbf{M}_{DL0} = m_{00}\begin{pmatrix} 1 & \cos\kappa & 0 & 0 \\ \cos\kappa & 1 & 0 & 0 \\ 0 & 0 & \sin\kappa & 0 \\ 0 & 0 & 0 & \sin\kappa \end{pmatrix} \tag{3.45}$$

The *counter-diattenuation*, given by the quantity $\sin\kappa \equiv \sqrt{1-D^2}$ will also appear frequently in calculations involving Mueller matrices of diattenuators.

It is worth noting that $\mathbf{M}_J$ can be diagonalized as (Collett 1971)

$$\begin{aligned}\mathbf{M}_{R2}^T\mathbf{C}_D^T\mathbf{M}_J\mathbf{C}_D\mathbf{M}_{R1}^T &= \mathrm{diag}\left(p_1^2, p_2^2, p_1p_2, p_1p_2\right) \\ &= m_{00}\,\mathrm{diag}\left(1+\cos\kappa, 1-\cos\kappa, \sin\kappa, \sin\kappa\right)\end{aligned} \tag{3.46}$$

where $\mathbf{C}_D$ is the following improper ($\det \mathbf{C}_D = -1$) orthogonal non-Mueller matrix, with

$$\mathbf{C}_D = \mathbf{C}_D^{-1} = \frac{1}{\sqrt{2}}\begin{pmatrix} 1 & 1 & 0 & 0 \\ 1 & -1 & 0 & 0 \\ 0 & 0 & \sqrt{2} & 0 \\ 0 & 0 & 0 & \sqrt{2} \end{pmatrix} \tag{3.47}$$

so that the ordered singular values of $\mathbf{M}_J$ (and of $\mathbf{M}_{DL0}$) are precisely $(p_1^2, p_2^2, p_1 p_2, p_1 p_2)$.

Given a pure Mueller matrix $\mathbf{M}_J$ and an arbitrary incident Stokes vector $\mathbf{s}$ with intensity I, the maximal intensity $I'_{\max}$ for the emerging Stokes vector is achieved for $\mathbf{s}_{\hat{D}+} = (1, \hat{\mathbf{D}}^T)^T$ ($\hat{\mathbf{D}} \equiv \mathbf{D}/D$ being the *normalized diattenuation vector* of $\mathbf{M}_J$). Therefore, the passivity condition for $\mathbf{M}_J$ is given by

$$I'_{\max}/I = m_{00}(1+D) \le 1 \tag{3.48}$$

and can also be written as

$$p_1^2 = m_{00}(1+D) = m_{00}(1+P) \le 1 \tag{3.49}$$

in agreement with the passivity condition (3.12) for the Jones matrix associated with $\mathbf{M}_J$. Except in very special and uncommon cases (Opatrný and Perina 1993), passivity is a physical requisite for polarimetric linear interactions, so that hereafter, unless otherwise indicated, we refer to passive Mueller matrices as simply Mueller matrices.

Considering all passive Mueller matrices proportional to $\mathbf{M}_J$, the one with maximum mean intensity coefficient (i.e., with minimum attenuation of intensity) is given by the *passive representative* Mueller matrix

$$\tilde{\mathbf{M}}_J = \frac{1}{1+D}\begin{pmatrix} 1 & \mathbf{D}^T \\ \mathbf{P} & \mathbf{m} \end{pmatrix} \tag{3.50}$$

Observe finally that, since pure Mueller matrices satisfy $P = D$, the passivity condition for $\mathbf{M}_J$ coincides with the passivity condition for $\mathbf{M}_J^T$ and for $\mathbf{M}_J^r$.

3.3.5 Algebraic Operations with Pure Mueller Matrices and their Physical Interpretation

As with Jones matrices, in the next paragraphs we summarize the mathematical operations that can be realized on pure Mueller matrices and outline their physical interpretation.

3.3.5.1 Product of Pure Mueller Matrices

Given a pair of pure Mueller matrices $\mathbf{M}_{J1}$ and $\mathbf{M}_{J2}$, the product $\mathbf{M}_{J2}\mathbf{M}_{J1}$ represents the successive serial actions of the media associated with them.

$$\mathbf{M}_{J2}\mathbf{M}_{J1}\mathbf{s} = \mathbf{M}_{J2}\mathbf{s}' = \mathbf{s}'' \tag{3.51}$$

The passivity of the product matrix $\mathbf{M}_J \equiv \mathbf{M}_{J2}\mathbf{M}_{J1}$ is ensured by the fact that both matrix factors are passive. Thus, as it is the case with Jones matrices, the matrix product of pure

Mueller matrices is an internal operation in the space of pure Mueller matrices. Moreover, like the product of Jones matrices, the product of Mueller matrices is not commutative. Recall that two Mueller matrices (pure or not) commute if and only if they can be diagonalized simultaneously.

3.3.5.2 Product of a Pure Mueller Matrix and a Non-negative Scalar

Given a pure Mueller matrix $\mathbf{M}_J$ and an arbitrary real non-negative number c, the product $c\mathbf{M}_J$ can be expressed as the matrix product

$$c\,\mathbf{M}_J = (c\,\mathbf{I}_4)\mathbf{M}_J = \mathbf{M}_{R2}\,\mathbf{M}_{DL0}\left(cp_1^2, cp_2^2\right)\mathbf{M}_{R1} \tag{3.52}$$

where $\mathbf{I}_4$ is the 4×4 identity matrix, so that the multiplication is equivalent to the effect of the neutral attenuating Mueller matrix $\mathbf{M}_{Jc} \equiv c\,\mathbf{I}_4$. Thus, provided $c \le 1$, $\mathbf{M}_{Jc}$ and hence $c\,\mathbf{M}_J$, are passive Mueller matrices, and $c\,\mathbf{M}_J$ only differs from $\mathbf{M}_J$ by the overall attenuation factor c. For values $c > 1$, $\mathbf{M}_{Jc}$ is not passive.

3.3.5.3 Determinant, Trace and Norms of a Pure Mueller Matrix

The determinant of a pure Mueller matrix is given by

$$\det\mathbf{M}_J = |\det\mathbf{T}|^4 = (p_1 p_2)^4 = m_{00}^4\left(1-D^2\right)^2 = \left(m_{00}\sin\kappa\right)^4 \tag{3.53}$$

so that, in general

$$0 \le \det\mathbf{M}_J \le m_{00}^4 \tag{3.54}$$

$\det\mathbf{M}_J = 0$ being satisfied if and only if $D = 1$ (i.e., $p_2 = 0$, or, equivalently, $\sin\kappa = 0$). As we shall see in Chapter 5, unlike in the case of pure Mueller matrices whose determinants are always non-negative, the determinants of depolarizing Mueller matrices can take any value (positive, negative or zero).

By using the polar decomposition of $\mathbf{M}_J$ (see Section 4.4), it can be proven that the trace of $\mathbf{M}_J$ can be expressed as

$$\operatorname{tr}\mathbf{M}_J = m_{00}\left[1+\sqrt{1-D^2}\,(1+2\cos\Delta)+\left(1-\sqrt{1-D^2}\right)\cos\psi\right] \tag{3.55}$$

where D and Δ are the diattenuation and retardance of $\mathbf{M}_J$ respectively, and ψ is the angle subtended by the vectors $\mathbf{D}$ and $\mathbf{P}$ (Gil and San José, 2016c).

The *Frobenius norm* of $\mathbf{M}_J$ is defined as

$$\left\|\mathbf{M}_J\right\|_2 = \sqrt{\sum_{i,j=0}^{3}\left|m_{ij}\right|^2} = \sqrt{\operatorname{tr}\left(\mathbf{M}_J^T\,\mathbf{M}_J\right)} = \|\mathbf{T}\|_2^2 = p_1^2 + p_2^2 = 2m_{00} \tag{3.56}$$

The *nuclear norm* of $\mathbf{M}_J$ is defined as the sum of its singular values

$$\left\|\mathbf{M}_J\right\|_* = \operatorname{tr}\sqrt{\mathbf{M}_J^T\mathbf{M}_J} = \|\mathbf{T}\|_*^2 = (p_1+p_2)^2 = 2m_{00}\,(1+\sin\kappa) \tag{3.57}$$

so that the following relations hold between the determinant and the norms of $\mathbf{M}_J$:

$$\|\mathbf{M}_J\|_* = \|\mathbf{M}_J\|_2 + 2\sqrt[4]{\det \mathbf{M}_J} \tag{3.58}$$

Furthermore, in analogy to the max-norm $\|\mathbf{s}\|_0$ defined for Stokes vectors, the *max-norm* of $\mathbf{M}_J$ is defined as follows (Gil 2007):

$$\|\mathbf{M}\|_0 \equiv m_{00} \tag{3.59}$$

Consequently, for any pure Mueller matrix $\mathbf{M}_J$,

$$m_{00} = \|\mathbf{M}_J\|_0 = \frac{1}{2}\|\mathbf{T}\|_2^2 = \frac{1}{2}\|\mathbf{M}_J\|_2 = \frac{1}{2}\|\mathbf{M}_J\|_* - \sqrt[4]{\det \mathbf{M}_J} \tag{3.60}$$

3.3.5.4 The Inverse of a Pure Mueller Matrix

Given a nonsingular pure Mueller matrix $\mathbf{M}_J = \mathbf{M}_{R2}\mathbf{M}_{DL0}(p_1,p_2)\mathbf{M}_{R1}$ $(0 < p_2 \leq p_1 \leq 1)$, from the singular value decomposition (3.46) and in accordance with Eq. (3.21), we get immediately the inverse of $\mathbf{M}_J$ from

$$\mathbf{M}_J^{-1} = \mathbf{M}_{R1}^T\,\mathbf{M}_{DL0}\left(1/p_1, 1/p_2\right)\mathbf{M}_{R2}^T \tag{3.61}$$

The inverse is not a passive Mueller matrix, because it produces certain amplification for some or all incident states. Nevertheless, given a real positive number c such that $c \leq p_1^2$, $c\mathbf{M}^{-1}$ is a passive Mueller matrix.

Another interesting and useful expression for $\mathbf{M}_J^{-1}$ is obtained through right-multiplying by $\mathbf{M}_J^{-1}$ the N-matrix of $\mathbf{M}_J$ in Eq. (3.37),

$$\mathbf{M}_J^{-1} \equiv \mathbf{G}\,\mathbf{M}_J^T\,\mathbf{G}\big/\sqrt{\det \mathbf{M}_J} \tag{3.62}$$

In the case of $\mathbf{M}_J$ being singular $(0 = p_2 < p_1)$ a *pseudoinverse* matrix can be defined as

$$\mathbf{M}_J^- = \mathbf{M}_{R1}^T\,\mathbf{M}_{DL0}\left(1/p_1, 0\right)\mathbf{M}_{R2}^T \tag{3.63}$$

so that

$$\mathbf{M}^-\mathbf{M} = \mathbf{M}\mathbf{M}^- = \mathbf{M}_{DL0}(1,0) \tag{3.64}$$

Like the inverse of a nonsingular pure Mueller matrix, the pseudoinverse of a singular pure Mueller matrix is not passive, but passivity can be achieved through the multiplication of $\mathbf{M}^-$ by a real positive number c that satisfies $c \leq p_1^2$.

3.3.5.5 Additive Composition of Mueller Matrices

Unlike the sum of Stokes vectors which corresponds to the superposition of their corresponding electromagnetic beams at a given point in space, the physical interpretation of the sum of Mueller matrices requires a specific analysis derived from the physically realizable linear transformations of the state of polarization of an electromagnetic beam. Additive compositions (also called *parallel compositions*) of Mueller matrices apply when a uniform light beam with a Stokes vector **s** interacts with a spatially inhomogeneous sample that can be considered as composed of a number n of elements (or tiles) characterized by pure Mueller matrices $\mathbf{M}_{Jk}$ $(k=1,\dots,n)$ with size ratios (relative cross-sections) $p_k\left(\sum_{k=1}^{n}p_k=1\right)$ $(p_k=A_k/A$ is the ratio of the area A_k of the element k and the total area A covered by the light beam). The incident light beam is shared among n pencils with Stokes vectors $\mathbf{s}_k=p_k\mathbf{s}$ that are transformed as $\mathbf{s}'_k=\mathbf{M}_{Jk}p_k\mathbf{s}$ (Figure 3.5) and consequently, provided the emerging pencils are incoherently superimposed, the Stokes vector of the emerging beam is given by

$$\mathbf{s}'=\sum_{k=1}^{n}\mathbf{s}'_k=\sum_{k=1}^{n}\mathbf{M}_{Jk}p_k\mathbf{s}=\left(\sum_{k=1}^{n}p_k\mathbf{M}_{Jk}\right)\mathbf{s} \tag{3.65}$$

where the composed Mueller matrix of the sample is

$$\mathbf{M}=\sum_{k=1}^{n}p_k\mathbf{M}_{Jk}\qquad\left(\sum_{k=1}^{n}p_k=1\right) \tag{3.66}$$

Note that this matrix is necessarily depolarizing (except for the trivial case where all $\mathbf{M}_{Jk}$ are mutually proportional).

This result shows that physically realizable sums of Mueller matrices should be performed in the form of convex linear combinations of Mueller matrices; the latter are only applicable when the emerging polarization states are incoherently superimposed. The passivity of the composed depolarizing Mueller matrix follows directly from the passivity of its components (Gil 2000a; San José and Gil 2020b).

3.3.6 Changes of Coordinate System and Rotated Mueller Matrices

The transformations of (pure or not) Mueller matrices represented with respect to different coordinate systems are derived from the rule (1.42) for Stokes vectors. Consider the action of a medium represented by a Mueller matrix **M** on a pure state **s** observed with respect to a coordinate system XY: $\mathbf{M}\mathbf{s}=\mathbf{s}'$, and write the analogous equation $\mathbf{M}_{X'Y'}\,\mathbf{s}_{X'Y'}=\mathbf{s}'_{X'Y'}$ referenced with respect to a coordinate system $X'Y'$ rotated through a counterclockwise angle θ with respect to XY. Then

$$\begin{gathered}\mathbf{s}'_{X'Y'}=\mathbf{M}_G(\theta)\,\mathbf{s}'=\mathbf{M}_G(\theta)\,\mathbf{M}\mathbf{s}=\mathbf{M}_G(\theta)\,\mathbf{M}\mathbf{M}_G(-\theta)\mathbf{s}_{X'Y'}=\mathbf{M}_{X'Y'}\,\mathbf{s}_{X'Y'}\\ \mathbf{M}_{X'Y'}=\mathbf{M}_G(\theta)\,\mathbf{M}\mathbf{M}_G(-\theta)\\ \mathbf{M}_G(\theta)\equiv\begin{pmatrix}1&0&0&0\\0&\cos 2\theta&\sin 2\theta&0\\0&-\sin 2\theta&\cos 2\theta&0\\0&0&0&1\end{pmatrix}\qquad\left[\mathbf{M}_G(-\theta)=\mathbf{M}_G^T(\theta)\right]\end{gathered} \tag{3.67}$$

When the coordinate system XYZ is fixed and the medium represented by **M** is rotated through an angle θ counterclockwise about the Z axis (the direction of propagation of the electromagnetic wave), the Mueller matrix of the rotated medium is given by

$$\mathbf{M}(\theta)=\mathbf{M}_G(-\theta)\,\mathbf{M}\mathbf{M}_G(\theta) \tag{3.68}$$

$\mathbf{M}_G(\theta)$ is a particular form of an orthogonal Mueller matrix, which is called a *rotator*. As we shall see in later sections, a rotator can be physically realized by means of a circular retarder.

3.4 Other Mathematical Representations of the Polarimetric Properties of Nondepolarizing Systems

Depending on the context or the convenience, the polarimetric action of a nondepolarizing system can be represented by means of its Jones matrix **T**, its pure Mueller matrix (or Mueller-Jones matrix) $\mathbf{M}_J(\mathbf{T})$, or through other alternative (and equivalent) mathematical forms like the ones presented in the next subsections.

3.4.1 Covariance Matrix

Given a nondepolarizing system characterized by $\mathbf{M}_J(\mathbf{T})$, the corresponding seven independent parameters are also contained in the *pure covariance matrix* $\mathbf{H}_J(\mathbf{M}_J) = \mathbf{H}_J(\mathbf{T})$ (Simon 1982), defined as

$$\mathbf{H}_J\left(\mathbf{M}_J\right) = \frac{1}{4}\sum_{i,j=0}^{3} m_{ij}\left(\boldsymbol{\sigma}_i \otimes \boldsymbol{\sigma}_j^*\right) \tag{3.69}$$

where $m_{kl}\ (k,l = 0,1,2,3)$ are the elements of **M**, $\otimes$ stands for Kronecker product and $\boldsymbol{\sigma}_k$ are the Pauli matrices [Eq. (3.34)]. $\mathbf{H}_J$ is Hermitian and positive semidefinite, with a unique nonzero eigenvalue ($\mathrm{rank}\,\mathbf{H}_J = 1$). Leaving aside passivity, which is characterized by the condition (3.49), $m_{00}(1+D) \le 1$, a 4×4 real matrix is a pure Mueller matrix if and only if its associated Hermitian matrix defined by Eq. (3.69) has an eigenvalue spectrum of the form $(\lambda_1, 0, 0, 0)$ with $\lambda_1 > 0$.

The representation through the covariance matrix is of particular relevance when applied to the general case of depolarizing systems which can always be considered as additive compositions of pure systems (with corresponding associated pure covariance matrices) (Cloude 1986).

By denoting the elements of the Jones matrix **T** in the following alternative manner

$$\mathbf{T} \equiv \begin{pmatrix} t_0 & t_1 \\ t_2 & t_3 \end{pmatrix} \tag{3.70}$$

$\mathbf{H}_J(\mathbf{T})$ can be expressed in terms of the elements of the *Jones covariance vector* $\mathbf{t} \equiv (t_0, t_1, t_2, t_3)^T$ as (Simon 1982; Girgel 1991)

$$\mathbf{H}_J = \frac{1}{2}\left(\mathbf{t} \otimes \mathbf{t}^\dagger\right) \qquad \mathbf{t} \equiv \left(t_0, t_1, t_2, t_3\right)^T \tag{3.71}$$

3.4.2 Covariance Vector

The Hermitian pure covariance matrix $\mathbf{H}_J$ can be diagonalized as

$$\mathbf{H}_J = \mathbf{U}\,\mathrm{diag}\left(\mathrm{tr}\,\mathbf{H}_J, 0, 0, 0\right)\mathbf{U}^\dagger = \mathbf{w} \otimes \mathbf{w}^\dagger \qquad |\mathbf{w}| = \sqrt{\mathrm{tr}\,\mathbf{H}_J} = \sqrt{m_{00}} \tag{3.72}$$

where the *covariance* vector $\mathbf{w} \equiv (w_0, w_1, w_2, w_3)^T$ is the single eigenvector with nonzero eigenvalue, constituting the first column of the unitary matrix **U** that diagonalizes $\mathbf{H}_J$.

The relation between **w** and the corresponding Jones covariance vector **t** (and hence, to the Jones matrix **T**) is simply (Gil and San José 2013)

$$\mathbf{w} = \sqrt{m_{00}}\,\hat{\mathbf{w}} = \mathbf{t}/\sqrt{2} \qquad \hat{\mathbf{w}} \equiv \mathbf{w}/\sqrt{m_{00}} \qquad |\hat{\mathbf{w}}| = 1 \tag{3.73}$$

Recall that global phase factors in **T**, **t** and **w** are physically irrelevant in polarimetric intensity measurements and thus, a pure system is characterized by up to seven independent real parameters.

3.4.3 Coherency Vector and Coherency Matrix

If the matrix **T** is expanded on the basis of the Pauli matrices, $\mathbf{T} = c_0\boldsymbol{\sigma}_0 + c_1\boldsymbol{\sigma}_1 + c_2\boldsymbol{\sigma}_2 + c_3\boldsymbol{\sigma}_3$, then the complex coefficients c_i of the expansion can be looked upon as the components of the so called *coherency vector* $\mathbf{c} = (c_0, c_1, c_2, c_3)^T$. The respective *pure coherency matrix* $\mathbf{C}_J$ (associated with **T**, $\mathbf{M}_J$, and $\mathbf{H}_J$) is defined as $\mathbf{C}_J = \mathbf{c} \otimes \mathbf{c}^\dagger = \mathcal{L}\mathbf{H}_J\mathcal{L}^{-1}$, where $\mathcal{L}$ is the transfer matrix introduced in Eq. (3.30b). $\mathbf{C}_J$ has the peculiar property that it is diagonal if and only if $\mathbf{M}_J$ is diagonal, so that the use of $\hat{\mathbf{c}}$ and $\mathbf{C}_J$ is advantageous when dealing with certain properties of Mueller matrices.

The coherency vector **w** of a serial combination of two pure media with respective coherency vectors **u** and **v** (the light passing through the medium represented by **u** and then, through that represented by **v**) is given by the *coherency product* (San José and Gil 2020a)

$$\mathbf{w} = \mathbf{u} \bullet \mathbf{v} = \begin{pmatrix} \mathbf{u}^T\mathbf{G}_0\mathbf{v} \\ \mathbf{u}^T\mathbf{G}_1\mathbf{v} \\ \mathbf{u}^T\mathbf{G}_2\mathbf{v} \\ \mathbf{u}^T\mathbf{G}_3\mathbf{v} \end{pmatrix}, \quad \mathbf{G}_0 \equiv \mathbf{I}_4, \quad \mathbf{G}_1 \equiv \begin{pmatrix} 0 & 1 & 0 & 0 \\ 1 & 0 & 0 & 0 \\ 0 & 0 & 0 & i \\ 0 & 0 & -i & 0 \end{pmatrix}, \quad \mathbf{G}_2 \equiv \begin{pmatrix} 0 & 0 & 1 & 0 \\ 0 & 0 & 0 & -i \\ 1 & 0 & 0 & 0 \\ 0 & i & 0 & 0 \end{pmatrix},$$
$$\mathbf{G}_3 \equiv \begin{pmatrix} 0 & 0 & 0 & 1 \\ 0 & 0 & i & 0 \\ 0 & -i & 0 & 0 \\ 1 & 0 & 0 & 0 \end{pmatrix}, \tag{3.74}$$

where $\mathbf{G}_0$ is the 4×4 identity matrix $\mathbf{I}_4$, while $\mathbf{G}_i$ $(i = 1,2,3)$ satisfy properties analogous to those of the Pauli matrices (Cloude, 1986; San José and Gil 2020a). The components w_i of the coherency vector $\mathbf{w} = \mathbf{u} \bullet \mathbf{v}$ are given by $w_i = \mathbf{u}^T\mathbf{G}_i\mathbf{v}$.

The elements m_{ij} of the Mueller matrix $\mathbf{M}_J$ associated with a given coherency vector **c** can be expressed as $m_{ij} = \mathbf{c}^\dagger(\mathbf{G}_i^*\mathbf{G}_j)\mathbf{c}$, which evidences the fact that each element m_{ij} is given by a norm (or measure) of the coherency vector **c**, through the corresponding metric $\mathbf{G}_i^*\mathbf{G}_j$. Moreover, $m_{ij} = \mathrm{tr}(\mathbf{G}_i^*\mathbf{G}_j\mathbf{C})$ and $c_{ij} = [\mathrm{tr}(\mathbf{G}_i^*\mathbf{G}_j\mathbf{M})]/4$.

The transformation $\mathbf{s}' = \mathbf{M}_J\mathbf{s}$ of an incident Stokes vector **s** into an emerging one $\mathbf{s}'$ is expressed in terms of the coherency vector **c** associated with $\mathbf{M}_J$ in the compact form $\mathbf{s}' = \mathbf{c} \bullet \mathbf{s} \bullet \mathbf{c}^*$. A comprehensive analysis of the coherency vector formalism can be found in (San José and Gil 2020a).

3.4.4 The Complex Mueller Matrix

As seen from Eq. (3.32), the transformation of a Jones matrix $\mathbf{T}$ to its associated pure Mueller matrix $\mathbf{M}_J$ is given by $\mathbf{M}_J \equiv \mathcal{L}\left(\mathbf{T}\otimes\mathbf{T}^*\right)\mathcal{L}^+$. In accordance with Parke (1948), the matrix $\mathbf{W}\equiv\mathbf{T}\otimes\mathbf{T}^*$ is sometimes referred to as the *complex pure Mueller matrix*. It is evident that the entire Stokes-Mueller algebra for pure systems, including the transformation $\mathbf{s}' = \mathbf{M}\mathbf{s}$, can alternatively be formulated in terms of (1) complex Mueller matrices $\mathbf{W}$ and (2) polarization covariance vectors defined as $\boldsymbol{\varphi} = \mathcal{L}^+\mathbf{s}/\sqrt{2}$ from the Stokes vectors $\mathbf{s}$, so that $\boldsymbol{\varphi}' = \mathbf{W}\boldsymbol{\varphi}$. As with Jones and Mueller matrices, the complex Mueller matrix of a serial combination of nondepolarizing media is given by the ordered product of their associated complex Mueller matrices.

3.4.5 Matrix States

Starting from the expression of the Jones matrix in terms of the elements of the coherency vector (Cloude 1986; Chipman 1987; Sheppard 2016a; Kuntman et al. 2017a; San José and Gil 2020a)

$$\mathbf{T}=\begin{pmatrix} c_0+c_1 & c_2-ic_3 \\ c_2+ic_3 & c_0-c_1 \end{pmatrix} \tag{3.75}$$

The corresponding *matrix state generator* is defined as (Chipman 1987; Kuntman 2017a; 2017b)

$$\mathbf{Z}=\begin{pmatrix} c_0 & c_1 & c_2 & c_3 \\ c_1 & c_0 & -ic_3 & ic_2 \\ c_2 & ic_3 & c_0 & -ic_1 \\ c_3 & -ic_2 & ic_1 & c_0 \end{pmatrix} \tag{3.76}$$

The corresponding Mueller matrix can be expressed as (Kuntman 2017a; 2017b)

$$\mathbf{M}_J = \mathbf{Z}\mathbf{Z}^* = \mathbf{Z}^*\mathbf{Z} = \mathcal{L}\left(\mathbf{T}\otimes\mathbf{I}_2\right)\mathcal{L}^{-1} = \mathcal{L}\left(\mathbf{I}_2\otimes\mathbf{T}\right)\mathcal{L}^{-1} \tag{3.77}$$

Note that, like $\mathbf{T}$, the matrix $\mathbf{Z}$ (called the Z-matrix) carries information on the absolute phase introduced by the medium. The Z-matrix of a serial composition of pure media is given by the ordered product of the Z-matrices of the components.

3.4.6 Quaternion States

The *quaternion state* characterizing the polarimetric properties of a nondepolarizing medium is defined by arranging the coherency vector in the following quaternion form (Kuntman et al. 2019):

$$Q \equiv c_0\mathbf{1} + ic_1\mathbf{i} + ic_2\mathbf{j} + ic_3\mathbf{k} \tag{3.78}$$

where $(\mathbf{1},\mathbf{i},\mathbf{j},\mathbf{k})$ constitute the quaternion basis.

The transformation of the quaternion form $s \equiv s_0 \mathbf{1} + i s_1 \mathbf{i} + i s_2 \mathbf{j} + i s_3 \mathbf{k}$ of an incident Stokes vector $\mathbf{s}$ into the quaternion form s' of an emerging Stokes vector $\mathbf{s}'$ through the action of a pure medium with quaternion state Q is given by $s' = Q\, s'\, Q^\dagger$ (Kuntman et al. 2019).

Like $\mathbf{T}$, $\mathbf{M}_J$ and $\mathbf{Z}$, the quaternion state of a serial composition of a set of pure media is given by the ordered product of their respective quaternions (Kuntman et al. 2019).

3.4.7 The Jones Operator

An additional quantum-like way for the representation of pure systems is the one based on the so-called *Pauli algebraic approach* which is a formulation consisting in the combined use of the Jones matrix (for the medium) and the polarization matrix (for the polarization state). The Pauli algebraic approach has been dealt with by some authors in the past (Whitney 1971) and was explicitly formulated by Tudor (2008) for both partially polarized states and nondepolarizing devices.

A linear operator A can be represented as $A = a_0\,\sigma_0 + \mathbf{a}\boldsymbol{\sigma}$ in the basis of the Pauli operators where σ_0 is the 2×2 identity operator, and $\boldsymbol{\sigma}(\sigma_1,\sigma_2,\sigma_3)$ is the Pauli vectorial operator. This representation is applicable to nondepolarizing devices transforming partially polarized states of electromagnetic radiation. The operator Φ, corresponding to a state of polarization, is written as $\Phi = s_0\hat{\Phi} = s_0(\sigma_0 + \mathcal{P}\,\hat{\mathbf{u}}\,\boldsymbol{\sigma})/2$, where $\hat{\Phi}$ is the *polarization density operator*, s_0 is the intensity (power density flux of the wave), $\mathcal{P}$ is the degree of polarization, and the *Poincaré vector* $\hat{\mathbf{u}}$ of the operator Φ is the real unit vector that characterizes the polarization ellipse (i.e., the state of polarization of the totally polarized component). The action on Φ by a linear pure system characterized by the linear *Jones operator* T is expressed as $\Phi' = T\,\Phi\,T^\dagger$, where Φ' is the linear operator representing the emerging state of polarization.

3.4.8 The Scattering Matrix: Sinclair Matrix and Kennaugh Matrix

Due to historical and practical reasons, the conventions for the matrix representation of the scattering medium (or "target") sometimes differ between optical and radar communities. Thus, two alternative conventions are used in radar polarimetry for the reference frame used for the scattered wave, namely the forward scatter alignment (FSA), and the backscatter alignment (BSA). The FSA convention coincides with that commonly used in Optics, where the reference axes XYZ form a right handed orthogonal frame (the positive Z axis being the direction of propagation) for both the incident and scattered wave. However, in the BSA convention the positive Z axis points towards the target for both incident and scattered waves (Figure 3.7). Therefore, FSA and BSA differ only with respect to the convention for the scattered wave. The BSA convention is less intuitive, but it is the common choice in synthetic aperture radar (SAR) imaging because in the case of monostatic radar configuration the BSA coordinate systems for the incident and backscattered waves coincide.

When using the FSA convention (as is the choice in all formulations throughout this book) the *scattering matrix* that describes the linear transformations of Jones vectors is the Jones matrix $\mathbf{T}$, while it is usually termed *Sinclair matrix* $\mathbf{S}$ when using the BSA convention. Therefore, $\mathbf{T}$ and $\mathbf{S}$ are linked through the simple relations $\mathbf{S} = \mathrm{diag}\,(1,-1)\,\mathbf{T}$ and $\mathbf{T} = \mathrm{diag}\,(1,-1)\,\mathbf{S}$.

Typically, a radar polarimeter transmits a linearly polarized wave and receives echoes along two orthogonal polarizations simultaneously; then, it transmits another wave whose polarization state is orthogonal with respect to the initially transmitted one and, again, receives echoes along the two orthogonal polarizations simultaneously.

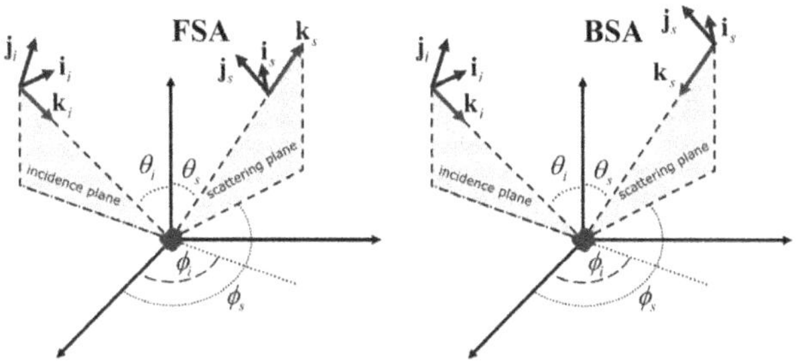

FIGURE 3.7
(a) FSA convention. (b) BSA convention. In both conventions, the incident wave is referred with respect to a right-handed orthogonal reference system ($\mathbf{i}_i$, $\mathbf{j}_i$, $\mathbf{k}_i$). However, unlike in the FSA, in the BSA convention the reference vector $\mathbf{k}_s$, taken along the scattered direction of propagation, points towards the scatterer.

Microwave remote sensing by means of SAR polarimetry differs from optical polarimetry. While optical waves have very short coherence lengths so that they are frequently depolarized by the effect of the sample (or the target, in the remote sensing terminology), microwaves and millimeter-waves have large coherence lengths so that coherence is preserved after their interaction with the target.

Since the vertical–horizontal (V-H) basis is commonly used in radar polarimetry (because of the antenna configuration of the airborne polarimeter), the scattering matrix is usually denoted as

$$\mathbf{S} \equiv \begin{pmatrix} S_{HH} & S_{HV} \\ S_{VH} & S_{VV} \end{pmatrix} \tag{3.79}$$

The elements of the scattering matrix (Jones or Sinclair) can be arranged in a vector form like the covariance vector $\mathbf{w}$ defined above (but generally not normalized) which, in SAR polarimetry context, is called *covariance vector*,

$$\mathbf{k}_C \equiv \left(S_{HH}, S_{HV}, S_{VH}, S_{VV}\right)^T \tag{3.80}$$

Due to the general reciprocity properties that hold in monostatic radar polarimetry, the matrix $\mathbf{S}$ is a symmetric matrix, so that $S_{HV} = S_{VH}$ and, thus, a reduced version of the covariance vector is formulated as $\mathbf{k}_C \equiv (S_{HH}, \sqrt{2}S_{HV}, S_{VV})^T$. Furthermore, the radar *covariance matrix* is defined in terms of measurable quantities in the power domain as the statistical average $\mathbf{C} \equiv \langle \mathbf{k}_C \otimes \mathbf{k}_C^{\dagger} \rangle$, which is employed either in the complete 4×4 formulation or in the reduced 3×3 version, depending on whether reciprocity is assumed or not. An alternative vector formulation commonly used in SAR polarimetry is that of the *scattering vector*, defined as

$$\mathbf{k}_T \equiv \frac{1}{\sqrt{2}}\left[S_{HH} + S_{VV}, S_{HH} - S_{VV}, S_{HV} + S_{VH}, i\left(S_{HV} - S_{VH}\right)\right]^T \tag{3.81}$$

or, assuming reciprocity (monostatic configuration), in its reduced version,

$$\mathbf{k}_T \equiv \frac{1}{\sqrt{2}}\left[S_{HH} + S_{VV}, S_{HH} - S_{VV}, 2S_{HV}\right]^T \tag{3.82}$$

The so-called coherency *matrix*, is defined in terms of measurable quantities in the power domain as the statistical average $\mathbf{T} \equiv \langle \mathbf{k}_T \otimes \mathbf{k}_T^{\dagger} \rangle$. The averaging can be either spatial, in order to reduce the speckle noise that affects the SAR images due to coherent interferences from adjacent scatterers, or temporal, over multiple exposures of the target. It should be stressed that $\mathbf{k}_C$ and $\mathbf{k}_T$ are alternative, but completely equivalent formulations. The same is true for the covariance and coherency matrices.

Moreover, when using the BSA convention, the *Kennaugh matrix* **K** is used instead of the Mueller matrix **M**. The two matrix representations are linked through the simple relations $\mathbf{K} = \mathrm{diag}(1,1,-1,1)\mathbf{M}$ and $\mathbf{M} = \mathrm{diag}(1,1,-1,1)\mathbf{K}$. It is straightforward to check that the matrix $\mathrm{diag}(1,1,-1,1)$, is not a Mueller matrix (its associated coherency matrix **C** has eigenvalues $(1/2,1/2,-1/2,0)$, in contrast with requirement that **C** should have a single nonzero and positive eigenvalue), and consequently, **K** is not a Mueller matrix either. Note also that, provided reciprocity holds, **K** is symmetric under the BSA convention.

In summary, leaving aside the variety of normalization criteria used, as well as the fact that the microwave pulses generated by the antenna of SAR polarimeters usually interact coherently with the target, there exists a parallelism between the mathematical treatments of optical polarimetry and SAR polarimetry. Note that in SAR polarimetry the covariance matrix is denoted by **C** rather than by **H**, and the coherency matrix is denoted by **T** (notation which we have reserved for Jones matrices, as common in related works in the optical field). To maintain a consistent notation, in conformity to that used in Optics, hereafter we will use **H** to refer to the covariance matrix and **C** to refer to the coherency matrix.

For detailed descriptions of the formulation and the conventions used in SAR polarimetry (remote sensing), as well as their relations to those used in Optics, we refer the reader to the treatises by (Lüneburg and Cloude 1997; Boerner 2006; Cloude 2009; Jin and Xu 2013; Sheppard et al. 2020b).

3.5 Singular States of Polarization

From the singular value decomposition (3.8) of the Jones matrix **T**, $\mathbf{T} = \mathbf{T}_{R2}\,\mathrm{diag}(p_1,p_2)\mathbf{T}_{R1}$, we deduce that it is always possible to find a pair of mutually orthogonal pure incident Jones vectors $\hat{\boldsymbol{\eta}}_1$ and $\hat{\boldsymbol{\eta}}_2$ (with unit intensities), such that their corresponding emerging Jones vectors $\hat{\boldsymbol{\eta}}'_1$ and $\hat{\boldsymbol{\eta}}'_2$ are mutually orthogonal. In fact, $\hat{\boldsymbol{\eta}}_1$ and $\hat{\boldsymbol{\eta}}_2$ are the orthonormal eigenvectors of the Jones matrix $\sqrt{\mathbf{T}^{\dagger}\mathbf{T}} = \mathbf{T}_{R1}^{\dagger}\,\mathrm{diag}(p_1,p_2)\mathbf{T}_{R1}$ that are given by the column vectors of the unitary matrix $\mathbf{T}_{R1}^{\dagger}$ (i.e., the row vectors of $\mathbf{T}_{R1}$),

$$\hat{\boldsymbol{\eta}}_1 \equiv \mathbf{T}_{R1}^{\dagger}\begin{pmatrix}1\\0\end{pmatrix} \qquad \hat{\boldsymbol{\eta}}_2 \equiv \mathbf{T}_{R1}^{\dagger}\begin{pmatrix}0\\1\end{pmatrix}$$

$$\hat{\boldsymbol{\eta}}'_1 = \mathbf{T}\hat{\boldsymbol{\eta}}_1 = p_1\left(\mathbf{T}_{R2}\mathbf{T}_{R1}\right)\hat{\boldsymbol{\eta}}_1 = p_1\mathbf{T}_{R2}\begin{pmatrix}1\\0\end{pmatrix} \qquad \hat{\boldsymbol{\eta}}'_2 = \mathbf{T}\hat{\boldsymbol{\eta}}_2 = p_2\left(\mathbf{T}_{R2}\mathbf{T}_{R1}\right)\hat{\boldsymbol{\eta}}_2 = p_2\mathbf{T}_{R2}\begin{pmatrix}0\\1\end{pmatrix} \tag{3.83}$$

wherefrom

$$\hat{\eta}_1'^{\dagger}\,\hat{\eta}_1' = p_1^2 \qquad \hat{\eta}_2'^{\dagger}\,\hat{\eta}_2' = p_2^2 \qquad \hat{\eta}_1'^{\dagger}\,\hat{\eta}_2' = 0 \tag{3.84}$$

It is straightforward to prove that $\hat{\eta}_1'$ and $\hat{\eta}_2'$ are the emerging states with highest and lowest intensities respectively (with respect to arbitrary incident states of unit intensity). Therefore, the *singular states of polarization* $\hat{\eta}_1$ and $\hat{\eta}_2$ correspond to the so-called *maximum and minimum transmittance states* (Lu and Chipman 1994).

The *emerging singular* states $\hat{\eta}_1' \equiv \eta_1'/p_1$ and $\hat{\eta}_2' \equiv \eta_2'/p_2$ are respectively given by the first and second columns of the unitary matrix $\mathbf{T}_{R2}$. Moreover, $\mathbf{T}$ can be expressed as $\mathbf{T} = (\mathbf{T}_{R2}\,\mathbf{T}_{R1})[\mathbf{T}_{R1}^{\dagger}\,\mathrm{diag}(p_1,p_2)\mathbf{T}_{R1}]$, evidencing that the singular states are precisely the eigenstates of $\mathbf{T}_{R1}^{\dagger}\,\mathrm{diag}(p_1,p_2)\mathbf{T}_{R1}$, with respective eigenvalues p_1 and p_2, and that these are transformed into the emerging states η_1' and η_2' by the action of the unitary matrix $\mathbf{T}_{R2}\,\mathbf{T}_{R1}$ (Figure 3.8).

Observe that, when $p_2 = 0$ (i.e., $\det\mathbf{T} = 0$), the second emerging singular state vanishes and the medium behaves as a polarizer (this kind of media will be studied in detail in later sections). It should also be noted that the pair $(\hat{\eta}_1, \hat{\eta}_1')$, together with the singular values (p_1, p_2) and the phase shift between $\hat{\eta}_1$ and $\hat{\eta}_2$, provides the seven independent physical parameters, completely determining $\mathbf{T}$.

If $\mathbf{T}$ is invertible (i.e., $p_2 > 0$), then the orthonormal vectors $\hat{\eta}_1'$ and $\hat{\eta}_2'$ are the singular states of polarization of the Jones matrix

$$\bar{\mathbf{T}} \equiv p_1 p_2 \mathbf{T}^{-1} = \mathbf{T}_{R1}^{\dagger}\,\mathrm{diag}(p_2, p_1)\,\mathbf{T}_{R2}^{\dagger} \tag{3.85}$$

with respective singular values p_2 and p_1

$$\bar{\mathbf{T}}\hat{\eta}_1' = p_2\hat{\eta}_1 \qquad \bar{\mathbf{T}}\hat{\eta}_2' = p_1\hat{\eta}_2 \tag{3.86}$$

If $p_2 = 0$, then the Jones matrix $\bar{\mathbf{T}}$ is defined as $\bar{\mathbf{T}} \equiv \mathbf{T}_{R1}^{\dagger}\,\mathrm{diag}(1,0)\,\mathbf{T}_{R2}^{\dagger}$, so that $\bar{\mathbf{T}}\hat{\eta}_1' = \hat{\eta}_1$. Since the Jones matrix $\bar{\mathbf{T}}$ defined in this manner is passive, the transformations (3.85) and (3.86) are physically realizable.

The singular states of polarization of a pure medium are pure states that can be represented by their Jones vectors $\hat{\eta}_1$ and $\hat{\eta}_2$ or by their associated pure Stokes vectors.

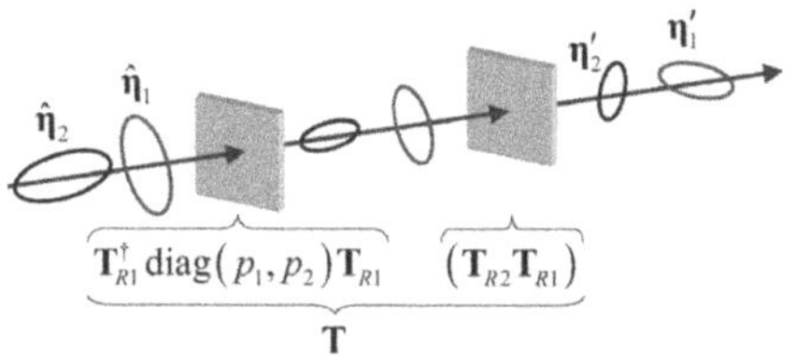

FIGURE 3.8
The singular states of polarization $\hat{\eta}_1$ and $\hat{\eta}_2$ of a Jones matrix $\mathbf{T}$ can be conceived as the eigenstates of the Hermitian matrix $\mathbf{T}_{R1}^{\dagger}\,\mathrm{diag}(p_1,p_2)\mathbf{T}_{R1}$, while the emerging singular states $\hat{\eta}_1'$ and $\hat{\eta}_2'$ can be conceived as the eigenstates of the unitary matrix $\mathbf{T}_{R2}$.

Thus, all previous considerations can be transferred to pure Mueller matrices in a straightforward manner. In fact, from the serial decomposition (3.44) of $\mathbf{M}_J \equiv \mathbf{M}(\mathbf{T})$, we get

$$\mathbf{M}_J = \left(\mathbf{M}_{R2}\,\mathbf{M}_{R1}\right)\left[\mathbf{M}_{R1}^T\,\mathbf{M}_{DL0}\left(p_1,p_2\right)\mathbf{M}_{R1}\right] \tag{3.87}$$

so that

$$\begin{gathered}
\hat{\mathbf{s}}_{\hat{D}+} \equiv \mathbf{M}_{R1}^T\,\mathbf{s}_{1+} \qquad \hat{\mathbf{s}}_{\hat{D}-} \equiv \mathbf{M}_{R1}^T\,\mathbf{s}_{1-} \\
\mathbf{s}'_{\hat{D}+} = \mathbf{M}_J\,\hat{\mathbf{s}}_{\hat{D}+} = p_1^2\,\mathbf{M}_{R2}\mathbf{M}_{R1}\,\hat{\mathbf{s}}_{\hat{D}+} = p_1^2\,\mathbf{M}_{R2}\,\mathbf{s}_{1+} = p_1^2\,\hat{\mathbf{s}}_{\hat{P}+} \\
\mathbf{s}'_{\hat{D}-} = \mathbf{M}_J\,\hat{\mathbf{s}}_{\hat{D}-} = p_2^2\,\mathbf{M}_{R2}\mathbf{M}_{R1}\,\hat{\mathbf{s}}_{\hat{D}-} = p_2^2\,\mathbf{M}_{R2}\,\mathbf{s}_{1-} = p_2^2\,\hat{\mathbf{s}}_{\hat{P}-} \\
\left\{\mathbf{s}_{1+} \equiv (1,1,0,0)^T \quad \mathbf{s}_{1-} \equiv (1,-1,0,0)^T \quad \hat{\mathbf{s}}_{\hat{D}\pm} \equiv \begin{pmatrix} 1 \\ \pm\mathbf{D}/D \end{pmatrix} \quad \hat{\mathbf{s}}_{\hat{P}\pm} \equiv \begin{pmatrix} 1 \\ \pm\mathbf{P}/D \end{pmatrix}\right\}
\end{gathered} \tag{3.88}$$

where $\mathbf{D}$ and $\mathbf{P}$ are the respective diattenuation and polarizance vectors of $\mathbf{M}_J$, with equal absolute values $D = P$.

Therefore, the Stokes vectors $\hat{\mathbf{s}}_{\hat{D}+}$ and $\hat{\mathbf{s}}_{\hat{D}-}$ represent the singular states and are the eigenvectors of the Mueller matrix $\mathbf{M}_{R1}^T\,\mathbf{M}_{DL0}(p_1,p_2)\,\mathbf{M}_{R1}$ with respective eigenvalues p_1^2 and p_2^2. The remaining two eigenvectors are unphysical states represented by vectors that are not Stokes vectors.

The pair $(\hat{\mathbf{s}}_{\hat{D}+}, \hat{\mathbf{s}}_{\hat{P}+})$, together with (p_1^2, p_2^2) and the phase shift between $\hat{\mathbf{s}}_{\hat{D}+}$ and $\hat{\mathbf{s}}_{\hat{D}-}$, provide the seven independent physical parameters necessary for the complete characterization of $\mathbf{M}_J$.

If $\mathbf{M}_J$ is invertible (i.e., $p_2 > 0$), then the orthonormal vectors $\hat{\mathbf{s}}_{\hat{P}+}$ and $\hat{\mathbf{s}}_{\hat{P}-}$ are the singular states of polarization of the Mueller matrix

$$\bar{\mathbf{M}}_J \equiv p_1^2\,p_2^2\,\mathbf{M}_J^{-1} = \mathbf{M}_{R1}^T\,\mathbf{M}_{DL0}\left(p_2,p_1\right)\mathbf{M}_{R2}^T \tag{3.89}$$

with respective singular values p_2^2 and p_1^2 and

$$\bar{\mathbf{M}}_J\,\hat{\mathbf{s}}_{\hat{P}+} = p_2^2\,\hat{\mathbf{s}}_{\hat{D}+} \qquad \bar{\mathbf{M}}_J\,\hat{\mathbf{s}}_{\hat{P}-} = p_1^2\,\hat{\mathbf{s}}_{\hat{D}-} \tag{3.90}$$

If $p_2 = 0$ (i.e., $\det\mathbf{M}_J = 0$) the second emerging singular state vanishes and the pure Mueller matrix $\bar{\mathbf{M}}_J$ is defined as $\bar{\mathbf{M}}_J = \mathbf{M}_{R1}^T\,\mathbf{M}_{DL0}(1,0)\,\mathbf{M}_{R2}^T$, so that $\bar{\mathbf{M}}_J\,\hat{\mathbf{s}}_{\hat{P}+} = \hat{\mathbf{s}}_{\hat{D}+}$. Since $\bar{\mathbf{M}}_J$ has been defined as passive, the transformations (3.89) and (3.90) are physically realizable.

3.6 Normality and Degeneracy of Jones and Mueller Matrices

Since pure Mueller matrices and Jones matrices represent, in two different ways, the same physical polarimetric transformation, a given pure Mueller matrix $\mathbf{M}_J$ has the same eigenstates (either one or two) as its corresponding Jones matrix $\mathbf{T}$. Therefore, two types of Jones and Mueller matrices can be distinguished according to the geometrical relation between their eigenstates (i.e., the polarization states which remain invariant under their action), namely: *normal* (or *homogeneous*) having two mutually orthogonal eigenstates, and *non-normal* (or *inhomogeneous*) having two mutually nonorthogonal eigenstates. The non-normal family of Jones and Mueller matrices includes the so-called *degenerate* matrices, which have only one eigenstate (with doubly degenerate eigenvalue) (Lu and Chipman 1994; Sudha and Gopala Rao 2001).

3.6.1 Normal Operators

In accordance with the usual concept of normal matrix, a Jones matrix $\mathbf{T}$ is said to be *normal* if it satisfies $\mathbf{T}\mathbf{T}^{\dagger} = \mathbf{T}^{\dagger}\mathbf{T}$ or, equivalently, if it is unitarily similar to a diagonal matrix. Normal Jones matrices are also characterized by the fact that their two eigenstates are mutually orthogonal. It should be noted that normal Jones and Mueller matrices are also called *homogeneous* (Lu and Chipman 1994). In spite of the common use of this term, in order to avoid confusion with further references to spatially homogeneous systems, we consider preferable using the term *normal* to refer to Jones matrices (as well as to the media represented by them) satisfying the above property (Tudor 2003a).

The singular values of a normal Jones matrix coincide with the absolute values of its eigenvalues. Furthermore, a Jones matrix is normal if and only if its singular states coincide with its eigenstates (eigenvectors of $\mathbf{T}$).

As will be shown in Section 4.9.1, a normal Jones matrix depends on up to five independent parameters, and its most general form corresponds to that of a *normal elliptic diattenuating retarder* whose Jones matrix can be written as

$$\begin{gathered}\mathbf{T}_{RD} = \mathbf{T}_{RL}\,\mathrm{diag}\left(p_1 e^{i\Delta/2}, p_2 e^{-i\Delta/2}\right)\mathbf{T}_{RL}^{\dagger} = \mathbf{T}_{RL}\left(\mathbf{T}_{DL0}\mathbf{T}_{RL0}\right)\mathbf{T}_{RL}^{\dagger} = \mathbf{T}_{RL}\left(\mathbf{T}_{RL0}\mathbf{T}_{DL0}\right)\mathbf{T}_{RL}^{\dagger}\\ \mathbf{T}_{DL0} \equiv \mathrm{diag}\left(p_1, p_2\right) \quad \mathbf{T}_{RL0} \equiv \mathrm{diag}\left(e^{i\Delta/2}, e^{-i\Delta/2}\right)\end{gathered} \tag{3.91}$$

where $\mathbf{T}_{RL}$ is the Jones matrix of a linear retarder (see Section 4.2.1.4). Moreover, a pure Mueller matrix is *normal* when $\mathbf{M}_J\,\mathbf{M}_J^T = \mathbf{M}_J^T\,\mathbf{M}_J$ and, in accordance with Equations (3.44c) and (3.45), has the general form

$$\begin{gathered}\mathbf{M}_{RD} = \mathbf{M}_{RL}\left[\mathbf{M}_{DL0}(p_1,p_2)\,\mathbf{M}_{RL0}(\Delta)\right]\mathbf{M}_{RL}^T = \mathbf{M}_{RL}\left[\mathbf{M}_{RL0}(\Delta)\,\mathbf{M}_{DL0}(p_1,p_2)\right]\mathbf{M}_{RL}^T\\ \mathbf{M}_{DL0}(p_1,p_2) \equiv m_{00}\begin{pmatrix} 1 & D & 0 & 0\\ D & 1 & 0 & 0\\ 0 & 0 & \sqrt{1-D^2} & 0\\ 0 & 0 & 0 & \sqrt{1-D^2}\end{pmatrix}, \quad \mathbf{M}_{RL0}(\Delta) \equiv \begin{pmatrix} 1 & 0 & 0 & 0\\ 0 & 1 & 0 & 0\\ 0 & 0 & \cos\Delta & \sin\Delta\\ 0 & 0 & -\sin\Delta & \cos\Delta\end{pmatrix}\\ \left[m_{00} = \left(p_1^2 + p_2^2\right)/2,\ \ D = \left(p_1^2 - p_2^2\right)/\left(p_1^2 + p_2^2\right)\right]\end{gathered} \tag{3.92}$$

Consequently, normal pure Mueller matrices have necessarily the following partitioned structure:

$$\mathbf{M}_{RD} = m_{00}\begin{pmatrix} 1 & \mathbf{D}^T\\ \mathbf{D} & \mathbf{m}_{RL}\mathbf{m}_{RD0}\mathbf{m}_{RL}^T\end{pmatrix} \quad \left[\mathbf{m}_{RD0} \equiv \mathbf{m}_{RL0}\mathbf{m}_{DL0} = \mathbf{m}_{DL0}\mathbf{m}_{RL0}\right] \tag{3.93}$$

and therefore, a necessary and sufficient condition for a pure system to be represented by a normal Mueller (or Jones) matrix is that the diattenuation and polarizance vectors be equal $(\mathbf{D} = \mathbf{P})$.

Recall that the singular value decomposition (3.46) of $\mathbf{M}_J$ is realized through the non-Mueller matrix $\mathbf{C}_D$, so that it is preferable to avoid the use of $\mathbf{C}_D$ in serial decompositions of Mueller matrices but rather express these in terms of physically realizable Mueller matrices like $\mathbf{M}_{DL0}$ (*horizontal linear diattenuator*), $\mathbf{M}_{RL0}$ (*horizontal linear retarder*) and $\mathbf{M}_R$ (*elliptic retarder*).

The key role played by normal matrices in Jones and Mueller algebras is evidenced by the fact that decompositions (3.8) and (3.44) are formulated in terms of normal matrices, namely the Jones matrices $\mathbf{T}_R$ (unitary, corresponding to generally elliptic retarders) and $\mathbf{T}_{DL0}$ (real and diagonal, corresponding to horizontal linear diattenuators), as well as the pure Mueller matrices $\mathbf{M}_R$ (orthogonal, corresponding to retarders) and $\mathbf{M}_{DL0}$ (symmetric, with non-negative diagonal elements, corresponding to horizontal linear diattenuators).

Particularly useful normal operators are the Jones and Mueller matrices $\mathbf{T}_R$ and $\mathbf{M}_R$ of a retarder (in general, elliptic), as well as the Jones and Mueller matrices $\mathbf{T}_D$ and $\mathbf{M}_D$ of a diattenuator (in general, elliptic).

3.6.2 Non-normal Operators

A non-normal pure Mueller matrix corresponds to a non-normal Jones matrix and vice versa. Given a Jones matrix $\mathbf{T}$, the *non-normality parameter* (or *inhomogeneity parameter*) $\eta(\mathbf{T})$ which provides a measure of the lack of normality of $\mathbf{T}$, is defined as the absolute value of the scalar product of the normalized Jones eigenvectors $\boldsymbol{\varepsilon}_q$ and $\boldsymbol{\varepsilon}_r$ of $\mathbf{T}$ (Lu and Chipman 1994),

$$\eta \equiv \left| \boldsymbol{\varepsilon}_q^{\dagger}\, \boldsymbol{\varepsilon}_r \right| \tag{3.94}$$

so that $0 \leq \eta \leq 1$. The minimum $\eta = 0$ is achieved when $\mathbf{T}$ (i.e., $\mathbf{M}(\mathbf{T})$) is normal, while the maximum non-normality $\eta = 1$ is reached when $\boldsymbol{\varepsilon}_q = \boldsymbol{\varepsilon}_r$ (corresponding to degenerate $\mathbf{T}$ and $\mathbf{M}(\mathbf{T})$).

As seen in the previous section, any pure Mueller matrix with $\mathbf{D} \neq \mathbf{P}$ is non-normal. An example of system with $\eta = 1$ (degenerate) is the *black sandwich*, i.e., the serial combination of a retarder placed between two mutually crossed total polarizers (Berry and Dennis 2004).

The quantity η^2 can be expressed as follows in terms of the corresponding Jones matrix $\mathbf{T}$ (Lu and Chipman 1994),

$$\eta^2 = \frac{2\|\mathbf{T}\|_2^2 - |\mathrm{tr}\,\mathbf{T}|^2 - \left|(\mathrm{tr}\,\mathbf{T})^2 - 4\det\mathbf{T}\right|}{2\|\mathbf{T}\|_2^2 - |\mathrm{tr}\,\mathbf{T}|^2 + \left|(\mathrm{tr}\,\mathbf{T})^2 - 4\det\mathbf{T}\right|} \tag{3.95}$$

or, in terms of the corresponding pure Mueller matrix $\mathbf{M}_J \equiv \mathbf{M}(\mathbf{T})$

$$\eta^2 = \frac{4m_{00} - |T|^2 - \left|T^2 - 4\sqrt[4]{\det\mathbf{M}_J}\right|}{4m_{00} - |T|^2 + \left|T^2 - 4\sqrt[4]{\det\mathbf{M}_J}\right|} \qquad T \equiv \frac{1}{\sqrt{2}} \frac{\mathrm{tr}\,\mathbf{M}_J + m_{01} + m_{10} + i(m_{23} + m_{32})}{\sqrt{m_{00} + m_{11} + m_{01} + m_{10}}} \tag{3.96}$$

An interesting physical and geometric interpretation of η comes from the fact that

$$\eta = \cos(\Theta/2) \tag{3.97}$$

where Θ is the angle subtended by the Poincaré vectors of the eigenstates in the Poincaré sphere representation.

3.6.3 Degenerate Operators

When the Jones matrix $\mathbf{T}$ has a single linearly independent eigenvector $\boldsymbol{\varepsilon}_q$, it has a doubly degenerate eigenvalue ξ_q. The associated pure Mueller matrix $\mathbf{M}_J$ has the same unique eigenstate, whose Stokes vector $\mathbf{s}_q$ provides the representation of the pure state $\boldsymbol{\varepsilon}_q$ on the Poincaré sphere. These kinds of Jones and pure Mueller matrices, as well as the devices represented by them, are called *degenerate* (Lu and Chipman 1994; Ossikovski and Gil 2017). The *degeneracy* defined as

$$\gamma = |t_{12}| + |t_{21}| = \frac{1}{\sqrt{2}}\left(\sqrt{m_{00} - m_{01} + m_{10} - m_{11}} + \sqrt{m_{00} + m_{01} - m_{10} - m_{11}}\right) \tag{3.98}$$

provides a measure for a degenerate $\mathbf{T}$ and its associated degenerate $\mathbf{M}_J$.

The degeneracy γ is invariant under unitary similarity transformations of $\mathbf{T}$, and, equivalently, under orthogonal transformations of $\mathbf{M}_J$,

$$\begin{gathered}\gamma(\mathbf{T}) = \gamma\left(\mathbf{T}_R^{\dagger}\mathbf{T}\,\mathbf{T}_R\right) \qquad \gamma(\mathbf{M}_J) = \gamma\left(\mathbf{M}_R^{T}\mathbf{M}_J\mathbf{M}_R\right) \\ \left[\mathbf{T}_R^{\dagger} = \mathbf{T}_R^{-1} \quad \mathbf{M}_R \equiv \begin{pmatrix} 1 & \mathbf{0}^T \\ \mathbf{0} & \mathbf{m}_R \end{pmatrix} \quad \mathbf{m}_R^{T} = \mathbf{m}_R^{-1}\right]\end{gathered} \tag{3.99}$$

A degenerate Jones matrix $\mathbf{T}_g$ is not diagonalizable through unitary similarity transformations. It can always be expressed as (Lu and Chipman 1994)

$$\mathbf{T}_g = \mathbf{T}_R\mathbf{T}'_{g\gamma}\mathbf{T}_R^{\dagger} \qquad \mathbf{T}'_{g\gamma} \equiv \begin{pmatrix} \xi_q & \gamma \\ 0 & \xi_q \end{pmatrix} \qquad \begin{bmatrix} \xi_q \in \mathbb{R},\ \gamma \in \mathbb{R},\ \gamma > 0 \\ \mathbf{T}_R^{\dagger} = \mathbf{T}_R^{-1} \end{bmatrix} \tag{3.100}$$

where the *Jones canonical degenerate form* $\mathbf{T}'_{g\gamma}$ depends on the doubly degenerate eigenvalue ξ_q and the degeneracy γ. Note that ξ_q is complex in general and can be expressed in the polar form $\xi_q \equiv q e^{i\alpha}$. Since a global phase shift does not affect the physical meaning of a Jones matrix, the representative Jones matrix $\mathbf{T}_{g\gamma} \equiv e^{-i\alpha}\mathbf{T}'_{g\gamma}$ of $\mathbf{T}'_{g\gamma}$ that satisfies the criterion $\det\mathbf{T}_{g\gamma} \in \mathbb{R}$ is

$$\mathbf{T}_{g\gamma} \equiv \begin{pmatrix} q & \gamma e^{-i\alpha} \\ 0 & q \end{pmatrix} \qquad [q, \gamma, \alpha \in \mathbb{R},\ q, \gamma > 0] \tag{3.101}$$

The Jones vector of the eigenstate of $\mathbf{T}_{g\gamma}$ is $\boldsymbol{\varepsilon}_q = (1,0)^T$.

The singular value decomposition (or general serial decomposition) of $\mathbf{T}_{g\gamma}$ is given by

$$\mathbf{T}_{g\gamma} = \mathbf{T}_{R2}\,\mathrm{diag}\left(p_{g1}, p_{g2}\right)\mathbf{T}_{R1}$$

$$\mathbf{T}_{R2} \equiv \frac{1}{\sqrt{\Gamma}}\begin{pmatrix} \dfrac{\sqrt{2}q}{\sqrt{\Gamma-\gamma}}e^{-i\alpha/2} & \dfrac{\sqrt{2}q}{\sqrt{\Gamma+\gamma}}e^{-i\alpha/2} \\ \dfrac{\sqrt{\Gamma-\gamma}}{\sqrt{2}}e^{i\alpha/2} & -\dfrac{\sqrt{\Gamma+\gamma}}{\sqrt{2}}e^{i\alpha/2} \end{pmatrix} \qquad \mathbf{T}_{R1} \equiv \frac{1}{\sqrt{\Gamma}}\begin{pmatrix} \dfrac{\sqrt{2}q}{\sqrt{\Gamma+\gamma}}e^{i\alpha/2} & \dfrac{\sqrt{\Gamma+\gamma}}{\sqrt{2}}e^{-i\alpha/2} \\ \dfrac{\sqrt{2}q}{\sqrt{\Gamma-\gamma}}e^{i\alpha/2} & -\dfrac{\sqrt{\Gamma-\gamma}}{\sqrt{2}}e^{-i\alpha/2} \end{pmatrix}$$

$$\left(\Gamma \equiv \sqrt{4q^2+\gamma^2}\right) \tag{3.102}$$

This representation allows for the characterization of degenerate Jones matrices as those that can be written in the form

$$\mathbf{T}_g = \mathbf{T}_R\,\mathbf{T}_{g\gamma}\,\mathbf{T}_R^{\dagger} = \mathbf{T}_R\left[\mathbf{T}_{R2}\,\mathrm{diag}\left(p_{g1}, p_{g2}\right)\mathbf{T}_{R1}\right]\mathbf{T}_R^{\dagger} \tag{3.103}$$

where $\mathbf{T}_{R1}$ and $\mathbf{T}_{R2}$ are unitary, with the particular forms shown in Eq. (3.102), while $\mathbf{T}_R$ is unitary, with arbitrary form. The square of the singular values, or principal intensity coefficients, p_{g1}^2 and p_{g2}^2, of $\mathbf{T}_g$ (which coincide with those of $\mathbf{T}_{g\gamma}$) are

$$\begin{aligned} p_{g1}^2 &= \frac{1}{4}(\Gamma+\gamma)^2 = \frac{1}{2}\left(2q^2+\gamma^2+\gamma\sqrt{4q^2+\gamma^2}\right) \\ p_{g2}^2 &= \frac{1}{4}(\Gamma-\gamma)^2 = \frac{1}{2}\left(2q^2+\gamma^2-\gamma\sqrt{4q^2+\gamma^2}\right) \end{aligned} \tag{3.104}$$

Note that the inequality $p_{g1} \neq p_{g2}$ necessarily holds for degenerate systems $(\gamma > 0)$.

The degenerate pure Mueller matrix $\mathbf{M}_{g\gamma}$ corresponding to $\mathbf{T}_{g\gamma}$ has the *Mueller canonical degenerate form*

$$\mathbf{M}_{g\gamma} = \frac{1}{2}\begin{pmatrix} 2q^2+\gamma^2 & -\gamma^2 & 2q\gamma\cos\alpha & -2q\gamma\sin\alpha \\ \gamma^2 & 2q^2-\gamma^2 & 2q\gamma\cos\alpha & -2q\gamma\sin\alpha \\ 2q\gamma\cos\alpha & -2q\gamma\cos\alpha & 2q^2 & 0 \\ -2q\gamma\sin\alpha & 2q\gamma\sin\alpha & 0 & 2q^2 \end{pmatrix} \quad (\gamma>0) \tag{3.105}$$

The Stokes vector of the eigenstate of $\mathbf{M}_{g\gamma}$ is $\mathbf{s}_q = (1,1,0,0)^T$, which obviously corresponds to the Jones vector $\boldsymbol{\varepsilon}_q$.

Observe that this particular representation, which corresponds to the Jones canonical degenerate form (3.101), has the property

$$\mathrm{diag}(-1,1,1)\mathbf{P} = \mathbf{D} \tag{3.106}$$

Thus, in accordance with the transformation (3.100) of an arbitrary degenerate Jones matrix $\mathbf{T}_g$ to its canonical degenerate form, any degenerate pure Mueller matrix $\mathbf{M}_g$

can be transformed to its canonical degenerate form $\mathbf{M}_{g\gamma}$ through a Mueller orthogonal transformation

$$\mathbf{M}_{g\gamma} = \mathbf{M}_R^T \, \mathbf{M}_g \, \mathbf{M}_R \tag{3.107}$$

in such a manner that the polarizance and diattenuation vectors of $\mathbf{M}_{g\gamma}$ satisfy the condition (3.106).

Furthermore, by means of the singular value decomposition $\mathbf{m}_{g\gamma} = \mathbf{m}'_{R2}\,\mathrm{diag}(a_1, a_2, a_3)\,\mathbf{m}'_{R1}$ of the submatrix $\mathbf{m}_{g\gamma}$ of $\mathbf{M}_{g\gamma}$, we find that any degenerate Mueller matrix can be expressed as

$$\mathbf{M}_g = \mathbf{M}_R \left(\mathbf{M}_{R2}\,\mathbf{M}_{gDL0}\,\mathbf{M}_{R1}\right)\mathbf{M}_R^T$$

$$\mathbf{M}_{gDL0} \equiv \frac{1}{2}\begin{pmatrix} 2q^2+\gamma^2 & \gamma\sqrt{4q^2+\gamma^2} & 0 & 0 \\ \gamma\sqrt{4q^2+\gamma^2} & 2q^2+\gamma^2 & 0 & 0 \\ 0 & 0 & 2q^2 & 0 \\ 0 & 0 & 0 & 2q^2 \end{pmatrix} \quad \begin{bmatrix} \mathbf{M}_{R1} \equiv \mathbf{M}(\mathbf{T}_{R1}) \\ \mathbf{M}_{R2} \equiv \mathbf{M}(\mathbf{T}_{R2}) \end{bmatrix} \tag{3.108}$$

It should be stressed that, while $\mathbf{M}_R$ is a general orthogonal Mueller matrix, $\mathbf{M}_{R1}$ and $\mathbf{M}_{R2}$ are the orthogonal Mueller matrices corresponding to the unitary Jones matrices $\mathbf{T}_{R1}$ and $\mathbf{T}_{R2}$ which have the particular forms expressed in terms of q, α and the degeneracy γ shown in Eq. (3.102). In other words, unlike the Jones and Mueller degenerate forms $\mathbf{T}_{g\gamma}$ and $\mathbf{M}_{g\gamma}$, $\mathbf{T}_{gDL0}$ and $\mathbf{M}_{gDL0}$ are not degenerate and therefore, the transformations $\mathbf{T}_{R2}\,\mathbf{T}_{gDL0}\,\mathbf{T}_{R1}$ and $\mathbf{M}_{R2}\,\mathbf{M}_{gDL0}\,\mathbf{M}_{R1}$ are precisely those that convert the nondegenerate $\mathbf{T}_{gDL0}$ and $\mathbf{M}_{gDL0}$ into the respective Jones and Mueller canonical degenerate forms.

As we shall see in Chapter 4, the polarizance-diattenuation D of a pure Mueller matrix is an intrinsic quantity that remains invariant under Mueller orthogonal transformations, so that $D(\mathbf{M}_g) = D(\mathbf{M}_{gDL0}) \equiv D_g$ and thus, the polarizance-diattenuation of a degenerate Mueller matrix $\mathbf{M}_g$ is given by

$$D_g = \gamma\frac{\sqrt{4q^2+\gamma^2}}{2q^2+\gamma^2} \tag{3.109}$$

From the above results, it is straightforward to show that the passivity condition for $\mathbf{M}_g$ is

$$\sqrt{4q^2+\gamma^2} + \gamma \leq 2 \tag{3.110}$$

4

Nondepolarizing Media: Retarders, Diattenuators and Serial Decompositions

4.1 Introduction

When dealing with the effects of material media on the polarization states of electromagnetic beams interacting with them, it is particularly interesting to identify the basic polarimetric phenomena and to represent them through appropriate matrix formalisms. As we shall see, the linear effect of a nondepolarizing medium can always be represented by means of an equivalent system constituted by a serial combination of two canonical components, namely a *retarder* and a *diattenuator*, whose respective physical interpretation and matrix representation is simple and complementary. The nature and characterization of nondepolarizing media can therefore be analyzed through serial combinations of the said kinds of simple media. The present chapter is devoted to the study and representation (in both Jones and Mueller formalisms) of retarders, diattenuators and any sort of serial combination of them. Moreover, concepts like normality, degeneracy, rotational invariance, as well as properties like polarizance, diattenuation and retardance are discussed. Furthermore, the contents of this chapter, together with those of Chapter 3, constitute the necessary framework to address, in later chapters, the characterization and study of depolarizing media.

4.2 Retarders

The term *retarder* refers to those media whose polarimetric action exhibits the phenomenon of *birefringence* (anisotropy of the refractive index). Any incident polarization state impinging on a retarder can be decomposed into two mutually orthogonal states (called *fast* and *slow* components) that "see" two different refractive indexes (sometimes, the difference in refractive indexes alone is referred to as birefringence). As a result, the retarder introduces a phase shift Δ between the fast and slow components, which constitute the two mutually orthogonal eigenstates of the retarder. Provided the thickness of the retarder is sufficiently small not to destroy the coherence between the two components, the emerging state is a coherent superposition of the fast and slow emerging components, resulting in a state of polarization different from the incident one (Figure 4.1). Unlike diattenuators we shall study later, ideal retarders preserve intensity, i.e., the intensity of the emerging state is equal to that of the incident one.

DOI: 10.1201/9780367815578-4

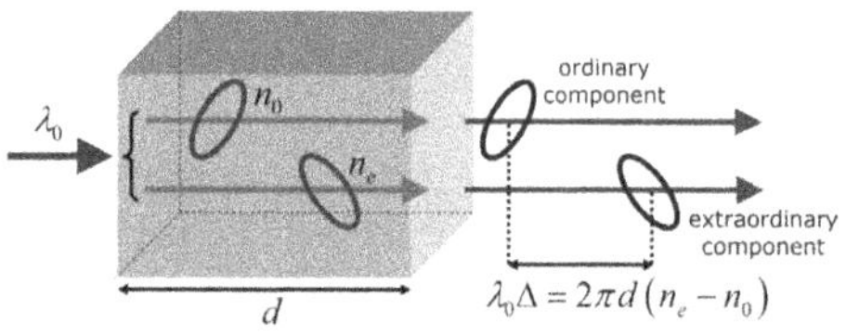

FIGURE 4.1
The incident state (polarized or not) is split into two beams with mutually orthogonal polarization states for which the retarder exhibits different refraction indexes n_o and n_e at the central wavelength λ_o. The emerging state results from the superposition of the respective exiting states. The superposition is coherent when the longitudinal displacement $\lambda_o\Delta$ is smaller than the coherence length of light (as in waveplates) and is partially coherent or incoherent otherwise (as in long optical fibers).

A great variety of media behaving as retarders can be found in nature and industry. Birefringence phenomena are produced e.g. by calcite and other anisotropic crystals; glucose solutions (optical activity); transparent plastic objects and sheets; cellophane; liquid crystal devices; optical fibers; Kerr, Pockels and Faraday cells (electrooptic and magnetooptic effects), etc. (Pye 2001).

Ideal (in the sense of pure, i.e. nondepolarizing, and intensity-preserving) retarders are represented either by unitary Jones matrices $\mathbf{T}_R$ (satisfying $\mathbf{T}_R^{-1} = \mathbf{T}_R^{\dagger}$) or by orthogonal Mueller matrices $\mathbf{M}_R$ (satisfying $\mathbf{M}_R^{-1} = \mathbf{M}_R^{T}$) which have zero diattenuation-polarizance and thus, have the form

$$\mathbf{M}_R = \begin{pmatrix} 1 & \mathbf{0}^T \\ \mathbf{0} & \mathbf{m}_R \end{pmatrix} \qquad \left[\mathbf{m}_R^T = \mathbf{m}_R^{-1} \qquad \det \mathbf{m}_R = +1\right] \tag{4.1}$$

where $\mathbf{m}_R$ is a 3×3 proper orthogonal matrix.

The previous definition of a retarder is an ideal limiting case of real situations where

- The intensity is reduced in an isotropic manner, in which case the medium is represented by $c\,\mathbf{M}_R$, with an isotropic attenuation coefficient c in the range $0 < c \le 1$.
- The retarder is thick enough to break the coherence between the ordinary and extraordinary modes, in which case the retarder produces depolarization. This effect can usually be neglected for thin slabs of birefringent media, while it may become important when the medium is relatively thick, as is especially the case in optical fibers. The depolarization caused by the dispersive effects of birefringent media will be considered in later chapters devoted to depolarizing systems.
- The internal multiple reflections between the two faces of a birefringent plate result in a diattenuating effect (Clarke and Grainger 1974), so that the fast and slow components are affected by different attenuation factors. When this phenomenon is appreciable, or requires to be accounted for, the medium is called a *non-ideal retarder* (or a *diattenuating retarder*) and is represented as a product of a diattenuator and a retarder whose optical axes have coincident orientations. This kind of medium will be considered in later sections, once the diattenuators have been studied.

As usual, hereafter, unless otherwise indicated, we shall use the term *retarder* to refer to an ideal pure retarder.

The physical phenomenon responsible for retarding effects is the birefringence and it is worth stressing that birefringent systems can likewise depolarize quasi-monochromatic polarization states. For instance, this is the case with long haul optical fiber systems where the difference between the path-length traveled by the modes can be larger than the coherence length. As we have indicated previously, these cases are not covered by the Jones formalism which only applies when the accumulated difference between the path-lengths of the modes is smaller than the coherence length of the electromagnetic wave. Thus, the treatment of linear interactions involving depolarizing effects requires the use of the Mueller formalism.

The wavelength dependence of the effective retardance of retarders is used in the design of Lyot and Solc filters, which are constituted of appropriate serial combinations of retarders placed between polarizers. The spectral calculations can be performed by using the Jones formalism whereas the study of the accompanying depolarization effects requires the more general Mueller formalism (Dlugnikov 1984). These considerations apply also to liquid crystal optical filters, electro-optical polarization controllers, polarization-maintaining fibers, ring lasers, birefringent tuning in dye lasers, thin film filters for optical sensing of gas concentration, active filters for color imaging, optical flip-flop systems and many other industrial and scientific devices.

4.2.1 Jones Matrices of Retarders

The most general case of a retarder is the elliptical retarder whose eigenstates are mutually orthogonal pure states characterized by the following unit Jones vectors:

$$\begin{gathered}\boldsymbol{\varepsilon}_1(\alpha,\delta)\equiv\begin{pmatrix}\cos\alpha\, e^{-i\delta/2}\\ \sin\alpha\, e^{i\delta/2}\end{pmatrix}=\begin{pmatrix}\cos\chi\cos\varphi-i\sin\chi\sin\varphi\\ \cos\chi\sin\varphi+i\sin\chi\cos\varphi\end{pmatrix}\\ \boldsymbol{\varepsilon}_2(\pi/2-\alpha,\delta+\pi)\equiv\begin{pmatrix}-i\sin\alpha\, e^{-i\delta/2}\\ i\cos\alpha\, e^{i\delta/2}\end{pmatrix}=\begin{pmatrix}-\cos\chi\sin\varphi+i\sin\chi\cos\varphi\\ \cos\chi\cos\varphi+i\sin\chi\sin\varphi\end{pmatrix}\end{gathered}\tag{4.2}$$

where, as introduced in Section 1.2, α is the angle whose tangent is the ratio between the amplitudes of the components of the electric field ($\tan\alpha\equiv a_2/a_1$) and δ is the phase shift between the said components. Thus, the pair of angles (α,δ) determines the shape and the orientation of the polarization ellipses of the states $\boldsymbol{\varepsilon}_1(\alpha,\delta)$ and $\boldsymbol{\varepsilon}_2(\pi/2-\alpha,\delta+\pi)$ and therefore, defines the corresponding azimuth φ and ellipticity angle χ (see Eq. (1.6)): $\tan 2\varphi=\tan 2\alpha\cos\delta$, $\sin 2\chi=\sin 2\alpha\sin\delta$.

4.2.1.1 *Elliptical Retarder*

The general form of the Jones matrix $\mathbf{T}_R(\alpha,\delta,\Delta)$ of an elliptical retarder with fast eigenstate $\boldsymbol{\varepsilon}_1(\alpha,\delta)$ introducing the phase shift Δ between the two eigenstates $\boldsymbol{\varepsilon}_1(\alpha,\delta)$ and $\boldsymbol{\varepsilon}_2(\pi/2-\alpha,\delta+\pi)$ can be expressed as

$$\mathbf{T}_R(\alpha,\delta,\Delta)=\begin{pmatrix}c_\alpha^2 e^{i\Delta/2}+s_\alpha^2 e^{-i\Delta/2} & i s_{2\alpha} s_{\Delta/2} e^{-i\delta}\\ i s_{2\alpha} s_{\Delta/2} e^{i\delta} & s_\alpha^2 e^{i\Delta/2}+c_\alpha^2 e^{-i\Delta/2}\end{pmatrix}\qquad\begin{bmatrix}s_x\equiv\sin x\\ c_x\equiv\cos x\end{bmatrix}\tag{4.3}$$

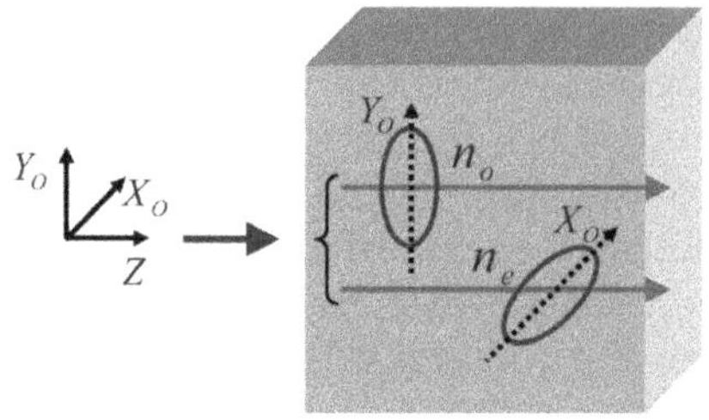

FIGURE 4.2
Reference axes taken so as to represent an elliptical retarder oriented at 0°.

which in turn can be considered as a general expression of a 2×2 unitary matrix $\mathbf{T}_R$ with $\det\mathbf{T} = +1$ (note that the most general expression of the Jones matrix of a retarder is obtained by multiplying $\mathbf{T}_R(\alpha,\delta,\Delta)$ by an arbitrary global phase factor $e^{i\phi}$, in which case $\det\mathbf{T} = e^{i2\phi}$).

4.2.1.2 Elliptical Retarder Oriented at 0°

In particular, the Jones matrix of the elliptical retarder can be represented with respect to its intrinsic reference axes $X_O Y_O$ defined in such a manner that X_O is aligned with the major axis of the polarization ellipse of the fast eigenstate $\boldsymbol{\varepsilon}_1$ $(\varphi=0 \Rightarrow \delta=\pi/2,\ \alpha=\chi)$ and consequently, Y_O is along the minor axis (or, equivalently, the major axis of the slow eigenstate). With this choice, the Jones matrix of the elliptical retarder oriented at 0° is expressed as (Figure 4.2)

$$\mathbf{T}_R(\chi,\pi/2,\Delta)=\begin{pmatrix} c_\chi^2 e^{i\Delta/2}+s_\chi^2 e^{-i\Delta/2} & s_{2\chi}s_{\Delta/2} \\ -s_{2\chi}s_{\Delta/2} & s_\chi^2 e^{i\Delta/2}+c_\chi^2 e^{-i\Delta/2} \end{pmatrix} \tag{4.4}$$

4.2.1.3 Circular Retarder and Rotator

When the retarder eigenstates are right- and left-circularly polarized states $(\chi=\pi/4 \Rightarrow \delta=\pi/2,\ \alpha=\pi/4)$ the Jones matrix of the *circular retarder* is

$$\mathbf{T}_{RC}(\Delta)\equiv\mathbf{T}_R(\pi/4,\pi/2,\Delta)=\begin{pmatrix} \cos(\Delta/2) & \sin(\Delta/2) \\ -\sin(\Delta/2) & \cos(\Delta/2) \end{pmatrix} \tag{4.5}$$

It is remarkable that $\mathbf{T}_{RC}(\Delta)$ has the form of a *rotator* (or *rotation matrix*, as in Eq. (3.26)) with counterclockwise rotation angle $\theta=\Delta/2$ (Figure 4.3) of the coordinate axes XY. Nevertheless, the action of a medium behaving as a rotator should be understood as a clockwise rotation through the angle θ of the electric field components of an incident polarization state, which is performed without change of the coordinate system.

Since a rotator is represented by a particular type of Jones matrix, rotators will be denoted as

$$\mathbf{T}_G(\theta)\equiv\mathbf{T}_{RC}(2\theta)=\begin{pmatrix} \cos\theta & \sin\theta \\ -\sin\theta & \cos\theta \end{pmatrix} \tag{4.6}$$

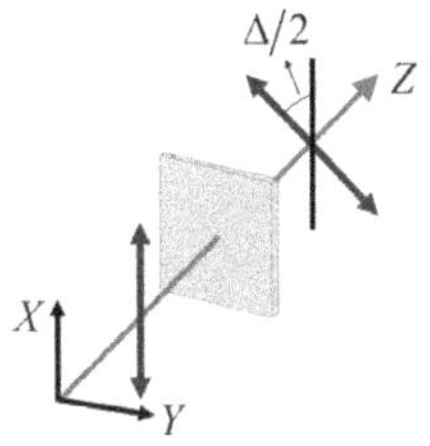

FIGURE 4.3
A circular retarder of retardance Δ behaves as a rotator of angle $\theta = \Delta/2$.

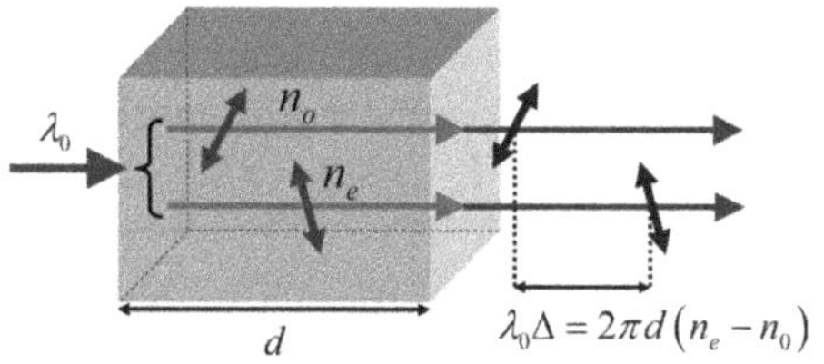

FIGURE 4.4
Linear retarder: the mutually orthogonal eigenstates are linearly polarized.

4.2.1.4 *Linear Retarder*

When the eigenstates of $\mathbf{T}_R$ are linearly polarized states ($\chi = 0 \Rightarrow \delta = 0,\ \alpha = \varphi$), the Jones matrix of the *linear retarder* adopts the form (Figure 4.4)

$$\mathbf{T}_{RL}(\varphi,\Delta) \equiv \mathbf{T}_R(\varphi,0,\Delta) = \begin{pmatrix} c_\varphi^2 e^{i\Delta/2} + s_\varphi^2 e^{-i\Delta/2} & i s_{2\varphi} s_{\Delta/2} \\ i s_{2\varphi} s_{\Delta/2} & s_\varphi^2 e^{i\Delta/2} + c_\varphi^2 e^{-i\Delta/2} \end{pmatrix} \tag{4.7}$$

4.2.1.5 *Horizontal Linear Retarder*

When the linear retarder is represented with respect to its intrinsic axes $X_O Y_O$, then

$$\mathbf{T}_{RL0}(\Delta) \equiv \mathbf{T}_{RL}(0,\Delta) \equiv \mathbf{T}_R(0,0,\Delta) = \begin{pmatrix} e^{i\Delta/2} & 0 \\ 0 & e^{-i\Delta/2} \end{pmatrix} \tag{4.8}$$

4.2.1.6 *Pseudorotator*

Let us consider a linear retarder with $\Delta = \pi$, usually called a *half-wave retarder*, whose Jones matrix is

$$\mathbf{T}_{RL}(\varphi,\pi) \equiv \mathbf{T}_R(\varphi,0,\pi) = \begin{pmatrix} \cos 2\varphi & \sin 2\varphi \\ \sin 2\varphi & -\cos 2\varphi \end{pmatrix} \tag{4.9}$$

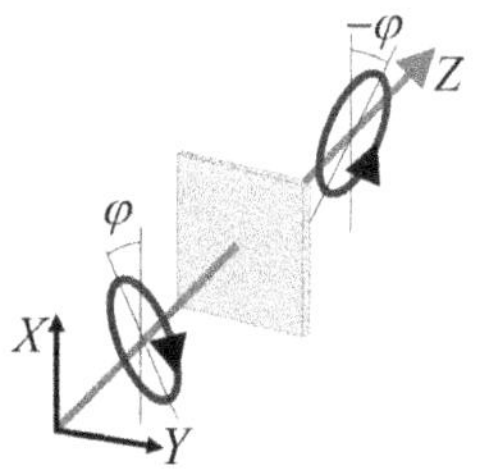

FIGURE 4.5
A half-wave plate or pseudorotator produces an inversion of the handedness of the polarization ellipse, combined with a rotation by the angle -2φ, φ being the azimuth of the polarization ellipse of the incident state.

and observe that $\mathbf{T}_{RL}(\varphi,\pi)$ is equivalent to a rotator of angle -2φ combined with an inversion of the y-component, represented by $\mathbf{T}_{RL}(0,\pi)=\mathrm{diag}(1,-1)$

$$\mathbf{T}_{RL}(\varphi,\pi)=\mathbf{T}_{RL}(0,\pi)\,\mathbf{T}_{G}(-2\varphi) \tag{4.10}$$

$\mathbf{T}_{RL}(\varphi,\pi)$ produces an improper rotation that reverses the handedness of the polarization ellipse while rotating it through the angle -2φ. Therefore, the action of $\mathbf{T}_{RL}(\varphi,\pi)$ differs from that of the rotator $\mathbf{T}_{G}(-2\varphi)$ (Figure 4.5), and this is the reason why $\mathbf{T}_{RL}(\varphi,\pi)$ is called a *pseudorotator*. Pseudorotators exhibit interesting properties that make them very useful in optics and photonics industry; for example, the fact that the action of a pseudorotator converts right-handed circularly polarized light into left-handed one and vice versa:

$$\mathbf{T}_{RL}(\varphi,\pi)\frac{1}{\sqrt{2}}\begin{pmatrix}1\\ i\end{pmatrix}=e^{i2\varphi}\frac{1}{\sqrt{2}}\begin{pmatrix}1\\ -i\end{pmatrix}\qquad \mathbf{T}_{RL}(\varphi,\pi)\frac{1}{\sqrt{2}}\begin{pmatrix}1\\ -i\end{pmatrix}=e^{-i2\varphi}\frac{1}{\sqrt{2}}\begin{pmatrix}1\\ i\end{pmatrix} \tag{4.11}$$

4.2.1.7 Operational Form of the Jones Matrix of a Retarder

The Jones matrix of a rotator $\mathbf{T}_{G}(\theta)$, together with the Jones matrix of a *horizontal linear retarder* $\mathbf{T}_{RL0}(\Delta)$, allows for expressing the Jones matrix of an arbitrary retarder in the following simple and useful manner

$$\mathbf{T}_{RL}(\varphi,\Delta)=\mathbf{T}_{RC}(-2\varphi)\,\mathbf{T}_{RL0}(\Delta)\,\mathbf{T}_{RC}(2\varphi)=\mathbf{T}_{G}(-\varphi)\,\mathbf{T}_{RL0}(\Delta)\,\mathbf{T}_{G}(\varphi) \tag{4.12}$$

which shows that a linear retarder is polarimetrically equivalent to a horizontal linear retarder $\mathbf{T}_{RL0}(\Delta)$ sandwiched between a circular retarder $\mathbf{T}_{RC}(-2\varphi)$ and its inverse circular retarder $\mathbf{T}_{RC}(2\varphi)$.

Furthermore, the Jones matrix of a general elliptical retarder can be expressed as

$$\mathbf{T}_{R}(\alpha,\delta,\Delta)=\mathbf{T}_{RL0}(-\delta)\,\mathbf{T}_{G}(-\alpha)\,\mathbf{T}_{RL0}(\Delta)\,\mathbf{T}_{G}(\alpha)\,\mathbf{T}_{RL0}(\delta) \tag{4.13}$$

which shows that an elliptical retarder $\mathbf{T}_{R}(\alpha,\delta,\Delta)$ is polarimetrically equivalent to a linear retarder $\mathbf{T}_{RL}(\alpha,\Delta)=\mathbf{T}_{G}(-\alpha)\,\mathbf{T}_{RL0}(\Delta)\,\mathbf{T}_{G}(\alpha)$ sandwiched between two identical, mutually crossed, linear retarders, $\mathbf{T}_{RL}(0,-\delta)$ (vertical) and $\mathbf{T}_{RL}(0,\delta)$ (horizontal).

The inverse matrix of $\mathbf{T}_R\left(\alpha,\delta,\Delta\right)$ is another Jones matrix given by

$$\mathbf{T}_R^{-1}\left(\alpha,\delta,\Delta\right)=\mathbf{T}_R\left(\left|\alpha-\pi/2\right|,\delta,\Delta\right) \tag{4.14}$$

producing an effect equivalent to a rotation through the angle $-\Delta$ about the Poincaré axis given by the angles (α,δ) determining the fast eigenstate. However, in virtue of the common convention $0\le\Delta\le\pi$, negative values of Δ are avoided and the inverse is formulated like in Eq. (4.14). Note also that the replacement of (α,δ,Δ) by $(\left|\alpha-\pi/2\right|,\delta,\Delta)$ corresponds to replacing the azimuth φ by $\varphi+\pi/2$ and the ellipticity angle χ by $-\chi$.

4.2.1.8 Exponential Form of the Jones Matrix of a Retarder

Once the main types of Jones matrices $\mathbf{T}_R$ of retarders have been considered, it is worth recalling the exponential form of $\mathbf{T}_R$. A unitary matrix $\mathbf{T}_R$ can always be expressed in terms of an associated Hermitian matrix $\mathbf{H}$ in the form $\mathbf{T}_R=\exp(i\mathbf{H})$; in particular (Whitney 1971; Takenaka 1973; Lu and Chipman 1994)

$$\mathbf{T}_R\left(\alpha,\delta,\Delta\right)=\boldsymbol{\sigma}_0\cos\left(\Delta/2\right)+i\left(\mathbf{u}_R\cdot\boldsymbol{\Sigma}\right)\sin\left(\Delta/2\right)=\exp\left[\left(-i\Delta/2\right)\left(\mathbf{u}_R\cdot\boldsymbol{\Sigma}\right)\right] \tag{4.15}$$

where $\mathbf{u}_R$ is the Poincaré vector of the fast eigenstate

$$\mathbf{u}_R\equiv\begin{pmatrix}u_{R1}\\u_{R2}\\u_{R3}\end{pmatrix}=\begin{pmatrix}\cos2\alpha\\\sin2\alpha\cos\delta\\\sin2\alpha\sin\delta\end{pmatrix}=\begin{pmatrix}\cos2\varphi\cos2\chi\\\sin2\varphi\cos2\chi\\\sin2\chi\end{pmatrix} \tag{4.16}$$

$\boldsymbol{\Sigma}\equiv\left(\boldsymbol{\sigma}_1,\boldsymbol{\sigma}_2,\boldsymbol{\sigma}_3\right)$ is the Pauli vectorial matrix ($\boldsymbol{\sigma}_i$, $i=1,2,3$, being the Pauli matrices), Δ is the phase shift introduced by the retarder between both eigenstates and the dot product stands for

$$\mathbf{u}_R\cdot\Sigma\equiv u_{R1}\boldsymbol{\sigma}_1+u_{R2}\boldsymbol{\sigma}_2+u_{R3}\boldsymbol{\sigma}_3 \tag{4.17}$$

Conversely, the associated Hermitian matrix is given by

$$\mathbf{H}=\left(\Delta/2\right)\left(\mathbf{u}_R\cdot\boldsymbol{\Sigma}\right)=i\left[\mathbf{I}_2-\mathbf{T}_R\right]\left[\mathbf{I}_2+\mathbf{T}_R\right]^{-1} \tag{4.18}$$

where $\mathbf{I}_2\equiv\boldsymbol{\sigma}_0$ is the 2×2 identity matrix.

The exponential representation (4.15) provides a convenient formulation of retarder matrices within the framework of the operational Pauli algebraic approach. As indicated in Section 3.4.7, the matrices in Eq. (4.15) are considered as linear operators without the need for an explicit reference to any particular coordinate frame (Tudor 2008). Furthermore, the exponential form (4.15) is useful in calculations dealing with the evolution of polarization states upon propagation within thick retarders (as, for example, in optical fibers). The exponential forms of Jones matrices are closely related to the concept of differential Jones matrices (Jones 1948, Barakat 1996a), which will be studied in Chapter 9.

4.2.1.9 Jones Matrix of a Serial Combination of Retarders

Since retarders are represented by unitary Jones matrices and the product of two unitary matrices is another unitary matrix, then any serial combination of retarders is polarimetrically equivalent to a retarder (in general, elliptical). In accordance with this fact, the Jones matrix $\mathbf{T}_R$ of a retarder can be expressed as a product of the form

$$\mathbf{T}_R = \mathbf{T}_R \mathbf{I}_4 = \left(\mathbf{T}_R \mathbf{T}_R'^T\right)\left(\mathbf{T}_R' \mathbf{T}_R''^T\right)\left(\mathbf{T}_R'' \mathbf{T}_R'''^T\right)\left(\mathbf{T}_R'''\ldots\right) \tag{4.19}$$

so that there is an infinite number of ways to decompose $\mathbf{T}_R$ into a product of Jones matrices of retarders.

4.2.2 Mueller Matrices of Retarders

Although pure Mueller matrices replicate the physical properties represented by the corresponding Jones matrices, a general formulation of polarization algebra including depolarization phenomena, requires the use of the Mueller-Stokes formalism, so that it is important to study the properties of pure Mueller matrices as a preliminary step to addressing the depolarizing case. Moreover, as we shall see, the structure of pure Mueller matrices has a direct relation to some interesting polarimetric properties of nondepolarizing media.

The Mueller matrix $\mathbf{M}_R$ of a retarder (in general, elliptical) is an orthogonal matrix that has the general form shown in Eq. (4.1)

$$\mathbf{M}_R = \begin{pmatrix} 1 & \mathbf{0}^T \\ \mathbf{0} & \mathbf{m}_R \end{pmatrix} \qquad \mathbf{M}_R^T = \mathbf{M}_R^{-1} \qquad \det \mathbf{M}_R = +1 \tag{4.20}$$

It is characterized by the following properties:

- Diattenuation and polarizance are zero: $D = P = 0$.
- The mean intensity coefficient (transmittance or reflectance) is unity, $m_{00}(\mathbf{M}_R) = 1$, for all incident states.
- The 3x3 submatrix $\mathbf{m}$ is orthogonal: $\mathbf{m}_R^T = \mathbf{m}_R^{-1}$ with $\det \mathbf{m}_R = +1$ (hence, improper orthogonal matrices are excluded). Note that $\mathbf{M}_R$ itself is a particular type of an orthogonal matrix.

As with the Jones matrix of a retarder, $\mathbf{M}_R$ depends on three parameters, namely the two angles that determine the polarization ellipse of one of its eigenstates (the pair α and δ, or the pair φ and χ, of the fast eigenstate), together with the *retardance* Δ (i.e., the phase shift introduced between the fast and slow eigenstates of the retarder, measured in radians, $0 \le \Delta \le \pi$).

4.2.2.1 Retardance Vector and Components of Retardance

The matrix $\mathbf{m}_R$ can be expressed in terms of the elements of the corresponding Poincaré vector $\mathbf{u}_R$ (Eq. (4.16)) and the retardance Δ as follows (Lu and Chipman 1996):

$$m_{Rij} = \delta_{ij} \cos\Delta + u_{Ri} u_{Rj} (1 - \cos\Delta) + \sum_{k=1}^{3} \left(\epsilon_{ijk}\, u_{Rk}\right) \sin\Delta \qquad (i,j = 1,2,3) \tag{4.21}$$

where δ_{ij} is the Kronecker delta and ϵ_{ijk} is the Levi-Cività permutation symbol. (Recall that $\epsilon_{ijk}=1$ if (i, j, k) is a cyclic (even) permutation of (1,2,3); $\epsilon_{ijk}=-1$ if it is an odd permutation, and $\epsilon_{ijk}=0$ if any index is repeated.)

The retardance of a retarder can be expressed as follows in terms of $\mathbf{m}_R$ or $\mathbf{M}_R$:

$$\cos\Delta = (\operatorname{tr}\mathbf{m}_R - 1)/2 = (\operatorname{tr}\mathbf{M}_R)/2 - 1 \tag{4.22}$$

Prior to introducing an alternative way for calculating $\mathbf{u}_R$, let us consider the particular cases where $\sin\Delta = 0$, namely

a) $\Delta = 0$, in which case $\mathbf{M}_R = \mathbf{I}_4$ ($\mathbf{I}_4$ being the 4×4 identity matrix). This limiting case of zero retardance corresponds to a transparent medium without polarimetric effects. Any pair of orthogonal polarization states can be considered as eigenstates. Both α and δ are undetermined, so that the values $\alpha = \delta = 0$ can be chosen without loss of generality.
b) $\Delta = \pi$, which implies that $\mathbf{M}_R$ is a symmetric matrix ($\mathbf{M}_R = \mathbf{M}_R^T$) and corresponds to a half-wave plate (or pseudorotator). The particular case where $\mathbf{M}_R = \operatorname{diag}(1,1,-1,-1)$ corresponds to $\Delta = \pi$, $\alpha = 0$, $\pi/2$ and δ is undetermined, so that the value $\delta = 0$ can be chosen without loss of generality. The case where $\mathbf{M}_R = \operatorname{diag}(1,-1,1,-1)$ corresponds to $\delta = 0$, $\alpha = \varphi = \pi/4$, $\Delta = \pi$, and $\mathbf{M}_R = \operatorname{diag}(1,-1,-1,1)$ corresponds to a half-wave circular retarder $(\alpha = \pi/4,\ \delta\pi/2,\ \Delta = \pi)$.

Except for the cases where $\mathbf{M}_R$ is symmetric (which implies $\sin\Delta = 0$), $\mathbf{u}_R$ can also be expressed as (Goldstein, 2011)

$$\mathbf{u}_R = \frac{1}{2\sin\Delta}\begin{pmatrix} m_{R23} - m_{R32} \\ m_{R31} - m_{R13} \\ m_{R12} - m_{R21} \end{pmatrix} \quad (0 < \Delta < \pi) \qquad \left[m_{Rij} \equiv (\mathbf{M}_R)_{ij} = (\mathbf{m}_R)_{ij} \quad i,j = 1,2,3 \right] \tag{4.23}$$

or equivalently (Lu and Chipman 1996)

$$u_{Ri} = \frac{1}{2\sin\Delta}\sum_{j,k=1}^{3} \epsilon_{ijk}\, m_{Rjk} \tag{4.24}$$

Therefore, despite the fact that the above expressions (4.23) and (4.24) are commonly used, we stress that they are not valid when $\Delta = 0, \pi$. Nevertheless, the form of $\mathbf{u}_R$ in Eq. (4.16), which is expressed as a function of the angular parameters of the fast eigenstate is always valid, regardless the value of the retardance Δ.

The *retardance vector* $\mathbf{R}$ is defined as (Lu and Chipman 1996)

$$\mathbf{R} \equiv \Delta\,\mathbf{u}_R \tag{4.25}$$

whose absolute value is the retardance $R \equiv |\mathbf{R}| = \Delta$ $(0 \le \Delta \le \pi)$, measured in radians.

A *Poincaré retardance vector* $\bar{\mathbf{R}}$ can be defined as the unit vector $\mathbf{u}_R$ scaled by Δ/π

$$\bar{\mathbf{R}} \equiv (\Delta/\pi)\mathbf{u}_R = \mathbf{R}/\pi \quad \bar{R} \equiv |\bar{\mathbf{R}}| = \Delta/\pi \quad 0 \le \bar{R} \le 1 \tag{4.26}$$

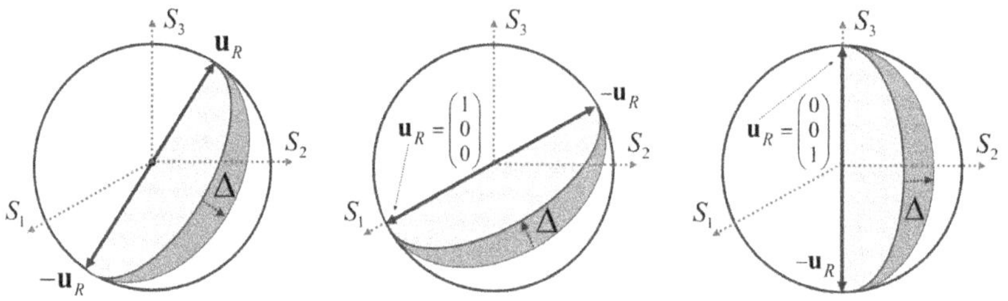

FIGURE 4.6
A retarder produces a rotation by the angle Δ about the Poincaré axis $\mathbf{u}_R$ of its fast eigenstate. (a) Elliptical retarder; (b) horizontal linear retarder, and (c) circular retarder.

so that the maximum amount of retardance $\Delta = \pi$ corresponds to the maximum achievable magnitude $\bar{R} = 1$ of $\bar{\mathbf{R}}$, while $\bar{R} = 0$ corresponds to zero retardance. Therefore, unlike what happens with $\mathbf{R}$, whose absolute value Δ (in radians) may be larger than 1, $\bar{\mathbf{R}}$ can be represented in the Poincaré sphere.

The action of $\mathbf{M}_R$ can be represented on the Poincaré sphere as a rotation by the angle Δ about the axis defined by $\mathbf{u}_R$; that is to say, any incident state (regardless of the value of its degree of polarization) is transformed into an emerging state obtained through this rotation (Figure 4.6). The two eigenstates of $\mathbf{M}_R$ are given by the totally polarized Stokes vectors

$$\mathbf{s}_{R+} \equiv \begin{pmatrix} 1 \\ \mathbf{u}_R \end{pmatrix} \quad \mathbf{s}_{R-} \equiv \begin{pmatrix} 1 \\ -\mathbf{u}_R \end{pmatrix} \tag{4.27}$$

while the two remaining eigenvectors of $\mathbf{M}_R$ are unphysical.

Let us now define the partially polarized Stokes vector $\mathbf{s}_{\bar{R}} \equiv (1, \bar{\mathbf{R}}^T)^T$, so that the *Poincaré linear retardance* $\bar{R}_L$ and *Poincaré circular retardance* $\bar{R}_C$ are defined as the respective *degree of linear polarization* and *degree of circular polarization* of the Stokes vector $\mathbf{s}_{\bar{R}}$

$$\bar{R}_L \equiv (\Delta/\pi)\sqrt{u_{R1}^2 + u_{R2}^2} \qquad \bar{R}_C \equiv (\Delta/\pi)|u_{R3}| \tag{4.28}$$

When $\bar{R}_C = 0$ all retardance is due to linear birefringence; conversely, when $\bar{R}_L = 0$ all retardance is due to circular birefringence. Note that, as with the definitions of the linear and circular degrees of polarization in Eq. (1.142), $\bar{R}_L$ and $\bar{R}_C$ are defined as non-negaive parameters.

It is also worth observing that from Eq. (3.67) it is straightforward to prove that the *components of retardance* $\bar{R}_L$ and $\bar{R}_C$ are invariant with respect to arbitrary rotations of the laboratory reference frame *XYZ* about the axis *Z* that is orthogonal to the plane *XY* containing the polarization ellipse of the incident and emerging states.

4.2.2.2 Mueller Matrix of a Rotator

Since the particular forms of Mueller matrices of retarders can be obtained through their respective operational forms in terms of horizontal linear retarders and rotators, we consider first the Mueller matrices of these two basic components.

As seen in Section 1.5.3, a change of the reference frame, from XY to $X'Y'$, for the representation of Stokes vectors is performed by means of a rotation transformation of the form $\mathbf{s}' = \mathbf{M}_G(\theta)\,\mathbf{s}$ (see Eq. (1.42)), with

$$\mathbf{M}_G(\theta) = \begin{pmatrix} 1 & 0 & 0 & 0 \\ 0 & \cos 2\theta & \sin 2\theta & 0 \\ 0 & -\sin 2\theta & \cos 2\theta & 0 \\ 0 & 0 & 0 & 1 \end{pmatrix} \tag{4.29}$$

which performs a counterclockwise rotation by the angle θ about the axis Z, from the original X reference axis to X'. $\mathbf{M}_G(\theta)$ formally coincides with the Mueller matrix of a circular retarder with retardance 2θ, $\mathbf{M}_G(\theta) = \mathbf{M}_{RC}(2\theta) = \mathbf{M}_R(\pi/4, \pi/2, 2\theta)$ and is usually called a rotator. Since the action of a rotator (a circular retarder) on an incident polarization state is performed without change of the coordinate system, $\mathbf{M}_G(\theta)$ represents a clockwise rotation, through the angle θ, of the Cartesian components of the electric field vector of the incident polarization state.

4.2.2.3 Horizontal Linear Retarder

The Mueller matrix of a linear retarder oriented at 0° with respect to the X axis of the reference frame is

$$\mathbf{M}_{RL0}(\Delta) = \begin{pmatrix} 1 & 0 & 0 & 0 \\ 0 & 1 & 0 & 0 \\ 0 & 0 & \cos\Delta & \sin\Delta \\ 0 & 0 & -\sin\Delta & \cos\Delta \end{pmatrix} \tag{4.30}$$

4.2.2.4 Operational Form of the Mueller Matrix of a Retarder

The serial decomposition (4.13) of the Jones matrix of a retarder (in general, elliptical) has the following counterpart in terms of Mueller matrices of rotators and linear retarders:

$$\mathbf{M}_R(\alpha, \delta, \Delta) = \mathbf{M}_{RL0}(-\delta)\,\mathbf{M}_G(-\alpha)\,\mathbf{M}_{RL0}(\Delta)\,\mathbf{M}_G(\alpha)\,\mathbf{M}_{RL0}(\delta) \tag{4.31}$$

The inverse matrix of $\mathbf{M}_R(\alpha, \delta, \Delta)$ is another orthogonal Mueller matrix given by

$$\mathbf{M}_R^{-1}(\alpha, \delta, \Delta) = \mathbf{M}_R\left(|\alpha - \pi/2|, \delta, \Delta\right) \tag{4.32}$$

producing an effect equivalent to a rotation by the angle $-\Delta$ about the Poincaré axis defined by the angles (α, δ) determining the fast eigenstate (recall that, in virtue of the common convention $0 \le \Delta \le \pi$, negative values of Δ are avoided in the formulation of equivalences). Note also that the replacement of (α, δ, Δ) by $(|\alpha - \pi/2|, \delta, \Delta)$ corresponds to replacing the azimuth φ by $\varphi + \pi/2$ and the ellipticity angle χ by $-\chi$.

4.2.2.5 Eigenvalues and Eigenstates of the Mueller Matrix of a Retarder

The Mueller matrix of a horizontal linear retarder can be diagonalized through the unitary similarity transformation (Marathay 1965)

$$\mathbf{M}_{RL0}(\Delta)=\mathbf{C}_R\,\mathrm{diag}\left(1,1,e^{i\Delta},e^{-i\Delta}\right)\mathbf{C}_R^{\dagger} \tag{4.33.a}$$

where the unitary (non-Mueller) matrix $\mathbf{C}_R$ is

$$\mathbf{C}_R\equiv\frac{1}{\sqrt{2}}\begin{pmatrix}1&1&0&0\\1&-1&0&0\\0&0&1&1\\0&0&-i&i\end{pmatrix} \tag{4.33.b}$$

Since the first two eigenvalues are equal, $\mathbf{C}_R$ is not the only unitary matrix that diagonalizes $\mathbf{M}_{RL0}(\Delta)$. Nevertheless, $\mathbf{C}_R$ has the peculiarity that it is the only option with two Stokes eigenvectors, so that, from a physical point of view, $\mathbf{C}_R$ is the most appropriate choice. The eigenvectors of $\mathbf{M}_{RL0}(\Delta)$ are given by the columns of $\mathbf{C}_R$ (expressed without the normalization factor $1/\sqrt{2}$, which has no physical significance in this context),

$$\mathbf{s}_{1+}\equiv(1,1,0,0)^T\qquad \mathbf{s}_{1-}\equiv(1,-1,0,0)^T\qquad \mathbf{t}_3\equiv(0,0,1,-i)^T\qquad \mathbf{t}_4\equiv(0,0,1,i)^T \tag{4.34}$$

$\mathbf{s}_{1+}$ and $\mathbf{s}_{1-}$ are the Stokes vectors of the eigenstates of $\mathbf{M}_{RL0}(\Delta)$, and the eigenvectors $\mathbf{t}_3$ and $\mathbf{t}_4$ are unphysical (i.e., are not Stokes vectors). Since $\mathbf{M}_{RL0}(\Delta)$ represents the same medium as the Jones matrix $\mathbf{T}_{RL0}(\Delta)$ having two physical eigenstates, namely horizontal and vertical linearly polarized pure states, the additional eigenvectors $\mathbf{t}_3$ and $\mathbf{t}_4$ of $\mathbf{M}_{RL0}(\Delta)$ arise as a result of $\mathbf{M}_{RL0}(\Delta)$ being a 4×4 matrix, but they do not bear any additional physical significance.

From the decomposition (4.31), the diagonalization of $\mathbf{M}_R(\alpha,\delta,\Delta)$ is obtained as

$$\mathbf{M}_R(\alpha,\delta,\Delta)=\mathbf{M}_{RL0}(-\delta)\,\mathbf{M}_G(-\alpha)\left[\mathbf{C}_R\,\mathrm{diag}\left(1,1,e^{i\Delta},e^{-i\Delta}\right)\mathbf{C}_R^{\dagger}\right]\mathbf{M}_G(\alpha)\,\mathbf{M}_{RL0}(\delta) \tag{4.35}$$

so that the Stokes vectors of the two eigenstates of a general retarder (or elliptical retarder) are

$$\mathbf{s}_{R+}\equiv\begin{pmatrix}1\\\hat{\mathbf{R}}\end{pmatrix}\qquad \mathbf{s}_{R-}\equiv\begin{pmatrix}1\\-\hat{\mathbf{R}}\end{pmatrix}\qquad \hat{\mathbf{R}}\equiv\begin{pmatrix}\cos 2\alpha\\ \sin 2\alpha\cos\delta\\ \sin 2\alpha\sin\delta\end{pmatrix}=\begin{pmatrix}\cos 2\chi\cos 2\varphi\\ \cos 2\chi\sin 2\varphi\\ \sin 2\chi\end{pmatrix} \tag{4.36}$$

where ϕ and χ are the corresponding azimuth and ellipticity angle of the polarization ellipse of the pure state $\mathbf{s}_{R+}$. Observe also that the two remaining eigenvectors of $\mathbf{M}_R(\alpha,\delta,\Delta)$ are unphysical.

As expected, the eigenstates of $\mathbf{M}_R(\alpha,\delta,\Delta)$ correspond to the mutually orthogonal pure states $\mathbf{s}_{R+}$ and $\mathbf{s}_{R-}$ (represented by antipodal points in the Poincaré sphere) whose polarization ellipses are determined by (α,δ) and $(\pi/2-\alpha,\delta+\pi)$, or equivalently, by (φ,χ) and $(\varphi+\pi/4,-\chi)$.

4.2.2.6 Elliptical Retarder Oriented at 0°

The Mueller matrix of an elliptical retarder can be represented with respect to its intrinsic reference axes $X_O Y_O$ defined in such a manner that X_O lies along the major axis of the polarization ellipse of the fast eigenstate $\mathbf{s}_+$. In this case α coincides with the ellipticity angle χ and $\delta = \pi/2$, so that $\mathbf{s}_+ = (\cos 2\chi, 0, \sin 2\chi)^T$, and therefore the Mueller matrix of the elliptical retarder oriented at 0° is expressed as

$$\mathbf{M}_R(\chi,\pi/2,\Delta) = \mathbf{M}_{RL0}(-\pi/2)\,\mathbf{M}_G(-\chi)\,\mathbf{M}_{RL0}(\Delta)\,\mathbf{M}_G(\chi)\,\mathbf{M}_{RL0}(\pi/2)$$
$$= \begin{pmatrix} 1 & 0 & 0 & 0 \\ 0 & \cos^2 2\chi + \cos\Delta \sin^2 2\chi & \sin 2\chi \sin\Delta & \sin^2(\Delta/2)\sin 4\chi \\ 0 & -\sin 2\chi \sin\Delta & \cos\Delta & \cos 2\chi \sin\Delta \\ 0 & \sin^2(\Delta/2)\sin 4\chi & -\cos 2\chi \sin\Delta & \sin^2 2\chi + \cos\Delta \cos^2 2\chi \end{pmatrix} \tag{4.37}$$

4.2.2.7 Circular Retarder

When the eigenstates are right- and left-circularly polarized ones ($\chi = \pi/4 \Rightarrow \delta = \pi/2,\ \alpha = \pi/4$), the Mueller matrix of the circular retarder has the form

$$\mathbf{M}_{RC}(\Delta) \equiv \mathbf{M}_R(\pi/4,\pi/2,\Delta) = \begin{pmatrix} 1 & 0 & 0 & 0 \\ 0 & \cos\Delta & \sin\Delta & 0 \\ 0 & -\sin\Delta & \cos\Delta & 0 \\ 0 & 0 & 0 & 1 \end{pmatrix} \tag{4.38}$$

4.2.2.8 Linear Retarder

When the eigenstates are linearly polarized ($\chi = 0 \Rightarrow \delta = 0,\ \alpha = \varphi$), the Mueller matrix of the *linear retarder* takes the form

$$\mathbf{M}_{RL}(\varphi,\Delta) \equiv \mathbf{M}_R(\varphi,0,\Delta) = \mathbf{M}_G(-\varphi)\,\mathbf{M}_{RL0}(\Delta)\,\mathbf{M}_G(\varphi)$$
$$= \begin{pmatrix} 1 & 0 & 0 & 0 \\ 0 & \cos^2 2\varphi + \cos\Delta \sin^2 2\varphi & (1-\cos\Delta)\sin 2\varphi \cos 2\varphi & -\sin 2\varphi \sin\Delta \\ 0 & (1-\cos\Delta)\sin 2\varphi \cos 2\varphi & \sin^2 2\varphi + \cos\Delta \cos^2 2\varphi & \cos 2\varphi \sin\Delta \\ 0 & \sin 2\varphi \sin\Delta & -\cos 2\varphi \sin\Delta & \cos\Delta \end{pmatrix} \tag{4.39}$$

The Mueller matrix of a linear retarder oriented at 0° ($\chi = 0 \Rightarrow \delta = 0,\ \alpha = \varphi = 0$) has the general form shown in Eq. (4.30)

$$\mathbf{M}_{RL0}(\Delta) \equiv \mathbf{M}_{RL}(0,\Delta) \equiv \mathbf{M}_R(0,0,\Delta) = \begin{pmatrix} 1 & 0 & 0 & 0 \\ 0 & 1 & 0 & 0 \\ 0 & 0 & \cos\Delta & \sin\Delta \\ 0 & 0 & -\sin\Delta & \cos\Delta \end{pmatrix} \tag{4.40}$$

4.2.2.9 Pseudorotator

The Mueller matrix of a half-wave retarder, or pseudorotator, is

$$\mathbf{M}_{RL}(\varphi,\pi) \equiv \mathbf{M}_G(-2\varphi)\,\mathbf{M}_{RL}(0,\pi) = \begin{pmatrix} 1 & 0 & 0 & 0 \\ 0 & \cos 4\varphi & \sin 4\varphi & 0 \\ 0 & \sin 4\varphi & -\cos 4\varphi & 0 \\ 0 & 0 & 0 & -1 \end{pmatrix} \tag{4.41}$$

Some characteristics of this interesting device have been considered in the previous subsection where the Jones matrices of pseudorotators were dealt with. Let us only recall that $\mathbf{M}_{RL}(\varphi,\pi)$ converts right-handed into left-handed circularly polarized states and vice versa,

$$\mathbf{M}_{RL}(\varphi,\pi)\,(1,0,0,1)^T = (1,0,0,-1)^T \qquad \mathbf{M}_{RL}(\varphi,\pi)\,(1,0,0,-1)^T = (1,0,0,1)^T \tag{4.42}$$

4.2.2.10 Mueller Matrix of a Serial Combination of Retarders

Since retarders are represented by orthogonal Mueller matrices and the product of two orthogonal matrices is another orthogonal matrix, then any serial combination of retarders is polarimetrically equivalent to a retarder (in general, elliptical). Furthermore, in agreement with Eq. (4.19), there is an infinite number of ways to decompose the Mueller matrix $\mathbf{M}_R$ of a given retarder into a product of Mueller matrices of retarders. In other words, an arbitrary rotation on the Poincaré sphere can be performed through arbitrary appropriate sets of consecutive rotations.

4.2.2.11 Euler Parameterization of the Mueller Matrix of a Retarder

Even though the parameterization of the Mueller matrix $\mathbf{M}_R$ of a retarder in terms of the angular parameters (α,δ,Δ) or (φ,χ,Δ) is the most common one, the 3x3 orthogonal submatrix $\mathbf{m}_R$ (and hence $\mathbf{M}_R$) can be further parameterized in terms of the Euler angles $(\Delta_1,\Delta_2,\Delta_3)$. In this case $\mathbf{M}_R$ can be expressed as the serial combination of a horizontal linear retarder with retardance Δ_2 sandwiched between two circular retarders with respective retardances Δ_1 and Δ_3 (Arnal 1990),

$$\mathbf{M}_R = \mathbf{M}_{RC}(\Delta_3)\,\mathbf{M}_{RL0}(\Delta_2)\,\mathbf{M}_{RC}(\Delta_1) \tag{4.43}$$

4.2.2.12 Symmetric Retarders

From the operational form of a linear retarder in Eq. (4.39) and of a general retarder in Eq. (4.31) it follows that the Mueller matrix of a retarder is symmetric if either $\Delta = 0$, in which case $\mathbf{M}_R = \mathbf{I}_4$, or $\Delta = \pi$. Thus, from Eq. (4.31), the Mueller matrix of a *symmetric retarder* with $\Delta = \pi$ can always be expressed as $\mathbf{M}_R(\alpha,\delta,\pi) = \mathbf{M}_R^T\,\mathrm{diag}(1,1,-1,-1)\,\mathbf{M}_R$, $\mathbf{M}_R$ being the

Mueller matrix of a retarder. Leaving aside the trivial case $\Delta = 0$, the explicit general form of the Mueller matrix of a symmetric retarder is

$$\mathbf{M}_R(\alpha,\delta,\pi) = \mathbf{M}_{RL0}(-\delta)\,\mathbf{M}_G(-\alpha)\,\mathbf{M}_{RL0}(\pi)\,\mathbf{M}_G(\alpha)\,\mathbf{M}_{RL0}(\delta)$$
$$= \begin{pmatrix} 1 & 0 & 0 & 0 \\ 0 & \cos 4\alpha & \sin 4\alpha \cos\delta & \sin 4\alpha \sin\delta \\ 0 & \sin 4\alpha \cos\delta & -\sin^2\delta - \cos 4\alpha \cos^2\delta & \sin\delta\cos\delta(1-\cos 4\alpha) \\ 0 & \sin 4\alpha \sin\delta & \sin\delta\cos\delta(1-\cos 4\alpha) & -\cos^2\delta - \cos 4\alpha \sin^2\delta \end{pmatrix} \quad (4.44)$$

4.2.2.13 Diagonal Retarders

An interesting subclass of symmetric retarders is that composed of *diagonal retarders* (i.e., retarders whose Mueller matrix is diagonal). The Mueller matrices of diagonal retarders can be obtained from Eq. (4.44), and take one of the following four *canonical diagonal forms*:

$$\begin{array}{ll} \mathbf{M}_{Rd0} \equiv \mathrm{diag}(1,1,1,1) & \mathbf{M}_{Rd1} \equiv \mathrm{diag}(1,1,-1,-1) \\ \mathbf{M}_{Rd2} \equiv \mathrm{diag}(1,-1,1,-1) & \mathbf{M}_{Rd3} \equiv \mathrm{diag}(1,-1,-1,1) \end{array} \quad (4.45)$$

where $\mathbf{M}_{Rd0}$ is the identity matrix, while the other three diagonal retarders $\mathbf{M}_{Rdi}$ $(i = 1,2,3)$ correspond to retarders with $\Delta = \pi$ and linear-horizontal and -vertical eigenstates ($\mathbf{M}_{Rd1}$), linear eigenstates at ±45° ($\mathbf{M}_{Rd2}$) and circular eigenstates ($\mathbf{M}_{Rd3}$) respectively.

Diagonal Mueller matrices play a key role in the interpretation of the information contained in a Mueller matrix (see Section 5.13). It is also remarkable that any pure (nondepolarizing) diagonal Mueller matrix takes necessarily one of the above canonical diagonal forms.

4.2.3 Equivalence Theorems for Serial Combinations of Retarders

Regarding the composition of media exhibiting polarimetric properties, we distinguish between *serial combinations* where the light passes through successive elements arranged along the direction of propagation, and *parallel combinations* where the incoming electromagnetic beam falls simultaneously on different parts of the material target and the wave pencils emerging from the different components are recombined into an emerging beam. The Mueller (or Jones) matrix of a serial combination is given by the ordered product of the Mueller (or Jones) matrices corresponding to the different components. The Jones matrix of a coherent parallel composition is given by the weighted sum of the Jones matrices corresponding to the different components (see Section 3.2.2.5). The Mueller matrix of an incoherent parallel combination is given by the convex sum of the Mueller matrices corresponding to the different components (Gil 2007; Gil and San José 2019) (see Section 3.3.5.5).

Since the Mueller or Jones matrix of an elliptical retarder constitutes a generic representation for all kinds of retarders, it is worth listing a number of interesting equivalence properties, summarized in Table 4.1, whose proofs are straightforward from the analysis performed in the previous subsections.

TABLE 4.1
Equivalence Theorems of Jones and Mueller–Jones Matrices of Retarders

Jones formulation	Mueller formulation
TR1	A system composed of a serial combination of any number of retarders (linear, circular and elliptical) is equivalent to an elliptical retarder (Whitney 1971).
TR2	An elliptical retarder is equivalent to a serial combination of a linear retarder and a circular retarder (rotator) (Hurwitz and Jones 1941)
TR3	Any elliptical retarder is equivalent to a serial combination of two linear retarders (not in a unique form) (Whitney 1971)
TR4	Any elliptical retarder is equivalent to a horizontal linear retarder sandwiched between two circular retarders $\mathbf{T}_R = \mathbf{T}_{RC}(\Delta_3)\,\mathbf{T}_{RL0}(\Delta_2)\,\mathbf{T}_{RC}(\Delta_1)$ $\quad$ $\mathbf{M}_R = \mathbf{M}_{RC}(\Delta_3)\,\mathbf{M}_{RL0}(\Delta_2)\,\mathbf{M}_{RC}(\Delta_1)$
TR5	Linear retarder $\mathbf{T}_{RL}(\varphi,\Delta) = \mathbf{T}_G(-\varphi)\,\mathbf{T}_{RL0}(\Delta)\,\mathbf{T}_G(\varphi)$ $\quad$ $\mathbf{M}_{RL}(\varphi,\Delta) = \mathbf{M}_G(-\varphi)\,\mathbf{M}_{RL0}(\Delta)\,\mathbf{M}_G(\varphi)$
TR6	Inverse retarder $\mathbf{T}_R^{-1}(\alpha,\delta,\Delta) = \mathbf{T}_R^{\dagger}(\alpha,\delta,\Delta) = \mathbf{T}_R(\lvert\alpha-\pi/2\rvert,\delta,\Delta)$ $\quad$ $\mathbf{M}_R^{-1}(\alpha,\delta,\Delta) = \mathbf{M}_R^{T}(\alpha,\delta,\Delta) = \mathbf{M}_R(\lvert\alpha-\pi/2\rvert,\delta,\Delta)$
TR7	Neutral retarder $\mathbf{T}_R(0,0,0) = \mathbf{I}_2\,[\mathbf{I}_2 \equiv \mathrm{diag}(1,1)]$ $\quad$ $\mathbf{M}_R(0,0,0) = \mathbf{I}_4\,[\mathbf{I}_4 \equiv \mathrm{diag}(1,1,1,1)]$
TR8	Horizontal half-wave plate $\mathbf{T}_R(0,0,\pi) = \mathrm{diag}(1,-1)$ $\quad$ $\mathbf{M}_R(0,0,\pi) = \mathrm{diag}(1,1,-1,-1)$
TR9	When $\alpha = \pi/2$ the retarder is a vertical linear retarder (δ is undetermined) $\mathbf{T}_R(\pi/2,\delta,\Delta) = \mathbf{T}_{RL}(\pi/2,\Delta)$ $\quad$ $\mathbf{M}_R(\pi/2,\delta,\Delta) = \mathbf{M}_{RL}(\pi/2,\Delta)$
TR10	A rotation matrix is equivalent to the matrix of a circular retarder with $\Delta = 2\theta$ $\mathbf{T}_G(\theta) = \mathbf{T}_R(\pi/4,\pi/2,2\theta)$ $\quad$ $\mathbf{M}_G(\theta) = \mathbf{M}_R(\pi/4,\pi/2,2\theta)$
TR11	An elliptical retarder is equivalent to a linear retarder placed between two mutually crossed identical linear retarders (horizontal and vertical, respectively) $\mathbf{T}_R(\alpha,\delta,\Delta) = \mathbf{T}_{RL0}(-\delta)\,\mathbf{T}_{RL}(\alpha,\Delta)\,\mathbf{T}_{RL0}(\delta)$ $\quad$ $\mathbf{M}_R(\alpha,\delta,\Delta) = \mathbf{M}_{RL0}(-\delta)\,\mathbf{M}_{RL}(\alpha,\Delta)\,\mathbf{M}_{RL0}(\delta)$
TR12	A linear retarder sandwiched between two collinear identical linear retarders is equivalent to a linear retarder (Gil 1982) $\mathbf{T}_{RL}(\varphi,\Delta) = \mathbf{T}_{RL0}(\delta_1)\,\mathbf{T}_{RL}(\theta,\delta_2)\,\mathbf{T}_{RL0}(\delta_1)$ $\quad$ $\mathbf{M}_{RL}(\varphi,\Delta) = \mathbf{M}_{RL0}(\delta_1)\,\mathbf{M}_{RL}(\theta,\delta_2)\,\mathbf{M}_{RL0}(\delta_1)$ The parameters φ,Δ of the resulting equivalent linear retarder are $\cos(\Delta/2) = \cos\delta_1\cos(\delta_2/2) - \sin\delta_1\sin(\delta_2/2)\cos 2\theta \quad \tan 2\varphi = \dfrac{\sin 2\theta}{\sin\delta_1\cot(\delta_2/2) + \cos\delta_1\cos 2\theta}$ This serial combination of three linear retarders provides a simple method for designing tunable compensators by adjusting the angle of the intermediate retarder

4.3 Diattenuators

A diattenuator designates a pure medium that exhibits anisotropic intensity attenuation (also called *dichroism*), i.e., the intensity coefficient (also called *transmittance* or *gain*) of the medium depends on the state of polarization of the incident electromagnetic beam. Obviously, this behavior is not exclusive of pure media, but unless otherwise indicated, the term diattenuator is commonly reserved in the literature to pure media.

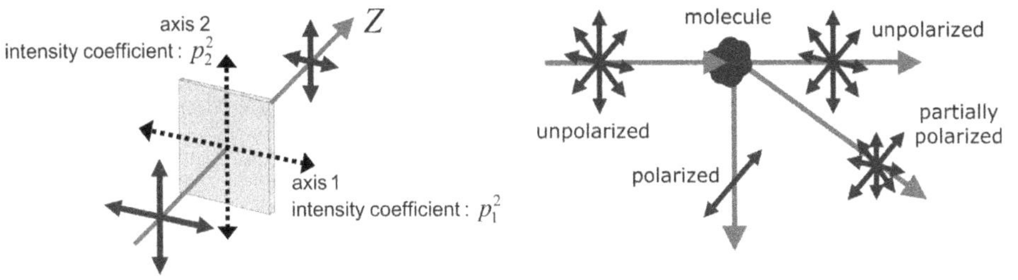

FIGURE 4.7
(a) Diattenuation by a partially polarizing slab and (b) by Rayleigh scattering.

Diattenuation arises in a great variety of natural and artificial situations, including light passing through a tourmaline crystal or a synthetic polarizing sheet, reflection by smooth surfaces, refraction, scattering and many other phenomena where electromagnetic waves propagate through a medium (Pye 2001; Bennet 1995) (Figure 4.7). Even birefringent media usually exhibit a certain amount of diattenuation affecting their eigenstates (Clarke and Grainger 1974).

In general, any pure medium whose Jones matrix has different singular values ($p_2 \neq p_1$) can be considered as a diattenuator (Savenkov et al. 2005). Nevertheless, a particularly interesting class of diattenuators is the one constituted by normal (or homogeneous) diattenuators, characterized by the fact that their Jones matrices are Hermitian (and hence, their Mueller matrices are symmetric). Thus, by considering the general serial decompositions (3.8) and (3.44), the Jones and Mueller matrices of a normal diattenuator can respectively be written as

$$\begin{gathered} \mathbf{T}_D = \mathbf{T}_D^\dagger = \mathbf{T}_R\,\mathbf{T}_{DL0}\,\mathbf{T}_R^\dagger \qquad \mathbf{T}_{DL0} \equiv \mathrm{diag}(p_1, p_2) \\ \mathbf{M}_D \equiv \mathbf{M}(\mathbf{T}_D) = \mathbf{M}_R\,\mathbf{M}_{DL0}\,\mathbf{M}_R^T \\ \left[\mathbf{M}_R = \mathbf{M}(\mathbf{T}_R),\ \mathbf{M}_{DL0} = \mathbf{M}(\mathbf{T}_{DL0}),\ \mathbf{M}_R^T = \mathbf{M}(\mathbf{T}_R^\dagger)\right] \end{gathered} \tag{4.46}$$

From these representations we see that any incident state can be decomposed into two mutually orthogonal pure states (whose Jones vectors are the columns of $\mathbf{T}_R$) for which the principal intensity coefficients are p_1^2 and p_2^2 and whose emerging state is a coherent composition of the transformed components.

The *diattenuation* of a nondepolarizing system is defined as the relative difference between the maximum p_1^2 and minimum p_2^2 intensity coefficients

$$D \equiv \left(p_1^2 - p_2^2\right)/\left(p_1^2 + p_2^2\right) \tag{4.47}$$

which in turn coincides with the *polarizance* P of the nondepolarizing system, defined as the degree of polarization of the exiting state corresponding to an unpolarized incident state.

In the case of a *normal* diattenuator (also called *partial polarizer*) one has $\mathbf{M}_D = \mathbf{M}_D^T$ and thus, the diattenuation vector $\mathbf{D}$ of $\mathbf{M}_D$ coincides with its polarizance vector $\mathbf{P}$, so that in this case the vector $\mathbf{D} = \mathbf{P}$ is called the *diattenuation-polarizance vector*.

4.3.1 Jones Matrices of Diattenuators

The most general case of a normal diattenuator is the *elliptical diattenuator* (or *elliptical partial polarizer*), whose eigenstates are given by a pair of mutually orthonormal pure states with Jones vectors (Azzam and Bashara 1987),

$$\begin{aligned}\boldsymbol{\varepsilon}_1(\alpha,\delta) &\equiv \begin{pmatrix}\cos\alpha\, e^{-i\,\delta/2}\\ \sin\alpha\, e^{i\,\delta/2}\end{pmatrix} = \begin{pmatrix}\cos\chi\cos\varphi - i\sin\chi\sin\varphi\\ \cos\chi\sin\varphi + i\sin\chi\cos\varphi\end{pmatrix}\\ \boldsymbol{\varepsilon}_2(\pi/2-\alpha,\delta+\pi) &\equiv \begin{pmatrix}-i\sin\alpha e^{-i\,\delta/2}\\ i\cos\alpha e^{i\,\delta/2}\end{pmatrix} = \begin{pmatrix}-\cos\chi\sin\varphi + i\sin\chi\cos\varphi\\ \cos\chi\cos\varphi + i\sin\chi\sin\varphi\end{pmatrix}\end{aligned} \tag{4.48}$$

Here, as introduced in Section 1.2, α is the angle whose tangent is the ratio between the amplitudes of the components of the electric field ($\tan\alpha \equiv a_2/a_1$) and δ is the phase shift between these components, so that the pair of angles (α,δ) determines the shape of the polarization ellipses of the states $\boldsymbol{\varepsilon}_1(\alpha,\delta)$ and $\boldsymbol{\varepsilon}_2(\pi/2-\alpha,\delta+\pi)$ and therefore, defines the corresponding azimuth φ and ellipticity angle χ (see Eq. (1.6)), with $\tan 2\varphi = \tan 2\alpha\cos\delta$ and $\sin 2\chi = \sin 2\alpha \sin\delta$.

4.3.1.1 Elliptical Diattenuator

A generic normal diattenuator is represented by a Hermitian Jones matrix $\mathbf{T}_D$ ($\mathbf{T}_D^{\dagger} = \mathbf{T}_D$), which can always be written as follows in terms of the characteristic angles of the eigenstates and the corresponding principal amplitude coefficients p_1 and p_2 (with the choice $0 \le p_2 \le p_1$, assumed without loss of generality):

$$\mathbf{T}_D(\alpha,\delta,p_1,p_2) \equiv \begin{pmatrix} p_1 c_\alpha^2 + p_2 s_\alpha^2 & s_\alpha c_\alpha (p_1-p_2) e^{-i\delta}\\ s_\alpha c_\alpha (p_1-p_2) e^{i\delta} & p_1 s_\alpha^2 + p_2 c_\alpha^2\end{pmatrix} \quad \begin{bmatrix} s_\alpha \equiv \sin\alpha\\ c_\alpha \equiv \cos\alpha\end{bmatrix} \tag{4.49}$$

Provided $p_2 > 0$, then $\mathbf{T}_D(\alpha,\delta,p_1,p_2)$ is nonsingular and its inverse matrix is given by

$$\mathbf{T}_D^{-1}(\alpha,\delta,p_1,p_2) = \mathbf{T}_D(\alpha,\delta,1/p_1,1/p_2) \tag{4.50}$$

which is not a passive Jones matrix because it produces intensity amplification for some or all incident states. Nevertheless, a passive Jones matrix $\bar{\mathbf{T}}_D$, proportional to the inverse of $\mathbf{T}_D$, can be defined as

$$\bar{\mathbf{T}}_D(\alpha,\delta,p_1,p_2) = p_1 p_2 \mathbf{T}_D^{-1}(\alpha,\delta,p_1,p_2) = \mathbf{T}_D(\alpha,\delta,p_2,p_1) \tag{4.51}$$

so that $\bar{\mathbf{T}}_D\mathbf{T}_D = p_1 p_2 \mathbf{I}_2$ ($\mathbf{I}_2$ being the 2x2 identity matrix).

When $p_2 = 0$, $\mathbf{T}_D(\alpha,\delta,p_1,0)$ is singular and corresponds to an elliptical polarizer. In accordance with Eq. (3.22) the pseudoinverse of $\mathbf{T}_D(\alpha,\delta,p_1,0)$ is given by the non-passive Hermitian matrix

$$\mathbf{T}_D^{-} = \mathbf{T}_D(\alpha,\delta,1/p_1,0) \tag{4.52}$$

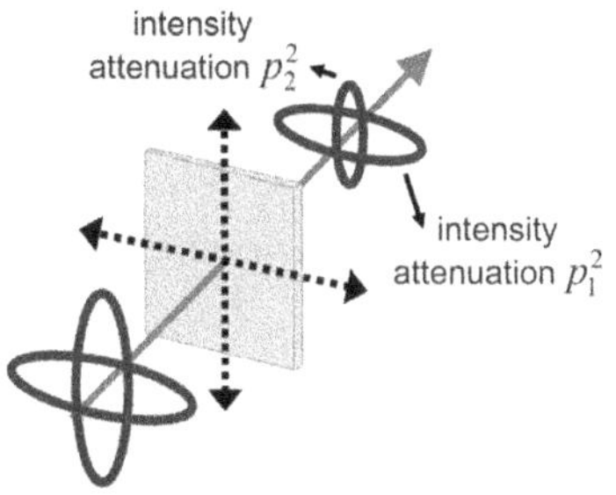

FIGURE 4.8
Elliptical diattenuator oriented at 0°. The symmetry axes of the polarization ellipses of the two mutually orthogonal eigenstates lie along the reference axes.

When $p_2 = 0$ the diattenuator totally extinguishes the eigenstate $\boldsymbol{\varepsilon}_2(\pi/2-\alpha, \delta+\pi)$ and is called an *elliptical polarizer* (or an *elliptical total polarizer*). In general, its Jones matrix is

$$\mathbf{T}_D(\alpha, \delta, p_1, 0) = p_1 \begin{pmatrix} c_\alpha^2 & s_\alpha c_\alpha e^{-i\delta} \\ s_\alpha c_\alpha e^{i\delta} & s_\alpha^2 \end{pmatrix} \quad \begin{bmatrix} s_\alpha \equiv \sin\alpha \\ c_\alpha \equiv \cos\alpha \end{bmatrix} \tag{4.53}$$

The passive representative of a normal diattenuator $\mathbf{T}_D(\alpha, \delta, p_1, p_2)$ (i.e., the Jones matrix proportional to $\mathbf{T}_D$ having the maximal intensity coefficient compatible with passivity) is $\mathbf{T}_D(\alpha, \delta, 1, p_2/p_1)$.

4.3.1.2 Elliptical Diattenuator Oriented at 0°

In particular, the Jones matrix of the elliptical diattenuator can be represented with respect to its intrinsic reference axes $X_O Y_O$, defined in such a manner that X_O is aligned with the major axis of the polarization ellipse of the eigenstate $\boldsymbol{\varepsilon}_1$ ($\varphi = 0 \Rightarrow \delta = \pi/2,\ \alpha = \chi$) and consequently, Y_O is along the minor axis. With this choice, the Jones matrix of the elliptical diattenuator oriented at 0° is expressed as (Figure 4.8)

$$\mathbf{T}_D(\chi, \pi/2, p_1, p_2) \equiv \begin{pmatrix} p_1 c_\chi^2 + p_2 s_\chi^2 & -i s_\chi c_\chi (p_1 - p_2) \\ i s_\chi c_\chi (p_1 - p_2) & p_1 s_\chi^2 + p_2 c_\chi^2 \end{pmatrix} \quad \begin{bmatrix} s_\chi \equiv \sin\chi \\ c_\chi \equiv \cos\chi \end{bmatrix} \tag{4.54}$$

For the particular case of an *elliptical polarizer* $(p_2 = 0)$ oriented at 0°, the Jones matrix is

$$\mathbf{T}_D(\chi, \pi/2, p_1, 0) \equiv p_1 \begin{pmatrix} c_\chi^2 & -i s_\chi c_\chi \\ i s_\chi c_\chi & s_\chi^2 \end{pmatrix} \tag{4.55}$$

4.3.1.3 Circular Diattenuator

When the eigenstates are right- and left-circularly polarized ($\chi = \pi/4 \Rightarrow \delta = \pi/2,\ \alpha = \pi/4$), the Jones matrix of the *circular diattenuator* takes the form

$$\mathbf{T}_{DC}(p_1,p_2) \equiv \mathbf{T}_D(\pi/4,\pi/2,p_1,p_2) \equiv \frac{1}{2}\begin{pmatrix} p_1+p_2 & -i(p_1-p_2) \\ i(p_1-p_2) & p_1+p_2 \end{pmatrix} \tag{4.56}$$

Consequently, the Jones matrix of a *circular polarizer* ($p_2 = 0$) is

$$\mathbf{T}_{DC}(p_1,0) = \mathbf{T}_D(\pi/4,\pi/2,p_1,0) \equiv \frac{p_1}{2}\begin{pmatrix} 1 & -i \\ i & 1 \end{pmatrix} \tag{4.57}$$

4.3.1.4 Linear Diattenuator

When the eigenstates of the normal diattenuator are linearly polarized states (i.e., $\chi = 0 \Rightarrow \delta = 0,\ \alpha = \varphi$) the system is called a *linear diattenuator*, or a *linear partial polarizer* (Figure 4.7, left). Its corresponding Jones matrix has the form

$$\mathbf{T}_{DL}(\varphi,p_1,p_2) \equiv \mathbf{T}_D(\varphi,0,p_1,p_2) = \begin{pmatrix} p_1 c_\varphi^2 + p_2 s_\varphi^2 & s_\varphi c_\varphi (p_1-p_2) \\ s_\varphi c_\varphi (p_1-p_2) & p_1 s_\varphi^2 + p_2 c_\varphi^2 \end{pmatrix} \quad \begin{bmatrix} s_\varphi \equiv \sin\varphi \\ c_\varphi \equiv \cos\varphi \end{bmatrix} \tag{4.58}$$

The Jones matrix of a *linear polarizer* (or *linear total polarizer*) ($p_2 = 0$) is

$$\mathbf{T}_{DL}(\varphi,p_1,0) = \mathbf{T}_D(\varphi,0,p_1,0) = p_1\begin{pmatrix} c_\varphi^2 & s_\varphi c_\varphi \\ s_\varphi c_\varphi & s_\varphi^2 \end{pmatrix} \tag{4.59}$$

4.3.1.5 Horizontal Linear Diattenuator

The Jones matrix $\mathbf{T}_{DL0}(p_1,p_2)$ of a linear diattenuator oriented at 0° ($\varphi = 0$) has the diagonal form

$$\mathbf{T}_{DL0}(p_1,p_2) \equiv \mathbf{T}_{DL}(0,p_1,p_2) \equiv \mathbf{T}_D(0,0,p_1,p_2) = \text{diag}(p_1,p_2) \tag{4.60}$$

and that of a linear polarizer ($p_2 = 0$) oriented at 0° is

$$\mathbf{T}_{DL0}(p_1,0) = \text{diag}(p_1,0) \tag{4.61}$$

4.3.1.6 Operational Form of the Jones Matrix of a Normal Diattenuator

The Jones matrix of a rotator $\mathbf{T}_G(\theta)$, together with the Jones matrix of a *horizontal linear diattenuator* $\mathbf{T}_{DL0}(\Delta)$, allows for expressing the Jones matrix of any normal diattenuator in the following operational form:

$$\begin{aligned} \mathbf{T}_D(\alpha,\delta,p_1,p_2) &= \mathbf{T}_{RL0}(-\delta)\,\mathbf{T}_{DL}(\alpha,p_1,p_2)\,\mathbf{T}_{RL0}(\delta) \\ &= \mathbf{T}_{RL0}(-\delta)\,\mathbf{T}_G(-\alpha)\,\mathbf{T}_{DL0}(p_1,p_2)\,\mathbf{T}_G(\alpha)\,\mathbf{T}_{RL0}(\delta) \end{aligned} \tag{4.62}$$

which shows that an elliptical diattenuator $\mathbf{T}_D(\alpha,\delta,p_1,p_2)$ is polarimetrically equivalent to a linear diattenuator $\mathbf{T}_{DL}(\alpha,p_1,p_2)$ sandwiched between two mutually crossed linear diattenuators $\mathbf{T}_{RL0}(-\delta)$ and $\mathbf{T}_{RL0}(\delta)$.

4.3.1.7 Exponential Form of the Jones Matrix of a Diattenuator

Once the main types of Jones matrices $\mathbf{T}_D$ of diattenuators have been considered, it is worth recalling their exponential form. Let us parameterize the principal amplitude coefficients as

$$p_1 \equiv e^{\eta_1} \qquad p_2 \equiv e^{\eta_2} \qquad (\eta_2 \le \eta_1 \le 0) \tag{4.63}$$

where we maintain the common convention of η_1 and η_2 being negative in order to ensure passivity together with the choice $p_2 \le p_1 \le 1$ for the order of the singular values of $\mathbf{T}_D$. The magnitude D of the diattenuation-polarizance vector $\mathbf{D}$ of $\mathbf{T}_D$ is

$$D \equiv |\mathbf{D}| = \left(p_1^2 - p_2^2\right)/\left(p_1^2 + p_2^2\right) = \tanh\eta \qquad (\eta \equiv \eta_1 - \eta_2) \tag{4.64}$$

so that $\mathbf{T}_D$ can be expressed in the form (Whitney 1971; Takenaka 1973; Lu and Chipman 1996)

$$\begin{aligned}\mathbf{T}_D &= \frac{p_1+p_2}{2}\left[\boldsymbol{\sigma}_0 + \frac{(\mathbf{D}\cdot\boldsymbol{\Sigma})}{1+\sin\kappa}\right] = \exp(\rho)\left[\boldsymbol{\sigma}_0\cosh(\eta/2) + \left(\hat{\mathbf{D}}\cdot\boldsymbol{\Sigma}\right)\sinh(\eta/2)\right] \\ &= \exp(\rho)\exp\left[(\eta/2)\left(\hat{\mathbf{D}}\cdot\boldsymbol{\Sigma}\right)\right] \qquad \left[D \equiv \cos\kappa \quad \rho \equiv (\eta_1+\eta_2)/2 \quad \eta \equiv \eta_1 - \eta_2\right]\end{aligned} \tag{4.65}$$

where $\hat{\mathbf{D}} \equiv \mathbf{D}/D$, and $\boldsymbol{\Sigma} \equiv (\boldsymbol{\sigma}_1,\boldsymbol{\sigma}_2,\boldsymbol{\sigma}_3)$ is the Pauli vectorial matrix ($\boldsymbol{\sigma}_i$, $i = 1,2,3$, being the Pauli matrices).

Furthermore, from Eq. (4.62) the Jones matrix of a normal diattenuator $\mathbf{T}_D(\alpha,\delta,p_1,p_2)$ can be expressed as

$$\mathbf{T}_D(\alpha,\delta,p_1,p_2) = \mathbf{T}_R^\dagger\, \mathbf{T}_{DL0}(p_1,p_2)\,\mathbf{T}_R \tag{4.66}$$

where $\mathbf{T}_R = \mathbf{T}_G(\alpha)\mathbf{T}_{RL0}(\delta)$ or, equivalently, by using the exponential form $\mathbf{T}_R = e^{i\mathbf{H}}$ of the unitary matrix $\mathbf{T}_R$,

$$\mathbf{T}_D(\alpha,\delta,p_1,p_2) = e^{-i\mathbf{H}}\,\mathrm{diag}\left(e^{\eta_1}, e^{\eta_2}\right)e^{i\mathbf{H}} \tag{4.67}$$

4.3.1.8 Serial Combination of Diattenuators

Unlike retarders, which are represented by unitary Jones matrices (corresponding to proper rotations on the Poincaré sphere) and whose product is another retarder (elliptical, in general), normal diattenuators are represented by Hermitian Jones matrices whose product is not, in general, a Hermitian matrix.

Next, we analyze the properties of Jones matrices that can be represented as products of Jones matrices of normal diattenuators. To do so, let us consider the Jones matrix of the serial combination of a pair of normal diattenuators represented by $\mathbf{T}_D(\alpha,\delta,p_1,p_2)$ and

$\mathbf{T}'_D(\alpha',\delta',p'_1,p'_2)$, and observe that $\mathbf{T}_D\mathbf{T}'_D \neq \mathbf{T}'_D\mathbf{T}_D$. $\mathbf{T}_D$ and $\mathbf{T}'_D$ commute ($\mathbf{T}_D\mathbf{T}'_D = \mathbf{T}'_D\mathbf{T}_D$) if and only if they share a common set of eigenvectors, in which case $\mathbf{T}_D\mathbf{T}'_D$ is a Hermitian Jones matrix representing a normal diattenuator. The Jones matrix corresponding to a product of Hermitian Jones matrices (normal diattenuators) depends on up to seven independent parameters, whose values are constrained by certain inequalities more restrictive than those involving the seven parameters of a general Jones matrix. Consequently, not all Jones matrices exhibiting diattenuation can be expressed as products of Jones matrices of normal diattenuators.

However, a normal (passive) diattenuator can always be written as a product of normal (passive) diattenuators. To prove this property, let us first note that the Jones matrix $\mathbf{T}_{DL0}$ of a linear diattenuator oriented at 0° can be written as the product of two axis-aligned linear diattenuators

$$\mathbf{T}_{DL0}(p_1,p_2) = \mathbf{T}''_{DL0}(p''_1,p''_2)\,\mathbf{T}'_{DL0}(p'_1,p'_2) \tag{4.68}$$

where p'_1,p''_1,p'_2,p''_2 can be considered nonnegative without loss of generality and satisfy the relations $p_1 = p'_1p''_1$, $p_2 = p'_2p''_2$, as well as the following passivity conditions:

$$p'_1,p''_1,p'_2,p''_2 \leq 1 \tag{4.69}$$

Thus, from Eqs. (4.62) and (4.68), any given Jones matrix of a normal diattenuator can be expressed as

$$\mathbf{T}_D(\alpha,\delta,p_1,p_2) = \mathbf{T}_{RL0}(-\delta)\,\mathbf{T}_G(-\alpha)\,\mathbf{T}_{DL0}(p''_1,p''_2)\,\mathbf{T}_{DL0}(p'_1,p'_2)\,\mathbf{T}_G(\alpha)\,\mathbf{T}_{RL0}(\delta) \tag{4.70}$$

By inserting the identity matrix $\operatorname{diag}(1,1) = \mathbf{T}_G(\alpha)\,\mathbf{T}_{RL0}(\delta)\,\mathbf{T}_{RL0}(-\delta)\,\mathbf{T}_G(-\alpha)$ between the Jones matrices of the linear diattenuators $\mathbf{T}_{DL0}(p''_1,p''_2)$ and $\mathbf{T}_{DL0}(p'_1,p'_2)$ we get

$$\mathbf{T}_D(\alpha,\delta,p_1,p_2) = \mathbf{T}_D(\alpha,\delta,p''_1,p''_2)\,\mathbf{T}_D(\alpha,\delta,p'_1,p'_2) \tag{4.71}$$

Note that, in this specific case, the matrices of the two diattenuators forming the product commute.

4.3.1.9 Diattenuating Retarder

Due to multiple internal reflections or to their molecular structure, real linear retarders, like birefringent plates, frequently exhibit some diattenuating effects whose optical axes coincide with those of the birefringence (Clarke and Grainger 1974). This combined effect of birefringence and dichroism has to be sometimes considered for the complete characterization of certain devices, as well as for the accurate representation of the retarding components of a polarimeter. A linear *non-ideal retarder* or *diattenuating retarder* (Gil 1983; Ossikovski 2008) has an associated Jones matrix $\mathbf{T}_{RDL}$ that depends on (1) the angle ϕ between its fast axis X_O and the X-reference axis (X_O coincides with the direction corresponding to the maximum intensity coefficient p_1^2), (2) the principal intensity coefficients p_1^2 and p_2^2 and (3) the effective retardance Δ introduced between the orthogonal eigenstates. Consequently, $\mathbf{T}_{RDL}$ can be expressed as the normal matrix

$$\mathbf{T}_{RDL}(\varphi,\Delta,p_1,p_2)=\mathbf{T}_G(-\varphi)\,\mathbf{T}_{RL0}(\Delta)\,\mathbf{T}_{DL0}(p_1,p_2)\,\mathbf{T}_G(\varphi) \tag{4.72}$$

or explicitly

$$\mathbf{T}_{RDL}(\varphi,\Delta,p_1,p_2)=\begin{pmatrix} p_1e^{i\Delta/2}c_\varphi^2+p_2e^{-i\Delta/2}s_\varphi^2 & s_\varphi c_\varphi\left(p_1e^{i\Delta/2}-p_2e^{-i\Delta/2}\right) \\ s_\varphi c_\varphi\left(p_1e^{i\Delta/2}-p_2e^{-i\Delta/2}\right) & p_1e^{i\Delta/2}s_\varphi^2+p_2e^{-i\Delta/2}c_\varphi^2 \end{pmatrix} \quad \begin{bmatrix} s_\varphi\equiv\sin\varphi \\ c_\varphi\equiv\cos\varphi \end{bmatrix} \tag{4.73}$$

where the central horizontal linear diattenuating retarder (or retarding diattenuator) is represented by the commutative product of the matrices of the horizontal linear retarder $\mathbf{T}_{RL0}(\Delta)$ and the horizontal linear diattenuator $\mathbf{T}_{DL0}(p_1,p_2)$,

$$\mathbf{T}_{RDL0}(\Delta,p_1,p_2)=\mathbf{T}_{RL0}(\Delta)\,\mathbf{T}_{DL0}(p_1,p_2)=\mathbf{T}_{DL0}(p_1,p_2)\,\mathbf{T}_{RL0}(\Delta)=\begin{pmatrix} p_1e^{i\Delta/2} & 0 \\ 0 & p_2e^{-i\Delta/2} \end{pmatrix} \tag{4.74}$$

Up to a global phase factor $e^{i\phi}$, the most general form of a normal Jones matrix is that of an elliptical diattenuating retarder which can always be written as $\mathbf{T}_{RD}=\mathbf{T}_{RL}\,\mathbf{T}_{RDL0}\,\mathbf{T}_{RL}^{\dagger}$ (see Section 4.9.1).

4.3.2 Mueller Matrices of Diattenuators

The Mueller matrix $\mathbf{M}_D$ of a normal diattenuator (in general, elliptical) has the general form

$$\begin{gathered} \mathbf{M}_D(m_{00},\mathbf{D})\equiv\mathbf{M}_D(\alpha,\delta,p_1,p_2)=m_{00}\,\hat{\mathbf{M}}_D(\mathbf{D}) \\ m_{00}=\frac{1}{2}\left(p_1^2+p_2^2\right) \qquad \hat{\mathbf{M}}_D(\mathbf{D})\equiv\begin{pmatrix} 1 & \mathbf{D}^T \\ \mathbf{D} & \mathbf{m}_D \end{pmatrix} \\ \mathbf{D}\equiv D\begin{pmatrix} \cos 2\alpha \\ \sin 2\alpha\cos\delta \\ \sin 2\alpha\sin\delta \end{pmatrix}=D\begin{pmatrix} \cos 2\varphi\cos 2\chi \\ \sin 2\varphi\cos 2\chi \\ \sin 2\chi \end{pmatrix} \\ \mathbf{m}_D\equiv(\sin\kappa)\mathbf{I}_3+\frac{1-\sin\kappa}{\cos^2\kappa}\mathbf{D}\otimes\mathbf{D}^T \qquad \mathbf{I}_3\equiv\operatorname{diag}(1,1,1) \qquad \cos\kappa\equiv D=\frac{p_1^2-p_2^2}{p_1^2+p_2^2} \end{gathered} \tag{4.75}$$

where ⊗ stands for the Kronecker product, $\sin\kappa=\sqrt{1-D^2}$ is the *counter-diattenuation* and

- $\mathbf{M}_D$ is a symmetric matrix defined by the four parameters (α,δ,p_1,p_2) or, equivalently, by (φ,χ,m_{00},D),
- the eigenstates of $\mathbf{M}_D$ correspond to the Stokes vectors

$$\mathbf{s}_{\hat{D}+}\equiv\begin{pmatrix} 1 \\ \hat{\mathbf{D}} \end{pmatrix} \qquad \mathbf{s}_{\hat{D}-}\equiv\begin{pmatrix} 1 \\ -\hat{\mathbf{D}} \end{pmatrix} \tag{4.76}$$

with respective eigenvalues p_1^2 and p_2^2,
- $\mathbf{M}_D$ is completely determined by $\mathbf{D}$ together with m_{00} (or by $\mathbf{D}$ and p_1^2),
- the mean intensity coefficient (or transmittance for incident unpolarized states) is $m_{00} = p_1^2/(1+D) = (p_1^2 + p_2^2)/2$. Passivity is ensured by the condition $p_1 \leq 1$, with $p_1^2 = m_{00}(1+D)$,
- $D = (p_1^2 - p_2^2)/(p_1^2 + p_2^2)$,
- $\mathbf{m}_D^T = \mathbf{m}_D$,
- the passive representative of $\mathbf{M}_D$ is $\tilde{\mathbf{M}}_D = [1/(1+D)]\hat{\mathbf{M}}_D$.
- When $p_2 = 0$ $(D = 1)$, the system is called a *polarizer* (also *perfect polarizer* or *ideal polarizer*) and its block expression reduces to the form

$$\mathbf{M}_{\hat{D}}\left(m_{00}, \hat{\mathbf{D}}\right) = m_{00}\begin{pmatrix} 1 & \hat{\mathbf{D}}^T \\ \hat{\mathbf{D}} & \hat{\mathbf{D}} \otimes \hat{\mathbf{D}}^T \end{pmatrix} = m_{00}\left(\mathbf{s}_{\hat{D}+} \otimes \mathbf{s}_{\hat{D}+}^T\right) \quad \mathbf{s}_{\hat{D}+} \equiv \begin{pmatrix} 1 \\ \hat{\mathbf{D}} \end{pmatrix} \quad D = 1 \quad m_{00} \leq \frac{1}{2} \tag{4.77}$$

As in the case of the generic Jones matrix of a normal diattenuator, $\mathbf{M}_D$ depends on four parameters: the two angles that determine the polarization ellipse of the eigenstate corresponding to the largest eigenvalue p_1^2, namely the pair (α, δ) or the pair (φ, χ), together with the two principal intensity coefficients p_1^2 and p_2^2. The Poincaré vector determining the polarization ellipses of the eigenstates is given by the normalized diattenuation-polarizance vector $\hat{\mathbf{D}} = \mathbf{D}/D$ and the principal intensity coefficients can be calculated from $\mathbf{M}_D$ from the expressions

$$p_1^2 = m_{00}\left(1 + D\right) \qquad p_2^2 = m_{00}\left(1 - D\right) \tag{4.78}$$

4.3.2.1 Components of Diattenuation

The *linear diattenuation* D_L and circular *diattenuation* D_C of a normal diattenuator are defined as the respective *degree of linear polarization* and *degree of circular polarization* of the Stokes vector $\mathbf{s}_{D+}$,

$$D_L \equiv \sqrt{D_1^2 + D_2^2} \qquad D_C \equiv |D_3| \tag{4.79}$$

When $D_C = 0$ all diattenuation-polarizance is due to linear diattenuation. Conversely, when $D_L = 0$ all diattenuation is identified as circular diattenuation. Note that, as in the definitions of the linear and circular degrees of polarization in Eq. (1.142), D_L and D_C are defined as nonnegative parameters.

As in the case of the components of retardance, the *components of diattenuation* D_L and D_C (which, since $\mathbf{P} = \mathbf{D}$ in the case of a normal diattenuator, coincide with the *components of polarizance* P_L and P_C) are invariant with respect to arbitrary rotations of the laboratory reference frame XYZ about the axis Z that is orthogonal to the plane XY containing the polarization ellipse of the incident and emerging states.

4.3.2.2 Horizontal Linear Diattenuator

The Mueller matrix of a linear diattenuator oriented at 0° with respect to the X reference axis can be expressed in the following alternative manners:

$$\mathbf{M}_{DL0}(m_{00},D)\equiv\mathbf{M}_{DL0}(p_1,p_2)=m_{00}\begin{pmatrix}1 & D & 0 & 0\\ D & 1 & 0 & 0\\ 0 & 0 & \sqrt{1-D^2} & 0\\ 0 & 0 & 0 & \sqrt{1-D^2}\end{pmatrix}$$

$$=\frac{p_1^2}{1+\cos\kappa}\begin{pmatrix}1 & \cos\kappa & 0 & 0\\ \cos\kappa & 1 & 0 & 0\\ 0 & 0 & \sin\kappa & 0\\ 0 & 0 & 0 & \sin\kappa\end{pmatrix}=\frac{1}{2}\begin{pmatrix}p_1^2+p_2^2 & p_1^2-p_2^2 & 0 & 0\\ p_1^2-p_2^2 & p_1^2+p_2^2 & 0 & 0\\ 0 & 0 & 2p_1p_2 & 0\\ 0 & 0 & 0 & 2p_1p_2\end{pmatrix} \tag{4.80}$$

4.3.2.3 Operational Form of the Mueller Matrix of a Normal Diattenuator

The serial decomposition (4.62) of the Jones matrix of a normal diattenuator (in general, elliptical) has the following counterpart in terms of Mueller matrices of rotators, linear retarders and a central linear diattenuator oriented at 0°:

$$\begin{aligned}\mathbf{M}_D(\alpha,\delta,p_1,p_2)&=\mathbf{M}_{RL0}(-\delta)\,\mathbf{M}_{DL}(\alpha,p_1,p_2)\,\mathbf{M}_{RL0}(\delta)\\ &=\mathbf{M}_{RL0}(-\delta)\,\mathbf{M}_G(-\alpha)\,\mathbf{M}_{DL0}(p_1,p_2)\,\mathbf{M}_G(\alpha)\,\mathbf{M}_{RL0}(\delta)\end{aligned} \tag{4.81}$$

Provided $p_2>0$ (recall that $0\le p_2\le p_1$ is assumed without loss of generality), $\mathbf{M}_D(\alpha,\delta,p_1,p_2)$ is nonsingular, and its inverse matrix is given by

$$\mathbf{M}_D^{-1}(\alpha,\delta,p_1,p_2)=\mathbf{M}_D(\alpha,\delta,1/p_1,1/p_2) \tag{4.82}$$

which is not a passive Mueller matrix because it produces intensity amplification for some or all incident states. By considering the partitioned form (4.75) of $\mathbf{M}_D(\alpha,\delta,p_1,p_2)$ we observe that its inverse matrix satisfies all the properties of the Mueller matrix of a diattenuator except for passivity and can be expressed as

$$\mathbf{M}_D^{-1}(\alpha,\delta,p_1,p_2)=\frac{1}{p_1^2(1-\cos\kappa)}\begin{pmatrix}1 & -\mathbf{D}^T\\ -\mathbf{D} & \mathbf{m}_D\end{pmatrix} \tag{4.83}$$

or, equivalently,

$$\mathbf{M}_D^{-1}(\alpha,\delta,p_1,p_2)=\frac{1}{p_1^2(1-D)}\mathbf{G}\hat{\mathbf{M}}_D\mathbf{G}\qquad[\mathbf{G}\equiv\mathrm{diag}(1,-1,-1,-1)] \tag{4.84}$$

When $p_2=0$ ($D=1$), $\mathbf{M}_D(\alpha,\delta,p_1,0)$ is singular and corresponds to an elliptical polarizer. The pseudoinverse of $\mathbf{M}_D(\alpha,\delta,p_1,0)$ is the symmetric matrix

$$\mathbf{M}_D^{-}=\mathbf{M}_D(\alpha,\delta,1/p_1,0)=(1/p_1)\mathbf{M}_{RL0}(-\delta)\,\mathbf{M}_G(-\alpha)\,\mathbf{M}_{DL0}(1,0)\,\mathbf{M}_G(\alpha)\,\mathbf{M}_{RL0}(\delta) \tag{4.85}$$

which is non-passive except for the limiting case where $p_1=1$ (i.e., the pseudoinverse of the passive representative $\tilde{\mathbf{M}}_D(\alpha,\delta,1,0)$ of $\mathbf{M}_D(\alpha,\delta,p_1,0)$ is also passive).

4.3.2.4 Eigenvalues and Eigenstates of the Mueller Matrix of a Normal Diattenuator

The Mueller matrix of a linear diattenuator oriented at 0° can be diagonalized through the unitary similarity transformation (Collett 1971)

$$\mathbf{M}_{DL0}(p_1,p_2)=\mathbf{C}_R\,\mathrm{diag}\left(p_1^2,p_2^2,p_1p_2,p_1p_2\right)\mathbf{C}_R^\dagger \tag{4.86}$$

where $\mathbf{C}_R$ is the unitary matrix introduced in Eq. (4.33) to diagonalize a linear retarder. As seen in Eqs. (3.46–3.47), since the last two eigenvalues p_1p_2 are equal, $\mathbf{C}_R$ is not the only unitary matrix that diagonalizes $\mathbf{M}_{DL0}(p_1,p_2)$, and another possible choice is the orthogonal (non-Mueller) matrix $\mathbf{C}_D$ defined in Eq. (3.47), which shares with $\mathbf{C}_R$ the first two columns constituting the pair of physical eigenstates $\mathbf{s}_{1+}\equiv(1,1,0,0)$ and $\mathbf{s}_{1-}\equiv(1,-1,0,0)$. The remaining pair of unphysical eigenvectors of $\mathbf{M}_{DL0}(p_1,p_2)$ depends on the particular choice of the diagonalization matrix, but this fact is physically irrelevant.

From decomposition (4.81), the diagonalization of the Mueller matrix $\mathbf{M}_D(\alpha,\delta,p_1,p_2)$ of a generic normal diattenuator is obtained as

$$\begin{aligned}\mathbf{M}_D(\alpha,\delta,p_1,p_2)=&\left[\mathbf{M}_{RL0}(-\delta)\,\mathbf{M}_G(-\alpha)\,\mathbf{C}_R\right]\mathrm{diag}\left(p_1^2,p_2^2,p_1p_2,p_1p_2\right)\\&\left[\mathbf{C}_R^\dagger\,\mathbf{M}_G(\alpha)\,\mathbf{M}_{RL0}(\delta)\right]\end{aligned} \tag{4.87}$$

whose two eigenstates are given by the following mutually orthogonal Stokes vectors (represented by antipodal points on the surface of the Poincaré sphere):

$$\mathbf{s}_{\hat{D}+}\equiv\begin{pmatrix}1\\\hat{\mathbf{D}}\end{pmatrix}\quad \mathbf{s}_{\hat{D}-}\equiv\begin{pmatrix}1\\-\hat{\mathbf{D}}\end{pmatrix}\quad\left[\hat{\mathbf{D}}\equiv\frac{\mathbf{D}}{D}\right] \tag{4.88}$$

4.3.2.5 Elliptical Diattenuator Oriented at 0°

The Mueller matrix of an elliptical diattenuator can be represented with respect to its intrinsic reference axes X_OY_O, defined in such a manner that X_O lies along the major axis of the polarization ellipse of the eigenstate $\mathbf{s}_{\hat{D}+}$, in which case α coincides with the ellipticity angle χ and $\delta=\pi/2$, so that $\mathbf{s}_{\hat{D}+}=(1,\cos 2\chi,0,\sin 2\chi)^T$. With this choice, the Mueller matrix of the elliptical diattenuator oriented at 0° is expressed as

$$\begin{aligned}\mathbf{M}_D(\chi,\pi/2,p_1,p_2)&=\mathbf{M}_{RL0}(-\pi/2)\,\mathbf{M}_G(-\chi)\,\mathbf{M}_{DL0}(p_1,p_2)\,\mathbf{M}_G(\chi)\,\mathbf{M}_{RL0}(\pi/2)=\\&=\frac{p_1^2}{1+\cos\kappa}\begin{pmatrix}1&\cos\kappa\cos 2\chi&0&\cos\kappa\sin 2\chi\\\cos\kappa\cos 2\chi&\sin\kappa+(1-\sin\kappa)\cos^2 2\chi&0&(1-\sin\kappa)\sin 2\chi\cos 2\chi\\0&0&\sin\kappa&0\\\cos\kappa\sin 2\chi&(1-\sin\kappa)\sin 2\chi\cos 2\chi&0&\sin\kappa+(1-\sin\kappa)\sin^2 2\chi\end{pmatrix}\\&\left[\cos\kappa\equiv D=\left(p_1^2-p_2^2\right)/\left(p_1^2+p_2^2\right)\qquad\sin\kappa\equiv\sqrt{1-D^2}\right]\end{aligned} \tag{4.89}$$

and, for the particular case an elliptical polarizer oriented at 0° ($D=1$, $\alpha=\chi$, $\delta=\pi/2$),

$$\mathbf{M}_D(\chi,\pi/2,p_1,0)=\frac{p_1^2}{2}\begin{pmatrix}1 & \cos 2\chi & 0 & \sin 2\chi\\ \cos 2\chi & \cos^2 2\chi & 0 & \sin 2\chi\cos 2\chi\\ 0 & 0 & 0 & 0\\ \sin 2\chi & \sin 2\chi\cos 2\chi & 0 & \sin^2 2\chi\end{pmatrix} \tag{4.90}$$

4.3.2.6 Circular Diattenuator

When the eigenstates are right- and left-circularly polarized ($\chi=\pi/4 \Rightarrow \delta=\pi/2,\ \alpha=\pi/4$), the Mueller matrix of the circular diattenuator has the form

$$\mathbf{M}_{DC}(p_1,p_2)\equiv\mathbf{M}_D(\pi/4,\pi/2,p_1,p_2)=\frac{p_1^2}{1+\cos\kappa}\begin{pmatrix}1 & 0 & 0 & \cos\kappa\\ 0 & \sin\kappa & 0 & 0\\ 0 & 0 & \sin\kappa & 0\\ \cos\kappa & 0 & 0 & 1\end{pmatrix} \tag{4.91}$$
$$\left[D\equiv\cos\kappa=\left(p_1^2-p_2^2\right)/\left(p_1^2+p_2^2\right)\right]$$

and, consequently, the Mueller matrix of a circular polarizer is given by

$$\mathbf{M}_{DC}(p_1,0)\equiv\mathbf{M}_D(\pi/4,\pi/2,p_1,0)=\frac{p_1^2}{2}\begin{pmatrix}1 & 0 & 0 & 1\\ 0 & 0 & 0 & 0\\ 0 & 0 & 0 & 0\\ 1 & 0 & 0 & 1\end{pmatrix} \tag{4.92}$$

4.3.2.7 Linear Diattenuator

When the eigenstates of the normal diattenuator are linearly polarized (i.e., $\chi=0 \Rightarrow \delta=0$, $\alpha=\varphi$), the system is called a *linear diattenuator*, or a *linear partial polarizer*, and the corresponding Mueller matrix adopts the form

$$\mathbf{M}_{DL}(\varphi,p_1,p_2)\equiv\mathbf{M}_D(\varphi,0,p_1,p_2)=$$
$$=\frac{p_1^2}{1+\cos\kappa}\begin{pmatrix}1 & \cos\kappa\cos 2\varphi & \cos\kappa\sin 2\varphi & 0\\ \cos\kappa\cos 2\varphi & \cos^2 2\varphi+\sin\kappa\sin^2 2\varphi & (1-\sin\kappa)\sin 2\varphi\cos 2\varphi & 0\\ \cos\kappa\sin 2\varphi & (1-\sin\kappa)\sin 2\varphi\cos 2\varphi & \sin^2 2\varphi+\sin\kappa\cos^2 2\varphi & 0\\ 0 & 0 & 0 & \sin\kappa\end{pmatrix} \tag{4.93}$$
$$\left[D\equiv\cos\kappa=\left(p_1^2-p_2^2\right)/\left(p_1^2+p_2^2\right)\qquad \sin\kappa\equiv\sqrt{1-D^2}\right]$$

so that the Mueller matrix of a *linear polarizer* has the form

$$\mathbf{M}_{DL}(\varphi, p_1, 0) \equiv \mathbf{M}_D(\varphi, 0, p_1, 0) = \frac{p_1^2}{2}\begin{pmatrix} 1 & \cos 2\varphi & \sin 2\varphi & 0 \\ \cos 2\varphi & \cos^2 2\varphi & \sin 2\varphi \cos 2\varphi & 0 \\ \sin 2\varphi & \sin 2\varphi \cos 2\varphi & \sin^2 2\varphi & 0 \\ 0 & 0 & 0 & 0 \end{pmatrix} \quad (4.94)$$

4.3.2.8 Horizontal Linear Diattenuator

The Mueller matrix $\mathbf{M}_{DL0}(p_1, p_2)$ of a linear diattenuator oriented at 0° has the form shown in Eq. (4.80) and that of a linear polarizer oriented at 0° is

$$\mathbf{M}_{DL0}(p_1, 0) \equiv \mathbf{M}_D(0, 0, p_1, 0) = \frac{p_1^2}{2}\begin{pmatrix} 1 & 1 & 0 & 0 \\ 1 & 1 & 0 & 0 \\ 0 & 0 & 0 & 0 \\ 0 & 0 & 0 & 0 \end{pmatrix} \quad (4.95)$$

4.3.2.9 Serial Combination of Diattenuators

In agreement with the considerations on Jones matrices of serial combinations of normal diattenuators, the pure Mueller matrices of such combinations are not, in general, symmetric. The pure Mueller matrix resulting from the product of pure symmetric matrices (normal diattenuators) depends on up to seven independent parameters whose values are constrained by certain inequalities more restrictive than those of the seven parameters of a general pure Mueller matrix. Consequently, not all pure Mueller matrices exhibiting diattenuation can be expressed as products of Mueller matrices of normal diattenuators.

However, in accordance with the physical interpretation of Eq. (4.71), a horizontal linear diattenuator can always be decomposed as the product of two (or more) horizontal linear diattenuators

$$\mathbf{M}_{DL0}(p_1, p_2) = \mathbf{M}_{DL0}(p_1'', p_2'')\, \mathbf{M}_{DL0}(p_1', p_2') \quad (4.96)$$

where the two diattenuations $D' \equiv (p_1'^2 - p_2'^2)/(p_1'^2 + p_2'^2)$ and $D'' \equiv (p_1''^2 - p_2''^2)/(p_1''^2 + p_2''^2)$ satisfy $p_1^2(1 + D'D'') = 1$, with $p_1' \le 1, p_1'' \le 1$. Consequently, a normal passive diattenuator can always be represented as a product of two normal passive diattenuators

$$\mathbf{M}_D(\alpha, \delta, p_1, p_2) = \mathbf{M}_D(\alpha, \delta, p_1'', p_2'')\, \mathbf{M}_D(\alpha, \delta, p_1', p_2') \quad (4.97)$$

Notice that the two matrix factors in the above relation commute.

4.3.2.10 Diattenuating Retarder

In accordance with the definition of diattenuating retarder whose Jones matrix has been considered in Section 4.3.1.9, a linear *non-ideal retarder* or *diattenuating retarder* has an associated pure Mueller matrix $\mathbf{M}_{RDL}$ that depends on (1) the orientation angle ϕ between its fast axis X_O and the X reference axis (X_O coincides with the direction corresponding to the maximum intensity coefficient p_1^2), (2) the principal intensity coefficients p_1^2 and p_2^2, and (3) the effective retardance Δ introduced between the orthogonal eigenstates. Consequently, $\mathbf{M}_{RDL}$ can be expressed as

$$\mathbf{M}_{RDL}(\varphi,\Delta,p_1,p_2)=\mathbf{M}_G(-\varphi)\,\mathbf{M}_{RL0}(\Delta)\,\mathbf{M}_{DL0}(p_1,p_2)\,\mathbf{M}_G(\varphi) \tag{4.98}$$

or explicitly (Gil 1983)

$$\mathbf{M}_{RDL}(\varphi,\Delta,p_1,p_2)=\frac{p_1^2}{1+\cos\kappa}\begin{pmatrix} 1 & c_\kappa c_{2\varphi} & c_\kappa s_{2\varphi} & 0 \\ c_\kappa c_{2\varphi} & c_{2\varphi}^2+s_\kappa c_\Delta s_{2\varphi}^2 & s_{2\varphi}c_{2\varphi}(1-s_\kappa c_\Delta) & -s_\kappa s_\Delta s_{2\varphi} \\ c_\kappa s_{2\varphi} & s_{2\varphi}c_{2\varphi}(1-s_\kappa c_\Delta) & s_{2\varphi}^2+s_\kappa c_\Delta c_{2\varphi}^2 & s_\kappa s_\Delta c_{2\varphi} \\ 0 & s_\kappa s_\Delta s_{2\varphi} & -s_\kappa s_\Delta c_{2\varphi} & s_\kappa c_\Delta \end{pmatrix} \tag{4.99}$$

$$\left[s_\alpha\equiv\sin\alpha \quad c_\alpha\equiv\cos\alpha \quad c_\kappa\equiv D \quad s_\kappa\equiv\sqrt{1-D^2}\right]$$

where the central horizontal diattenuating retarder (or retarding diattenuator) is represented by the commutative product of the Mueller matrices of the horizontal linear retarder $\mathbf{M}_{RL0}(\Delta)$ and the horizontal linear diattenuator $\mathbf{M}_{DL0}(p_1,p_2)$ (Ossikovski 2008)

$$\begin{aligned}\mathbf{M}_{RDL0}(\Delta,m_{00},D)&\equiv\mathbf{M}_{RDL0}(\Delta,p_1,p_2)=\mathbf{M}_{RL0}(\Delta)\,\mathbf{M}_{DL0}(p_1,p_2)=\mathbf{M}_{DL0}(p_1,p_2)\,\mathbf{M}_{RL0}(\Delta)\\ &=m_{00}\begin{pmatrix} 1 & \cos\kappa & 0 & 0 \\ \cos\kappa & 1 & 0 & 0 \\ 0 & 0 & \sin\kappa\cos\Delta & \sin\kappa\sin\Delta \\ 0 & 0 & -\sin\kappa\sin\Delta & \sin\kappa\cos\Delta \end{pmatrix}\end{aligned} \tag{4.100}$$

As seen in Section 3.6.1, the most general form of a normal Mueller matrix is that of an elliptical diattenuating retarder, which can always be written as $\mathbf{M}_{RD}=\mathbf{M}_{RL}\,\mathbf{M}_{RDL0}\,\mathbf{M}_{RL}^T$.

4.3.3 Equivalence Theorems for Serial Decompositions of Normal Diattenuators

Since the Mueller or Jones matrix of an elliptical diattenuator constitutes a generic representation for normal diattenuators, it is worth summarizing in Table 4.2 two useful equivalence theorems brought up among a number of equivalences that can be easily derived from the previous subsections (Hurwitz and Jones 1941; Whitney 1971; Savenkov et al. 2006).

4.4 Polar Decomposition of a Nondepolarizing System

A way to perform the serial decomposition of a pure system is given by the polar decomposition of its corresponding Jones or Mueller matrix. Any Jones matrix can be written as

TABLE 4.2
Equivalence Theorems on Jones and Mueller–Jones Matrices of Normal Diattenuators

TD1	Any elliptical diattenuator is equivalent to a linear diattenuator sandwiched between two mutually-crossed identical linear retarders (horizontal and vertical respectively) (Whitney 1971) $$\mathbf{M}_D(\alpha,\delta,p_1,p_2)=\mathbf{M}_{RL0}(-\delta)\,\mathbf{M}_{DL}(\alpha,p_1,p_2)\,\mathbf{M}_{RL0}(\delta)$$
TD2	Any elliptical diattenuator is equivalent to a serial combination of an arbitrary number of elliptical diattenuators with the same eigenstates $$\mathbf{M}_D(\alpha,\delta,p_1,p_2)=\prod_{i=1}^{n}\mathbf{M}_D(\alpha,\delta,p_{1,i},p_{2,i})$$ $$\left(p_1=\prod_{i=1}^{n}p_{1,i}\quad p_2=\prod_{i=1}^{n}p_{2,i}\quad p_{1,i},p_{2,i}\leq 1\right)$$

Note: Operators **M** can be considered either Mueller or Jones matrices.

the product of a Hermitian matrix (normal diattenuator) and a unitary matrix (retarder) in either of the two possible relative positions (Whitney 1971; Gil and Bernabéu 1987; Lu and Chipman 1996),

$$\mathbf{T}=\mathbf{T}_P\,\mathbf{T}_R=\mathbf{T}_R\,\mathbf{T}_D \tag{4.101}$$

where $\mathbf{T}_P$ and $\mathbf{T}_D$ represent the Jones matrices of normal diattenuators defined by their respective diattenuation-polarizance vectors **P** and **D**, and $\mathbf{T}_R$ stands for the Jones matrix of a retarder. Note that $\mathbf{T}_R$ is the same matrix regardless the order considered and that the Jones matrices of the two equivalent diattenuators are related to each other by $\mathbf{T}_D=\mathbf{T}_R^T\,\mathbf{T}_P\,\mathbf{T}_R$.

Thus, any given nondepolarizing system is polarimetrically equivalent to a serial composition of a normal diattenuator and a retarder. Recall that the term nondepolarizing (or pure) refers to systems whose action preserves the degree of polarization $\mathcal{P}=1$ of totally polarized incident states.

Obviously, the polar decomposition (4.101) of a Jones matrix is directly translatable to the corresponding pure Mueller matrix (Gil 1983; Gil and Bernabéu 1987; Lu and Chipman 1996),

$$\mathbf{M}_J(\mathbf{T})=\mathbf{M}_P\mathbf{M}_R=\mathbf{M}_R\mathbf{M}_D\quad\left(\mathbf{M}_D=\mathbf{M}_R^T\mathbf{M}_P\mathbf{M}_R\right) \tag{4.102}$$

where $\mathbf{M}_P$ and $\mathbf{M}_D$ are the symmetric pure Mueller matrices associated with $\mathbf{T}_P$ and $\mathbf{T}_D$ respectively, and $\mathbf{M}_R$ is the orthogonal Mueller matrix associated with $\mathbf{T}_R$.

By considering the partitioned expression (3.41) of a pure Mueller matrix $\mathbf{M}_J$, the polar decomposition can be written as

$$\begin{aligned}\mathbf{M}_J&=m_{00}\begin{pmatrix}1&\mathbf{D}^T\\\mathbf{P}&\mathbf{m}\end{pmatrix}=m_{00}\begin{pmatrix}1&\mathbf{P}^T\\\mathbf{P}&\mathbf{m}_P\end{pmatrix}\begin{pmatrix}1&\mathbf{0}^T\\\mathbf{0}&\mathbf{m}_R\end{pmatrix}=m_{00}\begin{pmatrix}1&\left(\mathbf{m}_R^T\mathbf{P}\right)^T\\\mathbf{P}&\mathbf{m}_P\mathbf{m}_R\end{pmatrix}\\&=m_{00}\begin{pmatrix}1&\mathbf{0}^T\\\mathbf{0}&\mathbf{m}_R\end{pmatrix}\begin{pmatrix}1&\mathbf{D}^T\\\mathbf{D}&\mathbf{m}_D\end{pmatrix}=m_{00}\begin{pmatrix}1&\mathbf{D}^T\\\mathbf{m}_R\mathbf{D}&\mathbf{m}_R\mathbf{m}_D\end{pmatrix}\end{aligned} \tag{4.103}$$

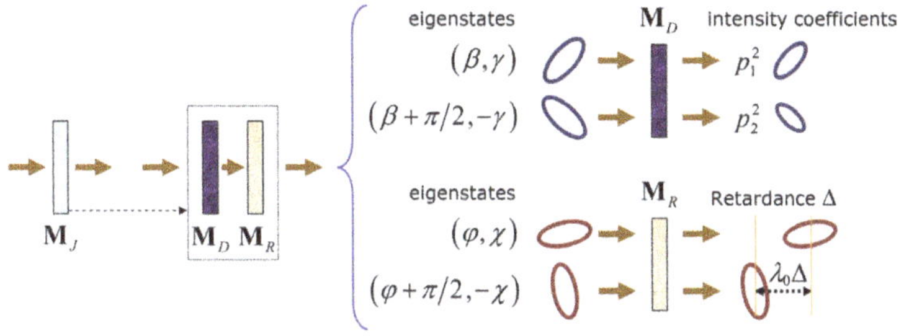

FIGURE 4.9
Polar decomposition of a pure Mueller matrix. Both pure components $\mathbf{M}_D$ and $\mathbf{M}_R$ are normal.

and therefore, the following interesting relations satisfied by pure Mueller matrices hold:

$$\begin{aligned} \mathbf{m}_R^T\mathbf{P} = \mathbf{D} \quad &\text{or} \quad \mathbf{m}_R\mathbf{D} = \mathbf{P} \\ \mathbf{m}_R\mathbf{m}_D\mathbf{m}_R^T = \mathbf{m}_P \quad &\text{or} \quad \mathbf{m}_R^T\mathbf{m}_P\mathbf{m}_R = \mathbf{m}_D \end{aligned} \tag{4.104}$$

These show, in particular, that $\mathbf{m}_R$ rotates the diattenuation vector $\mathbf{D}$ until it coincides with the polarizance vector $\mathbf{P}$. Conversely, $\mathbf{m}_R^T$ produces the inverse rotation carrying $\mathbf{P}$ into $\mathbf{D}$.

In general, $\mathbf{M}_J$ depends on up to seven independent parameters, namely up to four parameters β,γ,p_1,p_2 originating from $\mathbf{M}_D$, together with up to three parameters φ,χ,Δ from $\mathbf{M}_R$ (Figure 4.9).

In virtue of the polar decomposition, a given pure Mueller matrix $\mathbf{M}_J$ can be expressed as (Gil 2003)

$$\mathbf{M}_J = \mathbf{M}_P\left(\beta_P,\gamma_P,p_1,p_2\right)\mathbf{M}_R\left(\alpha,\delta,\Delta\right) = \mathbf{M}_R\left(\alpha,\delta,\Delta\right)\mathbf{M}_D\left(\beta_D,\gamma_D,p_1,p_2\right) \tag{4.105}$$

where the characteristic parameters of the polar components are restricted by the following limits:

$$0 \le \alpha \le \pi/2 \qquad 0 \le \beta_P,\beta_D \le \pi \qquad 0 \le \delta,\gamma_P,\gamma_D < 2\pi \qquad 0 \le \Delta \le \pi \qquad 0 \le p_2 \le p_1 \le 1 \tag{4.106}$$

and can be calculated through the procedure described below.

The diattenuation-polarizance is obtained from the following identities (regardless of whether $\mathbf{M}_J$ is singular or not):

$$D^2 = \frac{m_{01}^2 + m_{02}^2 + m_{03}^2}{m_{00}^2} = \frac{m_{10}^2 + m_{20}^2 + m_{30}^2}{m_{00}^2} = 1 - \frac{\sqrt{\det\mathbf{M}}}{m_{00}^2} \tag{4.107}$$

In terms of the Jones matrix $\mathbf{T}(\mathbf{M}_J)$ (Lu and Chipman 1994),

$$D = \sqrt{1 - \frac{4\left|\det\mathbf{T}\right|^2}{\mathrm{tr}^2\left(\mathbf{T}^\dagger\mathbf{T}\right)}} \tag{4.108}$$

From the diagonalization (3.46) of $\mathbf{M}_J$, the principal intensity coefficients p_1^2 and p_2^2 are given by

$$p_1^2 = m_{00}(1+D) \qquad p_2^2 = m_{00}(1-D) \tag{4.109}$$

The diattenuation can also be expressed as follows in terms of the eigenvalues $\xi_q \equiv q e^{i\alpha}$ and $\xi_r \equiv \varepsilon r e^{-i\alpha}$ (with $\varepsilon \equiv \det \mathbf{T}/|\det \mathbf{T}|$) of the Jones matrix $\mathbf{T}$ (which, in accordance to the common convention adopted in Section 3.2.1, has been chosen so as to satisfy $\det \mathbf{T} \in \mathbb{R}$) and the non-normality parameter η (Lu and Chipman 1994):

$$D = \sqrt{1 - \left[\frac{2(1-\eta^2)qr}{q^2 + r^2 - 2\varepsilon\eta^2 qr\cos 2\alpha}\right]^2} \qquad \varepsilon \equiv \frac{\det \mathbf{T}}{\det \mathbf{T}} \tag{4.110}$$

Further, the characteristic angles of the equivalent diattenuators are given by (Gil 1987)

$$\begin{aligned} \tan 2\beta_P &= \sqrt{m_{02}^2 + m_{03}^2}\Big/ m_{01} \qquad & \tan\gamma_P &= m_{03}/m_{02} \\ \tan 2\beta_D &= \sqrt{m_{20}^2 + m_{30}^2}\Big/ m_{10} \qquad & \tan\gamma_D &= m_{30}/m_{20} \end{aligned} \tag{4.111}$$

To obtain the parameters of the equivalent retarder, the two cases of $\mathbf{M}_J$ being singular or not should be distinguished. Note that $\det \mathbf{M}_J = 0$ if and only if $D = 1$ (i.e., $p_2 = 0$).

The explicit expressions for matrices $\mathbf{M}_D$ and $\mathbf{M}_P$ in terms of the elements of $\mathbf{M}_J$ are (Lamekin 2000)

$$\mathbf{M}_X = m_{00}\begin{pmatrix} 1 & \mathbf{X}^T \\ \mathbf{X} & \mathbf{m}_X \end{pmatrix} \qquad \mathbf{m}_X \equiv \frac{1}{1+d}\begin{pmatrix} x_1^2 + d^2 + d & x_1x_2 & x_1x_3 \\ x_1x_2 & x_2^2 + d^2 + d & x_2x_3 \\ x_1x_3 & x_2x_3 & x_3^2 + d^2 + d \end{pmatrix}$$
$$\left[\mathbf{X} \equiv (x_1, x_2, x_3) = \mathbf{D}, \mathbf{P} \qquad d \equiv \sqrt[4]{\det \hat{\mathbf{M}}_J}\,\right] \tag{4.112}$$

where the vector $\mathbf{X}$ represents either the diattenuation vector $\mathbf{D}$ or the polarizance vector $\mathbf{P}$; or in the canonical form as a function of m_{00} and $\mathbf{X}$

$$\mathbf{M}_X = m_{00}\begin{pmatrix} 1 & \mathbf{X}^T \\ \mathbf{X} & \sin\kappa\,\mathbf{I}_3 + (1-\sin\kappa)\hat{\mathbf{X}}\otimes\hat{\mathbf{X}}^T \end{pmatrix} \qquad \begin{bmatrix} |\mathbf{X}| \equiv D \equiv \cos\kappa, \; 0 \le \kappa < \pi/2 \\ \mathbf{X} \equiv \mathbf{D}, \mathbf{P} \quad \hat{\mathbf{X}} \equiv \mathbf{X}/|\mathbf{X}| \end{bmatrix} \tag{4.113}$$

4.4.1 Nonsingular Pure Mueller Matrices

By using the inverse matrices of the respective diattenuators, the matrix of the retarder is obtained as

$$\mathbf{M}_R(\alpha, \delta, \Delta) = \mathbf{M}_P(\beta_P, \gamma_P, 1/p_1, 1/p_2)\,\mathbf{M} = \mathbf{M}\,\mathbf{M}_D(\beta_D, \gamma_D, 1/p_1, 1/p_2) \tag{4.114}$$

and the azimuth φ, ellipticity angle χ and retardance Δ of $\mathbf{M}_R$ are calculated from the relations

$$\cos\Delta = \frac{\operatorname{tr}\mathbf{M}_R}{2} - 1 \quad \sin 2\chi = \frac{m_{R12} - m_{R21}}{2\sin(\Delta/2)} \quad \sin 2\varphi = \frac{m_{R31} - m_{R13}}{2\cos 2\chi \sin(\Delta/2)} \quad \left[m_{Rij} \equiv (\mathbf{M}_R)_{ij}\right] \tag{4.115}$$

which correspond to the following values for the angular parameters α and δ:

$$\cos 2\alpha = \cos 2\varphi \cos 2\chi \qquad \tan\delta = \tan 2\chi / \sin 2\varphi \tag{4.116}$$

The retardance Δ can also be obtained directly from the Jones matrix $\mathbf{T}(\mathbf{M}_J)$ as follows (Lu and Chipman 1994):

$$\cos(\Delta/2) = \frac{\left|\operatorname{tr}\mathbf{T} + \varepsilon \operatorname{tr}\mathbf{T}^{\dagger}\right|}{2\sqrt{\operatorname{tr}\left(\mathbf{T}^{\dagger}\mathbf{T}\right) + 2\left|\det\mathbf{T}\right|}} \qquad \varepsilon \equiv \frac{\det\mathbf{T}}{\left|\det\mathbf{T}\right|} \tag{4.117}$$

where the representative Jones matrix has been chosen so as to satisfy $\det\mathbf{T} \in \mathbb{R}$. Or, in terms of the eigenvalues $qe^{i\alpha}$ and $\varepsilon re^{-i\alpha}$ of $\mathbf{T}$ and the non-normality parameter η (Lu and Chipman 1994)

$$\cos(\Delta/2) = (q + r)\left|\cos\alpha\right| \sqrt{\frac{1 - \eta^2}{q^2 + r^2 + 2qr\left[1 - \eta^2 (1 + \varepsilon\cos 2\alpha)\right]}} \tag{4.118}$$

The explicit expression of matrix $\mathbf{M}_R$ in terms of the elements of $\mathbf{M}_J$ (assumed $D < 1$) is (Lamekin 2000)

$$\mathbf{M}_R = \begin{pmatrix} 1 & \mathbf{0}^T \\ \mathbf{0} & \mathbf{m}_R \end{pmatrix} \qquad \mathbf{m}_R \equiv \frac{1}{d}\begin{pmatrix} \hat{m}_{11} - \dfrac{P_1 D_1}{1+d} & \hat{m}_{12} - \dfrac{P_1 D_2}{1+d} & \hat{m}_{13} - \dfrac{P_1 D_3}{1+d} \\ \hat{m}_{21} - \dfrac{P_2 D_1}{1+d} & \hat{m}_{22} - \dfrac{P_2 D_2}{1+d} & \hat{m}_{23} - \dfrac{P_2 D_3}{1+d} \\ \hat{m}_{31} - \dfrac{P_3 D_1}{1+d} & \hat{m}_{32} - \dfrac{P_3 D_2}{1+d} & \hat{m}_{33} - \dfrac{P_3 D_3}{1+d} \end{pmatrix}$$

$$\left[\hat{m}_{kl} \equiv m_{kl}/m_{00} \ (k,l = 1,2,3) \quad \mathbf{D} \equiv (D_1, D_2, D_3) \quad \mathbf{P} \equiv (P_1, P_2, P_3) \quad d \equiv \sqrt[4]{\det\hat{\mathbf{M}}_J}\right] \tag{4.119}$$

or in terms of $\mathbf{m}$, $\mathbf{P}$ and $\mathbf{D}$ (with $D < 1$) (Goldstein 2011)

$$\mathbf{M}_R = \frac{1}{\sin\kappa}\begin{pmatrix} \sin\kappa & \mathbf{0}^T \\ \mathbf{0} & \mathbf{m} - (1 - \sin\kappa)\hat{\mathbf{P}} \otimes \hat{\mathbf{D}}^T \end{pmatrix} \qquad \begin{bmatrix} D \equiv \cos\kappa, \ 0 \le \kappa < \pi/2 \\ \hat{\mathbf{D}} \equiv \mathbf{D}/D \quad \hat{\mathbf{P}} \equiv \mathbf{P}/D \end{bmatrix} \tag{4.120}$$

The properties of nonsingular Jones and Mueller-Jones matrices can be studied by means of their representation in the SL(2C) group or in the proper orthochronous Lorentz group SO⁺(1, 3) respectively. In fact, SO⁺(1, 3) is constituted by real 4×4 matrices $\mathbf{L}$ satisfying $\mathbf{L}^T\mathbf{G}\mathbf{L}=\mathbf{G}$, $\det\mathbf{L}=+1$ and $(\mathbf{L})_{00}\geq 1$, so that any nonsingular pure Mueller matrix $\mathbf{M}_J$ can be normalized as $\mathbf{L}(\mathbf{M}_J)=\mathbf{M}_J/\det\mathbf{M}_J$ (with $\det\mathbf{M}_J=m_{00}^4(1-D^2)^2$, see Section 3.3.5.3), so that, as required for $\mathbf{L}(\mathbf{M}_J)$ to be considered an element of SO⁺(1, 3), $\det\mathbf{L}(\mathbf{M}_J)=+1$ and $l_{00}\equiv[\mathbf{L}(\mathbf{M}_J)]_{00}>1$. Note that the indicated normalization cannot be applied to singular Mueller matrices ($\det\mathbf{M}_J=0$), which prevents such $\mathbf{L}(\mathbf{M}_J)$ from being elements of SO⁺(1,3) (see Section 4.9.4). It should also be noted that the normalization $\mathbf{M}_J/\det\mathbf{M}_J$ is not compatible with the passivity condition $l_{00}(1+D)\leq 1$ except for pure retarders ($m_{00}=1$); otherwise $D>0$ implies $l_{00}>1$ (i.e., $l_{00}(1+D)>1$).

4.4.2 Singular Pure Mueller Matrices

When $\mathbf{M}_J$ is singular ($p_2=0\Leftrightarrow D=1\Leftrightarrow\det\mathbf{M}_J=(\det\mathbf{T})^4=0$) it is necessarily of the form (Lu and Chipman 1996)

$$\mathbf{M}_J=m_{00}\begin{pmatrix}1 & \hat{\mathbf{D}}^T\\ \hat{\mathbf{P}} & \hat{\mathbf{P}}\otimes\hat{\mathbf{D}}^T\end{pmatrix}=m_{00}\begin{pmatrix}1\\ \hat{\mathbf{P}}\end{pmatrix}\otimes\left(1,\hat{\mathbf{D}}^T\right)\qquad (P=D=1)\tag{4.121}$$

so that the equivalent normal diattenuators $\mathbf{M}_{\hat{D}}$ and $\mathbf{M}_{\hat{P}}$ are still determined by $\mathbf{D}$ and $\mathbf{P}$ vectors respectively

$$\begin{aligned}\mathbf{M}_{\hat{D}}&=m_{00}\begin{pmatrix}1 & \hat{\mathbf{D}}^T\\ \hat{\mathbf{D}} & \hat{\mathbf{D}}\otimes\hat{\mathbf{D}}^T\end{pmatrix}=m_{00}\begin{pmatrix}1\\ \hat{\mathbf{D}}\end{pmatrix}\otimes\left(1,\hat{\mathbf{D}}^T\right)\\ \mathbf{M}_{\hat{P}}&=m_{00}\begin{pmatrix}1 & \hat{\mathbf{P}}^T\\ \hat{\mathbf{P}} & \hat{\mathbf{P}}\otimes\hat{\mathbf{P}}^T\end{pmatrix}=m_{00}\begin{pmatrix}1\\ \hat{\mathbf{P}}\end{pmatrix}\otimes\left(1,\hat{\mathbf{P}}^T\right)[P=D=1]\end{aligned}\tag{4.122}$$

while the retarder is not unique.

The equivalent retarder can be chosen as the one having the minimum possible retardance Δ_m i.e., as the retarder that transforms $\mathbf{D}$ into $\mathbf{P}$ through the minimum rotation angle Δ_m on the Poincaré sphere. With this choice, and leaving aside the case where $\mathbf{P}=\mathbf{D}$ ($\Delta=0$), the normalized retardance vector $\hat{\mathbf{R}}$ is a unit vector orthogonal to the plane defined by $\mathbf{D}$ and $\mathbf{P}$ (Lu and Chipman 1996),

$$\hat{\mathbf{R}}=(\mathbf{P}\wedge\mathbf{D})/|\mathbf{P}\wedge\mathbf{D}|\qquad [P=D=1]\tag{4.123}$$

where $\wedge$ stands for the cross product (or vector product). The retardance and Poincaré retardance vectors are given by

$$\cos\Delta_m=\mathbf{P}^T\mathbf{D}\quad \mathbf{R}=\Delta_m(\mathbf{P}\wedge\mathbf{D})/|\mathbf{P}\wedge\mathbf{D}|\quad \bar{\mathbf{R}}=(\Delta_m\mathbf{P}\wedge\mathbf{D})/(\pi|\mathbf{P}\wedge\mathbf{D}|)\qquad [0\leq\Delta_m\leq\pi]\tag{4.124}$$

The expression for the retardance in Eq. (4.117), when applied to a singular Jones matrix $\mathbf{T}(\mathbf{M}_J)$, leads to the following formula for the minimum retardance (Lu and Chipman 1994):

$$\cos(\Delta_m/2) = |\mathrm{tr}\,\mathbf{T}| / \mathrm{tr}(\mathbf{T}^\dagger \mathbf{T}) \tag{4.125}$$

or, in terms of the non-normality parameter η (Lu and Chipman 1994)

$$\sin(\Delta_m/2) = \eta \tag{4.126}$$

Note that, from Eq. (4.117), the minimum retardance Δ_m corresponds to the maximum trace of $\mathbf{M}_R$.

4.4.3 Application of the Polar Decomposition to an Experimental Example

As an example, consider the matrix

$$\mathbf{M}_{LC} = \begin{pmatrix} 1 & -0.1392 & 0.1351 & 0.9217 \\ -0.1392 & 0.3495 & 0.0100 & -0.0997 \\ -0.1351 & -0.0100 & 0.3062 & -0.1929 \\ 0.9217 & -0.0997 & 0.1929 & 0.9567 \end{pmatrix}$$

which represents the calculated polarimetric response in reflection of a cholesteric liquid crystal (Bohley and Scharf 2004). It is easy to check that this Mueller matrix is both pure (or nondepolarizing), as well as non-singular. Its polar decomposition in the form $\mathbf{M}_R\mathbf{M}_D$ can be readily obtained by explicitly constructing the partitioned polar factors $\mathbf{M}_R$ and $\mathbf{M}_D$ reported in the previous sections. However, a direct algebraic approach without explicit partitioning is also possible. Indeed, because of the orthogonal nature of the retarder $\mathbf{M}_R$, it is straightforward to see that the diattenuator $\mathbf{M}_D$ is given by

$$\mathbf{M}_D = \sqrt{\mathbf{M}_{LC}^T \mathbf{M}_{LC}} = \begin{pmatrix} 1 & -0.1392 & 0.1351 & 0.9217 \\ -0.1392 & 0.3504 & -0.0141 & -0.0960 \\ 0.1351 & -0.0141 & 0.3496 & 0.0932 \\ 0.9217 & -0.0960 & 0.0932 & 0.9718 \end{pmatrix}$$

where the matrix square root is obtained through diagonalization, taking the square roots of the diagonal elements and subsequent remultiplication. The advantage of this method is that it performs on both non-singular and singular Mueller matrices. The retarder is finally determined by right-multiplication of $\mathbf{M}_{LC}$ by the inverse of $\mathbf{M}_D$,

$$\mathbf{M}_R = \mathbf{M}_{LC}\mathbf{M}_D^{-1} = \begin{pmatrix} 1 & 0 & 0 & 0 \\ 0 & 0.9974 & 0.0718 & -0.0109 \\ 0 & -0.0718 & 0.9522 & -0.2969 \\ 0 & -0.0109 & 0.2969 & 0.9548 \end{pmatrix}$$

(If $\mathbf{M}$ is singular, then so will be $\mathbf{M}_D$ and the methodology described in Section 4.4.2 has to be applied for the determination of $\mathbf{M}_R$ then.) Further, one finds for the magnitudes of the diattenuation and retardance vectors $D = 0.9419$ and $R = 0.3104$, the vectors themselves being oriented along the respective unit directions $\hat{\mathbf{D}} = (-0.1478, 0.1434, 0.9786)^T$ and $\hat{\mathbf{R}} = (-0.9720, 0, 0.2351)^T$ (the unit vectors $\hat{\mathbf{D}}$ and $\hat{\mathbf{R}}$ contain the cosine directors of **D** and **R** respectively).

We see that the cholesteric liquid crystal exhibits very important diattenuation, approximately directed along the S_3 axis of the Poincaré sphere, thus identifying its nature as being essentially circular. This is in perfect agreement with the well-known experimental fact that cholesteric liquid crystals behave as partial circular polarizers in reflection.

Next, the small but non-negligible retardance value (0.3104) is directed mostly along the S_1 axis pointing at linear birefringence; however, a smaller component is likewise oriented along the S_3 axis revealing the presence of (very) weak optical activity. This is again consistent with the polarimetric model of cholesteric liquid crystals as stratified structures made of twisted anisotropic layers (Bohley and Scharf 2004). Note that if the order of the polar factors were taken to be $\mathbf{M}_P\mathbf{M}_R$ (instead of $\mathbf{M}_R\mathbf{M}_D$) then the magnitudes D and R of the diattenuation and retardance vectors would not change, as well as the direction $\hat{\mathbf{R}}$ of the retardance one; but only the direction of the diattenuation vector would.

4.5 The General Serial Decomposition of a Nondepolarizing System

Because of the general non-commutativity of Mueller matrices, the polar decomposition of a pure Mueller matrix appears in two forms depending on the relative order of the diattenuator and the retarder. For some purposes, it is desirable to find an equivalent serial system with symmetric structure that reflects, in an explicit manner, the reciprocity properties inherent to Mueller matrices.

The equivalent model (retarder-diattenuator) of a pure system provided by the polar decomposition of its corresponding Jones or Mueller matrix, **T** and $\mathbf{M}_J$ respectively, is closely related to the singular value decomposition of **T** (see Section 3.2.1), called the *general serial decomposition* of **T**

$$\mathbf{T} = \mathbf{T}_{R2}\,\mathbf{T}_{DLO}(p_1, p_2)\,\mathbf{T}_{R1} \qquad \left[\mathbf{T}_{DLO}(p_1, p_2) = \mathrm{diag}(p_1, p_2)\right] \tag{4.127}$$

where $\mathbf{T}_{R1}$ and $\mathbf{T}_{R2}$ are unitary matrices and $\mathbf{T}_{DLO} \equiv \mathrm{diag}(p_1, p_2)$ is a diagonal matrix whose nonzero elements are the real nonnegative singular values p_1 and p_2, called the principal amplitude coefficients of **T**. Leaving aside a global phase factor $e^{i\phi}$, **T** depends on seven independent parameters and therefore, there exists a relation between the three angular parameters $(\alpha_2, \delta_2, \Delta_2)$ of $\mathbf{T}_{R2}$ and those of $\mathbf{T}_{R1}$, $(\alpha_1, \delta_1, \Delta_1)$. Thus, the information contained in the seven independent parameters of **T** can be considered as shared among the components of the equivalent system: two parameters (p_1, p_2) of the central linear horizontal diattenuator plus five additional parameters. The latter are shared among the left and right equivalent retarders so that, in virtue of the equivalence theorems for serial combinations of retarders (Section 4.2.3), $\mathbf{T}_{R1}$ and $\mathbf{T}_{R2}$ can be parameterized non-uniquely.

In particular, the components of the polar decomposition can be rearranged in different ways as, for instance,

$$\begin{gathered}\mathbf{T} = \mathbf{T}_{R2}\left[\mathbf{T}_G(-\beta_P)\,\mathbf{T}_{DLO}(p_1, p_2)\,\mathbf{T}_G(\beta_P)\right]\mathbf{T}_{R1} \\ \mathbf{T}_{R2} \equiv \mathbf{T}_{RLO}(-\gamma_P) \qquad \mathbf{T}_{R1} \equiv \mathbf{T}_{RLO}(\gamma_P)\,\mathbf{T}_R(\alpha, \delta, \Delta)\end{gathered} \tag{4.128}$$

where the left retarder (represented by $\mathbf{T}_{R2}$) has been chosen to be linear (while $\mathbf{T}_{R1}$ corresponds, in general, to an elliptical retarder) or alternatively,

$$\begin{gathered}\mathbf{T}=\mathbf{T}'_{R2}\left[\mathbf{T}_G(-\beta_D)\mathbf{T}_{DL0}(p_1,p_2)\mathbf{T}_G(\beta_D)\right]\mathbf{T}'_{R1}\\ \mathbf{T}'_{R2}\equiv\mathbf{T}_R(\alpha,\delta,\Delta)\mathbf{T}_{RL0}(-\gamma_D)\qquad \mathbf{T}'_{R1}\equiv\mathbf{T}_{RL0}(\gamma_D)\end{gathered}\tag{4.129}$$

where the right retarder (represented by $\mathbf{T}'_{R1}$) has been chosen to be linear (while $\mathbf{T}'_{R2}$ corresponds, in general, to an elliptical retarder). Obviously, the indicated equivalences can be formulated in terms of corresponding pure Mueller matrices and the explicit expressions for the elements of $\mathbf{M}_J$ in terms of the parameters of the serial components of the particular forms (4.128) and (4.129) can be found in (Gil, 1983).

By considering the partitioned expression (3.41) for a pure Mueller matrix $\mathbf{M}_J$, the *general serial decomposition* of $\mathbf{M}_J$ results in

$$\begin{aligned}\mathbf{M}_J&\equiv m_{00}\begin{pmatrix}1&\mathbf{D}^T\\ \mathbf{P}&\mathbf{m}\end{pmatrix}=\mathbf{M}_{R2}\,\mathbf{M}_{DL0}\,\mathbf{M}_{R1}\\ &=\begin{pmatrix}1&\mathbf{0}^T\\ \mathbf{0}&\mathbf{m}_{R2}\end{pmatrix}\left[m_{00}\begin{pmatrix}1&\mathbf{D}_0^T\\ \mathbf{D}_0&\mathbf{m}_{D_0}\end{pmatrix}\right]\begin{pmatrix}1&\mathbf{0}^T\\ \mathbf{0}&\mathbf{m}_{R1}\end{pmatrix}=m_{00}\begin{pmatrix}1&\left(\mathbf{m}_{R1}^T\mathbf{D}_0\right)^T\\ \mathbf{m}_{R2}\mathbf{D}_0&\mathbf{m}_{R2}\mathbf{m}_{D_0}\mathbf{m}_{R1}\end{pmatrix}\\ &\left[\mathbf{D}_0\equiv(D,0,0)^T\qquad \mathbf{m}_{D_0}\equiv\operatorname{diag}(1,\sin\kappa,\sin\kappa)\qquad \cos\kappa\equiv D\right]\end{aligned}\tag{4.130}$$

and therefore,

$$\mathbf{D}=\mathbf{m}_{R1}^T\mathbf{D}_0\qquad \mathbf{P}=\mathbf{m}_{R2}\mathbf{D}_0\qquad \mathbf{m}=\mathbf{m}_{R2}\mathbf{m}_{D_0}\mathbf{m}_{R1}\tag{4.131}$$

showing that $\mathbf{m}_{R1}$ and $\mathbf{m}_{R2}^T$ rotate respectively the diattenuation vector $\mathbf{D}$ and the polarizance vector $\mathbf{P}$ until they coincide with $\mathbf{D}_0\equiv(D,0,0)^T$.

4.6 The Symmetric Decomposition of a Nondepolarizing System

As we have seen, the general serial decomposition has a symmetric structure in the sense that the equivalent system is constituted of a horizontal linear diattenuator sandwiched between two retarders. A particularly interesting symmetrized decomposition of a pure Mueller matrix can be achieved through the following choice for the parameterization of $\mathbf{M}_{R1}$ and $\mathbf{M}_{R2}$:

$$\mathbf{M}_{R1}=\mathbf{M}_{RL1}(\varphi_1,\Delta_1)\mathbf{M}_{RL0}(\Delta/2)\qquad \mathbf{M}_{R2}=\mathbf{M}_{RL0}(\Delta/2)\mathbf{M}_{RL2}(\varphi_2,\Delta_2)\tag{4.132}$$

so that $\mathbf{M}_J$ can always be expressed as

$$\mathbf{M}_J=\left[\mathbf{M}_{RL2}(\varphi_2,\Delta_2)\mathbf{M}_{RL0}(\Delta/2)\right]\mathbf{M}_{DL0}(p_1,p_2)\left[\mathbf{M}_{RL0}(\Delta/2)\mathbf{M}_{RL1}(\varphi_1,\Delta_1)\right]\tag{4.133}$$

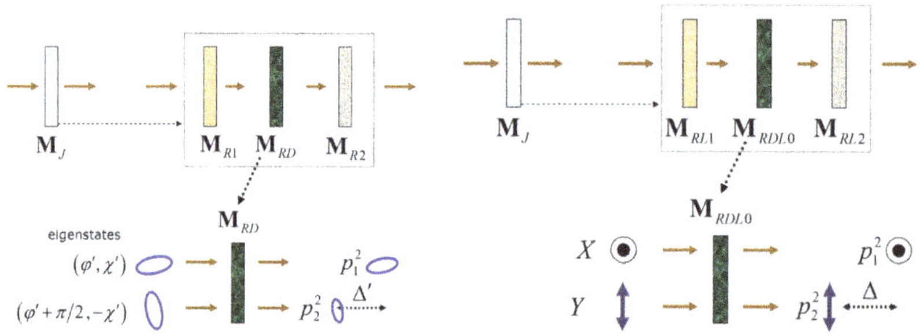

FIGURE 4.10
(a) The Mueller matrix $\mathbf{M}_J$ of any nondepolarizing system can be obtained through appropriate dual-retarder transformations of elliptical diattenuating retarders; (b) among such transformations, $\mathbf{M}_J$ can be represented by means of its symmetric decomposition, where the central diattenuating retarder is linear-horizontal while the entrance and exit retarders are linear (in general different).

in terms of the seven parameters $(\varphi_1, \Delta_1, \Delta, p_1, p_2, \varphi_2, \Delta_2)$. One thus obtains the *symmetric decomposition* (Ossikovski 2008),

$$\begin{gathered}\mathbf{M}_J = \mathbf{M}_{RL2}(\varphi_2, \Delta_2)\mathbf{M}_{RDL0}(\Delta, p_1, p_2)\,\mathbf{M}_{RL1}(\varphi_1, \Delta_1) \\ \left[\mathbf{M}_{RDL0}(\Delta, p_1, p_2) \equiv \mathbf{M}_{RL0}(\Delta)\mathbf{M}_{DL0}(p_1, p_2) = \mathbf{M}_{DL0}(p_1, p_2)\mathbf{M}_{RL0}(\Delta)\right]\end{gathered} \tag{4.134}$$

where $\mathbf{M}_{RL1}$ and $\mathbf{M}_{RL2}$ are *linear* retarders, while the central component, whose Mueller matrix $\mathbf{M}_{RDL0}$ is given by Eq. (4.100), is a serial combination of a linear diattenuator $\mathbf{M}_{DL0}$ and a linear retarder $\mathbf{M}_{RL0}$ whose eigenaxes are aligned with the laboratory references axes. Since $\mathbf{M}_{DL0}$ and $\mathbf{M}_{RL0}$ commute, the horizontal *linear diattenuating retarder* (or horizontal *linear retarding diattenuator*) represented by $\mathbf{M}_{RDL0}$ (Figure 4.10) takes the form

$$\mathbf{M}_{RDL0} = m_{00}\begin{pmatrix} 1 & \mathbf{D}_0^T \\ \mathbf{D}_0 & \mathbf{m}_{RDL0} \end{pmatrix} \quad \mathbf{D}_0 \equiv \begin{pmatrix} \cos\kappa \\ 0 \\ 0 \end{pmatrix} \quad \mathbf{m}_{RDL0} \equiv \begin{pmatrix} 1 & 0 & 0 \\ 0 & \sin\kappa\cos\Delta & \sin\kappa\sin\Delta \\ 0 & -\sin\kappa\sin\Delta & \sin\kappa\cos\Delta \end{pmatrix}$$
$$[\cos\kappa \equiv D] \tag{4.135}$$

The mean intensity coefficient m_{00}, as well as the diattenuation-polarizance D of the central horizontal linear diattenuating retarder are equal to those of the pure Mueller matrix $\mathbf{M}_J$,

$$D \equiv \cos\kappa = \sqrt{m_{01}^2 + m_{02}^2 + m_{03}^2}\,/m_{00} = \sqrt{m_{10}^2 + m_{20}^2 + m_{30}^2}\,/m_{00} \tag{4.136}$$

In virtue of Eq. (4.78), the principal intensity coefficients p_1^2 and p_2^2 of $\mathbf{M}_J$ (which are equal to those of $\mathbf{M}_{RDL0}$) are given by $p_1^2 = m_{00}(1+D)$ and $p_2^2 = m_{00}(1-D)$.

To calculate the parameters of the linear retarders, observe that, from Eq. (4.133),

$$\mathbf{M}_J \equiv m_{00}\begin{pmatrix} 1 & \mathbf{D}^T \\ \mathbf{P} & \mathbf{m} \end{pmatrix} = m_{00}\begin{pmatrix} 1 & \left(\mathbf{m}_{RL1}^T\mathbf{D}_0\right)^T \\ \mathbf{m}_{RL2}\mathbf{D}_0 & \mathbf{m}_{RL2}\mathbf{m}_{RDL0}\mathbf{m}_{RL1} \end{pmatrix} \tag{4.137}$$

so that the orthogonal 3×3 submatrices $\mathbf{m}_{RL1}$ and $\mathbf{m}_{RL2}$ of the respective linear retarders satisfy

$$\mathbf{m}_{RL1}^T \mathbf{D}_0 = \mathbf{D} \qquad \mathbf{m}_{RL2} \mathbf{D}_0 = \mathbf{P} \tag{4.138}$$

The first one implies that the unit vector defined by the first column $(\mathbf{m}_{RL1}^T)_1$ of $\mathbf{m}_{RL1}^T$ is given by the normalized diattenuation vector $\hat{\mathbf{D}} \equiv \mathbf{D}/D = (\mathbf{m}_{RL1}^T)_1$. From the second one, it follows that the first column-vector $(\mathbf{m}_{RL2})_1$ of $\mathbf{m}_{RL2}$ is given by the normalized polarizance vector $\hat{\mathbf{P}} \equiv \mathbf{P}/D = (\mathbf{m}_{RL2})_1$. The remaining column vectors of $\mathbf{m}_{RL1}^T$ and $\mathbf{m}_{RL2}$ are determined from the already obtained first columns $(\mathbf{m}_{RL1}^T)_1$ and $(\mathbf{m}_{RL2})_1$ with the help of the explicit expression (4.39) of a linear retarder. As a result, the respective retardances Δ_1, Δ_2 and azimuths φ_1, φ_2 of the left and right linear retarders are calculated from the following expressions:

$$\begin{matrix} \tan(\Delta_1/2) = (1-\hat{D}_1)/\hat{D}_3^2 & \sin 2\varphi_1 = \hat{D}_3/\sin\Delta_1 \\ \tan(\Delta_2/2) = (1-\hat{P}_1)/\hat{P}_3^2 & \sin 2\varphi_2 = \hat{P}_3/\sin\Delta_2 \end{matrix} \quad \begin{bmatrix} \hat{D}_i \equiv D_i/D \\ \hat{P}_i \equiv P_i/P \end{bmatrix} \tag{4.139}$$

so that

$$\mathbf{M}_{RDL0} = \mathbf{M}_{RL2}^T(\varphi_2, \Delta_2)\, \mathbf{M}_J\, \mathbf{M}_{RL1}^T(\varphi_1, \Delta_1) \tag{4.140}$$

and the retardance Δ of the diattenuating retarder is obtained from

$$\sin\Delta = (\mathbf{M}_{RDL0})_{23}/(m_{00}\sin\kappa) \qquad \cos\Delta = (\mathbf{M}_{RDL0})_{22}/m_{00}\sin\kappa \tag{4.141}$$

Once the seven parameters $(\varphi_1, \Delta_1, \Delta, p_1, p_2, \varphi_2, \Delta_2)$ of the equivalent system have been calculated, the decomposition (4.133) can be obtained by expanding $\mathbf{M}_{RDL0}$ in the form

$$\mathbf{M}_{RDL0} = \mathbf{M}_{RL0}(\Delta/2)\, \mathbf{M}_{DL0}(p_1, p_2)\, \mathbf{M}_{RL0}(\Delta/2) \tag{4.142}$$

A more detailed study of the symmetric serial decomposition of a pure Mueller matrix $\mathbf{M}_J$ is reported in Ossikovski 2008, where the option of a vertical retarding diattenuator, besides horizontal one, is also considered.

Let us finally note that an alternative form of generalized decomposition of a pure system as a serial combination of a linear diattenuator, a circular diattenuator, a linear retarder and a circular retarder was developed by Savenkov et al. (2006) by means of the Jones calculus.

4.7 Two-Vector Representation of a Nondepolarizing System

Let us consider again the polar decomposition $\mathbf{M}_J = \mathbf{M}_R\mathbf{M}_D = \mathbf{M}_P\mathbf{M}_R$ of a pure Mueller matrix (representing a nondepolarizing medium or device) and observe that $\mathbf{M}_J$ is completely characterized by its mean intensity coefficient m_{00} together with the Poincaré retardance vector $\bar{\mathbf{R}}$ of $\mathbf{M}_R$ and the diattenuation-polarizance vector $\mathbf{D}$ of $\mathbf{M}_D$. Equivalently, since $\mathbf{m}_R\mathbf{D} = \mathbf{P}$, $\mathbf{M}_J$ is also determined by m_{00}, $\bar{\mathbf{R}}$ and $\mathbf{P}$. Observe that $\bar{\mathbf{R}}$, $\mathbf{D}$ and $\mathbf{P}$ are the respective polarization vectors of the fast eigenstate of the retarder $\mathbf{M}_R$, the eigenstate with the largest eigenvalue of the normal diattenuator $\mathbf{M}_D$, and the eigenstate with the largest eigenvalue of the normal diattenuator $\mathbf{M}_P$.

Consequently, pure Mueller matrices can be classified into three types (Gil and San José 2016b):

(a) $D = 0$. Then $\mathbf{M}_J = m_{00}\,\mathbf{M}_R$, which corresponds to a pure retarder with intensity coefficient m_{00} that is completely determined by $\bar{\mathbf{R}}$ and m_{00}. In this case m_{00} is the intensity coefficient for all incident polarization states.
(b) $D \neq 0$. Then $\mathbf{M}_J$ is completely determined by the scalar quantity m_{00} together with the pair of vectors $\mathbf{D}$ and $\bar{\mathbf{R}}$. The direction of $\mathbf{P}$ is obtained by means of a rotation of $\mathbf{D}$ about the direction defined by $\bar{\mathbf{R}}$ by an angle (retardance) $\Delta = 2\arccos\left\{[\mathrm{tr}(\mathbf{M}_J\mathbf{M}_D^{-1})]/2-1\right\}$. When $\bar{\mathbf{R}}$ and $\mathbf{D}$ are parallel, then $\mathbf{P} = \mathbf{D}$, and the system is a diattenuating retarder (i.e., $\mathbf{M}_J$ is a normal matrix, and therefore, $\eta = 0$). If $D = 1$, then $\mathbf{M}_J$ is associated with a normal polarizer whose equivalent retarder $\mathbf{M}_R$ is the identity matrix (i.e., $\Delta = 0$).
(c) $D \neq 0$ and $\mathbf{D} = \mathbf{P}$. Then $\bar{\mathbf{R}}$ and $\mathbf{D}$ (and $\mathbf{P}$) are parallel, so that the system is a diattenuating retarder (i.e., $\mathbf{M}_J$ is a normal matrix, and therefore, $\eta = 0$ and $\gamma = 0$), with retardance given by $\Delta = 2\arccos\left\{[\mathrm{tr}(\mathbf{M}_J\mathbf{M}_D^{-1})]/2-1\right\}$. If $D = 1$, then $\mathbf{M}_J$ is associated with a normal polarizer whose equivalent retarder $\mathbf{M}_R$ is the identity matrix (i.e., $\Delta = 0$).

It is remarkable that any arbitrary pair of three-dimensional real vectors $\mathbf{v}$ and $\mathbf{w}$, with $|\mathbf{v}| < 1$ and $|\mathbf{w}| < 1$, can be respectively considered, without any additional constraint, as the diattenuation and Poincaré retardance vectors of a certain normalized pure Mueller matrix $\hat{\mathbf{M}}_J$ and therefore, an arbitrary pure Mueller matrix can be built from $\mathbf{v}$, $\mathbf{w}$, and a positive scalar m_{00} satisfying $m_{00}(1+|\mathbf{v}|) \leq 1$.

4.8 Invariant Quantities of a Nondepolarizing Mueller Matrix

As in the case of 2D polarization states whose intensity and degree of polarization are invariant with respect to any change of reference frame, as well as to any transformation by means of retarders (representing rotations on the Poincaré sphere), some properties of the nondepolarizing medium represented by $\mathbf{M}_J$ remain invariant with respect to certain transformations.

As in Section 3.3.6, let us first consider Eq. (3.67) representing an arbitrary rotation about the Z axis of the coordinate system XY containing the polarization ellipse of the incident and emerging polarization states, and study the physical quantities that remain unchanged when the Mueller matrix is submitted to a rotation transformation $\mathbf{M}_{JX'Y'} = \mathbf{M}_G(\theta)\,\mathbf{M}_J\,\mathbf{M}_G(-\theta)$. The transformation can be expressed as

$$\mathbf{M}_{JX'Y'} = \mathbf{M}_G(\theta)\, m_{00}\begin{pmatrix} 1 & \mathbf{D}^T \\ \mathbf{P} & \mathbf{m} \end{pmatrix}\mathbf{M}_G(-\theta) = m_{00}\begin{pmatrix} 1 & \mathbf{D}^T\mathbf{m}_G^T \\ \mathbf{m}_G\mathbf{P} & \mathbf{m}_G\mathbf{m}\mathbf{m}_G^T \end{pmatrix}$$
$$\mathbf{m}_G \equiv \begin{pmatrix} \cos 2\theta & \sin 2\theta & 0 \\ -\sin 2\theta & \cos 2\theta & 0 \\ 0 & 0 & 1 \end{pmatrix} \qquad (4.143)$$

showing that the following quantities remain invariant (Gil 2016b; Yakovlev and Yakovlev 2019):

- The mean intensity coefficient (average transmittance-reflectance or gain) m_{00}.
- The linear diattenuation D_L, defined as the degree of linear polarization of the Stokes vector $\mathbf{s}_D \equiv (1, \mathbf{D}^T)^T$, where $\mathbf{D}$ is the diattenuation vector of $\mathbf{M}_J$.
- The circular diattenuation D_C, defined as the degree of circular polarization of $\mathbf{s}_D$.
- The linear polarizance P_L, defined as the degree of linear polarization of the Stokes vector $\mathbf{s}_P \equiv (1, \mathbf{P}^T)^T$, where $\mathbf{P}$ is the polarizance vector of $\mathbf{M}_J$.
- The circular polarizance P_C, defined as the degree of circular polarization of $\mathbf{s}_P$.

The transformation of the components of retardance can be analyzed by means of the polar decomposition $\mathbf{M}_J = \mathbf{M}_R \mathbf{M}_D$, together with the decomposition of the equivalent retarder $\mathbf{M}_R$ as the serial combination $\mathbf{M}_R = \mathbf{M}_{RL} \mathbf{M}_{RC}$ of a linear retarder $\mathbf{M}_{RL}$ and a circular retarder $\mathbf{M}_{RC}$, so that

$$\begin{aligned} \mathbf{M}_{JX'Y'} &= \mathbf{M}_G \mathbf{M}_J \mathbf{M}_G^T = \mathbf{M}_G \mathbf{M}_R \mathbf{M}_D \mathbf{M}_G^T = \left(\mathbf{M}_G \mathbf{M}_R \mathbf{M}_G^T\right)\left(\mathbf{M}_G \mathbf{M}_D \mathbf{M}_G^T\right) \\ &= \left(\mathbf{M}_G \mathbf{M}_{RL} \mathbf{M}_G^T\right)\left(\mathbf{M}_G \mathbf{M}_{RC} \mathbf{M}_G^T\right)\left(\mathbf{M}_G \mathbf{M}_D \mathbf{M}_G^T\right) \end{aligned} \tag{4.144}$$

Now, by defining the retardance vector $\mathbf{R}$ of $\mathbf{M}_J$ as the retardance vector of $\mathbf{M}_R$, and observing that $\mathbf{M}_G \mathbf{M}_{RL} \mathbf{M}_G^T$ preserves the linear retardance R_L while $\mathbf{M}_G \mathbf{M}_{RC} \mathbf{M}_G^T$ preserves the circular retardance R_C, it is straightforward to show that the following quantities remain invariant under the transformation (4.143) (Gil 2016c):

- The Poincaré linear retardance $\bar{R}_L$, defined as the degree of linear polarization of the Stokes vector $\mathbf{s}_R \equiv (1, \bar{\mathbf{R}}^T)^T$, where $\bar{\mathbf{R}}$ is the Poincaré retardance vector of the retarder component $\mathbf{M}_R$ in the polar decomposition $\mathbf{M}_J = \mathbf{M}_R \mathbf{M}_D = \mathbf{M}_P \mathbf{M}_R$ of $\mathbf{M}_J$
- The Poincaré circular retardance $\bar{R}_C$, defined as the degree of circular polarization of $\mathbf{s}_R$.

Obviously, other derived quantities such as the diattenuation D; the polarizance P (with $P = D$); the principal intensity coefficients p_1^2 and p_2^2, and the retardance R are likewise invariant. Further, the degree of polarimetric purity (or depolarization index) P_Δ defined in Eq. (3.42) can be expressed as

$$P_\Delta^2 = 2P_P^2/3 + P_S^2 \tag{4.145}$$

in terms of the so-called *components of purity* of the corresponding Mueller matrix $\mathbf{M}$ (Gil 2011), namely the *degree of polarizance*,

$$P_P \equiv \sqrt{(P^2 + D^2)/2} \tag{4.146}$$

which, in the case of pure systems, satisfies $P_P = P = D$, and the *polarimetric dimension index*

$$P_S \equiv \|\mathbf{m}\|_2 / \sqrt{3} \tag{4.147}$$

For a pure Mueller matrix, $P_\Delta = 1$, which implies

$$P_S^2 = 1 - 2D^2/3 \tag{4.148}$$

and therefore, since D and P are invariant, P_S is necessarily invariant too.

In summary, regarding arbitrary rotations of the reference frame XY with respect to the Z axis, we have identified the following set of six independent invariant parameters: $m_{00}, D_L, P_L, D, \bar{R}_L, \bar{R}_C$ (which, together the rotation angle θ, constitute a specific set of the seven independent parameters of $\mathbf{M}_J$). Any other invariant quantity can be derived from this set in a straightforward manner, like

$$\begin{array}{lll} D_C = \sqrt{D^2 - D_L^2} & \bar{R} = \sqrt{\bar{R}_L^2 + \bar{R}_C^2} & p_1^2 = m_{00}(1+D) \\ P_C = \sqrt{D^2 - P_L^2} & P_S^2 = 1 - 2D^2/3 & p_2^2 = m_{00}(1-D) \end{array} \tag{4.149}$$

Obviously, different alternative sets of six independent parameters can be chosen as, for example, $m_{00}, D_L, D_C, P_L, \bar{R}_L, \bar{R}_C$, or $m_{00}, D_L, P_L, D, \bar{R}_L, \bar{R}$.

Finally, let us consider the invariants of a nondepolarizing system with respect to arbitrary changes of generalized bases for the independent representation of the incident and emerging polarization states, i.e., the invariants of $\mathbf{M}_J$ with respect to *dual retarder transformations*

$$\mathbf{M}'_J = \mathbf{M}_{R2}\mathbf{M}_J\mathbf{M}_{R1} \tag{4.150}$$

where $\mathbf{M}_{R1}$ and $\mathbf{M}_{R2}$ are arbitrary orthogonal Mueller matrices (i.e., representing respective retarders). Transformation (4.150) preserves the two independent scalar parameters p_1^2 and p_2^2, as well as other quantities that can be derived from them, like m_{00}, D and P (with $P = D$) (Gil 2016c).

4.9 Particular Forms of Nondepolarizing Mueller Matrices

Despite the fact that many particular forms of pure Mueller matrices have been considered in the previous sections of this chapter, it is worth considering and putting together a number of special cases frequently met in experimental polarimetry. Thus, we shall devote this section to the identification, comparison and interpretation of the main families of pure Mueller matrices by means of the variety of tools described in previous sections of this chapter.

4.9.1 Normal Pure Mueller Matrices

A normal (also called *homogeneous*) pure Mueller matrix $\mathbf{M}_J$ is characterized by the property that it has two orthogonal eigenstates, or equivalently, $\mathbf{M}_J\mathbf{M}_J^T = \mathbf{M}_J^T\mathbf{M}_J$. The general form of a normal pure Mueller matrix is that of a serial combination of a normal diattenuator $\mathbf{M}_D$ and a retarder $\mathbf{M}_R$ with coincident eigenstates (in general elliptical), so that the combination is called a *diattenuating retarder* (or *retarding diattenuator*). Its Mueller matrix $\mathbf{M}_{RD}$ can be expressed as

$$\begin{aligned}\mathbf{M}_{RD}(\alpha,\delta,\Delta,p_1,p_2) &= \mathbf{M}_D(\alpha,\delta,p_1,p_2)\,\mathbf{M}_R(\alpha,\delta,\Delta) = \mathbf{M}_R(\alpha,\delta,\Delta)\,\mathbf{M}_D(\alpha,\delta,p_1,p_2)\\ &= \mathbf{M}_{RL0}(-\delta)\,\mathbf{M}_G(-\alpha)\,\mathbf{M}_{RDL0}(\Delta,p_1,p_2)\mathbf{M}_G(\alpha)\,\mathbf{M}_{RL0}(\delta)\end{aligned}$$

$$\mathbf{M}_{RDL0}(\Delta,p_1,p_2) \equiv \mathbf{M}_{DL0}(p_1,p_2)\,\mathbf{M}_{RL0}(\Delta) = \mathbf{M}_{RL0}(\Delta)\,\mathbf{M}_{DL0}(p_1,p_2) \tag{4.151}$$

where the Mueller matrix of the central horizontal linear diattenuating retarder can be factored as the product of a horizontal linear diattenuator $\mathbf{M}_{DL0}(p_1,p_2)$ and a horizontal linear retarder $\mathbf{M}_{RL0}(\Delta)$ in either of the two possible relative positions.

When $\Delta = 0$ the system takes the form of the normal diattenuator $\mathbf{M}_D$, and when $D = 0$ (i.e., $p_1 = p_2$) $\mathbf{M}_{RD}$ takes the form $\mathbf{M}_{RD} = m_{00}\,\mathbf{M}_R$ which up to the intensity coefficient $m_{00} = p_1^2$ corresponds to a retarder.

It should be noted that, due to the particular form of $\mathbf{M}_{RD}$, which depends on up to five independent parameters, the matrix $\mathbf{M}_R$ satisfying $\mathbf{M}_{RD} = \mathbf{M}_R\,\mathbf{M}_{RDL0}\,\mathbf{M}_R^T$ is not unique, so that, in accordance with the symmetric decomposition (also called *dual linear retarder transformation*, see Section 4.6), $\mathbf{M}_R$ can be chosen as being a linear retarder. In fact, by taking advantage of the simplicity of the Jones formalism and by inserting appropriately the identity matrix, the transformation (4.151) can be converted into the following *single linear retarder transformation*:

$$\begin{aligned}\mathbf{T}_{RD}(\alpha,\delta,\Delta,p_1,p_2) &= \mathbf{T}_{RL0}(-\delta)\,\mathbf{T}_G(-\alpha)\,\mathbf{T}_{RDL0}(p_1,p_2,\Delta)\,\mathbf{T}_G(\alpha)\,\mathbf{T}_{RL0}(\delta)\\ &= \mathbf{T}_{RL0}(-\delta)\,\mathbf{T}_G(-\alpha)\left[\mathbf{T}_{R\delta}\,\mathbf{T}_{R\delta}^\dagger\right]\mathbf{T}_{RDL0}(p_1,p_2,\Delta)\left[\mathbf{T}_{R\delta}\,\mathbf{T}_{R\delta}^\dagger\right]\mathbf{T}_G(\alpha)\,\mathbf{T}_{RL0}(\delta)\\ &= \left[\mathbf{T}_{RL0}(-\delta)\,\mathbf{T}_G(-\alpha)\,\mathbf{T}_{R\delta}\right]\left[\mathbf{T}_{R\delta}^\dagger\,\mathbf{T}_{RDL0}(p_1,p_2,\Delta)\,\mathbf{T}_{R\delta}\right]\left[\mathbf{T}_{R\delta}^\dagger\,\mathbf{T}_G(\alpha)\,\mathbf{T}_{RL0}(\delta)\right]\\ &= \left[\mathbf{T}_{RL0}(-\delta)\,\mathbf{T}_G(-\alpha)\,\mathbf{T}_{R\delta}\right]\left[\mathbf{T}_{RDL0}(p_1,p_2,\Delta)\right]\left[\mathbf{T}_{R\delta}^\dagger\,\mathbf{T}_G(\alpha)\,\mathbf{T}_{RL0}(\delta)\right]\\ &= \mathbf{T}_{RL}\,\mathbf{T}_{RDL0}(p_1,p_2,\Delta)\,\mathbf{T}_{RL}^\dagger\end{aligned} \tag{4.152.a}$$

where

$$\mathbf{T}_{R\delta} \equiv \begin{pmatrix} ie^{-i\delta/2} & 0 \\ 0 & -ie^{i\delta/2}\end{pmatrix} \qquad \mathbf{T}_{RL} \equiv \mathbf{T}_{RL0}(-\delta)\mathbf{T}_G(-\alpha)\mathbf{T}_{R\delta} = \begin{pmatrix} \cos\alpha e^{i\delta/2} & i\sin\alpha \\ i\sin\alpha & \cos\alpha e^{-i\delta/2}\end{pmatrix} \tag{4.152.b}$$

Consequently, a normal pure Mueller matrix $\mathbf{M}_{RD}$ can always be written as $\mathbf{M}_{RD} = \mathbf{M}_{RL}\,\mathbf{M}_{RDL0}\,\mathbf{M}_{RL}^T$.

4.9.2 Non-normal Pure Mueller Matrices

Pure Mueller matrices are, in general, non-normal (also called inhomogeneous) so that the normal diattenuators $\mathbf{M}_P$ and $\mathbf{M}_D$ of the two forms of the polar decomposition

$$\mathbf{M}_J = \mathbf{M}_P(\beta_P,\gamma_P,p_1,p_2)\,\mathbf{M}_R(\alpha,\delta,\Delta) = \mathbf{M}_R(\alpha,\delta,\Delta)\,\mathbf{M}_D(\beta_D,\gamma_D,p_1,p_2) \tag{4.153}$$

are different ($\mathbf{M}_P \neq \mathbf{M}_D$), while the linear retarders $\mathbf{M}_{RL1}$ and $\mathbf{M}_{RL2}$ of the dual linear retarder transformation of $\mathbf{M}_J$ do not satisfy $\mathbf{M}_{RL2} \neq \mathbf{M}_{RL1}^T$.

4.9.3 Degenerate Pure Mueller Matrices

Degenerate pure Mueller matrices constitute a particular case of non-normal pure Mueller matrices characterized by the fact that they have only one eigenstate (whose corresponding eigenvalue is doubly degenerate). This kind of Mueller matrices has been considered in Section 3.6.3, where it has been shown that a given degenerate Mueller matrix $\mathbf{M}_g$ can always be expressed as

$$\mathbf{M}_g = \mathbf{M}_R \, \mathbf{M}_{g\gamma} \, \mathbf{M}_R^T$$
$$\mathbf{M}_{g\gamma} = \frac{1}{2}\begin{pmatrix} 2q^2+\gamma^2 & -\gamma^2 & 2q\gamma\cos\alpha & -2q\gamma\sin\alpha \\ \gamma^2 & 2q^2-\gamma^2 & 2q\gamma\cos\alpha & -2q\gamma\sin\alpha \\ 2q\gamma\cos\alpha & -2q\gamma\cos\alpha & 2q^2 & 0 \\ -2q\gamma\sin\alpha & 2q\gamma\sin\alpha & 0 & 2q^2 \end{pmatrix} \tag{4.154}$$

with $0 \le q \le 1$, $0 \le \gamma \le 1$ and the passivity constraint $\sqrt{4q^2+\gamma^2}+\gamma \le 2$.

4.9.4 Singular Pure Mueller Matrices

Singular Mueller matrices constitute a subset of Mueller matrices with genuine peculiarities. From Eq. (3.37), it follows that $\mathbf{M}_J^T \mathbf{G} \mathbf{M}_J = \sqrt{\det \mathbf{M}_J}\,\mathbf{G}$, with $\mathbf{G} \equiv \mathrm{diag}(1,-1,-1,-1)$. As seen in Section 4.4, unlike nonsingular pure Mueller matrices, a singular pure Mueller matrix cannot be considered as being proportional to an element of the proper orthochronous Lorentz group, SO+(1, 3), i.e., to a real 4×4 matrix $\mathbf{L}$ satisfying $\mathbf{L}^T\mathbf{G}\mathbf{L} = \mathbf{G}$, $\det \mathbf{L} = +1$ and $(\mathbf{L})_{00} \ge 1$. Therefore, leaving aside the fact that experimental measurements have always a limited accuracy (and consequently, experimental Mueller matrices are, in general, nonsingular), the properties of singular pure Mueller matrices have to be considered from a specific algebraic point of view in order to achieve a complete theoretical characterization.

Eq. (3.53) shows that $\det \mathbf{M}_J = 0$ if and only if $D = 1$, or equivalently, the counter-polarizance is zero ($\sin\kappa = 0$), the second intensity coefficient is zero ($p_2^2 = 0$), and the Frobenius and nuclear norms are equal ($\|\mathbf{M}_J\|_2 = \|\mathbf{M}_J\|_* = 2m_{00}$). Due to the peculiar structure of pure Mueller matrices, any singular pure Mueller matrix $\mathbf{M}_J$ can be written in the form shown in Eq. (4.121), so that

$$\mathbf{M}_J = \mathbf{M}_{\hat{P}\hat{D}} \equiv m_{00}\left(\mathbf{s}_{\hat{P}} \otimes \mathbf{s}_{\hat{D}}^T\right) \qquad \mathbf{s}_{\hat{P}} \equiv \begin{pmatrix} 1 \\ \hat{\mathbf{P}} \end{pmatrix} \quad \mathbf{s}_{\hat{D}} \equiv \begin{pmatrix} 1 \\ \hat{\mathbf{D}} \end{pmatrix} \tag{4.155}$$

where $\hat{\mathbf{D}}$ and $\hat{\mathbf{P}}$ are the unit diattenuation and polarizance vectors determining the two-vector representation of $\mathbf{M}_J$.

When $\hat{\mathbf{P}} = \hat{\mathbf{D}}$, then $\mathbf{M}_{\hat{P}\hat{D}}$ corresponds to a *normal polarizer-analyzer* of the form shown in Eq. (4.77). The two-vector representation of $\mathbf{M}_{\hat{P}\hat{D}}$ is thus given by the unique vector $\hat{\mathbf{P}} = \hat{\mathbf{D}}$.

When $\hat{\mathbf{P}} \neq \hat{\mathbf{D}}$, $\mathbf{M}_{\hat{P}\hat{D}}$ is non-normal. In the particular case of a degenerate polarizer-analyzer, the double eigenvalue q^2 (corresponding to the unique eigenstate) is necessarily equal to zero ($q = 0$) and the degeneracy reaches its maximum value $\gamma = 1$, so that the canonical form of the degenerate polarizer-analyzer takes the form

$$\mathbf{M}_{g\gamma} = \frac{1}{2}\begin{pmatrix} 1 & -1 & 0 & 0 \\ 1 & -1 & 0 & 0 \\ 0 & 0 & 0 & 0 \\ 0 & 0 & 0 & 0 \end{pmatrix} = \frac{1}{2}\begin{pmatrix} 1 \\ 1 \\ 0 \\ 0 \end{pmatrix} \otimes (1,-1,0,0) \tag{4.156}$$

The diattenuation and polarizance vectors are mutually orthogonal and correspond respectively to the points $(1,0,0)$ and $(-1,0,0)$ on the Poincaré sphere.

A serial combination composed of a retarder sandwiched between two pure normal polarizers whose transmission axes are mutually crossed is called a *black sandwich* (Berry and Dennis 2004). When expressed with respect to a laboratory coordinate system whose X axis coincides with the transmission axis of the first polarizer, its Mueller matrix takes precisely the form of the degenerate polarizer-analyzer shown in Eq. (4.156). Other forms of degenerate polarizers-analyzers can be obtained through *single retarder transformations* of the form $\mathbf{M}_g = \mathbf{M}_R^T \mathbf{M}_{g\gamma} \mathbf{M}_R$, where $\mathbf{M}_R$ is a retarder Mueller matrix. Observe that degenerate polarizers are square roots of the zero matrix (are nilpotent), that is to say, given a singular degenerate Mueller matrix $\mathbf{M}_g$, it satisfies $\mathbf{M}_g^2 = 0$. (Sudha and Gopala Rao 2001).

4.9.4.1 Non-Normal Elliptical Polarizer

Given an arbitrary normal polarizer $\mathbf{M}_{\hat{D}}$, any prescribed emerging pure state (denoted as $\mathbf{s}_{\hat{P}}$ for reasons that will become clear later) can be obtained (non-uniquely) by the action on any incident pure state of the serial combination of $\mathbf{M}_{\hat{D}}$ and an appropriate retarder $\mathbf{M}_R$. The Mueller matrix $\mathbf{M}_{\hat{P}\hat{D}}$ of this kind of devices has the form

$$\mathbf{M}_{\hat{P}\hat{D}} = \mathbf{M}_R \mathbf{M}_{\hat{D}} = m_{00}\begin{pmatrix} 1 & \mathbf{0}^T \\ \mathbf{0} & \mathbf{m}_R \end{pmatrix}\begin{pmatrix} 1 & \hat{\mathbf{D}}^T \\ \hat{\mathbf{D}} & \hat{\mathbf{D}}\otimes\hat{\mathbf{D}}^T \end{pmatrix} = m_{00}\begin{pmatrix} 1 & \hat{\mathbf{D}}^T \\ \mathbf{m}_R\hat{\mathbf{D}} & \mathbf{m}_R\hat{\mathbf{D}}\otimes\hat{\mathbf{D}}^T \end{pmatrix} = m_{00}\left(\mathbf{s}_{\hat{P}} \otimes \mathbf{s}_{\hat{D}}^T\right)$$
$$m_{00} = \frac{p_1^2}{2} \qquad \mathbf{s}_{\hat{P}} \equiv \begin{pmatrix} 1 \\ \mathbf{m}_R\hat{\mathbf{D}} \end{pmatrix} \qquad \mathbf{s}_{\hat{D}} \equiv \begin{pmatrix} 1 \\ \hat{\mathbf{D}} \end{pmatrix} \tag{4.157}$$

so that, up to an intensity coefficient $m_{00} \leq 1/2$, the emerging state is uniquely determined by the unit polarizance vector $\hat{\mathbf{P}} = \mathbf{m}_R \hat{\mathbf{D}}$ of the composed system (i.e., by the Stokes vector $\mathbf{s}_{\hat{P}}$) while the diattenuation vector $\hat{\mathbf{D}}$ controls the overall attenuation. Obviously, when $\mathbf{m}_R = \mathbf{I}_3$ then the device is a normal polarizer.

Since linear polarizers and linear retarders are easily achievable components in practice, an interesting illustrative example of a non-normal elliptical polarizer is the combination of a linear polarizer $\mathbf{M}_{\hat{D}L}$ followed by a linear retarder $\mathbf{M}_{RL}$. The emerging state is determined by the retardance Δ of $\mathbf{M}_{RL}$, as well as by the respective orientations ψ and φ of the transmission axis of $\mathbf{M}_{\hat{D}L}$ and of the fast axis of $\mathbf{M}_{RL}$,

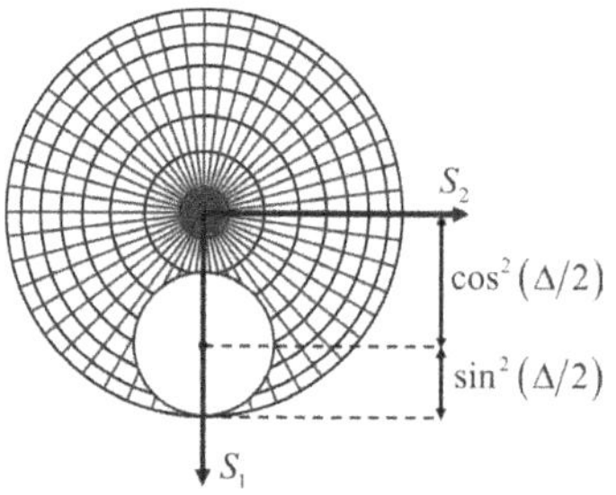

FIGURE 4.11
Action of a linear polarizer $\mathbf{M}_{\hat{D}L}(\psi, D)$ followed by a linear retarder $\mathbf{M}_{RL}(\varphi, \Delta)$. For fixed values of ψ and Δ, and variable values of φ (i.e., for a variable relative angle $\varphi - \psi$) the emerging state of polarization traces a "figure-eight" contour on the surface of the Poincaré sphere defined by the intersection of the sphere with a cylinder whose generatrix is parallel to the S_3 axis and touches the surface at the equatorial point defined by the linear eigenstate of $\mathbf{M}_{\hat{D}L}$. The radius of the cylinder is $\sin^2(\Delta/2)$, while the distance from the generatrix to the origin of the sphere is given by $\cos^2(\Delta/2)$ (Azzam 2000).

$$\mathbf{M}_{\hat{P}\hat{D}L} = m_{00}\,\mathbf{M}_{RL}(\varphi, \Delta)\,\hat{\mathbf{M}}_{\hat{D}L}(\psi, D) = m_{00}\left(\mathbf{s}_{\hat{P}} \otimes \mathbf{s}^T_{\hat{D}L}\right)$$
$$m_{00} = \frac{p_1^2}{2} \qquad \mathbf{s}_{\hat{P}} \equiv \begin{pmatrix} 1 \\ \mathbf{m}_{RL}(\varphi, \Delta)\,\hat{\mathbf{D}}_L(\psi, D) \end{pmatrix} \qquad \mathbf{s}_{\hat{D}L} \equiv \begin{pmatrix} 1 \\ \hat{\mathbf{D}}_L(\psi, D) \end{pmatrix} \quad (4.158)$$

Note that $\mathbf{s}_{\hat{P}}$ does not necessarily correspond to a linearly polarized state.

As shown by Azzam (2000), for fixed values of ψ and Δ and variable values of φ (i.e., for a variable relative angle $\varphi - \psi$), the emerging state of polarization traces a "figure-eight" contour in the surface of the Poincaré sphere. The latter is defined by its intersection with a cylinder whose generatrix is parallel to the S_3 axis of the Poincaré sphere and touches the sphere at the equatorial point defined by the linear eigenstate of $\mathbf{M}_{\hat{D}L}$ (Figure 4.11). The radius of the cylinder is $\sin^2(\Delta/2)$, while the distance from the generatrix to the origin of the sphere is given by $\cos^2(\Delta/2)$ (Azzam 2000). When $\Delta = \pi/2$ (quarter-wave retarder) the contour includes both right- and left-circular polarization states. This kind of combination is frequently used in polarimetry as a polarization state generator or analyzer (in the latter case, the normal linear polarizer $\mathbf{M}_{\hat{D}L}$ is placed after the linear retarder $\mathbf{M}_{RL}$). In the more general case where elliptical polarization states are impinging upon a rotating linear retarder, the emerging polarization states correspond to trajectories determined by the intersection of the surface of the Poincaré and a cone (Salazar-Ariza and Torres 2018).

5

The Concept of Mueller Matrix

5.1 Introduction

Mueller matrices play a central role in polarimetry because they represent the polarimetric effect of media in light-matter linear interactions. As seen in Chapters 3 and 4, under certain circumstances the medium behaves as polarimetrically *pure*, or *nondepolarizing*, because it preserves the degree of polarization for any totally polarized incident state. Nevertheless, many scientific and industrial applications of polarimetry involve depolarization phenomena. Depolarization is the property of transforming some, or all, totally polarized incident states into partially polarized emerging states. As seen in Chapters 1 and 2, depolarization is due to loss of coherence between the components of the electric field of light. This phenomenon can be produced by certain kinds of interaction between electromagnetic waves and matter, such as

- Scattering: as for instance scattering by particles with arbitrary shape or random orientation (Van de Hulst 1957; Bohren and Huffman 1983).
- Reflection or refraction by spatially inhomogeneous targets or devices.
- More generally, the spatial, spectral or temporal integration or averaging inherent to the measurement process, where the instrument features are determinant (Ossikovski and Hingerl 2016; Hingerl and Ossikovski 2016). A special situation is that of polychromatic light passing through dispersive birefringent media, in which case the frequency-dependent retardance leads to depolarization if the thickness of the device is larger than the coherence length of the wave (Dlugnikov, 1984).

In analogy to the concept of color of an object, which involves, in a combined manner, the properties of the object, the illuminating light and the observation conditions, a medium may exhibit different Mueller matrices for different geometric configurations of the probing light (of a given spectral profile), angles of incidence, observation (detection) direction, etc. The Mueller matrix likewise depends on the instrument specifications (spot size, sensitive area of the detector, measurement time…). Therefore, when the expression "Mueller matrix of the medium" is used, the measurement configuration and conditions are understood implicitly.

The main objective of the present chapter is to introduce the concept of Mueller matrix, not only for nondepolarizing media (as seen in Chapters 3 and 4) but also for any kind of linear polarimetric interaction, thus including depolarizing processes. For this purpose, the contents of Chapter 5 are organized as described below. In the remaining part of the

DOI: 10.1201/9780367815578-5

introduction, the main physical conceptions that underlie the polarimetric interactions and that lead to the construction of the physical concept of Mueller matrix are briefly summarized.

Section 5.2 is devoted to the construction of a physical Mueller matrix as the result of the synthesis of basic nondepolarizing transformations undergone by the polarization states of the probing light under interaction with the material medium. The concepts of covariance $\mathbf{H}(\mathbf{M})$ and coherency $\mathbf{C}(\mathbf{M})$ matrices, associated with the Mueller matrix $\mathbf{M}$, are required for a proper mathematical characterization of Mueller matrices and are introduced in Section 5.3.

As a direct consequence of its physical nature, as shown in Section 5.4, any Mueller matrix can be considered as a convex sum (or ensemble average) of pure (nondepolarizing) and passive (i.e., not increasing the intensity of the probing light) Mueller matrices. The *ensemble criterion* established in such section leads to the identification of the space of Mueller matrices with 4×4 real matrices that satisfy two sets of inequalities, namely four *covariance conditions* and two *passivity conditions*. The rules for the transformation of Mueller matrices under rotations of the Cartesian laboratory reference frame, or rotations of the medium itself, are briefly presented in Section 5.5.

Section 5.6 is devoted to the formulation of the covariance conditions, which, as shown in Section 5.7, are intimately linked to the properties of the two possible canonical types of depolarizing Mueller matrices resulting from their *normal form*. The transformation of a given Mueller matrix $\mathbf{M}$ upon interchanging the directions of propagation of the incident and emerging probing light is an important topic, considered in Section 5.8, which allows for the formulation of the passivity conditions in Section 5.9.

A pure Mueller matrix $\mathbf{M}_J$ has the genuine property that it does not admit sum decompositions into different pure Mueller matrices; the closeness of a given Mueller matrix $\mathbf{M}$ to a pure one is termed the *polarimetric purity* of $\mathbf{M}$ and can be quantified in several ways described in Section 5.10.

Among other specific features of the mathematical structure of Mueller matrices, five constitutive three-dimensional real vectors are identified in Section 5.11; these are particularly useful for the analysis of certain anisotropy properties studied in Chapter 6.

Section 5.12 includes the formulation of the *arrow decomposition* of a Mueller matrix $\mathbf{M}$, which is based on the fact that any linear polarimetric interaction is equivalent to the successive actions of an entrance retarder, a medium (the arrow component) that may be interpreted as free of retardance properties, and an exit retarder, in such a manner that $\mathbf{M}$ is given by the product of those of the indicated serial components. Section 5.13 is dedicated to the analysis of certain specific features of the arrow form of $\mathbf{M}$ regarding the structure of the information relative to enpolarizing and depolarizing properties that allow for a simple general characterization of Mueller matrices based on their statistical nature.

Section 5.14 deals with the different ways of decomposing a given a Mueller matrix $\mathbf{M}$ into respective sums of simpler Mueller matrices, among which the *spectral*, the *arbitrary* and the *characteristic* decompositions provide meaningful frameworks for the analysis and treatment of relevant polarimetric properties.

Then, the way in which Mueller matrices are generated by respective random Jones matrices, called the *Jones generators*, that allow for classifying and determining interesting geometric features of Mueller matrices is analyzed in Section 5.15.

Section 5.16 is devoted to the synthesis of Mueller matrices corresponding to two essential types of polarization transformations causing depolarization, namely partial spectral coherence and partial spatial coherence, where the effect of the measurement device plays a critical role.

Prior to addressing in detail the general mathematical formulation of Mueller algebra, it is necessary to recall that the interaction of a photon with a single atom or molecule is necessarily nondepolarizing. However, as it will be shown in further sections, from a macroscopic point of view, there are many circumstances under which the light-matter interaction and its measurement produce depolarization. In general, for an incident totally polarized electromagnetic beam, the interaction with a material system produces a superposition of totally polarized outgoing light states with different polarizations. Depending on the spectral distribution of the incident wave, on the nature of the material target and on the nature of the interaction, these exiting pencils can present different degrees of mutual coherence. Thus, the superposition of exiting pencils is, in general, partially coherent with two limiting physical situations: (1) *coherent*, so that their respective Jones vectors must be added in order to obtain the resulting Jones vector, and (2) *incoherent*, in which case their respective Stokes vectors must be added in order to obtain the resulting Stokes vector.

The above observations imply that the polarimetric effect of a material medium in general involves spatial, spectral or temporal averages of basic nondepolarizing interactions and is polarimetrically equivalent to that of a system composed of a *parallel combination* (or spatially inhomogeneous mixture) of a number of nondepolarizing optical elements distributed over the illuminated area (Figure 5.1).

Throughout this book we deal with physical (or physically realizable) Mueller matrices, i.e., 4×4 real matrices that represent the linear polarimetric behavior of material systems. In accordance with the above considerations, we will show that the Mueller matrix of a material system can always be obtained through a convex sum (a weighted sum whose coefficients are positive and sum to one) of a set of pure Mueller matrices associated with respective equivalent nondepolarizing elements constitutive of a certain parallel combination. This concept of general Mueller matrix is equivalent to saying that the Mueller matrix of the system can be considered as the result of an ensemble average of pure polarization transformations (Parke 1948; Kim et al. 1987).

Although in the literature the properties of matrices transforming Stokes vectors into Stokes vectors (i.e., satisfying the so-called *Stokes criterion*) have been extensively studied (Simon 1987; Van der Mee and Hovenier 1992; Van der Mee 1993; Kostinski et al. 1993; Givens and Kostinski 1993; Sridhar and Simon, 1994), it will be shown in Section 5.3 that this kind of matrices, hereafter named *Stokes matrices* (Nagirner (1993; Gil 2000b), also called *pre-Mueller matrices* (Simon et al. 2010a; 2010b)), does not coincide with the set of *Mueller matrices*, which is derived from the *ensemble criterion* mentioned in the above paragraph.

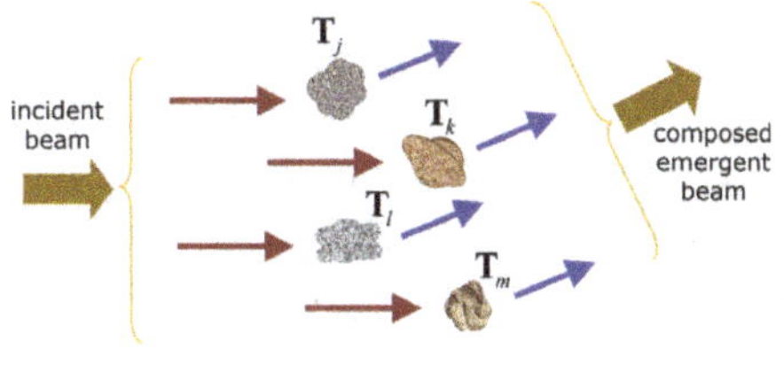

FIGURE 5.1
Any macroscopic linear polarimetric transformation of the polarization state of the probing electromagnetic wave can be considered as an average of the actions of a number of elementary deterministic (or nondepolarizing) elements with respective well-defined Jones matrices $\mathbf{T}_i$.

Obviously, any physical Mueller matrix is a Stokes matrix but, in general, the converse is not true. A straightforward example is given by $\mathbf{G} \equiv \mathrm{diag}(1,-1,-1,-1)$ which is clearly a Stokes matrix, however it is not a physical Mueller one (since it does not satisfy the ensemble criterion, i.e., it is not physically realizable). Consequently, no method to physically realize a Stokes matrix non-derivable from the ensemble criterion can be established. Moreover, it is obvious that only physical Mueller matrices (simply called Mueller matrices from now on) are the subject of practical interest. As some authors have pointed out (Parke 1948; Gil and Bernabéu 1985; Cloude 1986; Kim et al. 1987; Van der Mee 1993; Bolshakov et al. 1996; Gopala Rao et al. 1998b; Gil 2000a; Simon et al. 2010a; Ossikovski and Gil 2017), the ensemble-based model represents a suitable theoretical framework for the study of polarimetric phenomena.

5.2 Synthesis of Depolarizing Mueller Matrices

As mentioned in the introduction, in general a material sample can exhibit spatial inhomogeneity, as well as dispersive effects, over the area illuminated by the incident electromagnetic beam, producing depolarization. The emerging electromagnetic wave is composed of a number of contributions that superimpose incoherently and consequently, the polarimetric effect of the material system cannot be represented by means of a Jones matrix. However, since any macroscopic linear polarimetric effect is essentially due to a certain combination of elementary electronic excitations, the system can be considered as being composed of a number of deterministic nondepolarizing elements, each one with a well-defined Jones matrix, in such a manner that the emerging wave is in turn composed of a coherent, partially coherent or incoherent mixture of the pencils emerging from such elements.

As seen in Section 3.4.2, the covariance vector $\mathbf{h}$ of a pure medium is defined as $\mathbf{h}^T = (t_0, t_1, t_2, t_3)/\sqrt{2}$ where t_i are the elements of the corresponding Jones matrix $\mathbf{T}$. Thus, structures like $\mathbf{h}$ and $\mathbf{T}$ do not contain enough parameters to represent the 16 independent polarimetric quantities involved in a general depolarizing linear interaction. Nevertheless, unlike what happens with $\mathbf{h}$ and $\mathbf{T}$, which depend on up to eight real parameters (including a global phase factor), the concept of covariance matrix (see Section 3.4.1) can be extended to represent general depolarizing systems. Such generalization is obtained through an appropriate ensemble averaging of pure covariance matrices that results in a general covariance matrix, i.e., a 4×4 positive semidefinite Hermitian matrix that depends on up to 16 independent real parameters.

To perform the generalization, consider a uniform electromagnetic wave with fixed state of polarization represented by its corresponding Stokes vector $\mathbf{s}$, interacting linearly with a sample that exhibits inhomogeneous polarimetric structure along the transverse area crossed by the wave. The sample can be regarded as being composed of a number n of parallel components "i" with respective well-defined covariance vectors $\mathbf{h}_i$ (up to, possibly fluctuating, respective phase factors $e^{i\phi_i}$) and respective cross-sections $p_i = A_i/A$ (A and A_i, being the total transverse area covered by the incident wave and the area of the i element, respectively) (Gil 2007; 2000a; Kim et al. 1987; Gil and San José 2019).

To handle mathematically the possible statistical fluctuations of the relative phases among the pure components, it is necessary to establish a reference for the global phase shift introduced by each element, for instance, by writing $\mathbf{h}_i = e^{i\phi_i}\hat{\mathbf{h}}_i$, with $\hat{\mathbf{h}}_i$ satisfying

$\hat{h}_{i0}\hat{h}_{i3} - \hat{h}_{i1}\hat{h}_{i2} \in \mathbb{R}$ (i.e., the associated Jones matrix $\hat{\mathbf{T}}_i$ satisfies $\det \hat{\mathbf{T}}_i \in \mathbb{R}$). In case the field reaches each patch or tile *i* affected by different respective phases (like, for instance, in speckle patterns), those can be included in ϕ_i. Thus, for each realization (temporal or spectral) determined by a specific set of fixed covariance vectors $\mathbf{h}_i$, the covariance matrix $\mathbf{H}_J$ of the composed medium is pure and is generated by the covariance vector $\mathbf{h} \equiv \sum_{i=1}^{n} \sqrt{p_i}\, \mathbf{h}_i$ as follows (San José and Gil 2020a):

$$\mathbf{H}_J = \left(\sum_{i=1}^{n} \sqrt{p_i}\,\mathbf{h}_i\right) \otimes \left(\sum_{i=1}^{n} \sqrt{p_i}\,\mathbf{h}_i\right)^{\dagger} = \left(\sum_{i=1}^{n} \sqrt{p_i}\,\mathbf{h}_i\right) \otimes \left(\sum_{i=1}^{n} \sqrt{p_i}\,\mathbf{h}_i^{\dagger}\right) = \mathbf{X} + \mathbf{Y}$$

$$\mathbf{X} \equiv \sum_{i=1}^{n} p_i \mathbf{H}_{Ji} \quad \mathbf{Y} \equiv \sum_{\substack{i,j=1 \\ i \neq j}}^{n} \sqrt{p_i p_j}\left[\left(\mathbf{h}_i \otimes \mathbf{h}_j^{\dagger}\right) + \left(\mathbf{h}_j \otimes \mathbf{h}_i^{\dagger}\right)\right] \quad \left[\mathbf{H}_{Ji} = \mathbf{h}_i \otimes \mathbf{h}_i^{\dagger}\right] \tag{5.1}$$

where $\mathbf{X}$ is the convex sum of the covariance matrices $\mathbf{H}_{Ji}$ of the components. Notice that $\mathbf{X}$ is a nonpure covariance matrix with $1 < \operatorname{rank}\mathbf{X} \leq 4$ (the effective value of $\operatorname{rank}\mathbf{X}$ depends on the dimension of the subspace of $\mathbb{C}^4$ generated by the set of vectors $\mathbf{h}_i$ $(i = 1,\ldots,n)$), while the cross term $\mathbf{Y}$ is a Hermitian matrix that is not positive semidefinite (otherwise one would have $\operatorname{rank}\mathbf{H}_J > 1$, which violates the starting hypothesis that $\mathbf{H}_J$ is a pure covariance matrix). Despite these properties of $\mathbf{X}$ and $\mathbf{Y}$, their sum always results in the pure covariance matrix $\mathbf{H}_J = \mathbf{X} + \mathbf{Y}$.

As described in (Ossikovski and Hingerl 2016; Hingerl and Ossikovski 2016; Kuntman et al. 2017b; San José and Gil 2020a), together with the spatial statistical distribution considered, temporal or spectral fluctuations of the phase differences $\phi_{ij} = \phi_i - \phi_j$ among each pair of pure components can occur, so that the effective covariance matrix $\mathbf{H}$ results from the following ensemble average $\langle \mathbf{H}_J \rangle$:

$$\mathbf{H} = \langle \mathbf{H}_J \rangle = \langle \mathbf{X} \rangle + \langle \mathbf{Y} \rangle$$

$$\langle \mathbf{X} \rangle = \sum_{i=1}^{n} \langle p_i \mathbf{H}_{Ji} \rangle = \sum_{i=1}^{n} p_i \mathbf{H}_{Ji} = \mathbf{X} \quad \langle \mathbf{Y} \rangle = \sum_{\substack{i,j=1 \\ i \neq j}}^{n} \sqrt{p_i p_j}\left[\langle e^{i\phi_{ij}} \rangle \left(\hat{\mathbf{h}}_i \otimes \hat{\mathbf{h}}_j^{\dagger}\right) + \langle e^{-i\phi_{ij}} \rangle \left(\hat{\mathbf{h}}_i \otimes \hat{\mathbf{h}}_j^{\dagger}\right)^{\dagger}\right]. \tag{5.2}$$

This result involves several important remarks such as

(1) $\mathbf{H}$, obtained through an ensemble average of pure covariance matrices is necessarily a covariance matrix with $1 \leq \operatorname{rank}\mathbf{H} \leq 4$, the value of $r \equiv \operatorname{rank}\mathbf{H}$ depending on the nature of the fluctuations of the phase differences ϕ_{ij}.
(2) $\mathbf{H}_{Ji} = \mathbf{h}_i \otimes \mathbf{h}_i^{\dagger}$ are the pure covariance matrices of the components.
(3) $\mathbf{X}$ represents the convex sum of the covariance matrices of the components, and therefore, describes their incoherent composition.
(4) The Hermitian matrix $\mathbf{Y} = \mathbf{H} - \mathbf{X}$ is not positive semidefinite (in fact, it is necessarily indefinite), so that, except for the trivial limiting case where $\mathbf{Y} = \mathbf{0}$ (i.e., either all ϕ_{ij} are uniformly distributed or all covariance vectors $\mathbf{h}_i$ are mutually proportional), $\mathbf{Y}$ is not a covariance matrix.
(5) When all ϕ_{ij} are fixed, the sum $\mathbf{X} + \mathbf{Y}$ results in the pure covariance matrix $\mathbf{H}_J = \mathbf{h} \otimes \mathbf{h}^{\dagger}$ associated with the covariance vector $\mathbf{h} = \sum_{i=1}^{n} \sqrt{p_i}\, \mathbf{h}_i$ (this is the limiting case of fully coherent composition).

(6) When all fluctuating phase differences ϕ_{ij} are uniformly distributed, the resulting covariance matrix $\mathbf{H} = \mathbf{X}$ is given by the incoherent composition (convex sum) of the covariance matrices $\mathbf{H}_{Ji} = \mathbf{h}_i \otimes \mathbf{h}_i^{\dagger}$, so that $r = 2,3,4$ depending on the dimension of the subspace of $\mathbb{C}^4$ covered by the set of covariance vectors $\mathbf{h}_i$ $(i = 1,\ldots,n)$.

(7) In general, each ϕ_{ij} can have a specific probability distribution, resulting in intermediate cases where $\mathbf{Y}$ is nonzero while it takes forms different from that of the fully coherent composition. Therefore, the composition is partially coherent and results in a composed covariance matrix $\mathbf{C}$ with $r = 2,3,4$, depending on the nature of the probability density functions of the phase differences ϕ_{ij}.

8) The elements m_{ij} of the Mueller matrix $\mathbf{M}$ (pure or nonpure) and the elements h_{ij} of its associated covariance matrix $\mathbf{H}$ are linked through the following biunivocal linear relation (San José and Gil 2020a, see Section 3.4.1):

$$\mathrm{m}_{ij} = \mathbf{tr}\left(\mathbf{E}_{ij}\,\mathbf{H}\right) \qquad \mathrm{h}_{ij} = \left[\mathbf{tr}\left(\mathbf{F}_i^{*}\,\mathbf{F}_j\,\mathbf{M}\right)\right]/4$$

$$\mathbf{E}_{ij} \equiv \boldsymbol{\sigma}_i \otimes \boldsymbol{\sigma}_j^{*} = \boldsymbol{\mathcal{L}}^{\dagger}\mathbf{G}_i^{*}\mathbf{G}_j\boldsymbol{\mathcal{L}} \qquad \boldsymbol{\mathcal{L}} \equiv \frac{1}{\sqrt{2}}\begin{pmatrix} 1 & 0 & 0 & 1 \\ 1 & 0 & 0 & -1 \\ 0 & 1 & 1 & 0 \\ 0 & i & -i & 0 \end{pmatrix}$$

$$\mathbf{G}_0 \equiv \begin{pmatrix} 1 & 0 & 0 & 0 \\ 0 & 1 & 0 & 0 \\ 0 & 0 & 1 & 0 \\ 0 & 0 & 0 & 1 \end{pmatrix} \quad \mathbf{G}_1 \equiv \begin{pmatrix} 0 & 1 & 0 & 0 \\ 1 & 0 & 0 & 0 \\ 0 & 0 & 0 & i \\ 0 & 0 & -i & 0 \end{pmatrix} \quad \mathbf{G}_2 \equiv \begin{pmatrix} 0 & 0 & 1 & 0 \\ 0 & 0 & 0 & -i \\ 1 & 0 & 0 & 0 \\ 0 & i & 0 & 0 \end{pmatrix} \quad \mathbf{G}_3 \equiv \begin{pmatrix} 0 & 0 & 0 & 1 \\ 0 & 0 & i & 0 \\ 0 & -i & 0 & 0 \\ 1 & 0 & 0 & 0 \end{pmatrix} \tag{5.3}$$

$$\mathbf{F}_0 = \frac{\mathbf{G}_0 + \mathbf{G}_1}{\sqrt{2}} \quad \mathbf{F}_1 = \frac{\mathbf{G}_2 + i\mathbf{G}_3}{\sqrt{2}} \quad \mathbf{F}_2 = \frac{\mathbf{G}_2 - i\mathbf{G}_3}{\sqrt{2}} \quad \mathbf{F}_3 = \frac{\mathbf{G}_0 - \mathbf{G}_1}{\sqrt{2}}$$

where $\boldsymbol{\sigma}_i$ is the set composed of the three Pauli matrices and the identity matrix in Eq. (1.32) and $\mathbf{E}_{ij}$ constitutes the set of 16 *modified Dirac matrices* (San José and Gil 2011).
Matrices $\mathbf{G}_i$ and $\mathbf{F}_i$ will be useful to formulate certain further expressions. Note also that, because of the general property $\mathrm{tr}(\mathbf{AB}) = \mathrm{tr}(\mathbf{BA})$, the above expressions for both m_{ij} and h_{ij} admit respective different equivalent forms.

Therefore, the results from Eq. (5.2) can likewise be formulated in terms of Mueller matrices as follows:

$$\mathbf{M} = \langle \mathbf{M}_J \rangle = \sum_{i=1}^{n} p_i \mathbf{M}_{Ji} + \mathbf{Q}(\langle \mathbf{Y} \rangle) \tag{5.4}$$

where the elements q_{ij} of the non-Mueller matrix $\mathbf{Q}(\langle \mathbf{Y} \rangle)$ are given by $q_{ij} = \mathrm{tr}(\boldsymbol{\mathcal{L}}^{\dagger}\mathbf{G}_i^{*}\mathbf{G}_j\boldsymbol{\mathcal{L}}\langle \mathbf{Y} \rangle)$, $\mathbf{Q}(\langle \mathbf{Y} \rangle)$ adopting different forms depending on the probability distributions of the phase differences ϕ_{ij}. In particular, (1) when ϕ_{ij} do not fluctuate, $\mathbf{M}$ is a pure Mueller matrix; (2) when all ϕ_{ij} are uniformly distributed (i.e., $\phi_{ij} = 1/2\pi$), $\mathbf{M}$ is a depolarizing Mueller matrix given by the incoherent composition $\mathbf{M} = \sum_{i=1}^{n} p_i \mathbf{M}_{Ji}$, and (3) in general, $\mathbf{M}$ is a nonpure Mueller matrix that is determined by $\mathbf{H}_{Ji}$ and by the specific probability distributions of ϕ_{ij}.

Consequently, there are infinite ways to synthesize a given depolarizing Mueller matrix $\mathbf{M}$, the maximum number of parallel components of $\mathbf{M}$ with independent associated covariance vectors being $\mathrm{rank}\,\mathbf{H}(\mathbf{M}) \equiv r$, with $1 \le r \le 4$. When $\mathbf{M}$ is pure, then $r = 1$; otherwise $\mathbf{M}$

admits arbitrary parallel decompositions as convex sums of r pure Mueller matrices (Gil and San José 2019, see Section 5.14).

The synthesis of depolarizing Mueller matrices from partially coherent compositions of nondepolarizing parallel components has been algebraically formulated in some previous works devoted to the cases of partial spatial and spectral coherence polarimetry (Hingerl and Ossikovski 2016; Ossikovski and Hingerl 2016), as well as by means of $\mathbf{Z}$ -matrix states (Kuntman et al. 2017b) or coherency vectors (San José and Gil 2020a)

As an example, the synthesis of a Mueller matrix can be modeled by using the truncated normal probability density function

$$p(\phi_{ij}) = \frac{e^{-(\phi_{ij}-\phi_0)^2/(2\sigma^2)}}{\sqrt{2}\sigma\int_{-(\pi+\phi_0)/(\sqrt{2}\sigma)}^{(\pi-\phi_0)/(\sqrt{2}\sigma)} e^{-t^2}\,dt} \tag{5.5}$$

which tends to Dirac delta-function when $\sigma \approx 0$ (fixed ϕ_{ij}, corresponding to a coherent composition), while $p(\phi_{ij}) \approx 1/2\pi$ when $\sigma \gg \pi$ (fully random ϕ_{ij} in the interval $-\pi \leq \phi_{ij} \leq \pi$, thus corresponding to an incoherent composition) (San José and Gil 2020a). As expected, the smaller σ, the closer to a pure Mueller matrix $\mathbf{M}$ is, while the larger σ, the smaller the contribution of $\mathbf{Q}(\langle \mathbf{Y} \rangle)$, in which case the system tends to an incoherent parallel composition represented by $\mathbf{M} = \sum_{i=1}^{n} p_i \mathbf{M}_{Ji}$, where $\mathbf{M}_{Ji}$ are the pure Mueller matrices of the parallel components. In general, each phase difference can have a particular probability density function $p(\phi_{ij})$, so that the overall parallel composition may involve a sort of mixture of coherent, incoherent and partially coherent combinations, resulting in a Mueller matrix $\mathbf{M}$ that, except for the pure case (totally coherent composition), is equivalent to infinite different possible parallel combinations of pure components (by virtue of the *arbitrary decomposition*, Gil and San José 2019).

The above results on the statistical nature of the Mueller matrix have been formulated by using the covariance vector $\mathbf{h}$ for the representation of pure media and then, have been interpreted in terms of Mueller matrices. Let us now recall that, up to a common factor $1/\sqrt{2}$, the elements of the covariance vector $\mathbf{h}$ are precisely those of the Jones matrix $\mathbf{T}$, so that Jones and Mueller descriptions can be formally related to each other if the depolarization phenomenon is viewed as originating from a measurement-induced averaging process on an otherwise nondepolarizing, but random (or fluctuating) medium (Kim et al. 1987). This fundamental concept underlies the statistical essence of the definition of a Mueller matrix. Since the two types of polarization matrices (Jones and Mueller) are related through the statistical definition, the basic properties of Mueller matrices must likewise be closely related to those of Jones matrices. Such intimate relations provide a proper theoretical basis for a further classification of Mueller matrices based on their basic properties (Ossikovski and Gil 2017) (see Section 5.15).

From the relation $\mathbf{M}_J = \boldsymbol{\mathcal{L}}(\mathbf{T} \otimes \mathbf{T}^*)\boldsymbol{\mathcal{L}}^\dagger$ between a Jones matrix $\mathbf{T}$ and its associated pure Mueller matrix $\mathbf{M}_J$, the ensemble average in Eq. (5.4) can be expressed as

$$\mathbf{M} = \langle \mathbf{M}_J \rangle = \boldsymbol{\mathcal{L}}\langle \mathbf{T} \otimes \mathbf{T}^* \rangle \boldsymbol{\mathcal{L}}^\dagger, \tag{5.6}$$

where the fluctuating Jones matrix $\mathbf{T}$, termed the *Jones generator* of $\mathbf{M}$ (Ossikovski and Gil 2017) (see Section 5.15), denotes the ensemble of realizations $\mathbf{T}^{(s)}$ and their associated probabilities k_s, $s = 1,2,3,\ldots$ Since Eq. (5.6) only involves the second-order conjugate moments

$\langle t_{ij} t_{kl}^{*} \rangle = k_s t_{ij}^{(s)} t_{kl}^{(s)*}$ $(i,j,k,l = 1,2)$ of the elements $t_{ij}^{(s)}$ of the realizations $\mathbf{T}^{(s)}$ (summation over s being understood), any given depolarizing Mueller matrix $\mathbf{M}$ can have infinitely many Jones generators. If $\mathbf{T}$ is not fluctuating, i.e., $\mathbf{T}^{(s)} = \mathbf{T}$ for all values of s, then the averaging process can be omitted and, in virtue of Eq. (5.6), the one-to-one correspondence between $\mathbf{T}$ and $\mathbf{M}$ (except for a global phase factor affecting $\mathbf{T}$), genuine of nondepolarizing systems, is retrieved. Whether a given Mueller matrix is depolarizing or not can be established by using a simple scalar *purity criterion* (Gil and Bernabéu 1985) that will be dealt with in Section 5.10.

5.3 Covariance and Coherency Matrices Associated with a Mueller Matrix

From its statistical definition in Eq. (5.2), the matrix $\mathbf{H} = \langle \mathbf{H}_J \rangle$ has the mathematical structure of a covariance matrix and therefore, it is characterized by the fact that it is a positive semidefinite Hermitian matrix (i.e., its four eigenvalues are non-negative). Consequently, the minimum number r of pure parallel components of $\mathbf{H}$ is given by the integer descriptor $r = \operatorname{rank}\mathbf{H}$ (i.e., r is the number of nonzero eigenvalues of $\mathbf{H}$). Infinite parallel decompositions of a given $\mathbf{H}$ are possible as convex sums of n (with $n \geq r$) pure components (the formulation of the arbitrary decompositions of $\mathbf{H}$ and its associated $\mathbf{M}$ is described in Section 5.14.2). Also, by virtue of the biunivocal relation between $\mathbf{M}$ and its corresponding $\mathbf{H}$, any given Mueller matrix $\mathbf{M}$ can always be expressed in the form of a convex sum (or parallel composition) of r pure Mueller matrices $\mathbf{M}_{Ji}$.

As shown in Section 3.4.1, each pure component can be represented by its corresponding covariance matrix $\mathbf{H}_{Ji}(\mathbf{M}_{Ji})$ defined in Eq. (3.69) so that the *covariance matrix* $\mathbf{H}$ associated with $\mathbf{M}$ can be expressed as follows:

$$\mathbf{H}(\mathbf{M}) \equiv \langle \mathbf{H}_J \rangle = \frac{1}{2}\langle \mathbf{t} \otimes \mathbf{t}^{\dagger} \rangle = \sum_{i=1}^{n} p_i \mathbf{H}_{Ji} = \frac{1}{4}\sum_{i=1}^{n}\left\{ p_i \sum_{k,l=0}^{3}\left[\left(\mathbf{M}_{Ji}\right)_{kl} \left(\boldsymbol{\sigma}_k \otimes \boldsymbol{\sigma}_l^{*}\right)\right]\right\} \tag{5.7}$$

where $\mathbf{t}$ is the *Jones covariance* vector, defined in Section 3.4.1, whose components are the elements of the corresponding Jones matrix $\mathbf{T}$.

Thus, the covariance matrix $\mathbf{H}$ of an additive composition $\mathbf{M} \equiv \sum p_i \mathbf{M}_{Ji}$ of Mueller matrices is equal to the additive composition $\sum p_i \mathbf{H}_{Ji}$, with the same coefficients p_i, of the covariance matrices $\mathbf{H}_{Ji}$ associated with the components. Furthermore, the elements h_{kl} of $\mathbf{H}$ can be expressed as

$$h_{kl} \equiv \sum_{i=1}^{n} p_i \left(\mathbf{H}_{Ji}\right)_{kl} = \frac{1}{2}\langle t_k t_l^{*} \rangle_e \qquad (k,l = 0,1,2,3) \tag{5.8}$$

$\mathbf{H}_{Ji}(\mathbf{M}_{Ji})$ are pure covariance matrices (viz. $\operatorname{rank}\mathbf{H}_{Ji} = 1$), so that, depending on the particular statistical mixture, the rank $r \equiv \operatorname{rank}\mathbf{H}$ of their average $\mathbf{H}$ can take integer values in the interval $1 \leq \operatorname{rank}\mathbf{H} \leq 4$, and equals the number of nonzero (positive) eigenvalues of $\mathbf{H}$.

Since the additive properties are preserved in the transformation from $\mathbf{M}$ to $\mathbf{H}$ and vice versa, the elements h_{kl} of the covariance matrix $\mathbf{H}(\mathbf{M})$ can be expressed in terms of the elements m_{ij} of $\mathbf{M}$ as follows:

$$\mathbf{H} = \frac{1}{4}\sum_{k,l=0}^{3} m_{kl}\,\mathbf{E}_{kl} \quad \left(\mathbf{E}_{kl} \equiv \boldsymbol{\sigma}_k \otimes \boldsymbol{\sigma}_l^*\right) \tag{5.9}$$

where $\mathbf{E}_{kl}$ are the 16 *modified Dirac matrices* (Arfken 1970) defined from the direct products of the Pauli matrices (including the 2×2 identity matrix $\boldsymbol{\sigma}_0$). As is the case with the bases used for the expansions of the 2×2 and 3×3 coherency matrices, matrices $\mathbf{E}_{ij}$ are Hermitian ($\mathbf{E}_{ij} = \mathbf{E}_{ij}^{\dagger}$); trace-orthogonal ($\mathrm{tr}(\mathbf{E}_{ij}\mathbf{E}_{kl}) = 4\delta_{(ij)(kl)}$, $\delta_{(m)(n)}$ being the Kronecker delta), and satisfy $\mathbf{E}_{ij}^2 = \mathbf{I}_4$ ($\mathbf{I}_4$ being the 4×4 identity matrix). Thus, these matrices are unitary and, except for $\mathbf{E}_{00} = \mathbf{I}_4$, are traceless.

The explicit form of Eq. (5.9), useful for practical calculations, is

$$\mathbf{H}(\mathbf{M}) = \frac{1}{4}\begin{pmatrix} \begin{matrix} m_{00}+m_{01} \\ +m_{10}+m_{11} \end{matrix} & \begin{matrix} m_{02}+m_{12} \\ +i(m_{03}+m_{13}) \end{matrix} & \begin{matrix} m_{20}+m_{21} \\ -i(m_{30}+m_{31}) \end{matrix} & \begin{matrix} m_{22}+m_{33} \\ +i(m_{23}-m_{32}) \end{matrix} \\ \begin{matrix} m_{02}+m_{12} \\ -i(m_{03}+m_{13}) \end{matrix} & \begin{matrix} m_{00}-m_{01} \\ +m_{10}-m_{11} \end{matrix} & \begin{matrix} m_{22}-m_{33} \\ -i(m_{23}+m_{32}) \end{matrix} & \begin{matrix} m_{20}-m_{21} \\ -i(m_{30}-m_{31}) \end{matrix} \\ \begin{matrix} m_{20}+m_{21} \\ +i(m_{30}+m_{31}) \end{matrix} & \begin{matrix} m_{22}-m_{33} \\ +i(m_{23}+m_{32}) \end{matrix} & \begin{matrix} m_{00}+m_{01} \\ -m_{10}-m_{11} \end{matrix} & \begin{matrix} m_{02}-m_{12} \\ +i(m_{03}-m_{13}) \end{matrix} \\ \begin{matrix} m_{22}+m_{33} \\ -i(m_{23}-m_{32}) \end{matrix} & \begin{matrix} m_{20}-m_{21} \\ +i(m_{30}-m_{31}) \end{matrix} & \begin{matrix} m_{02}-m_{12} \\ -i(m_{03}-m_{13}) \end{matrix} & \begin{matrix} m_{00}-m_{01} \\ -m_{10}+m_{11} \end{matrix} \end{pmatrix} \tag{5.10}$$

Conversely, the elements m_{kl} of $\mathbf{M}$ can be expressed as

$$m_{kl} = \mathrm{tr}\left(\mathbf{E}_{kl}\,\mathbf{H}\right) \tag{5.11}$$

or, equivalently,

$$\mathbf{M}(\mathbf{H}) = \begin{pmatrix} \begin{matrix} h_{00}+h_{11} \\ +h_{22}+h_{33} \end{matrix} & \begin{matrix} h_{00}-h_{11} \\ +h_{22}-h_{33} \end{matrix} & \begin{matrix} h_{01}+h_{10} \\ +h_{23}+h_{32} \end{matrix} & \begin{matrix} -i(h_{01}-h_{10}) \\ -i(h_{23}-h_{32}) \end{matrix} \\ \begin{matrix} h_{00}+h_{11} \\ -h_{22}-h_{33} \end{matrix} & \begin{matrix} h_{00}-h_{11} \\ -h_{22}+h_{33} \end{matrix} & \begin{matrix} h_{01}+h_{10} \\ -h_{23}-h_{32} \end{matrix} & \begin{matrix} -i(h_{01}-h_{10}) \\ +i(h_{23}-h_{32}) \end{matrix} \\ \begin{matrix} h_{02}+h_{20} \\ +h_{13}+h_{31} \end{matrix} & \begin{matrix} h_{02}+h_{20} \\ -h_{13}-h_{31} \end{matrix} & \begin{matrix} h_{03}+h_{30} \\ +h_{12}+h_{21} \end{matrix} & \begin{matrix} -i(h_{03}-h_{30}) \\ +i(h_{12}-h_{21}) \end{matrix} \\ \begin{matrix} i(h_{02}-h_{20}) \\ +i(h_{13}-h_{31}) \end{matrix} & \begin{matrix} i(h_{02}-h_{20}) \\ -i(h_{13}-h_{31}) \end{matrix} & \begin{matrix} i(h_{03}-h_{30}) \\ +i(h_{12}-h_{21}) \end{matrix} & \begin{matrix} h_{03}+h_{30} \\ -h_{12}-h_{21} \end{matrix} \end{pmatrix} \tag{5.12}$$

While the general characterization of a pure Mueller matrix $\mathbf{M}_J$ is directly obtained from its mathematical relation to its corresponding Jones matrix $\mathbf{T}$, the general characterization of the complete set of physical Mueller matrices (both nondepolarizing and depolarizing) follows from the non-negativity of the eigenvalues of $\mathbf{H}$ (Cloude's criterion) (Cloude 1986; 1989).

As we have seen, Cloude's criterion has a statistical foundation that originates from the underlying argument that any physical Mueller matrix is the integral result of the additive combination (coherent, partially coherent or incoherent) of elementary deterministic linear transformations, each of them characterized by a well-defined Mueller-Jones matrix (Parke 1948; Gil 2000a). As pointed out by Parke (a student of Hans Mueller) a proper statistical average of deterministic Jones devices bears the same relation to the phenomenological Mueller approach that statistical mechanics bears to thermodynamics.

At this point, it is important to recall that the Cloude's criterion is more restrictive than the requirement that a Mueller matrix **M** be necessarily a Stokes matrix (i.e., **M** necessarily transforms all Stokes vectors into Stokes vectors). Obviously, any Mueller matrix is a Stokes matrix, but the converse statement is not true because all Stokes matrices whose associated Hermitian matrix **H** has negative eigenvalues are not Mueller matrices. Therefore, the numerous inequalities quoted in the literature for Stokes matrices are all derivable, as necessary conditions, from the Cloude's criterion.

In summary, the statistical concept of Mueller matrix entails more restrictive conditions (inequalities) than the very fact that a Mueller matrix is a Stokes matrix (Van der Mee and Hovenier 1992; Nagirner 1993; Kostinski et al. 1993). Only matrices satisfying the covariance (or statistical) criterion are physically realizable (leaving aside passivity conditions) and can strictly be called Mueller matrices. As pointed out by B. N. Simon et al., in agreement with the common terminology in mathematics and in quantum information theory, Stokes matrices correspond to *positive maps of the states of polarization (Poincaré sphere) and the subset of Stokes matrices which are Mueller matrices correspond to completely positive maps* (Simon et al. 2010a; Gamel and James 2011).

Different physical arguments have been considered to justify the statistical criterion (Parke 1948; Gil and Bernabéu 1985; Cloude 1986; Kim et al. 1987; Van der Mee 1993; Bolshakov et al. 1996; Gil 2000a; Sudha et al. 2008; Simon et al. 2010a; 2010b; Gamel and James 2011). Any alternative concept of physical Mueller matrix (as, for example, identifying Mueller matrices with Stokes matrices) fails due to the impossibility of supporting it by, at least, one way to physically realize a Mueller matrix not satisfying the covariance (or Cloude's) criterion.

Note that any unitary similarity transformation $\mathbf{A} = \mathbf{V}\mathbf{H}\mathbf{V}^{\dagger}$ $(\mathbf{V}^{\dagger} = \mathbf{V}^{-1})$ of **H** allows for defining an alternative positive semidefinite Hermitian matrix **A** that also contains all the polarimetric information of the medium, and therefore, can be alternatively used as representative of the same medium. A particularly interesting form of **A** is the so-called *coherency matrix* **C** of the medium, given by Cloude (1986):

$$\mathbf{C}(\mathbf{M}) = \mathcal{L}\left[\mathbf{H}(\mathbf{M})\right]\mathcal{L}^{\dagger} = \frac{1}{4}\begin{pmatrix} \begin{matrix} m_{00}+m_{11} \\ +m_{22}+m_{33} \end{matrix} & \begin{matrix} m_{01}+m_{10} \\ -i(m_{23}-m_{32}) \end{matrix} & \begin{matrix} m_{02}+m_{20} \\ +i(m_{13}-m_{31}) \end{matrix} & \begin{matrix} m_{03}+m_{30} \\ -i(m_{12}-m_{21}) \end{matrix} \\ \begin{matrix} m_{01}+m_{10} \\ +i(m_{23}-m_{32}) \end{matrix} & \begin{matrix} m_{00}+m_{11} \\ -m_{22}-m_{33} \end{matrix} & \begin{matrix} m_{12}+m_{21} \\ +i(m_{03}-m_{30}) \end{matrix} & \begin{matrix} m_{13}+m_{31} \\ -i(m_{02}-m_{20}) \end{matrix} \\ \begin{matrix} m_{02}+m_{20} \\ -i(m_{13}-m_{31}) \end{matrix} & \begin{matrix} m_{12}+m_{21} \\ -i(m_{03}-m_{30}) \end{matrix} & \begin{matrix} m_{00}-m_{11} \\ +m_{22}-m_{33} \end{matrix} & \begin{matrix} m_{23}+m_{32} \\ +i(m_{01}-m_{10}) \end{matrix} \\ \begin{matrix} m_{03}+m_{30} \\ +i(m_{12}-m_{21}) \end{matrix} & \begin{matrix} m_{13}+m_{31} \\ +i(m_{02}-m_{20}) \end{matrix} & \begin{matrix} m_{23}+m_{32} \\ -i(m_{01}-m_{10}) \end{matrix} & \begin{matrix} m_{00}-m_{11} \\ -m_{22}+m_{33} \end{matrix} \end{pmatrix} \tag{5.13}$$

Since the term *coherency matrix* has also been used in some works to refer to **H**, to avoid confusion hereafter we maintain the terms proposed originally by Cloude who assigned the name *covariance matrix* to **H** (Cloude 1989) and reserved the term *coherency matrix* for **C** (Cloude 1986).

The explicit expression of $\mathbf{M}(\mathbf{C})$ reads

$$\mathbf{M}(\mathbf{C}) = \begin{pmatrix} \begin{matrix} c_{00}+c_{11} \\ +c_{22}+c_{33} \end{matrix} & \begin{matrix} c_{01}+c_{01}^* \\ -i\left(c_{23}-c_{23}^*\right) \end{matrix} & \begin{matrix} c_{02}+c_{02}^* \\ +i\left(c_{13}-c_{13}^*\right) \end{matrix} & \begin{matrix} c_{03}+c_{03}^* \\ -i\left(c_{12}-c_{12}^*\right) \end{matrix} \\ \begin{matrix} c_{01}+c_{01}^* \\ +i\left(c_{23}-c_{23}^*\right) \end{matrix} & \begin{matrix} c_{00}+c_{11} \\ -c_{22}-c_{33} \end{matrix} & \begin{matrix} c_{12}+c_{12}^* \\ +i\left(c_{03}-c_{03}^*\right) \end{matrix} & \begin{matrix} c_{13}+c_{13}^* \\ -i\left(c_{02}-c_{02}^*\right) \end{matrix} \\ \begin{matrix} c_{02}+c_{02}^* \\ -i\left(c_{13}-c_{13}^*\right) \end{matrix} & \begin{matrix} c_{12}+c_{12}^* \\ -i\left(c_{03}-c_{03}^*\right) \end{matrix} & \begin{matrix} c_{00}-c_{11} \\ +c_{22}-c_{33} \end{matrix} & \begin{matrix} c_{23}+c_{23}^* \\ +i\left(c_{01}-c_{01}^*\right) \end{matrix} \\ \begin{matrix} c_{03}+c_{03}^* \\ +i\left(c_{12}-c_{12}^*\right) \end{matrix} & \begin{matrix} c_{13}+c_{13}^* \\ +i\left(c_{02}-c_{02}^*\right) \end{matrix} & \begin{matrix} c_{23}+c_{23}^* \\ -i\left(c_{01}-c_{01}^*\right) \end{matrix} & \begin{matrix} c_{00}-c_{11} \\ -c_{22}+c_{33} \end{matrix} \end{pmatrix} \quad \left[m_{ij} = \mathrm{tr}\left(\mathbf{G}_i^* \mathbf{G}_j \mathbf{C}\right)\right] \tag{5.14}$$

Among others, a remarkable feature of **C** is that its diagonal elements are determined by the diagonal elements of **M** and vice versa. This property leads to useful mathematical simplifications that will be exploited in future calculations.

Let us stress that Mueller, covariance and coherency matrices corresponding to the same medium contain identical physical information. $\mathbf{H}(\mathbf{M})$ and $\mathbf{C}(\mathbf{M})$ share the same non-negative eigenvalues, $\lambda_0, \lambda_1, \lambda_2, \lambda_3$ (hereafter we will use this notation for the ordered eigenvalues $\lambda_0 \geq \lambda_1 \geq \lambda_2 \geq \lambda_3$). A detailed review of the relations between **M**, **H** and **C** can be found in (Cloude 2009).

The statistical nature of the coherency matrix $\mathbf{C}(\mathbf{M})$ allows for a detailed analysis of the structure of the information contained in **M**, which will be dealt with in Section 5.13.4.

Due to the nature of the relations between **M**, $\mathbf{H}(\mathbf{M})$ and $\mathbf{C}(\mathbf{M})$, any parallel decomposition formulated in terms of covariance or coherency matrices can directly be translated into the corresponding expression in terms of Mueller matrices and vice-versa with identical coefficients.

The particular forms of Jones and Mueller matrices with respect to the bases commonly used in radar polarimetry for forward and backward scattering, as well as their respective relations with the coherent scattering matrix (Sinclair matrix), with the covariance and coherency matrices, and with the Kennaugh matrix, have been considered briefly in Section 3.4.8. Comprehensive descriptions can be found in (Lüneburg and Cloude 1997; Cloude 2009; Boerner 2006; Jin and Xu 2013; Sheppard et al. 2020b).

5.4 The Concept of Mueller Matrix

From the analysis performed in the previous sections, it follows that, given a 4×4 real matrix, it can be considered a physically realizable Mueller matrix (hereafter, simply, a Mueller matrix) if and only if it is obtainable as an ensemble average (or convex sum) of pure and passive pure Mueller matrices. This defining feature of Mueller matrices, called by us the *ensemble criterion*, leads to the general characterization of Mueller matrices through two sets of conditions, namely, (1) the four *covariance conditions*, which are embodied in the non-negativity of the four eigenvalues of $\mathbf{H}(\mathbf{M})$ (Cloude 1986; Arnal 1990) and (2) the

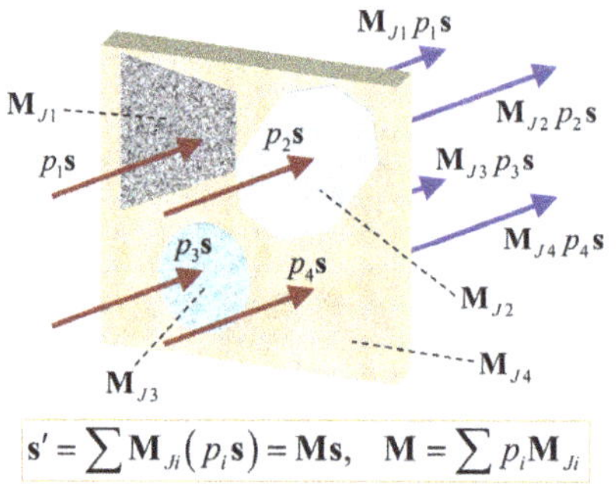

FIGURE 5.2
The incident beam, whose state of polarization is given by the Stokes vector **s**, is shared among a given number of transversely separated portions of the sample. The overall Mueller matrix **M** is equivalent to that of a convex sum of up to four pure Mueller matrices representing equivalent nondepolarizing components.

pair of necessary and sufficient conditions for **M** to represent a passive system, which are given by $m_{00}(1+D) \leq 1$ and $m_{00}(1+P) \leq 1$, where D and P are the diattenuation and the polarizance of **M** (Gil 2000a, San José and Gil 2020b) (see Section 5.9).

Given a depolarizing medium, its associated covariance, coherency and Mueller matrices, **H**, **C** and **M**, can respectively be decomposed in infinite ways in terms of pure components, the minimum possible number of them being $r = \operatorname{rank}\mathbf{H} = \operatorname{rank}\mathbf{C}$. Therefore, as illustrated in Figure 5.2, **M** can always be expressed as (Gil 2007; Gil and San José 2019)

$$\mathbf{M} = \sum_{i=1}^{r} p_i \mathbf{M}_{Ji} \quad \left[p_i > 0 \quad \sum_{i=1}^{r} p_i = 1 \quad r = \operatorname{rank}\mathbf{H} \right] \tag{5.15}$$

where $\mathbf{M}_{Ji}$ are pure and passive Mueller matrices whose associated covariance vectors are independent and constitute a basis of the image subspace of **H** (denoted as $\operatorname{range}\mathbf{H}$). This subject will be dealt with in detail in further sections.

Let us finish this section by recalling that, as indicated in Section 3.3.1, it was Jones (1947), who coined the name *Mueller matrices*, despite the fact that it was Soleillet (1929) who stated for the first time the linear polarimetric transformations of Stokes parameters, and that Perrin (1942) used this formulation to represent the action of various scattering media on polarized light (see also Arteaga and Nichols 2018).

5.5 Changes of Coordinate System and Rotated Mueller Matrices

The transformations for both nondepolarizing and depolarizing Mueller matrices when they are represented with respect to different coordinate systems are derived from the rule in Eq. (1.42) for Stokes vectors and have already been considered in Section 3.3.6. In particular, given the transformation $\mathbf{M}\mathbf{s} = \mathbf{s}'$ of a state of polarization **s** described with respect to a coordinate system XY, the same relation $\mathbf{M}_{X'Y'}\mathbf{s}_{X'Y'} = \mathbf{s}'_{X'Y'}$ with respect to a coordinate system $X'Y'$ rotated through a counterclockwise angle θ about the axis Z orthogonal to the plane XY (i.e., about the light propagation direction), then

$$\mathbf{M}_{X'Y'} = \mathbf{M}_G(\theta)\,\mathbf{M}\,\mathbf{M}_G(-\theta) \tag{5.16}$$

where $\mathbf{M}_G(\theta)$ is the rotation matrix (or *rotator*) shown in Eq. (3.67)

$$\mathbf{M}_G(\theta) \equiv \begin{pmatrix} 1 & 0 & 0 & 0 \\ 0 & \cos 2\theta & \sin 2\theta & 0 \\ 0 & -\sin 2\theta & \cos 2\theta & 0 \\ 0 & 0 & 0 & 1 \end{pmatrix} \tag{5.17}$$

If, instead of the coordinate system, it is the medium what is rotated through a counterclockwise angle θ about the Z axis, then its Mueller matrix is given by

$$\mathbf{M}(\theta) = \mathbf{M}_G(-\theta)\,\mathbf{M}\,\mathbf{M}_G(\theta) \tag{5.18}$$

Let us finally observe the transitivity of the transformation law (5.18) when applied to the components $\mathbf{M}_i$ of a parallel combination of Mueller matrices $\mathbf{M} = \sum p_i\,\mathbf{M}_i$

$$\sum_i p_i\,\mathbf{M}_G(-\theta)\,\mathbf{M}_i\,\mathbf{M}_G(\theta) = \mathbf{M}_G(-\theta)\left[\sum_i p_i\,\mathbf{M}_i\right]\mathbf{M}_G(\theta) = \mathbf{M}_G(-\theta)\,\mathbf{M}\,\mathbf{M}_G(\theta) \tag{5.19}$$

showing that the laws (5.16) and (5.18) hold for general Mueller matrices (either nondepolarizing or depolarizing).

5.6 Characterization of Mueller Matrices: Covariance Criterion

Due to the progressive increase of the applications of polarimetry, and the consequent necessity of appropriate mathematical tools for exploiting the measurements, the mathematical characterization of Mueller matrices has become a key issue dealt with by a number of authors. Moreover, the rigorous characterization of Mueller matrices is the appropriate basis for identifying representative physical quantities, as well as serial and parallel decompositions in terms of easily interpretable components.

As mentioned in the previous sections, well-established physical arguments support the concept of a general Mueller matrix as an ensemble average of pure, passive, Mueller matrices. This *ensemble criterion* has been formulated mathematically in Eq. (5.15) by stating that a given 4×4 real matrix $\mathbf{M}$ is a Mueller matrix if, and only if, it can be expressed as a convex sum of pure Mueller matrices. The ensemble criterion entails a set of necessary and sufficient conditions split into two complementary criteria dealt with in respective sections, namely, the *covariance criterion* and the *passivity criterion*.

A characteristic property of Mueller matrices is that their associated Hermitian matrices **H**, given by Eq. (5.10) are positive semidefinite. This is the so-called *Cloude's criterion* (Cloude 1986), or *covariance criterion*, which states that, for any Mueller matrix **M**, its associated matrix $\mathbf{H}(\mathbf{M})$ has the structure and properties of a covariance matrix, i.e., the four eigenvalues of $\mathbf{H}(\mathbf{M})$ are non-negative. This condition was obtained independently by Arnal (1990) and is also equivalent to the fact that the four leading principal minors of $\mathbf{H}(\mathbf{M})$ are non-negative. In fact, all the 15 principal minors of $\mathbf{H}(\mathbf{M})$ are necessarily non-negative, but the non-negativity of the four leading principal minors is sufficient to ensure the fulfillment of Cloude's criterion. This particular formulation leads to explicit expressions for a set of four necessary and sufficient *covariance conditions* in terms of the statistical parameterization of **H** (Gil 2000a).

Obviously, experimentally measured Mueller matrices have a limited precision and, to some extent, may fail to satisfy the non-negativity of **H**. That is to say, one or more eigenvalues of the measured $\mathbf{H}(\mathbf{M})$ can be negative, so that the non-negativity criterion provides a way for analyzing how "noisy" are the polarimetric measurements, together with certain procedures for filtering measured Mueller matrices (usually based on replacing by zero the negative eigenvalues, but preserving the unitary matrix **U** that diagonalizes **H**; see Chapter 11).

5.7 Normal Form of a Mueller Matrix

As shown in Section 4.3.2.4 the diagonalization of the Mueller matrix of a diattenuator through an orthogonal similarity transformation involves the non-Mueller matrix $\mathbf{C}_R$, and therefore it cannot be performed in terms of Mueller matrices. Consequently, in general, it is not possible to diagonalize pure and nonpure Mueller matrices through orthogonal similarity transformations where all serial components are Mueller matrices.

It seems to be Xing (1992) who first proposed a possible diagonalization of a Mueller matrix **M** through the serial decomposition $\mathbf{M} = \mathbf{M}_{J2}\,\mathbf{M}_{\Delta d}\,\mathbf{M}_{J1}$, where $\mathbf{M}_{J1}$ and $\mathbf{M}_{J2}$ are pure Mueller matrices, and $\mathbf{M}_{\Delta d}$ is a depolarizing diagonal Mueller matrix. This *normal form* of **M** was formally established by Sridhar and Simon (1994) for a large subset of depolarizing Mueller matrices, but it was Van der Mee (1993) who first pointed out the existence of a complementary subset of Mueller matrices for which the central matrix of the normal form cannot be expressed in the diagonal form. Together with Bolshakov et al. (1996; 1997), he completed the general characterization of Mueller matrices (except for passivity) based on the normal form. This result was independently obtained by Gopala Rao et al. (1998a; 1998b) in an equivalent form. A simplified and physically meaningful review of this problem was performed later by Ossikovski (2009; 2010a), so that it was established that any Mueller matrix **M** belongs to one of the two following categories:

(I) Type-I, where the auxiliary matrix $\mathbf{N} \equiv \mathbf{G}\mathbf{M}^T\mathbf{G}\mathbf{M}$ (with $\mathbf{G} \equiv \mathrm{diag}(1,-1,-1,-1)$), is diagonalizable (i.e., there exists an invertible matrix **A** such that $\mathbf{A}^{-1}\mathbf{N}\mathbf{A}$ is diagonal). In this case **M** can be written in the *type-I normal form*

$$\mathbf{M} = \mathbf{M}_{J2}\,\mathbf{M}_{\Delta d}\,\mathbf{M}_{J1} \tag{5.20}$$

where $\mathbf{M}_{J1}$ and $\mathbf{M}_{J2}$ are pure Mueller matrices, and the *type-I canonical Mueller matrix* $\mathbf{M}_{\Delta d}$ has the form

$$\mathbf{M}_{\Delta d} = \mathrm{diag}(d_0, d_1, d_2, \varepsilon d_3) \qquad [0 \le d_3 \le d_2 \le d_1 \le d_0 \quad \varepsilon \equiv \det \mathbf{M}/|\det \mathbf{M}|] \tag{5.21}$$

$\mathbf{M}_{\Delta d}$ is a diagonal Mueller matrix that, in general, represents an *intrinsic depolarizer*, or *pure depolarizer* (Lu and Chipman 1996) (note that here the term *pure* is used to indicate that the medium exclusively exhibits depolarization properties and is, furthermore, centered, that is, unrotated, with respect to the laboratory reference axes). The diagonal elements of $\mathbf{M}_{\Delta d}$ are the square roots of the eigenvalues of the auxiliary matrix $\mathbf{N}$. Observe also that d_k have been taken ordered on the diagonal so as to satisfy $d_3 \le d_2 \le d_1 \le d_0$, but provided $d_i \le d_0$ $(i = 1, 2, 3)$ any other choice of order can be performed through Mueller orthogonal transformations.

Since any pure Mueller matrix $\mathbf{M}_J$ can be written as $\mathbf{M}_J = \mathbf{M}_J \mathbf{I}_4$ ($\mathbf{I}_4$ being the 4×4 identity matrix), with $\mathbf{M}_{\Delta d} = \mathbf{I}_4$, it is evident that pure Mueller matrices are always of type-I.

(II) Type-II, when the auxiliary matrix $\mathbf{N}$ is not diagonalizable, in which case $\mathbf{M}$ can be written in the *type-II normal form*

$$\mathbf{M} = \mathbf{M}_{J2}\, \mathbf{M}_{\Delta nd}\, \mathbf{M}_{J1} \tag{5.22}$$

where $\mathbf{M}_{J1}$ and $\mathbf{M}_{J2}$ are nonsingular pure Mueller matrices, and $\mathbf{M}_{\Delta nd}$ is the *type-II canonical Mueller matrix* (Ossikovski 2010a)

$$\mathbf{M}_{\Delta nd} \equiv \begin{pmatrix} 2a_0 & -a_0 & 0 & 0 \\ a_0 & 0 & 0 & 0 \\ 0 & 0 & a_2 & 0 \\ 0 & 0 & 0 & a_2 \end{pmatrix} \qquad [0 \le a_2 \le a_0] \tag{5.23}$$

a_0, a_2 being the square roots of the eigenvalues of $\mathbf{N}$. Through suitable representations of the pure serial components $\mathbf{M}_{J1}$ and $\mathbf{M}_{J2}$ in Eq. (5.22), the central type-II depolarizer can adopt different alternative forms that will be considered in Section 5.7.2.

It should be noted that, when both N-matrices $\mathbf{N} \equiv \mathbf{G}\mathbf{M}^T\mathbf{G}\mathbf{M}$ and $\mathbf{N}' \equiv \mathbf{M}\mathbf{G}\mathbf{M}^T\mathbf{G}$ are different from the zero matrix, both pure components $\mathbf{M}_{J1}$ and $\mathbf{M}_{J2}$ of the normal form are nonsingular. Furthermore, when $\mathbf{N} = \mathbf{0}$ and $\mathbf{N}' \ne \mathbf{0}$, then $\mathbf{M}$ is a singular type-I matrix that corresponds to a *depolarizing analyzer* [$D(\mathbf{M}) = 1$]; when $\mathbf{N} \ne \mathbf{0}$ and $\mathbf{N}' = \mathbf{0}$, then $\mathbf{M}$ is a singular type-I matrix that corresponds to a *depolarizing polarizer* [$P(\mathbf{M}) = 1$], and when both $\mathbf{N} = \mathbf{0}$ and $\mathbf{N}' = \mathbf{0}$, then $\mathbf{M}$ is a *perfect polarizer* [$D(\mathbf{M}) = P(\mathbf{M}) = 1$] (recall that $D(\mathbf{M})$ and $P(\mathbf{M})$ are the diattenuation and polarizance of $\mathbf{M}$, given by the absolute values of the diattenuation and polarizance vectors, $\mathbf{D}$ and $\mathbf{P}$, of $\mathbf{M}$, respectively). Consequently, any serial combination containing a perfect polarizer is necessarily a type-I Mueller matrix (we use the term *perfect polarizer* or simply *polarizer*, as equivalent to *polarizer-analyzer*, or *total polarizer*).

For the sake of simplicity in classifying Mueller matrices, the above definitions for type-I and type-II matrices, where depolarizing analyzers and depolarizing polarizers belong to type-I category, differ slightly from the criterion used by Gopala Rao et al. (1998a; 1998b), where the said kinds of depolarizers, containing either a singular $\mathbf{M}_{J1}$ or a singular $\mathbf{M}_{J2}$, belong to type-II. The above established definitions will be used and applied throughout this book.

The procedure for the calculation of the pure components and the central depolarizer of the normal form or *symmetric decomposition* (Ossikovski 2009; 2010a, San José et al. 2020)

$$\mathbf{M} = \mathbf{M}_{J2}\mathbf{M}_{\Delta}\mathbf{M}_{J1} \tag{5.24}$$

for any given Mueller matrix $\mathbf{M}$, including the cases where $\mathbf{M}$ is singular, is dealt with in Section 8.3.

Due to the key role played by the canonical forms of the Mueller matrices, it is worth going deeper into their properties and structure through the next two subsections.

5.7.1 Type-I Canonical Mueller Matrix

Regarding the type-I canonical form $\mathbf{M}_{\Delta d}$, its associated coherency matrix $\mathbf{C}_{\Delta d}$ has the diagonal form

$$\mathbf{C}_{\Delta d} \equiv \mathbf{C}(\mathbf{M}_{\Delta d}) = \\ = \frac{1}{4}\begin{pmatrix} d_0+d_1+d_2+\varepsilon d_3 & 0 & 0 & 0 \\ 0 & d_0+d_1-d_2-\varepsilon d_3 & 0 & 0 \\ 0 & 0 & d_0-d_1+d_2-\varepsilon d_3 & 0 \\ 0 & 0 & 0 & d_0-d_1-d_2+\varepsilon d_3 \end{pmatrix} \tag{5.25}$$

whose diagonal entries are precisely the eigenvalues $\lambda_{\Delta di}$ $(i=0,1,2,3)$ of $\mathbf{C}_{\Delta d}$, which, since $d_3 \le d_2 \le d_1 \le d_0$, are ordered so as to satisfy

$$0 \le \lambda_{\Delta d3} \le \lambda_{\Delta d2} \le \lambda_{\Delta d1} \le \lambda_{\Delta d0} \tag{5.26}$$

(Note that, even though the covariance matrix $\mathbf{H}(\mathbf{M}_{\Delta d})$ is not diagonal, the present analysis can likewise be performed by using the $\mathbf{H}(\mathbf{M}_{\Delta d})$ instead of $\mathbf{C}(\mathbf{M}_{\Delta d})$.)

The non-negativity of $\lambda_{\Delta di}$ is equivalent to the fact that $\mathbf{M}_{\Delta d}$ satisfies the covariance conditions

$$d_0 \ge -d_1-d_2-\varepsilon d_3 \quad d_0 \ge -d_1+d_2+\varepsilon d_3 \quad d_0 \ge d_1-d_2+\varepsilon d_3 \quad d_0 \ge d_1+d_2-\varepsilon d_3 \tag{5.27}$$

which can also be formulated as (Gopala Rao el al. 1998b)

$$\begin{matrix} d_0 \ge d_i \quad (i=1,2,3) & d_k \ge 0 \quad (k=0,1,2,3) \\ d_0+d_1 \ge d_2+\varepsilon d_3 & d_0-d_1 \ge d_2-\varepsilon d_3 \end{matrix} \tag{5.28}$$

Note that the passivity of $\mathbf{M}_{\Delta d}$ holds if and only if $d_0 \le 1$.

Let us recall that, from its very definition, $\mathbf{M}_{\Delta d}$ represents either (*a*) a type-I canonical depolarizer, or (*b*), as a limiting case, $\mathbf{M}_{\Delta d} = \mathbf{I}_4$ (so that pure Mueller matrices are formally included in the normal form). Therefore, diagonal pure Mueller matrices other than $\mathbf{I}_4$ are not included in the form of $\mathbf{M}_{\Delta d}$ in Eq. (5.21). To avoid misunderstanding, and for the sake of comprehensiveness, note in passing that the general covariance conditions for a diagonal 4×4 real matrix $\mathbf{M} = \mathrm{diag}(m_{00}, m_{11}, m_{22}, m_{33})$ to be a Mueller matrix (pure or not) are the following:

$$\begin{aligned} &m_{00}+m_{11}+m_{22}+m_{33}\geq 0 \quad && m_{00}+m_{11}-m_{22}-m_{33}\geq 0\\ &m_{00}-m_{11}+m_{22}-m_{33}\geq 0 \quad && m_{00}-m_{11}-m_{22}+m_{33}\geq 0 \end{aligned} \qquad (5.29)$$

Once the eigenvalues $\lambda_{\Delta di}$ of $\mathbf{C}_{\Delta d}$ (coinciding with those of $\mathbf{H}_{\Delta d}$) have been obtained, let us analyze the possible cases according to the value of $\operatorname{rank}\mathbf{C}_{\Delta d}$. In order to make easier the interpretation of the following case study, hereafter we denote $\mathbf{C}_{\Delta d}$ and $\mathbf{M}_{\Delta d}$ as $\mathbf{C}_{\Delta dr}$ and $\mathbf{M}_{\Delta dr}$ respectively where $r \equiv \operatorname{rank}\mathbf{C}_{\Delta d}$:

(1) $\operatorname{rank}\mathbf{C}_{\Delta d}=1$
In this case $d_0=d_1=d_2=d_3$, so that $\mathbf{M}_{\Delta d1}=d_0\mathbf{I}_4$, therefore the Mueller matrix $\mathbf{M}$ in Eq. (5.20) is nondepolarizing (or pure), $\mathbf{M}_J \equiv d_0\,\mathbf{M}_{J2}\,\mathbf{M}_{J1}$. The properties and serial decompositions of pure Mueller matrices have been dealt with in Chapters 3 and 4.

(2) $\operatorname{rank}\mathbf{C}_{\Delta d}=2$
This case corresponds to $d_0=d_1\neq d_2=d_3$, so that $\mathbf{M}_{\Delta d2}=\operatorname{diag}(d_0,d_0,d_2,d_2)$.
As we shall see in Chapter 7, devoted to parallel decompositions of depolarizing Mueller matrices, $\mathbf{M}_{\Delta d2}$ admits an infinite number of parallel decompositions. Let us mention, in advance, the following example of decomposition:

$$\mathbf{M}_{\Delta d2}=\frac{d_0+d_2}{2d_0}\left[d_0\mathbf{I}_4\right]+\frac{d_0-d_2}{2d_0}\left[d_0\operatorname{diag}(1,1,-1,-1)\right] \qquad (5.30)$$

where $\mathbf{M}_{\Delta d2}$ is represented as a parallel combination of two pure components with the same mean intensity coefficient d_0, namely, an attenuating filter, with relative cross section $(d_0+d_2)/2d_0$, and a horizontal half-wave plate with relative cross section $(d_0-d_2)/2d_0$.

(3) $\operatorname{rank}\mathbf{C}_{\Delta d}=3$
This case corresponds to $d_0-d_1=d_2-d_3\equiv q>0$, so that $\mathbf{M}_{\Delta d3}=\operatorname{diag}(d_0,d_0-q,d_2,d_2-q)$ and a particular example of parallel decomposition of $\mathbf{M}_{\Delta d3}$ is given by

$$\mathbf{M}_{\Delta d3}=\frac{d_0+d_2-q}{2d_0}\left[d_0\mathbf{I}_4\right]+\frac{d_0-d_2}{2d_0}\left[d_0\operatorname{diag}(1,1,-1,-1)\right]+\frac{q}{2d_0}\left[d_0\operatorname{diag}(1,-1,1,-1)\right] \qquad (5.31)$$

where $\mathbf{M}_{\Delta d3}$ is represented as a convex combination of three pure components with the same mean intensity coefficient d_0, namely an attenuating filter with relative cross section $(d_0+d_2-q)/2d_0$, a horizontal half-wave plate with relative cross section $(d_0-d_2)/2d_0$, and a half-wave plate oriented at 45° with relative cross section $q/2d_0$.

(4) $\operatorname{rank}\mathbf{C}_{\Delta d}=4$
In this most general case $d_0-d_1\neq d_2-d_3$, and an example of particular parallel decomposition of $\mathbf{M}_{\Delta d4}$ is given by the spectral decomposition:

$$\begin{aligned}\mathbf{M}_{\Delta d4}=&\frac{\lambda_{\Delta d0}}{d_0}\left[d_0\mathbf{I}_4\right]+\frac{\lambda_{\Delta d1}}{d_0}\left[d_0\operatorname{diag}(1,1,-1,-1)\right]+\\ &+\frac{\lambda_{\Delta d2}}{d_0}\left[d_0\operatorname{diag}(1,-1,1,-1)\right]+\frac{\lambda_{\Delta d3}}{d_0}\left[d_0\operatorname{diag}(1,-1,-1,1)\right]\end{aligned} \qquad (5.32)$$

where $\mathbf{M}_{\Delta d4}$ is represented as a convex combination of four pure components with the same mean intensity coefficient d_0, namely an attenuating filter, with relative cross section $\lambda_{\Delta d0}/d_0$, a horizontal half-wave plate with relative cross section $\lambda_{\Delta d1}/d_0$, a half-wave plate oriented at 45° with relative cross section $\lambda_{\Delta d2}/d_0$, and a half-wave circular retarder with relative cross section $\lambda_{\Delta d3}/d_0$.

5.7.2 Type-II Canonical Mueller Matrix

The type-II canonical Mueller matrix $\mathbf{M}_{\Delta nd}$ has peculiar characteristics that make it very different from $\mathbf{M}_{\Delta d}$. Namely, $\mathbf{M}_{\Delta nd}$ exhibits residual diattenuation and polarizance given by the respective vectors $\mathbf{D}_{\Delta nd} = (-1/2,0,0)^T$ and $\mathbf{P}_{\Delta nd} = (1/2,0,0)^T$ As will be shown in Chapter 10, $\mathbf{M}_{\Delta nd}$ maps the Poincaré sphere into an ellipsoid that touches the unit sphere at a single point (unlike $\mathbf{M}_{\Delta d}$, whose image ellipsoid either does not touch the unit sphere, or touches it at two points, or coincides with the entire sphere) (Ossikovski et al. 2013; Ossikovski and Gil 2017).

In the case of type-II Mueller matrices, the normal form (5.24) can be expressed in the following manner (Ossikovski 2010a):

$$\begin{aligned}\mathbf{M} &= \mathbf{M}_{J2}\mathbf{M}_{\Delta nd}\mathbf{M}_{J1}\\ &= \mathbf{M}_{J2}\left[\mathbf{M}_{Da}\mathbf{M}_{Da}^{-1}\mathbf{M}_{\Delta nd}\mathbf{M}_{Da}\mathbf{M}_{Da}^{-1}\right]\mathbf{M}_{J1} = \left(\mathbf{M}_{J2}\mathbf{M}_{Da}\right)\left(\mathbf{M}_{Da}^{-1}\mathbf{M}_{\Delta nd}\mathbf{M}_{\mathbf{Da}}\right)\left(\mathbf{M}_{Da}^{-1}\mathbf{M}_{J1}\right)\\ &= \mathbf{M}_{JO}\mathbf{M}_{DII}\mathbf{M}_{JI} \qquad \left[\mathbf{M}_{DII} = \mathbf{M}_{Da}^{-1}\mathbf{M}_{\Delta nd}\mathbf{M}_{Da} \quad \mathbf{M}_{JO} \equiv \mathbf{M}_{\mathbf{J2}}\mathbf{M}_{\mathbf{Da}} \quad \mathbf{M}_{JI} \equiv \mathbf{M}_{Da}^{-1}\mathbf{M}_{J1}\right]\end{aligned} \tag{5.33}$$

where the Mueller matrix of the auxiliary diattenuator $\mathbf{M}_{Da}$ and its inverse matrix $\mathbf{M}_{Da}^{-1}$ are given by

$$\mathbf{M}_{Da} \equiv \frac{1}{1+c_{\kappa a}}\hat{\mathbf{M}}_{Da+} \qquad \mathbf{M}_{Da}^{-1} \equiv \frac{1}{1-c_{\kappa a}}\hat{\mathbf{M}}_{Da-}$$

$$\hat{\mathbf{M}}_{Da+} \equiv \begin{pmatrix} 1 & c_{\kappa a} & 0 & 0 \\ c_{\kappa a} & 1 & 0 & 0 \\ 0 & 0 & s_{\kappa a} & 0 \\ 0 & 0 & 0 & s_{\kappa a} \end{pmatrix} \qquad \hat{\mathbf{M}}_{Da-} \equiv \begin{pmatrix} 1 & -c_{\kappa a} & 0 & 0 \\ -c_{\kappa a} & 1 & 0 & 0 \\ 0 & 0 & s_{\kappa a} & 0 \\ 0 & 0 & 0 & s_{\kappa a} \end{pmatrix}$$
$$\left[c_{\kappa a} \equiv \cos\kappa_a = D_a = \left|a_0 - a/a_0 + a\right|, \quad s_{\kappa a} \equiv \sin\kappa_a\right] \tag{5.34}$$

so that the new entrance and exit pure Mueller matrices are respectively $\mathbf{M}_{JI} = \mathbf{M}_{Da}^{-1}\mathbf{M}_{J1}$ and $\mathbf{M}_{JO} = \mathbf{M}_{J2}\mathbf{M}_{Da}$, while the *second canonical type-II depolarizer* has the alternative form (Bolshakov et al. 1996; 1997)

$$\mathbf{M}_{\Delta II} = \begin{pmatrix} a_0+a & -a & 0 & 0 \\ a & a_0-a & 0 & 0 \\ 0 & 0 & a_2 & 0 \\ 0 & 0 & 0 & a_2 \end{pmatrix} \qquad [0 \le a_2 \le a_0 \quad 0 < a_0 \quad 0 < a] \tag{5.35}$$

a_0, a_2 being the square roots of the eigenvalues of $\mathbf{N}$, and a being an arbitrary positive parameter (which determines the diattenuation-polarizance of the diattenuator $\mathbf{M}_{Da}$). Observe that, as indicated in Eq. (5.33), the apparent extra parameter a in $\mathbf{M}_{\Delta II}$ is removed through the similarity transformation $\mathbf{M}_{Da}\mathbf{M}_{\Delta II}\mathbf{M}_{Da}^{-1} = \mathbf{M}_{\Delta nd}$.

Furthermore, both $\mathbf{M}_{\Delta II}$ and $\mathbf{M}_{\Delta nd}$ are degenerate Mueller matrices with the same eigenstate $(1,1,0,0)$ and doubly degenerate associated eigenvalue a_0. However, in general, the product of $\mathbf{M}_{\Delta II}$, or $\mathbf{M}_{\Delta nd}$, and another depolarizing Mueller matrix is not a degenerate matrix.

Expression (5.35) of $\mathbf{M}_{\Delta II}$ will be useful for the procedure of generating type-II matrices from respective reference pure Mueller matrices through the progressive increase of the depolarization descriptors (see Section 6.8).

Another interesting alternative form of the normal decomposition is (Bolshakov et al. 1996; 1997; Gopala Rao et al. 1998b; Ossikovski 2010a)

$$\mathbf{M} = \mathbf{M}_{JB}\,\mathbf{M}'_{\Delta II}\,\mathbf{M}_{JA}$$

$$\mathbf{M}'_{\Delta II} = \begin{pmatrix} x & y-x & 0 & 0 \\ 0 & y & 0 & 0 \\ 0 & 0 & a_2 & 0 \\ 0 & 0 & 0 & a_2 \end{pmatrix} \quad \begin{bmatrix} x \equiv \sqrt{a_0(2a+a_0)} \quad y \equiv a_0^2\Big/\sqrt{a_0(2a+a_0)} \\ 0 \le a_2 \le a_0 \quad 0 < a_0 \quad 0 < a \end{bmatrix} \tag{5.36}$$

where

$$\mathbf{M}'_{\Delta II} = \mathbf{M}_{Dw}^{-1}\mathbf{M}_{Dv}^{-1}\mathbf{M}_{\Delta nd}\mathbf{M}_{Dv} \quad \mathbf{M}_{JB} = \mathbf{M}_{J2}\mathbf{M}_{Dv}\mathbf{M}_{Dw} \quad \mathbf{M}_{JA} = \mathbf{M}_{Dv}^{-1}\mathbf{M}_{J1}$$

$$\mathbf{M}_{Dv} \equiv \frac{1}{1+c_{\kappa v}}\hat{\mathbf{M}}_{Dv+} \quad \mathbf{M}_{Dv}^{-1} \equiv \frac{1}{1-c_{\kappa v}}\hat{\mathbf{M}}_{Dv-} \quad \mathbf{M}_{Dw} \equiv \frac{1}{s_{\kappa w}}\hat{\mathbf{M}}_{Dw+} \quad \mathbf{M}_{Dw}^{-1} \equiv \frac{1}{s_{\kappa w}}\hat{\mathbf{M}}_{Dw-} \tag{5.37}$$
$$\left[c_{\kappa i} \equiv \cos\kappa_i \quad s_{\kappa i} \equiv \sin\kappa_i \quad (i = v, w), \quad \cos\kappa_v \equiv \frac{3x-y}{x+y} \quad \cos\kappa_w \equiv \frac{x-y}{x+y}\right]$$

the forms of matrices $\hat{\mathbf{M}}_{Di+}$ and $\hat{\mathbf{M}}_{Di-}$ $(i = v, w)$ being those of Eq. (5.34) but replacing the subscript a by i.

Thus, through appropriate choices of the left and right pure serial components of the normal form, the central depolarizer can adopt different forms so that the choice between $\mathbf{M}_{\Delta nd}$, $\mathbf{M}_{\Delta II}$ and $\mathbf{M}'_{\Delta II}$ can be done, without loss of generality, in accordance with which is the best suited one for the problem considered.

As for the type-II canonical form $\mathbf{M}_{\Delta nd}$, let us now consider its associated covariance matrix $\mathbf{C}_{\Delta nd}$

$$\mathbf{C}_{\Delta nd} \equiv \mathbf{C}(\mathbf{M}_{\Delta nd}) = \frac{1}{2}\begin{pmatrix} a_0+a_2 & 0 & 0 & 0 \\ 0 & a_0-a_{20} & 0 & 0 \\ 0 & 0 & a_0 & -ia_0 \\ 0 & 0 & ia_0 & a_0 \end{pmatrix} \tag{5.38}$$

whose eigenvalues are

$$\lambda_{\Delta nd0} = a_0 \quad \lambda_{\Delta nd1} = (a_0+a_2)/2 \quad \lambda_{\Delta nd2} = (a_0-a_2)/2 \quad \lambda_{\Delta nd3} = 0$$
$$\left[0 \le \lambda_{\Delta nd2} \le \lambda_{\Delta nd1} \le \lambda_{\Delta nd0}\right] \tag{5.39}$$

The non-negativity of $\lambda_{\Delta ndi}$ requires that $\mathbf{M}_{\Delta nd}$ satisfies the covariance conditions

$$0 \leq a_2 \leq a_0 \tag{5.40}$$

while the passivity condition is given by $a_0 \leq 1/3$

Let us next consider separately the two possible values for $r \equiv \mathbf{C}_{\Delta nd}$, namely, $r = 2$ and $r = 3$. Throughout the analysis that will be developed below, we will denote $\mathbf{C}_{\Delta nd}$ and $\mathbf{M}_{\Delta nd}$ as $\mathbf{C}_{\Delta ndr}$ and $\mathbf{M}_{\Delta ndr}$ respectively, where $r \equiv \operatorname{rank} \mathbf{C}_{\Delta nd}$.

(1) $\operatorname{rank} \mathbf{C}_{\Delta nd} = 2$
In this case $a_0 = a_2$ and thus,

$$\mathbf{M}_{\Delta nd2} = 2a_0 \begin{pmatrix} 1 & -1/2 & 0 & 0 \\ 1/2 & 0 & 0 & 0 \\ 0 & 0 & 1/2 & 0 \\ 0 & 0 & 0 & 1/2 \end{pmatrix} \quad [0 < a_0] \tag{5.41}$$

An example of the physical realizability of $\mathbf{M}_{\Delta nd2}$ in the form of a parallel combination of pure components is

$$\mathbf{M}_{\Delta nd2} = a_0 \mathbf{I}_4 + 2a_0 \left\{ \frac{1}{2} \begin{pmatrix} 1 & -1 & 0 & 0 \\ 1 & -1 & 0 & 0 \\ 0 & 0 & 0 & 0 \\ 0 & 0 & 0 & 0 \end{pmatrix} \right\} \tag{5.42}$$

in terms of a neutral filter $\mathbf{I}_4$ and a non-normal perfect polarizer.

(2) $\operatorname{rank} \mathbf{C}_{\Delta nd} = 3$
In this case $a_2 < a_0$. An example of parallel decomposition of $\mathbf{M}_{\Delta nd3}$ is given by

$$\mathbf{M}_{\Delta nd3} = \frac{a_0 + a_2}{2} \mathbf{I}_4 + \frac{a_0 - a_2}{2} \left\{ \begin{pmatrix} 1 & 0 & 0 & 0 \\ 0 & 1 & 0 & 0 \\ 0 & 0 & -1 & 0 \\ 0 & 0 & 0 & -1 \end{pmatrix} \right\} + 2a_0 \left\{ \frac{1}{2} \begin{pmatrix} 1 & -1 & 0 & 0 \\ 1 & -1 & 0 & 0 \\ 0 & 0 & 0 & 0 \\ 0 & 0 & 0 & 0 \end{pmatrix} \right\} \tag{5.43}$$

in terms of a neutral filter, a horizontal half wave-plate and a non-normal pure polarizer.

5.7.3 Covariance Characterization of Mueller Matrices Through their Normal Form

By taking into account the normal forms for type-I and type-II Mueller matrices, the covariance criterion can be formulated as follows (Bolshakov et al. 1996, Gopala Rao et al. 1998b, Ossikovski 2010a):

(1) Any Mueller matrix **M** is necessarily either of type-I or of type-II (in an exclusive manner, i.e., if **M** is type-I, it is not of type-II and vice versa).

(2) Any type-I Mueller matrix **M** can be expressed as $\mathbf{M} = \mathbf{M}_{J2}\mathbf{M}_{\Delta d}\mathbf{M}_{J1}$ where $\mathbf{M}_{J1}$ and $\mathbf{M}_{J2}$ are pure Mueller matrices and $\mathbf{M}_{\Delta d}$ is the diagonal Mueller matrix $\mathbf{M}_{\Delta d} = \mathrm{diag}(d_0, d_1, d_2, \varepsilon d_3)$ $[\varepsilon \equiv \det\mathbf{M}/|\det\mathbf{M}|]$ satisfying the following inequalities:

$$d_0 \geq -d_1 - d_2 - \varepsilon d_3 \quad d_0 \geq -d_1 + d_2 + \varepsilon d_3 \quad d_0 \geq d_1 - d_2 + \varepsilon d_3 \quad d_0 \geq d_1 + d_2 - \varepsilon d_3 \tag{5.44}$$

which are equivalent to (Gopala Rao el al. 1998b)

$$d_0 \geq d_i \ (i = 1,2,3) \quad d_k \geq 0 \ (k = 0,1,2,3) \quad d_0 + d_1 \geq d_2 + \varepsilon d_3 \quad d_0 - d_1 \geq d_2 - \varepsilon d_3 \tag{5.45}$$

Observe that, unlike pure Mueller matrices, whose determinants are necessarily non-negative, depolarizing Mueller matrices can have positive or negative determinants depending on whether $\varepsilon = 1$ or $\varepsilon = -1$ respectively.

(3) Any type-II Mueller matrix **M** can be expressed as $\mathbf{M} = \mathbf{M}_{J2}\mathbf{M}_{\Delta nd}\mathbf{M}_{J1}$ where $\mathbf{M}_{J1}$ and $\mathbf{M}_{J2}$ are nonsingular pure Mueller matrices and $\mathbf{M}_{\Delta nd}$ is the Mueller matrix in the form shown in Eq. (5.23), with the elements satisfying $0 \leq a_2 \leq a_0$.
The determinant of a type-II Mueller matrix can be written as follows:

$$\det\mathbf{M} = \det\mathbf{M}_{J2}\det\mathbf{M}_{\Delta nd}\det\mathbf{M}_{J1} = a_0^2 a_2^2 \det\mathbf{M}_{J2}\det\mathbf{M}_{J1} \tag{5.46}$$

and, since pure Mueller matrices have necessarily non-negative determinants, the determinant of a type-II Mueller matrix is non-negative. Furthermore, in Section 8.3.2 it will be demonstrated that, for a type-II Mueller matrix, the inequalities $\det\mathbf{M}_{J1} > 0$ and $\det\mathbf{M}_{J2} > 0$ must necessarily hold (otherwise **M** corresponds to a depolarizing analyzer or to a depolarizing polarizer, both being type-I matrices). Consequently, a type-II Mueller matrix is singular if and only if $a_2 = 0$.

Considering the structural similarities between the coherency matrices representing the polarimetric properties for 2D polarization states (2×2 coherency matrices), 3D polarization states (3×3 coherency matrices) and material media (4×4 covariance or coherency matrices), it is worth noting that the constraining inequalities between the elements m_{ij}, derived from the non-negativity of the eigenvalues of **H**, are of higher degree of complexity than those of the 2D and 3D Stokes parameters, and involve the invariants $\mathrm{tr}\,\mathbf{H}$, $\mathrm{tr}\,\mathbf{H}^2$, $\mathrm{tr}\,\mathbf{H}^3$ and $\mathrm{tr}\,\mathbf{H}^4$. Leaving aside the fact that the 4×4 covariance matrix **H** represents the polarimetric properties of material media, from a mathematical point of view, through appropriate restrictions of **H** we can reproduce easily the coherency matrices $\mathbf{\Phi}$ and **R** representing 2D and 3D polarization states. Therefore, by eliminating the corresponding extra components of the variables t_i in Eq. (5.8), the 2D model appears as the upper-left 2×2 matrix if one puts

$$s_0 = 2(m_{00} + m_{01}) \quad s_1 = 2(m_{10} + m_{11}) \quad s_2 = 2(m_{20} + m_{21}) \quad s_3 = 2(m_{30} + m_{31}) \tag{5.47}$$

The restriction to the 3D model can be performed in a straightforward way but, as it happens in the restriction from 3D to 2D, the explicit expressions for the 3D Stokes parameters in terms of m_{ij} are more complicated because of the mathematical asymmetry of the Gell-Mann matrices with respect to the modified Dirac matrices $\mathbf{E}_{ij}$.

The expansion given by Eq. (5.9) provides fundamental meaning to the elements of the Mueller matrix, deeper than their phenomenological role as simple coefficients of the linear transformation of Stokes vectors. We see that the relation between the covariance matrix **H** and its corresponding Mueller matrix is, in fact, analogous to the relation between the 2×2 coherency matrix and its corresponding Stokes parameters. This shows the symmetry of the operators representing polarization quantities for both electromagnetic waves and material media. In this way, Sudha et al. (2008) have emphasized the mathematical analogy between the formulation of the covariance matrix **H**, derived from the ensemble criterion, and the positive operator valued measures (POVM) of the quantum density matrix, and have proposed a way for realizing different types of Mueller devices.

5.8 Reciprocity Properties of Mueller Matrices

The reciprocity properties of pure Mueller matrices have been considered in Section 3.3.3 where, leaving aside systems exhibiting magnetooptic effects whose reciprocity property $\mathbf{M}_J^r = \mathbf{M}_J$ differs from the usual rule, it was found that, for a given pure Mueller matrix $\mathbf{M}_J$, the pure *reverse Mueller matrix* $\mathbf{M}_J^r$ that represents the same nondepolarizing medium as $\mathbf{M}_J$, but with the incident and emerging directions of propagation of the electromagnetic wave interchanged, is given by (Sekera 1966, Schönhofer and Kuball 1987)

$$\mathbf{M}_J^r = \text{diag}(1,1,-1,1)\, \mathbf{M}_J^T\, \text{diag}(1,1,-1,1) \tag{5.48}$$

or, in block form,

$$\mathbf{M}_J^r = \mathbf{X}\mathbf{M}_J^T\mathbf{X} = m_{00}\,\mathbf{X}\begin{pmatrix} 1 & \mathbf{P}^T \\ \mathbf{D} & \mathbf{m}^T \end{pmatrix}\mathbf{X} = m_{00}\begin{pmatrix} 1 & \mathbf{P}^T\mathbf{x} \\ \mathbf{x}\mathbf{D} & \mathbf{x}\mathbf{m}^T\mathbf{x} \end{pmatrix} \tag{5.49}$$

$$\mathbf{X} \equiv \begin{pmatrix} 1 & \mathbf{0}^T \\ \mathbf{0} & \mathbf{x} \end{pmatrix} \qquad \mathbf{x} \equiv \text{diag}(1,-1,1)$$

By virtue of the covariance criterion, any depolarizing Mueller matrix results from the spatial or temporal averaging of a number of elementary scattering processes, each of which is characterized by its respective pure Mueller matrix. Taking into account that the property (5.48) is preserved under addition,

$$\mathbf{M}^r = \sum_i p_i\,\mathbf{M}_{Ji}^r = \sum_i p_i\,\mathbf{X}\mathbf{M}_{Ji}^T\,\mathbf{X} = \mathbf{X}\left(\sum_i p_i\,\mathbf{M}_{Ji}^T\right)\mathbf{X} = \mathbf{X}\mathbf{M}^T\mathbf{X} \tag{5.50}$$

we deduce that the same rule

$$\mathbf{M}^r = \text{diag}(1,1,-1,1)\, \mathbf{M}^T\, \text{diag}(1,1,-1,1) \tag{5.51}$$

holds for depolarizing Mueller matrices, and shows that the knowledge of a depolarizing Mueller matrix **M** fully determines the knowledge of the reverse Mueller matrix $\mathbf{M}^r$. In consequence, any requisite for **M** to be physically realizable must also be satisfied by $\mathbf{M}^r$. Further, the constraining inequalities that characterize Mueller matrices are invariant with respect to the replacement of $\mathbf{M}^r$ by $\mathbf{M}^T$. Thus, we conclude that all the physical conditions for a matrix **M** to be a Mueller matrix must also be satisfied by $\mathbf{M}^T$; in particular, it can be easily shown that $\mathbf{M}^T$ satisfies the Cloude's criterion if and only if **M** does.

From the definition of diattenuation and polarizance as the absolute values of the respective vectors **D** and **P**, we observe that

$$D(\mathbf{M}) = P(\mathbf{M}^T) = P(\mathbf{M}^r) \qquad P(\mathbf{M}) = D(\mathbf{M}^T) = D(\mathbf{M}^r) \tag{5.52}$$

and therefore, the diattenuation of **M** can be called the *reverse polarizance* of **M** and, conversely, the polarizance of **M** can be called the *reverse diattenuation* of **M** (Gil 2007). Thus, diattenuation and polarizance share a common physical nature and refer to complementary aspects of the enpolarizing properties of the medium represented by **M**.

5.9 Passivity Constraints for Mueller Matrices

The covariance conditions satisfied by Mueller matrices have been obtained from the construction of the Mueller matrix as the average given by Eq. (5.6) where the Jones matrices $\mathbf{T}_i$ of the pure parallel components are just 2×2 complex matrices, without considering the fact that $\mathbf{T}_i$ must furthermore represent *passive* pure media. Thus, in order to complete the set of conditions for a 4×4 real matrix to be a Mueller matrix, it is necessary to consider the constraints derived from the fact that all pure Mueller matrices involved in the average must be passive.

The maximum intensity coefficient of a given Mueller matrix **M** is achieved for an input pure Stokes vector $\hat{\mathbf{s}}_{\hat{D}}$ whose polarization vector is precisely the normalized diattenuation vector $\hat{\mathbf{D}} \equiv \mathbf{D}/D$ of **M**

$$\mathbf{M}\hat{\mathbf{s}}_{\hat{D}} = m_{00}\begin{pmatrix} 1 & \mathbf{D}^T \\ \mathbf{P} & \mathbf{m} \end{pmatrix}\begin{pmatrix} 1 \\ \hat{\mathbf{D}} \end{pmatrix} \tag{5.53}$$

so that the forward intensity coefficient $g_f \equiv s'_0/s_0$ (with $\mathbf{s}' = \mathbf{M}\mathbf{s}$) is limited by

$$g_f \le m_{00}(1+D) \tag{5.54}$$

and therefore, a necessary and sufficient condition for **M** to be passive with respect to incident polarization states in the forward direction is (Barakat 1987a, Kostinski and Givens 1992)

$$m_{00}(1+D) \le 1 \tag{5.55}$$

Moreover, from the two premises of the ensemble criterion, namely (1) the system is considered as an ensemble (the Mueller matrix is given by a convex sum of pure Mueller matrices) and (2) each single realization is a passive system (i.e., its pure Mueller matrix satisfies the passivity condition); the following additional passivity condition follows (Gil 2000a):

$$m_{00}(1+P) \leq 1 \tag{5.56}$$

To show that Eq. (5.56) is satisfied, as a necessary condition, in the general case (i.e., not only for pure Mueller matrices), let us denote by P_{il} the components of the diattenuation vector $\mathbf{P}_i$ (with absolute value P_i) of each pure component $\mathbf{M}_{Ji}$ and expand Eq. (5.56) with the use of the parallel decomposition (5.15) as

$$\begin{aligned} m_{00}(1+P) &= m_{00}\left\{\sum_i p_i + \sqrt{\sum_{l=1}^{3}\left[\sum_i (p_i P_{il})\right]^2}\right\} \\ &= m_{00}\left[\sum_i p_i + \left|\sum_i p_i \mathbf{P}_i\right|\right] \leq m_{00}\left(\sum_i p_i + \sum_i p_i P_i\right) = \sum_i p_i\left[m_{00}(1+P_i)\right] \end{aligned} \tag{5.57}$$

By considering the assumed passivity of each pure component, expressed as $m_{00}(1+P_i) \leq 1$, together with the convexity condition $\sum p_i = 1$ from Eq. (5.15), the above expression leads to $m_{00}(1+P) \leq 1$. Furthermore, it has been demonstrated that the pair of inequalities $m_{00}(1+D) \leq 1$ and $m_{00}(1+P) \leq 1$ is not only necessary, but also sufficient for $\mathbf{M}$ to represent a passive system (San José and Gil 2020b) and constitute a complete set of conditions characterizing passivity.

In summary, the procedure followed to construct a general Mueller matrix as a convex sum of passive pure Mueller matrices, leads to the fact that any Mueller matrix must satisfy both the *forward passivity condition* (5.55) and the *reverse passivity condition* (5.56)

$$m_{00}(1+D) \leq 1 \qquad m_{00}(1+P) \leq 1 \tag{5.58}$$

whose expressions in terms of the elements of the covariance and coherency matrices $\mathbf{H}$ and $\mathbf{C}$ are the following (Gil 2000a, San José and Gil 2020b):

$$\begin{aligned} &\operatorname{tr}\mathbf{H} + \sqrt{\sum_{j=1}^{3}\operatorname{tr}^2\left(\mathcal{L}^{\dagger}\mathbf{G}_j\mathcal{L}\mathbf{H}\right)} \leq 1 \quad \operatorname{tr}\mathbf{H} + \sqrt{\sum_{i=1}^{3}\operatorname{tr}^2\left(\mathcal{L}^{\dagger}\mathbf{G}_i^{*}\mathcal{L}\mathbf{H}\right)} \leq 1 \\ &\operatorname{tr}\mathbf{C} + \sqrt{\sum_{j=1}^{3}\operatorname{tr}^2\left(\mathbf{G}_j\mathbf{C}\right)} \leq 1 \quad \operatorname{tr}\mathbf{C} + \sqrt{\sum_{i=1}^{3}\operatorname{tr}^2\left(\mathbf{G}_i^{*}\mathbf{C}\right)} \leq 1 \end{aligned} \tag{5.59}$$

where $\mathcal{L}$ and $\mathbf{G}_i$ are the matrices introduced in Eq. (5.3).

A good example for analyzing the nature and scope of this pair of conditions is the following matrix

$$\begin{pmatrix} 1 & \mathbf{0}^T \\ \mathbf{P} & \mathbf{0} \end{pmatrix} \quad [0 < |\mathbf{P}| \leq 1] \tag{5.60}$$

whose only nonzero column is the first one. It is easy to check that this matrix satisfies the covariance conditions as well as the forward passivity condition, but it does not satisfy the reverse passivity condition. This is consistent with the fact that this matrix cannot be obtained as an average of pure passive Mueller matrices and, hence, it is not possible to physically realize it.

As we have seen, in the case of pure Mueller matrices, the equality $P = D$ necessarily holds and the above pair of passivity conditions reduces to a unique one. Let us now denote $D = Q$ if $D \geq P$ or $P = Q$ if $P > D$, and observe that, among the Mueller matrices proportional to a given one (denoted by $\mathbf{M}$), the largest mean intensity coefficient compatible with passivity is given by $\tilde{m}_{00} = 1/(1+Q)$. Consequently, the *passive form* (or *passive representative*) $\tilde{\mathbf{M}}$ of $\mathbf{M}$ is defined as (San José and Gil 2020b)

$$\tilde{\mathbf{M}} \equiv \left[1/(1+Q)\right]\hat{\mathbf{M}} \quad Q \equiv \max(D,P). \tag{5.61}$$

Experimental polarimetry is called *absolute* (or *complete*) when all the 16 elements of the Mueller matrix are determined without any arbitrary scale factor, i.e., when m_{00} is measured. Typically, Mueller polarimeters provide the 16 elements of the Mueller matrix $\mathbf{M}$ of the sample only within a positive common scale factor (*relative Mueller polarimetry*).

Absolute polarimetry can be carried out by means of any kind of complete Mueller polarimeter whose scale factor is calculated by placing, instead of the sample, a system whose absolute Mueller matrix is previously known (Compain et al. 1999, De Martino et al. 2004). Particularly simple is the case where the polarimeter can be arranged in direct transmission (i.e., the polarization state generator and the polarization state analyzer are disposed along a common optical axis) so that the test sample is the ambient (or, in different words, there is no sample) whose Mueller matrix is the identity matrix (Gil 1983, Goldstein 1992).

As with pure Mueller matrices, given a matrix $\mathbf{M}$ that does not satisfy the passivity conditions, it is always possible to find a real positive coefficient c such that $c\,\mathbf{M}$ fulfils them. Obviously, this is a reason why many experimentalists work with relative measurements instead of absolute Mueller polarimetry. Nevertheless, this fact does not reduce the interest of the study of the mathematical and physical constraints derived from the passivity condition.

5.10 Polarimetric Purity of a Mueller Matrix

5.10.1 Norms of the Covariance, Coherency and Mueller Matrices

In order to find a simple criterion for distinguishing between pure and nonpure Mueller matrices, as well as to define a parameter that provides an overall measure of how pure a Mueller matrix is, let us first consider the Frobenius norms of $\mathbf{C}$ and $\mathbf{M}$, defined as

$$\|\mathbf{C}\|_2 \equiv \sqrt{\sum_{k,l=0}^{3} |c_{kl}|^2} = \sqrt{\operatorname{tr}\left(\mathbf{C}^{\dagger}\mathbf{C}\right)} = \sqrt{\operatorname{tr}\mathbf{C}^2} = \sqrt{\sum_{k=0}^{3} \lambda_k^2} \tag{5.62}$$

where λ_k are the eigenvalues of **C**, and

$$\|\mathbf{M}\|_2 \equiv \sqrt{\sum_{k,l=0}^{3} m_{kl}^2} = \sqrt{\operatorname{tr}\left(\mathbf{M}^T\mathbf{M}\right)} = \sqrt{\operatorname{tr}\left(\mathbf{M}\mathbf{M}^T\right)} \tag{5.63}$$

Let us also consider the *trace norm* of **C**

$$\|\mathbf{C}\|_{tr} \equiv \operatorname{tr}\mathbf{C} = \sum_{k=0}^{3} \lambda_k \tag{5.64}$$

as well as the *max* norm of **M** as

$$\|\mathbf{M}\|_0 \equiv m_{00} \tag{5.65}$$

Note that, since **C** (the coherency matrix) and **H** (the covariance matrix) have the same spectrum of eigenvalues, their norms are equal

$$\|\mathbf{H}\|_2 = \|\mathbf{C}\|_2 \qquad \|\mathbf{H}\|_{tr} = \|\mathbf{C}\|_{tr} \tag{5.66}$$

It is straightforward to show that the above norms satisfy the following relations:

$$\|\mathbf{C}\|_2 = \|\mathbf{M}\|_2/2 \tag{5.67}$$

$$\|\mathbf{C}\|_{tr} = m_{00} = \|\mathbf{M}\|_0 \tag{5.68}$$

$$\|\mathbf{C}\|_{tr} \le 2\|\mathbf{C}\|_2 \le 2\|\mathbf{C}\|_{tr} \tag{5.69}$$

5.10.2 Purity criterion

Relations (5.69) between $\|\mathbf{C}\|_{tr}$ and $\|\mathbf{C}\|_2$ can also be expressed as follows in terms of **M**:

$$\|\mathbf{M}\|_0 \le \|\mathbf{M}\|_2 \le 2\|\mathbf{M}\|_0 \tag{5.70}$$

and become evident when written in terms of the eigenvalues λ_k, showing that the equality

$$\|\mathbf{C}\|_2 = \|\mathbf{C}\|_{tr} \tag{5.71}$$

or, equivalently

$$\sum_{i,j=0}^{3} m_{ij}^2 = 4m_{00}^2 \qquad \left(\|\mathbf{M}\|_2^2 = 4\|\mathbf{M}\|_0^2\right) \tag{5.72}$$

is a necessary and sufficient condition for **C** to have a single nonzero eigenvalue. In consequence, it is also a necessary and sufficient condition for a 4×4 coherency matrix **C** to correspond to a pure Mueller matrix (*purity criterion*) (Gil and Bernabéu 1985).

The lower limit $2\|\mathbf{C}\|_2 = \|\mathbf{C}\|_{tr}$ is reached when all the eigenvalues are equal, so that the system is composed of an equiprobable mixture of pure parallel components resulting in a perfect depolarizer represented by a Mueller matrix $\mathbf{M}_{\Delta 0}$ whose elements are zero except for m_{00}.

Therefore, under the premise that a Mueller matrix is expressible as a convex sum of pure Mueller matrices, we can state that "*Given a Mueller matrix* $\mathbf{M}$ *(hence satisfying the covariance conditions), the equality* $\|\mathbf{M}\|_2 = 2\|\mathbf{M}\|_0$ *is a necessary and sufficient condition for* $\mathbf{M}$ *to be a pure Mueller matrix.*" All arguments contrary to this statement quoted in the literature arise from identifying Mueller matrices with Stokes matrices (obeying the Stokes criterion), which do not satisfy the covariance conditions in general.

5.10.3 Depolarization Index and Depolarizance

A measure of the degree of polarimetric purity of a given Mueller matrix $\mathbf{M}$ is provided by the *depolarization index* P_Δ, defined as (Gil and Bernabéu 1986)

$$P_\Delta = \sqrt{\frac{1}{3}\left(4\|\hat{\mathbf{C}}\|_2^2 - 1\right)} = \sqrt{\frac{1}{3}\left(\|\hat{\mathbf{M}}\|_2^2 - 1\right)} = \sqrt{\frac{1}{3}\left(\operatorname{tr}\left(\hat{\mathbf{M}}^T\hat{\mathbf{M}}\right) - 1\right)} = \sqrt{\frac{1}{3m_{00}^2}\left(\sum_{i,j=0}^{3} m_{ij}^2 - m_{00}^2\right)} \tag{5.73}$$

$$\left(\hat{\mathbf{C}} \equiv \mathbf{C}/\|\mathbf{C}\|_{tr} = \mathbf{C}/m_{00} \qquad \hat{\mathbf{M}} \equiv \mathbf{M}/\|\mathbf{M}\|_0 = \mathbf{M}/m_{00}\right)$$

or, in terms of the eigenvalues of $\mathbf{C}$

$$P_\Delta = \frac{1}{\sqrt{3}}\sqrt{4\sum_{k=0}^{3}\hat{\lambda}_k^2 - 1} \qquad \left[\hat{\lambda}_k \equiv \frac{\lambda_k}{\operatorname{tr}\mathbf{C}}\right] \tag{5.74}$$

This parameter is invariant with respect to changes of the laboratory coordinate system for the representation of $\mathbf{M}$, as well as with respect to the pre- and post- multiplication of $\mathbf{M}$ by Mueller matrices of retarders (Gil 2011). Furthermore, P_Δ is restricted to the interval

$$0 \le P_\Delta \le 1 \tag{5.75}$$

The lower limit $P_\Delta = 0$ corresponds to an ideal depolarizer, characterized by the fact that all eigenvalues of $\mathbf{C}$ are equal, i.e., the medium does not exhibit any polarimetric preference

$$P_\Delta(\mathbf{M}) = 0 \Leftrightarrow \mathbf{M} = m_{00}\begin{pmatrix} 1 & \mathbf{0}^T \\ \mathbf{0} & \mathbf{0} \end{pmatrix} \tag{5.76}$$

The maximum achievable value $P_\Delta = 1$ corresponds to a pure system ($\lambda_0 > 0$, $\lambda_1 = \lambda_2 = \lambda_3 = 0$). Therefore, the purity criterion can be stated as: *given a Mueller matrix* $\mathbf{M}$, *it is pure if and only if* $P_\Delta(\mathbf{M}) = 1$ (Gil and Bernabéu 1986). From its very definition, P_Δ constitutes an objective measure of the global polarimetric purity of the system, and provides criteria for the analysis of measured Mueller matrices (Le Roy-Brehonnet et al. 1997).

As in the case of the 2D and 3D degrees of polarimetric purity (studied in Chapters 1 and 2 respectively), the *4D degree of polarimetric purity* P_Δ (which refers to a medium instead of to a state of polarization of an electromagnetic wave, unlike the 2D and 3D degrees of polarimetric purity) has a simple geometric interpretation as the distance of the normalized coherency matrix $\hat{\mathbf{C}} \equiv \mathbf{C}/\mathrm{tr}\,\mathbf{C} = \mathbf{C}\mathbf{H}/m_{00}$ from the coherency covariance matrix of the ideal depolarizer, $\hat{\mathbf{C}}_{\Delta 0} = \mathbf{I}_4/4$,

$$P_\Delta^2 = \frac{4}{3}\mathrm{tr}\left[\left(\frac{1}{4}\mathbf{I}_4 - \frac{\mathbf{C}}{\mathrm{tr}\mathbf{C}}\right)^2\right] = \frac{4}{3}\mathrm{tr}\left[\left(\hat{\mathbf{C}}_{\Delta 0} - \hat{\mathbf{C}}\right)^2\right] \tag{5.77}$$

where $\mathbf{I}_4$ is the 4×4 identity matrix. Consequently, P_Δ can also be considered as a distance of the normalized Mueller matrix $\hat{\mathbf{M}} \equiv \mathbf{M}/m_{00}$ from the normalized Mueller matrix of the ideal depolarizer $\hat{\mathbf{M}}_{\Delta 0}$.

Eq. (5.77) can be considered a particular case of the polarimetric distance $P_{\Delta(a,b)}$ between two normalized coherency matrices $\hat{\mathbf{C}}_a$ and $\hat{\mathbf{C}}_b$ defined as

$$P_{\Delta(a,b)}^2 = \frac{4}{3}\mathrm{tr}\left[\left(\hat{\mathbf{C}}_a - \hat{\mathbf{C}}_b\right)^2\right] \tag{5.78}$$

It is worth noting that P_Δ can also be expressed as follows in terms of the so-called *components of purity* of $\mathbf{M}$, namely, the diattenuation D, the polarizance P and the polarimetric dimension index $P_S(\mathbf{M}) \equiv \sqrt{\|\mathbf{m}\|_2^2/3}$ ($\|\mathbf{m}\|_2$ being the Frobenius norm of the submatrix $\mathbf{m}$ of $\mathbf{M}$) (Gil 2011):

$$P_\Delta = \sqrt{D^2 + P^2 + 3P_S^2}\Big/\sqrt{3} \tag{5.79}$$

In order to go deeper into the physical meaning of P_Δ let us revisit the procedure followed originally for its definition (Gil and Bernabéu 1986).

Given a state of polarization represented by a Stokes vector $\mathbf{s}$, consider the associated quantity

$$\mathcal{D}^2(\mathbf{s}) \equiv \left(\mathbf{s}^T\mathbf{G}\,\mathbf{s}\right)/s_0^2 = \left(s_0^2 - s_1^2 - s_2^2 - s_3^2\right)/s_0^2 = 1 - \mathcal{P}^2(\mathbf{s}) \tag{5.80}$$

where $\mathcal{D}$ is the *randomness density* (or *degree of depolarization*) defined in Section 1.5.3, and

$$\mathcal{P}(\mathbf{s}) \equiv \sqrt{s_1^2 + s_2^2 + s_3^2}\Big/s_0 = \sqrt{1 - \mathcal{D}^2(\mathbf{s})} \tag{5.81}$$

is the *degree of polarization* of $\mathbf{s}$. Let us recall that the quantity $s_0^2\,\mathcal{D}^2(\mathbf{s})$ was called the *mean randomness* of $\mathbf{s}$ by Barakat (1987a).

Now, given a medium represented by a Mueller matrix $\mathbf{M}$, let us calculate the average $D_{\Delta f}^2$ of the quadratic quantities $\mathcal{D}^2(\mathbf{s}_i^f)$ associated with the emerging states $\mathbf{s}_i^f = \mathbf{M}\mathbf{s}_i$ corresponding to the input canonical Stokes vectors (which constitute a set of evenly distributed states on the Poincaré sphere, as drawn in Figure 5.3)

$$\begin{array}{lll} \mathbf{s}_{1+} \equiv (1,1,0,0)^T & \mathbf{s}_{2+} \equiv (1,0,1,0)^T & \mathbf{s}_{3+} \equiv (1,0,0,1)^T \\ \mathbf{s}_{1-} \equiv (1,-1,0,0)^T & \mathbf{s}_{2-} \equiv (1,0,-1,0)^T & \mathbf{s}_{3-} \equiv (1,0,0,-1)^T \end{array} \tag{5.82}$$

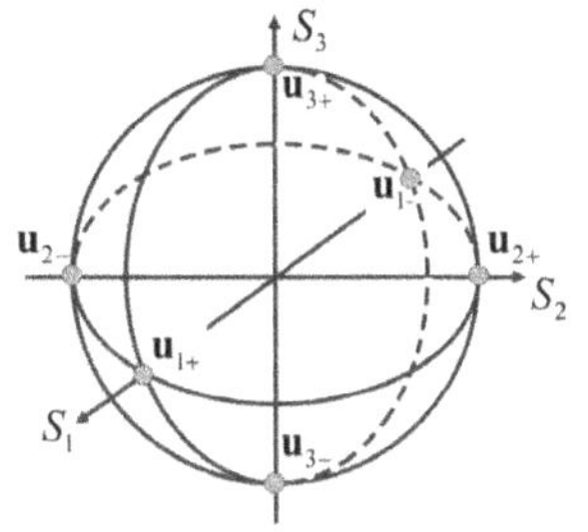

FIGURE 5.3
Poincaré vectors $\mathbf{u}_{i\pm}$ associated with the canonical Stokes vectors $\mathbf{s}_{i\pm}$, which constitute a sufficient set of pure states to determine a normalized Mueller matrix $\hat{\mathbf{M}}$ through the transformations $\mathbf{s}^f_{i\pm} = \hat{\mathbf{M}}\mathbf{s}_{i\pm}$.

so that $D^2_{\Delta f}$ is given by

$$D^2_{\Delta f} = \frac{1}{6}\sum_{1=1}^{3}\left[\mathcal{D}^2\left(\mathbf{s}^f_{i+}\right)+\mathcal{D}^2\left(\mathbf{s}^f_{i-}\right)\right]=\left(1+\frac{1}{3}D^2-P^2-P_s^2\right) \qquad \left[\mathbf{s}^f_{i+}\equiv\mathbf{M}\mathbf{s}_{i+} \quad \mathbf{s}^f_{i-}\equiv\mathbf{M}\mathbf{s}_{i-}\right] \quad (5.83)$$

Let us now recall that a Mueller matrix that is depolarizing for incident polarization states in the forward direction can be nondepolarizing for states incoming in the reverse direction and vice versa. This occurs, for instance, for a *depolarizing polarizer* constituted by a perfect polarizer followed by a perfect depolarizer

$$\mathbf{M}=\begin{pmatrix}1 & \mathbf{0}^T\\ 0 & 0\end{pmatrix}\frac{1}{2}\begin{pmatrix}1 & \hat{\mathbf{D}}^T\\ \hat{\mathbf{D}} & \hat{\mathbf{D}}\otimes\hat{\mathbf{D}}^T\end{pmatrix}=\frac{1}{2}\begin{pmatrix}1 & \hat{\mathbf{D}}^T\\ 0 & 0\end{pmatrix} \qquad \mathbf{M}^r=\mathbf{X}\mathbf{M}^T\mathbf{X}=\frac{1}{2}\begin{pmatrix}1 & \mathbf{0}^T\\ \hat{\mathbf{P}} & 0\end{pmatrix} \quad (5.84)$$
$$\left[\mathbf{X}\equiv\mathrm{diag}(1,1,-1,1) \qquad \hat{\mathbf{P}}\equiv\mathrm{diag}(1,-1,1)\hat{\mathbf{D}} \qquad |\hat{\mathbf{P}}|=|\hat{\mathbf{D}}|=1\right]$$

Thus, a proper global measure of the depolarizing properties of $\mathbf{M}$ requires the consideration of the emerging states $\mathbf{M}^r\mathbf{s}$ too. To do so, let us now consider the average $D^2_{\Delta r}$ of the quadratic forms $\mathcal{D}^2\left(\mathbf{s}^r_i\right)$ associated with the emerging states $\mathbf{s}^r_{i\pm}=\mathbf{M}^r\mathbf{s}_{i\pm}=\mathbf{X}\mathbf{M}^T\mathbf{X}$ corresponding to the incident canonical states $\mathbf{s}_{i\pm}$. Through the same procedure as that followed for $D^2_{\Delta f}$ while taking into account that the contributions of m_{kl} in the expressions of $D^2_{\Delta f}$ and $D^2_{\Delta r}$ always appear as quadratic terms m^2_{kl}, we obtain

$$D^2_{\Delta r}(\mathbf{M})=D^2_{\Delta f}\left(\mathbf{M}^T\right)=\frac{1}{6}\sum_{1=1}^{3}\left[\mathcal{D}^2\left(\mathbf{s}^r_{i+}\right)+\mathcal{D}^2\left(\mathbf{s}^r_{i-}\right)\right]=\left(1-D^2+\frac{1}{3}P^2-P_s^2\right) \quad (5.85)$$
$$\left[\mathbf{s}^r_{i+}\equiv\mathbf{M}^r\mathbf{s}_{i+} \quad \mathbf{s}^r_{i-}\equiv\mathbf{M}^r\mathbf{s}_{i-}\right]$$

Thus, an overall measure of the depolarization produced by $\mathbf{M}$ for both forward and reverse incident states can be defined as the average

$$D^2_{\Delta}\equiv\frac{1}{2}\left[D^2_{\Delta f}(\mathbf{M})+D^2_{\Delta r}(\mathbf{M})\right]=\frac{1}{2}\left[D^2_{\Delta r}\left(\mathbf{M}^T\right)+D^2_{\Delta f}\left(\mathbf{M}^T\right)\right] \quad (5.86)$$
$$=\frac{1}{3m^2_{00}}\left(4m^2_{00}-\sum_{k,l=0}^{3}m^2_{kl}\right)=\frac{1}{3}\left(4\left\|\hat{\mathbf{M}}\right\|_0^2-\left\|\hat{\mathbf{M}}\right\|_2^2\right)$$

and, in analogy to the quadratic relation (5.81) between the degree of polarization $\mathcal{P}(\mathbf{s})$ and the randomness density $\mathcal{D}(\mathbf{s})$ of a given state of polarization $\mathbf{s}$, the *degree of polarimetric purity* (or *depolarization index*) P_Δ is defined as

$$P_\Delta = \sqrt{1-D_\Delta^2} = \sqrt{\frac{D^2+P^2+3P_s^2}{3}} \tag{5.87}$$

which, as expected, coincides with the definition of P_Δ in Eq. (5.73).

It is remarkable that, while D_Δ^2 has been obtained from an average of the randomness densities of the twelve emerging states $\mathbf{s}_{i\pm}^f$ and $\mathbf{s}_{i\pm}^r$, the same result is obtained when $D_{\Delta f}^2$ and $D_{\Delta r}^2$ are calculated through respective integration over the pure states covering the entire Poincaré sphere, that is,

$$D_\Delta^2 = \frac{1}{2\pi}\int_{\varphi=0}^{\varphi=\pi}\int_{\chi=-\pi/4}^{\chi=\pi/4}\left[D_{\Delta f}^2(\varphi,\chi)+D_{\Delta r}^2(\varphi,\chi)\right]\cos 2\chi \, d\varphi \, d\chi \tag{5.88}$$

Therefore, in accordance with the physical meaning of P_Δ as the degree of polarimetric purity, an objective and global measure of how depolarizing is the medium represented by $\mathbf{M}$ is given by the *depolarizance* or *depolarizing degree* $D_\Delta(\mathbf{M})$

$$D_\Delta \equiv \sqrt{1-P_\Delta^2} \tag{5.89}$$

satisfying $P_\Delta^2 + D_\Delta^2 = 1$. The values of D_Δ run from 0 (pure system) to 1 (ideal depolarizer). The expressions of the depolarizance in terms of the norms of $\mathbf{M}$, $\mathbf{C}$ and $\mathbf{H}$ are the following:

$$D_\Delta = \sqrt{\frac{4}{3}\left(1-\|\mathbf{C}\|_2^2/\|\mathbf{C}\|_{tr}^2\right)} = \sqrt{\frac{1}{3}\left(4-\|\mathbf{M}\|_2^2/\|\mathbf{M}\|_0^2\right)} \quad \left[\|\mathbf{C}\|_2 = \|\mathbf{H}\|_2 \quad \|\mathbf{C}\|_{tr} = \|\mathbf{H}\|_{tr}\right] \tag{5.90}$$

or, equivalently, in terms of the normalized eigenvalues $\hat{\lambda}_k$ of $\mathbf{C}$

$$D_\Delta = \frac{2}{\sqrt{3}}\sqrt{1-\sum_{k=0}^{3}\hat{\lambda}_k^2} \tag{5.91}$$

Let us finally observe that the depolarizance was defined previously by us as $D_\Delta = 1 - P_\Delta$ (Gil 2007), but we consider preferable the above definition because its mathematical expression is more akin to other quantities related to polarimetric purity.

5.10.4 Polarization Entropy

In analogy to the polarization density matrices $\hat{\mathbf{\Phi}} = \mathbf{\Phi}/(\operatorname{tr}\mathbf{\Phi})$ and $\hat{\mathbf{R}} = \mathbf{R}/(\operatorname{tr}\mathbf{R})$ representing 2D and 3D states of polarization respectively, the normalized coherency matrix $\hat{\mathbf{C}} = \mathbf{C}/\operatorname{tr}\mathbf{C}$, as well as the normalized covariance matrix $\hat{\mathbf{H}} = \mathbf{H}/\operatorname{tr}\mathbf{H}$, have the mathematical structure and properties of a density matrix (Gil 2020). This fact motivates the definition of the polarization entropy $S(\mathbf{C})$ of a medium represented by $\mathbf{C}$ as the Von Neumann entropy of $\hat{\mathbf{C}}$ (Cloude and Pottier 1995),

$$S(\mathbf{C}) \equiv -\mathrm{tr}\left(\hat{\mathbf{C}}\log_4\hat{\mathbf{C}}\right) = -\mathrm{tr}\left(\hat{\mathbf{H}}\log_4\hat{\mathbf{H}}\right) = -\sum_{i=0}^{3}\left(\hat{\lambda}_i \log_4 \hat{\lambda}_i\right) \tag{5.92}$$

where $\hat{\lambda}_i \equiv \lambda_i/\mathrm{tr}\,\mathbf{C}$ are the normalized eigenvalues of **C** (which coincide with the normalized eigenvalues of **H**). A base-4 logarithm has been taken, instead of the Napierian logarithm, in order to restrict the values of S to the interval $0 \le S \le 1$. The maximum achievable value $S = 1$ corresponds to a perfect depolarizer ($P_\Delta = 0$, i.e., $\lambda_0 = \lambda_1 = \lambda_2 = \lambda_3$: fully random polarimetric properties), while the minimum one, $S = 0$, is reached for pure systems ($P_\Delta = 1$, i.e., only one nonzero eigenvalue).

The relation between P_Δ and S leads to interesting physical interpretations that have been analyzed by several authors like (Aiello and Woerdman 2005, Puentes et al. 2005, Pires and Monken 2008, Tariq et al. 2017) and Ossikovski and Vizet, 2019, where a comprehensive study of the (P_Δ, S) depolarization diagram and its relation to other depolarization metrics was presented. Both S and P_Δ (or D_Δ) constitute alternative measures of the overall depolarizing power of the system considered. The interdependence between S and P_Δ is complicated, in general; it will be described in Section 6.3 on the basis of the representation of the polarimetric purity through a set of three indices of purity, defined in a way similar to that of the pair of 3D indices of polarimetric purity for 3D polarization states (see Section 2.10.4).

The measurement of S as a representative quantity of the spatial inhomogeneity of a great variety of terrestrial and oceanic surface targets is commonly used in remote sensing, as well as in problems concerning light scattering, SAR (*synthetic aperture radar*) polarimetry, imaging polarimetry, lidar detection, etc. (Cloude 2009).

5.10.5 Lorentz Depolarization Indices

Some overall measures of depolarization, alternative to P_Δ (or D_Δ), have been defined to emphasize particular aspects of the depolarization properties of a medium.

Let us first consider two indices of depolarization whose definition is motivated by the fact that, as with the eigenvalues of the coherency matrix $\mathbf{C}(\mathbf{M})$, the eigenvalues $\rho_0, \rho_1, \rho_2, \rho_3$ (with $\rho_0 \ge \rho_i$) of the N-matrix $\mathbf{N} \equiv \mathbf{G}\mathbf{M}^T\mathbf{G}\mathbf{M}$ associated with the Mueller matrix **M** are non-negative.

As shown in Eqs. (5.21) and (5.23), the elements of the Mueller matrix $\mathbf{M}_\Delta$ of the central depolarizer of the normal form $\mathbf{M} = \mathbf{M}_{J2}\mathbf{M}_\Delta\mathbf{M}_{J1}$ of **M** (Eqs. (5.20) and (5.22)) are determined by $\sqrt{\rho_i}$. Note also that $d_i \equiv \sqrt{\rho_i}$ are the absolute values of the eigenvalues of the type-I canonical depolarizer $\mathbf{M}_{\Delta d}$ and that $(a_0, a_0, a_2, a_2) \equiv (\sqrt{\rho_0}, \sqrt{\rho_0}, \sqrt{\rho_2}, \sqrt{\rho_2})$ are precisely the eigenvalues of the type-II canonical depolarizer $\mathbf{M}_{\Delta nd}$.

The *first Lorentz depolarization index* is defined as (Ossikovski 2010b)

$$L_1(\mathbf{M}) \equiv \sqrt{(\rho_1 + \rho_2 + \rho_3)/3\rho_0} \tag{5.93}$$

so that L_1 is a measure of the *intrinsic* depolarizing power of the system represented by **M**, free from the potential diattenuating-polarizing effect of the pure components of the normal form.

In the case of a type-I Mueller matrix $\mathbf{M}_I$ the central depolarizer $\mathbf{M}_{\Delta d}$ is diagonal, so that L_1 is precisely the depolarization index P_Δ of $\mathbf{M}_{\Delta d}$, $L_1(\mathbf{M}_I) = P_\Delta(\mathbf{M}_{\Delta d})$

$$L_1(\mathbf{M}_I) = P_\Delta(\mathbf{M}_{\Delta d}) = \sqrt{(d_1^2 + d_2^2 + d_3^2)/3d_0^2} \tag{5.94}$$

and, thus, the range of $L_1(\mathbf{M}_I)$ runs from $L_1(\mathbf{M}_I)=0$ in which case $\mathbf{M}_{\Delta d}$ represents a perfect depolarizer, to $L_1(\mathbf{M}_I)=1$ in which case $\mathbf{M}_{\Delta d}=\mathbf{I}_4$ and $\mathbf{M}_I$ corresponds to a nondepolarizing system.

For a type-II Mueller matrix $\mathbf{M}_{II}$

$$L_1(\mathbf{M}_{II})=\sqrt{\left(a_0^2+a_2^2+a_2^2\right)/3a_0^2} \tag{5.95}$$

and, unlike $L_1(\mathbf{M}_I)$, $L_1(\mathbf{M}_{II})\neq P_\Delta(\mathbf{M}_{\Delta nd})$ because the central depolarizer $\mathbf{M}_{\Delta nd}$ of the normal form of a type-II Mueller matrix $\mathbf{M}_{II}$ retains certain amount of residual diattenuation-polarizance $D(\mathbf{M}_{\Delta nd})=P(\mathbf{M}_{\Delta nd})=1/2$. Observe that $1/\sqrt{3}\leq L_1(\mathbf{M}_{II})\leq 1$, where the lower limit is reached when $a_2=0$, whereas $L_1(\mathbf{M}_{II})=1$ when $a_2=a_0$.

Another depolarization measure is provided by the *second Lorentz depolarization index* (Ossikovski 2010b) defined as

$$L_2(\mathbf{M})\equiv\frac{1}{\sqrt{3}\sum_{i=0}^{3}\rho_i}\sqrt{4\sum_{i=0}^{3}\rho_i^2-\left(\sum_{i=0}^{3}\rho_i\right)^2} \tag{5.96}$$

which reaches its minimum $L_2(\mathbf{M})=0$ when $\mathbf{M}_\Delta$ is proportional to the identity matrix, in which case $\mathbf{M}$ is a pure Mueller matrix, and reaches its maximum $L_2(\mathbf{M})=1$ when $\mathbf{M}_\Delta=\mathrm{diag}(\rho_0,0,0,0)$, in which case $\mathbf{M}_\Delta$ represents a perfect depolarizer. The computational advantage of the second depolarization index over the first one is that it does not require the diagonalization of the N-matrix $\mathbf{N}\equiv\mathbf{G}\mathbf{M}^T\mathbf{G}\mathbf{M}$ but can be rather computed from the trace of $\mathbf{N}$ and of its square.

5.10.6 Other Overall Measures of Depolarization

5.10.6.1 Average Degree of Depolarization

The *average degree of depolarization* A_f (Chipman 2005) is defined by integrating the degree of polarization $\mathcal{P}_f$ of the emerging state of polarization for incident states (in the forward direction) over the entire surface of the Poincaré sphere and normalizing the result by $1/4\pi$ (i.e., by the area of the Poincaré sphere),

$$A_f(\mathbf{M})\equiv\frac{1}{\pi}\int_{\varphi=0}^{\varphi=\pi}\int_{\chi=-\pi/4}^{\chi=\pi/4}\mathcal{P}_f(\varphi,\chi)\cos 2\chi\,\mathrm{d}\varphi\,\mathrm{d}\chi \tag{5.97}$$

where

$$\mathcal{P}_f(\varphi,\chi)\equiv\mathcal{P}\left(\mathbf{s}^f\right)\qquad \mathbf{s}^f(\varphi,\chi)\equiv\mathbf{M}\left(1,\cos 2\varphi\cos 2\chi,\sin 2\varphi\cos 2\chi,\sin 2\chi\right)^T \tag{5.98}$$

A_f has a complicated expression in terms of the elements m_{kl} of $\mathbf{M}$. Moreover, while both measures P_Δ and A_f are defined from certain averages of the degree of polarization of the exiting states, the difference between these two quantities is particularly significant for systems whose structure is polarimetrically equivalent to an asymmetric serial arrangement of diattenuators and depolarizers. This is a direct consequence of the fact that, while P_Δ is an invariant quantity defined from a symmetric average involving emerging states for both forward and reverse directions of the incident electromagnetic beam, A_f only takes into account the effects of forward interaction.

Obviously, an *average degree of reverse depolarization* can be defined as $A_r(\mathbf{M}) \equiv A_f(\mathbf{M}^r) = A_f(\mathbf{M}^T)$, whose value is, in general, different from that of $A_f(\mathbf{M})$. An *overall average degree of depolarization A* can be defined as

$$A(\mathbf{M}) \equiv \frac{1}{2}\left[A_f(\mathbf{M}) + A_r(\mathbf{M})\right] \tag{5.99}$$

$A(\mathbf{M})$ is a positive parameter restricted by $0 \le A(\mathbf{M}) \le 1$, with $A(\mathbf{M}) = 1$ if and only if $\mathbf{M}$ is pure, and $A(\mathbf{M}) = 0$ for a perfect depolarizer. Nevertheless, despite the apparent similitude between $A(\mathbf{M})$ and $P_\Delta(\mathbf{M})$, it should be stressed that $A(\mathbf{M})$ (defined from the average of the degree of polarization of the emerging polarization states) differs from $P_\Delta(\mathbf{M})$ (defined from the average of the square of the degree of polarization of the emerging polarization states).

5.10.6.2 Depolarization Power

The *depolarization power* of a Mueller matrix $\mathbf{M}$ is defined as (Ossikovski 2010b)

$$\Delta \equiv 1 - \frac{\sqrt{\rho_1} + \sqrt{\rho_2} + \sqrt{\rho_3}}{3\sqrt{\rho_0}} \tag{5.100}$$

which is a generalization of the quantity (Lu and Chipman 1996)

$$\Delta = 1 - \frac{|d_1| + |d_2| + |d_3|}{3d_0} \tag{5.101}$$

defined originally for *intrinsic depolarizers* (also *diagonal depolarizers*) of the form $\mathbf{M}_d = \operatorname{diag}(d_0, a, b, c)$ or for *general depolarizers* with a symmetric Mueller matrix of the form

$$\mathbf{M}_{\Delta s} = \mathbf{M}_R^T\, \mathbf{M}_d\, \mathbf{M}_R \tag{5.102}$$

The depolarization power reaches its lower limit $\Delta = 0$ for pure Mueller matrices, while the maximum $\Delta = 1$ corresponds to an ideal depolarizer.

5.10.6.3 The Overall Purity Index

The so-called *overall purity index* (Tariq et al. 2017) *PI* is a measure of polarimetric purity defined as

$$PI \equiv \sqrt{P_1^2 + P_2^2 + P_3^2}\,/3 \tag{5.103}$$

in terms of the three *indices of polarimetric purity* (IPP) (San José and Gil 2011), P_1, P_2, P_3, which in turn are defined from the eigenvalues λ_i of the coherency matrix $\mathbf{C}(\mathbf{M})$ and that, as described in Section 6.3, constitute a minimum and complete set of parameters determining the quantitative polarimetric randomness of the medium represented by **M**. The particular features of PI, as well as its relation to other depolarization descriptors, have been studied by Tariq et al. (2017) and Ossikovski and Vizet (2019)

5.11 Constitutive Vectors of a Mueller Matrix

By considering the block form of a Mueller matrix and the expression of the retardance vector of a pure retarder in Eq. (4.20), let us write a generic Mueller matrix as

$$\mathbf{M} \equiv m_{00} \begin{pmatrix} 1 & D_1 & D_2 & D_3 \\ P_1 & k_1 & r_3 & r_2 \\ P_2 & q_3 & k_2 & r_1 \\ P_3 & q_2 & q_1 & k_3 \end{pmatrix} \tag{5.104}$$

in terms of the mean intensity coefficient m_{00} and the five vectors

$$\mathbf{D} \equiv \begin{pmatrix} D_1 \\ D_2 \\ D_3 \end{pmatrix} \quad \mathbf{P} \equiv \begin{pmatrix} P_1 \\ P_2 \\ P_3 \end{pmatrix} \quad \mathbf{k} \equiv \frac{1}{\sqrt{3}} \begin{pmatrix} k_1 \\ k_2 \\ k_3 \end{pmatrix} \quad \mathbf{r} \equiv \begin{pmatrix} r_1 \\ r_2 \\ r_3 \end{pmatrix} \quad \mathbf{q} \equiv \begin{pmatrix} q_1 \\ q_2 \\ q_3 \end{pmatrix} \tag{5.105}$$

whose respective absolute values D, P, k, r, q are limited by $D \le 1, P \le 1, k \le 1, r \le 1$, and $q \le 1$ (see Section 5.13.4).

D and **P** are the diattenuation and polarizance vectors, **k** is built from the diagonal elements of the submatrix **m**, and **r** and **q** are two complementary vectors defined from the off-diagonal elements of **m**. These five constitutive vectors are closely related to statistical structure of **M**, which will be analyzed in detail in Section 5.13 in the light of the so-called *arrow form* of **M**. Consequently, the above parameterization of **M** is advantageous for the analysis of some interesting properties of **M**, as is the case of its anisotropy coefficients, which will be considered in Section 6.7.

5.12 The Arrow Form of a Mueller Matrix

Even though the serial decompositions of depolarizing Mueller matrices are studied in Chapter 8, while the invariance properties of the dual-retarder transformations $\mathbf{M}_{Ra}\mathbf{M}\mathbf{M}_{Rb}$ of a Mueller matrix (where $\mathbf{M}_{Ra}$ and $\mathbf{M}_{Rb}$ represent respective retarders) are discussed in Section 6.4, the so-called *arrow form* of **M** is considered in this section because it provides both a meaningful view of the structure of the information contained in **M** and a general characterization of Mueller matrices linked to their statistical nature.

Consider the following transformation of the submatrix **m** of **M**, based on its singular value decomposition:

$$\mathbf{m} = \mathbf{m}_{RO}\,\mathbf{m}_A\,\mathbf{m}_{RI} \qquad \left[\begin{array}{ll} \mathbf{m}_{Ri}^{-1} = \mathbf{m}_{Ri}^{T} \quad \det \mathbf{m}_{Ri} = +1 & (i = I, O) \\ \mathbf{m}_A \equiv \operatorname{diag}(a_1, a_2, \varepsilon a_3) & \varepsilon \equiv \det \mathbf{m}/|\det \mathbf{m}| \end{array}\right] \tag{5.106}$$

where the non-negative parameters (a_1, a_2, a_3) are the singular values of **m**, so that the following orthogonal Mueller matrices (pure transparent retarders) are built as

$$\mathbf{M}_{Ri} = \begin{pmatrix} 1 & \mathbf{0}^T \\ \mathbf{0} & \mathbf{m}_{Ri} \end{pmatrix} \quad (i = I, O) \tag{5.107}$$

and the arrow matrix $\mathbf{M}_A(\mathbf{M})$ is then defined as

$$\mathbf{M}_A(\mathbf{M}) \equiv \mathbf{M}_{RO}^T \mathbf{M} \mathbf{M}_{RI}^T = m_{00} \begin{pmatrix} 1 & \mathbf{D}_A^T \\ \mathbf{P}_A & \mathbf{m}_A \end{pmatrix} \qquad \left[\begin{array}{l} \mathbf{m}_A \equiv \mathbf{m}_{RO}^T \mathbf{m} \mathbf{m}_{RI}^T = \operatorname{diag}(a_1, a_2, \varepsilon a_3) \\ a_1 \ge a_2 \ge a_3 \ge 0 \quad \varepsilon \equiv \det \mathbf{m}/|\det \mathbf{m}| \\ \mathbf{D}_A = \mathbf{m}_{RI} \mathbf{D} \quad \mathbf{P}_A = \mathbf{m}_{RO}^T \mathbf{P} \end{array}\right] \tag{5.108}$$

The *arrow decomposition* of **M** is formulated as

$$\mathbf{M} = \mathbf{M}_{RO}\,\mathbf{M}_A\,\mathbf{M}_{RI} \tag{5.109}$$

(see Figure 5.4). To avoid ambiguity, in the definition of $\mathbf{M}_A(\mathbf{M})$, the retarders $\mathbf{M}_{RI}$ and $\mathbf{M}_{RO}$ have been chosen so as to satisfy $a_1 \ge a_2 \ge a_3$ (with, $1 \ge a_1 \ge a_2 \ge a_3 \ge 0$) with the sign of εa_3 equal to the sign of $\det \mathbf{m}$ (thus ensuring that $\det \mathbf{M}_{RI} = \det \mathbf{M}_{RO} = +1$, as required for $\mathbf{M}_{RI}$ and $\mathbf{M}_{RO}$ to represent Mueller matrices of retarders).

The diattenuation and polarizance vectors of **M** are recovered from those of $\mathbf{M}_A$ through the respective rotations $\mathbf{D}_A = \mathbf{m}_{RI}^T \mathbf{D}_A$ and $\mathbf{P} = \mathbf{m}_{RO} \mathbf{P}_A$, which are directly determined from the entrance and exit retarders $\mathbf{M}_{RI}$ and $\mathbf{M}_{RO}$ of **M**. Note that such rotations are performed with respect to the abstract Poincaré coordinate system.

The constitutive vectors **r** and **q**, defined in Section 5.11 for general Mueller matrices, vanish for $\mathbf{M}_A(\mathbf{M})$. Moreover, the polarimetric dimension index of $\mathbf{M}_A(\mathbf{M})$, defined in Section 6.2, is given by $P_S(\mathbf{M}) = P_S(\mathbf{M}_A) = \sqrt{(a_1^2 + a_2^2 + a_3^2)/3}$. Some interesting and genuine properties of $\mathbf{M}_A(\mathbf{M})$ are studied in Section 8.6.

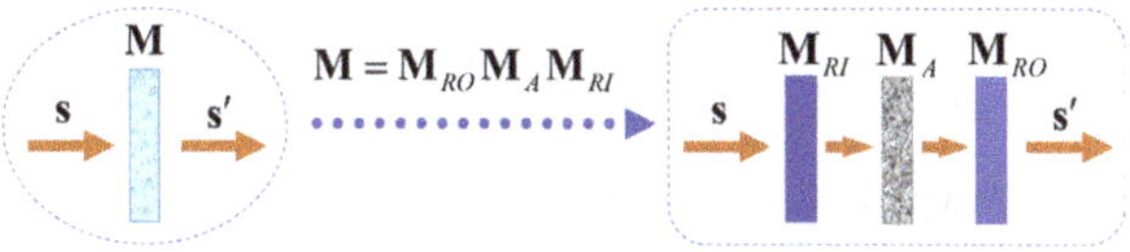

FIGURE 5.4
Arrow decomposition: The effect of a Mueller matrix **M** is equivalent to the consecutive actions of the *entrance retarder* $\mathbf{M}_{RI}$, the *arrow form* $\mathbf{M}_A(\mathbf{M})$ and the *exit retarder* $\mathbf{M}_{RO}$.

5.13 Characterization of Mueller Matrices Through the Arrow Form

As seen in the previous sections, the structure of Mueller matrices relies on the non-negativity of the four eigenvalues of the coherency matrix **C** associated with the Mueller matrix **M**. As a result of the need for a solution to an eigenvalue problem, the mathematical characterization of Mueller matrices through explicit analytic expressions becomes very intricate. Therefore, the problem of obtaining relatively simple analytic formulations for the characteristic properties of Mueller matrices, clearly revealing the general structural features of Mueller matrices, becomes relevant to the theoretical corpus of polarimetry and its wide scope of applications. The arrow form provides the framework for solving the indicated problem through specific procedures shown throughout this section.

5.13.1 Covariance Characterization of Mueller Matrices

As will be shown in Section 6.4.1, dual-retarder transformations of a Mueller matrix **M** (among which $\mathbf{M}_A = \mathbf{M}_{RI}^T \mathbf{M} \mathbf{M}_{RO}^T$ is a particular type) preserve all the characteristic properties of **M** as being a Mueller matrix (i.e., both the covariance conditions and the passivity conditions). Consequently, the arrow form $\mathbf{M}_A(\mathbf{M})$ is a Mueller matrix if and only if **M** is a Mueller matrix, and therefore the conditions for **M** to be a Mueller matrix can be checked through the corresponding conditions for $\mathbf{M}_A(\mathbf{M})$ (Gil and San José 2014).

Furthermore, **M** satisfies the covariance conditions if and only if the coherency matrix $\mathbf{C}_A(\mathbf{M}) \equiv \mathbf{C}(\mathbf{M}_A)$ associated with $\mathbf{M}_A(\mathbf{M})$

$$\mathbf{C}_A(\mathbf{M}) = m_{00}\frac{1}{4}\begin{pmatrix} 4c_0 & D_{A1}+P_{A1} & D_{A2}+P_{A2} & D_{A3}+P_{A3} \\ D_{A1}+P_{A1} & 4c_1 & i(D_{A3}-P_{A3}) & -i(D_{A2}-P_{A2}) \\ D_{A2}+P_{A2} & -i(D_{A3}-P_{A3}) & 4c_2 & i(D_{A1}-P_{A1}) \\ D_{A3}+P_{A3} & i(D_{A2}-P_{A2}) & -i(D_{A1}-P_{A1}) & 4c_3 \end{pmatrix}$$

$$\begin{bmatrix} 4c_0 \equiv 1+a_1+a_2+\varepsilon a_3 & 4c_1 \equiv 1+a_1-a_2-\varepsilon a_3 \\ 4c_2 \equiv 1-a_1+a_2-\varepsilon a_3 & 4c_3 \equiv 1-a_1-a_2+\varepsilon a_3 \end{bmatrix} \quad \left[\varepsilon \equiv \frac{\det \mathbf{m}}{|\det \mathbf{m}|}\right] \tag{5.110}$$

is positive semidefinite, which in turn holds if and only if the four leading (nested) principal minors of $\mathbf{C}_A(\mathbf{M})$ are non-negative (see Section 5.6). These conditions constitute an alternative set of constraints that is equivalent to the non-negativity of the four eigenvalues of $\mathbf{C}_A(\mathbf{M})$, but formulated through the following set of four analytic and much more concise expressions in terms of the physical parameters of $\mathbf{M}_A(\mathbf{M})$ (Gil and San José 2014):

$$\begin{aligned}
& m_{00} \geq 0 \\
& 16c_0c_1 \geq (D_{A1}+P_{A1})^2 \\
& 16c_0c_1c_2 \geq c_2(D_{A1}+P_{A1})^2 + c_1(D_{A2}+P_{A2})^2 + c_3(D_{A3}-P_{A3})^2 \\
& 16c_0c_1c_2c_3 + \frac{1}{16}(D^2-P^2)^2 \geq c_0c_1(D_{A1}-P_{A1})^2 + c_2c_3(D_{A1}+P_{A1})^2 \\
& \qquad + c_0c_2(D_{A2}-P_{A2})^2 + c_1c_3(D_{A2}+P_{A2})^2 \\
& \qquad + c_0c_3(D_{A3}-P_{A3})^2 + c_1c_2(D_{A3}+P_{A3})^2
\end{aligned} \tag{5.111}$$

The first condition, $m_{00} \geq 0$, is obviously very well-known, but it is remarkable that it appears here as one of the set of four covariance conditions for a 4×4 real matrix to be a Mueller matrix.

In accordance with the reciprocity properties of Mueller matrices, the above characteristic inequalities are equally valid for the transposed matrix $\mathbf{M}^T$, and thus they reflect the fact that, given a Mueller matrix $\mathbf{M}$, $\mathbf{M}^T$ is also a Mueller matrix.

5.13.2 Statistical Parameterization of the Arrow Form

An alternative way to express explicitly the covariance conditions is achieved by noticing that $\mathbf{C}_A$ is a positive semidefinite Hermitian matrix and observing that certain off-diagonal elements of $\mathbf{C}_A$ in Eq. (5.110) are real, while other are pure imaginary. As a result, $\mathbf{C}_A$ can always be expressed through the following statistical parameterization in terms of the four equivalent *intrinsic variances* σ_i^2 and the six equivalent real *intrinsic correlation parameters* $\mu_{01}, \mu_{02}, \mu_{03}, \nu_{23}, \nu_{13}, \nu_{12}$ (Gil and San José, 2022):

$$\mathbf{C}_A = \begin{pmatrix} \sigma_0^2 & \mu_{01}\sigma_0\sigma_1 & \mu_{02}\sigma_0\sigma_2 & \mu_{03}\sigma_0\sigma_3 \\ \mu_{01}\sigma_0\sigma_1 & \sigma_1^2 & i\nu_{12}\sigma_1\sigma_2 & -i\nu_{13}\sigma_1\sigma_3 \\ \mu_{02}\sigma_0\sigma_2 & -i\nu_{12}\sigma_1\sigma_2 & \sigma_2^2 & i\nu_{23}\sigma_2\sigma_3 \\ \mu_{03}\sigma_0\sigma_3 & i\nu_{13}\sigma_1\sigma_3 & -i\nu_{23}\sigma_2\sigma_3 & \sigma_3^2 \end{pmatrix} \quad [\sigma_k \geq 0] \tag{5.112}$$

where σ_k have been taken as non-negative without loss of generality, while, by construction, the absolute values of the correlation parameters are less than 1. Furthermore, since $m_{00} = \operatorname{tr}\mathbf{C}_A$, the passivity condition implies that $\sigma_0^2 + \sigma_1^2 + \sigma_2^2 + \sigma_3^2 \leq 1/(1+Q)$, where $Q = \max(D, P)$.

To get a detailed view of the structure of the statistical information supported by the coherency matrix $\mathbf{C}_A$, let us consider its following factorization where σ_i are decoupled from the correlation parameters:

$$\mathbf{C}_A = \boldsymbol{\Sigma}\boldsymbol{\Gamma}\boldsymbol{\Sigma} \quad \boldsymbol{\Sigma} \equiv 2\begin{pmatrix} \sigma_0 & 0 & 0 & 0 \\ 0 & \sigma_1 & 0 & 0 \\ 0 & 0 & \sigma_2 & 0 \\ 0 & 0 & 0 & \sigma_3 \end{pmatrix} \quad \boldsymbol{\Gamma} \equiv \frac{1}{4}\begin{pmatrix} 1 & \mu_{01} & \mu_{02} & \mu_{03} \\ \mu_{01} & 1 & i\nu_{12} & -i\nu_{13} \\ \mu_{02} & -i\nu_{12} & 1 & i\nu_{23} \\ \mu_{03} & i\nu_{13} & -i\nu_{23} & 1 \end{pmatrix} \tag{5.113}$$

The respective coefficients 2 and 1/4 in the definitions of $\boldsymbol{\Sigma}$ and $\boldsymbol{\Gamma}$ have been chosen so as to ensure that $\boldsymbol{\Gamma}$, being proportional to a correlation matrix, also takes the form of a particular type of coherency matrix with an associated normalized Mueller matrix $\hat{\mathbf{M}}_\Gamma$.

It can be readily seen that the convention taken for the ordering of the diagonal elements of $\mathbf{M}_A$ $(a_3 \leq a_2 \leq a_1)$ leads to $\sigma_3 \leq \sigma_2 \leq \sigma_1 \leq \sigma_0$, and vice versa. When $\sigma_3 = 0$, the correlation parameters μ_{03}, ν_{23} and ν_{13} are undetermined and can be taken to be zero-valued $\mu_{03} = \nu_{23} = \nu_{13} = 0$ without loss of generality. Likewise, when $\sigma_3 = \sigma_2 = 0$, we adopt the convention that all but μ_{01} correlation parameters are zero. In addition, since the number of zero-valued diagonal elements of matrix $\boldsymbol{\Sigma}$ determines the corresponding *active submatrix* of $\boldsymbol{\Gamma}$, we define it as in Eq. (5.113) for the case where $\sigma_3 > 0$ (i.e., $\det\boldsymbol{\Sigma} > 0$, in which case, wherever appropriate, $\boldsymbol{\Gamma}$ is denoted as $\boldsymbol{\Gamma}_4$). In the remaining cases, $\boldsymbol{\Gamma}$ is defined as

$$\overbrace{\mathbf{\Gamma}\equiv\mathbf{\Gamma}_3\equiv\frac{1}{4}\begin{pmatrix}1 & \mu_{01} & \mu_{02} & 0\\ \mu_{01} & 1 & i\nu_{12} & 0\\ \mu_{02} & -i\nu_{12} & 1 & 0\\ 0 & 0 & 0 & 0\end{pmatrix}}^{\sigma_3=0,\ \sigma_2>0},\quad \overbrace{\mathbf{\Gamma}\equiv\mathbf{\Gamma}_2\equiv\frac{1}{4}\begin{pmatrix}1 & \mu_{01} & 0 & 0\\ \mu_{01} & 1 & 0 & 0\\ 0 & 0 & 0 & 0\\ 0 & 0 & 0 & 0\end{pmatrix}}^{\sigma_2=0,\ \sigma_1>0},\quad \overbrace{\mathbf{\Gamma}\equiv\mathbf{\Gamma}_1\equiv\frac{1}{4}\begin{pmatrix}1 & 0 & 0 & 0\\ 0 & 0 & 0 & 0\\ 0 & 0 & 0 & 0\\ 0 & 0 & 0 & 0\end{pmatrix}}^{\sigma_1=0,\ \sigma_0>0} \tag{5.114}$$

The above convention for the definition of $\mathbf{\Gamma}$ implies that the equality $\operatorname{rank}\mathbf{\Gamma}=\operatorname{rank}\mathbf{C}_A$ is always satisfied (recall that and $\operatorname{rank}\mathbf{C}_A=\operatorname{rank}\mathbf{C}$). The physical significance of the integer parameter $\operatorname{rank}\mathbf{C}$ is that it equals the minimal number of parallel pure components of $\mathbf{M}$ (see Section 7.3, Gil and San José 2022).

From the expression of $\mathbf{C}_A$ in terms of the corresponding $\mathbf{M}_A$, it follows that the diagonal elements $c_k=\sigma_k^2$ of $\mathbf{C}_A$ are given by the following non-negative quantities, which in turn are determined from the diagonal elements of $\mathbf{M}_A$:

$$\begin{aligned}\sigma_0^2&=\operatorname{tr}(\mathbf{M}_A\mathbf{M}_{Rd0})=m_{00}(1+a_1+a_2+\varepsilon a_3)/4\\ \sigma_1^2&=\operatorname{tr}(\mathbf{M}_A\mathbf{M}_{Rd1})=m_{00}(1+a_1-a_2-\varepsilon a_3)/4\\ \sigma_2^2&=\operatorname{tr}(\mathbf{M}_A\mathbf{M}_{Rd2})=m_{00}(1-a_1+a_2-\varepsilon a_3)/4\\ \sigma_3^2&=\operatorname{tr}(\mathbf{M}_A\mathbf{M}_{Rd3})=m_{00}(1-a_1-a_2+\varepsilon a_3)/4\end{aligned}\quad\begin{bmatrix}\mathbf{M}_{Rd0}\equiv\operatorname{diag}(1,1,1,1)\\ \mathbf{M}_{Rd1}\equiv\operatorname{diag}(1,1,-1,-1)\\ \mathbf{M}_{Rd2}\equiv\operatorname{diag}(1,-1,1,-1)\\ \mathbf{M}_{Rd3}\equiv\operatorname{diag}(1,-1,-1,1)\end{bmatrix} \tag{5.115}$$

where $\mathbf{M}_{Rd0}$ is the identity matrix. The other three diagonal matrices $\mathbf{M}_{Rdi}\,(i=1,2,3)$ represent retarders with $\Delta=\pi$ and respectively, linear-horizontal and linear-vertical eigenstates ($\mathbf{M}_{Rd1}$), linear eigenstates at ±45° ($\mathbf{M}_{Rd2}$) and circular eigenstates ($\mathbf{M}_{Rd3}$) (see Section 4.2.2).

Note that, as shown in Section 6.4.2, $\sigma_0^2\le\lambda_0$ and $\sigma_3^2\ge\lambda_3$, where λ_0 and λ_3 are the largest and the smallest eigenvalues of $\mathbf{C}_A$ respectively, so that the maximal values for the differences $\lambda_0-\sigma_0^2$, $\lambda_3-\sigma_3^2$ are reached when $\sigma_0^2=\sigma_1^2=\sigma_2^2=\sigma_3^2$. Observe also that in the limiting case of a perfect depolarizer ($\mathbf{M}_{\Delta 0}=\operatorname{diag}(1,0,0,0)$), $\lambda_k=\sigma_k^2\ (k=0,1,2,3)$.

Since the matrices $\mathbf{\Sigma}$ and $\mathbf{\Gamma}$ are formally similar to typical statistical matrices of standard deviations ($\mathbf{\Sigma}$) and correlations ($\mathbf{\Gamma}$), hereafter they will be called the *intrinsic standard deviations matrix*, and the *intrinsic correlation matrix* of $\mathbf{M}$, respectively. The Mueller matrix $\hat{\mathbf{M}}_\Gamma$ associated with $\mathbf{\Gamma}$ (which is obviously different from $\mathbf{M}_A$) will be termed the *correlation Mueller matrix* (CMM) of both $\mathbf{M}_A$ and $\mathbf{M}$ ($\mathbf{M}$ being the Mueller matrix whose arrow form is $\mathbf{M}_A$).

The mathematical characterization of $\mathbf{\Gamma}$, defined by one of the forms shown in Eqs. (5.113) and (5.114), is determined by the following conditions: (a) the absolute values of the correlation parameters are less than 1 (by the construction of $\mathbf{\Gamma}$), and (b) the eigenvalues of $\mathbf{\Gamma}$ are non-negative (because $\mathbf{\Gamma}$ has the mathematical form of a particular type of covariance matrix). The eigenvalues $\gamma_i\,(i=0,1,2,3)$ of $\mathbf{\Gamma}$ are given by (Gil and San José 2022)

$$\mathbf{\Gamma}_4\begin{cases}\gamma_0=(1+D_\Gamma+P_\Gamma)/4\\ \gamma_1=(1+D_\Gamma-P_\Gamma)/4\\ \gamma_2=(1-D_\Gamma+P_\Gamma)/4\\ \gamma_3=(1-D_\Gamma-P_\Gamma)/4\end{cases}\quad\begin{bmatrix}D_\Gamma\equiv|\mathbf{D}_\Gamma|\quad P_\Gamma\equiv|\mathbf{P}_\Gamma|\\ \mathbf{D}_\Gamma\equiv(\boldsymbol{\mu}+\boldsymbol{\nu})/2\quad \mathbf{P}_\Gamma\equiv(\boldsymbol{\mu}-\boldsymbol{\nu})/2\\ \boldsymbol{\mu}\equiv(\mu_{01},\mu_{02},\mu_{03})^T\quad \boldsymbol{\nu}\equiv(\nu_{23},\nu_{13},\nu_{12})^T\\ (\gamma_3\le\gamma_2\le\gamma_1\le\gamma_0,\ \text{or}\ \gamma_3\le\gamma_1\le\gamma_2\le\gamma_0)\end{bmatrix}$$

$$\mathbf{\Gamma}_3\begin{cases}\gamma_0=(1+2D_\Gamma)/4\\ \gamma_1=1/4\\ \gamma_2=(1-2D_\Gamma)/4\\ \gamma_3=0\end{cases}\quad\begin{bmatrix}D_\Gamma\equiv|\mathbf{D}_\Gamma|=|\mathbf{P}_\Gamma|=|\boldsymbol{\mu}+\boldsymbol{\nu}|/2=|\boldsymbol{\mu}-\boldsymbol{\nu}|/2\\ \boldsymbol{\mu}=(\mu_{01},\mu_{02},0)^T\quad \boldsymbol{\nu}=(0,0,\nu_{12})^T\\ (\gamma_2\le\gamma_1\le\gamma_0)\end{bmatrix}$$

$$\mathbf{\Gamma}_2\begin{cases}\gamma_0=(1+2D_\Gamma)/4\\ \gamma_1=(1-2D_\Gamma)/4\\ \gamma_2=0\\ \gamma_3=0\end{cases}\quad\begin{bmatrix}D_\Gamma=P_\Gamma=|\mu_{01}|/2\quad \boldsymbol{\mu}=(\mu_{01},0,0)^T\quad \boldsymbol{\nu}=(0,0,0)^T\\ (\gamma_1\le\gamma_0)\end{bmatrix}\tag{5.116}$$

$$\mathbf{\Gamma}_1\;\{\gamma_0=1/4,\;\gamma_1=\gamma_2=\gamma_3=0$$

Leaving aside the trivial case of $\mathbf{\Gamma}_1$, for each respective $\mathbf{\Gamma}_i$ $(i=2,3,4)$ the necessary and sufficient conditions for the non-negativity of γ_i are the following: $\mathbf{\Gamma}_4$: $D_\Gamma+P_\Gamma\le 1$; $\mathbf{\Gamma}_3$: $\psi\le 1$, with $\psi\equiv|\boldsymbol{\psi}|\equiv|(\mu_{01},\mu_{02},\nu_{12})^T|$; $\mathbf{\Gamma}_2$: $|\mu_{01}|\le 1$

The CMM, $\hat{\mathbf{M}}_\Gamma$, associated with $\mathbf{M}_A$ (and with $\mathbf{M}$) has the simple form

$$\hat{\mathbf{M}}_\Gamma=\begin{pmatrix}1 & \mathbf{D}_\Gamma^T\\ \mathbf{P}_\Gamma & \mathbf{0}\end{pmatrix}\tag{5.117}$$

so that $\mathbf{D}_\Gamma=(\boldsymbol{\mu}+\boldsymbol{\nu})/2$ and $\mathbf{P}_\Gamma=(\boldsymbol{\mu}-\boldsymbol{\nu})/2$ are the diattenuation and polarizance vectors of $\hat{\mathbf{M}}_\Gamma$ respectively, while $Ps(\hat{\mathbf{M}}_\Gamma)=0$. Therefore, the degree of polarimetric purity of $\hat{\mathbf{M}}_\Gamma$ is determined by $P_\Delta^2(\hat{\mathbf{M}}_\Gamma)=(\mathbf{D}_\Gamma^2+\mathbf{P}_\Gamma^2)/3$, which implies that $P_\Delta^2(\hat{\mathbf{M}}_\Gamma)\le 1/3$, the limit value $P_\Delta^2(\hat{\mathbf{M}}_\Gamma)=1/3$ being reached when either $D_\Gamma=1$ and $P_\Gamma=0$, or $D_\Gamma=0$ and $P_\Gamma=1$ (because of the condition $D_\Gamma+P_\Gamma\le 1$).

The corresponding expression of $\mathbf{M}_A$ in terms of the statistical parameters is

$$\mathbf{M}_A=\begin{pmatrix}\sigma_0^2+\sigma_1^2+\sigma_2^2+\sigma_3^2 & 2\sigma_0\sigma_1\mu_{01}+2\sigma_2\sigma_3\nu_{23} & 2\sigma_0\sigma_2\mu_{02}+2\sigma_1\sigma_3\nu_{13} & 2\sigma_0\sigma_3\mu_{03}+2\sigma_1\sigma_2\nu_{12}\\ 2\sigma_0\sigma_1\mu_{01}-2\sigma_2\sigma_3\nu_{23} & \sigma_0^2+\sigma_1^2-\sigma_2^2-\sigma_3^2 & 0 & 0\\ 2\sigma_0\sigma_2\mu_{02}-2\sigma_1\sigma_3\nu_{13} & 0 & \sigma_0^2-\sigma_1^2+\sigma_2^2-\sigma_3^2 & 0\\ 2\sigma_0\sigma_3\mu_{03}-2\sigma_1\sigma_2\nu_{12} & 0 & 0 & \sigma_0^2-\sigma_1^2-\sigma_2^2+\sigma_3^2\end{pmatrix}\tag{5.118}$$

In spite of the obvious fact that $\hat{\mathbf{M}}_\Gamma\ne\hat{\mathbf{M}}_A$ and that arrow Mueller matrices with different associated $\mathbf{\Sigma}$ can share the same associated $\hat{\mathbf{M}}_\Gamma$, $\hat{\mathbf{M}}_\Gamma$ encodes all information on the two *intrinsic correlation vectors*, $\boldsymbol{\mu}$ and $\boldsymbol{\nu}$, while the remaining information on the four σ_k is contained in the matrix $\mathbf{\Sigma}$ in a decoupled and exclusive manner.

It is remarkable that the degree of polarizance $P_P(\mathbf{M})$ of $\mathbf{M}$ (which always satisfies $P_P(\mathbf{M})=P_P(\mathbf{M}_A)$) is zero if and only if $P_P(\hat{\mathbf{M}}_\Gamma)=0$, showing how the information on $P_P(\mathbf{M})$ and $P_S(\mathbf{M})$ encoded in $\mathbf{M}_A$ is decoupled in a peculiar manner through the transformation $\mathbf{C}_A=\mathbf{\Sigma}\mathbf{\Gamma}\mathbf{\Sigma}$ in Eq. (5.113).

5.13.3 Characterization of Mueller Matrices Through the Correlation Criterion

The powerful decoupling features of the statistical approach formulated above allows for a succinct general characterization of Mueller matrices, which can be stated as follows (Gil and San José 2022):

Characterization theorem: A given 4×4 real matrix **M** is a Mueller matrix if and only if the following conditions hold:

(1) *Variance conditions.* $\text{tr}(\mathbf{M}_A \mathbf{M}_{Rdi}) \geq 0$ ($i = 0,1,2,3$, see Eq. (5.115)).
(2) *Correlation condition.* $\hat{\mathbf{M}}_\Gamma(\mathbf{M})$ satisfies $D_\Gamma + P_\Gamma \leq 1$, where D_Γ and P_Γ are the diattenuation and polarizance of $\hat{\mathbf{M}}_\Gamma$.
(3) *Passivity condition.* $m_{00}(1+Q) \leq 1$, where $Q \equiv \max(D,P)$ (see Section 5.9).

It is remarkable that the four covariance conditions established by Cloude through the non-negativity of the eigenvalues of $\mathbf{C}(\mathbf{M})$ (Cloude 1986) are fully equivalent to the set of five conditions composed of the combination of the four variance inequalities and the correlation condition. Even though the latter set involves five conditions (plus the passivity one) instead of the four Cloude's conditions (the non-negativity of the four eigenvalues of $\mathbf{C}$ or $\mathbf{C}_A$), the advantage of the above approach is that the characterization is made in an explicit and simple manner in terms of intrinsic properties that are directly expressed through the Mueller matrices $\mathbf{M}_A$ and $\hat{\mathbf{M}}_\Gamma$ associated with **M** (Gil and San José 2022).

Note that condition $D_\Gamma + P_\Gamma \leq 1$ can also be expressed as $|\boldsymbol{\mu}+\mathbf{v}| + |\boldsymbol{\mu}-\mathbf{v}| \leq 2$, while other required inequalities are directly satisfied because of the very definitions of the matrices $\boldsymbol{\Sigma}$ and $\boldsymbol{\Gamma}$.

Therefore, any Mueller matrix **M** can be synthesized through the following steps:

(1) Take four arbitrary non-negative parameters $\sigma_0 \geq \sigma_1 \geq \sigma_2 \geq \sigma_3 \geq 0$ and build the matrix $\boldsymbol{\Sigma} \equiv 2\,\text{diag}(\sigma_0, \sigma_1, \sigma_2, \sigma_3)$;
(2) take a pair of arbitrary real 3D vectors $\boldsymbol{\mu}$ and $\mathbf{v}$ that satisfy the inequalities $|\boldsymbol{\mu}| \leq 1$, $|\mathbf{v}| \leq 1$, and $|\boldsymbol{\mu}+\mathbf{v}|/2 + |\boldsymbol{\mu}-\mathbf{v}|/2 \leq 1$;
(3) calculate the corresponding arrow Mueller matrix $\mathbf{M}_A(\boldsymbol{\Sigma},\boldsymbol{\mu},\mathbf{v})$ by means of Eq. (5.118);
(4) assign a physically realizable MIC $m_{00} \leq 1/(1+Q)$ ($Q \equiv \min(D,P)$) to $\mathbf{M}_A = m_{00}\hat{\mathbf{M}}_A$;
(5) take two Mueller matrices $\mathbf{M}_{RI}$ and $\mathbf{M}_{RO}$ representing (arbitrary) retarders; and
(6) build the corresponding Mueller matrix through the dual-retarder transformation $\mathbf{M} = \mathbf{M}_{RO}\mathbf{M}_A\mathbf{M}_{RI}$.

The inspection of the particular forms of $\boldsymbol{\Gamma}$ in Eqs. (5.113, 5.114) together with the explicit expressions of their eigenvalues in Eq. (5.116) is sufficient to analyze the achievable values of $r \equiv \text{rank}\boldsymbol{\Gamma} = \text{rank}\,\mathbf{C}_A = \text{rank}\,\mathbf{C}$,

$$2 \leq \text{rank}\,\boldsymbol{\Gamma}_4 \leq 4 \qquad 2 \leq \text{rank}\,\boldsymbol{\Gamma}_3 \leq 3 \qquad 1 \leq \text{rank}\,\boldsymbol{\Gamma}_2 \leq 2 \qquad \text{rank}\,\boldsymbol{\Gamma}_1 = 1 \tag{5.119}$$

so that the following cases can be distinguished:

$r = 4$. All variances are nonzero ($\sigma_3 > 0$, i.e., $\det\boldsymbol{\Sigma} > 0$): The intrinsic correlation matrix necessarily takes the form $\boldsymbol{\Gamma} = \boldsymbol{\Gamma}_4$ with $D_\Gamma + P_\Gamma < 1$.

$r = 3$. One of the following two alternatives holds:

(a) $\sigma_3 > 0$: $\boldsymbol{\Gamma} = \boldsymbol{\Gamma}_4$ with $D_\Gamma + P_\Gamma = 1$ and $0 < D_\Gamma < 1$.
(b) $\sigma_3 = 0$ and $\sigma_2 > 0$: $\boldsymbol{\Gamma} = \boldsymbol{\Gamma}_3$ with $|\boldsymbol{\psi}| < 1$, $[\boldsymbol{\psi} \equiv (\mu_{01}, \mu_{02}, \nu_{12})^T]$.

$r = 2$. One of the following three alternatives holds:

(a) $\sigma_3 > 0$: $\mathbf{\Gamma}=\mathbf{\Gamma}_4$ with $\mathbf{v}=\pm\boldsymbol{\mu}$ and $|\boldsymbol{\mu}|=1$, so that the eigenvalues of $\mathbf{\Gamma}_4$ are given by $\gamma_0=\gamma_1=1/2$ and $\gamma_2=\gamma_3=0$.
(b) $\sigma_3 = 0$ and $\sigma_2 > 0$: $\mathbf{\Gamma}=\mathbf{\Gamma}_3$ with $|\boldsymbol{\psi}|=1$, $[\boldsymbol{\psi}\equiv(\mu_{01},\mu_{02},\nu_{12})^T]$.
(c) $\sigma_3=\sigma_2=0$ and $\sigma_1>0$: $\mathbf{\Gamma}=\mathbf{\Gamma}_2$ and $\mu_{01}^2<1$.

$r = 1$. One of the following two alternatives holds:

(a) $\sigma_2=0$ and $\sigma_1>0$: $\mathbf{\Gamma}=\mathbf{\Gamma}_2$ with $\mu_{01}^2=1$. $\mathbf{M}_A$ has the form of the canonical horizontal diattenuator $\mathbf{M}_{DL0}(m_{00},D)$ given by Eq. (4.80) and consequently, $\mathbf{M}$ is a pure Mueller matrix with nonzero diattenuation.
(b) $\sigma_1=0$ and $\sigma_0>0$: $\mathbf{\Gamma}=\mathbf{\Gamma}_1$. $\hat{\mathbf{M}}_A$ is the identity matrix, so that $\mathbf{M}$ has the form of a pure attenuating retarder $m_{00}\,\mathbf{M}_R$

With regard to the arrow form $\mathbf{M}_A$ of a general Mueller matrix $\mathbf{M}$ in Eq. (5.118), the nature of the pair of vectors ($\boldsymbol{\mu}$,$\mathbf{v}$) is closely related to the dichroism-type anisotropy of the medium (i.e., enpolarizing properties encoded into the diattenuation and polarizance vectors of $\mathbf{M}_A$ and $\mathbf{M}$); see Section 6.7 dealing with the anisotropy coefficients of a Mueller matrix. Nonenpolarizing media are characterized by $\boldsymbol{\mu}=\mathbf{v}=\mathbf{0}$.

Furthermore, all birefringence anisotropy parameters are zero-valued for $\mathbf{M}_A$, which suggests that the dual-retarder transformation from $\mathbf{M}$ to $\mathbf{M}_A$ removes all birefringence-type anisotropy, allocating it to the entrance and exit retarders $\mathbf{M}_{RI}$ and $\mathbf{M}_{RO}$. Thus, except for certain particular cases, it is not correct to attribute the birefringence properties of a depolarizing medium to a single retarding serial component.

5.13.4 Structure of the Polarimetric Information Contained in a General Mueller Matrix

Once the statistical structure of the coherency matrix $\mathbf{C}_A$ associated with a given arrow Mueller matrix $\mathbf{M}_A$ has been studied and characterized through the matrix $\mathbf{\Sigma}$ together with the intrinsic correlation vectors $\boldsymbol{\mu}$ and $\mathbf{v}$, it is worth reconsidering the statistical structure of the coherency matrix $\mathbf{C}$ associated with a general Mueller matrix $\mathbf{M}$, which can always be expressed as (Gil and San José 2022)

$$\mathbf{C}=\mathbf{\Sigma}\mathbf{\Omega}\mathbf{\Sigma}$$

$$\mathbf{\Sigma}\equiv 2\begin{pmatrix}\sigma_0 & 0 & 0 & 0\\ 0 & \sigma_1 & 0 & 0\\ 0 & 0 & \sigma_2 & 0\\ 0 & 0 & 0 & \sigma_3\end{pmatrix}\quad \mathbf{\Omega}\equiv\frac{1}{4}\begin{pmatrix}1 & \mu_{01}+i\,\eta_{01} & \mu_{02}-i\,\eta_{02} & \mu_{03}+i\,\eta_{03}\\ \mu_{01}-i\,\eta_{01} & 1 & \tau_{12}+i\,\nu_{12} & \tau_{13}-i\,\nu_{13}\\ \mu_{02}+i\,\eta_{02} & \tau_{12}-i\,\nu_{12} & 1 & \tau_{23}+i\,\nu_{23}\\ \mu_{03}-i\,\eta_{03} & \tau_{13}+i\,\nu_{13} & \tau_{23}-i\,\nu_{23} & 1\end{pmatrix} \tag{5.120}$$

$$\left[\sigma_k,\mu_{0k},\nu_{lk},\eta_{0k},\tau_{lk}\in\mathbb{R}\quad \sigma_k\geq 0\quad l,k=1,2,3\quad l<k\right]$$

In general, σ_k not necessarily satisfy the order $\sigma_0\geq\sigma_1\geq\sigma_2\geq\sigma_3$ taken for the canonical arrow form, and there arise two additional correlation real vectors, namely $\boldsymbol{\eta}\equiv(\eta_{01},\eta_{02},\eta_{03})^T$ and $\boldsymbol{\tau}\equiv(\tau_{23},\tau_{13},\tau_{12})^T$.

It should also kept in mind that, despite the fact that, for simplicity, the same notation for the statistical parameters of **C** and $\mathbf{C}_A$ has been used, $\boldsymbol{\Sigma}$, $\boldsymbol{\mu}$ and $\mathbf{v}$ of **M** are different from those of $\mathbf{M}_A$. As with the matrix $\boldsymbol{\Gamma}$, and without loss of generality, the convention that when one or more σ_k are zero-valued the corresponding k rows and columns of Ω are zero is adopted.

The role played by σ_k and the correlation vectors $\boldsymbol{\mu}$, $\mathbf{v}$, $\boldsymbol{\eta}$, and $\boldsymbol{\tau}$ in the information structure of **M**, becomes explicit when **M** is expressed in terms of them

$$\mathbf{M}=\begin{pmatrix} \sigma_0^2+\sigma_1^2+\sigma_2^2+\sigma_3^2 & 2\sigma_0\sigma_1\mu_{01}+2\sigma_2\sigma_3\nu_{23} & 2\sigma_0\sigma_2\mu_{02}+2\sigma_1\sigma_3\nu_{13} & 2\sigma_0\sigma_3\mu_{03}+2\sigma_1\sigma_2\nu_{12} \\ 2\sigma_0\sigma_1\mu_{01}-2\sigma_2\sigma_3\nu_{23} & \sigma_0^2+\sigma_1^2-\sigma_2^2-\sigma_3^2 & -2\sigma_0\sigma_3\eta_{03}+2\sigma_1\sigma_2\tau_{12} & -2\sigma_0\sigma_2\eta_{02}+2\sigma_1\sigma_3\tau_{13} \\ 2\sigma_0\sigma_2\mu_{02}-2\sigma_1\sigma_3\nu_{13} & 2\sigma_0\sigma_3\eta_{03}+2\sigma_1\sigma_2\tau_{12} & \sigma_0^2-\sigma_1^2+\sigma_2^2-\sigma_3^2 & -2\sigma_0\sigma_1\eta_{01}+2\sigma_2\sigma_3\tau_{23} \\ 2\sigma_0\sigma_3\mu_{03}-2\sigma_1\sigma_2\nu_{12} & 2\sigma_0\sigma_2\eta_{02}+2\sigma_1\sigma_3\tau_{13} & 2\sigma_0\sigma_1\eta_{01}+2\sigma_2\sigma_3\tau_{23} & \sigma_0^2-\sigma_1^2-\sigma_2^2+\sigma_3^2 \end{pmatrix} \tag{5.121}$$

where the statistical nature and the structure of the constitutive vectors **D**, **P**, **k**, **r**, and **q** defined in Section 5.11 is also revealed through the relations

$$\mathbf{P}_+\equiv\frac{\mathbf{D}+\mathbf{P}}{2}=\frac{2}{m_{00}}\bar{\boldsymbol{\Sigma}}_0\bar{\boldsymbol{\Sigma}}_1\boldsymbol{\mu}\quad \mathbf{P}_-\equiv\frac{\mathbf{D}-\mathbf{P}}{2}=\frac{2}{m_{00}}\bar{\boldsymbol{\Sigma}}_2\bar{\boldsymbol{\Sigma}}_3\mathbf{v}$$

$$\bar{\mathbf{R}}_+\equiv\frac{\mathbf{q}+\mathbf{r}}{2}=\frac{2}{m_{00}}\bar{\boldsymbol{\Sigma}}_2\bar{\boldsymbol{\Sigma}}_3\boldsymbol{\tau}\quad \bar{\mathbf{R}}_-\equiv\frac{\mathbf{q}-\mathbf{r}}{2}=\frac{2}{m_{00}}\bar{\boldsymbol{\Sigma}}_0\bar{\boldsymbol{\Sigma}}_1\boldsymbol{\eta}$$

$$\mathbf{k}=\frac{1}{\sqrt{3}m_{00}}\begin{pmatrix}\sigma_0^2+\sigma_1^2-\sigma_2^2-\sigma_3^2\\ \sigma_0^2-\sigma_1^2+\sigma_2^2-\sigma_3^2\\ \sigma_0^2-\sigma_1^2-\sigma_2^2+\sigma_3^2\end{pmatrix}\qquad\left[m_{00}=\sigma_0^2+\sigma_1^2+\sigma_2^2+\sigma_3^2\right]$$

$$\bar{\boldsymbol{\Sigma}}_0\equiv\begin{pmatrix}\sigma_0&0&0\\0&\sigma_0&0\\0&0&\sigma_0\end{pmatrix}\quad \bar{\boldsymbol{\Sigma}}_1\equiv\begin{pmatrix}\sigma_1&0&0\\0&\sigma_2&0\\0&0&\sigma_3\end{pmatrix}\quad \bar{\boldsymbol{\Sigma}}_2\equiv\begin{pmatrix}\sigma_2&0&0\\0&\sigma_3&0\\0&0&\sigma_1\end{pmatrix}\quad \bar{\boldsymbol{\Sigma}}_3\equiv\begin{pmatrix}\sigma_3&0&0\\0&\sigma_1&0\\0&0&\sigma_2\end{pmatrix} \tag{5.122}$$

It is straightforward to check that the *correlation retarding vectors* $\boldsymbol{\eta}$ and $\boldsymbol{\tau}$ have the property that the entrance and exit retardances of the arrow decomposition of **M** are zero (i.e., $\mathbf{M}_{RI}=\mathbf{M}_{RO}=\mathbf{I}$) if and only if $\boldsymbol{\eta}=\boldsymbol{\tau}=\mathbf{0}$. Therefore, the transformation $\mathbf{M}_A=\mathbf{M}_{RO}^T\mathbf{M}\mathbf{M}_{RI}^T$ leads to the cancellation of vectors $\boldsymbol{\eta}$ and $\boldsymbol{\tau}$. Furthermore, as seen above, the medium lacks both diattenuation and polarizance if and only if both *correlation enpolarizing vectors*, $\boldsymbol{\mu}$ and $\mathbf{v}$, are zero (Gil and San José 2022). In fact, the *dichroic anisotropy coefficients* $(\alpha_{1D},\alpha_{2D},\alpha_{3D})$ and the *birefringent anisotropy coefficients* $(\alpha_{1R},\alpha_{2R},\alpha_{3R})$ (Arteaga et al. 2011) defined in Section 6.7 are none other than the components of vectors $2\mathbf{P}_+$ and $-2\bar{\mathbf{R}}_-$.

The arrow form and statistical approaches considered above provide a privileged framework to analyze, in the next subsections, the peculiar properties of two particular, but important, cases of (1) *nonenpolarizing* Mueller matrices $\mathbf{M}_O$ defined as Mueller matrices with *zero degree of polarizance*, $0=P_P(\mathbf{M}_O)=\sqrt{[P^2(\mathbf{M}_O)+D^2(\mathbf{M}_O)]/2}$ and (2) *symmetric* Mueller matrices $\mathbf{M}_S$ $(\mathbf{M}_S^T=\mathbf{M}_S)$.

5.13.5 Characterization of Nonenpolarizing Mueller Matrices

We define *nonenpolarizing Mueller matrices* as Mueller matrices $\mathbf{M}_O$ with zero diattenuation and polarizance (as in the case of parallel compositions of retarders, for instance). As a result, $\mathbf{M}_O$ has a diagonal arrow form, $\mathbf{M}_A(\mathbf{M}_O) = m_{00}\,\mathrm{diag}(1, a_1, a_2, \varepsilon a_3)$ that is, $\mathbf{M}_A(\mathbf{M}_O)$ is a type-I canonical depolarizer $\mathbf{M}_{\Delta d}$ and consequently, the decomposition $\mathbf{M}_O = \mathbf{M}_{RO}\,\mathbf{M}_A\,\mathbf{M}_{RI}$ coincides with both the normal form and the symmetric decomposition of $\mathbf{M}_O$. The only nonzero component of purity is $P_S(\mathbf{M}_O) = P_\Delta(\mathbf{M}_O)$, and thus $\mathbf{C}_A(\mathbf{M}_O) = \mathrm{diag}(\sigma_0^2, \sigma_1^2, \sigma_2^2, \sigma_3^2)$, where σ_k^2 coincide with the eigenvalues of $\mathbf{C}_A(\mathbf{M}_O)$. Therefore, $\hat{\mathbf{M}}_\Gamma = \mathrm{diag}(1,0,0,0)$, and, in terms of the statistical parameters, the covariance conditions trivially require that σ_k^2 be non-negative, while all correlation coefficients are zero (Gil and San José 2022).

5.13.6 Characterization of Symmetric Mueller Matrices

In the case of symmetric Mueller matrices $\mathbf{M}_S$ (which includes, for instance, any parallel combination of normal diattenuators) the entrance and exit retarders are inverse to each other ($\mathbf{M}_{RO} = \mathbf{M}_{RI}^T$, because the submatrix $\mathbf{m}$ is symmetric). Furthermore, the arrow form $\mathbf{M}_{AS}$ of $\mathbf{M}_S$ is also a symmetric matrix, so that $\mathbf{M}_S = \mathbf{M}_R^T\,\mathbf{M}_{AS}\,\mathbf{M}_R$. Note also that the coherency matrix $\mathbf{C}_A(\mathbf{M}_S)$ is symmetric too (i.e., $\boldsymbol{\nu} = \mathbf{0}$), and consequently, the set of covariance conditions reduces to the trivial inequalities $\sigma_k^2 \geq 0$ (with the convention $\sigma_k \geq 0$) together with $D_\Gamma \leq 1/2$ (Gil and San José 2022)

By applying to $\mathbf{M}_S$ the characterization criterion described in Section 5.13.3, it is straightforward to get the specific structure of $\mathbf{M}_{AS}$ depending on the value of $\mathrm{rank}\,\mathbf{C}_A(\mathbf{M}_S)$.

5.14 Spectral, Arbitrary and Characteristic Decompositions of a Mueller Matrix

The decomposition of a depolarizing Mueller matrix $\mathbf{M}$ into a convex sum of Mueller matrices $\mathbf{M}_i$, called a *parallel decomposition* of $\mathbf{M}$, is formulated as

$$\mathbf{M} = m_{00}\hat{\mathbf{M}} = \sum_{i=1}^{n} k_i\left(m_{00i}\hat{\mathbf{M}}_i\right) \quad \left(k_i \geq 0 \quad \sum_{i=1}^{n} k_i = 1\right) \tag{5.123}$$

Parallel decompositions of a given Mueller matrix can be performed in different ways, in terms of only pure, of only nonpure, or of a mixture of pure and nonpure Mueller matrices. This is an important subject with very interesting applications in both experimental and theoretical fields that will be studied in detail in Chapter 7. Nevertheless, it is worth introducing here three particular types of parallel decompositions of $\mathbf{M}$ that will be useful for the interpretation of some descriptors of the polarimetric purity, introduced in the next sections.

5.14.1 Spectral Decomposition

Since $\mathbf{C}$ is a positive semi-definite Hermitian matrix (Cloude 1986; Arnal 1990), it can be diagonalized through a unitary transformation

$$\mathbf{C} = \mathbf{U}\,\mathrm{diag}(\lambda_0, \lambda_1, \lambda_2, \lambda_3)\,\mathbf{U}^\dagger \tag{5.124}$$

where λ_i are the four ordered $(0 \leq \lambda_3 \leq \lambda_2 \leq \lambda_1 \leq \lambda_0)$ non-negative eigenvalues of $\mathbf{C}$. The columns $\mathbf{u}_i$ $(i = 0,1,2,3)$ of the 4×4 unitary matrix $\mathbf{U}$ are the orthonormal eigenvectors of

C. Thus, C can be expressed as the following convex linear combination of four rank-one coherency matrices respectively representing pure systems:

$$\mathbf{C}=\sum_{i=0}^{3}\frac{\lambda_i}{\operatorname{tr}\mathbf{C}}\mathbf{C}_i,\quad \mathbf{C}_i\equiv(\operatorname{tr}\mathbf{C})\left(\mathbf{u}_i\otimes\mathbf{u}_i^{\dagger}\right) \tag{5.125}$$

Note that, from the relation (5.13) that links matrices **C** and **H** associated with a given **M**, it follows that the above expressions are also valid when **C** is replaced by **H**.

Due to the biunivocal relation between **C**, **H** and **M**, and taking into account that $\operatorname{tr}\mathbf{C}=\operatorname{tr}\mathbf{H}=m_{00}$, the above expression can be written as follows in terms of the corresponding Mueller matrices $\mathbf{M}_{Ji}(\mathbf{C}_i)$:

$$\mathbf{M}=\sum_{i=0}^{3}\frac{\lambda_i}{m_{00}}\left(m_{00}\hat{\mathbf{M}}_{Ji}\right)=\sum_{i=0}^{3}\hat{\lambda}_i\left(m_{00}\hat{\mathbf{M}}_{Ji}\right)\qquad\left[\hat{\lambda}_i\equiv\lambda_i/\operatorname{tr}\mathbf{C}=\lambda_i/m_{00}\right] \tag{5.126}$$

This parallel decomposition is called *spectral decomposition* (or *Cloude's decomposition*) because its coefficients are given by the normalized eigenvalues of **C** (the spectrum of **C**). It shows that any linear system can be considered as a parallel combination of up to four pure independent systems with weights equal to the normalized eigenvalues $\hat{\lambda}_i$ of **C**.

Let us note that when an eigenvalue λ_i has a multiplicity m higher than one $(1<m\leq 4)$, the eigenvectors of the corresponding m-dimensional image subspace of **C** (denoted as $\operatorname{im}\mathbf{C}$) are not unique and can be chosen arbitrarily as a set of orthonormal basis vectors covering $\operatorname{im}\mathbf{C}$. Moreover, the statistical nature of **C** leads to a probabilistic interpretation of its eigenvalues. This fact has direct consequences on the physical interpretation of quantities such as the *depolarization index*, the *polarization entropy* (see Section 5.10) and the *indices polarimetric of purity*, which will be defined in Section 6.3 in terms of $\hat{\lambda}_i$.

5.14.2 Arbitrary Decomposition

The use of the synthetic procedure described in detail in Section 7.3, allows for expressing **C** as a convex sum of r (with $r\equiv\operatorname{rank}\mathbf{C}=\operatorname{rank}\mathbf{H}$) pure coherency matrices $\mathbf{C}_{Ji}=(\operatorname{tr}\mathbf{C})(\hat{\mathbf{c}}_i\otimes\hat{\mathbf{c}}_i^{\dagger})$ $(i=0,\ldots,r-1)$ ($\otimes$ stands for the Kronecker product), where the set of r unit coherency vectors $\hat{\mathbf{c}}_i$ can appropriately be chosen in infinitely many different manners leading to the *homogeneous arbitrary decomposition* (Gil 2007; Gil and San José 2013; 2019)

$$\mathbf{C}=(\operatorname{tr}\mathbf{C})\sum_{i=0}^{r-1}p_i\,\hat{\mathbf{C}}_{Ji}\quad p_i=\frac{1}{(\operatorname{tr}\mathbf{C})\sum_{q=0}^{r-1}\frac{1}{\lambda_q}\left|\left(\mathbf{U}^{\dagger}\hat{\mathbf{c}}_i\right)_q\right|^2}=\frac{1}{\hat{\mathbf{c}}_i^{\dagger}\mathbf{C}^{-}\hat{\mathbf{c}}_i}\qquad\left[\sum_{i=0}^{r-1}p_i=1\right] \tag{5.127}$$

$\mathbf{C}^{-}$ represents the pseudoinverse of $\hat{\mathbf{C}}\equiv\mathbf{C}/\operatorname{tr}\mathbf{C}$ defined as $\mathbf{C}^{-}=\mathbf{U}\boldsymbol{\Lambda}^{-}\mathbf{U}^{\dagger}$, $\boldsymbol{\Lambda}^{-}$ being the diagonal matrix whose r first diagonal elements are $1/\lambda_0, 1/\lambda_1,\ldots,1/\lambda_{r-1}$ and the last $4-r$ elements are zero (Gil and San José 2019).

The Mueller counterpart of the above decomposition reads

$$\mathbf{M}=\sum_{i=9}^{r-1}p_i\mathbf{M}_{Ji}\quad \mathbf{M}_{Ji}=m_{00}\hat{\mathbf{M}}_{Ji}\quad \hat{\mathbf{M}}_{Ji}\equiv\mathbf{M}_J(\hat{\mathbf{c}}_i)\quad p_i=\frac{1}{\hat{\mathbf{c}}_i^{\dagger}\mathbf{C}^{-}\hat{\mathbf{c}}_i}\qquad\left[\sum_{i=0}^{r-1}p_i=1\right] \tag{5.128}$$

where all pure components have equal mean intensity coefficient $m_{00i} = m_{00}$, which coincides with that of the Mueller matrix **M**; this is the origin of the adjective *homogeneous* for decomposition (5.127) and (5.128). Nevertheless, the condition $m_{00i} = m_{00}$ is too restrictive and, in general, it is not required for physically realizable arbitrary decompositions. The ensemble criterion (i.e., the combination of Cloude's and passivity criteria) leads to the following generalized expression for the arbitrary decomposition of a Mueller matrix (Gil and San José 2019):

$$\begin{gathered}\mathbf{M}=\sum_{i=0}^{r-1} k_i \mathbf{M}_{Ji} \quad \mathbf{M}_{Ji}=m_{00i}\hat{\mathbf{M}}_{Ji} \quad \hat{\mathbf{M}}_{Ji}\equiv\begin{pmatrix}1 & \mathbf{D}_i^T \\ \mathbf{P}_i & \mathbf{m}_i\end{pmatrix}\equiv\mathbf{M}_J(\hat{\mathbf{c}}_i) \\ m_{00i}(1+D_i)\leq 1 \quad k_i=\frac{m_{00}}{m_{00i}}p_i=\frac{m_{00}}{m_{00i}}\frac{1}{\hat{\mathbf{c}}_i^{\dagger}\mathbf{C}^{-}\hat{\mathbf{c}}_i} \quad \left[\sum_{i=0}^{r-1} k_i=1\right]\end{gathered} \tag{5.129}$$

5.14.3 Characteristic Decomposition

The pure components of the spectral decomposition can be regrouped into different sets of components, some of which are depolarizing. A particularly interesting case is the so-called *characteristic decomposition*, or *trivial decomposition* (Gil 2007), which is formulated as an extension of the characteristic decomposition of 3D coherency matrices (Samson 1973), which in turn is an extension of the well-known decomposition of 2D polarization matrices into a sum of a totally polarized state and an unpolarized state.

The characteristic decomposition of the 4×4 coherency matrix **C** associated with a Mueller matrix **M** is defined as the following convex sum of covariance matrices having the same trace:

$$\mathbf{C}=(\hat{\lambda}_0-\hat{\lambda}_1)\mathbf{C}_{J0}+2(\hat{\lambda}_1-\hat{\lambda}_2)\mathbf{C}_1+3(\hat{\lambda}_2-\hat{\lambda}_3)\mathbf{C}_2+4\hat{\lambda}_3\mathbf{C}_{\Delta 0} \tag{5.130.a}$$

with

$$\begin{aligned}\mathbf{C}_{J0} &\equiv (\mathrm{tr}\,\mathbf{C})\left[\mathbf{U}\mathrm{diag}(1,0,0,0)\mathbf{U}^{\dagger}\right]=(\mathrm{tr}\,\mathbf{C})(\mathbf{u}_0\otimes\mathbf{u}_0^{\dagger}) \\ \mathbf{C}_1 &\equiv \frac{1}{2}(\mathrm{tr}\,\mathbf{C})\left[\mathbf{U}\mathrm{diag}(1,1,0,0)\mathbf{U}^{\dagger}\right]=\frac{1}{2}(\mathrm{tr}\,\mathbf{C})\sum_{i=0}^{1}\mathbf{u}_i\otimes\mathbf{u}_i^{\dagger} \\ \mathbf{C}_2 &\equiv \frac{1}{3}(\mathrm{tr}\,\mathbf{C})\left[\mathbf{U}\mathrm{diag}(1,1,1,0)\mathbf{U}^{\dagger}\right]=\frac{1}{3}(\mathrm{tr}\,\mathbf{C})\sum_{i=0}^{2}\mathbf{u}_i\otimes\mathbf{u}_i^{\dagger} \\ \mathbf{C}_{\Delta 0} &\equiv \frac{1}{4}(\mathrm{tr}\,\mathbf{C})\left[\mathbf{U}\mathrm{diag}(1,1,1,1)\mathbf{U}^{\dagger}\right]=\frac{1}{4}(\mathrm{tr}\,\mathbf{C})\sum_{i=0}^{3}\mathbf{u}_i\otimes\mathbf{u}_i^{\dagger}=\frac{1}{4}(\mathrm{tr}\,\mathbf{C})\mathbf{I}_4\end{aligned} \tag{5.130.b}$$

where $\mathbf{I}_4$ is the 4×4 identity matrix, $\mathbf{u}_i$ are the eigenvectors of **C** (i.e., the column vectors of the unitary matrix **U** that diagonalizes **C**), $\mathbf{C}_{J0}$ is a pure component, $\mathbf{C}_i$ $(i=1,2)$ are particular nonpure components with $\mathrm{rank}\,\mathbf{C}_i=i+1$ and with $i+1$ degenerate nonzero eigenvalues, and $\mathbf{C}_{\Delta 0}$ is proportional to the identity matrix and represents a perfect depolarizer.

The characteristic decomposition can be expressed in terms of Mueller matrices as

$$\mathbf{M}=\left[\hat{\lambda}_0-\hat{\lambda}_1\right]m_{00}\hat{\mathbf{M}}_{J0}+\left[2\left(\hat{\lambda}_1-\hat{\lambda}_2\right)\right]m_{00}\hat{\mathbf{M}}_1+\left[3\left(\hat{\lambda}_2-\hat{\lambda}_3\right)\right]m_{00}\hat{\mathbf{M}}_2+\left[4\hat{\lambda}_3\right]m_{00}\hat{\mathbf{M}}_{\Delta 0} \quad (5.131)$$
$$\left[m_{00}\hat{\mathbf{M}}_{J0}=\mathbf{M}\left(\mathbf{C}_{J0}\right) \qquad m_{00}\hat{\mathbf{M}}_i=\mathbf{M}\left(\mathbf{C}_i\right)\ (i=1,2) \qquad m_{00}\hat{\mathbf{M}}_{\Delta 0}=\mathbf{M}\left(\mathbf{C}_{\Delta 0}\right)\right]$$

Thus, the characteristic decomposition of a Mueller matrix $\mathbf{M}$ is formulated as a convex sum of (1) a pure component $\mathbf{M}_{J0}$ determined by $\mathbf{u}_0$, (2) a nonpure component $\mathbf{M}_1$ containing an equiprobable mixture of two pure components (namely, the first two spectral components), (3) a nonpure component $\mathbf{M}_2$ containing an equiprobable mixture of three pure components (namely, the first three spectral components) and (4) a perfect depolarizer $\mathbf{M}_{\Delta 0}$ (fully random component constituted by an equiprobable mixture of the four spectral components).

5.15 Jones Generators of Mueller Matrices

Throughout this section, the statistical definition of the Mueller matrix is considered as the basis for the analysis of the relations between certain fundamental properties, such as the number of contact points c of the characteristic ellipsoid E associated with a given Mueller matrix $\mathbf{M}$ with the Poincaré sphere (Ossikovski et al. 2013) and the rank r of the corresponding covariance matrix $\mathbf{H}$.

The characteristic ellipsoid E (or the P-image of $\mathbf{M}$) is described by the end points of the intensity-normalized polarization vectors $\mathbf{p}_l\equiv(s'_1,s'_2,s'_3)^T/s'_0$ of the Stokes vectors $\mathbf{s}'\equiv\left(s'_0,s'_1,s'_2,s'_3\right)^T=\mathbf{M}\hat{\mathbf{s}}_p$, where $\hat{\mathbf{s}}_p$ are the intensity-normalized pure Stokes vectors sweeping the entire surface of the Poincaré sphere,

$$\mathbf{p}_l\equiv\frac{1}{\left(\mathbf{M}\hat{\mathbf{s}}_p\right)_0}\left[\left(\mathbf{M}\hat{\mathbf{s}}_p\right)_1,\left(\mathbf{M}\hat{\mathbf{s}}_p\right)_2,\left(\mathbf{M}\hat{\mathbf{s}}_p\right)_3\right]^T \quad (5.132)$$

E is studied in Chapter 10, where it is termed the *forward ellipsoid* $E_{\Delta P}$ in order to distinguish it from the *reverse ellipsoid* $E_{\Delta D}$ associated with $\mathbf{M}^T$. Provided $\mathbf{M}$ is a physical Mueller matrix, both $E_{\Delta P}$ and $E_{\Delta D}$ do not protrude the Poincaré sphere. Throughout this section, we will use E as synonym of $E_{\Delta P}$.

The combined analysis of c and r enables the comprehensive classification of any experimental depolarizing Mueller matrix into one of six possible classes, thus making possible a meaningful phenomenological interpretation in terms of a specific fluctuating Jones generator or, equivalently, of a finite sum of nondepolarizing Mueller matrices.

The rank $r=\operatorname{rank}\mathbf{H}$ of the covariance matrix $\mathbf{H}$ plays a key role in characterizing the randomness of the fluctuating process that generates the depolarizing Mueller matrix $\mathbf{M}$. When $r=1$, then the covariance matrix has pure character and corresponds to the matrix $\mathbf{H}_J$ introduced in Section 3.4.1. Indeed, if $r=1$, then $\mathbf{H}$ is generated by a single covariance vector $\mathbf{h}$ and takes the particular form $\mathbf{H}_J=\mathbf{h}\otimes\mathbf{h}^\dagger$. Therefore, $\mathbf{H}_J$ is a projector so that the Jones generator $\mathbf{T}$ is not fluctuating and consequently, $\mathbf{M}$ is nondepolarizing (Simon 1982).

The rank r of $\mathbf{H}$ is likewise intimately related to the notion of linear dependence between the elements of the Jones generator $\mathbf{T}$ of $\mathbf{M}$. Thus, when $r=2,3,4$, $\mathbf{M}$ is depolarizing and its Jones generator $\mathbf{T}$ contains respectively two, three or four linearly independent fluctuating elements. Observe that having a single independently fluctuating Jones generator element, say t_{11}, is polarimetrically equivalent to the absence of fluctuations. Indeed, the three remaining matrix elements must be proportional to t_{11} in virtue of their dependence on it; t_{11} can be factored out of the covariance vector and its only effect is that the covariance matrix takes the form $\mathbf{H} = \langle |t_{11}|^2 \rangle \mathbf{H}_J$.

A Mueller matrix $\mathbf{M}$ transforms incident polarization states into emerging ones in accordance with the general relation $\mathbf{s}' = \mathbf{M}\mathbf{s}$ where $\mathbf{s}$ and $\mathbf{s}'$ are the Stokes vectors corresponding to the incident and emerging polarization states respectively. It has been shown that if $\mathbf{s}$ sweeps the Poincaré sphere representing all totally polarized light states, then $\mathbf{s}'$ describes an ellipsoid E lying within the sphere (Lu and Chipman 1998; see Chapter 10). From a topological point of view, the following situations can occur (see Chapter 10): (1) E coincides with the Poincaré sphere (only if $\mathbf{M}$ is nondepolarizing), (2) E touches the Poincaré sphere at two points or at one point, or (3) E does not touch the Poincaré sphere at all (Ossikovski et al. 2013). The number c of contact points ($c = \infty$, 2, 1 or 0) constitutes an important descriptor for the interpretation and classification of Mueller matrices.

Both parameters, c and r, are intimately related (Ossikovski et al. 2013, Ossikovski and Gil 2017). In what follows, we shall analyze their interrelations from the point of view of the statistical definitions of $\mathbf{M}$ and $\mathbf{H}$ given by Eqs. (5.6) and $\mathbf{H} = \langle \mathbf{h} \otimes \mathbf{h}^\dagger \rangle$, rather than by using exclusively the algebraic approach based on the symmetric decomposition (Ossikovski 2009); see Section 8.3) and the canonical forms of $\mathbf{M}$ (Ossikovski 2010a; see Section 5.7) employed in Ossikovski et al. (2013).

5.15.1 Relation between Mapping and Statistical Properties of Mueller Matrices

Since the correspondence between the number of contact points c of the forward ellipsoid E of the Mueller matrix $\mathbf{M}$ with the Poincaré sphere and the rank r of the covariance matrix $\mathbf{H}$ associated with $\mathbf{M}$ is not of one-to-one nature, we shall use the statistical definition of $\mathbf{M}$ to establish the corresponding values of r given those of c. Eventually, this will allow us to draw a diagram summarizing the c-to-r correspondence in a compact visual form as well as to classify all depolarizing Mueller matrices into six classes. The topological aspects will be deal with in more detail in Chapter 10.

5.15.1.1 Nondepolarizing Transformation of a Mueller Matrix

Let $\mathbf{M}_{J1}$ and $\mathbf{M}_{J2}$ be two nonsingular pure Mueller matrices (and consequently, their Jones counterparts, $\mathbf{T}_1$ and $\mathbf{T}_2$, are also nonsingular). We call the transformation $\mathbf{M}_{J2}\mathbf{M}\mathbf{M}_{J1}$ the *nondepolarizing transformation* of the (generally depolarizing) Mueller matrix $\mathbf{M}$ (Xing 1992). Within the statistical definition of $\mathbf{M}$, the Jones counterpart of $\mathbf{M}_{J2}\mathbf{M}\mathbf{M}_{J1}$ is $\mathbf{T}_2\mathbf{T}\mathbf{T}_1$, where the Jones generator $\mathbf{T}$ of $\mathbf{M}$ is generally fluctuating, whereas the Jones matrices $\mathbf{T}_1$ and $\mathbf{T}_2$ are fixed (i.e., are not fluctuating). Since $\mathbf{T}_1$ and $\mathbf{T}_2$ are assumed nonsingular, the transformed Jones generator $\mathbf{T}_2\mathbf{T}\mathbf{T}_1$ is singular if and only if $\mathbf{T}$ is.

It is straightforward to see that the nondepolarizing transformation $\mathbf{M}_{J2}\mathbf{M}\mathbf{M}_{J1}$ preserves the number of contact points c of $\mathbf{M}$. Indeed, both $\mathbf{M}_{J1}$ and $\mathbf{M}_{J2}$ transform totally polarized incident states into totally polarized emerging ones, that is, they map the Poincaré sphere onto itself. Consequently, the value of c will not be modified by right- and left-multiplication of $\mathbf{M}$ by $\mathbf{M}_{J1}$ and $\mathbf{M}_{J2}$ respectively, and will be determined uniquely by the properties of $\mathbf{M}$. It can be likewise shown (Ossikovski and Gil 2017, Appendix A) that the nondepolarizing transformation preserves the rank r of $\mathbf{H}(\mathbf{M})$. The nondepolarizing transformation $\mathbf{M}_{J2}\mathbf{M}\mathbf{M}_{J1}$ thus appears as a useful tool for reducing a general Mueller matrix (as well as its Jones generator) to a simpler (canonical) form having the same values of c and r as the original one.

5.15.1.2 From the Number of Contact Points c of E to the Rank r of H

(1) $c=\infty$. Condition $c=\infty$ implies $r=1$ and a nonsingular, non-fluctuating, Jones generator $\mathbf{T}$ of $\mathbf{M}_J$.

(2) $c=2$. It has been proven that $c=2$ necessarily entails $r=2$ (Ossikovski and Gil 2017). The corresponding canonical Jones generator has the diagonal form $\mathbf{T}_c^{(2)}=\operatorname{diag}(t_{c1},t_{c2})$ and is both nonsingular and non-degenerate (the notion of degeneracy is discussed in Section 3.6).

(3) $c=1$. When E touches the Poincaré sphere at a single point, then exactly one totally polarized state still remains totally polarized after the light-matter interaction. It has been demonstrated that condition $c=1$ entails $r=3,2,1$. $r=3$ is achieved for a nonsingular and non-degenerate canonical generator

$$\mathbf{T}_c^{(1)}=\begin{pmatrix} t_{c0} & t_{c1} \\ 0 & t_{c3} \end{pmatrix} \tag{5.133}$$

If $\mathbf{T}_c^{(1)}$represents a (singular) depolarizing polarizer or if $\mathbf{T}_c^{(1)}$ is degenerate, then $r=2$. If $\mathbf{T}_c^{(1)}$ represents a polarizer, then $r=1$.

(4) $c=0$. E is located entirely inside the Poincaré sphere: $\mathbf{M}$ transforms all totally polarized incident states into partially polarized ones. The canonical Jones generator of $\mathbf{M}$ has the most general form

$$\mathbf{T}_c^{(0)}=\begin{pmatrix} t_{c0} & t_{c1} \\ t_{c2} & t_{c3} \end{pmatrix} \tag{5.134}$$

where none of its elements is identically zero. The Jones generator $\mathbf{T}_c^{(0)}$ is compatible with the following types of medium: (1) $\mathbf{T}_c^{(0)}$ is a depolarizing analyzer (hence singular) with $r=2$; (2) $\mathbf{T}_c^{(0)}$ is nonsingular, with $r=3$, and (3) $\mathbf{T}_c^{(0)}$ is nonsingular, with $r=4$. It is remarkable that the depolarizing analyzer (DA) (whose associated Mueller matrix is depolarizing but with $D=1$) is the only device whose singular Jones generator produces $r=2$. Degenerate Jones generators result in either $r=3$ or $r=4$ depending on whether they are singular or not.

Here, we list the Jones generators discussed so far, labeled with a subscript "c,", together with other types of Jones matrices to which we will refer to in the next paragraphs,

$$\begin{aligned} &\mathbf{T}_c^{(0)}=\begin{pmatrix} t_{c0} & t_{c1} \\ t_{c2} & t_{c3} \end{pmatrix} \quad \mathbf{T}_c^{(1)}=\begin{pmatrix} t_{c0} & t_{c1} \\ 0 & t_{c3} \end{pmatrix} \quad \mathbf{T}_c^{(2)}=\begin{pmatrix} t_{c0} & 0 \\ 0 & t_{c3} \end{pmatrix} \\ &\mathbf{T}_{cDP}=\begin{pmatrix} a_0 & a_1 \\ 0 & 0 \end{pmatrix} \quad \mathbf{T}_{cDA}=\begin{pmatrix} a_0 & 0 \\ a_2 & 0 \end{pmatrix} \quad \mathbf{T}_{ds}=\begin{pmatrix} a_2 a_3 & a_2^2 \\ -a_3^2 & -a_2 a_3 \end{pmatrix} \end{aligned} \tag{5.135}$$

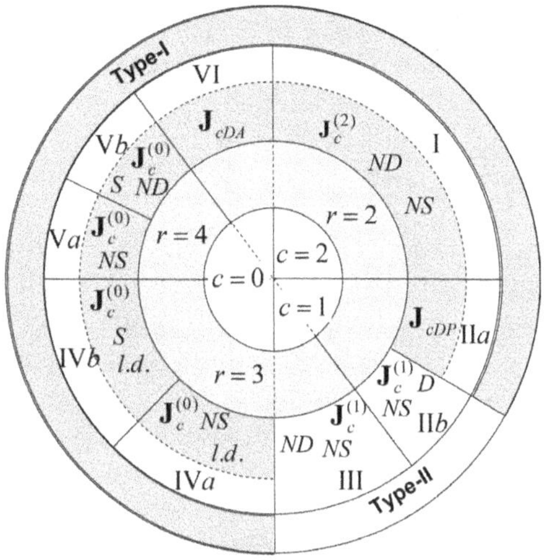

FIGURE 5.5
Circular diagram showing the correspondence between the number of contact points c of the characteristic ellipsoid E of $\mathbf{M}$ and the rank r of its covariance matrix $\mathbf{H}$. *S*: singular; *NS*: nonsingular; *D*: degenerate; *ND*: non-degenerate; *l.d.*: linear dependence (Ossikovski and Gil 2017).

5.15.1.3 Diagram Representation of Depolarizing Mueller Matrices

Figure 5.5 represents schematically the interrelations between the number of contact points c of the characteristic ellipsoids of depolarizing Mueller matrices and the rank r of their respective covariance matrices, established above. The trivial nondepolarizing case ($c = 1$ or ∞; $r = 1$) is not represented for simplicity.

The diagram consists of four concentric rings and six sectors. The innermost ring contains the values of c; the second one, those of r, whereas the third one specifies the nature of the Jones generators providing a given combination of c and r. The possible combinations of c and r numbers form six radial sectors, labeled from I to VI. Some of the sectors (II, IV and V) are subdivided into two subsectors (*a* and *b*) to account for different Jones generators producing the same (c; r) pair (c-r degeneracy). Finally, the outermost ring of the diagram contains two subrings: that of type-II (comprising sectors II*b* and III) and type-I (comprising the rest of the sectors) Mueller matrices.

Sector I ($c = 2; r = 2$) features a diagonal nonsingular canonical Jones generator $\mathbf{T}_c^{(2)}$.

Sector II ($c = 1; r = 2$) is subdivided into II*a* and II*b*. Subsector II*a* is that of the depolarizing polarizer *DP* (whose associated singular Mueller matrix is depolarizing but with $P = 1$) whose canonical Jones generator $\mathbf{T}_{cDP}$ is singular, whereas the degenerate nonsingular canonical Jones generator of subsector II*b* is given by $\mathbf{T}_c^{(1)}$ with $t_{c0} = t_{c3}$. Sector II is the only one whose two subsectors are described by two Jones generators having two different forms. Notice that the essential geometrical difference between subsectors II*a* and II*b* is that E degenerates into a point in the first subsector.

Sector III ($c = 1; r = 3$) is that of the non-degenerate nonsingular canonical Jones generator $\mathbf{T}_c^{(1)}$.

Sector IV ($c = 0; r = 3$) features the Jones generator $\mathbf{T}_c^{(0)}$ for which a linear dependence between Jones matrix elements exists, $\sum_0^3 \alpha_i t_{ci} = 0$, where α_i are constant (i.e., non-fluctuating) coefficients. Such linear dependence reduces the rank of the fluctuating process associated with $\mathbf{T}_c^{(0)}$ from 4 to 3. This sector is further divided into two subsectors, IV*b* and IV*a*, depending on whether the Jones generator $\mathbf{T}_c^{(0)}$ (with the additional linear dependence indicated above) is singular or not. The special case where $\mathbf{T}_c^{(0)}$ is both singular and degenerate can be parameterized like $\mathbf{T}_{ds}$ (see Eq. (5.135)) and the condition $t_0 + t_3 = a_2 a_3 - a_2 a_3 = 0$ is fulfilled identically, so that $\mathbf{T}_{ds}$ belongs to subsector IV*b*.

Sector V ($c = 0; r = 4$) is specified by the general Jones generator $\mathbf{T}_c^{(0)}$ whose matrix elements are linearly independent. This sector can likewise be divided into two subsectors, V*a* and V*b*, depending on whether $\mathbf{T}_c^{(0)}$ is nonsingular (degenerate or not) or singular and non-degenerate (if $\mathbf{T}_c^{(0)}$ were both singular and degenerate it would belong to subsector IV*b*, as we have seen before).

Sector VI ($c = 0$; $r = 2$) is that of the depolarizing analyzer *DA* whose canonical Jones generator is $\mathbf{T}_{cDA}$.

Since a given physically realizable (Cloude 1989) depolarizing Mueller matrix necessarily belongs to one of the six sectors, every experimental Mueller matrix can be classified into a subsector specific to it.

Recall that, unlike in the original definition of type-I and type-II Mueller matrices by Gopala Rao et al. (1998a; 1998b), and in Ossikovski and Gil (2017), the definition introduced in Section 5.7 identifies type-II depolarizing Mueller matrices as those that cannot be reduced to diagonal form by using the nondepolarizing transformation $\mathbf{M}_{J2}\mathbf{M}\mathbf{M}_{J1}$, even when $\mathbf{M}_{J1}$ or $\mathbf{M}_{J2}$ are singular. Therefore, according to this definition, used throughout the book, depolarizing polarizers and depolarizing analyzers belong to the type-I category.

Type-II Mueller matrices belong to sector III or to subsector II*b*, thus featuring $c = 1$ and $r = 3$ or $r = 2$, but excluding subsector II*a*. On the other hand, the depolarizing polarizer *DP* generating the subsector II*a* with $c = 1$ and $r = 2$ is represented by a type-I Mueller matrix for which $\mathbf{M}_{J2}$ is singular (while $\mathbf{M}_{J1}$ is nonsingular) (Ossikovski 2010a). Consequently, type-II Mueller matrices generate sectors III and II*b* whereas type-I ones cover the rest of the sectors, as shown in Figure 5.5.

5.15.2 Synthesis of Depolarizing Mueller Matrices with Given Basic Properties

Most generally, there are two ways of synthesizing a depolarizing Mueller matrix with given c and r numbers. The first one is based on a random fluctuating medium with specific Jones generator whereas the second one results from the incoherent superposition of a small number $n = r$ of nondepolarizing Mueller matrix responses. The first method realizes a depolarizing Mueller matrix with predefined basic properties from its statistical definition (5.6) while the second approach (Cloude 2013) synthesizes it from its arbitrary (Gil and San José 2019) or Cloude (1986) parallel decomposition. We will refer to these two methods as *statistical* and *deterministic* respectively.

5.15.2.1 Statistical Generation of a Depolarizing Mueller Matrix

Each sector with given c and r numbers from the diagram shown in Figure 5.5 can be put into direct correspondence with a specific canonical Jones generator identified in Eq. (5.135). Table 5.1 summarizes the possible combinations of c and r numbers and their associated Jones generators.

To every (c, r) entry of Table 5.1 there corresponds a specific Jones generator, with the exception of the (1;2) entry that is subdivided into two different Jones generators, the first

TABLE 5.1
Canonical Jones Generators for the Statistical Synthesis of Depolarizing Mueller Matrices with Given c and r Numbers

c \ r	1	3	4
0	$\begin{pmatrix} t_0 & 0 \\ t_2 & 0 \end{pmatrix}$	$\begin{pmatrix} t_0 & t_1 \\ t_2 & t_3 \end{pmatrix} \left[\sum_{i=0}^{3} \alpha_i t_i = 0\right]$	$\begin{pmatrix} t_0 & t_1 \\ t_2 & t_3 \end{pmatrix}$
1	$\begin{pmatrix} t_0 & t_1 \\ 0 & 0 \end{pmatrix}$	$\begin{pmatrix} t_0 & t_1 \\ 0 & t_3 \end{pmatrix}$	
	$\begin{pmatrix} t_0 & t_1 \\ 0 & t_0 \end{pmatrix}$		
2	$\begin{pmatrix} t_0 & 0 \\ 0 & t_3 \end{pmatrix}$		

Source: Ossikovski and Gil 2017.

generator being that of the depolarizing polarizer *DP*, whereas the second one, being degenerate and nonsingular, corresponds to a non-diagonalizable Mueller matrix. Jones generators from different table entries may have identical forms (e.g. those from (0;3) and (0;4)), the difference residing in the additional constraints imposed on the matrix elements. The Jones generators of entries (0;2) and (1; 2) (upper half) differ only by a matrix transposition. The entries (1; 2) (lower part) and (1; 3) generate non-diagonalizable Mueller matrices.

5.15.2.2 Deterministic Generation of a Depolarizing Mueller Matrix

The arbitrary decomposition (see Section 7.3) represents a given depolarizing Mueller matrix **M** as the convex sum of a number $n = r$ of nondepolarizing Mueller matrices that correspond to respective covariance matrix projectors. As mentioned, sum decompositions (or *parallel decompositions*) represent physically the incoherent superposition of a finite number of nondepolarizing responses and can be realized experimentally in a finite spot size (Ossikovski et al. 2014) or in aperture division (Ossikovski, R. et al. 2009b) measurement configurations over several nondepolarizing media.

Table 5.2 presents the different nondepolarizing matrix sums leading to depolarizing Mueller matrices with given c and r values.

Like in Table 5.1, the entry (1;2) is subdivided into two since it can be achieved experimentally in two different ways. The (nondepolarizing) matrices $\mathbf{M}_i$ $(i = 1,2,3,4)$ in all entries are arbitrary (but different from one another) while the matrices $\mathbf{M}_{IPi}$ $(i = 1,2)$ and $\mathbf{M}_{IP}$ appearing in entries (0;2), (1;2) and (1;3) are those of non-normal elliptical polarizers. The latter entries likewise contain additional constraints imposed on the matrix terms. The details on the matrix sums and constraints appearing in these three entries can be found in (Ossikovski and Gil 2017, Appendices C and D). If a matrix $\mathbf{M}_i$ from the remaining entries coincides occasionally with a non-normal polarizer matrix $\mathbf{M}_{IP}$, then it should not satisfy the additional constraints. For instance, if $\mathbf{M}_2$ from (2;2) equals accidentally $\mathbf{M}_{IP}$ from (1;2) (lower half) then there must exist no Stokes vectors $\mathbf{s}_i$ and $\mathbf{s}_o$ such that $\mathbf{M}_{IP}\,\mathbf{s}_i = \mathbf{0}$, $\mathbf{s}_o^T\mathbf{M}_{IP} = \mathbf{0}$ and $\mathbf{s}_o^T\mathbf{M}_{IP}\,\mathbf{s}_i = 0$.

TABLE 5.2
Sums of Nondepolarizing Mueller Matrices for the Deterministic Synthesis of Depolarizing Mueller Matrices with Given c and r Numbers

<table>
<tr><th>c \ r</th><th>2</th><th>3</th><th>4</th></tr>
<tr><td>0</td><td>$\mathbf{M}_{IP1}+\mathbf{M}_{IP2}$
$\mathbf{M}_{IP1}\mathbf{s}_i=\mathbf{M}_{IP2}\mathbf{s}_i=0$</td><td>$\mathbf{M}_1+\mathbf{M}_2+\mathbf{M}_3$</td><td>$\mathbf{M}_1+\mathbf{M}_2+\mathbf{M}_3+\mathbf{M}_4$</td></tr>
<tr><td rowspan="2">1</td><td>$\mathbf{M}_{IP1}+\mathbf{M}_{IP2}$
$\mathbf{s}_o^T\mathbf{M}_{IP1}=\mathbf{s}_o^T\mathbf{M}_{IP2}=0$</td><td rowspan="2">$\mathbf{M}_1+\mathbf{M}_2+\mathbf{M}_{IP}$
$\mathbf{M}_{IP}\mathbf{s}_i=\mathbf{s}_o^T\mathbf{M}_{IP}=\mathbf{0}$
$\mathbf{s}_o^T(\mathbf{M}_1+\mathbf{M}_2)\mathbf{s}_i=0$</td><td rowspan="2"></td></tr>
<tr><td>$\mathbf{M}_1+\mathbf{M}_{IP1}$
$\mathbf{M}_{IP}\mathbf{s}_i=0,\mathbf{s}_o^T\mathbf{M}_1\mathbf{s}_i=0$
$\mathbf{s}_o^T\mathbf{M}_1\mathbf{s}_i=0$</td></tr>
<tr><td>2</td><td>$\mathbf{M}_1+\mathbf{M}_2$</td><td></td><td></td></tr>
</table>

Source: Ossikovski and Gil 2017.

Experimental examples for the different types of physical situations represented in Tables 5.1 and 5.2 can be found in (Ossikovski and Gil 2017, Appendices C and D).

5.16 Depolarization Generated by Partial Spectral and Spatial Coherence

As discussed in the previous sections, basic light-matter interactions are intrinsically nondepolarizing. It is the spatial and spectral averaging involved in the measurement procedure, together with the spectral properties of the light source, the interaction configuration and conditions (reflection, refraction, scattering, spot size and its location on the material sample), the nature and polarimetric properties of the sample itself, and the nature, time and frequency resolution of the measurement instrument, that produce depolarization. Thus, depolarization is not an intrinsic property of a given material sample (under fixed interaction conditions) but also depends essentially on the measurement process. The fact that polarimeters typically employ sufficient measurement times and feature high spectral resolutions, as well as small spot sizes, makes this condition implicitly assumed in most cases, so that the role of the instrument is not always apparent.

Because of the importance of depolarization in the polarimetric characterization of material media and of the variety of physical phenomena involved, this section summarizes the main results of several works dealing with this problem (Ossikovski and Hingerl, 2016; Hingerl and Ossikovski 2016; Miranda-Medina et al. 2017). It is therefore devoted to the detailed study of depolarization generated by partial spectral and spatial coherence.

5.16.1 Depolarization Generated by Partial Spectral Coherence

The present subsection deals with a general formalism for modeling partial coherence in spectroscopic Mueller matrix measurements. The approach is based on the statistical definition of a Mueller matrix, as well as on the fundamental representation of the measurement process as the convolution of the sample response with a specific instrumental function. The formalism can readily be extended to describe other measurement imperfections, occurring jointly with or separately from partial coherence, and resulting in depolarizing experimental Mueller matrices.

As mentioned above, finite spectral resolution, imperfectly collimated probing beam, or inhomogeneous sample thickness under the spot, result in the presence of depolarization in spectroscopic ellipsometry or Mueller matrix polarimetry measurements. Physically, the depolarization phenomenon originates from the partial or total loss of coherence of the probing beam upon its interaction with the material sample and generally depends on the optical properties of both the sample and the measurement equipment.

The Mueller matrix **M** (depolarizing, in general) can be expressed as in Eq. (5.6), $\mathbf{M} = \langle \mathbf{M}_J \rangle = \boldsymbol{\mathcal{L}} \langle \mathbf{T} \otimes \mathbf{T}^* \rangle \boldsymbol{\mathcal{L}}^{\dagger}$, in terms of the corresponding Jones generator **T**, where the average is performed by the measurement instrument on the various Jones matrix realizations $\mathbf{T}^{(s)}$ of the medium (Kim et al. 1987, Ossikovski and Hingerl 2016). The elements m_{ij} of **M** admit the following expression involving explicit averaging of the second order conjugate products of the Jones elements $t^{(1)} \equiv t_{11}, t^{(2)} \equiv t_{22}, t^{(3)} \equiv t_{12}, t^{(4)} \equiv t_{21}$:

$$m_{ij} = c_{ijkl} \langle t^{(k)} t^{(l)*} \rangle \tag{5.136}$$

in which the brackets $\langle \ldots \rangle$ denote ensemble averaging and c_{ijkl} are the respective linear combination coefficients (summation over repeated indices being understood).

Generally, different depolarization mechanisms entail different ensemble averages. If we assume that it is the finite spectral resolution that is responsible for the depolarization observed in a polarimetric experiment, the averaged second-order moments entering Eq. (5.136) can be written as

$$\begin{gathered} \langle t^{(k)} t^{(l)*} \rangle = \int t^{(k)}(\omega')\, t^{(l)*}(\omega')\, w(\omega - \omega'; \Delta\omega)\, d\omega' \\ \left[\int w(\omega - \omega'; \Delta\omega)\, d\omega' = \int w(\omega; \Delta\omega)\, d\omega = 1 \right] \end{gathered} \tag{5.137}$$

where ω is the angular frequency of the probing light, Δw is the spectral resolution (or width, or uncertainty) of the measurement equipment and ω is its instrumental (broadening) function satisfying the indicated normalization condition (integration over infinite bounds being assumed). Eq. (5.137) constitutes a classic representation of the measurement process through the convolution of the sample response (given by its Jones matrix **T**) and the properties of the measurement device (described by its instrumental function w and its spectral resolution $\Delta\omega$). It makes evident the fact that the depolarization observed experimentally is not inherent to the sample in an exclusive manner but is rather a function of the optical properties of both the sample and the measurement equipment. In particular, the spectral variations of $t^{(i)}$ are partially or totally smoothed by the finite resolution of the instrument, while the instrumental resolution is directly related to the spectral coherence of the probing light (Gabriel and Nedoluha 1971; Mandel and Wolf 1995).

By definition, the instrumental function w is real and non-negative and plays the role of a weighting function. The spectral resolution (or uncertainty) $\Delta\omega$ determines its width (i.e., the length of the interval $\omega - \Delta\omega/2 \leq \omega' \leq \omega + \Delta\omega/2$ over which it takes non-negligible values). When $\Delta\omega \to 0$, then ω tends to Dirac's delta function $\delta(\omega - \omega')$ and the convolution in Eq. (5.137) tends to the non-averaged conjugate product of the respective Jones matrix elements, which corresponds to the limiting case of *total coherence* (there is no averaging introduced by the measurement). Conversely, when $\Delta w \to \infty$, then w tends to the infinitesimal constant $1/\Delta\omega$ and the convolution tends to the mean value of the conjugate product of Jones matrix elements, which corresponds to the limiting case of *total incoherence*

$$\begin{aligned} \langle t^{(k)} t^{(l)*} \rangle &\xrightarrow{\Delta\omega \to 0} t^{(k)}(\omega)\, t^{(l)*}(\omega) \\ \langle t^{(k)} t^{(l)*} \rangle &\xrightarrow{\Delta\omega \to \infty} \frac{1}{\Delta\omega} \int_{-\Delta\omega/2}^{\Delta\omega/2} t^{(k)}(\omega') t^{(l)*}(\omega')\, d\omega' \end{aligned} \tag{5.138}$$

Relations (5.137) and (5.138) can be formulated in terms of Mueller matrices as

$$\begin{aligned} &\mathbf{M}(\omega) = \int \mathbf{M}(\omega')\, w(\omega - \omega'; \Delta\omega)\, d\omega' . \\ &\mathbf{M}(\omega) \xrightarrow{\Delta\omega \to 0} \mathbf{M}(\omega_0) \\ &\mathbf{M}(\omega) \xrightarrow{\Delta\omega \to \infty} \frac{1}{\Delta\omega} \int_{-\Delta\omega/2}^{\Delta\omega/2} \mathbf{M}(\omega')\, d\omega' \\ &\left[\int w(\omega - \omega'; \Delta\omega)\, d\omega' = \int w(\omega; \Delta\omega)\, d\omega = 1 \right] \end{aligned} \tag{5.139}$$

Specific details on the application of this approach to stratified media can be found in (Hingerl and Ossikovski, 2016) where Gaussian, Lorentzian, rectangular and triangular forms for the instrumental function w are considered to describe real measurement conditions.

5.16.2 Depolarization Generated by Partial Spatial Coherence

In analogy to the previous section, the expressions governing partially coherent Mueller matrix reflection polarimetry on spatially inhomogeneous samples are now developed from first principles (Ossikovski and Hingerl 2016).

Spatial inhomogeneity is one of several experimental situations whereby the measured polarimetric response is depolarizing due to the averaging process introduced by the inherent coupling (or interaction) between the measuring instrument and the measured medium or system.

The procedure followed in (Ossikovski and Hingerl 2016) for the derivation of the expressions for the second-order moments $\langle t^{(k)} t^{(l)*} \rangle$, which is summarized here, resembles that of the van Cittert-Zernike theorem from optical coherence theory (Born and Wolf 2005); therefore, the quasi-monochromatic approximation is assumed, i.e., the variations of the sample surface profile (e.g. roughness) are much smaller than the coherence length of the probing electromagnetic wave.

Also, some natural additional assumptions are considered, like (1) both the light spot and the light source areas A and S are much smaller than the squares of the source-sample and sample-detector distances, (2) typical ellipsometer and polarimeter sources consist of halogen or glow-discharge lamps whose individual atomic emissions are spatially uncorrelated and (3) the light source exhibits rotational symmetry. Observe that a light source of uniform brightness is characterized by its coherence area $A_C = \bar{\lambda}^2 D^2 / S$ at the source-sample distance D, where $\bar{\lambda}$ is the mean emission wavelength (Goodman 2015), which allows for introducing the weighting function

$$w(\mathbf{r} - \mathbf{r}') \equiv C F_s\left(|x - x'| / \sqrt{A_C}\, , |y - y'| \cos\varphi / \sqrt{A_C} \right) / A_C \tag{5.140.a}$$

where $\mathbf{r} = (x, y)$, $\mathbf{r}' = (x', y')$ refer to two points of the sample, φ is the incidence angle, F_s is the spatial Fourier transform of the light source intensity I_s, and the constant C is determined by the normalization condition

$$\int_A \int_A w(\mathbf{r} - \mathbf{r}')\, d^2\mathbf{r}\, d^2\mathbf{r}' = 1. \tag{5.140.b}$$

where A denotes the area of the probing light spot. With the above definition, the spatial average of the conjugate product of Jones matrix elements is given by

$$\left\langle t^{(m)}\, t^{(n)*} \right\rangle = \int_A \int_A t^{(m)}(\mathbf{r})\, t^{(n)*}(\mathbf{r}')\, w(\mathbf{r} - \mathbf{r}')\, d^2\mathbf{r}\, d^2\mathbf{r}'. \tag{5.141}$$

Eqs. (5.140) and (5.141) describe the measurement-induced averaging process as the correlation of the (inhomogeneous) sample response in two points, $\mathbf{r}$ and $\mathbf{r}'$, with the point-like detector instrument-dependent weighting function. To generalize such equations to a finite-extent detector, it is enough the use in Eq. (5.141) an effective weighting function $w = w_S w_D$ which is the product of those of the light source, w_S, and the detector, w_D, where w_D is parameterized by the detector coherence area A_D, in full analogy with Eq. (5.140) for the light source. Note that, as expected, the resulting expression is "time reversible" (source $\leftrightarrow$ detector).

There are two limiting cases of Eq. (5.141), namely the coherent and the incoherent one. When $A_C \to \infty$ (A_C the coherence area of the source) which corresponds to a point-like source, then $w \to 1/A^2$ and consequently,

$$\begin{aligned} \left\langle t^{(m)}\, t^{(n)*} \right\rangle^{coh} &= \frac{1}{A^2} \int_A \int_A t^{(m)}(\mathbf{r})\, t^{(n)*}(\mathbf{r}')\, d^2\mathbf{r}\, d^2\mathbf{r}' \\ &= \left[\frac{1}{A} \int_A t^{(m)}(\mathbf{r})\, d^2\mathbf{r} \right] \left[\frac{1}{A} \int_A t^{(n)*}(\mathbf{r}')\, d^2\mathbf{r}' \right] = \left\langle t^{(m)} \right\rangle_A \left\langle t^{(n)*} \right\rangle_A \end{aligned} \tag{5.142}$$

where $\langle \ldots \rangle_A$ denotes averaging over the spot area A. In this case of *total coherence* the spatial average of the conjugate product is equal to the product of the averages of the two factors over the spot area. Conversely, when $A_C \to 0$ then $w \to \delta(\mathbf{x} - \mathbf{x}')/A$, where δ is Dirac's delta function, so that

$$\left\langle t^{(m)}\, t^{(n)*} \right\rangle^{incoh} = \frac{1}{A} \int_A t^{(m)}(\mathbf{r})\, t^{(n)*}(\mathbf{r})\, d^2\mathbf{r} = \left\langle t^{(m)} t^{(n)*} \right\rangle_A \tag{5.143}$$

i.e., in the case of *total incoherence* the spatial average of the conjugate product tends to the average of the same product over the spot area.

In practice, a particularly interesting case of the general relation (5.141) arises when the spatially inhomogeneous sample comprises two juxtaposed media (labeled '*a*' and '*b*') with different, but spatially invariant, Jones matrices $\mathbf{T}_a$ and $\mathbf{T}_b$, that is, $\mathbf{T} = \mathbf{T}_a$ over the spot area A_a and $\mathbf{T} = \mathbf{T}_b$ over the spot area A_b with$A_a + A_b = A$. In this special case

$$\left\langle t^{(m)}\, t^{(n)*} \right\rangle_{ab} = W_{aa}\, t_a^{(m)}\, t_a^{(n)*} + W_{bb}\, t_b^{(m)}\, t_b^{(n)*} + W_{ab}\, t_a^{(m)}\, t_b^{(n)*} + W_{ba}\, t_b^{(m)}\, t_a^{(n)*} \tag{5.144.a}$$

where

$$W_{ab} = \int_{Aa}\int_{Ab} w(\mathbf{r}-\mathbf{r}')d^2\mathbf{r}\, d^2\mathbf{r}' \quad [W_{ab} = W_{ba},\ W_{aa} + W_{bb} + 2W_{ab} = 1] \tag{5.144.b}$$

The coherent and incoherent limiting cases become

$$\begin{gathered} \left\langle t^{(m)}\, t^{(n)*}\right\rangle_{ab}^{coh} = \left[w_a\, t_a^{(m)} + w_b\, t_b^{(m)}\right]\left[w_a\, t_a^{(n)*} + w_b\, t_b^{(n)*}\right] = \left\langle t^{(m)}\right\rangle_A \left\langle t^{(n)*}\right\rangle_A \\ \left\langle t^{(m)}\, t^{(n)*}\right\rangle_{ab}^{incoh} = w_a\, t_a^{(m)}\, t_a^{(n)*} + w_b\, t_b^{(m)}\, t_b^{(n)*} = \left\langle t^{(m)} t^{(n)*}\right\rangle_A \\ \left[w_a = A_a/A,\ w_b = A_b/A,\ w_a + w_b = 1\right] \end{gathered} \tag{5.145}$$

These relations suggest the introduction of the *coherence index* $CI = W_{ab}/\sqrt{W_{aa}W_{bb}}$ as a measure of the spatial coherence of the *binary mixture*. Obviously, $0 \le CI \le 1$, with $CI = 1$ and $CI = 0$ for the coherent and incoherent limiting cases respectively.

The above relations can be readily reformulated by appropriately extending the definition of the Mueller matrix in terms of its Jones generator. In this way, the *cross-Mueller matrix* $\mathbf{K}$ is defined as $\mathbf{K}(\mathbf{T}_1,\mathbf{T}_2) = \mathcal{L}\left(\mathbf{T}_1 \otimes \mathbf{T}_2^*\right)\mathcal{L}^\dagger$. Note that, unlike $\mathbf{M}_J = \mathcal{L}\left(\mathbf{T} \otimes \mathbf{T}^*\right)\mathcal{L}^\dagger$ $\mathbf{K}$ is generally not a Mueller matrix, $\mathbf{K}(\mathbf{T},\mathbf{T})$ is, while $\mathbf{K}(\mathbf{T}_1,\mathbf{T}_2) = \mathbf{K}^*(\mathbf{T}_1,\mathbf{T}_2)$, so that the matrix $\mathbf{K}(\mathbf{T}_1,\mathbf{T}_2) + \mathbf{K}(\mathbf{T}_2,\mathbf{T}_1) = 2\,\mathrm{Re}\left[\mathbf{K}(\mathbf{T}_1,\mathbf{T}_2)\right]$ is real (without still being a Mueller matrix). In virtue of the element-wise definition of $\mathbf{M}$ and the linearity property, the fundamental relation (5.141) can now be rewritten as

$$\mathbf{M} = \int_A\int_A \mathbf{K}(\mathbf{r},\mathbf{r}')\, w(\mathbf{r}-\mathbf{r}')\, d^2\mathbf{r}\, d^2\mathbf{r}' \tag{5.146}$$

with the coherent and incoherent limiting cases given by

$$\mathbf{M}^{coh} = \mathbf{M}\left[\frac{1}{A}\int_A \mathbf{T}(\mathbf{r})\, d^2\mathbf{r}\right] \equiv \mathbf{M}(\langle\mathbf{T}\rangle_A) \qquad \mathbf{M}^{incoh} = \frac{1}{A}\int_A \mathbf{M}(\mathbf{r})\, d^2\mathbf{r} \equiv \langle\mathbf{M}\rangle_A \tag{5.147}$$

Similarly, in the case of the binary mixture,

$$\begin{gathered} \mathbf{M}_{ab} = W_{aa}\,\mathbf{M}(\mathbf{T}_a) + W_{bb}\,\mathbf{M}(\mathbf{T}_b) + 2W_{ab}\,\mathrm{Re}\left[\mathbf{K}(\mathbf{T}_a,\mathbf{T}_b)\right] \\ \mathbf{M}_{ab}^{coh} = \mathbf{M}(w_a\mathbf{T}_a + w_b\mathbf{T}_b) \quad \mathbf{M}_{ab}^{incoh} = w_a\,\mathbf{M}(\mathbf{T}_a) + w_b\,\mathbf{M}(\mathbf{T}_b) \end{gathered} \tag{5.148}$$

In accordance with Section 5.2, the relations obtained for the coherent case show that the result of a polarimetric experiment in the limit of total coherence is the pure Mueller matrix corresponding to the averaged Jones matrices of the different constituents of the inhomogeneous sample. Conversely, the measurement result in the totally incoherent case is the average of the Mueller-Jones matrices of the constituents, resulting in a depolarizing $\mathbf{M}$. In particular, the above formulation for the incoherent case has been widely used in interpreting finite spot-size measurements on patterned samples performed with extended thermal light source polarimeters whose coherence areas satisfy $A_C \ll A$.

In the general case, the Mueller matrix of a spatially inhomogeneous medium is represented as an average of the cross-Mueller matrices of all possible pairs of its constituents. When the measured medium is spatially homogeneous the cross-Mueller matrices are spatially independent (i.e., $\mathbf{K}(\mathbf{r}_0, \mathbf{r}_0) = \mathbf{M}(\mathbf{r}_0)$, with $\mathbf{r} = \mathbf{r}' = \mathbf{r}_0$) so that the averaging process operates over constant quantities and consequently, the experimental Mueller matrix is nondepolarizing. Therefore, the polarimetric response is nondepolarizing either (1) when the medium is spatially homogeneous, or (2) when the measurement is totally coherent (formally $A_C \to \infty$; in practice $A_C >> A$). Conversely, the measured Mueller matrix is depolarizing when both the medium is inhomogeneous and the coherence is partial (formally $A_C \to \infty$, in practice $A_C >> A$). It is remarkable that partial coherence alone is a necessary, but not a sufficient, condition for the presence of depolarization.

5.16.3 Generalized Formulation of Depolarization Generated by Partial Spectral and Spatial Coherence

It is instructive to compare Eq. (5.146) to the fundamental relation (5.139) for a partially spectrally coherent polarimetric experiment. The analogies between instrumental function and correlation function (both denoted by w) and spectral resolution ($\Delta\omega$) – coherence area (A_C) are evident. However, Eq. (5.139) mathematically represents a convolution over a one-dimensional variable (the frequency ω), whereas Eq. (5.146) is a correlation (or double convolution) over two two-dimensional variables (the positions $\mathbf{r}$ and $\mathbf{r}'$). This is so for the following reasons: (a) spectral (or temporal) and spatial coherence are respectively one and two-dimensional phenomena, and (b) the detector has a spectroscopic capability in the first case while it is position-insensitive in the second. If a spatially resolved detector is used with uniform sensitivity (e.g., an imaging CCD camera) instead, it is easy to use the above derivations to generalize Eq. (5.146) as

$$\begin{aligned}\mathbf{M}(\mathbf{r}_D) &= \int_A \int_A \mathbf{K}(\mathbf{r}, \mathbf{r}')\, w(\mathbf{r} - \mathbf{r}')\, e^{-ik(\mathbf{r}-\mathbf{r}')^T \gamma \mathbf{r}_D}\, d^2\mathbf{r}\, d^2\mathbf{r}' \\ &= \int_A \int_A \mathbf{K}(\mathbf{r}, \mathbf{r}')\, \tilde{w}(\mathbf{r} - \mathbf{r}', \mathbf{r}_D)\, d^2\mathbf{r}\, d^2\mathbf{r}'\end{aligned} \tag{5.149}$$

where $\mathbf{r}_D$ is the position on the detector and $\gamma \equiv \mathrm{diag}(1/r_S, \cos\varphi/r_S)$ is a 2×2 diagonal matrix in which r_S is the sample-to-detector distance and φ is the incidence angle. Thus, the measured Mueller matrix image $\mathbf{M}(\mathbf{r}_D)$ is interpretable as either (a) the double Fourier transform of the product of the cross-Mueller matrix $\mathbf{K}$ of the spatially inhomogeneous medium and the light source correlation function w (first integral), or (b) the double convolution of $\mathbf{K}$ with complex instrumental function $\tilde{w}$ given by the product of the (real) w and the (complex) detector phase factor (second integral).

6

Physical Quantities in a Mueller Matrix

6.1 Introduction

The concept and basic properties (both depolarizing and nondepolarizing) of Mueller matrices having been introduced in the previous chapters, the contents of the present chapter are devoted to a deeper analysis of the main physical quantities involved in polarimetric phenomena. In particular, the *sources of polarimetric purity* are studied from the point of view of certain complementary attributes like *diattenuation, polarizance* and *polarimetric dimension index*. The depolarization of a medium represented by a Mueller matrix **M** is also intimately linked to the randomness of the covariance matrix **H** (or, equivalently, to the randomness of the coherency matrix **C**) associated with **M**. Therefore, the analysis of the eigenvalue spectrum of **H** provides information about the polarimetric purity in terms of the number and the relative weights of the canonical pure constituents of the medium and leads to the definition of a set of representative parameters called *indices of polarimetric purity*. Other possible alternative sets of three scalar purity-randomness descriptors are also studied in a general framework for the depolarization metric spaces. Overall measures of the closeness of **M** to a pure Mueller matrix (polarimetric purity), providing global information of the depolarizing features of the interaction represented by **M**, like the degree of polarimetric purity P_Δ and others, have been introduced in Section 5.10 and thus, are not replicated in this chapter.

All of the above-mentioned quantities are examples of certain parameters contained in the Mueller matrix which are invariant with respect to certain transformations and therefore bear polarimetric information in an independent and objective manner. Thus, the anisotropies exhibited by a given medium can be likewise characterized through a set of *anisotropy coefficients*, which are also defined and studied in this chapter.

The last section deals with a synthesis procedure for generating any depolarizing Mueller matrix **M** from an associated reference pure Mueller matrix $\mathbf{M}_J(\mathbf{M})$ through appropriate adjustment of a set of indices of polarimetric purity.

6.2 Components of Purity of a Mueller Matrix

This section is devoted to the analysis of how the diattenuation D, the polarizance P and an additional quantity, the so-called *polarimetric dimension index* P_S, contribute separately to the overall degree of polarimetric purity P_Δ.

DOI: 10.1201/9780367815578-6

Next, we extend to Mueller matrices (with elements m_{kl}) associated with media the concept of components of purity introduced in Chapters 1 and 2 for 2D and 3D polarization states respectively. Obviously, the components of purity of a Mueller matrix have a peculiar physical meaning, different from that of polarization states.

Let us consider the block form of a generic Mueller matrix **M**, and recall the definitions of the *components of purity* D, P and P_S (Gil 2011)

$$\mathbf{M} = m_{00}\begin{pmatrix} 1 & \mathbf{D}^T \\ \mathbf{P} & \mathbf{m} \end{pmatrix}$$

$$D \equiv |\mathbf{D}| = \frac{\sqrt{m_{01}^2 + m_{02}^2 + m_{03}^2}}{m_{00}} \qquad P \equiv |\mathbf{P}| = \frac{\sqrt{m_{10}^2 + m_{20}^2 + m_{30}^2}}{m_{00}} \qquad P_S \equiv \frac{\|\mathbf{m}\|_2}{\sqrt{3}} = \frac{\sqrt{\sum_{k,l=1}^{3} m_{kl}^2}}{\sqrt{3}\,m_{00}} \tag{6.1}$$

The square of the Frobenius norm of **M** can be expressed as

$$\|\mathbf{M}\|_2^2 = m_{00}^2\,\|\hat{\mathbf{M}}\|_2^2 = m_{00}^2\left(1 + D^2 + P^2 + 3P_S^2\right) \tag{6.2}$$

where D, P and P_S provide separate contributions to $\|\hat{\mathbf{M}}\|_2$, and, as shown in Eq. (5.79), the square of the degree of polarimetric purity of **M** can be expressed as (Gil 2011)

$$P_\Delta^2 = \frac{1}{3}\left(D^2 + P^2 + 3P_s^2\right) \tag{6.3}$$

where D, P and P_S represent separate contributions to P_Δ.

To inspect the physical information provided by each of the components of purity, as well as their role played as sources of polarimetric purity, the following sections are devoted to the transformations, through **M** and $\mathbf{M}^T$, of an incident state $\hat{\mathbf{s}}$ (with unit intensity) into an emerging state $\mathbf{s}$,^{f}

$$\mathbf{s}^f = \mathbf{M}\hat{\mathbf{s}} = m_{00}\begin{pmatrix} 1 & \mathbf{D}^T \\ \mathbf{P} & \mathbf{m} \end{pmatrix}\begin{pmatrix} 1 \\ \mathcal{P}\mathbf{u} \end{pmatrix} = m_{00}\begin{pmatrix} 1 + \mathcal{P}\mathbf{D}^T\mathbf{u} \\ \mathbf{P} + \mathcal{P}\mathbf{m}\mathbf{u} \end{pmatrix} \tag{6.4}$$

6.2.1 Mean Intensity Coefficient

Let us first observe that for any given incident state $\hat{\mathbf{s}} = (1, \mathcal{P}\mathbf{u}^T)^T$, there exists a corresponding orthogonal incident state $\hat{\mathbf{s}}_\perp = (1, -\mathcal{P}\mathbf{u}^T)^T$, so that the average of the output intensities s_0^f and $s_{0\perp}^f$ for each pair of mutually orthogonal incident states is given by

$$\frac{1}{2}\left(s_0^f + s_{0\perp}^f\right) = \frac{1}{2}m_{00}\left(1 + \mathcal{P}\mathbf{D}^T\mathbf{u} + 1 - \mathcal{P}\mathbf{D}^T\mathbf{u}\right) = m_{00} \tag{6.5}$$

Therefore, the overall average of the intensity coefficient for all the possible incident polarization states (i.e., the entire Poincaré ball) is precisely m_{00}, which in turn coincides with the intensity coefficient corresponding to an unpolarized incident state,

$$\mathbf{M}\begin{pmatrix}1\\ \mathbf{0}\end{pmatrix}=m_{00}\begin{pmatrix}1 & \mathbf{D}^T\\ \mathbf{P} & \mathbf{m}\end{pmatrix}\begin{pmatrix}1\\ \mathbf{0}\end{pmatrix}=m_{00}\begin{pmatrix}1\\ \mathbf{P}\end{pmatrix} \tag{6.6}$$

It is straightforward to show that an analogous result is obtained when the reciprocal Mueller matrix $\mathbf{M}^r = \text{diag}(1,1,-1,1)\,\mathbf{M}^T\,\text{diag}(1,1,-1,1)$ is considered, and consequently, m_{00} is the *mean intensity coefficient* for both the forward and the reverse (reciprocal) directions.

6.2.2 Diattenuation

From Eq. (6.4) we deduce that the maximal $s^f_{0\,\text{max}}$ and minimal $s^f_{0\,\text{min}}$ intensities for the emerging state $\mathbf{s}^f = \mathbf{M}\hat{\mathbf{s}}$ correspond respectively to the following mutually orthogonal incident states:

$$\hat{\mathbf{s}}_{\hat{D}} \equiv \begin{pmatrix}1\\ \hat{\mathbf{D}}\end{pmatrix} \qquad \hat{\mathbf{s}}_{-\hat{D}} \equiv \begin{pmatrix}1\\ -\hat{\mathbf{D}}\end{pmatrix} \qquad \hat{\mathbf{D}} \equiv \mathbf{D}/D \tag{6.7}$$

where $\mathbf{D}$ is the diattenuation vector of $\mathbf{M}$, so that

$$s^f_{0\,\text{max}} = m_{00}(1+D) \qquad s^f_{0\,\text{min}} = m_{00}(1-D) \tag{6.8}$$

and therefore

$$D = \left(s^f_{0\,\text{max}} - s^f_{0\,\text{min}}\right)\Big/\left(s^f_{0\,\text{max}} + s^f_{0\,\text{min}}\right) = \left(s^f_{0\,\text{max}} - s^f_{0\,\text{min}}\right)\Big/2m_{00} \tag{6.9}$$

Thus, the diattenuation D of $\mathbf{M}$ is a measure of the relative difference between the maximal and minimal intensities of the emerging states with respect to all possible incident polarization states.

Due to its very definition, D is non-negative and takes values in the interval $0 \le D \le 1$. The value $D = 0$ corresponds to media that do not exhibit diattenuation, while $D = 1$ implies that there exists an incident state $\mathbf{s}_{-\hat{D}}$ for which the output intensity is zero, whereas the output intensity is maximal for the incident state $\mathbf{s}_{\hat{D}}$ (orthogonal to $\mathbf{s}_{-\hat{D}}$, i.e., $\mathbf{s}_{\hat{D}}$ and $\mathbf{s}_{-\hat{D}}$ are represented by antipodal points on the Poincaré sphere). When $D = 1$ the medium is called an *analyzer* (Lu and Chipman 1996). The projection property of Mueller matrices of analyzers canceling the intensity of the image of $\mathbf{s}_{-\hat{D}}$ entails necessarily $\det\mathbf{M} = 0$. Moreover, the Mueller matrix $\mathbf{M}_{P\hat{D}}$ of any analyzer can always be expressed as

$$\mathbf{M}_{P\hat{D}} = m_{00}\left(\mathbf{s}_P \otimes \mathbf{s}^T_{\hat{D}}\right) = m_{00}\begin{pmatrix}1 & \hat{\mathbf{D}}^T\\ \mathbf{P} & \mathbf{P}\otimes\hat{\mathbf{D}}^T\end{pmatrix}$$
$$\left[\mathbf{s}_{\hat{D}} \equiv \begin{pmatrix}1\\ \hat{\mathbf{D}}\end{pmatrix} \qquad \mathbf{s}_P \equiv \begin{pmatrix}1\\ \mathbf{P}\end{pmatrix} \qquad D \equiv |\hat{\mathbf{D}}| = 1 \qquad P \equiv |\mathbf{P}| \le 1\right] \tag{6.10}$$

where $\mathbf{s}_{\hat{D}}$ is a totally polarized Stokes vector defined from the unit diattenuation vector $\hat{\mathbf{D}}$ of $\mathbf{M}_{P\hat{D}}$, and $\mathbf{s}_P$ is a Stokes vector with arbitrary degree of polarization. When $D = 1$ and

$P = 1$, then $\mathbf{M}_{\hat{P}\hat{D}}$ is the pure Mueller matrix of a *perfect polarizer* (also called *pure polarizer, nondepolarizing polarizer* or *ideal polarizer*) and is non-normal, in general. When $D = 1$ and $P < 1$, $\mathbf{M}_{P\hat{D}}$ is nonpure and corresponds to a *depolarizing analyzer* (Lu and Chipman 1996), in which case $\mathbf{M}_{P\hat{D}}$ can be expressed as the serial combination of a pure polarizer followed by an *intrinsic depolarizer* (or *diagonal depolarizer*)

$$\mathbf{M}_{P\hat{D}} = m_{00}\begin{pmatrix} 1 & \hat{\mathbf{D}}^T \\ \mathbf{P} & \mathbf{P}\otimes\hat{\mathbf{D}}^T \end{pmatrix} = \left\{2m_{00}\begin{pmatrix} 1 & \mathbf{0}^T \\ \mathbf{0} & \mathbf{m}_d \end{pmatrix}\right\}\left\{\frac{1}{2}\begin{pmatrix} 1 & \hat{\mathbf{D}}^T \\ \hat{\mathbf{D}} & \hat{\mathbf{D}}\otimes\hat{\mathbf{D}}^T \end{pmatrix}\right\}$$
$$\begin{bmatrix} \mathbf{m}_d \equiv \mathrm{diag}(d_1, d_2, d_3) \quad \hat{\mathbf{D}} \equiv (D_1, D_2, D_3)^T \quad \mathbf{P} \equiv (P_1, P_2, P_3)^T \\ D \equiv \sqrt{D_1^2 + D_2^2 + D_3^2} = 1 \quad |d_i| \le 1 \quad d_i D_i = P_i \quad (i = 1,2,3) \end{bmatrix} \tag{6.11}$$

Note that the passivity of $\mathbf{M}_{P\hat{D}}$ entails $m_{00} \le 1/2$, and therefore, both the pure polarizer and the intrinsic depolarizer of the above serial combination can be chosen to be passive. Moreover, the diagonal elements of the depolarizer are calculated from the components of the diattenuation and polarizance vectors $\hat{\mathbf{D}}$ and $\mathbf{P}$ of $\mathbf{M}_{P\hat{D}}$.

As seen in Chapter 3, for the sake of simplicity and clarity of many expressions, it is sometimes useful to express the diattenuation D in the form $D \equiv \cos\kappa_D$, where κ_D is the *diattenuation angle* (with $0 \le \kappa_D \le \pi/2$), which also determines the *counter-diattenuation* defined as $\sin\kappa_D \equiv \sqrt{1-D^2}$.

6.2.3 Reciprocal Diattenuation

Let us now consider incident states with unit intensity in the reverse incidence direction, so that the medium is represented by $\mathbf{M}^r$ instead of $\mathbf{M}$, and the emerging states are given by

$$\mathbf{s}^r = \mathbf{M}^r\hat{\mathbf{s}} = \mathbf{X}\mathbf{M}^T\mathbf{X}\hat{\mathbf{s}} = m_{00}\,\mathbf{X}\begin{pmatrix} 1 & \mathbf{P}^T \\ \mathbf{D} & \mathbf{m}^T \end{pmatrix}\mathbf{X}\begin{pmatrix} 1 \\ \mathcal{P}\mathbf{u} \end{pmatrix} = m_{00}\begin{pmatrix} 1+\mathcal{P}\,\mathbf{P}^T\mathbf{x}\mathbf{u} \\ \mathbf{x}\mathbf{D}+\mathcal{P}\,\mathbf{x}\mathbf{m}\mathbf{x}\mathbf{u} \end{pmatrix}$$
$$[\mathbf{X} \equiv \mathrm{diag}(1,1,-1,1) \quad \mathbf{x} \equiv \mathrm{diag}(1,-1,1)] \tag{6.12}$$

where the maximal $s^r_{0\max}$ and minimal $s^r_{0\min}$ intensities of the emerging state $\mathbf{s}^r = \mathbf{M}^r\hat{\mathbf{s}}$

$$s^f_{0\max} = m_{00}(1+P) \qquad s^f_{0\max} = m_{00}(1-P) \tag{6.13}$$

correspond respectively to the incident states

$$\hat{\mathbf{s}}_{rP} \equiv \begin{pmatrix} 1 \\ \mathbf{x}\hat{\mathbf{P}} \end{pmatrix} \qquad \hat{\mathbf{s}}_{-rP} \equiv \begin{pmatrix} 1 \\ -\mathbf{x}\hat{\mathbf{P}} \end{pmatrix} \qquad [\hat{\mathbf{P}} \equiv \mathbf{P}/P] \tag{6.14}$$

and therefore

$$P = \left(s^r_{0\max} - s^r_{0\min}\right)/\left(s^r_{0\max} + s^r_{0\min}\right) = \left(s^r_{0\max} - s^r_{0\min}\right)/2m_{00} \tag{6.15}$$

As indicated in Section 5.8, the diattenuation of $\mathbf{M}^T$ is equal to the diattenuation of the reverse Mueller matrix $\mathbf{M}^r$and is equal to the polarizance P of $\mathbf{M}$. Thus, the polarizance P of $\mathbf{M}$ can be considered as a measure of the relative difference between the maximal and minimal intensities of the emerging states with respect to all possible polarization states incoming in the reverse direction. Obviously, since P is the diattenuation of $\mathbf{M}^T$, it takes non-negative values in the interval $0 \le P \le 1$. When the reverse incidence direction is considered, $P = 0$ corresponds to media that do not exhibit diattenuation, while $P = 1$ implies that there exists an incident state $\mathbf{s}_{-rP}$ for which the output intensity is zero, whereas the output intensity is maximal for the incident state $\mathbf{s}_{rP}$ (orthogonal to $\mathbf{s}_{-rP}$).

6.2.4 Polarizance

Consider an unpolarized incident state $\hat{\mathbf{s}}_u$ interacting in the forward direction, so that

$$\mathbf{s}^f = \mathbf{M}\hat{\mathbf{s}}_u = m_{00}\begin{pmatrix} 1 & \mathbf{D}^T \\ \mathbf{P} & \mathbf{m} \end{pmatrix}\begin{pmatrix} 1 \\ \mathbf{0} \end{pmatrix} = m_{00}\begin{pmatrix} 1 \\ \mathbf{P} \end{pmatrix} = m_{00}\begin{pmatrix} 1 \\ P\hat{\mathbf{P}} \end{pmatrix} \tag{6.16}$$

and therefore, the polarizance P of $\mathbf{M}$ is the degree of polarization of the emerging state. Consequently, P is a measure of the ability of $\mathbf{M}$ to *enpolarize* (i.e., to increase the degree of polarization of) an incident unpolarized state in the forward direction. It follows from the above sections that the diattenuation of the reciprocal Mueller matrix $\mathbf{M}^r$ is precisely the polarizance P of $\mathbf{M}$.

As with diattenuation, when appropriate, P will be expressed in the form $P \equiv \cos\kappa_P$, where κ_P is the *polarizance angle* (with $0 \le \kappa_P \le \pi/2$), which in turn determines the *counter-polarizance*, defined as $\sin\kappa_P \equiv \sqrt{1-P^2}$.

When $P = 1$ the medium is called a *polarizer* and, regardless of the incident state of polarization, the emerging state of polarization is totally polarized with Stokes vector proportional to $\mathbf{s}_P$. The projection property of polarizers consisting in converting any incident state into a fixed emerging pure state implies that necessarily $\det\mathbf{M} = 0$. Moreover, in accordance with the fact that the transposed Mueller matrix of an analyzer is a polarizer, and by considering Eq. (6.11), we deduce that the Mueller matrix $\mathbf{M}_{\hat{P}D}$ of a polarizer can always be expressed as

$$\begin{gathered} \mathbf{M}_{\hat{P}D} = m_{00}\left(\mathbf{s}_{\hat{P}} \otimes \mathbf{s}_D^T\right) = m_{00}\begin{pmatrix} 1 & \mathbf{D}^T \\ \hat{\mathbf{P}} & \hat{\mathbf{P}} \otimes \mathbf{D}^T \end{pmatrix} \\ \left[\mathbf{s}_D \equiv \begin{pmatrix} 1 \\ \mathbf{D} \end{pmatrix} \quad \mathbf{s}_{\hat{P}} \equiv \begin{pmatrix} 1 \\ \hat{\mathbf{P}} \end{pmatrix} \quad D \equiv |\mathbf{D}| \le 1 \quad P \equiv |\hat{\mathbf{P}}| = 1\right] \end{gathered} \tag{6.17}$$

where $\mathbf{s}_{\hat{P}}$ is a totally polarized Stokes vector defined from the unit polarizance vector $\hat{\mathbf{P}}$ of $\mathbf{M}_{\hat{P}D}$, and $\mathbf{s}_D$ is a Stokes vector defined from the diattenuation vector $\mathbf{D}$ of $\mathbf{M}_{\hat{P}D}$ so that the absolute value D of $\mathbf{D}$ determines the degree of polarization of $\mathbf{s}_D$. As we have noted above, when $D = 1$ and $P = 1$, then $\mathbf{M}_{\hat{P}\hat{D}}$ is the pure Mueller matrix of a *perfect polarizer* (also termed *nondepolarizing polarizer* or *ideal polarizer*), and is non-normal, in general. When $P = 1$ and $D < 1$, $\mathbf{M}_{\hat{P}D}$ is the nonpure Mueller matrix of a *depolarizing polarizer* (Lu and Chipman 1996), in which case $\mathbf{M}_{\hat{P}D}$ can be expressed as the serial combination of an intrinsic (or diagonal) depolarizer, followed by a pure polarizer

$$\mathbf{M}_{\hat{P}D} = m_{00}\begin{pmatrix} 1 & \mathbf{D}^T \\ \hat{\mathbf{P}} & \hat{\mathbf{P}}\otimes\mathbf{D}^T \end{pmatrix} = \left\{\frac{1}{2}\begin{pmatrix} 1 & \hat{\mathbf{P}}^T \\ \hat{\mathbf{P}} & \hat{\mathbf{P}}\otimes\hat{\mathbf{P}}^T \end{pmatrix}\right\}\left\{2m_{00}\begin{pmatrix} 1 & \mathbf{0}^T \\ \mathbf{0} & \mathbf{m}_d \end{pmatrix}\right\}$$
$$\left[\begin{matrix} \mathbf{m}_d \equiv \mathrm{diag}(d_1,d_2,d_3) \quad \mathbf{D}\equiv(D_1,D_2,D_3)^T \quad \mathbf{P}\equiv(P_1,P_2,P_3)^T \\ P\equiv\sqrt{P_1^2+P_2^2+P_3^2}=1 \quad D\equiv\sqrt{D_1^2+D_2^2+D_3^2}<1 \quad |d_i|\le 1 \quad D_i=d_iP_i \quad (i=1,2,3) \end{matrix}\right] \tag{6.18}$$

Note that the passivity of $\mathbf{M}_{\hat{P}D}$ entails $m_{00}\le 1/2$, and therefore, both the pure polarizer and the intrinsic depolarizer can be chosen to be passive. Notice also that the diagonal elements of the depolarizer are calculated from the components of the diattenuation and polarizance vectors $\mathbf{D}$ and $\hat{\mathbf{P}}$ of $\mathbf{M}_{\hat{P}D}$.

6.2.5 Reciprocal Polarizance

For an incident unpolarized state in the reverse direction the emerging state is

$$\mathbf{s}^r = \mathbf{M}^r\hat{\mathbf{s}}_u = m_{00}\begin{pmatrix} 1 & \mathbf{P}^T\mathbf{x} \\ \mathbf{x}\mathbf{D} & \mathbf{x}\mathbf{m}^T\mathbf{x} \end{pmatrix}\begin{pmatrix} 1 \\ 0 \end{pmatrix} = m_{00}\begin{pmatrix} 1 \\ \mathbf{x}\mathbf{D} \end{pmatrix} = m_{00}\begin{pmatrix} 1 \\ D\mathbf{x}\hat{\mathbf{D}} \end{pmatrix} \qquad [\mathbf{x}\equiv\mathrm{diag}(1,-1,1)] \tag{6.19}$$

where $\mathbf{x}\hat{\mathbf{D}}$ is a unit vector, and the diattenuation D is precisely the degree of polarization of the emerging state image of $\hat{\mathbf{s}}_u$ through $\mathbf{M}^r$, that is to say, D is the polarizance of $\mathbf{M}^r$ and $\mathbf{M}^T$.

6.2.6 Degree of Polarizance

The above considerations about the physical meaning of the polarizance P and the diattenuation D of a Mueller matrix show their dual role depending on the direction of propagation of light (forward or reverse). Thus, D is both the diattenuation of $\mathbf{M}$ and the polarizance of $\mathbf{M}^r$, while P is both the polarizance of $\mathbf{M}$ and the diattenuation of $\mathbf{M}^r$. Regardless of the fact that P and D have respective specific and well-defined physical meanings, their conceptual relationship motivates the introduction of the *degree of polarizance* P_P measuring their joint contribution to polarimetric purity (Gil 2011)

$$P_P \equiv \sqrt{\left(P^2+D^2\right)/2} \tag{6.20}$$

Thus, P_P constitutes a global measure of the diattenuation-polarizance of a given Mueller matrix. Since P and D are measures of the enpolarizing power of the medium for forward and reverse interactions respectively, media exhibiting $P_p > 0$ will be hereafter referred to as *enpolarizing*, while the term *nonenpolarizing* will refer to media with zero degree of polarizance ($P_p = 0$).

The value of P_P is restricted to $0 \le P_P \le 1$, so that $P_P = 1$ corresponds to a perfect polarizer. The value $P_P = 0$ is attained when the corresponding Mueller matrix **M** does not exhibit any enpolarizing effect $(P = D = 0)$ and, consequently, **M** has the form of a *nonenpolarizing Mueller matrix*

$$\mathbf{M}_O = m_{00}\begin{pmatrix} 1 & \mathbf{0}^T \\ \mathbf{0} & \mathbf{m} \end{pmatrix} \tag{6.21}$$

which can always be represented as a parallel mixture of pure retarders (Gil 2007). In the particular case where $\mathbf{m}_O$ is orthogonal, $\mathbf{M}_O$ corresponds to a pure retarder.

The degree of polarizance has been shown to be useful for certain applications, for example in simple graphic representation and physical interpretation of Mueller matrices with respect to their values of P_P and P_S (see Section 6.5).

Given a Mueller matrix **M**, it is interesting to consider now the *I*-image (or *intensity image*), defined as the surface swept by the end-point of the three-dimensional real vectors $\mathbf{i}'_p$ whose directions (in the Poincaré sphere) are those of the polarization vectors $\mathbf{p}'$ of the transformed Stokes vectors $\mathbf{M}\hat{\mathbf{s}}_p$ ($\hat{\mathbf{s}}_p$ being the intensity-normalized pure Stokes vectors sweeping the entire surface of the Poincaré sphere) and whose magnitudes are those of the intensities $(\mathbf{M}\hat{\mathbf{s}}_p)_0$ of $\mathbf{M}\hat{\mathbf{s}}_p$. Thus,

$$\mathbf{i}'_p \equiv \left(\mathbf{M}\hat{\mathbf{s}}_p\right)_0 \hat{\mathbf{p}}'$$
$$\left[\hat{\mathbf{p}}' \equiv \frac{\mathbf{p}'}{|\mathbf{p}'|} \quad \mathbf{p}' \equiv \left[\left(\mathbf{M}\hat{\mathbf{s}}_p\right)_1, \left(\mathbf{M}\hat{\mathbf{s}}_p\right)_2, \left(\mathbf{M}\hat{\mathbf{s}}_p\right)_3\right]^T \quad \hat{\mathbf{s}}_p \equiv \begin{pmatrix} 1 \\ \hat{\mathbf{p}} \end{pmatrix} \quad |\hat{\mathbf{p}}| = 1\right] \tag{6.22}$$

Note that, provided $P < 1$, the *I*-image of **M** is a sphere if and only if $D = 0$. When $P = 1$, the *I*-image is a single point determined by the end point of the polarizance vector **P** of **M**. When $D \ne 0$, the *I*-image exhibits an anisotropic deformation with respect to the unit sphere. This indicates selective transmittance associated with the diattenuation. In the particular case where $\mathbf{P} = \mathbf{0}$ and $\mathbf{m} = \mathbf{0}_3$, the *I*-image is not determined.

Note that the *I*-image is different from the *P*-image (determining the characteristic ellipsoid *E* associated with **M** that will be studied in detail in Chapter 10). While the distances of the points of the *P*-image to the origin are the respective degrees of polarization, the distances of the points of the *I*-image to the origin are the respective intensities. Both images provide interesting complementary geometric views.

From the reciprocity properties of **M** we see that the *I*-image of $\mathbf{M}^r$ is a sphere if and only if $P = 0$. When $P \ne 0$, the *I*-image of $\mathbf{M}^r$exhibits non-homothetic (non-isotropic) deformation with respect to the unit sphere. This indicates selective transmittance associated with the polarizance. Thus, values of P_P in the interval $0 < P_P \le 1$ indicate dependence of transmittance with respect to the state of polarization of the incoming light in the forward or/and the reverse direction. Furthermore, the *I*-image of $\mathbf{M}^r$ is akin to that of $\mathbf{M}^T$ in the sense that the transformation connecting these does not affect the values of *D* and *P* (hence of P_P). Therefore, we conclude that the *I*-images of **M** and of $\mathbf{M}^T$ are both spheres if and only if $P_P = 0$. When $P_P \ne 0$, at least one of these two *I*-images is not a sphere and, thus, nonzero values of P_P are associated with certain anisotropic distribution of the intensity in the *I*-images.

6.2.7 Components of Diattenuation and Polarizance

The linear diattenuation D_L and *circular diattenuation* D_C of a Mueller matrix $\mathbf{M}$ with a diattenuation vector $\mathbf{D} \equiv (D_1, D_2, D_3)^T$ are defined as the respective *degree of linear polarization* and *degree of circular polarization* of the Stokes vector $\mathbf{s}_D \equiv (1, \mathbf{D}^T)^T$

$$D_L \equiv \sqrt{D_1^2 + D_2^2} \qquad D_C \equiv |D_3| \tag{6.23}$$

When $D_C = 0$ the diattenuation affects exclusively the s_1 and s_2 components (linear components) of the Stokes vectors of the incident states. Conversely, when $D_L = 0$ the diattenuation affects only the s_3 component (the circular component) of the Stokes vectors of the incident states.

The linear polarizance P_L and *circular polarizance* P_C of a Mueller matrix $\mathbf{M}$ with a polarizance vector $\mathbf{P} \equiv (P_1, P_2, P_3)^T$ are defined as the respective *degree of linear polarization* and *degree of circular polarization* of the Stokes vector $\mathbf{s}_P \equiv \mathbf{M}\,\mathbf{s}_u$ that is the image of the incident unpolarized state $\mathbf{s}_u \equiv (1, \mathbf{0}^T)^T$

$$P_L \equiv \sqrt{P_1^2 + P_2^2} \qquad P_C \equiv P_3 \tag{6.24}$$

When $P_C = 0$ the polarizance affects exclusively the linear components of the Stokes vector $\mathbf{s}_P$. Conversely, when $P_L = 0$ the polarizance affects only the circular component of $\mathbf{s}_P$.

6.2.8 Polarimetric Dimension Index

The quantity $P_S(\mathbf{M}) = \|\mathbf{m}\|_2/\sqrt{3}$ introduced in Eq. (6.1) and satisfying $0 \le P_S \le 1$, can be expressed in terms of the absolute values of the constitutive vectors $\mathbf{k}$, $\mathbf{r}$ and $\mathbf{q}$ of the partitioned form of $\mathbf{m}$ in Eqs. (5.104–5.105)

$$P_S = \sqrt{k^2 + \left(r^2 + q^2\right)/3} \tag{6.25}$$

or, in virtue of the invariance properties of the arrow form $\mathbf{M}_A$ of $\mathbf{M}$ (see Section 5.12),

$$P_S(\mathbf{M}) = P_S(\mathbf{M}_A) = \sqrt{(a_1^2 + a_2^2 + a_3^2)/3} \tag{6.26}$$

or, equivalently, in terms of the statistical parameters σ_i^2 $(i = 0,1,2,3)$ (see Section 5.13.2) (Gil 2022)

$$P_S(\mathbf{M}) = \sqrt{\left(\hat{\sigma}_0^2 - \hat{\sigma}_1^2\right)^2 + \left(\hat{\sigma}_0^2 - \hat{\sigma}_2^2\right)^2 + \left(\hat{\sigma}_0^2 - \hat{\sigma}_3^2\right)^2 + \left(\hat{\sigma}_1^2 - \hat{\sigma}_2^2\right)^2 + \left(\hat{\sigma}_1^2 - \hat{\sigma}_3^2\right)^2 + \left(\hat{\sigma}_2^2 - \hat{\sigma}_3^2\right)^2} \Big/ \sqrt{3}$$
$$\left[\hat{\sigma}_i^2 \equiv \sigma_i^2 / \left(\sigma_0^2 + \sigma_1^2 + \sigma_2^2 + \sigma_3^2\right)\right] \tag{6.27}$$

so that $P_S = 1$ when $\sigma_3 = \sigma_2 = \sigma_1 = 0$ (with $\sigma_0 \neq 0$) and $P_S = 0$ when $\sigma_3 = \sigma_2 = \sigma_1 = \sigma_0$.

Thus, the definition of P_S formally constitutes the four-dimensional version of the dimensionality index d defined by Norrman et al. (2017), see Eq. (2.88), for three-dimensional

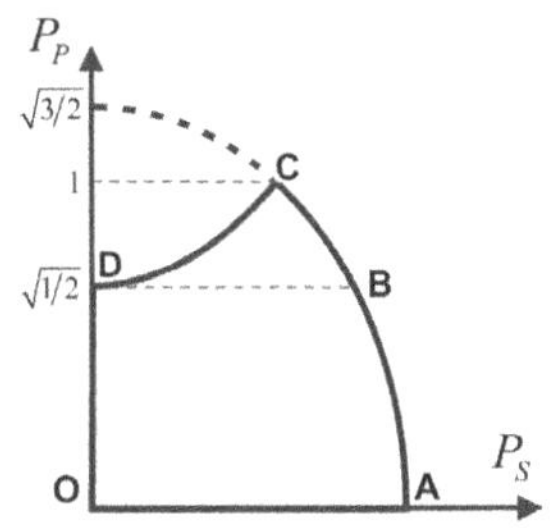

FIGURE 6.1
Feasible region for the components of purity P_P and P_S. Elliptic branch AC: pure Mueller matrices; segment OA: zero degree of polarizance; segment OD: zero polarimetric dimension index, and hyperbolic branch CD: maximum achievable degree of polarizance (Gil 2011).

polarization states. The *polarimetric dimension* of **M**, a measure of the effective statistical dimensions involved in $\mathbf{C}(\mathbf{M})$, can be defined as (Gil 2022)

$$D_\Sigma \equiv 4 - 3P_S \tag{6.28}$$

in such a manner that the maximum $D_\Sigma = 4$ corresponds to equiprobable spectral mixtures of four components (perfect depolarizer, $\mathbf{M} = \mathbf{M}_{\Delta 0}$); D_Σ decreases as the differences between the eigenvalues of $\mathbf{C}(\mathbf{M})$ increase, and the minimal value $D_\Sigma = 1$ corresponds to pure states (single spectral component).

When introduced for the first time (Gil 2011), P_S was termed the *degree of spherical purity* of **M** because the equality $P_S = 1$ (which implies $P_P = 0$) occurs uniquely when both the *I*-image and the *P*-image of **M** are spheres. Nevertheless, in the light of the statistical formulation of **M** studied in Section 5.13.2, P_S can also be referred to as the *polarimetric dimension index.*

From Eqs. (5.79) and (6.20), it follows that

$$P_\Delta^2 = 2P_P^2/3 + P_S^2 \tag{6.29}$$

so that P_Δ can be considered as the result of the separate contributions of P_P and P_S. When $P_S < 1$, the departure from sphericity of the *I*-images of **M** and $\mathbf{M}^r$ depends on the values of *D* and *P* respectively. Observe also that the minimum value of P_S compatible with total polarimetric purity of the system ($P_\Delta = 1$) is $P_S = 1/\sqrt{3}$, while the minimal value $P_S = 0$ is reached when all elements of the submatrix **m** are zero ($m_{kl} = 0;\ k,l = 1,2,3$).

The feasible region for the values of the components of purity P_P and P_S is shown in Figure 6.1 and a detailed case analysis is included in Section 6.5 (Gil 2011).

6.2.9 Physical Significance of the Components of Purity

The degree of polarizance of a Mueller matrix **M** contributes to purity in a peculiar way which may mask a certain lack of purity due to depolarization effects. To analyze these phenomena, we consider next some illustrative cases.

6.2.9.1 Components of Purity of Polarizers and Analyzers

(a) Consider a system composed of a transparent perfect depolarizer followed by a perfect polarizer. The Mueller matrix $\mathbf{M}_{\hat{P}0}$ of the system is

$$\mathbf{M}_{\hat{P}0} = m_{00}\begin{pmatrix} 1 & \hat{\mathbf{P}}^T \\ \hat{\mathbf{P}} & \mathbf{m}_{\hat{P}} \end{pmatrix}\begin{pmatrix} 1 & \mathbf{0}^T \\ \mathbf{0} & \mathbf{0} \end{pmatrix} = m_{00}\begin{pmatrix} 1 & \mathbf{0}^T \\ \hat{\mathbf{P}} & \mathbf{0} \end{pmatrix} = m_{00}\left(\mathbf{s}_{\hat{P}} \otimes \mathbf{s}_u^T\right)$$
$$\left[\mathbf{s}_{\hat{P}} \equiv \begin{pmatrix} 1 \\ \hat{\mathbf{P}} \end{pmatrix} \quad \mathbf{s}_u \equiv \begin{pmatrix} 1 \\ \mathbf{0} \end{pmatrix} \quad P = |\hat{\mathbf{P}}| = 1\right] \tag{6.30}$$

For any state of polarization incoming in the forward direction, the emerging state is totally polarized. The system behaves as a depolarizing polarizer without diattenuation. The *I*-image of $\mathbf{M}_{\hat{P}0}$ is given by a single point corresponding to the end point of vector $\hat{\mathbf{P}}$. In this case, $P_\Delta = 1/\sqrt{3}$, $P = 1$, $D = 0$ and $P_S = 0$, so that the only contribution to purity is due to forward polarizance. We also observe that, despite the fact that the system behaves as a perfect polarizer for incident states in the forward direction, P_S and P_Δ correctly reflect the lack of purity resulting from the presence of the perfect depolarizer.

(b) When the Mueller matrix $\mathbf{M}$ corresponds to the reverse case of the previous one, i.e., $\mathbf{M}$ is the Mueller matrix $\mathbf{M}_{0\hat{D}}$ of a pure polarizer followed by a perfect depolarizer, it adopts the form

$$\mathbf{M}_{0\hat{D}} = m_{00}\begin{pmatrix} 1 & \mathbf{0}^T \\ \mathbf{0} & \mathbf{0} \end{pmatrix}\begin{pmatrix} 1 & \hat{\mathbf{D}}^T \\ \hat{\mathbf{D}} & \mathbf{m}_{\hat{D}} \end{pmatrix} = m_{00}\begin{pmatrix} 1 & \hat{\mathbf{D}}^T \\ \mathbf{0} & \mathbf{0} \end{pmatrix} = m_{00}\left(\mathbf{s}_u \otimes \mathbf{s}_{\hat{D}}^T\right)$$
$$\left[\mathbf{s}_u \equiv \begin{pmatrix} 1 \\ \mathbf{0} \end{pmatrix} \quad \mathbf{s}_{\hat{D}} \equiv \begin{pmatrix} 1 \\ \hat{\mathbf{D}} \end{pmatrix} \quad D = |\hat{\mathbf{D}}| = 1\right] \tag{6.31}$$

For any state of polarization incoming in the forward direction, the emerging sate is totally depolarized. The system behaves as a total depolarizer, but maintaining diattenuation properties. As seen in Section 6.2.6, the *I*-image of $\mathbf{M}_{0\hat{D}}$ is undetermined.

(c) To complete this set of cases involving polarizers and perfect depolarizers, let us consider the Mueller matrix of a serial composition of a perfect depolarizer sandwiched between two perfect polarizers

$$\mathbf{M}_{\hat{D}\hat{P}} = m_{00}\begin{pmatrix} 1 & \hat{\mathbf{P}}^T \\ \hat{\mathbf{P}} & \mathbf{m}_{\hat{P}} \end{pmatrix}\begin{pmatrix} 1 & \mathbf{0}^T \\ \mathbf{0} & \mathbf{0} \end{pmatrix}\begin{pmatrix} 1 & \hat{\mathbf{D}}^T \\ \hat{\mathbf{D}} & \mathbf{m}_{\hat{D}} \end{pmatrix} = m_{00}\begin{pmatrix} 1 & \hat{\mathbf{D}}^T \\ \hat{\mathbf{P}} & \hat{\mathbf{P}} \otimes \hat{\mathbf{D}}^T \end{pmatrix} = m_{00}\left(\mathbf{s}_{\hat{P}} \otimes \mathbf{s}_{\hat{D}}^T\right)$$
$$\left[\mathbf{s}_{\hat{P}} \equiv \begin{pmatrix} 1 \\ \hat{\mathbf{P}} \end{pmatrix} \quad \mathbf{s}_{\hat{D}} \equiv \begin{pmatrix} 1 \\ \hat{\mathbf{D}} \end{pmatrix} \quad |\hat{\mathbf{D}}| = |\hat{\mathbf{P}}| = 1\right] \tag{6.32}$$

The resulting system is a pure, in general non-normal, polarizer-analyzer (viz., a perfect polarizer). The value of P_S is $P_S = 1/\sqrt{3}$, which is the necessary condition for a polarizer to be pure. When, in particular, $\mathbf{P} = \mathbf{D}$, then the Mueller matrix corresponds to a normal perfect polarizer.

We have observed that polarizance and diattenuation can mask, to some extent, the intrinsic depolarization properties of the system. As a limiting case, the sandwich considered in (c) results in a total masking of the central depolarizer. Moreover, serial combinations of polarizers, diattenuators and depolarizers can produce critical situations where P_Δ can reach values close to 1, in spite of the fact that some incident polarized states are converted into highly depolarized emerging states (Ossikovski 2010b). In general, the analysis of the components of purity provides a way to distinguish the sources of the polarimetric purity (Gil 2016a).

6.2.9.2 Components of Purity of the Canonical Depolarizers

A complementary view of the allocation of purity is obtained from the normal form of $\mathbf{M}$, introduced in Section 5.7. In the case of type-I Mueller matrices, which is the most extensive subgroup of the whole set of Mueller matrices, the corresponding canonical form (central intrinsic depolarizer) satisfies

$$P_P\left(\mathbf{M}_{\Delta d}\right)=0 \qquad P_S\left(\mathbf{M}_{\Delta d}\right)=P_\Delta\left(\mathbf{M}_{\Delta d}\right)=L_1\left(\mathbf{M}\right) \tag{6.33}$$

where $L_1(\mathbf{M})$ is the first Lorentz depolarization index (Ossikovski 2010b). Obviously, $L_1(\mathbf{M})$ and $P_\Delta(\mathbf{M})$ are different in general, and $L_1(\mathbf{M}) = P_\Delta(\mathbf{M})$ if and only if $P_P(\mathbf{M}) = 0$.

Particularly interesting is the case of type-II Mueller matrices where $\mathbf{M}_{\Delta nd}$ not only exhibits depolarizing properties, but also retains certain amounts of diattenuation and polarizance, which result in the following components of purity:

$$P_P\left(\mathbf{M}_{\Delta nd}\right)=P\left(\mathbf{M}_{\Delta nd}\right)=D\left(\mathbf{M}_{\Delta nd}\right)=1/2 \qquad P_S\left(\mathbf{M}_{\Delta nd}\right)=a_2\big/\left(\sqrt{6}\,a_0\right) \tag{6.34}$$

wherefrom

$$P_\Delta\left(\mathbf{M}_{\Delta nd}\right)=\sqrt{\frac{2}{3}P_P^2\left(\mathbf{M}_{\Delta nd}\right)+P_S^2\left(\mathbf{M}_{\Delta nd}\right)}=\frac{1}{\sqrt{6}}\sqrt{1+\frac{a_2^2}{a_0^2}} \tag{6.35}$$

In order to quantify the relative weight of the components of purity, we define the ratio c between the polarimetric dimension index and the weighted degree of polarizance of $\mathbf{M}_{\Delta nd}$ (Gil 2011)

$$c\left(\mathbf{M}_{\Delta nd}\right)\equiv\sqrt{\frac{3}{2}}\frac{P_S\left(\mathbf{M}_{\Delta nd}\right)}{P_P\left(\mathbf{M}_{\Delta nd}\right)}=\frac{a_2}{a_0}\le 1 \tag{6.36}$$

For $\mathbf{M}_{\Delta nd}$, the contribution of P_S to the total purity is less than or equal to the contribution of the degree of polarizance. The equality holds when $a_2 = a_0$. Thus, the feasible values of $P_\Delta\left(\mathbf{M}_{\Delta nd}\right)$ are restricted to the interval $\sqrt{1/6} \le P_\Delta\left(\mathbf{M}_{\Delta nd}\right) \le \sqrt{1/3}$, whose minimum and maximum correspond respectively to $P_S\left(\mathbf{M}_{\Delta nd}\right) = 0$ and $P_S\left(\mathbf{M}_{\Delta nd}\right) = \sqrt{1/6}$.

6.3 Indices of Polarimetric Purity

Depolarization properties are intrinsically related to polarimetric purity, which in turn can be characterized through the knowledge of the normalized eigenvalues $\hat{\lambda}_i$ of the covariance matrix **H**. Therefore, as with 3×3 coherency matrices representing 3D polarization states, a complete quantitative description of the polarimetric purity of a medium requires to consider three independent parameters derived from $\hat{\lambda}_i$. Even though the depolarization index P_Δ, the Lorentz depolarization indices L_1 and L_2, the polarization entropy S, the average degree of depolarization A_f, and other parameters considered in Section 5.10, provide different overall measures of the depolarizing properties of a medium, they do not contain information sufficient to deduce the $\hat{\lambda}_i$. Moreover, the components of purity provide specific knowledge of the sources of purity, but are not sufficient to derive $\hat{\lambda}_i$.

As a generalization to four-dimensional coherency matrices of the two 3D indices of polarimetric purity (Section 2.10.4), a set of three invariant *indices of polarimetric purity* (IPP) of a given medium represented by a covariance matrix **H** is defined as (San José and Gil 2011)

$$P_1 \equiv \hat{\lambda}_0 - \hat{\lambda}_1 \qquad P_2 \equiv \hat{\lambda}_0 + \hat{\lambda}_1 - 2\hat{\lambda}_2 \qquad P_3 \equiv \hat{\lambda}_0 + \hat{\lambda}_1 + \hat{\lambda}_2 - 3\hat{\lambda}_3 = 1 - 4\hat{\lambda}_3 \tag{6.37}$$

or, equivalently, through the generic expression (Ossikovski and Vizet 2019)

$$P_k \equiv \sum_{q=1}^{k} q\Delta_q \qquad \Delta_q \equiv \hat{\lambda}_{q-1} - \hat{\lambda}_q \qquad (k, q = 1,2,3) \tag{6.38}$$

where the four non-negative normalized eigenvalues $\hat{\lambda}_i \equiv \lambda_i/(\mathrm{tr}\,\mathbf{H})$ of **H** (which are also the normalized eigenvalues of the coherency matrix **C**) have been taken ordered in decreasing magnitude $\hat{\lambda}_0 \geq \hat{\lambda}_1 \geq \hat{\lambda}_2 \geq \hat{\lambda}_3$. Let us stress that this choice of ordering must be preserved in order to maintain the mathematical and physical meaning of the IPP.

Conversely, the expressions of the normalized eigenvalues of **H** in terms of the IPP are given by

$$\begin{aligned} &\hat{\lambda}_0 = \frac{1}{4}\left(1 + 2P_1 + \frac{2}{3}P_2 + \frac{1}{3}P_3\right) \quad \hat{\lambda}_1 = \frac{1}{4}\left(1 - 2P_1 + \frac{2}{3}P_2 + \frac{1}{3}P_3\right) \quad \hat{\lambda}_2 = \frac{1}{4}\left(1 - \frac{4}{3}P_2 + \frac{1}{3}P_3\right) \\ &\hat{\lambda}_3 = \frac{1}{4}(1 - P_3) \end{aligned} \tag{6.39}$$

The IPP constitute a particularly interesting triplet among the infinite possible choices whose general formal analysis is dealt with in Section 6.6.

The depolarization index P_Δ (also called the *degree of polarimetric purity*) can be calculated from the IPP as follows (San José and Gil 2011):

$$P_\Delta^2 = \frac{1}{3}\left(2P_1^2 + \frac{2}{3}P_2^2 + \frac{1}{3}P_3^2\right) \tag{6.40}$$

Another interesting expression of P_Δ as a symmetric quadratic average of all relative differences between pairs of eigenvalues of **H** is given by

$$P_\Delta^2 = \frac{1}{3}\sum_{\substack{i,j=0\\i<j}}^{3} p_{ij}^2 \qquad p_{ij} \equiv \hat{\lambda}_i - \hat{\lambda}_j \tag{6.41}$$

Obviously, since $\hat{\lambda}_0 + \hat{\lambda}_1 + \hat{\lambda}_2 + \hat{\lambda}_3 = 1$, three relative differences, like those defined in Eq. (6.37), are sufficient to determine the parameters p_{ij} entering Eq. (6.41).

Pure systems are characterized by $P_\Delta = P_1 = P_2 = P_3 = 1$, while the lower limiting values $P_\Delta = P_1 = P_2 = P_3 = 0$ correspond to a perfect depolarizer (i.e., an equiprobable parallel mixture of four, or more, components).

By applying the ordering choice for the eigenvalues, $0 \le \lambda_3 \le \lambda_2 \le \lambda_1 \le \lambda_0$, we find that the indices of purity are restricted by the following nested inequalities:

$$0 \le P_1 \le P_2 \le P_3 \le 1 \tag{6.42}$$

Thus, if $P_1 = 1$, then $P_2 = P_3 = P_\Delta = 1$ (pure system). Moreover, if $P_3 = 0$, then $P_1 = P_2 = P_\Delta = 0$ (perfect depolarizer). Total purity is characterized by $P_1 = 1 (P_\Delta = 1)$, while perfect depolarizers are characterized by $P_3 = 0$ $(P_\Delta = 0)$.

The characteristic decomposition (5.131) of a given Mueller matrix **M** can be expressed in terms of its IPP P_i $(i = 1, 2, 3)$ as follows:

$$\mathbf{M} = P_1\left(m_{00}\hat{\mathbf{M}}_{J0}\right) + (P_2 - P_1)\left(m_{00}\hat{\mathbf{M}}_1\right) + (P_3 - P_2)\left(m_{00}\hat{\mathbf{M}}_2\right) + (1 - P_3)\left(m_{00}\hat{\mathbf{M}}_3\right) \tag{6.43}$$

It highlights the physical meaning and the role played by P_i concerning polarimetric purity:

- P_1 is the relative portion of the medium that behaves as pure in the characteristic decomposition and whose pure Mueller matrix $\mathbf{M}_{J0}$ coincides with that of the spectral decomposition with maximum weight $(\hat{\lambda}_0)$, so that $\mathbf{M}_{J0}$ is a proper *pure representative* of **M** (see Section 6.8).
- $P_2 - P_1$ is the relative portion of the medium that behaves as a *2D-depolarizer*, i.e., as a medium composed of an equiprobable mixture of the first two spectral pure components. Therefore, P_2 quantifies the portion of the medium that can be represented as a parallel combination of the pure component $\mathbf{M}_{J0}$ and the 2D-depolarizer $\mathbf{M}_1(\mathbf{H}_1)$, with $\mathbf{H}_1 = m_{00}\,\mathrm{diag}(1,1,0,0)$.
- $P_3 - P_2$ is the relative portion of the medium that behaves as a *3D-depolarizer*, i.e., a medium composed of an equiprobable mixture of the first three spectral pure components. Therefore, P_3 quantifies the portion of the medium that can be represented as a parallel combination of a pure component $\mathbf{M}_{J0}$; a 2D depolarizer $\mathbf{M}_1(\mathbf{H}_1)$, with $\mathbf{H}_1 = (m_{00}/2)\,\mathrm{diag}(1,1,0,0)$, and a *3D depolarizer* $\mathbf{M}_2(\mathbf{H}_2)$, with $\mathbf{H}_2 = (m_{00}/3)\,\mathrm{diag}(1,1,1,0)$. In other words, P_3 is the weight (or cross section) of the portion of the medium that is not a perfect depolarizer.
- In agreement with the previous interpretation, $1 - P_3$ is the portion of the medium that behaves as a *perfect depolarizer* (or *4D-depolarizer*).

When $P_3 = 1$, then $\mathrm{rank}\,\mathbf{H} < 4$ and the perfect depolarizer (the fourth component) is absent from the characteristic decomposition. When $P_2 = 1$ $(\mathrm{rank}\,\mathbf{H} < 3)$, the perfect depolarizer and the 3D depolarizer are both absent from the characteristic decomposition, so that the system is composed of a pure component and a 2D depolarizer. When $P_1 = 1$ $(\mathrm{rank}\,\mathbf{H} = 1)$, the system is necessarily nondepolarizing, so that the equalities $1 = P_\Delta = \hat{\lambda}_0 = P_1 = P_2 = P_3$ are satisfied. The system cannot be considered a parallel composition of more than one pure component.

As shown above, the IPP constitute a particularly appropriate set of parameters for the quantitative description of the polarimetric purity of **M**. The IPP contain complete information on the polarimetric purity (randomness) of the medium (under given interaction and measurement conditions); such information on the depolarizing features of **M** is complementary to that provided by the components of purity D, P, and P_S. Unlike the components of purity (abbreviated as CP), the IPP are insensitive to the nature (diattenuation, retardance...) of the spectral or characteristic components of **M**.

The knowledge of P_1, P_2, P_3 alone does not imply necessarily the knowledge of the components of purity; nevertheless, the set of five quantities (P_1, P_2, P_3, P, D) is sufficient to calculate P_Δ and P_S and therefore constitutes a complete set of descriptors characterizing the intrinsic depolarizing properties of **M** from both quantitative and qualitative points of view. The feasible region for the IPP is represented in Figure 6.2.

Once the IPP of a Mueller matrix have been considered, it is worth emphasizing the role played by n-dimensional coherency matrices (positive semi-definite Hermitian matrices) in polarization algebra. The polarimetric purity of an electromagnetic beam whose polarization ellipse lies in a fixed plane (2D polarization states, represented by their associated two-dimensional coherency matrices) is given by a single index of purity, namely the *degree of polarization*. For polarization states whose polarization ellipse lies in a fluctuating plane (3D polarized states, represented by three-dimensional coherency matrices), a pair of indices of purity is required to characterize completely the polarimetric purity (Gil 2007). Therefore, the IPP, P_1, P_2 and P_3, of a Mueller matrix appear as a natural extension of the concept of indices of purity defined for polarization states of electromagnetic beams.

To complete this section, let us consider again the concept of polarization entropy $S(\mathbf{M})$ and observe that, despite the complicated relation between S and the depolarization index P_Δ (Ossikovski and Vizet 2019), S admits the following analytic expression in terms of the IPP:

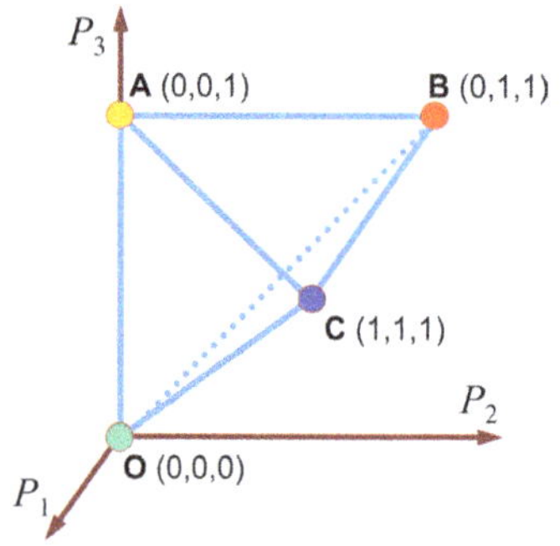

FIGURE 6.2
The purity space Σ_{IPP} is defined as the tetrahedron determining the feasible region for the indices of polarimetric purity (P_1, P_2, P_2) (San José and Gil 2011).

$$\begin{aligned} S(\mathbf{H}) &= -\sum_{i=0}^{3}\left(\hat{\lambda}_i \log_4 \hat{\lambda}_i\right) \\ &= -\frac{1}{4}\left(1+2P_1+\frac{2}{3}P_2+\frac{1}{3}P_3\right)\left[\log_4\left(1+2P_1+\frac{2}{3}P_2+\frac{1}{3}P_3\right)-1\right] \\ &\quad -\frac{1}{4}\left(1-2P_1+\frac{2}{3}P_2+\frac{1}{3}P_3\right)\left[\log_4\left(1-2P_1+\frac{2}{3}P_2+\frac{1}{3}P_3\right)-1\right] \\ &\quad -\frac{1}{4}\left(1-\frac{4}{3}P_2+\frac{1}{3}P_3\right)\left[\log_4\left(1-\frac{4}{3}P_2+\frac{1}{3}P_3\right)-1\right] \\ &\quad -\frac{1}{4}(1-P_3)\left[\log_4(1-P_3)-1\right] \end{aligned} \tag{6.44}$$

so that S represents a surface in the *purity space* with reference frame axes P_1, P_2, P_3.

Furthermore, following the idea introduced in Section 2.10.7 devoted to the entropy of 3×3 coherency matrices, it is possible to extend to four dimensions the concept of partial entropies by defining the following three partial entropies of **M**:

$$S(P_i) \equiv -\left\{\frac{1}{2}(1+P_i)\log_2\left[\frac{1}{2}(1+P_i)\right]+\frac{1}{2}(1-P_i)\log_2\left[\frac{1}{2}(1-P_i)\right]\right\} \ldots (i=1,2,3) \tag{6.45}$$

These invariant quantities contain objective information equivalent to that provided by the IPP. The condition $0 \le P_1 \le P_2 \le P_3 \le 1$ on the indices of purity has its counterpart in the partial entropies $0 \le S(P_3) \le S(P_2) \le S(P_1) \le 1$.

The thermodynamic view of the 2D and 3D coherency matrices representing polarization states (Brosseau and Bicout 1994) can be generalized to 4D covariance (or coherency) matrices representing the polarimetric properties of media by means of the concept of 4D *temperature* of *polarization*, defined as

$$\tau_\Delta \equiv \frac{2}{\ln(1+P_\Delta)-\ln(1-P_\Delta)} \tag{6.46}$$

The temperature of polarization is zero for nondepolarizing systems and increases monotonically as the polarimetric purity decreases.

6.4 Invariant Quantities of a Mueller Matrix

The analysis of the physical properties that remain unchanged under certain transformations has proven to be very useful for the interpretation of 2D and 3D coherency matrices associated with polarization states. The present section is devoted to the study of the quantities involved in a Mueller matrix that are invariant with respect to physically realizable transformations.

6.4.1 Dual Retarder Transformation

Let us consider the *dual retarder transformation* of a Mueller matrix $\mathbf{M}$ through its pre- and post-multiplication by respective orthogonal Mueller matrices $\mathbf{M}_{R2}$ and $\mathbf{M}_{R1}$ (Gil 2016c)

$$\mathbf{M}' = \mathbf{M}_{R2}\mathbf{M}\mathbf{M}_{R1} = \begin{pmatrix} 1 & \mathbf{0}^T \\ \mathbf{0} & \mathbf{m}_{R2} \end{pmatrix} m_{00} \begin{pmatrix} 1 & \mathbf{D}^T \\ \mathbf{P} & \mathbf{m} \end{pmatrix} \begin{pmatrix} 1 & \mathbf{0}^T \\ \mathbf{0} & \mathbf{m}_{R1} \end{pmatrix} = m_{00} \begin{pmatrix} 1 & \mathbf{D}^T\mathbf{m}_{R1} \\ \mathbf{m}_{R2}\mathbf{P} & \mathbf{m}_{R2}\mathbf{m}\mathbf{m}_{R1} \end{pmatrix} \tag{6.47}$$

and observe that

$$\begin{aligned}
&(\mathbf{M}')_{00} = m_{00} \\
&|\mathbf{D}'| = \sqrt{\mathbf{D}'^T\mathbf{D}'} = \sqrt{\mathbf{D}^T\mathbf{m}_{R1}\mathbf{m}_{R1}^T\mathbf{D}} = \sqrt{\mathbf{D}^T\mathbf{D}} = |\mathbf{D}| \\
&|\mathbf{P}'| = \sqrt{\mathbf{P}'^T\mathbf{P}'} = \sqrt{\mathbf{P}^T\mathbf{m}_{R2}\mathbf{m}_{R2}^T\mathbf{P}} = \sqrt{\mathbf{P}^T\mathbf{P}} = |\mathbf{P}| \\
&P'_S = \frac{1}{\sqrt{3}}\|\mathbf{m}'\|_2 = \frac{1}{\sqrt{3}}\sqrt{\mathrm{tr}\left(\mathbf{m}'^T\mathbf{m}'\right)} = \frac{1}{\sqrt{3}}\sqrt{\mathrm{tr}\left(\mathbf{m}_{R1}^T\mathbf{m}^T\mathbf{m}_{R2}^T\mathbf{m}_{R2}\mathbf{m}\mathbf{m}_{R1}\right)} = \frac{1}{\sqrt{3}}\sqrt{\mathrm{tr}\left(\mathbf{m}^T\mathbf{m}\right)} = P_S \\
&\mathbf{P}'^T\mathbf{m}'\mathbf{D}' = \mathbf{P}^T\mathbf{m}_{R2}^T\mathbf{m}_{R2}\mathbf{m}\mathbf{m}_{R1}\mathbf{m}_{R1}^T\mathbf{D} = \mathbf{P}^T\mathbf{m}\mathbf{D} \\
&\det\mathbf{M}' = \det\mathbf{M}_{R2}\det\mathbf{M}\det\mathbf{M}_{R1} = \det\mathbf{M}
\end{aligned} \tag{6.48}$$

These series of equalities show that the following quantities of the transformed Mueller matrix $\mathbf{M}'$ are equal to their respective counterparts from $\mathbf{M}$, namely the mean intensity coefficient m_{00}, the diattenuation D, the polarizance P, the polarimetric dimension index P_S, the scalar quantity $\Upsilon \equiv \mathbf{P}^T\mathbf{m}\mathbf{D} = \mathbf{D}^T\mathbf{m}^T\mathbf{P}$ and $\det\mathbf{M}$. As seen below, other quantities that are not affected by the multiplication of $\mathbf{M}$ by retarders are the depolarization descriptors, like the IPP (P_1, P_2, P_3), the depolarization index P_Δ, the first and second Lorentz depolarization indices L_1, L_2, etc.

Moreover, by considering the symmetric decomposition of $\mathbf{M}$ in the form $\mathbf{M} = \mathbf{M}_{R2}\mathbf{M}_{D2}\mathbf{M}_\Delta\mathbf{M}_{D1}\mathbf{M}_{R1}$, it follows that

$$\begin{gathered}
\mathbf{M}' = \mathbf{M}'_{R2}\left(\mathbf{M}_{R2}\mathbf{M}_{D2}\mathbf{M}_\Delta\mathbf{M}_{D1}\mathbf{M}_{R1}\right)\mathbf{M}'_{R1} = \mathbf{M}''_{R2}\mathbf{M}_{D2}\mathbf{M}_\Delta\mathbf{M}_{D1}\mathbf{M}''_{R1} \\
\left[\mathbf{M}''_{R2} = \mathbf{M}'_{R2}\mathbf{M}_{R2} \qquad \mathbf{M}''_{R1} = \mathbf{M}'_{R1}\mathbf{M}_{R1}\right]
\end{gathered} \tag{6.49}$$

and therefore, $\mathbf{D}_1$, $\mathbf{D}_2$ and $\mathbf{M}_\Delta$ (featuring ten independent parameters when $\mathbf{M}$ is type-I and eight independent parameters when $\mathbf{M}$ is type-II) are invariant under dual retarder transformations. The same holds for parameters like L_1, L_2 and the eigenvalues ρ_i of the N-matrix $\mathbf{N} = \mathbf{G}\mathbf{M}^T\mathbf{G}\mathbf{M}$ from whose square roots are defined the elements of $\mathbf{M}_\Delta$.

6.4.1.1 *Interpretation of* Υ, *and* $\det\mathbf{M}$

The parameter Υ is given by the scalar product of vectors $\mathbf{P}$ and $\mathbf{m}\mathbf{D}$. It takes its maximal value $\Upsilon = 1$ when $\mathbf{M}$ corresponds to a normal perfect polarizer, while $\Upsilon = 0$ when $\mathbf{P} = \mathbf{0}$, or $\mathbf{m}\mathbf{D} = \mathbf{0}$, or $\mathbf{P}$ and $\mathbf{m}\mathbf{D}$ are mutually orthogonal vectors. By considering the arrow form $\mathbf{M}_A$ of $\mathbf{M}$ defined through the dual retarder transformation in Eq. (5.108), Υ can be interpreted as the scalar product of $\mathbf{D}_A = \mathbf{m}_{RI}\mathbf{D}$ and $\mathbf{P}_A = \mathbf{m}_{RO}^T\mathbf{P}$ through the diagonal metric determined by $\mathbf{m}_A$, i.e., $\Upsilon = \mathbf{P}_A^T\mathbf{m}_A\mathbf{D}_A$.

The determinant $\det \mathbf{M}$ can be expressed as

$$\det \mathbf{M} = m_{00}^4 \det \widehat{\mathbf{M}}_{J2} \det \widehat{\mathbf{M}}_{\Delta} \det \widehat{\mathbf{M}}_{J1} = m_{00}^4\, V(1-D_1^2)^2(1-D_2^2)^2 \quad (6.50)$$

where D_1 abd D_2 are the (invariant) diattenuations of $\mathbf{M}_{J1}$ and $\mathbf{M}_{J2}$ while V is the (invariant) product of the square roots of the eigenvalues of $\mathbf{G}\hat{\mathbf{M}}^T\mathbf{G}\hat{\mathbf{M}}$ (see Section 5.7). V, termed *volume coefficient*, takes the following expressions for type-I and type-II matrices respectively (Gil et al. 2022b):

$$\begin{aligned} &V = \varepsilon \hat{d}_1\hat{d}_2\hat{d}_3 \quad \left[\hat{\mathbf{M}}_\Delta = \hat{\mathbf{M}}_{\Delta d} = diag\left(1,\hat{d}_1,\hat{d}_2,\varepsilon\hat{d}_3\right) \quad \left(0 \le \hat{d}_3 \le \hat{d}_2 \le \hat{d}_1\right)\right] \\ &V = \hat{a}_2^2/16 \quad \left[\hat{\mathbf{M}}_\Delta = \hat{\mathbf{M}}_{\Delta nd} = \begin{pmatrix} 1 & -1/2 & 0 & 0 \\ 1/2 & 0 & 0 & 0 \\ 0 & 0 & \hat{a}_2/2 & 0 \\ 0 & 0 & 0 & \hat{a}_2/2 \end{pmatrix} \quad \left(0 \le a_2 \le a_0\right)\right] \end{aligned} \quad (6.51)$$

Leaving aside its sign, V provides a scaled measure of the volume ($4\pi\hat{d}_1\hat{d}_2\hat{d}_3/3$ or $4\pi\hat{a}_2^2/27$) of the ellipsoid E_Δ associated with the central depolarizer $\hat{\mathbf{M}}_\Delta$ (see Chapter 10; Gil et al. 2022b). Therefore, the invariance of $\det \mathbf{M}$ is a mere consequence of the invariance of the four parameters entering it in Eq. (6.50), namely m_{00}, D_1, D_2 and V.

Regarding the sign of $\det \mathbf{M}$, recall that it is positive for pure Mueller matrices (see Section 3.3.1). Therefore, Eq. (6.50) shows that $\det \mathbf{M}$ is positive for all type-II matrices, as well as for those type-I Mueller matrices for which $\varepsilon = 1$, and is negative only for depolarizing type-I matrices with $\varepsilon = -1$. A deeper insight into the physical significance of the sign and other properties of $\det \mathbf{M}$ is provided by the parameterization of $\hat{d}_i$ in terms of the indices of polarimetric purity (P_1, P_2, P_3) of $\mathbf{M}_{\Delta d}$ (see Section 6.8.1): $\hat{d}_1 = (2P_2 + P_3)/3$, $\hat{d}_2 = P_1 + (P_3 - P_2)/3$, $\varepsilon\hat{d}_3 = P_1 - (P_3 - P_2)$, $[0 \le P_1 \le P_2 \le P_3 \le 1]$; after substitution in Eq. (6.50), it follows that $\det \mathbf{M} < 0$ if and only if $P_3 > 3P_1 + P_2$, i.e., if the combined amount of purity $3P_1 + P_2$ is smaller than P_3. It can be readily seen that this condition cannot be fulfilled when $P_3 = P_2$, which includes the cases where $r \equiv \operatorname{rank} \mathbf{C}(\mathbf{M}) < 3$ (or, equivalently $P_3 = P_2 = 1$). Therefore, only certain type-I depolarizers with $r = 3$ or $r = 4$ can exhibit negative determinants (Gil et al. 2022b).

6.4.1.2 Independent Invariant Quantities under Dual Retarder Transformations

Obviously, other quantities derived from the set indicated in Eq. (6.48), as is the case of the degree of polarizance P_P or the degree of polarimetric purity P_Δ, are also invariant with respect to dual retarder transformations. Moreover, by considering the spectral decomposition of $\mathbf{M}$ in Eq. (5.126), we see that $\mathbf{M}'$ can be expressed as

$$\mathbf{M}' = \mathbf{M}_{R2}\mathbf{M}\mathbf{M}_{R1} = \mathbf{M}_{R2}\left[\sum_{i=0}^{3}\hat{\lambda}_i\left(m_{00}\hat{\mathbf{M}}_{Ji}\right)\right]\mathbf{M}_{R1} = \sum_{i=0}^{3}\hat{\lambda}_i\left(m_{00}\hat{\mathbf{M}}'_{Ji}\right) \quad \left(\hat{\mathbf{M}}'_{Ji} \equiv \mathbf{M}_{R2}\hat{\mathbf{M}}'_{Ji}\mathbf{M}_{R1}\right) \quad (6.52)$$

so that $\mathbf{M}'$ is obtained as a convex sum of pure components $\mathbf{M}'_{Ji}$ with the same coefficients $\hat{\lambda}_i$, i.e., the spectrum of eigenvalues of the covariance matrix, as well as the IPP, are preserved under the dual retarder transformation (6.47).

Furthermore, observe that the singular values (a_1, a_2, a_3) of $\mathbf{m}$ are not affected by a dual retarder transformation. By taking into account Eq. (6.26) we see that P_S is entirely determined by the set (a_1, a_2, a_3), and, in addition, P_Δ can be obtained from D, P and P_S, whereas one of the IPP (say, P_3) can be obtained from P_Δ and the two remaining IPP.

In summary, a set of independent parameters remaining invariant under dual retarder transformations is given by the following ten quantities (Gil 2016c):

$$m_{00},\ P,\ D,\ a_1,\ a_2,\ a_3,\ P_1,\ P_2,\ \mathbf{P}^T\mathbf{m}\,\mathbf{D},\ \det\mathbf{M} \tag{6.53}$$

Other invariant parameters obtainable from the above set are, for instance, P_P, P_S, P_Δ, P_3, ρ_0, ρ_1, ρ_2, ρ_3, D_1 and D_2

Since orthogonal Mueller matrices represent pure retarders, the transformation (6.47) can be physically realized by sandwiching the medium represented by $\mathbf{M}$ between the respective retarders. Therefore, any serial combination of a Mueller matrix $\mathbf{M}$ and an arbitrary number of retarders preserves the values of the above-mentioned set of quantities, which can be considered as *intrinsic* to $\mathbf{M}$. Given a Mueller matrix $\mathbf{M}$ and an arbitrary pair of orthogonal Mueller matrices $\mathbf{M}_{R1}, \mathbf{M}_{R2}$, the matrix $\mathbf{M}'(\mathbf{M}) = \mathbf{M}_{R2}\mathbf{M}\mathbf{M}_{R1}$ is said to be *invariant-equivalent* to $\mathbf{M}$.

6.4.2 Single Retarder Transformation

A particular type of the above dual retarder transformation is the case of an orthogonal similarity transformation of $\mathbf{M}$

$$\mathbf{M}' = \mathbf{M}_R\mathbf{M}\mathbf{M}_R^T = \begin{pmatrix} 1 & \mathbf{0}^T \\ \mathbf{0} & \mathbf{m}_R \end{pmatrix} m_{00}\begin{pmatrix} 1 & \mathbf{D}^T \\ \mathbf{P} & \mathbf{m} \end{pmatrix}\begin{pmatrix} 1 & \mathbf{0}^T \\ \mathbf{0} & \mathbf{m}_R^T \end{pmatrix} = m_{00}\begin{pmatrix} 1 & \mathbf{D}^T\mathbf{m}_R^T \\ \mathbf{m}_R\mathbf{P} & \mathbf{m}_R\mathbf{m}\mathbf{m}_R^T \end{pmatrix} \tag{6.54}$$

This *single retarder transformation* of $\mathbf{M}$ can physically be performed by sandwiching the medium represented by $\mathbf{M}$ between two identical retarders whose fast eigenstates are mutually crossed.

In this case, in addition to relations (6.48), the following equalities are satisfied:

$$\begin{aligned} &\mathbf{P}'^T\mathbf{D}' = \mathbf{P}^T\mathbf{m}_R^T\mathbf{m}_R\mathbf{D} = \mathbf{P}^T\mathbf{D} \\ &\mathbf{P}'^T\mathbf{m}'^T\mathbf{D}' = \mathbf{P}^T\mathbf{m}_R^T\mathbf{m}_R\mathbf{m}^T\mathbf{m}_R^T\mathbf{m}_R\mathbf{D} = \mathbf{P}^T\mathbf{m}^T\mathbf{D} \\ &\operatorname{tr}\mathbf{M}' = \operatorname{tr}\mathbf{M} \end{aligned} \tag{6.55}$$

The invariant parameter $\mathbf{P}^T\mathbf{D}$ represents the scalar product of the diattenuation, $\mathbf{D}$, and polarizance, $\mathbf{P}$, vectors and gives a measure of the projection of $\mathbf{D}$ on $\mathbf{P}$ (and vice versa). Analogously, $\mathbf{P}^T\mathbf{m}^T\mathbf{D} = \mathbf{D}^T\mathbf{m}\mathbf{P}$ is the scalar product of $\mathbf{D}$ and $\mathbf{m}\mathbf{P}$ (i.e., the polarizance vector transformed through $\mathbf{m}$).

Moreover, $\operatorname{tr}\mathbf{M} = m_{00}\operatorname{tr}\hat{\mathbf{M}}$, with $m_{00} > 0$ and $0 \le \operatorname{tr}\hat{\mathbf{M}} \le 4$, where the maximal achievable value $\operatorname{tr}\hat{\mathbf{M}} = 4$ corresponds exclusively to the identity matrix $\hat{\mathbf{M}} = \mathbf{I}_4$ (i.e., $\hat{\mathbf{M}}$ does not produce any polarimetric effects), while the minimal achievable value $\operatorname{tr}\hat{\mathbf{M}} = 0$ corresponds to different kinds of depolarizing and nondepolarizing systems. Now, by recalling that (1) the maximal and minimal diagonal elements of a coherency matrix $\mathbf{C}$ with respect to arbitrary unitary transformations $\mathbf{U}\mathbf{C}\mathbf{U}^\dagger$ ($\mathbf{U}$ unitary) coincide with the largest and smallest eigenvalues λ_0 and λ_3 of $\mathbf{C}$ respectively, and (2) the element c_{00} of $\mathbf{C}(\mathbf{M})$ is given by $c_{00} = \mathrm{tr}\,\mathbf{M}$ (see Eq. (5.13)), an additional property is that $\lambda_3 \leq \mathrm{tr}\,\mathbf{M} \leq \lambda_0$, λ_0 and λ_3 being the largest and smallest eigenvalues of the $\mathbf{C}(\mathbf{M})$ (i.e., λ_0 and λ_3 are the largest and smallest coefficients in the spectral decomposition of $\mathbf{C}(\mathbf{M})$ or $\mathbf{H}(\mathbf{M})$). Therefore, $\mathrm{tr}\,\mathbf{M}$ is always non-negative and combines, in an intricate manner, information on m_{00}, diattenuation-polarizance, retardance and depolarization.

In summary, a set of independent parameters which remain invariant under single retarder transformations is given by the following 13 quantities (Gil 2016c):

$$m_{00},\ P,\ D,\ a_1,\ a_2,\ a_3,\ P_1\ ,P_2\ ,\mathbf{P}^T\mathbf{m}\,\mathbf{D},\ \mathbf{P}^T\mathbf{m}^T\mathbf{D},\ \mathbf{P}^T\mathbf{D},\ \mathrm{tr}\,\mathbf{M},\ \det\mathbf{M} \tag{6.56}$$

6.4.3 Dual Rotation Transformation

A strict subset of the dual retarder transformations can be realized through physical rotations of the laboratory reference frames of the incident and emerging polarization states about the respective axes Z and Z′ defining the directions of propagation of the incident and emerging electromagnetic waves (Figure 6.3).

This *dual rotation transformation* corresponds to the case where the orthogonal Mueller matrices $\mathbf{M}_{R2}$ and $\mathbf{M}_{R1}$ in Eq. (6.47) have the form of respective rotation matrices $\mathbf{M}_{R2} = \mathbf{M}_G(\theta_2), \mathbf{M}_{R1} = \mathbf{M}_G(\theta_1)$

$$\mathbf{M}_{Gi} \equiv \begin{pmatrix} 1 & 0 & 0 & 0 \\ 0 & c_i & s_i & 0 \\ 0 & -s_i & c_i & 0 \\ 0 & 0 & 0 & 1 \end{pmatrix} \qquad \begin{bmatrix} s_i \equiv \sin 2\theta_i & \\ & i = 1,2 \\ c_i \equiv \cos 2\theta_i & \end{bmatrix} \tag{6.57}$$

so that, by taking advantage of the vectorial partitioned notation (5.104)

$$\mathbf{M} \equiv m_{00} \begin{pmatrix} 1 & D_1 & D_2 & D_3 \\ P_1 & k_1 & r_3 & r_2 \\ P_2 & q_3 & k_2 & r_1 \\ P_3 & q_2 & q_1 & k_3 \end{pmatrix} \tag{6.58}$$

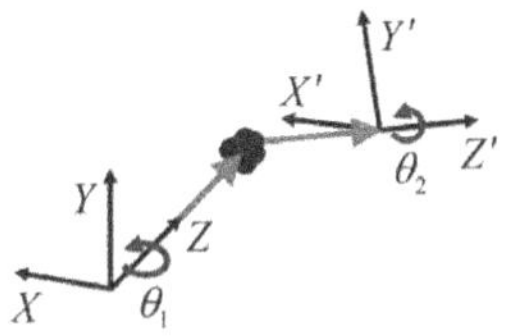

FIGURE 6.3
The *dual rotation transformation* consists of respective rotations θ_1 and θ_2 of the reference frames of the incident and emerging polarization states about the respective propagation axes Z and Z′.

the transformed Mueller matrix $\mathbf{M}'$ is given by

$$\mathbf{M}' = \mathbf{M}_{G2}\mathbf{M}\mathbf{M}_{G1} = \begin{pmatrix} 1 & \mathbf{0}^T \\ \mathbf{0} & \mathbf{m}_{G2} \end{pmatrix} m_{00} \begin{pmatrix} 1 & \mathbf{D}^T \\ \mathbf{P} & \mathbf{m} \end{pmatrix} \begin{pmatrix} 1 & \mathbf{0}^T \\ \mathbf{0} & \mathbf{m}_{G1} \end{pmatrix} = m_{00} \begin{pmatrix} 1 & \mathbf{D}^T\mathbf{m}_{G1} \\ \mathbf{m}_{G2}\mathbf{P} & \mathbf{m}_{G2}\mathbf{m}\mathbf{m}_{G1} \end{pmatrix}$$

$$= m_{00} \begin{pmatrix} 1 & c_1D_1 + s_1D_2 & -s_1D_1 + c_1D_2 & D_3 \\ c_2P_1 + s_2P_2 & c_2(c_1k_1 + s_1r_3) + s_2(c_1q_3 + s_1k_2) & c_2(-s_1k_1 + c_1r_3) + s_2(-s_1q_3 + c_1k_2) & c_2r_2 + s_2r_1 \\ -s_2P_1 + c_2P_2 & -s_2(c_1k_1 + s_1r_3) + c_2(c_1q_3 + s_1k_2) & s_2(s_1k_1 - c_1r_3) + c_2(-s_1q_3 + c_1k_2) & -s_2r_2 + c_2r_1 \\ P_3 & c_1q_2 + s_1q_1 & -s_1q_2 + c_1q_1 & k_3 \end{pmatrix} \tag{6.59}$$

Therefore, the following parameters (not all mutually independent), which include the set (6.53), remain invariant under dual rotation transformations (Gil 2016c)

$$\begin{gathered} m_{00},\, D_L,\, D_C,\, D,\, P_L,\, P_C,\, P,\, P_P,\, r_L,\, r_C,\, r,\, q_L,\, q_C,\, q,\, a_1,\, a_2,\, a_3, \\ k_L^2 + r_C^2 + q_C^2,\, k_C,\, P_S,\, P_1,\, P_2,\, P_3,\, P_\Delta,\, \mathbf{P}^T\mathbf{m}\mathbf{D},\, \det\mathbf{M}, \\ \rho_0,\, \rho_1,\, \rho_2,\, \rho_3,\, L_1,\, L_2 \end{gathered} \tag{6.60}$$

where the linear and circular components of $\mathbf{k}$, $\mathbf{r}$ and $\mathbf{q}$ have been defined as

$$k_L^2 \equiv k_1^2 + k_2^2 \qquad r_L^2 \equiv r_1^2 + r_2^2 \qquad q_L^2 \equiv q_1^2 + q_2^2 \qquad k_C \equiv k_3 \qquad r_C \equiv r_3 \qquad q_C \equiv q_3 \tag{6.61}$$

Different sets of fourteen mutually independent invariant parameters can be chosen from (6.60), as for instance

$$m_{00},\, D_L,\, D_C,\, P_L,\, P_C,\, r_L,\, r_C,\, q_L,\, q_C,\, k_L^2 + r_C^2 + q_C^2,\, k_C,\, P_1,\, P_2,\, \det\mathbf{M} \tag{6.62}$$

6.4.4 Single Rotation Transformation

A particular, but very common, type of dual rotation transformation occurs when $\theta_2 = -\theta_1$, which corresponds to a joint rotation of the laboratory reference frames of the incident and emerging polarization states (Figure 6.4). In this case, the *single rotation transformation* is formulated as

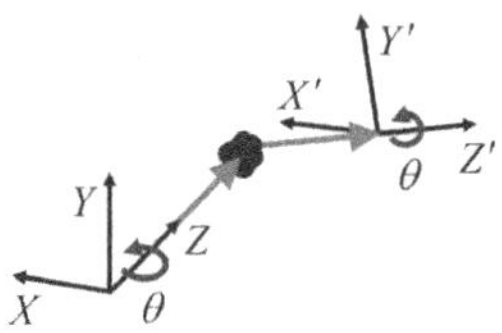

FIGURE 6.4
The *single rotation transformation* consists of respective rotations of a same angle θ of the reference frames of the incident and emerging polarization states about the respective propagation axes Z and Z'.

$$\mathbf{M}' = \mathbf{M}_G\,\mathbf{M}\mathbf{M}_G^T = \begin{pmatrix} 1 & \mathbf{0}^T \\ \mathbf{0} & \mathbf{m}_G \end{pmatrix} m_{00}\begin{pmatrix} 1 & \mathbf{D}^T \\ \mathbf{P} & \mathbf{m} \end{pmatrix}\begin{pmatrix} 1 & \mathbf{0}^T \\ \mathbf{0} & \mathbf{m}_G^T \end{pmatrix} = m_{00}\begin{pmatrix} 1 & \mathbf{D}^T\mathbf{m}_G^T \\ \mathbf{m}_G\mathbf{P} & \mathbf{m}_G\mathbf{m}\mathbf{m}_G^T \end{pmatrix}$$
$$= m_{00}\begin{pmatrix} 1 & cD_1+sD_2 & -sD_1+cD_2 & D_3 \\ cP_1+sP_2 & c^2k_1+s^2k_2+sc(r_3+q_3) & c^2r_3-s^2q_3+sc(k_2-k_1) & cr_2+sr_1 \\ -sP_1+cP_2 & -s^2r_3+c^2q_3+sc(k_2-k_1) & s^2k_1+c^2k_2-sc(r_3+q_3) & -sr_2+cr_1 \\ P_3 & cq_2+sq_1 & -sq_2+cq_1 & k_3 \end{pmatrix} \quad (6.63)$$

where $s \equiv \sin\theta$, $c \equiv \cos\theta\,(\theta \equiv \theta_2 = -\theta_1)$.

Therefore, the following parameters (not all mutually independent) that include the sets (6.56) and (6.60) remain invariant under single rotation transformations (Gil 2016c):

$$\begin{gathered} m_{00},\, D_L,\, D_C,\, D,\, P_L,\, P_C,\, P,\, P_P,\, r_L,\, r_C,\, r,\, q_L,\, q_C,\, q,\, k_L^2+r_C^2+q_C^2,\, r_C-q_C,\, k_C, \\ a_1,\, a_2,\, a_3,\, P_S,\, P_1,\, P_2,\, P_3,\, P_\Delta,\, \mathbf{P}^T\mathbf{m}\mathbf{D},\, \mathbf{P}^T\mathbf{m}^T\mathbf{D},\, \mathbf{P}^T\mathbf{D},\, \operatorname{tr}\mathbf{M},\, \det\mathbf{M}, \\ \rho_0,\, \rho_1,\, \rho_2,\, \rho_3,\, L_1,\, L_2 \end{gathered} \quad (6.64)$$

Different sets of 15 mutually independent invariant parameters can be chosen from (6.64) as for instance

$$m_{00},\, D_L,\, D_C,\, P_L,\, P_C,\, r_L,\, r_C,\, q_L,\, q_C,\, k_L^2+r_C^2+q_C^2,\, k_C,\, P_1,\, P_2,\, \operatorname{tr}\mathbf{M},\, \det\mathbf{M} \quad (6.65)$$

or, alternatively, the following combination of the set of invariants (6.56) of the single retarder transformation with the set (6.62) of invariants of the dual rotation transformation:

$$m_{00},\, D_L,\, D_C,\, P_L,\, P_C,\, a_1,\, a_2,\, a_3,\, \mathbf{P}^T\mathbf{D},\, \mathbf{P}^T\mathbf{m}^T\mathbf{D},\, \mathbf{P}^T\mathbf{m}\mathbf{D},\, P_1,\, P_2,\, \operatorname{tr}\mathbf{M},\, \det\mathbf{M} \quad (6.66)$$

6.5 Purity Space

Since the components of purity (P, D, P_S) of a Mueller matrix are mutually related by certain constraining inequalities, it is worth representing them graphically in order to get a pictorial view of the relations involved. Moreover, the indices of polarimetric purity (IPP) (P_1, P_2, P_3) can be considered as reference frame axes of a *purity space* where each point corresponds to matrices with the same IPP.

6.5.1 Purity Space for the Components of Purity

In order to simplify the graphic representation of the components of purity while taking into account the intimate relation between diattenuation and polarizance analyzed in Section 6.2, let us consider P_P and P_S as the reference axes for the *purity figure* of the components of purity.

As pointed out previously, the nature of a Mueller matrix imposes restrictions on the values of P_P and P_S. If we consider the case of $P_S = 0$, the value $P_P = 1$ is compatible with the expression $P_\Delta^2 = 2P_P^2/3 + P_S^2$ of the degree of polarimetric purity in terms of P_P and P_S, but this possibility is unphysical because $P_P = 1$ implies $P_\Delta = 1$, which is incompatible with $P_S = 0$. In consequence, it is desirable to inspect exhaustively the constraining inequalities involving P_P and P_S. Let us first note that values $P_S < 1/\sqrt{3}$ are incompatible with total purity; furthermore, in that case P_P cannot reach the value $P_P = 1$. In fact, it has been shown that P_P and P_S of a Mueller matrix **M** must satisfy the following inequality (Gil 2011):

$$P_P^2 \leq \left(1 + 3P_S^2\right)/2 \tag{6.67}$$

To prove it, let us recall that for any Mueller matrix **M**, the eigenvalues of its N-matrix $\mathbf{N} \equiv \mathbf{GM}^T\mathbf{GM}$ are non-negative and therefore $\operatorname{tr}\mathbf{N} \geq 0$. Moreover, it is straightforward to show that

$$\operatorname{tr}\mathbf{N} = m_{00}^2\left[\left(1 + 3P_S^2\right) - 2P_P^2\right] \geq 0 \tag{6.68}$$

which leads to inequality (6.67). Therefore, the feasible region for P_P and P_S is limited by the positive branch of the centered hyperbola defined by the maximum value $P_{P\max}$ of P_P for a given value of P_S

$$2P_{P\max}^2 - 3P_S^2 = 1 \tag{6.69}$$

An additional limit is determined by the positive quadrant of the centered ellipse defined by (P_Δ: parameter; P_P,P_S: variables)

$$1 = \frac{2}{3}\frac{P_P^2}{P_\Delta^2} + \frac{P_S^2}{P_\Delta^2} \tag{6.70}$$

Figure 6.5 shows the feasible region (*purity figure*) for P_P and P_S, as well as the feasible elliptical segments $P_P(P_S)$ for certain values of P_Δ. The feasible points for each given value of P_Δ are indicated, as well as the corresponding limiting values for P_P and P_S. When $P_\Delta > 1/\sqrt{3}$ the constraining inequality (6.67) must be considered because the hyperbola intersects the elliptical curves of the function $P_P(P_S)$; nevertheless, when $P_\Delta \leq 1/\sqrt{3}$ the ellipses in the entire (positive) quadrant are feasible. In summary, the resulting feasible region is given by the area OACDO in Figure 6.5.

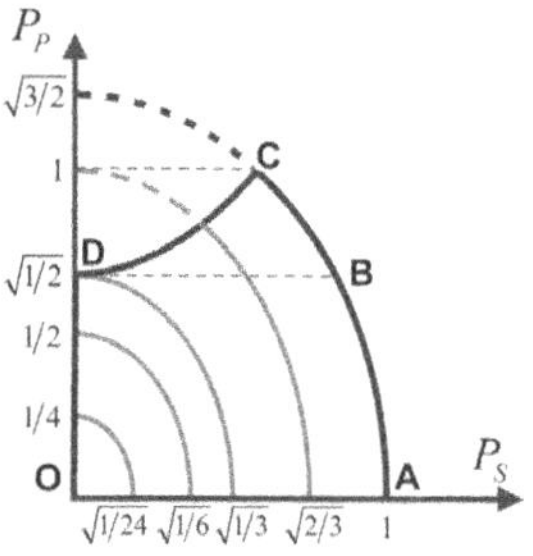

FIGURE 6.5
Purity figure of **M**. The feasible region for the components of purity P_P and P_S is limited by the edges OACDO. Iso-purity curves for some values of the degree of polarimetric purity P_Δ are represented (Gil 2011).

6.5.2 Classification of Mueller Matrices According to the Values of the Components of Purity

Next, we analyze the different types of Mueller matrices according to the distribution of polarimetric purity between P_P and P_S.

(a) Edge AC: $1 = P_\Delta^2 = 2P_P^2/3 + P_S^2$. Pure system, whose interpretation in terms of the CP is described in Table 6.1

(b) Edge OA (vertices O and A excluded): $P_P = 0\ (0 < P_\Delta < 1)$. Depolarizing nonenpolarizing system.

Nonpure Mueller matrix with $P_P = D = P = 0$. The only contribution to purity comes from P_S so that the Mueller matrix has the form

$$\mathbf{M}_O = m_{00}\begin{pmatrix} 1 & \mathbf{0}^T \\ \mathbf{0} & \mathbf{m} \end{pmatrix} \tag{6.71}$$

(c) Vertex O: $P_P = 0,\ P_S = 0\ (P_\Delta = 0)$. Perfect depolarizer.

The submatrix $\mathbf{m}$ is the 3×3 zero matrix, so that the only nonzero element of $\mathbf{M}$ is m_{00}. The system behaves as a perfect depolarizer with fixed intensity coefficient m_{00} for all incident states.

(d) Edge OD (vertex O excluded): $0 < P_P \le 1/\sqrt{2},\ P_S = 0\left(0 < P_\Delta \le \sqrt{1/3}\right)$. Partial polarizer-analyzer with $P_S = 0$.

The only contribution to purity comes from P_P and $\mathbf{M}$ has the particular form

$$\mathbf{M} = m_{00}\begin{pmatrix} 1 & \mathbf{D}^T \\ \mathbf{P} & \mathbf{0} \end{pmatrix} \tag{6.72}$$

Point D corresponds to a Mueller matrix with $P_P = 1/\sqrt{2}$, which is the maximum value of P_P compatible with $P_S = 0$. The general restrictions on $\mathbf{D}$ and $\mathbf{P}$ for matrices

TABLE 6.1
Interpretation of Pure Mueller Matrices in Terms of the Components of Purity

Region	Values for P_P and P_S	Interpretation
Vertex A	$P_S = P_\Delta = 1$ $(P_P = 0)$	*Pure retarder*. This is the only point compatible with $P_S = 1$. Submatrix $\mathbf{m}$ is orthogonal, and the system behaves as a pure retarder with a fixed value of the intensity coefficient m_{00} for all incoming polarization states: $\mathbf{M} = m_{00}\,\mathbf{M}_R$.
Edge AC (points A and C excluded)	$0 \le P_P < 1\ P_\Delta = 1$ $1/\sqrt{3} < P_S < 1$	*Pure system with nonzero diattenuation*. Pure Mueller matrix, in general non-normal, with nonzero diattenuation-polarizance $P_P = D = P$.
Vertex C	$P_P = 1$ $P_\Delta = 1$ $P_S = 1/\sqrt{3}$	*Perfect polarizer* (also called polarizer-analyzer or nondepolarizing polarizer). Pure Mueller matrix with maximum diattenuation-polarizance $P_P = D = P = 1$, in general non-normal. P_S reaches its minimum value compatible with total purity $P_S = 1/\sqrt{3}$. Point C is the only physically feasible point compatible with $P_P = 1$. The extremal values for the intensity coefficient are $t_{\max}(\mathbf{M}) = t_{\max}(\mathbf{M}^r) = 2m_{00}$, $t_{\min}(\mathbf{M}) = t_{\min}(\mathbf{M}^r) = 0$.

TABLE 6.2
Interpretation of Mueller Matrices with $P_S = 0$ in Terms of the Components of Purity

Values for the CP	Interpretation
$0 < P_P \le 1/\sqrt{2}$ $P_S = 0,\ D = 0$	**M** does not exhibit diattenuation, the *I*-image of **M** is a sphere of radius m_{00}, whereas the *I*-image of $\mathbf{M}^T$ is not a sphere. The system exhibits simultaneously polarizing and depolarizing properties and its Mueller matrix is of the form (6.72) with $\mathbf{D} = \mathbf{0}$ and $\mathbf{P} \neq \mathbf{0}$. The forward intensity coefficient has the fixed value m_{00}, whereas the extremal values for the reverse intensity coefficient are $t_{max}(\mathbf{M}^r) = m_{00}(1+P)$, $t_{min}(\mathbf{M}^r) = m_{00}(1-P)$. In the particular case where $P = 1$, the matrix corresponds to a depolarizing polarizer with zero diattenuation.
$0 < P_P \le 1/\sqrt{2}$ $P_S = 0,\ P = 0$	**M** does not exhibit forward polarizance, the *I*-image of **M** is not a sphere, whereas the *I*-image of $\mathbf{M}^T$ is a sphere of radius m_{00}. The system exhibits simultaneously diattenuating and depolarizing properties and its Mueller matrix is of the form (6.72) with $\mathbf{D} \neq \mathbf{0}$ and $\mathbf{P} = \mathbf{0}$. The extremal values for the forward intensity coefficient are $t_{max}(\mathbf{M}) = m_{00}(1+D)$, $t_{min}(\mathbf{M}) = m_{00}(1-D)$, whereas the reverse intensity coefficient has the fixed value m_{00}. In the particular case where $D = 1$, the matrix corresponds to a depolarizing analyzer with zero polarizance.
$0 < P_P \le 1/\sqrt{2}$ $P_S = 0$ $P \neq 0,\ D \neq 0$	The system, whose matrix has the form (6.72), exhibits polarizance $0 < P < 1$ and diattenuation $0 < D < 1$. Polarizance and diattenuation contribute to polarimetric purity. Both *I*-images of **M** and $\mathbf{M}^T$ are not spherical.

with the above structure to be Mueller matrices can be found in (Gil and San José 2021a).

Although P_P provides a global measure of the purity linked to polarizing and diattenuation properties, when $P_P \neq 0$ it is also interesting to analyze the individual contributions of the polarizance P and the diattenuation D to P_Δ. The interpretation of Mueller matrices with $P_S = 0$ in terms of the CP is described in Table 6.2

(e) Edge DC (vertices D and C excluded): $2P_P^2 - 3P_S^2 = 1,\ 0 < P_S < \sqrt{1/3}$.

This hyperbolic segment corresponds to nonpure systems with $P_S \neq 0$ and whose degree of polarizance P_P has its maximum value compatible with P_S. P_P takes values in the interval $1/\sqrt{2} < P_P < 1$.

(f) Region OABD (edges OA, AB and OD excluded): $0 < P_P < 1/\sqrt{2},\ 0 < P_\Delta < 1\ (0 < P_S < 1)$.

As shown in Figure 6.5, all the values of P_S in the interval $0 < P_S < 1$ are achievable within this region. The limit $0 < P_P < 1/\sqrt{2}$ ensures that P_P and P_S satisfy the restriction (6.67), while both P_P and P_S contribute to purity. Values of P_P are nonzero ($0 < P_P < 1/\sqrt{2}$), but are compatible with a zero value for either D or P. At least one of the *I*-images of **M** and $\mathbf{M}^T$ is not spherical. The extremal values for the forward intensity coefficient are $t_{max}(\mathbf{M}) = m_{00}(1+D)$, $t_{min}(\mathbf{M}) = m_{00}(1-D)$, whereas the extremal values for the reverse intensity coefficient are $t_{max}(\mathbf{M}^r) = m_{00}(1+P)$, $t_{min}(\mathbf{M}^r) = m_{00}(1-P)$.

(g) Region BCD (edges BC and CD excluded): $1/\sqrt{2} < P_P < \sqrt{(1+3P_S^2)/2}$, $0 < P_S < \sqrt{2/3}\ (\sqrt{1/2} < P_\Delta \le 1)$.

For all the points inside this region, nonzero contributions of P, D and P_S to the total purity of the system are realized.

Tables 6.3 and 6.4 list some particular forms of Mueller matrices according to the values of P and D, and P_P and P_S respectively

TABLE 6.3
Particular Forms of Mueller Matrices According to the Values of Diattenuation and Polarizance

	$D=0$	$0<D<1$	$D=1$
$P=0$	Nonenpolarizing $\mathbf{M}_O=m_{00}\begin{pmatrix}1 & \mathbf{0}^T\\ \mathbf{0} & \mathbf{m}\end{pmatrix}$	Depolarizing partial analyzer $\mathbf{M}=m_{00}\begin{pmatrix}1 & \mathbf{D}^T\\ \mathbf{0} & \mathbf{m}\end{pmatrix}$	Depolarizing analyzer, $P=0$ $\mathbf{M}_{0\hat{D}}=m_{00}\begin{pmatrix}1 & \hat{\mathbf{D}}^T\\ \mathbf{0} & \mathbf{0}_3\end{pmatrix}$
$0<P<1$	Depolarizing partial polarizer $\mathbf{M}=m_{00}\begin{pmatrix}1 & \mathbf{0}^T\\ \mathbf{P} & \mathbf{m}\end{pmatrix}$	Depolarizing partial polarizer-analyzer $\mathbf{M}=m_{00}\begin{pmatrix}1 & \mathbf{D}^T\\ \mathbf{P} & \mathbf{m}\end{pmatrix}$	Depolarizing analyzer, $0<P<1$ $\mathbf{M}_{P\hat{D}}=m_{00}\begin{pmatrix}1 & \hat{\mathbf{D}}^T\\ \mathbf{P} & \mathbf{P}\otimes\hat{\mathbf{D}}^T\end{pmatrix}$
$P=1$	Depolarizing polarizer, $D=0$ $\mathbf{M}_{\hat{P}0}=m_{00}\begin{pmatrix}1 & \mathbf{0}^T\\ \hat{\mathbf{P}} & \mathbf{0}_3\end{pmatrix}$	Depolarizing polarizer, $0<D<1$ $\mathbf{M}_{\hat{P}D}=m_{00}\begin{pmatrix}1 & \mathbf{D}^T\\ \hat{\mathbf{P}} & \hat{\mathbf{P}}\otimes\mathbf{D}^T\end{pmatrix}$	Perfect polarizer $\mathbf{M}_{\hat{P}\hat{D}}=m_{00}\begin{pmatrix}1 & \hat{\mathbf{D}}^T\\ \hat{\mathbf{P}} & \hat{\mathbf{P}}\otimes\hat{\mathbf{D}}^T\end{pmatrix}$

TABLE 6.4
Particular Forms of Mueller Matrices According to the Values of the Polarimetric Dimension Index and the Degree of Polarizance

	$P_S=0$	$0<P_S<1$	$P_S=1$
$P_P=0$	Perfect depolarizer -point O- $\mathbf{M}_{\Delta 0}=m_{00}\begin{pmatrix}1 & \mathbf{0}^T\\ \mathbf{0} & \mathbf{0}\end{pmatrix}$	Nonenpolarizing depolarizer -segment]OA[- $\mathbf{M}_O=m_{00}\begin{pmatrix}1 & \mathbf{0}^T\\ \mathbf{0} & \mathbf{m}\end{pmatrix}$	Retarder -point A- $\mathbf{M}=m_{00}\begin{pmatrix}1 & \mathbf{0}^T\\ \mathbf{0} & \mathbf{m}_R\end{pmatrix}$
$0<P_P\le 1/\sqrt{2}$	Partial polarizer-analyzer with $P_S=0$ -segment]OD]- $\mathbf{M}=m_{00}\begin{pmatrix}1 & \mathbf{D}^T\\ \mathbf{P} & \mathbf{0}\end{pmatrix}$	Depolarizing partial polarizer-analyzer-region inside]OABD[- $\mathbf{M}=m_{00}\begin{pmatrix}1 & \mathbf{D}^T\\ \mathbf{P} & \mathbf{m}\end{pmatrix}$	Physically not achievable
$1/\sqrt{2}<P_P<1$	Physically not achievable	(a) Depolarizing partial polarizer-analyzer-region BCD-(edge BC excluded) $\mathbf{M}=m_{00}\begin{pmatrix}1 & \mathbf{D}^T\\ \mathbf{P} & \mathbf{m}\end{pmatrix}$ (b) Pure system-edge [BC[- (point C excluded) $\mathbf{M}_J=m_{00}\begin{pmatrix}1 & \mathbf{D}^T\\ \mathbf{m}_R\mathbf{D} & \mathbf{m}_R\mathbf{m}_D\end{pmatrix}$	Physically not achievable
$P_P=1$	Physically not achievable	Perfect polarizer-point C- $\mathbf{M}=m_{00}\begin{pmatrix}1 & \hat{\mathbf{D}}^T\\ \hat{\mathbf{P}} & \hat{\mathbf{P}}\otimes\hat{\mathbf{D}}^T\end{pmatrix}$	Physically not achievable

6.5.3 Purity Space for the Indices of Polarimetric Purity

From the arbitrary decomposition of **H**, it follows that rank**H** is equal to the minimum number r of pure components of the system (Gil 2007). The IPP can be obtained from the normalized eigenvalues $\hat{\lambda}_i \equiv \lambda_i/(\mathrm{tr}\,\mathbf{H})$ of **H** by means of Eq. (6.37); they are directly related to the weights of the components of the characteristic expansion (6.43).

Figure 6.2 shows the tetrahedron that constitutes the feasible region of the Mueller matrices with respect to a reference frame whose axes are defined by the IPP. Each point in the tetrahedron corresponds to Mueller matrices with equal IPP, regardless of the different possible values of the components of purity and other parameters. The restriction to the plane $P_3 = 1$ reproduces the feasible region ABC for P_1 and P_2 corresponding formally to 3×3 coherency matrices (Section 2.10.5), and the feasible region for the degree of polarization of 2×2 coherency matrices corresponds to the segment BC ($P_2 = 1,\ 0 \le P_1 \le 1$). Next, we analyze the different types of Mueller matrices with respect to the values of the IPP. A comprehensive case analysis can be found in (San José and Gil 2011).

(a) Face ABC corresponds to states with $P_3 = 1$, so that $0 \le P_1 \le P_2 \le 1$ and therefore $1/3 \le P_\Delta \le 1$.
    The spectral decomposition contains one, two, or three pure elements (the possibility of four non-vanishing pure components is excluded because it requires $\lambda_3 \ne 0$, which is not compatible with $P_3 = 1$). This case is mathematically equivalent to that obtained for 3×3 coherency matrices (3D polarization states). Vertex C corresponds to a pure system; vertex B corresponds to a system composed of two pure spectral elements with equal relative cross-sections, and vertex A corresponds to a system composed of three pure spectral elements with equal relative cross-sections.
(b) Face OAB (edges OA and AB excluded): $P_1 = 0, 0 < P_2 \le P_3 < 1 \quad (0 < P_\Delta < 1/\sqrt{3})$.
    The spectral decomposition contains four pure elements $(0 < \lambda_3 \le \lambda_2 < \lambda_1 = \lambda_0)$. The two most significant ones have equal relative cross-sections $(\lambda_0 = \lambda_1)$.
(c) Face OAC (edges OA and AC excluded): $0 < P_1 = P_2 \le P_3 < 1 \quad (0 < P_\Delta < 1)$.
    The spectral decomposition contains four pure elements $(0 < \lambda_3 \le \lambda_2 = \lambda_1 < \lambda_0)$. The second and the third most significant components have equal relative cross-sections $(\lambda_1 = \lambda_2)$, different from that of the first most significant one $(\lambda_2 = \lambda_1 < \lambda_0)$.
(d) Face OBC (edges OB, OC and CB excluded): $0 < P_1 < P_2 = P_3 < 1 \quad (0 < P_\Delta < 1)$.
    The spectral decomposition contains four pure elements. The two most significant components have different relative cross-sections and the two less significant ones have equal relative cross-sections $(0 < \lambda_3 = \lambda_2 < \lambda_1 < \lambda_0)$.
(e) Edge OA (vertices O and A excluded): $P_1 = P_2 = 0,\ 0 < P_3 < 1 \quad (0 < P_\Delta < 1/3)$.
    The spectral decomposition contains four pure elements, where the three most significant ones have equal relative cross-sections, different from that of the less significant component $(0 < \lambda_3 < \lambda_2 = \lambda_1 = \lambda_0)$.
(f) Vertex O: $P_1 = P_2 = P_3 = 0 \quad (P_\Delta = 0)$. Perfect depolarizer.
    The spectral decomposition of this system, equivalent to a perfect depolarizer, contains four pure elements with equal relative cross-sections $(0 < \lambda_3 = \lambda_2 = \lambda_1 = \lambda_0)$. All the elements of the Mueller matrix are zero except for m_{00}.
(g) Tetrahedron OABC (faces excluded): $0 < P_1 < P_2 < P_3 < 1 \quad (0 < P_\Delta < 1)$.
    The system can be considered as composed of four pure spectral components with different relative cross-sections $(0 < \lambda_3 < \lambda_2 < \lambda_1 < \lambda_0)$.

6.5.4 Purity Regions in the Space of Components of Purity

Once the two complementary approaches for the representation of the purity structure of a Mueller matrix, namely the components of purity and the indices of polarimetric purity, have been studied, we are in a position to analyze the feasible regions of the space P_P, P_S in terms of $r \equiv \text{rank}\mathbf{H}$ (Figure 6.6; Gil 2016a).

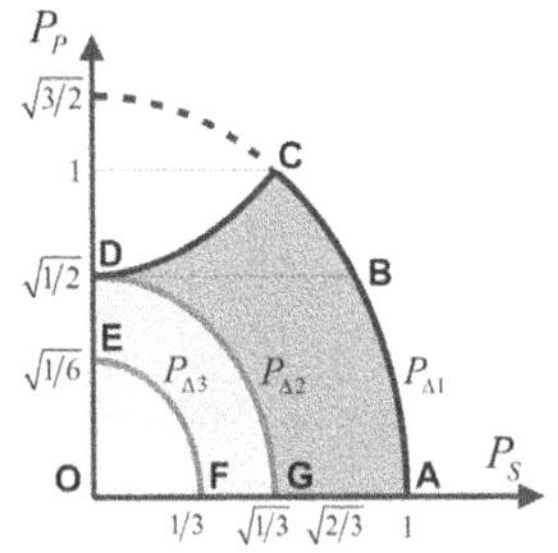

FIGURE 6.6
Purity regions in terms of $r \equiv \text{rank}[\mathbf{H}(\mathbf{M})]$: region OFEO (curve FE excluded) is exclusive of Mueller matrices with $r = 4$; region FGDEF (curve GD excluded) corresponds to Mueller matrices with $r = 4$ or $r = 3$; region GACDG (curve AC excluded) corresponds to Mueller matrices with $r = 4$, $r = 3$ or $r = 2$, and curve AC corresponds to pure Mueller matrices ($r = 1$). (From Gil 2016a).

(a) $\text{rank}\mathbf{H} = 1$. In this case, $P_\Delta = 1, P_P^2 = 3(1 - P_S^2)/2$ and $P_1 = P_2 = P_3 = 1$. The corresponding Mueller matrices are pure and are represented exclusively in the elliptic branch AC. The system is composed of a unique pure component.
(b) $\text{rank}\mathbf{H} = 2$. In this case, $1/\sqrt{3} \le P_\Delta < 1$ and $0 \le P_1 < P_2 = P_3 = 1$. The corresponding Mueller matrices can be considered as composed of two pure components and are represented by points in the region ACDGA (edge AC excluded), which in turn correspond to points in edge BC (vertex C excluded) in the IPP space (Figure 6.2).
(c) $\text{rank}\mathbf{H} = 3$. In this case, $1/3 \le P_\Delta < 1$ and $0 \le P_1 \le P_2 < P_3 = 1$. The corresponding Mueller matrices can be considered as composed of three pure components and are represented by points in the region ACDEFA (edge AC excluded), which in turn correspond to points in face ABC (edge BC excluded) in the IPP space (Figure 6.2).
(d) $\text{rank}\mathbf{H} = 4$. In this case, $0 \le P_\Delta < 1 \; 0 \le P_1 \le P_2 \le P_3 < 1$. The corresponding Mueller matrices can be considered as composed of four pure components and are represented by points in the region ACDOA (edge AC excluded), which in turn correspond to points in the tetrahedron (face ABC excluded) in the IPP space (Figure 6.2).

In summary, region OFEO (curve FE excluded) is exclusive for Mueller matrices with $r = 4$ ($P_3 < 1$); region FGDEF (curve GD excluded) corresponds to Mueller matrices with $r = 4$ or $r = 3$ ($P_3 \le 1, P_2 < 1$); region GACDG (curve AC excluded) corresponds to Mueller matrices with $r = 4$, $r = 3$ or $r = 2$ ($P_3 \le 1, P_2 \le P_3, P_1 < 1$), and curve AC corresponds to pure Mueller matrices ($r = 1$) ($P_1 = 1$).

6.6 General Formulation of Eigenvalue-Based Depolarization Metric Spaces

As seen in previous sections, the quantitative information on the polarimetric randomness-purity encoded in a Mueller matrix is provided by a set of three independent parameters that can be defined from the eigenvalues of the associated covariance matrix $\mathbf{H}$. Among the (infinite) possible choices for the parameters constitutive of such a set, the IPP described in Section 6.3 constitute a representative example. The present section, mainly based on the study performed by Ossikovski and Vizet (2019), is devoted to the development of a unified formal description and analysis of the depolarization spaces defined by depolarization metrics based on the eigenvalues of **H**.

The spectral decomposition in Eq. (5.126) can be rewritten in the equivalent form

$$\mathbf{M} \equiv m_{00}\hat{\mathbf{M}} = \lambda_0 \mathbf{M}_{J0} + \sum_{k=1}^{3} \lambda_k \mathbf{M}_k = \mathbf{M}_0 + \Delta\mathbf{M}$$
$$\left[\mathbf{M}_0 \equiv \lambda_0 \mathbf{M}_{J0} = m_{00}\hat{\lambda}_0 \hat{\mathbf{M}}_{J0} \quad \Delta\mathbf{M} \equiv \sum_{k=1}^{3} \lambda_k \mathbf{M}_k = m_{00} \sum_{k=1}^{3} \hat{\lambda}_k \hat{\mathbf{M}}_k \quad \lambda_0 \geq \lambda_1 \geq \lambda_2 \geq \lambda_3 \right] \tag{6.73}$$

which, in accordance with the discussions in Chapter 11, allows for an immediate statistical interpretation within the fluctuating medium model (Brosseau 1998; Ossikovski and Arteaga 2015). Thus, $\mathbf{M}_0$ can be identified as the nondepolarizing estimate of **M** and physically interpreted as the mean value of **M**, while $\Delta\mathbf{M}$ can be interpreted as the depolarizing part (or residual) due to the spatial, spectral and temporal variations or fluctuations exhibited by the medium during the measurement process.

The mean intensity coefficient m_{00} of **M** plays the mathematical role of a scaling coefficient of **M** and therefore, as already seen with $P_\Delta, L_1, L_2, PI, P_1, P_2, P_3$ and S, it does not take place in the depolarization descriptors, whose values are insensitive to the effective value of m_{00}. By considering the normalized eigenvalues $\hat{\lambda}_i \equiv \lambda_i / m_{00}$ $(i = 0,1,2,3)$, with $m_{00} = \operatorname{tr}\mathbf{H} = \operatorname{tr}\mathbf{C}$, it follows that $\hat{\lambda}_0 + \hat{\lambda}_1 + \hat{\lambda}_2 + \hat{\lambda}_3 = 1$, so that, as with the IPP (see Section 6.3), any set of three independent parameters derived from $\hat{\lambda}_i$ provides complete quantitative information on the polarimetric purity-randomness of the interactions represented by **M**. Obviously, there are different ways to define such kinds of sets (Gil 2007; San José and Gil 2011; Sheppard et al. 2019; Ossikovski and Vizet 2019). Among them, let us consider the triplet $(\hat{\lambda}_1, \hat{\lambda}_2, \hat{\lambda}_3,)$ with the assumed convention $\hat{\lambda}_0 \geq \hat{\lambda}_1 \geq \hat{\lambda}_2 \geq \hat{\lambda}_3$, which forms a three-dimensional space Σ_λ, called by Ossikovski and Vizet (2019) the *natural depolarization space*. In analogy to the purity space Σ_{IPP} defined by the IPP and represented by the tetrahedron in Figure 6.2, every Mueller matrix **M** is represented by a point in Σ_λ; however, to every point in Σ_λ there correspond infinitely many Mueller matrices in general (those whose associated covariance matrices have equal eigenvalue spectra).

Despite the fact that there also exist alternative definitions of depolarization parameters, appropriate for homogeneous depolarizing media (Noble and Chipman 2012; Ossikovski and Arteaga 2014), most of the commonly used depolarization metrics are derived from $\hat{\lambda}_i$ (Gil 2007; San José and Gil 2011; Sheppard et al. 2019; Ossikovski and Vizet 2019). These metrics can be regarded as linear or nonlinear functions of the points from Σ_λ and, if taken as triplets, they themselves define their proper depolarization spaces.

Regarding the natural depolarization space Σ_λ, its geometry is determined by the constraints

$$0 \leq \lambda_3 \leq \lambda_2 \leq \lambda_1 \leq 1 - \lambda_3 - \lambda_2 - \lambda_1 \tag{6.74}$$

Consequently, Σ_λ is bounded by the irregular tetrahedron represented in Figure 6.7, which is delimited by four planes defining its four faces: $\lambda_3 = 0$ (face $A_\lambda B_\lambda C_\lambda$), $\lambda_2 = \lambda_3$ (face $O_\lambda B_\lambda C_\lambda$), $\lambda_1 = \lambda_2$ (face $O_\lambda A_\lambda C_\lambda$) and $2\lambda_1 + \lambda_2 + \lambda_3 = 1$ (face $O_\lambda A_\lambda B_\lambda$). The coordinates of the vertices of the tetrahedron are $O_\lambda(1/4, 1/4, 1/4)$, $A_\lambda(0, 1/3, 1/3)$, $B_\lambda(0,0,1/2)$ and $C_\lambda(0,0,0)$.

In particular, all nondepolarizing Mueller matrices correspond to the vertex C_λ, whereas the depolarizing ones lie within the tetrahedron $O_\lambda A_\lambda B_\lambda C_\lambda$ including its faces, edges and

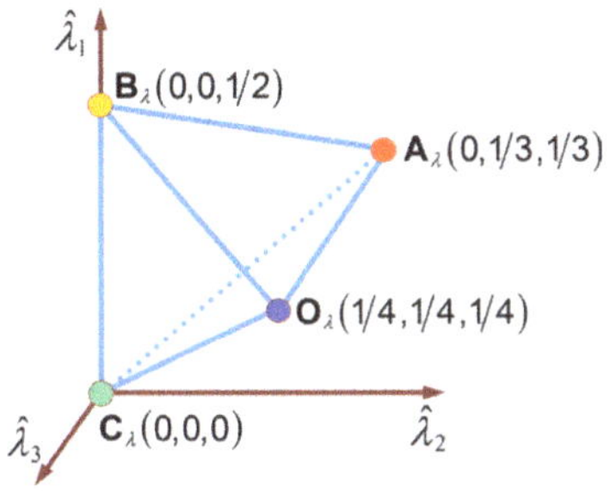

FIGURE 6.7
Natural depolarization space Σ_λ.

vertices, except for vertex C_λ. The perfect depolarizer $\mathbf{M}_{\Delta 0} = \mathrm{diag}(1,0,0,0)$ corresponds, in an exclusive manner, to vertex O_λ.

As with the purity space Σ_{IPP} for the IPP, different regions of Σ_λ can be identified for the different values of the rank r of $\mathbf{H}$. As a result, $\mathbf{M}$ is mapped onto the following respective regions: within the tetrahedron (faces, edges and vertices excluded) when $r = 4$; on the face ABC ($\lambda_3 = 0$) when $r = 3$; on the edge BC ($\lambda_2 = \lambda_3 = 0$) when $r = 2$, and on the vertex C ($\lambda_1 = \lambda_2 = \lambda_3 = 0$) when $r = 1$.

Other conventions for the definition of respective purity spaces can be formulated by considering triplets p_k $(k = 1,2,3)$ determined by $p_k = f_k\left(\hat{\lambda}_1, \hat{\lambda}_2, \hat{\lambda}_3\right)$, where f_k represents a functional relation (linear or not). Geometrically, the triplet (p_1, p_2, p_3) maps the depolarization space Σ_λ into the new corresponding depolarization space Σ_p, in such a manner that the irregular tetrahedron $O_\lambda A_\lambda B_\lambda C_\lambda$ from Σ_λ is mapped into a new tetrahedron O′A′B′C′ in Σ_p.

When the functions f_k are linear, the tetrahedron O′A′B′C′ is *rectilinear* (and irregular, like $O_\lambda A_\lambda B_\lambda C_\lambda$); while it is *curvilinear* if f_k are nonlinear. In all cases there is a one-to-one correspondence between tetrahedrons and, in particular, between respective vertices.

If, in addition, the triplet (p_1, p_2, p_3) is ordered so as to satisfy the nested inequalities $0 \le p_1 \le p_2 \le p_3 \le 1$, then the tetrahedron O′A′B′C′ of Σ_p lies entirely within the (rectilinear) irregular tetrahedron OABC defined by the four planes $p_1 = 0$, $p_1 = p_2$, $p_2 = p_3$ and $p_3 = 1$ and whose vertices are $O(0,0,0)$, $A(0,0,1)$, $B(0,1,1)$ and $C(1,1,1)$.

Let us now consider the three differences $\Delta_q \equiv \hat{\lambda}_{q-1} - \hat{\lambda}_q$ $(q = 1,2,3)$ and define the two vectors $\mathbf{p} \equiv (p_3, p_2, p_1)^T$ (not to be confused with the polarization vector of a polarization state, which is also denoted by $\mathbf{p}$) and $\mathbf{\Delta} \equiv (\Delta_3, \Delta_2, \Delta_1)^T$. Then, the linear relation between both vectors can be expressed as $\mathbf{p} = \mathbf{\Omega}\,\mathbf{\Delta}$, where $\mathbf{\Omega}$ is the corresponding transformation 3×3 matrix.

Then, the elements of $\mathbf{\Omega}$ are determined from the one-to-one mapping correspondence between the vertices $O_\lambda A_\lambda B_\lambda C_\lambda$ of Σ_λ and O′A′B′C′ of Σ_p. Such four simultaneous linear equations lead to $\Omega_{kq} = q$ if $q \le k$ and $\Omega_{kq} = 0$ if $q > k$ for the elements Ω_{kq} of $\mathbf{\Omega}$, which is strictly equivalent to the definition of the IPP in Eq. (6.38). Thus, among the depolarization spaces Σ_p defined through linear depolarization metrics and obeying the inequalities $0 \le p_1 \le p_2 \le p_3 \le 1$, the one whose tetrahedron has the largest possible volume is Σ_{IPP} (see Figure 6.2), so that the separation among points representing Mueller matrices with different values of the metric triplet is maximized, which constitutes a peculiar extremal feature of the IPP.

An interesting triplet is that constituted by $1 - S$, where S is the polarization entropy S (Cloude 1986), the depolarization index P_Δ (Gil and Bernabéu 1986), and the overall purity index *PI* (Tariq et al. 2017), each of them constituting a different global measure of the polarimetric purity of a Mueller matrix **M**, as described in Section 5.10. It has been shown that these parameters satisfy the inequalities (Ossikovski and Vizet 2019, Appendix A)

$$0 \le 1 - S \le P_\Delta \le PI \le 1 \tag{6.75}$$

Therefore, the ordered triplet $(1-S, P_\Delta, PI)$ defines a nonlinear depolarization space Σ_{SPI} delimited by a curvilinear tetrahedron whose vertices, edges and faces remain in a one-to-one correspondence with those of the tetrahedrons associated with Σ_λ and Σ_{IPP}.

In analogy to the polarization metrics considered above based on the eigenvalues of the covariance matrix **H** associated with **M**, the corresponding intrinsic polarization metrics can be defined from the eigenvalues of the covariance matrix $\mathbf{H}_\Delta$ associated with the central depolarizer $\mathbf{M}_\Delta$ of the normal form of **M** (see Section 5.7). As a result, respective canonical depolarization spaces are defined for type-I and type-II canonical Mueller matrices.

If the depolarization indices are grouped in pairs instead of in triplets, the respective depolarization spaces reduce to two-dimensional geometric representations, called depolarization diagrams (Ossikovski and Vizet 2019). For instance, the depolarization diagram (P_Δ, S) was considered by Le Roy-Brehonnet et al. (1997) as well as by Aiello and Woerdman (2005). More recently, the depolarization diagram (P_Δ, PI) has been studied by Tariq. et al. (2017). It turns out (if abstraction is made of permuting coordinates and complementing *S* to the unit) that these depolarization diagrams are nothing but the two-dimensional projections of the three-dimensional $(1-S, P_\Delta, PI)$ depolarization space onto two of its coordinate planes, *PI*=0 and $1-S=0$, respectively, as pointed out by Ossikovski and Vizet (2019).

From a general point of view, the depolarization space Σ_p defined by the triplet (p_1, p_2, p_3) is completely described by the set of its three depolarization diagrams (p_1, p_2), (p_1, p_3) and (p_2, p_3), given by the respective projections onto the three coordinate planes $p_3 = 0$, $p_2 = 0$ and $p_1 = 0$, respectively. Despite the appealing two-dimensional representation of the depolarization diagrams, it should be stressed that they constitute only partial views of the quantitative structure of polarimetric purity of Mueller matrices, and three-dimensional space representations are required for complete description.

It is remarkable that, despite the apparent difficulty to find Mueller matrices represented by points belonging to certain "forbidden" or "hidden" areas of specific depolarization diagrams, all points of a given depolarization space Σ_p are physically achievable and the associated tetrahedron does not contain "empty" regions (Ossikovski and Vizet 2019).

As indicated in the introduction of this section, a comprehensive study of depolarization metric spaces can be found in Ossikovski and Vizet (2019), while the characterization of invariant depolarizing properties through the combination of the IPP and the CP has been studied by Van Eeckhout et al. (2021a).

6.7 Anisotropy Coefficients of a Mueller Matrix

The structure of a Mueller matrix **M** is conditioned by the polarimetric anisotropies of the medium represented by **M**. So, diattenuation-polarizance (dichroism), as well as birefringence, are relative to certain symmetry axes, and the properties of **M** reflect the corresponding anisotropy properties.

By considering the canonical Stokes vectors (see Figure 5.3)

$$\begin{array}{lll} \mathbf{s}_{1+} \equiv (1,1,0,0)^T & \mathbf{s}_{2+} \equiv (1,0,1,0)^T & \mathbf{s}_{3+} \equiv (1,0,0,1)^T \\ \mathbf{s}_{1-} \equiv (1,-1,0,0)^T & \mathbf{s}_{2-} \equiv (1,0,-1,0)^T & \mathbf{s}_{3-} \equiv (1,0,0,-1)^T \end{array} \tag{6.76}$$

representing the respective states of polarization: linear horizontal/vertical; linear 45°/ -45°, and circular right/left, it is observed that the following nondepolarizing *linear horizontal anisotropy Mueller matrix* $\hat{\mathbf{M}}_L$, *linear 45° anisotropy Mueller matrix* $\hat{\mathbf{M}}_{L'}$ and *circular anisotropy Mueller matrix* $\hat{\mathbf{M}}_C$, leave unchanged $\mathbf{s}_{1\pm}$, $\mathbf{s}_{2\pm}$ and $\mathbf{s}_{3\pm}$ respectively (except for a positive coefficient) (Arteaga et al. 2011):

$$\hat{\mathbf{M}}_L = \begin{pmatrix} 1 & c_{\kappa_L} & 0 & 0 \\ c_{\kappa_L} & 1 & 0 & 0 \\ 0 & 0 & s_{\kappa_L} c_{\Delta_L} & s_{\kappa_L} s_{\Delta_L} \\ 0 & 0 & -s_{\kappa_L} s_{\Delta_L} & s_{\kappa_L} c_{\Delta_L} \end{pmatrix} = \begin{pmatrix} 1 & c_{\kappa_L} & 0 & 0 \\ c_{\kappa_L} & 1 & 0 & 0 \\ 0 & 0 & s_{\kappa_L} & 0 \\ 0 & 0 & 0 & s_{\kappa_L} \end{pmatrix} \begin{pmatrix} 1 & 0 & 0 & 0 \\ 0 & 1 & 0 & 0 \\ 0 & 0 & c_{\Delta_L} & s_{\Delta_L} \\ 0 & 0 & -s_{\Delta_L} & c_{\Delta_L} \end{pmatrix}$$

$$\hat{\mathbf{M}}_{L'} = \begin{pmatrix} 1 & 0 & c_{\kappa_{L'}} & 0 \\ 0 & s_{\kappa_{L'}} c_{\Delta_{L'}} & 0 & s_{\kappa_{L'}} s_{\Delta_{L'}} \\ c_{\kappa_{L'}} & 0 & 1 & 0 \\ 0 & -s_{\kappa_{L'}} s_{\Delta_{L'}} & 0 & s_{\kappa_{L'}} c_{\Delta_{L'}} \end{pmatrix} = \begin{pmatrix} 1 & 0 & c_{\kappa_{L'}} & 0 \\ 0 & s_{\kappa_{L'}} & 0 & 0 \\ c_{\kappa_{L'}} & 0 & 1 & 0 \\ 0 & 0 & 0 & s_{\kappa_{L'}} \end{pmatrix} \begin{pmatrix} 1 & 0 & 0 & 0 \\ 0 & c_{\Delta_{L'}} & 0 & s_{\Delta_{L'}} \\ 0 & 0 & 1 & 0 \\ 0 & -s_{\Delta_{L'}} & 0 & c_{\Delta_{L'}} \end{pmatrix} \tag{6.77}$$

$$\hat{\mathbf{M}}_C = \begin{pmatrix} 1 & 0 & 0 & c_{\kappa_C} \\ 0 & s_{\kappa_C} c_{\Delta_C} & s_{\kappa_C} s_{\Delta_C} & 0 \\ 0 & -s_{\kappa_C} s_{\Delta_C} & s_{\kappa_C} c_{\Delta_C} & 0 \\ c_{\kappa_C} & 0 & 0 & 1 \end{pmatrix} = \begin{pmatrix} 1 & 0 & 0 & c_{\kappa_C} \\ 0 & s_{\kappa_C} & 0 & 0 \\ 0 & 0 & s_{\kappa_C} & 0 \\ c_{\kappa_C} & 0 & 0 & 1 \end{pmatrix} \begin{pmatrix} 1 & 0 & 0 & 0 \\ 0 & c_{\Delta_C} & s_{\Delta_C} & 0 \\ 0 & -s_{\Delta_C} & c_{\Delta_C} & 0 \\ 0 & 0 & 0 & 1 \end{pmatrix}$$

where the abbreviated notations $s_x \equiv \sin x$ and $c_x \equiv \cos x$ have been used.

Each of these pure Mueller matrices corresponds to a diattenuating retarder (or retarding diattenuator) which can respectively be considered as a serial combination of a diattenuator (linear, linear-45°, circular) and a retarder (linear, linear-45°, circular). Furthermore, in virtue of the commutativity of their respective matrices, the diattenuator and the retarder can be placed in either of the two possible relative positions. Thus, dichroism and birefringence are uncoupled in each of these particular pure Mueller matrices.

Let us recall the vectorial partitioned expression (5.104) of a generic Mueller matrix as

$$\mathbf{M} \equiv m_{00} \begin{pmatrix} 1 & D_1 & D_2 & D_3 \\ P_1 & k_1 & r_3 & r_2 \\ P_2 & q_3 & k_2 & r_1 \\ P_3 & q_2 & q_1 & k_3 \end{pmatrix} \quad \mathbf{D} \equiv \begin{pmatrix} D_1 \\ D_2 \\ D_3 \end{pmatrix} \quad \mathbf{P} \equiv \begin{pmatrix} P_1 \\ P_2 \\ P_3 \end{pmatrix} \quad \mathbf{k} \equiv \frac{1}{\sqrt{3}} \begin{pmatrix} k_1 \\ k_2 \\ k_3 \end{pmatrix} \quad \mathbf{r} \equiv \begin{pmatrix} r_1 \\ r_2 \\ r_3 \end{pmatrix} \quad \mathbf{q} \equiv \begin{pmatrix} q_1 \\ q_2 \\ q_3 \end{pmatrix} \tag{6.78}$$

where $\mathbf{D}$ and $\mathbf{P}$ are the diattenuation and polarizance vectors, $\mathbf{k}$ is built from the diagonal elements of the submatrix $\mathbf{m}$, and $\mathbf{r}$ and $\mathbf{q}$ are complementary vectors embedded into $\mathbf{m}$. The absolute values of vectors $\mathbf{k}$, $\mathbf{r}$ and $\mathbf{q}$ are denoted as $|\mathbf{k}| \equiv k$, $|\mathbf{r}| \equiv r$, and $|\mathbf{q}| \equiv q$.

TABLE 6.5
Quantities Associated with Different Types of Polarimetric Anisotropies

	Enpolarizing anisotropy	Retarding anisotropy	Anisotropy coefficients
Linear-horizontal	$\alpha_{1D} \equiv D_1 + P_1$	$\alpha_{1R} \equiv r_1 - q_1$	$\alpha_1 \equiv \sqrt{\left(\alpha_{1D}^2 + \alpha_{1R}^2\right)/\Sigma}$
Linear-45°	$\alpha_{2D} \equiv D_2 + P_2$	$\alpha_{2R} \equiv r_2 - q_2$	$\alpha_2 \equiv \sqrt{\left(\alpha_{2D}^2 + \alpha_{2R}^2\right)/\Sigma}$
Linear	$\alpha_{LD} \equiv \sqrt{\left(\alpha_{1D}^2 + \alpha_{2D}^2\right)/\Sigma}$	$\alpha_{LR} \equiv \sqrt{\left(\alpha_{1R}^2 + \alpha_{2R}^2\right)/\Sigma}$	$\alpha_L \equiv \sqrt{\left(\alpha_{1D}^2 + \alpha_{1R}^2 + \alpha_{2D}^2 + \alpha_{2R}^2\right)/\Sigma}$
Circular	$\alpha_{3D} \equiv D_3 + P_3$	$\alpha_{3R} \equiv r_3 - q_3$	$\alpha_C \equiv \alpha_3 \equiv \sqrt{\left(\alpha_{3D}^2 + \alpha_{3R}^2\right)/\Sigma}$

Note: The right column summarizes the combination of enpolarizing and retarding anisotropies.

From the particular cases reported in Eq. (6.77), one may define a set of quantities expressing the different types of polarimetric anisotropies. The set, based on the anisotropy coefficients by Arteaga et al. (2011), is reported in Table 6.5, where the auxiliary parameter

$$\Sigma \equiv 3\left(1 - k^2\right) + 2\mathbf{D}^T\mathbf{P} - 2\mathbf{r}^T\mathbf{q} \tag{6.79}$$

has been used.

The parameters in Table 6.5 measure different types of polarimetric anisotropies, which can be grouped in different ways. For instance, linear anisotropies α_{LD} and α_{LR} are given by respective quadratic averages of linear-horizontal and linear-45° ones, while the enpolarizing (dichroism) and retardance (birefringence) anisotropies can be combined in the so-called *anisotropy coefficients* defined as (Arteaga et al. 2011)

$$\alpha_1 \equiv \sqrt{\left(\alpha_{1D}^2 + \alpha_{1R}^2\right)/\Sigma} \quad \alpha_2 \equiv \sqrt{\left(\alpha_{2D}^2 + \alpha_{2R}^2\right)/\Sigma} \quad \alpha_3 \equiv \sqrt{\left(\alpha_{3D}^2 + \alpha_{3R}^2\right)/\Sigma} \tag{6.80}$$

Moreover, as in the case of the components of diattenuation and polarizance, it is particularly interesting to distinguish between linear and circular contributions, so that linear and circular anisotropy coefficients are defined as

$$\alpha_L \equiv \sqrt{\left(\alpha_{1D}^2 + \alpha_{1R}^2 + \alpha_{2D}^2 + \alpha_{2R}^2\right)/\Sigma} \quad \alpha_C \equiv \alpha_3 \equiv \sqrt{\left(\alpha_{3D}^2 + \alpha_{3R}^2\right)/\Sigma} \tag{6.81}$$

The anisotropy coefficients are independent of the value of m_{00}, and can be grouped into an *anisotropy vector* $\boldsymbol{\alpha} \equiv (\alpha_1, \alpha_2, \alpha_3)^T$, whose absolute value P_α, called the *degree of anisotropy* of $\mathbf{M}$, is given by

$$P_\alpha^2 \equiv |\boldsymbol{\alpha}|^2 = \alpha_1^2 + \alpha_2^2 + \alpha_3^2 = \alpha_L^2 + \alpha_C^2 = \frac{D^2 + P^2 + r^2 + q^2 + 2\mathbf{D}^T\mathbf{P} - 2\mathbf{r}^T\mathbf{q}}{3 - 3k^2 + 2\mathbf{D}^T\mathbf{P} - 2\mathbf{r}^T\mathbf{q}} \tag{6.82}$$

and is limited by the depolarization index P_Δ of $\mathbf{M}$

$$P_\alpha \le P_\Delta \le 1 \tag{6.83}$$

Note that $\boldsymbol{\alpha}\left(\hat{\mathbf{M}}_L\right) = (1,0,0)^T$; $\boldsymbol{\alpha}\left(\hat{\mathbf{M}}_{L'}\right) = (0,1,0)^T$, and $\boldsymbol{\alpha}\left(\hat{\mathbf{M}}_C\right) = (0,0,1)^T$, so that (Arteaga et al. 2011)

$$P_\alpha\left(\hat{\mathbf{M}}_L\right)=P_\Delta\left(\hat{\mathbf{M}}_L\right)=P_\alpha\left(\hat{\mathbf{M}}_{L'}\right)=P_\Delta\left(\hat{\mathbf{M}}_{L'}\right)=P_\alpha\left(\hat{\mathbf{M}}_C\right)=P_\Delta\left(\hat{\mathbf{M}}_C\right)=1 \tag{6.84}$$

The maximal value $P_\alpha = P_\Delta$ is reached if and only if $\mathbf{M}$ is a pure Mueller matrix ($P_\Delta = 1$). Moreover, it is enough to consider the identity Mueller matrix to show that $P_\alpha < 1$ does not imply necessarily the existence of depolarization ($P_\Delta < 1$). Observe also that it is possible to define an *anisotropy Stokes vector* of $\mathbf{M}$ as $\mathbf{s}_\alpha(\mathbf{M}) \equiv (1, \alpha_1, \alpha_2, \alpha_3)^T$, which can be represented in the positive octant of the Poincaré sphere, providing an interesting analogy to Stokes vectors of polarized states.

To illustrate the values of the anisotropy coefficients for different types of Mueller matrices, some representative examples are considered in Table 6.6.

To go deeper into the physical information provided by the anisotropy coefficients, let us consider the arrow form $\mathbf{M}_A$ of $\mathbf{M}$ (see Section 5.12), which is calculated through the dual retarder transformation in Eq. (5.108) based on the singular value decomposition of the submatrix $\mathbf{m}$ of $\mathbf{M}$. The anisotropy coefficients of $\mathbf{M}_A$ are given by

$$\alpha_1=\alpha_{1D}=\sqrt{\left(D_{A1}+P_{A1}\right)^2/\Sigma_A} \qquad \alpha_2=\alpha_{2D}=\sqrt{\left(D_{A2}+P_{A2}\right)^2/\Sigma_A} \qquad \alpha_3=\alpha_{3D}=\sqrt{\left(D_{A3}+P_{A3}\right)^2/\Sigma_A}$$
$$\left[\Sigma_A \equiv 3\left(1-P_S^2\right)+2\mathbf{D}_A^T\mathbf{P}_A \qquad P_S^2\left(\mathbf{M}_A\right)=\left(a_1^2+a_2^2+a_3^2\right)/3=P_S^2(\mathbf{M})\right] \tag{6.85}$$

or, grouped into linear and circular coefficients

$$\alpha_L=\sqrt{\frac{\left(D_{A1}+P_{A1}\right)^2+\left(D_{A2}+P_{A2}\right)^2}{3\left(1-P_S^2\right)+2\mathbf{D}_A^T\mathbf{P}_A}} \qquad \alpha_C=\sqrt{\frac{\left(D_{A3}+P_{A3}\right)^2}{3\left(1-P_S^2\right)+2\mathbf{D}_A^T\mathbf{P}_A}} \tag{6.86}$$

Note that they contain no birefringence anisotropy contribution ($\alpha_{1R} = \alpha_{2R} = \alpha_{3R} = 0$).

Thus, the only remaining anisotropies after the transformation are due to the transformed diattenuation vector $\mathbf{D}_A$, the transformed polarizance vector $\mathbf{P}_A$, and $P_S(\mathbf{M}_A) = P_S(\mathbf{M})$. Observe that the linear, α_L, and circular, α_C, anisotropy coefficients are invariant under single rotation transformations. The degree of anisotropy of $\mathbf{M}_A$, which in general is different from that of $\mathbf{M}$ because $\mathbf{M}_A$ lacks birefringence anisotropies, is given by

$$P_\alpha^2\left(\mathbf{M}_A\right)=\frac{D^2+P^2+2\mathbf{D}_A^T\mathbf{P}_A}{3\left(1-P_S^2\right)+2\mathbf{D}_A^T\mathbf{P}_A} \qquad \left(3P_S^2=a_1^2+a_2^2+a_3^2\right) \tag{6.87}$$

By denoting by ψ the angle subtended between vectors $\mathbf{D}_A$ and $\mathbf{P}_A$, their scalar product is expressed as $\mathbf{D}_A^T\mathbf{P}_A = D_A P_A \cos\psi = DP\cos\psi$. Since the quantities D, P and P_S are invariant under dual retarder transformations, the values of the anisotropy coefficients only change as ψ changes. Moreover, as seen in Section 6.4.2, $\mathbf{D}^T\mathbf{P}$ is invariant under single retarder transformations (including single rotation transformations), and therefore the anisotropy coefficients, as well as the degree of anisotropy, are invariant under any kind of single retarder (or rotation) transformation.

TABLE 6.6
Anisotropy Coefficients for Various Particular Forms of Mueller Matrices

Mueller matrix		Anisotropy
Symmetric **m** and $D \neq 0, P \neq 0, r \neq 0$	$\mathbf{M} \equiv m_{00}\begin{pmatrix} 1 & D_1 & D_2 & D_3 \\ P_1 & k_1 & r_3 & r_2 \\ P_2 & r_3 & k_2 & r_1 \\ P_3 & r_2 & r_1 & k_3 \end{pmatrix}$	The only contribution to P_α is due to diattenuation and polarizance. When $P_\alpha = 1$, **M** corresponds to a perfect polarizer.
Symmetric nonenpolarizing depolarizer Symmetric **m** and $D = P = 0$	$\mathbf{M} \equiv m_{00}\begin{pmatrix} 1 & 0 & 0 & 0 \\ 0 & k_1 & r_3 & r_2 \\ 0 & r_3 & k_2 & r_1 \\ 0 & r_2 & r_1 & k_3 \end{pmatrix}$	$P_\alpha = 0$. The medium is depolarizing without exhibiting apparent anisotropies. When $r = 0$, the medium is called an *intrinsic depolarizer* (or *diagonal depolarizer*). When $k = 1$, **M** is a diagonal pure Mueller matrix
Symmetric **M** and $D \neq 0, r \neq 0$	$\mathbf{M} \equiv m_{00}\begin{pmatrix} 1 & D_1 & D_2 & D_3 \\ D_1 & k_1 & r_3 & r_2 \\ D_2 & r_3 & k_2 & r_1 \\ D_3 & r_2 & r_1 & k_3 \end{pmatrix}$	The only contribution to P_α is due to diattenuation and polarizance. This case covers normal diattenuators and parallel mixtures of normal diattenuators.
Arrow Mueller matrix $r = q = 0$	$\mathbf{M} \equiv m_{00}\begin{pmatrix} 1 & D_1 & D_2 & D_3 \\ P_1 & k_1 & 0 & 0 \\ P_2 & 0 & k_2 & 0 \\ P_3 & 0 & 0 & k_3 \end{pmatrix}$	The only contribution to P_α is due to diattenuation and polarizance. Recall that any Mueller matrix can be transformed to its arrow form through the dual retarder transformation defined from the singular value decomposition of the submatrix **m**.
Intrinsic depolarizer	$\mathbf{M} \equiv m_{00}\begin{pmatrix} 1 & 0 & 0 & 0 \\ 0 & k_1 & 0 & 0 \\ 0 & 0 & k_2 & 0 \\ 0 & 0 & 0 & k_3 \end{pmatrix}$	$P_\alpha = 0$. All the anisotropy coefficients are zero. This case covers intrinsic depolarizers (diagonal, including perfect depolarizers), half-wave retarders and attenuators (**M** proportional to the identity matrix).
Nonenpolarizing depolarizer $\begin{pmatrix} P = D = 0 \\ k^2 + q^2 + r^2 < 3 \end{pmatrix}$	$\mathbf{M} \equiv m_{00}\begin{pmatrix} 1 & 0 & 0 & 0 \\ 0 & k_1 & r_3 & r_2 \\ 0 & q_3 & k_2 & r_1 \\ 0 & q_2 & q_1 & k_3 \end{pmatrix}$	The only contribution to P_α is due to birefringence. **M** is equivalent to that of a parallel mixture of retarders.
Retarder $\begin{pmatrix} P = D = 0 \\ k^2 + q^2 + r^2 = 3 \end{pmatrix}$	$\mathbf{M} \equiv m_{00}\begin{pmatrix} 1 & 0 & 0 & 0 \\ 0 & k_1 & r_3 & r_2 \\ 0 & q_3 & k_2 & r_1 \\ 0 & q_2 & q_1 & k_3 \end{pmatrix}$	$P_\alpha = 1$, and the only contribution to P_α is due to birefringence. Since $1 = (k^2 + q^2 + r^2)/3 = P_\Delta$, **M** corresponds to a pure retarder.
Linear anisotropic $P_3 = D_3 = k_3 = q_3 = r_3 = 0$	$\mathbf{M} \equiv m_{00}\begin{pmatrix} 1 & D_1 & D_2 & 0 \\ P_1 & k_1 & 0 & r_2 \\ P_2 & 0 & k_2 & r_1 \\ 0 & q_2 & q_1 & k_3 \end{pmatrix}$	$\alpha_C = 0$. The only contribution to P_α is due to linear anisotropies. **M** is equivalent to that of a parallel mixture of linear diattenuators (in general non-normal) and linear retarders.
Circular anisotropic $P_1 = P_2 = D_1 = D_2 = 0$ $k_1 = k_2 = r_1 = r_2 = q_1 = q_2 = 0$	$\mathbf{M} \equiv m_{00}\begin{pmatrix} 1 & 0 & 0 & D_3 \\ 0 & k_1 & r_3 & 0 \\ 0 & q_3 & k_2 & 0 \\ P_3 & 0 & 0 & k_3 \end{pmatrix}$	$\alpha_L = 0$. The only contribution to P_α is due to circular anisotropies. **M** is equivalent to that of a parallel mixture of circular diattenuators (in general non-normal) and circular retarders.

6.8 From a Nondepolarizing to a Depolarizing Mueller Matrix

For a deeper knowledge of the mathematical structure of Mueller matrices, as well as of the role played by some relevant polarimetric quantities, it is important to understand how depolarization can be synthesized from a given reference pure Mueller matrix. This section is devoted to addressing the problem of how to generate all possible kinds of Mueller matrices from the appropriate choice of a reference pure Mueller matrix through the adjustment of certain parameters in an objective and systematic way.

The need for a procedure for the synthesis of depolarization by the progressive modification of a reference nondepolarizing system was emphasized by Cloude (2013), who developed a synthesis method based on the successive addition of nondepolarizing elements represented by pure Mueller matrices. Later, an alternative procedure outlined in the next paragraphs, was developed (Gil 2014c) by taking advantage of the general serial decomposition of a pure Mueller matrix (Section 4.5), and of the normal form (Section 5.7), and the symmetric decomposition (5.24) of a depolarizing Mueller matrix. This procedure allows one to achieve a constructive synthesis of a given arbitrary depolarizing Mueller matrix **M** by identifying a reference pure Mueller matrix $\mathbf{M}_J$ and by varying, in a consistent manner, its depolarizing properties through the gradual variation of the indices of polarimetric purity.

The steps of this procedure are the following:

(1) take a generic pure Mueller matrix $\mathbf{M}_J$ as starting reference;
(2) split $\mathbf{M}_J$ into a product (or serial composition) of two pure Mueller matrices $\mathbf{M}_J = \mathbf{M}_{J2}\,\mathbf{M}_{J1}$ (this step can be performed in an infinite number of ways, so that the choice of $\mathbf{M}_{J1}$ and $\mathbf{M}_{J2}$ determines the pure components of the normal form of the depolarizing Mueller matrix that will be synthesized);
(3) insert, in the middle of the initial serial decomposition, a canonical depolarizing Mueller matrix $\mathbf{M}_\Delta$, expressed in terms of its own IPP (P_1, P_2, P_3) taken at their maximum values; and
(4) gradually reduce the values of the IPP, in a correlated way (Figure 6.8).

From the general serial decomposition dealt with in Section 4.5, any arbitrary pure passive Mueller matrix $\mathbf{M}_J$ can be expressed as

$$\mathbf{M}_J = d_0\,\mathbf{M}_{R2}\,\tilde{\mathbf{M}}_{DL0}(1, p_2)\,\mathbf{M}_{R1} \tag{6.88}$$

where $\tilde{\mathbf{M}}_{DL0}(1, p_2)$ is the passive form of a horizontal diattenuator

$$\tilde{\mathbf{M}}_{DL0}(1, p_2) \equiv \frac{1}{1+\cos\kappa}\begin{pmatrix} 1 & \cos\kappa & 0 & 0 \\ \cos\kappa & 1 & 0 & 0 \\ 0 & 0 & \sin\kappa & 0 \\ 0 & 0 & 0 & \sin\kappa \end{pmatrix} \quad \begin{bmatrix} p_2 \equiv \sin\kappa \\ \cos\kappa = D \end{bmatrix} \tag{6.89}$$

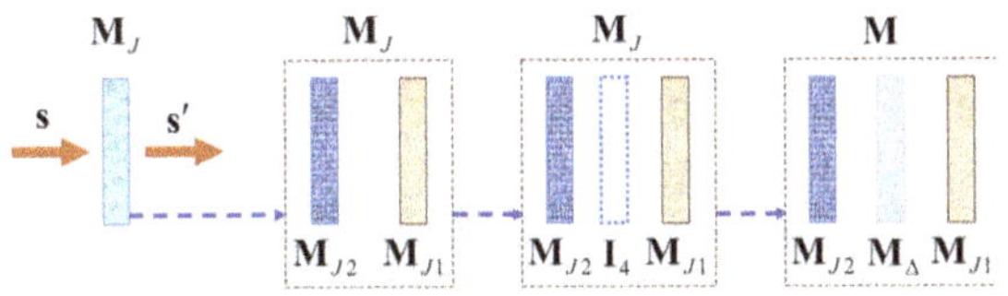

FIGURE 6.8
A reference pure Mueller matrix is decomposed into a product of two pure Mueller matrices. By reducing the values of the indices of polarimetric purity of the starting central identity matrix, it becomes a canonical depolarizer, which allows for synthesizing any depolarizing Mueller matrix.

$\mathbf{M}_{R2}$ and $\mathbf{M}_{R1}$ are pure retarders and $0 < d_0 \leq 1$. Moreover, as shown in Section 4.3.2.9, $\mathbf{M}_{DL0}(1, p_2)$ can be written as the product of two passive axis-aligned linear diattenuators

$$\tilde{\mathbf{M}}_{DL0}(1,p_2)=\tilde{\mathbf{M}}_{DL0}(1,p_2'')\,\tilde{\mathbf{M}}_{DL0}(1,p_2')\quad [p_2=p_2'p_2''\quad 0<p_2'\leq 1\quad 0<p_2''\leq 1] \tag{6.90}$$

By inserting in the middle the identity matrix in the form $\mathbf{I}_4 = \mathbf{M}_R\,\mathbf{M}_R^T$, $\mathbf{M}_R$ representing an arbitrary retarder,

$$\tilde{\mathbf{M}}_{DL0}(1,p_2)=\tilde{\mathbf{M}}_{DL0}(1,p_2'')\,\mathbf{M}_R\,\mathbf{M}_R^T\,\tilde{\mathbf{M}}_{DL0}(1,p_2') \tag{6.91}$$

and considering Eq. (6.87), we conclude that $\mathbf{M}_J$ can always be factored as a serial product of two pure passive Mueller matrices $\tilde{\mathbf{M}}_{J2}$ and $\tilde{\mathbf{M}}_{J1}$ in the following manner:

$$\mathbf{M}_J = d_0\,\mathbf{M}_{R2}\,\tilde{\mathbf{M}}_{DL0}(1,p_2'')\,\mathbf{M}_R\,\mathbf{M}_R^T\,\tilde{\mathbf{M}}_{DL0}(1,p_2')\,\mathbf{M}_{R1} = d_0\,\tilde{\mathbf{M}}_{J2}\,\tilde{\mathbf{M}}_{J1} \tag{6.92.a}$$

where

$$\tilde{\mathbf{M}}_{J2} \equiv \mathbf{M}_{R2}\,\mathbf{M}_{DL0}(1,p_2'')\,\mathbf{M}_R \qquad \tilde{\mathbf{M}}_{J1} \equiv \mathbf{M}_R^T\,\mathbf{M}_{DL0}(1,p_2')\,\mathbf{M}_{R1} \tag{6.92.b}$$

This procedure provides a general method for representing a given pure passive Mueller matrix as a product of two pure passive Mueller matrices.

In order to prepare an appropriate framework for the synthesis of type-I and type-II Mueller matrices from given reference pure Mueller matrices, let us observe that, by inserting the identity matrix $\mathbf{I}_4$ between $\tilde{\mathbf{M}}_{J2}$ and $\tilde{\mathbf{M}}_{J1}$ in Eq. (6.91), we can write $\mathbf{M}_J$ as

$$\mathbf{M}_J = \tilde{\mathbf{M}}_{J2}\,(d_0\mathbf{I}_4)\,\tilde{\mathbf{M}}_{J1} \tag{6.93}$$

6.8.1 Synthesis of a Type-I Mueller Matrix

Let us consider the normal form (or symmetric decomposition) $\mathbf{M} = \mathbf{M}_{J2}\mathbf{M}_{\Delta d}\mathbf{M}_{J1}$ of a type-I depolarizing Mueller matrix $\mathbf{M}$. The eigenvalues λ_i (with $0 \leq \lambda_3 \leq \lambda_2 \leq \lambda_1 \leq \lambda_0$) of the coherency matrix $\mathbf{C}_{\Delta d}$ associated with the Mueller matrix of the canonical type-I depolarizer $\mathbf{M}_{\Delta d} \equiv \mathrm{diag}(d_0, d_1, d_2, \varepsilon d_3)$ are precisely the diagonal elements of $\mathbf{C}_{\Delta d}$ in Eq. (5.25). By additionally applying to $\mathbf{M}_{\Delta d}$ the expressions (6.37) for the indices of polarimetric purity as functions of $\hat{\lambda}_i$, we get

$$d_1 = d_0(2P_2+P_3)/3 \qquad d_2 = d_0\left[P_1+(P_3-P_2)/3\right] \qquad \varepsilon d_3 = d_0\left[P_1-(P_3-P_2)/3\right] \tag{6.94}$$

or, conversely,

$$P_1(\mathbf{M}_{\Delta d}) = \frac{d_2-\varepsilon d_3}{2d_0} \qquad P_2(\mathbf{M}_{\Delta d}) = \frac{2d_1-d_2+\varepsilon d_3}{2d_0} \qquad P_3(\mathbf{M}_{\Delta d}) = \frac{d_1+d_2-\varepsilon d_3}{d_0} \tag{6.95}$$

Let us now replace $d_0\mathbf{I}_4$ in Eq. (6.93) by $\mathbf{M}_{\Delta d}$ expressed in terms of its IPP

$$\mathbf{M}_{\Delta d} = d_0 \operatorname{diag}\left(1, \frac{2P_2+P_3}{3}, P_1+\frac{P_3-P_2}{3}, P_1-\frac{P_3-P_2}{3}\right) \tag{6.96}$$

It should be emphasized that the previous considerations allows us to state that a 4×4 real matrix $\mathbf{M}$ is a type-I Mueller matrix if and only if it can be expressed as $\mathbf{M} = \mathbf{M}_{J2}\,\mathbf{M}_{\Delta d}\,\mathbf{M}_{J1}$ where $\mathbf{M}_{J2}$ and $\mathbf{M}_{J1}$ are pure Mueller matrices and $\mathbf{M}_{\Delta d}$ has the form (6.96) with $0 \le P_1 \le P_2 \le P_3 \le 1$ (Gil 2014c).

From the starting values $P_1 = P_2 = P_3 = 1$ corresponding to the pure reference Mueller matrix, the first level of deviation from complete purity occurs when $\operatorname{rank}\mathbf{C}_{\Delta d} = 2$ $(1 = P_3 = P_2 > P_1)$, so that

$$\mathbf{M}_{\Delta d}(P_2 = 1) = d_0 \operatorname{diag}(1,1,P_1,P_1) \tag{6.97}$$

and the symmetric composition takes the form

$$\mathbf{M} = \mathbf{M}_{J2}\, d_0 \operatorname{diag}(1,1,P_1,P_1)\, \mathbf{M}_{J1} \tag{6.98}$$

whose lowest degree of polarimetric purity $P_\Delta = 1/\sqrt{3}$ corresponds to $P_1 = 0$.

$\operatorname{rank}\mathbf{C}_{\Delta d} = 3$ is achieved by allowing $P_2 < 1$, but maintaining $P_3 = 1$, so that

$$\mathbf{M}_{\Delta d}(P_3 = 1) = d_0 \operatorname{diag}\left(1, \frac{1+2P_2}{3}, P_1+\frac{1-P_2}{3}, P_1-\frac{1-P_2}{3}\right) \tag{6.99}$$

whose lowest polarimetric purity is reached for $P_1 = P_2 = 0$ $(P_\Delta = 1/3)$.

Finally, $\operatorname{rank}\mathbf{C}_{\Delta d} = 4$, is achieved when $P_3 < 1$ $(1 > P_3 \ge P_2 \ge P_1 \ge 0)$, and $\mathbf{M}_{\Delta d}$ has the form shown in Eq. (6.96), whose lowest polarimetric purity corresponds to a perfect depolarizer $\mathbf{M}_{\Delta d} = d_0 \operatorname{diag}(1,0,0,0)$.

It is worth emphasizing that the starting central depolarizer $\mathbf{M}_{\Delta d}(P_1 = 1) = d_0\mathbf{I}_4$ is proportional to the identity matrix, and therefore the breakdown and reduction of polarimetric purity occurs smoothly and continuously.

6.8.2 Synthesis of a Type-II Mueller Matrix

Now $d_0\mathbf{I}_4$ in Eq. (6.93) should be replaced by $\mathbf{M}_{\Delta nd}$, which can be expressed as

$$\mathbf{M}_{\Delta nd} = a_0\frac{2}{3}\tilde{\mathbf{M}}_p + a_0\frac{1}{3}\tilde{\mathbf{M}}_{\Delta d2}$$
$$\tilde{\mathbf{M}}_p \equiv \frac{1}{2}\begin{pmatrix} 1 & -1 & 0 & 0 \\ 1 & -1 & 0 & 0 \\ 0 & 0 & 0 & 0 \\ 0 & 0 & 0 & 0 \end{pmatrix} \quad \tilde{\mathbf{M}}_{\Delta d2} \equiv \begin{pmatrix} 1 & 0 & 0 & 0 \\ 0 & 1 & 0 & 0 \\ 0 & 0 & P_1' & 0 \\ 0 & 0 & 0 & P_1' \end{pmatrix} \quad \left[P_1' \equiv \frac{a_2}{a_0}\right] \tag{6.100}$$

in terms of the fixed non-normal perfect polarizer $\tilde{\mathbf{M}}_p$ and the nonenpolarizing depolarizer $\tilde{\mathbf{M}}_{\Delta d2}$ (with $\operatorname{rank}\mathbf{C}(\tilde{\mathbf{M}}_{\Delta d2})=2$), which only depends on its first index of polarimetric purity P_1'. Note that $\operatorname{rank}\mathbf{C}(\tilde{\mathbf{M}}_{\Delta d2})=2 \Leftrightarrow P_1'=1$, while $\operatorname{rank}\mathbf{C}(\tilde{\mathbf{M}}_{\Delta d2})=3 \Leftrightarrow P_1'<1$. This simple procedure for generating type-II Mueller matrices does not allow for a smooth transition from a pure Mueller matrix to a type-II one (because of the abrupt insertion of the fixed component $\tilde{\mathbf{M}}_p$).

In order to achieve a smooth and continuous transition from total polarimetric purity to decreasing levels of polarimetric purity, let us observe that, as mentioned in Section 5.7.2, any type-II Mueller matrix can be expressed as

$$\mathbf{M}=\mathbf{M}_{J2}\,\mathbf{M}_{\Delta II}\,\mathbf{M}_{J1} \tag{6.101}$$

without loss of generality, where $\mathbf{M}_{J2}$ and $\mathbf{M}_{J1}$ are pure Mueller matrices and the *second canonical type-II depolarizer* $\mathbf{M}_{\Delta II}$ has the form

$$\mathbf{M}_{\Delta II}=\begin{pmatrix} a_0+a & -a & 0 & 0 \\ a & a_0-a & 0 & 0 \\ 0 & 0 & a_2 & 0 \\ 0 & 0 & 0 & a_2 \end{pmatrix} \quad \left(0\le a_2\le a_0 \quad 0<a_0 \quad 0<a\right) \tag{6.102}$$

On the other hand, the eigenvalues of the coherency matrix $\mathbf{C}_{\Delta II}$ associated with $\mathbf{M}_{\Delta II}$ are given by

$$\lambda_\alpha=\left(a_0+a_2\right)/2,\ \ \lambda_\beta=\left(a_0-a_2\right)/2,\ \ \lambda_\gamma=a,\ \ \lambda_3=0 \tag{6.103}$$

where Greek subscripts α,β,γ are used in order not to predefine the hierarchy of their values. (Obviously $\lambda_\alpha\ge\lambda_\beta$, but λ_γ can take any non-negative value.) $\mathbf{M}_{\Delta II}$ can be expressed as the convex combination

$$\mathbf{M}_{\Delta II}=\frac{1}{2}\left(a_0+a_2\right)\mathbf{I}_4+\frac{1}{2}\left(a_0-a_2\right)\mathbf{M}_{Rd1}+2a\tilde{\mathbf{M}}_p \tag{6.104}$$

where $\mathbf{I}_4$ is the 4×4 identity matrix (transparent neutral filter), $\mathbf{M}_{Rd1}=\operatorname{diag}(1,1,-1,-1)$ represents a horizontal half wave plate (or mirror, in reflection) and $\tilde{\mathbf{M}}_p$ is the Mueller matrix of the horizontal non-normal linear polarizer, considered in the beginning of the present subsection.

Therefore, in analogy to what happens in the case of type-I Mueller matrices, we deduce that a 4×4 real matrix $\mathbf{M}$ is a type-II Mueller matrix if and only if it can be expressed as $\mathbf{M}=\mathbf{M}_{J2}\,\mathbf{M}_{\Delta II}\,\mathbf{M}_{J1}$ where $\mathbf{M}_{J2}$ and $\mathbf{M}_{J1}$ are pure Mueller matrices and $\mathbf{M}_{\Delta II}$ has the form (6.104) with $(0\le a_2\le a_0,\ 0<a)$ (Gil 2014c).

Let us now recall that, while the values 1, 2, 3, 4 are achievable for $\operatorname{rank}\mathbf{C}_{\Delta d}$, only the values 2, 3 are achievable for $\operatorname{rank}\mathbf{C}_{\Delta II}$, so that the following three different cases can be considered depending on the relative values of the parameters a, a_0 and a_2: (a) $\lambda_\beta\ge\lambda_\gamma$; (b) $\lambda_\alpha\ge\lambda_\gamma\ge\lambda_\beta$; and (c) $\lambda_\gamma\ge\lambda_\alpha$. Thus, in contrast to the situation with $\mathbf{M}_{\Delta d}$, different families of type-II Mueller matrices can be generated depending on the relative value of λ_γ with respect to the other two eigenvalues of $\mathbf{C}_{\Delta II}$.

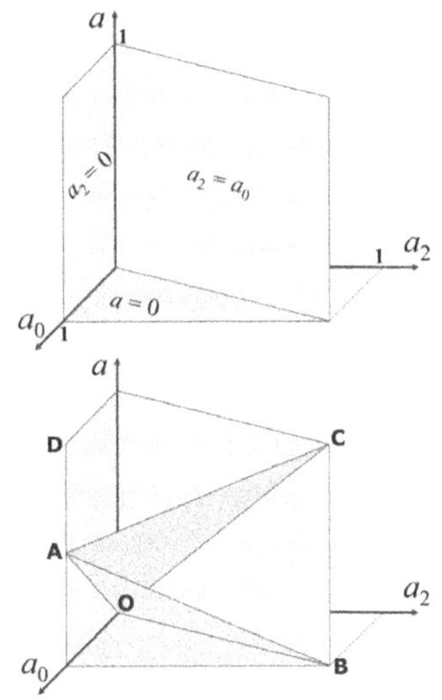

FIGURE 6.9
(a) Feasible region for the second canonical type-II depolarizers $\mathbf{M}_{\Delta II}$. Excluding points lying in the plane $a=0$ (which correspond to type-I matrices of the form $\mathrm{diag}(a_0,a_0,a_2,a_2)$) and points lying on the axis a (which correspond to the type-I pure matrix $\tilde{\mathbf{M}}_p$), the region limited by the planes $a=0$, $a_2=0$ and $a_0=a_2$ contains all possible second canonical type-II depolarizers $\mathbf{M}_{\Delta II}$. The scale indications marked with "1" have been placed along the axes in order to show the relative scales and do not designate any limit for the values of a_0, a_2 and a (Gil 2014c). (b) The three families of the second canonical type-II depolarizers $\mathbf{M}_{\Delta II}$ are represented by points lying in the respective regions (a), (b) and (c). Region (a) is limited by the planes $a=0$, $a_2=0$ and $a=(a_0-a_2)/2$ (OAB, with $a\le(a_0-a_2)/2$); region (b) is limited by the planes $a=(a_0-a_2)/2$ (OAB), $a=(a_0+a_2)/2$ (OAC), and $a_0=a_2$ (OBC). Region (c) is limited by the planes $a=(a_0+a_2)/2$ (OAC, with $a\ge(a_0+a_2)/2$), $a_2=0$ and $a_0=a_2$ (Gil 2014c).

The corresponding feasible regions for the cases (a), (b) and (c) can be identified through the representation of $\mathbf{M}_{\Delta II}$ in a three-dimensional space with the parameters (a_0,a_2,a) taken along mutually orthogonal coordinate axes. Excluding points lying in the plane $a=0$ (which would correspond to diagonal type-I matrices) and points lying on the axis a (which would correspond to the type-I pure matrix $\tilde{\mathbf{M}}_p$), the region limited by the planes $a=0$, $a_2=0$ and $a_0=a_2$ contains all the sets of second canonical type-II central depolarizers $\mathbf{M}_{\Delta II}$ (Figure 6.9a). The Mueller matrices generated in each case are represented by points in the regions limited by the following respective planes, (a) $a=0$, $a_2=0$ and $a=(a_0-a_2)/2$ (with $a\le(a_0-a_2)/2$); (b) $a=(a_0-a_2)/2$, $a=(a_0+a_2)/2$, and $a_0=a_2$; and (c) $a=(a_0+a_2)/2$, $a_2=0$ and $a_0=a_2$ (with $a\ge(a_0+a_2)/2$) (Figure 6.9b).

The decomposition (6.104) evidences the additional peculiarity that, leaving aside the limiting points lying along the edges OA ($a_2=0, a=a_0/2$) and OC ($a=a_2=a_0$), and unlike cases (a) and (b), the whole family of second canonical type-II depolarizers $\mathbf{M}_{\Delta IIc}$ corresponding to case (c) has a common reference pure Mueller matrix built from $\tilde{\mathbf{M}}_p$ and not from $\mathbf{I}_4$. Next, we analyze the indicated three regions separately.

(a) When $\lambda_\beta\ge\lambda_\gamma$, the IPP are expressed as follows in terms of a,a_0,a_2:

$$P_1=a_2/(a+a_0)\quad P_2=(a_0-2a)/(a+a_0)\quad P_3=1 \tag{6.105}$$

so that

$$\mathbf{M}_{\Delta IIa} = \frac{1}{6}(a_0 + a)\left[(2+P_2+3P_1)\mathbf{I}_4 + (2+P_2-3P_1)\mathbf{M}_{Rd1} + 2(1-P_2)\tilde{\mathbf{M}}_p\right] \tag{6.106}$$

where it is possible to increase the parameter a continuously from its smaller value $a = 0$ to its maximum $a = (a_0 - a_2)/2$. The combined values $a_2 = a_0$ and $a = 0$ (segment OB) correspond to the identity matrix. Regardless of the value of a_2, $a = 0$ corresponds to type-I diagonal Mueller matrices of the form $\operatorname{diag}(a_0, a_0, a_2, a_2)$, while type-II Mueller matrices are generated as long as a differs from zero. Thus, in this present case (a), the condition $a > 0$ entails $P_2 < 1$, so that type-II Mueller matrices with $\operatorname{rank}\mathbf{C}_{\Delta IIa} = 2$ are not achievable and the following family of type-II central depolarizers with $\operatorname{rank}\mathbf{C}_{\Delta IIa} = 3$ is generated:

$$\mathbf{M}_{\Delta IIa} = (a_0 + a)\begin{pmatrix} 1 & -(1-P_2)/3 & 0 & 0 \\ (1-P_2)/3 & (1+2P_2)/3 & 0 & 0 \\ 0 & 0 & P_1 & 0 \\ 0 & 0 & 0 & P_1 \end{pmatrix} \tag{6.107}$$

When, in particular, $P_1 = 0$, and hence $P_\Delta = \sqrt{2P_2^2 + 1}/3$, $\mathbf{M}_{\Delta IIa}$ takes the form

$$\mathbf{M}_{\Delta IIa}(P_1 = 0) = (a_0 + a)\begin{pmatrix} 1 & -(1-P_2)/3 & 0 & 0 \\ (1-P_2)/3 & (1+2P_2)/3 & 0 & 0 \\ 0 & 0 & 0 & 0 \\ 0 & 0 & 0 & 0 \end{pmatrix} \tag{6.108}$$

and the lowest polarimetric purity ($P_1 = P_2 = 0$, segment OA, $P_\Delta = 1/3$) is achieved for

$$\mathbf{M}_{\Delta IIa}(P_2 = 0) = (a_0 + a)\begin{pmatrix} 1 & -1/3 & 0 & 0 \\ 1/3 & 1/3 & 0 & 0 \\ 0 & 0 & 0 & 0 \\ 0 & 0 & 0 & 0 \end{pmatrix} \tag{6.109}$$

(b) When $\lambda_\alpha \geq \lambda_\gamma \geq \lambda_\beta$, the IPP are given by

$$P_1 = \frac{(a_0 + a_2)/2 - a}{a + a_0} \qquad P_2 = \frac{a - (a_0 - 3a_2)/2}{a + a_0} \qquad P_3 = 1 \tag{6.110}$$

and the expression of $\mathbf{M}_{\Delta II}$ in terms of the IPP is

$$\mathbf{M}_{\Delta IIb} = (a_0 + a)\begin{pmatrix} 1 & -(2+P_2-3P_1)/6 & 0 & 0 \\ (2+P_2-3P_1)/6 & P_1 + (1-P_2)/3 & 0 & 0 \\ 0 & 0 & (P_1+P_2)/2 & 0 \\ 0 & 0 & 0 & (P_1+P_2)/2 \end{pmatrix} \tag{6.111}$$

Unlike case (a), now the value $P_2 = 1$ (corresponding to $a_2 = a_0$) is physically achievable and leads to the following family of type-II Mueller matrices with rank $\mathbf{C}_{\Delta IIb} = 2$:

$$\mathbf{M}_{\Delta IIb}(P_2 = 1) = (a_0 + a)\begin{pmatrix} 1 & -(1-P_1)/2 & 0 & 0 \\ (1-P_1)/2 & P_1 & 0 & 0 \\ 0 & 0 & (1+P_1)/2 & 0 \\ 0 & 0 & 0 & (1+P_1)/2 \end{pmatrix} \tag{6.112}$$

with $P_1 < 1$. We observe that, as occurs in case (a), $P_1 = 1$ corresponds to the identity matrix (edge OB). The lowest polarimetric purity ($P_1 = 0$) within the family $P_2 = 1$ is achieved for points in the edge OC representing matrices

$$\mathbf{M}_{\Delta IIb}(P_2 = 1, P_1 = 0) = (a_0 + a)\begin{pmatrix} 1 & -1/2 & 0 & 0 \\ 1/2 & 0 & 0 & 0 \\ 0 & 0 & 1/2 & 0 \\ 0 & 0 & 0 & 1/2 \end{pmatrix} \tag{6.113}$$

Values $P_2 < 1$ correspond to type-II Mueller matrices with rank $\mathbf{C}_{\Delta IIb} = 3$. The lowest polarimetric purity, $P_2 = P_1 = 0$ ($P_\Delta = 1/3$), is reached for matrices $\mathbf{M}_{\Delta IIb}(P_2 = 0) = \mathbf{M}_{\Delta IIa}(P_2 = 0)$ (Eq. (6.109)) represented along the edge OA where the region (b) touches the region (a). Note that regions (a) and (b) share also the limiting edge OB ($P_1 = 1$) which corresponds to the identity matrix, i.e., $\mathbf{I}_4 = \mathbf{M}_{\Delta IIb}(P_1 = 1) = \mathbf{M}_{\Delta IIa}(P_1 = 1)$.

We have seen that the general expression of $\mathbf{M}_{\Delta IIb}$ in Eq. (6.111) allows us to generate the entire family of type-II matrices corresponding to case (b) starting from the identity matrix $\mathbf{I}_4 = \mathbf{M}_{\Delta IIb}(P_1 = 1)$ without the necessity of using $\tilde{\mathbf{M}}_p$ as a starting pure Mueller matrix. Consequently, type-II Mueller matrices like

$$\begin{pmatrix} a_0 + a & -a & 0 & 0 \\ a & a_0 - a & 0 & 0 \\ 0 & 0 & -|a_2| & 0 \\ 0 & 0 & 0 & -|a_2| \end{pmatrix} \tag{6.114}$$

can be generated from the symmetric composition $\tilde{\mathbf{M}}_p\,\mathbf{M}_{\Delta IIb}\,\mathbf{I}_4$.

(c) Let us finally consider the case $\lambda_\gamma \geq \lambda_\alpha$, where the IPP are given by

$$P_1 = \frac{a - (a_0 + a_2)/2}{a + a_0} \qquad P_2 = \frac{a - (a_0 - 3a_2)/2}{a + a_0} \qquad P_3 = 1 \tag{6.115}$$

and therefore the weight of the component $\tilde{\mathbf{M}}_p$ in the convex sum (6.104) is greater than or equal to that of $\mathbf{I}_4$. The corresponding expression for $\mathbf{M}_{\Delta II}$ in terms of the IPP is given by

$$\mathbf{M}_{\Delta IIc} = (a_0 + a)\begin{pmatrix} 1 & -(2+P_2+3P_1)/6 & 0 & 0 \\ (2+P_2+3P_1)/6 & -P_1 + (1-P_2)/3 & 0 & 0 \\ 0 & 0 & (P_2 - P_1)/2 & 0 \\ 0 & 0 & 0 & (P_2 - P_1)/2 \end{pmatrix} \tag{6.116}$$

The value $P_2 = 1$ is achievable for points above the edge OC ($P_1 = 0$) in the plane $a_2 = a_0$ and leads to the following family of type-II Mueller matrices with rank $\mathbf{C}_{\Delta IIc} = 2$:

$$\mathbf{M}_{\Delta IIc}(P_2 = 1) = (a_0 + a)\begin{pmatrix} 1 & -(1+P_1)/2 & 0 & 0 \\ (1+P_1)/2 & -P_1 & 0 & 0 \\ 0 & 0 & (1-P_1)/2 & 0 \\ 0 & 0 & 0 & (1-P_1)/2 \end{pmatrix} \quad (6.117)$$

with $P_1 < 1$. Unlike the previous cases (a) and (b), the limiting value $P_1 = 1$ corresponds to the perfect polarizer represented by $\tilde{\mathbf{M}}_p$ and not to the identity matrix. This result implies that the reference pure Mueller matrix corresponding to the entire type-II family of matrices generated in case (c) can be written in the form $\mathbf{M}_{Jc} = \mathbf{M}_{J2}\,\tilde{\mathbf{M}}_p\,\mathbf{M}_{J1}$. Since $\tilde{\mathbf{M}}_p$ is a perfect polarizer then $D(\mathbf{M}_{Jc}) = P(\mathbf{M}_{Jc}) = 1$ and therefore, the polar decomposition of $\mathbf{M}_{Jc}$ always contains a perfect polarizer (in general, elliptical).

The lowest polarimetric purity ($P_1 = 0$) within the family $P_2 = 1$ corresponds to the edge OC, associated with the Mueller matrix given by Eq. (6.113). Moreover, values $P_2 < 1$ correspond to type-II Mueller matrices with rank $\mathbf{C}_{\Delta IIc} = 3$. The lowest polarimetric purity $P_2 = 0$ ($P_\Delta = 1/3$) is reached for matrices of the form given by Eq. (6.109), which are represented by points lying in the edge OA which, in turn, is shared by the three families (a), (b) and (c).

Once the different families of type-II central depolarizers $\mathbf{M}_{\Delta II}$ have been analyzed and geometrically represented, let us observe that the type-II canonical Mueller matrix $\mathbf{M}_{\Delta nd}$ (Eq. (5.23)), which corresponds to the particular form adopted by $\mathbf{M}_{\Delta II}$ when $a = a_0$, is represented by points in the plane OBC contained inside the feasible region relative to case (c). Hence, it is generated exclusively from singular reference Mueller matrices of the form $\mathbf{M}_{Jc}$ given above.

6.8.3 On the Reference Pure Mueller Matrix

As shown above, any Mueller matrix $\mathbf{M}$ is can be synthesized from a respective reference pure Mueller matrix $\mathbf{M}_J(\mathbf{M})$ by its factorization into two pure components, $\mathbf{M}_J(\mathbf{M}) = \mathbf{M}_{J2}\mathbf{M}_{J1}$ and by reducing, in a smooth and continuous way, the nested values of the starting unit-valued indices of polarimetric purity of the central matrix $\mathbf{M}_\Delta$.

The case of type-II Mueller matrices requires the particular analysis performed in Section 6.8.2 because of its peculiar nature (Ossikovski 2010a; Ossikovski et al. 2013; Gil 2014c) resulting in its incompatibility with complete purity. A smooth and continuous procedure for generating any kind of type-II Mueller matrix has been developed by using the generic form $\mathbf{M}_{\Delta II}$ of the second canonical type-II depolarizer instead of the form $\mathbf{M}_{\Delta nd}$. The generation of the entire achievable set of type-II matrices has been performed by splitting $\mathbf{M}_{\Delta II}$ into three complementary families and by identifying the respective IPP (Gil 2014c).

Moreover, given a depolarizing Mueller matrix $\mathbf{M}$, its *reference pure Mueller matrix* $\mathbf{M}_J(\mathbf{M})$ is calculated as the product $\mathbf{M}_{J2}\mathbf{M}_{J1} = \mathbf{M}_J(\mathbf{M})$ of the left and right pure components of the normal form $\mathbf{M} = \mathbf{M}_{J2}\mathbf{M}_\Delta\mathbf{M}_{J1}$. A notable exception to this procedure is the case (c) where the parameters (a_0, a_2, a) of the second canonical type-II depolarizer satisfy $a \geq (a_0 + a_2)/2$ in such a manner that the reference Mueller matrix $\mathbf{M}_J(\mathbf{M})$ can be written as $\mathbf{M}_J(\mathbf{M}) = \mathbf{M}_{J2}\tilde{\mathbf{M}}_p\mathbf{M}_{J1}$, $\tilde{\mathbf{M}}_p$ being the non-normal perfect polarizer in Eq. (6.100).

It is remarkable that, from its very definition, the above mentioned reference pure Mueller matrix $\mathbf{M}_J(\mathbf{M})$ coincides with the pure representative $\mathbf{M}_{J0}$ defined from the characteristic decomposition of **M** in Eq. (6.43), and also equals the pure spectral component of **M** associated with the largest eigenvalue of **H**(**M**).

6.8.4 Depolarization Synthesis

The synthesis procedure described above constitutes a way for generating any physical Mueller matrix and provides a constructive view of the concept of depolarization, which arises from the existence of two or more pure parallel components that are combined incoherently and that can be parameterized through the set of three indices of polarimetric purity.

Observe that depolarization is synthesized by tuning the IPP of $\mathbf{M}_\Delta$ instead of those of **M**. The only possible effect of $\mathbf{M}_{J1}$ and $\mathbf{M}_{J2}$ on the depolarization, beyond that due to $\mathbf{M}_\Delta$, is the increase of the polarimetric purity produced by the enpolarizing properties of $\mathbf{M}_{J1}$ and $\mathbf{M}_{J2}$. Therefore, when the degree of polarizance of **M** is nonzero, the overall depolarizing properties, which are quantitatively characterized through the IPP of **M**, arise as a sort of balance among polarizance, diattenuation and the depolarization introduced by $\mathbf{M}_\Delta$, so that $P_\Delta(\mathbf{M}_\Delta) \le P_\Delta(\mathbf{M})$ (Ossikovski 2010a; Gil 2011). However, since $\text{rank}[\mathbf{C}(\mathbf{M}_\Delta)] = \text{rank}[\mathbf{C}(\mathbf{M})]$ (Gil et al. 2013), $\mathbf{M}_\Delta$ and **M** have the same number μ of unit-valued IPP and thus, there exists a parallel between the structure of polarimetric purity of $\mathbf{M}_\Delta$ and **M**.

It is important to recall at this point that even pure media produce depolarization for some incident *partially* polarized states (Simon 1990). This is a direct consequence of the fact that polarizance, diattenuation and depolarization are essentially entangled (Gil 2011). Moreover, when $D(\mathbf{M}) = P(\mathbf{M}) = 0$, **M** is type-I and thus, while $\mathbf{M}_{\Delta II}$ always exhibits a residual polarizance-diattenuation, the type-I canonical Mueller matrix $\mathbf{M}_{\Delta d}$ lacks both diattenuation and polarizance. Consequently, unlike type-II Mueller matrices, in the case of type-I Mueller matrices the effect of nonzero values of D and P on the polarimetric purity of **M** is fully described by the set of parameters $P_1(\mathbf{M}), P_2(\mathbf{M}), P_3(\mathbf{M})$. However, the quantities D and P are not derivable from the sole IPP of **M** and hence, D and P add independent and complementary information that allows for analyzing and interpreting the nature and origin of the depolarization. When $D = P = 0$ the IPP of $\mathbf{M}_{\Delta d}$ are equal to those of **M**; when the medium exhibits nonzero values of either D or P, then $P_\Delta(\mathbf{M}_{\Delta d}) < P_\Delta(\mathbf{M})$, so that the depolarizing properties of the canonical component $\mathbf{M}_\Delta$ are masked to some extent by the diattenuation-polarizance of $\mathbf{M}_{J1}$ and $\mathbf{M}_{J2}$. The case of a serial composition of a perfect depolarizer and a perfect polarizer, provides a clear example of this phenomenon.

Therefore, a complete description of the enpolarizing-depolarizing properties of the medium represented by **M**, requires the consideration of the five quantities D, P, $P_1(\mathbf{M})$, $P_2(\mathbf{M})$ and $P_3(\mathbf{M})$. In the particular case of pure media, all the IPP reach their maximal values $1 = P_\Delta(\mathbf{M}) = P_1(\mathbf{M}) = P_2(\mathbf{M}) = P_3(\mathbf{M})$, and necessarily $D = P$. Furthermore, as the medium becomes depolarizing, then $P_1(\mathbf{M}) < 1$ (and hence, $P_\Delta(\mathbf{M}) < 1$), while D and P are not necessarily equal.

7

Parallel Decompositions of Mueller Matrices

7.1 Introduction

Even though a measured Mueller matrix **M** contains a wealth of information about the polarimetric properties and structure of the sample, this information is enclosed in an implicit way. This fact stresses the importance of developing adequate procedures for the decomposition of Mueller matrices, allowing for the detailed physical interpretation of the polarimetric behavior of the corresponding samples. Therefore, from both the theoretical and experimental points of view, it is of great interest to analyze the various polarimetric properties that can be identified through appropriate approaches for the decomposition of the Mueller matrix describing the sample under measurement. Among the wide scope of applications we can mention fields like radar polarimetry (Cloude 2009), polarimetric diagnosis of biological tissues (Gosh and Vitkin 2011; Sun et al. 2014; Van Eeckhout et al. 2021b), and polarimetric characterization of optical grating devices (Foldyna et al. 2009).

In general, two kinds of decompositions of a Mueller matrix can be performed, namely *serial decompositions* (through product of Mueller matrices) and *parallel decompositions* (through weighted sums of Mueller matrices). Furthermore, both decompositions can be combined leading to *serial-parallel decompositions.*

Parallel decompositions consist of representing a Mueller matrix as a convex sum of Mueller matrices. The physical meaning of parallel decompositions is that the incoming electromagnetic wave splits into a set of pencils that interact, without overlapping, with a number of components that are spatially distributed in the illuminated area, and the emerging pencils are incoherently recombined into the emerging beam (Figure 7.1). The so-called *spectral, arbitrary* and *characteristic* decompositions of a Mueller matrix have been briefly introduced in Chapter 5 in order to provide the appropriate support to the descriptors of polarimetric purity, as well as to the synthesis of depolarizing Mueller matrices with given basic properties.

The present chapter is devoted to the analysis of all possible parallel decompositions of Mueller matrices. One of the procedures, called the *arbitrary decomposition*, consists of finding a set of pure Mueller matrices whose convex sum (i.e., a linear combination of matrix terms with positive coefficients that sum to one) is equal to the original Mueller matrix. A particular realization of the arbitrary decomposition is the spectral one, while the characteristic decomposition is constituted by the convex sum of the elements $\mathbf{M}_{J0}$, $\mathbf{M}_1$, $\mathbf{M}_2$ and $\mathbf{M}_{\Delta 0}$ whose respective degrees of polarimetric purity have the scaled values: $P_\Delta(\mathbf{M}_{J0}) = 1$ (nondepolarizing component), $P_\Delta(\mathbf{M}_1) = 1/\sqrt{3}$ (2D depolarizer), $P_\Delta(\mathbf{M}_2) = 1/3$ (3D depolarizer), and $P_\Delta(\mathbf{M}_{\Delta 0}) = 0$ (perfect depolarizer).

DOI: 10.1201/9780367815578-7

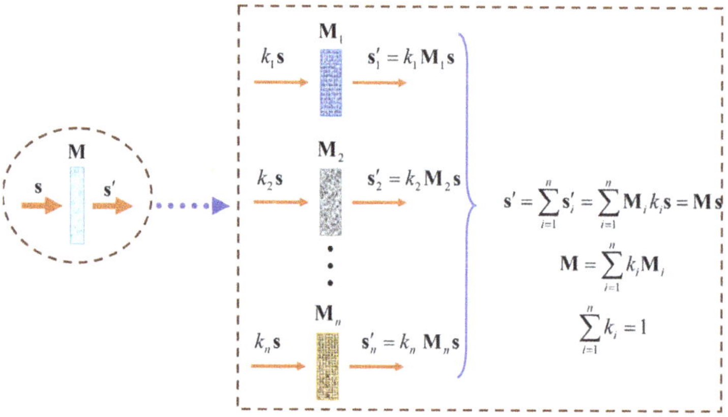

FIGURE 7.1
Parallel decomposition of a Mueller matrix. The Mueller matrix **M** of the system is expressed as a convex linear combination of the Mueller matrices of the parallel components.

The theory of parallel decompositions of Mueller matrices can be developed by means of the covariance (or the coherency) matrix representation. It also allows for finding the solution to the inverse problem by subtracting (from the measured depolarizing Mueller matrix **M**) the Mueller matrix component of a given constituent whose polarimetric effect is desired to be decoupled from the overall polarimetric response. As a first step for the analysis of the parallel decompositions, it is necessary to consider the formulation and the physical interpretation of the additive composition of Mueller matrices.

7.2 Additive Composition of Mueller Matrices

As dealt with in Section 5.2, the notion of additive (or parallel) composition of Mueller matrices underlies the very concept of Mueller matrix. Therefore, all considerations and developments of the subsequent sections are based on this notion, which in turn relies on the concept of additive composition of Stokes vectors (considered in Section 1.12). Nevertheless, we should recall that there exists an important difference: while any sum of Stokes vectors, i.e., any incoherent combination of polarization states, is physically admissible regardless of the values of their respective intensities, the additive composition of Mueller matrices obeys certain rules originating from the fact that a Mueller matrix represents a linear transformation of the input Stokes vectors into the output ones. In fact, in accordance with the arguments set forth in Section 3.3.5.5, the mere sum operation of Mueller matrices is not consistent from a physical point of view, in general. To prove this, it is enough to recall that, given a Mueller matrix **M** with mean intensity coefficient (MIC) m_{00}, the MIC of the sum $\mathbf{M}+\mathbf{M}$ equals $2m_{00}$ and therefore, through a sufficient number of added sums, the resulting MIC can reach a value higher than any predetermined positive number, thus violating the passivity condition.

Considering the Mueller matrix component $\mathbf{M}_i$ from Figure 7.1, it is understood that $\mathbf{M}_i$ represents the interaction of a given medium with a certain cross section of the input electromagnetic beam. Thus, the physical meaning of the additive composition of a given number of Mueller matrix components $\mathbf{M}_i$ is that the input electromagnetic beam, represented by a

Stokes vector **s**, is spatially shared among the respective media represented by $\mathbf{M}_i$, so that **s** is transformed as (Figure 7.1)

$$\mathbf{s}' = \sum_{i=1}^{n}\left[(k_i\mathbf{M}_i)\mathbf{s}\right] = \left(\sum_{i=1}^{n} k_i\mathbf{M}_i\right)\mathbf{s} \qquad \sum_{i=1}^{n} k_i = 1 \tag{7.1}$$

where k_i are the relative cross sections, or ratios $k_i = I_i/I$ between the intensity falling upon respective elements represented by $\mathbf{M}_i$ and $I = \sum I_i$. That is, the cross section of a beam incident on the combined medium is equal to the sum of the cross sections of the beams incident on the spatially distributed constituent media. To complete the interpretation of the additive composition (7.1) it is necessary to identify the criterion for an appropriate normalization of the addend Mueller matrices.

To identify a proper normalization criterion for the Mueller matrices entering the additive composition, consider polarization states with intensity equal to the unit, covering the entire Poincaré sphere and recall that the maximum output intensity, for a given Mueller matrix **M**, is $m_{00}(1+D)$ for forward interaction, and $m_{00}(1+P)$ for reverse interaction (see Section 6.2). Furthermore, as dealt with in Section 5.9, the passivity condition is given by $m_{00}(1+Q) \leq 1$, with $Q \equiv \max(D,P)$, that is, the maximum value for the MIC (mean intensity coefficient, m_{00}) of **M** compatible with passivity is $\tilde{m}_{00} = 1/(1+Q)$. Therefore, an adequate procedure for the normalization of the components $\mathbf{M}_i$ consists in taking their respective *passive forms* $\tilde{\mathbf{M}}_i \equiv \left[1/(1+Q_i)\right]\hat{\mathbf{M}}_i$, with $Q_i \equiv \max(D_i, P_i)$. It should be stressed that arbitrary MICs $m_{00i} \leq 1/(1+Q_i)$ are physically realizable and admissible in parallel compositions or decompositions (San José and Gil 2020b).

Since for any pure Mueller matrix $\mathbf{M}_{Ji}$ polarizance and diattenuation are equal ($P_i = D_i$) the additive composition of a number of pure Mueller matrices $\mathbf{M}_{Ji}$ can be formulated as follows in terms of their passive forms (San José and Gil 2020b):

$$\mathbf{M} = \sum_{i=1}^{n} k_i\tilde{\mathbf{M}}_i \qquad \left[\mathbf{M} = m_{00}\hat{\mathbf{M}} \quad \tilde{\mathbf{M}}_i = \tilde{m}_{00i}\hat{\mathbf{M}}_i \quad \sum_{i=1}^{n} k_i \leq 1\right] \tag{7.2}$$

Note that, as seen in Section 5.9, the passivity of the parallel components $\mathbf{M}_{Ji}$ ensures the passivity of the composed Mueller matrix **M**. Furthermore, it has been proven that when either the diattenuation vectors $\mathbf{D}_i$ or the polarizance vectors $\mathbf{P}_i$ are mutually parallel, then **M** takes its passive form $\tilde{\mathbf{M}}$ (San José and Gil 2020b).

7.3 Arbitrary Decomposition of a Mueller Matrix

In accordance with the definition of the covariance, **H**, and coherency, **C**, matrices associated with **M**, the spectrum of eigenvalues of **H** and **C** coincides. Therefore, all developments performed in this and other chapters in terms of **H** can entirely be formulated in terms of **C**, provided covariance vectors and matrices are reinterpreted as respective coherency vectors and matrices.

As indicated in Section 5.14.1, a straightforward way to achieve a parallel decomposition of a Mueller matrix **M** in terms of pure Mueller matrices is given by the *spectral decomposition* (or *Cloude's decomposition*) of the covariance matrix **H** (or the coherency matrix **C**) associated with **M**, determined by the diagonalization of **H** (or **C**)

$$\left[\begin{array}{c}\mathbf{H}=\mathbf{U}\,\boldsymbol{\Lambda}\,\mathbf{U}^{\dagger}=\sum_{i=0}^{r-1}p_i\left(m_{00}\,\hat{\mathbf{H}}_{Ji}\right)\\ \boldsymbol{\Lambda}=\operatorname{diag}\left(\lambda_0,\lambda_1,\lambda_2,\lambda_3\right)\\ r=\operatorname{rank}\mathbf{H}\quad m_{00}=\operatorname{tr}\mathbf{H}\quad p_i=\lambda_i/m_{00}\quad \sum_{i=0}^{r-1}p_i=1\quad \hat{\mathbf{H}}_{Ji}\equiv\mathbf{U}\mathbf{D}_i\mathbf{U}^{\dagger}\\ \mathbf{D}_0\equiv\operatorname{diag}(1,0,0,0)\quad \mathbf{D}_1\equiv\operatorname{diag}(0,1,0,0)\quad \mathbf{D}_2\equiv\operatorname{diag}(0,0,1,0)\quad \mathbf{D}_3\equiv\operatorname{diag}(0,0,0,1)\end{array}\right] \tag{7.3}$$

where the positive coefficients p_i of the convex sum are the normalized nonzero eigenvalues $p_i=\hat{\lambda}_i\equiv\lambda_i/m_{00}$ of **H**.

The above formulation, where the mean intensity coefficients of all components are equal to that of **M** (denoted by m_{00}), is called the *homogeneous spectral decomposition*, and can be generalized as follows in order that the components have different mean intensity coefficients m_{00i}:

$$\mathbf{H}=\sum_{i=0}^{r-1}k_i\left(m_{00i}\,\hat{\mathbf{H}}_{Ji}\right)\quad\left[k_i=m_{00}\,p_i/m_{00i}\quad \sum_{i=0}^{r-1}k_i=\sum_{i=0}^{r-1}p_i=1\right] \tag{7.4}$$

Since the additive properties are preserved by the biunivocal relation between **M** and its associated **H** (or **C**, see Section 5.3), the spectral decomposition can be expressed as follows (see Eq. (5.126)) in terms of Mueller matrices:

$$\begin{gathered}\mathbf{M}=\sum_{i=0}^{r-1}p_i\left(m_{00}\,\hat{\mathbf{M}}_{Ji}\right)=\sum_{i=0}^{r-1}k_i\left(m_{00i}\,\hat{\mathbf{M}}_{Ji}\right)\\ \left[\hat{\mathbf{M}}_{Ji}=\mathbf{M}\left(\hat{\mathbf{H}}_{Ji}\right)\quad k_i=m_{00}p_i/m_{00i}\quad \sum_{i=0}^{r-1}k_i=\sum_{i=0}^{r-1}p_i=1\right]\end{gathered} \tag{7.5}$$

Nevertheless, as shown in previous works (Gil 2007; Gil and San José 2013) the spectral decomposition is not the only way for performing a parallel decomposition of **M**, but is a particular case of the *arbitrary decomposition*, which can be applied to $n\times n$ covariance matrices and, in particular, to **H**. Moreover, the existence of decompositions other than the spectral one is evident from the physical possibility of experimentally synthesizing nonpure systems by means of arbitrary parallel combinations of pure systems which do not necessarily represent spectral components.

Let us consider an arbitrary set of r four-component orthonormal complex vectors $\hat{\mathbf{v}}_0,\hat{\mathbf{v}}_1\ldots,\hat{\mathbf{v}}_{r-1}$ (with $r=\operatorname{rank}\mathbf{H}$) whose last $4-r$ components are zero. Then the 4×4 diagonal matrix $\mathbf{I}_r$ whose first r diagonal elements are equal to one and whose last $4-r$ diagonal elements are zero can be expressed as

$$\mathbf{I}_r=\sum_{i=0}^{r-1}\hat{\mathbf{v}}_i\otimes\hat{\mathbf{v}}_i^{\dagger} \tag{7.6}$$

The above expression of $\mathbf{I}_r$ in terms of an arbitrary set of r orthonormal vectors $\hat{\mathbf{v}}_i$ allows for transforming the spectral decomposition of **H** as follows (Gil 2007; Gil and San José 2013; Gil and San José 2019):

$$\mathbf{H}=\mathbf{U}\boldsymbol{\Lambda}\mathbf{U}^{\dagger}=\mathbf{U}\sqrt{\boldsymbol{\Lambda}}\,\mathbf{I}_r\sqrt{\boldsymbol{\Lambda}}\,\mathbf{U}^{\dagger}=\mathbf{U}\sqrt{\boldsymbol{\Lambda}}\left(\sum_{i=0}^{r-1}\hat{\mathbf{v}}_i\otimes\hat{\mathbf{v}}_i^{\dagger}\right)\sqrt{\boldsymbol{\Lambda}}\mathbf{U}^{\dagger}=\sum_{i=0}^{r-1}\left(\mathbf{U}\sqrt{\boldsymbol{\Lambda}}\hat{\mathbf{v}}_i\right)\otimes\left(\mathbf{U}\sqrt{\boldsymbol{\Lambda}}\hat{\mathbf{v}}_i\right)^{\dagger} \tag{7.7}$$

where the r covariance vectors $\mathbf{w}_i = \mathbf{U}\sqrt{\boldsymbol{\Lambda}}\hat{\mathbf{v}}_i$ $(i = 0,\dots,r-1)$ are linearly independent vectors belonging to the image subspace of $\mathbf{H}$ (i.e., the subspace generated by all possible linear combinations of the r eigenvectors associated with nonzero eigenvalues of $\mathbf{H}$; such an *image subspace* is also called the *range* of $\mathbf{H}$ and will be denoted as $\operatorname{im}\mathbf{H}$). The covariance vectors generate their respective pure covariance matrices $\mathbf{H}_{Ji} = \mathbf{w}_i \otimes \mathbf{w}_i^{\dagger} = k_i(\hat{\mathbf{w}}_i \otimes \hat{\mathbf{w}}_i^{\dagger})$, with $k_i = |\mathbf{w}_i|^2$ and $\hat{\mathbf{w}}_i = \mathbf{w}_i/|\mathbf{w}_i|$, so that $\mathbf{H}$ can be expressed as a convex linear combination of pure covariance matrices $m_{00i}\hat{\mathbf{H}}_{Ji}$ with respective mean intensity coefficients m_{00i} (Gil and San José 2019)

$$\mathbf{H} = \sum_{i=0}^{r-1} k_i\left(m_{00i}\hat{\mathbf{H}}_{Ji}\right)$$
$$\left[r = \operatorname{rank}\mathbf{H} \quad \sum_{i=0}^{r-1} k_i = 1 \quad \sum_{i=0}^{r-1} m_{00i}k_i = m_{00} = \operatorname{tr}\mathbf{H}\right] \tag{7.8a}$$

The positive coefficients k_i (relative cross sections) are given by (Gil 2007; Gil and San José 2013; Gil and San José 2019)

$$k_i = \frac{1}{m_{00i}\sum_{j=0}^{r-1}\frac{1}{\lambda_j}\left|\left(\mathbf{U}^{\dagger}\hat{\mathbf{w}}_i\right)_j\right|^2} = \frac{1}{m_{00i}\left(\hat{\mathbf{w}}_i^{\dagger}\mathbf{H}^{-}\hat{\mathbf{w}}_i\right)} \qquad \left[\sum_{i=0}^{r-1} m_{00i}k_i = m_{00} \quad \sum_{i=0}^{r-1} k_i = 1\right] \tag{7.8b}$$

where λ_j are the r nonzero eigenvalues of $\mathbf{H}$, $\mathbf{H}^{-}$ is the pseudoinverse of $\mathbf{H}$ defined as $\mathbf{H}^{-} = \mathbf{U}\mathbf{D}^{-}\mathbf{U}^{\dagger}$, $\mathbf{D}^{-}$ being the diagonal matrix whose r first diagonal elements are $1/\lambda_0, 1/\lambda_1, \dots, 1/\lambda_{r-1}$ and the last $4-r$ elements are zero (Gil and San José 2019). Each pure component $\mathbf{H}_{Ji} = m_{00i}\hat{\mathbf{H}}_{Ji}$ (with $\operatorname{rank}\mathbf{H}_{Ji} = 1$) is determined by its corresponding covariance vector $\sqrt{m_{00i}}\,\hat{\mathbf{w}}_i$. In the particular case where $m_{00i} = m_{00}$, the corresponding form of the arbitrary decomposition is called *homogeneous*.

The minimum number r of pure parallel components of the equivalent system is given by $r = \operatorname{rank}\mathbf{H}$. In accordance with the iterative procedure described below to perform the arbitrary decomposition, when $1 < \operatorname{rank}\mathbf{H} \le 4$ different sets of pure components are physically realizable (Gil and San José 2013; 2019). When $\operatorname{rank}\mathbf{H} = 1$, the system is pure and therefore it cannot be expressed as a linear combination of different pure components. Obviously, due to the biunivocal relation between $\mathbf{H}$ and its associated coherency matrix $\mathbf{C}$, all of the above considerations (as well as those below) are also valid if one replaces $\mathbf{H}$ by $\mathbf{C}$.

The arbitrary decomposition (7.8) can be expressed as follows in terms of its corresponding Mueller matrix components:

$$\mathbf{M} = \sum_{i=0}^{r-1} k_i\mathbf{M}_{Ji} = \sum_{i=0}^{r-1} k_i m_{00i}\hat{\mathbf{M}}_{Ji} \qquad k_i = \frac{1}{m_{00i}\left(\hat{\mathbf{w}}_i^{\dagger}\mathbf{H}^{-}\hat{\mathbf{w}}_i\right)}$$
$$\left[\mathbf{M} \equiv \mathbf{M}(\mathbf{H}) = m_{00}\hat{\mathbf{M}} \quad \mathbf{M}_{Ji} \equiv \mathbf{M}\left(\mathbf{H}_{Ji}\right) = m_{00i}\hat{\mathbf{M}}_{Ji} \quad \sum_{i=0}^{r-1} k_i = 1\right] \tag{7.9}$$

where the Mueller matrix $\mathbf{M}$ of the system is obtained as a convex linear combination of a number of r pure matrices $\mathbf{M}_{Ji}$ with respective relative cross sections given by k_i.

It is remarkable that, by assigning appropriate MICs m_{00i} to a number of $r-1$ components (denoted $m_{00,0}, \dots, m_{00,r-2}$), any arbitrary decomposition can be built through a simple transformation of its homogeneous form ($m_{00i} = m_{00}$, $i = 0,\dots,r-1$):

$$\mathbf{M} = m_{00}\hat{\mathbf{M}} = \sum_{i=0}^{r-1} p_i \left(m_{00}\hat{\mathbf{M}}_{Ji}\right) \quad p_i = \frac{1}{m_{00}\left(\hat{\mathbf{w}}_i^{\dagger}\mathbf{H}^{-}\hat{\mathbf{w}}_i\right)} \quad \left[\sum_{i=0}^{r-1} p_i = 1\right] \tag{7.10}$$

The new coefficients k_i and the remaining MIC $m_{00,r-1}$ of $\mathbf{M}_{J,r-1}$ are obtained as

$$k_i = \frac{p_i m_{00}}{m_{00i}} \quad (i = 0,\ldots,r-2) \qquad k_{r-1} = 1 - \sum_{i=0}^{r-2} k_i \qquad m_{00,r-1} = \frac{m_{00}\, p_{r-1}}{1 - m_{00}\sum_{i=0}^{r-2} p_i / m_{00i}} \tag{7.11}$$

which leads to the general form in Eq. (7.9)

It is important to note that the fulfilment by $\mathbf{M}$ of the passivity condition $m_{00}(1+Q) \le 1$ (with $Q \equiv \max(D,P)$), does not ensure that all components of all arbitrary decompositions of $\mathbf{M}$ are passive, so that, in general, the passive-realizability of a given parallel decomposition of $\mathbf{M}$ requires a subsequent specific test. Thus, different practical situations can be considered, as for instance:

(1) When $m_{00} \le 1/2$, then any arbitrary decomposition of $\mathbf{M}$ can be achieved with passive components (for a detailed demonstration, see San José and Gil 2020b).
(2) Most polarimeters provide measured Mueller matrices up to a scaling factor. These are represented in the normalized form $\hat{\mathbf{M}}$, in which case passivity is disregarded and thus, the MIC is usually taken to be $m_{00} = 1$. This means that the information on m_{00} is, in fact, unknown, and it belongs to the experimentalist to assign the value for m_{00}, like $m_{00} = 1/2$ (in order to ensure that all applied arbitrary decompositions are passive-realizable) or $m_{00} = 1$ and disregard passivity constraints. In the case of absolute polarimetry (where the measurements of all 16 Mueller elements is provided, and thus, m_{00} satisfies $m_{00}(1+Q) \le 1$, within the measurement uncertainty), the use of the passivity criterion constitutes a useful tool that allows for discarding those parallel decompositions that are not compatible with the passivity of the components.
(3) when $1/2 < m_{00} \le 1/(1+Q)$, then there always exist arbitrary decompositions with passive components; in particular, it is always possible to decompose $\mathbf{M}$ into a set of incoherently combined components $\mathbf{M}_{Ji}$ such that either their diattenuation vectors $\mathbf{D}_i$ are parallel to $\mathbf{D}$ (if $Q = D$), or their polarizance vectors $\mathbf{P}_i$ are parallel to $\mathbf{P}$ (if $Q = P$).
(4) It has been proven by San José and Gil (2020b; 2020c) that, given an experimentally obtained $\mathbf{M}$, two situations can be distinguished (within the measurement uncertainty). In the case where $P = D$ it is always possible to express $\mathbf{M}$ as a convex sum of a passive $\mathbf{M}_{J0}$ whose diattenuation and polarizance vectors are parallel to $\mathbf{D}$ and $\mathbf{P}$, respectively, and a number of $r-1$ (with $r = \operatorname{rank}\mathbf{H}$) retarders. In the alternative case ($P \ne D$) it is always possible to express $\mathbf{M}$ as a convex sum of r components, namely two passive components $\mathbf{M}_{J0}$ and $\mathbf{M}_{J1}$ such that either their nonzero diattenuation vectors $\mathbf{D}_i$ are parallel to $\mathbf{D}$ (if $Q = D$), or their nonzero polarizance vectors $\mathbf{P}_i$ are parallel to $\mathbf{P}$ (if $Q = P$) and a number of $r-2$ retarders.

To illustrate the above considerations, some examples are discussed below.

(a) Consider the Mueller matrix $\mathbf{M}_{\Delta 0} = \operatorname{diag}(1,0,0,0)$ of the perfect depolarizer. A possible parallel decomposition of $\mathbf{M}_{\Delta 0}$ is

$$\mathbf{M}_{\Delta 0} = \frac{1}{4}\mathbf{M}_{P1} + \frac{1}{4}\mathbf{M}_{P2} + \frac{1}{4}\mathbf{M}_{P3} + \frac{1}{4}\mathbf{M}_{P4} \tag{7.12}$$

with

$$\mathbf{M}_{P1}=\begin{pmatrix}1&1&0&0\\1&1&0&0\\0&0&0&0\\0&0&0&0\end{pmatrix}\quad \mathbf{M}_{P2}=\begin{pmatrix}1&-1&0&0\\1&-1&0&0\\0&0&0&0\\0&0&0&0\end{pmatrix}\quad \mathbf{M}_{P3}=\begin{pmatrix}1&1&0&0\\-1&-1&0&0\\0&0&0&0\\0&0&0&0\end{pmatrix}$$

$$\mathbf{M}_{P4}=\begin{pmatrix}1&-1&0&0\\-1&1&0&0\\0&0&0&0\\0&0&0&0\end{pmatrix}\tag{7.13}$$

where $k_i = 1/4$ $(i=0,1,2,3)$ and the components $\mathbf{M}_{Pi}$ (normalized perfect polarizers) are not passive (their passivity would require $m_{00,i} \le 1/2$ instead of $m_{00,i} = 1$). This incompatibility with passivity is a natural consequence of the fact that $\mathbf{M}_{\Delta 0}$ exhibits $m_{00} = 1$ with zero degree of polarizance of (i.e., $\mathbf{M}_{\Delta 0}$ does not produce loss of intensity for any incident polarization state in either forward or reverse incidence), while the enpolarizing nature of the components $\mathbf{M}_{Pi}$ implies necessarily respective losses of intensity. Nevertheless, infinitely many passive-realizable decompositions of $\mathbf{M}_{\Delta 0}$ can be realized, provided the matrix components are transparent retarders as, for instance

$$\mathbf{M}_{\Delta 0}=\frac{1}{4}\begin{pmatrix}1&0&0&0\\0&1&0&0\\0&0&1&0\\0&0&0&1\end{pmatrix}+\frac{1}{4}\begin{pmatrix}1&0&0&0\\0&1&0&0\\0&0&-1&0\\0&0&0&-1\end{pmatrix}+\frac{1}{4}\begin{pmatrix}1&0&0&0\\0&-1&0&0\\0&0&1&0\\0&0&0&-1\end{pmatrix}+\frac{1}{4}\begin{pmatrix}1&0&0&0\\0&-1&0&0\\0&0&-1&0\\0&0&0&1\end{pmatrix}\tag{7.14}$$

Moreover, observe that, if a variable value for m_{00} is considered, then $\mathbf{M} = m_{00}\mathbf{M}_{\Delta 0}$ can be decomposed as

$$\mathbf{M}=m_{00}\hat{\mathbf{M}}_{\Delta 0}=\frac{1}{4}\left(m_{00}\hat{\mathbf{M}}_{P1}\right)+\frac{1}{4}\left(m_{00}\hat{\mathbf{M}}_{P2}\right)+\frac{1}{4}\left(m_{00}\hat{\mathbf{M}}_{P3}\right)+\frac{1}{4}\left(m_{00}\hat{\mathbf{M}}_{P4}\right)\qquad (7.15))$$

which is passive-realizable provided that $m_{00} \le 1/2$. This result shows how the value of m_{00} is critical for the analysis of the passive realizability of the arbitrary decompositions of a given Mueller matrix, so that the measurement of m_{00} is not a negligible matter in polarimetry.

(b) Consider the Mueller matrix

$$\tilde{\mathbf{M}}=\begin{pmatrix}3/4&1/4&0&0\\1/12&1/4&0&0\\0&0&-1/24&-1/12\\0&0&1/12&1/24\end{pmatrix}=\frac{3}{4}\begin{pmatrix}1&1/3&0&0\\1/9&1/3&0&0\\0&0&-1/18&-1/9\\0&0&1/9&1/18\end{pmatrix}\tag{7.16}$$

for which $r=4$, $D=1/3$ and $P=1/9$, so that $Q=D$. Since $m_{00}(\tilde{\mathbf{M}})=1/(1+Q)=3/4$, $\tilde{\mathbf{M}}$ has the passive form defined in Section 5.9, so that any arbitrary decomposition contains at least two diattenuators (recall that pure Mueller matrices always satisfy $D = P$). A particular parallel decomposition is given by

$$\tilde{\mathbf{M}} = \frac{1}{6}\left[\frac{1}{2}\begin{pmatrix} 1 & 1 & 0 & 0 \\ -1 & -1 & 0 & 0 \\ 0 & 0 & 0 & 0 \\ 0 & 0 & 0 & 0 \end{pmatrix}\right] + \frac{2}{6}\left[\frac{1}{2}\begin{pmatrix} 1 & 1 & 0 & 0 \\ 1 & 1 & 0 & 0 \\ 0 & 0 & 0 & 0 \\ 0 & 0 & 0 & 0 \end{pmatrix}\right] + \frac{2}{6}\begin{pmatrix} 1 & 0 & 0 & 0 \\ 0 & 1 & 0 & 0 \\ 0 & 0 & 0 & -1 \\ 0 & 0 & 1 & 0 \end{pmatrix} + \frac{1}{6}\begin{pmatrix} 1 & 0 & 0 & 0 \\ 0 & -1 & 0 & 0 \\ 0 & 0 & -1 & 0 \\ 0 & 0 & 0 & 1 \end{pmatrix} \tag{7.17}$$

where all the components are passive (in fact, they coincide with their respective passive forms), and, since m_{00} has the maximal value compatible with the passivity of $\tilde{\mathbf{M}}$, its enpolarizing components necessarily exhibit diattenuation vectors that are parallel to that of $\tilde{\mathbf{M}}$. In accordance with the mathematical formulation for arbitrary decompositions of Mueller matrices, other infinite parallel decompositions of $\tilde{\mathbf{M}}$ are obtainable, as for instance

$$\tilde{\mathbf{M}} = k_1 m_{00,1}\,\hat{\mathbf{M}}_{J1} + k_2 m_{00,2}\,\hat{\mathbf{M}}_{J2} + k_3 m_{00,3}\,\hat{\mathbf{M}}_{J3} + k_4 m_{00,4}\,\hat{\mathbf{M}}_{J4} \tag{7.18}$$

with

$$k_1 = 0.3242 \quad k_2 = 0.3538 \quad k_3 = 0.1900 \quad k_4 = 0.1321 \quad \left[\sum_{i=1}^{4} k_i = 1\right]$$

$$m_{00,1} = 0.6504 \quad m_{00,2} = 0.6137 \quad m_{00,3} = 1 \quad m_{00,4} = 1$$

$$\hat{\mathbf{M}}_{J1} = \begin{pmatrix} 1 & 0.5375 & 0 & 0 \\ 0.0316 & 0.0588 & -0.7952 & 0.2762 \\ 0.3794 & 0.7059 & -0.1625 & -0.5748 \\ 0.3392 & 0.7059 & 0.2288 & 0.5517 \end{pmatrix} \quad \hat{\mathbf{M}}_{J3} = \begin{pmatrix} 1 & 0 & 0 & 0 \\ 0 & 3/5 & 4/5 & 0 \\ 0 & -4/5 & 3/5 & 0 \\ 0 & 0 & 0 & 1 \end{pmatrix}$$

$$\hat{\mathbf{M}}_{J2} = \begin{pmatrix} 1 & 0.6294 & 0 & 0 \\ 0.3531 & 0.5610 & -0.5276 & -0.3681 \\ -0.3684 & -0.5854 & -0.5673 & 0.2742 \\ -0.3684 & -0.5854 & 0.0617 & -0.6270 \end{pmatrix} \quad \hat{\mathbf{M}}_{J4} = \begin{pmatrix} 1 & 0 & 0 & 0 \\ 0 & 1/73 & 72/73 & 12/73 \\ 0 & 72/73 & 1/73 & -12/73 \\ 0 & -12/73 & 12/73 & -71/73 \end{pmatrix} \tag{7.19}$$

Note that, as frequently occurs for the passive forms of depolarizing Mueller matrices with nonzero degree of polarizance, the spectral decomposition of the above $\tilde{\mathbf{M}}$ is not passive-realizable. Nevertheless, for values $m_{00} < \tilde{m}_{00}$, there always exist sets of passive components that allow the physical realizability of the corresponding arbitrary decompositions of $m_{00}\hat{\mathbf{M}}$. As the value taken for m_{00} decreases from $\tilde{m}_{00}$, the number of sets of components compatible with passivity increases, and when $m_{00} \leq 1/2$ all arbitrary decompositions become realizable in terms of passive components. Therefore, for those arbitrary decompositions of $\tilde{\mathbf{M}}$ that are not passive-realizable, there always exists a value for m_{00} in the range $1/2 \leq m_{00} < \tilde{m}_{00}$ for which such decompositions become passive-realizable.

In summary, a *synthetic procedure* for generating arbitrary decompositions of a given Mueller matrix $\mathbf{M}$ can be performed as follows (Gil and San José 2013):

(1) Calculate the covariance matrix $\mathbf{H}$ associated with $\mathbf{M}$.
(2) Calculate the unitary matrix $\mathbf{U}$ and the eigenvalue diagonal matrix $\mathbf{\Lambda}=\operatorname{diag}(\lambda_0,\lambda_1,\lambda_2,\lambda_3)$ (with $\lambda_0 \geq \lambda_1 \geq \lambda_2 \geq \lambda_3$) entering the diagonalization $\mathbf{H}=\mathbf{U}\,\mathbf{\Lambda}\,\mathbf{U}^{\dagger}$.
(3) Take the unit eigenvectors $\hat{\mathbf{u}}_i$ of $\mathbf{H}$ with nonzero eigenvalues λ_i, $i=0,\ldots,r-1$ ($r \equiv \operatorname{rank}\mathbf{H}$) as well as λ_i.
(4) Build a set of r orthonormal unit vectors $\hat{\mathbf{v}}_i$ as linear combinations of $\hat{\mathbf{u}}_k$ (note that such sets of vectors $\hat{\mathbf{v}}_i$ can be constructed in infinitely many ways), or equivalently, take an arbitrary set of r orthonormal unit complex vectors $\hat{\mathbf{v}}_i$ whose last $4-r$ components are zero.
(5) Build the set of r independent covariance vectors $\mathbf{w}_i=\mathbf{U}\sqrt{\mathbf{\Lambda}}\hat{\mathbf{v}}_i$.
(6) Calculate the coefficients $p_i=|\mathbf{w}_i|^2$ corresponding to the respective pure components of the homogeneous arbitrary decomposition.
(7) Assign appropriate values for the MICs m_{00i} in such a manner that $m_{00}\sum p_i/m_{00i}=1$ (see Eq. (7.8a)), so that the coefficients of the respective components $m_{00i}\hat{\mathbf{M}}_{Ji}(\hat{\mathbf{w}}_i)$ are precisely $k_i=m_{00}\,p_i/m_{00i}$. Recall that the choice of sets of m_{00i} compatible with the passivity of the parallel components is not always possible and that, in general, such compatibility depends on the value of m_{00}.
(8) The corresponding arbitrary decomposition of $\mathbf{M}$ is given by Eq. (7.9).

An *iterative procedure* for the arbitrary decomposition of a given Mueller matrix $\mathbf{M}$ is the following (Gil 2007):

(1) Calculate the covariance matrix $\mathbf{H}$ associated with $\mathbf{M}$.
(2) Calculate the eigenvalues λ_i and unit eigenvectors $\hat{\mathbf{u}}_i$ of $\mathbf{H}$ and build the diagonalization unitary matrix $\mathbf{U}$, whose columns are $\hat{\mathbf{u}}_i$.
(3) Take an arbitrary covariance vector $\mathbf{w}_0$ constructed as a linear combination $\mathbf{w}_0=\sum_{i=0}^{r-1}c_{0i}\hat{\mathbf{u}}_i$ ($r \equiv \operatorname{rank}\mathbf{H}$) where $\hat{\mathbf{u}}_i$ ($i=1,\ldots,r-1$) are the eigenvectors of $\mathbf{H}$ with nonzero eigenvalues, and c_{0i} are arbitrary complex coefficients; then normalize $\mathbf{w}_0$ as $\hat{\mathbf{w}}_0=\mathbf{w}_0/|\mathbf{w}_0|$.
    Alternatively, build $\mathbf{w}_0 \equiv (w_{00},w_{01},w_{02},w_{023})^T$ from the components $t_{0i}=\sqrt{2}w_{0i}$ of a given Jones matrix $\mathbf{T}_0$ (see Section 3.4.2); if $\operatorname{rank}[\mathbf{H}+(\mathbf{w}_0\otimes\mathbf{w}_0^{\dagger})]=\operatorname{rank}\mathbf{H}$, then continue; if not, the pure medium represented by $\mathbf{w}_0$ cannot be considered as a parallel component of the medium represented by $\mathbf{M}$ and other appropriate $\mathbf{w}_0$ should be chosen.
(4) Build the normalized pure covariance matrix $\hat{\mathbf{H}}_{J0}=\hat{\mathbf{w}}_0\otimes\hat{\mathbf{w}}_0^{\dagger}$, and then calculate the corresponding normalized pure Mueller matrix ($\hat{\mathbf{M}}_{J0}$ see Eq. (5.12)).
(5) Calculate $\tilde{m}_{00,0}=1/(1+D_0)$, where $D_0=P_0$ is the diattenuation-polarizance of $\hat{\mathbf{M}}_{J0}$, and build the passive form $\tilde{\mathbf{M}}_{J0}=\tilde{m}_{00,0}\,\hat{\mathbf{M}}_{J0}$ (note that other MICs $m_{00,0}<1/(1+D_0)$ can also be assigned, but then the passivity of further components becomes more constrained, see Eq. (5.61)).
(6) Calculate the coefficient k_0 corresponding to $\tilde{\mathbf{M}}_{J0}$ by means of $k_0=1/[\tilde{m}_{00,0}(\hat{\mathbf{w}}_0^{\dagger}\mathbf{H}^{-}\hat{\mathbf{w}}_0)]$.
(7) Calculate the remainder covariance matrix $\mathbf{H}_{r1}=[\mathbf{H}-k_0\,\tilde{m}_{00,0}(\hat{\mathbf{w}}_0\times\hat{\mathbf{w}}_0^{\dagger})]/(1-k_0)$, and its associated Mueller matrix $\mathbf{M}_{r1}(\mathbf{H}_{r1})$.
(8) If the MIC $m_{00,1}$ of $\mathbf{M}_{r1}$ does not satisfy the passivity condition $m_{00,1}(1+Q_1)\leq 1$ [$Q_1=\max(D_1,P_1)$], then the intended decomposition is not passive-realizable; the closer $m_{00,0}$ to $\tilde{m}_{00,0}$ is (the choice of $m_{00,0}$ has been made in step 5), the less strict the achievement of the passivity condition for $\mathbf{M}_{r1}$ is.

(9) If $\text{rank}\,\mathbf{H}=2$, then $\mathbf{M}_{r1}$ is a pure Mueller matrix $\mathbf{M}_{r1}\equiv\mathbf{M}_{J1}$ and its corresponding coefficient in the convex sum is $k_1=1-k_0$. The arbitrary decomposition is therefore completed as $\mathbf{M}=k_0\tilde{\mathbf{M}}_{J0}+k_1\mathbf{M}_{J1}$.

(10) If $\text{rank}\,\mathbf{H}>2$, then take an arbitrary covariance vector $\mathbf{w}_1$ constructed as a linear combination $\mathbf{w}_1=\sum_{i=0}^{r-1}c_{1i}\,\hat{\mathbf{u}}_k$ where c_{1i} are arbitrary complex coefficients, provided that $\mathbf{w}_1$ is linearly independent of $\mathbf{w}_0$, and normalize it as $\hat{\mathbf{w}}_1=\mathbf{w}_1/|\mathbf{w}_1|$. Alternatively, as done in step 3, build $\mathbf{w}_1$ from the components of a given Jones matrix $\mathbf{T}_1$, if $\text{rank}[\mathbf{H}_{r1}+(\mathbf{w}_1\otimes\mathbf{w}_1^{\dagger})]=\text{rank}\,\mathbf{H}_{r1}$, then continue; if not, other appropriate $\mathbf{w}_1$ should be chosen.

(11) Build the pure and normalized covariance matrix $\hat{\mathbf{H}}_{J1}=\hat{\mathbf{w}}_1\otimes\hat{\mathbf{w}}_1^{\dagger}$, and then use Eq. (5.12) to calculate the corresponding normalized pure Mueller matrix $\hat{\mathbf{M}}_{J1}$ of the second arbitrary component.

(12) Calculate the MIC $\tilde{m}_{00,1}=1/(1+D_1)$, where $D_1=P_1$ is the diattenuation-polarizance of $\hat{\mathbf{M}}_{J1}$, and build the passive form $\tilde{\mathbf{M}}_{J1}=\tilde{m}_{00,1}\hat{\mathbf{M}}_{J1}$ (other MICs satisfying $m_{00,1}<1/(1+D_1)$ can also be assigned if appropriate).

(13) Calculate the coefficient k_1 corresponding to $\tilde{\mathbf{M}}_{J1}$ by means of $k_1=1/[\tilde{m}_{00,1}(\hat{\mathbf{w}}_1^{\dagger}\mathbf{H}^{-}\hat{\mathbf{w}}_1)]$.

(14) Calculate the remainder covariance matrix $\mathbf{H}_{r2}=[\mathbf{H}_{r1}-k_1\tilde{m}_{00,1}(\hat{\mathbf{w}}_1\times\hat{\mathbf{w}}_1^{\dagger})]/(1-k_1)$, and its associated Mueller matrix $\mathbf{M}_{r2}(\mathbf{H}_{r2})$.

(15) If the MIC $m_{00,2}$ of $\mathbf{M}_{r2}$ does not satisfy the passivity condition $m_{00,2}(1+Q_2)\leq 1$ $[Q_2=\max(D_2,P_2)]$, then the intended decomposition is not passive-realizable; the closer to their respective canonical passive representatives $\mathbf{M}_{J0}$ and $\mathbf{M}_{J1}$ are, the less strict the achievement of the passivity condition for $\mathbf{M}_{r2}$ is.

(16) If $\text{rank}\,\mathbf{H}=3$, then $\mathbf{M}_{r2}$ is a pure Mueller matrix $\mathbf{M}_{r2}\equiv\mathbf{M}_{J2}$ and its corresponding coefficient in the convex sum is $k_2=1-k_0-k_1$. The arbitrary decomposition is therefore completed as $\mathbf{M}=k_0\tilde{\mathbf{M}}_{J0}+k_1\tilde{\mathbf{M}}_{J1}+k_2\mathbf{M}_{J2}$.

(17) If $\text{rank}\,\mathbf{H}=4$, then take an arbitrary covariance vector $\mathbf{w}_2$ constructed as a linear combination $\mathbf{w}_2=\sum_{k=0}^{r-1}c_{2i}\,\hat{\mathbf{u}}_i$ where c_{2i} are arbitrary complex coefficients, provided that $\mathbf{w}_2$ is linearly independent of $\mathbf{w}_0$ and $\mathbf{w}_1$, and normalize it as $\hat{\mathbf{w}}_2=\mathbf{w}_2/|\mathbf{w}_2|$. Alternatively, as done in steps 3 and 10, build $\mathbf{w}_2$ from the components of a given Jones matrix $\mathbf{T}_2$, if $\text{rank}[\mathbf{H}_{r2}+(\mathbf{w}_{02}\otimes\mathbf{w}_2^{\dagger})]=\text{rank}\,\mathbf{H}_{r2}$, then continue; if not, other appropriate $\mathbf{w}_2$ should be chosen.

(18) Build the pure and normalized covariance matrix $\hat{\mathbf{H}}_{J2}=\hat{\mathbf{w}}_2\otimes\hat{\mathbf{w}}_2^{\dagger}$, and then use Eq. (5.12) to calculate the corresponding normalized pure Mueller matrix $\hat{\mathbf{M}}_{J2}$ of the third arbitrary component.

(19) Calculate the MIC $\tilde{m}_{00,2}=1/(1+D_2)$, where $D_2=P_2$ is the diattenuation-polarizance of $\hat{\mathbf{M}}_{J2}$, and build the passive form $\tilde{\mathbf{M}}_{J2}=\tilde{m}_{00,2}\hat{\mathbf{M}}_{J2}$ (note that other MICs $m_{00,2}<1/(1+D_2)$ can also be assigned).

(20) Calculate the coefficient k_2 corresponding to $\tilde{\mathbf{M}}_{J2}$ by means of $k_2=1/[\tilde{m}_{00,2}(\hat{\mathbf{w}}_2^{\dagger}\mathbf{H}^{-}\hat{\mathbf{w}}_2)]$.

(21) Calculate the coefficient of the fourth pure component as $k_3=1-k_0-k_1-k_2$.

(22) Calculate the fourth pure component as $\mathbf{M}_{J3}=(\mathbf{M}-k_0\mathbf{M}_{J0}-k_1\mathbf{M}_{J1}-k_2\mathbf{M}_{J2})/k_3$.

(23) If the MIC $m_{00,3}$ of the obtained $\mathbf{M}_{J3}$ does not satisfy the passivity condition $m_{00,3}(1+D_3)\leq 1$ $(D_3=P_3)$, then the intended decomposition is not passive-realizable; the closer to their respective canonical passive representatives $\mathbf{M}_{J0}$, $\mathbf{M}_{J1}$ and $\mathbf{M}_{J2}$ are, the less strict the achievement of the passivity condition for $\mathbf{M}_{J3}$ is.

(24) The parallel decomposition of $\mathbf{M}$ is finally achieved as $\mathbf{M}=k_0\tilde{\mathbf{M}}_{J0}+k_1\tilde{\mathbf{M}}_{J1}+k_2\tilde{\mathbf{M}}_{J2}+k_3\mathbf{M}_{J3}$.

7.3.1 Application of the Arbitrary Decomposition to an Experimental Example

As an example of the above developments, consider the normalized Mueller matrix

$$\hat{\mathbf{M}}_{BP} = \begin{pmatrix} 1 & 0.072 & 0.093 & 0.921 \\ 0.072 & 0.014 & 0.001 & 0.108 \\ -0.093 & -0.001 & -0.026 & -0.068 \\ 0.922 & 0.108 & 0.068 & 0.960 \end{pmatrix}$$

representing a simulated response of averaged Bragg reflection on the so-called *blue phase* of a cholesteric liquid crystal (Bohley and Scharf 2004). Since, as with many polarimeters, $\hat{\mathbf{M}}_{BP}$ has not been obtained through absolute polarimetry, its MIC m_{00} is unknown. The values for the diattenuation and polarizance are $D = 0.862$ and $P = 0.864$, so that the MIC $\tilde{m}_{00}$ of the passive form of $\hat{\mathbf{M}}_{BP}$ takes the value $\tilde{m}_{00} = 1/(1+P) = 0.537$ (maximum MIC compatible with passivity). Taking into account that effects of loss of intensity usually occur in addition to those due to the mere diattenuation-polarizance, it is reasonable to hypothesize that the sample exhibits an effective MIC $m_{00} \le 1/2$, and consequently assume that $m_{00} = 1/2$. The covariance matrix of $\mathbf{M}_{BP} = m_{00}\hat{\mathbf{M}}_{BP}$ is found to be

$$\mathbf{H}_{BP} = \frac{1}{2}\begin{pmatrix} 0.2895 & 0.0235+i0.2573 & -0.0235-i0.2575 & 0.2335-i0.0340 \\ 0.0235-i0.2573 & 0.2465 & -0.2465 & -0.0230-i0.2035 \\ -0.0235+i0.2575 & -0.2465 & 0.2465 & 0.0230+i0.2033 \\ 0.2335+i0.0340 & -0.0230+i0.2035 & 0.0230-i0.2033 & 0.2175 \end{pmatrix}$$

and its eigenvalues are 0.9585, 0.0381, 0.0034 and zero. These eigenvalues are nothing but the weights p_k of the pure components $\mathbf{M}_k$ in the spectral (or Cloude's) decomposition of $\mathbf{M}_{BP}$. The Mueller matrices of the three pure components, obtained from the respective eigenvectors of $\mathbf{H}_{BP}$ with nonzero eigenvalues, are

$$\mathbf{M}_0 = \frac{1}{2}\begin{pmatrix} 1 & 0.0948 & 0.0856 & 0.9917 \\ 0.0948 & -0.0012 & 0.0007 & 0.0956 \\ -0.0855 & -0.0007 & -0.0176 & -0.0847 \\ 0.9917 & 0.0956 & 0.0847 & 0.9836 \end{pmatrix} \quad \mathbf{M}_1 = \frac{1}{2}\begin{pmatrix} 1 & -0.5298 & 0.3219 & -0.7452 \\ -0.5360 & 0.3386 & 0.0096 & 0.4828 \\ -0.3267 & -0.0096 & -0.1677 & 0.3728 \\ -0.7386 & 0.4758 & -0.3686 & 0.4937 \end{pmatrix}$$

$$\mathbf{M}_2 = \frac{1}{2}\begin{pmatrix} 1 & 0.3939 & -0.3870 & -0.3585 \\ 0.4791 & 0.6556 & -0.0102 & -0.6051 \\ 0.4295 & 0.0102 & -0.8071 & -0.3153 \\ -0.1395 & -0.5402 & 0.2545 & -0.4790 \end{pmatrix}$$

and therefore,

$$\mathbf{M}_{BP} = p_0\mathbf{M}_0 + p_1\mathbf{M}_1 + p_2\mathbf{M}_2$$

is the spectral decomposition of $\mathbf{M}_{BP}$. We see that the Bragg reflection appears as a rank-three phenomenon ($r = 3$) which is consistent with the backscattering configuration assumed in the model (backscattering is a well-known rank-three process; see, for instance, Cloude and Pottier 1996). A closer look at the normalized eigenvalues shows that the third one (0.0034) is comparable to the experimental uncertainty (of the order of several 0.001 typically) and, therefore, in practice, the Bragg scattering on cholesteric blue phases turns out to be rather a rank-two process ($r = 2$). This observation is likewise consistent with the phenomenological scattering theory on cholesteric liquid crystals (Hornreich and Shtrikman 1983). Therefore, within the typical experimental accuracy we can represent the polarimetric response of the blue phase as

$$\mathbf{M}_{BP2} = p_0\mathbf{M}_0 + p_1\mathbf{M}_1 = \frac{1}{2}\begin{pmatrix} 1 & 0.0709 & 0.0946 & 0.9253 \\ 0.0706 & 0.0118 & 0.0010 & 0.1104 \\ -0.0948 & -0.0010 & -0.0234 & -0.0672 \\ 0.9256 & 0.1102 & 0.0674 & 0.9648 \end{pmatrix}$$

to be compared to the original matrix $\mathbf{M}_{BP}$. As we shall see in Chapter 10, because of $r = 2$, the mapping ellipsoid of the Mueller matrix $\mathbf{M}$ of a cholesteric blue phase touches the Poincaré sphere in either two or in a single point; that is, $\mathbf{M}$ can be either a type-I or a type-II Mueller matrix (Ossikovski et al. 2009). In the specific case here, $\mathbf{M}_{BP2}$ features two contact points, i.e., it is a (special) type-I matrix. Returning to the eigenvalues, we further see that the second normalized eigenvalue (0.0381), although measurable, is by more than an order of magnitude smaller than the first one (0.9585), so that the polarimetric behavior of the blue phase is essentially governed by the first nondepolarizing component $\mathbf{M}_0$ of the spectral decomposition of $\mathbf{M}_{BP}$. Geometrically, this means that the surface of the mapping ellipsoid of $\mathbf{M}_{BP}$ (see Chapter 10), although not coinciding with the Poincaré sphere, is lying very close to it. If we approximate $\mathbf{M}_{BP}$ by $\mathbf{M}_0$ alone, and polar-decompose $\mathbf{M}_0$ as $\mathbf{M}_0 = \mathbf{M}_R\mathbf{M}_D$ to determine its diattenuation, $\mathbf{D}$, and retardance, $\mathbf{R}$, vectors (see Chapter 4), we find $D = 0.9999$ and $R = 2.525$ for their magnitudes, and $\hat{\mathbf{D}} = (0.0948, 0.0856, 0.9918)^T$ and $\hat{\mathbf{R}} = (-0.1223, 0, -0.9918)^T$ for their direction unit-vectors respectively. Recall that, unlike $\hat{\mathbf{R}}$, the direction vector $\hat{\mathbf{D}}$ is order-dependent, i.e., the two polar factorizations $\mathbf{M}_R\mathbf{M}_D$ and $\mathbf{M}_P\mathbf{M}_R$ generally produce different $\hat{\mathbf{D}}$ unless $\mathbf{M}_D$ and $\mathbf{M}_R$ commute or equivalently, $\hat{\mathbf{R}}$ and $\hat{\mathbf{D}}$ are parallel (see Section 4.7). In our case, here, the last condition is largely met, that is, we have practically $\hat{\mathbf{D}} \,||\, \hat{\mathbf{R}}$. We thus see that both diattenuation and retardance vectors are oriented almost perfectly along the S_3 axis of the Poincaré sphere and that the diattenuation is practically equal to unity. This observation is in perfect agreement with the well-known experimental fact that cholesteric liquid crystals, as well as some similarly constituted natural structures such as beetle cuticles, roughly behave as circular polarizers in reflection.

7.4 On the Rank of the Covariance Matrix of a Parallel Composition

Let us now analyze the value of the rank of a covariance matrix $\mathbf{H}$ that is synthesized through an additive composition of covariance matrices. In agreement with the arbitrary decomposition considered in Section 7.3, $r \equiv \operatorname{rank}\mathbf{H}$ is the minimum number of pure parallel components of $\mathbf{M}(\mathbf{H})$. The additive composition of two nondepolarizing elements with corresponding Mueller-Jones matrices $\mathbf{M}_{Ji}\,(i = 0,1)$ and associated covariance matrices $\mathbf{H}_{Ji} = m_{00i}\left(\hat{\mathbf{w}}_i \otimes \hat{\mathbf{w}}_i^{\dagger}\right)$ is given by

$$\mathbf{H} = k\mathbf{H}_{J0} + (1-k)\mathbf{H}_{J1} = k m_{00,0}(\hat{\mathbf{w}}_0 \otimes \hat{\mathbf{w}}_0^{\dagger}) + (1-k) m_{00,1}(\hat{\mathbf{w}}_1 \otimes \hat{\mathbf{w}}_1^{\dagger}) \tag{7.20}$$

Except for the trivial case where $\hat{\mathbf{w}}_1 = \hat{\mathbf{w}}_0$ (i.e., $\hat{\mathbf{M}}_{J1} = \hat{\mathbf{M}}_{J0}$), one has $\operatorname{rank}\mathbf{H} = 2$.

The additive composition of n (with $n > 2$) nondepolarizing systems is expressed as follows in terms of the corresponding covariance matrices:

$$\mathbf{H} = \sum_{i=0}^{n-1} k_i \mathbf{H}_{Ji} = \sum_{i=0}^{n-1} k_i\, m_{00i} \left(\hat{\mathbf{w}}_i \otimes \hat{\mathbf{w}}_i^{\dagger}\right) \qquad \left[|\hat{\mathbf{w}}_i| = 1 \quad \sum_{i=0}^{n-1} k_i = 1 \right] \tag{7.21}$$

It is straightforward to show that $\operatorname{rank}\mathbf{H}$ is equal to the number r of vectors $\hat{\mathbf{w}}_i\,(i = 0,\ldots,r-1)$ that are linearly independent, while the remaining $n-r$ vectors $\hat{\mathbf{w}}_l\,(l = r,\ldots,n-1)$ are linearly dependent of $\hat{\mathbf{w}}_i$. That is to say, $\operatorname{im}\mathbf{H}$ is the image subspace generated by the r independent vectors $\hat{\mathbf{w}}_i$. Therefore, given a pair of covariance matrices $\mathbf{H}_1$ and $\mathbf{H}_2$, with $r_1 \equiv \operatorname{rank}\mathbf{H}_1$ and $r_2 \equiv \operatorname{rank}\mathbf{H}_2$, the rank of a parallel combination $\mathbf{H} = k\mathbf{H}_1 + (1-k)\mathbf{H}_2$ $(0 < k < 1)$, is equal to the dimension of $\operatorname{im}\mathbf{H} = \operatorname{im}\mathbf{H}_1 \cup \operatorname{im}\mathbf{H}_2$. This statement is equivalent to saying that $\operatorname{rank}\mathbf{H}$ is equal to the number of independent vectors within the set composed of the union of 1) the eigenvectors of $\mathbf{H}_1$ with nonzero eigenvalues, and 2) the eigenvectors of $\mathbf{H}_2$ with nonzero eigenvalues.

Consequently, $r_1 \le \operatorname{rank}\mathbf{H}$, $r_2 \le \operatorname{rank}\mathbf{H}$, and $\operatorname{rank}\mathbf{H} \le r_1 + r_2$. When all vectors of the said set are independent, then $\operatorname{rank}\mathbf{H} = r_1 + r_2$. These results agree with the physical fact that a Mueller matrix obtained by means of the parallel composition of a number $n > \operatorname{rank}\mathbf{H}$ of pure Mueller matrices can always be expressed as an *equivalent* convex combination of r pure Mueller matrices.

7.5 Retarders and Diattenuators as Parallel Components of a Given Mueller Matrix

The fact that the arbitrary decomposition offers infinitely many ways to represent a depolarizing Mueller matrix $\mathbf{M}$ as a parallel composition of pure ones leads naturally to questions like how many retarders and how many diattenuating (i.e., enpolarizing) pure components can be included in such compositions. The answers to such questions are directly related to (1) the diattenuation D and polarizance P of $\mathbf{M}$; (2) the value r of $\operatorname{rank}\mathbf{H}(\mathbf{M})$, and (3) the value of the MIC m_{00} of $\mathbf{M}$.

For instance, if $\mathbf{M}$ exhibits nonzero degree of polarizance (i.e., $D > 0$ or $P > 0$), it turns out that necessarily some of its parallel components also have nonzero degree of polarizance, because of the vectors $\mathbf{D}$ and $\mathbf{P}$ of $\mathbf{M}$ being given by the respective weighted sums of the diattenuation and polarizance vectors $\mathbf{D}_i$ and $\mathbf{P}_i$ of the components. Moreover, as seen in Section 7.3, the pure parallel components of a transparent perfect depolarizer $\mathbf{M}_{\Delta 0}$ are necessarily retarders, otherwise the loss of intensity associated with the diattenuation-polarizance of certain components cannot be recovered by the action of others. Therefore, the knowledge of the restrictions on the nature of the pure components of physically realizable arbitrary decompositions of a given $\mathbf{M}$ should be taken into account in the analysis of measured Mueller matrices and appears as a relevant matter in polarimetry.

This problem has been studied by San José and Gil (2020c). The results are summarized in Table 7.1 in terms of the possible values of D, P, $\operatorname{rank}\mathbf{H}(\mathbf{M})$ (denoted by r), and m_{00}. Obviously, when $r = 1$ then $\mathbf{M}$ is itself a pure Mueller matrix, so that only the cases $r = 2, 3, 4$ require analysis. As in other sections, we use the notation $Q \equiv \max(D, P)$, i.e., $Q = D$ when $D \geq P$ and $Q = P$ when $P \geq D$, while q and t denote the number of pure parallel components without diattenuation (retarders) and the number of pure parallel components with nonzero diattenuation, respectively (the total number of components $q + t = r$ having independent associated covariance vectors).

Some important remarks, which are closely related to the classification shown in Table 7.1, are given below.

If $Q = 0$ (i.e., $D = P = 0$), then the r components can always be retarders (affected by equal or different respective MICs $m_{00,i}$). The physical realizability of arbitrary decompositions including components with nonzero polarizance-diattenuation depends on the value of m_{00}. When $m_{00} = 1$ only retarders are physically admissible as components; when $1/2 < m_{00} < 1$, the family of passive realizable decompositions including diattenuating components gets broader the closer to $1/2$ m_{00} is; all mathematically achievable arbitrary decompositions are passive-realizable when $m_{00} \leq 1/2$. Note that the possibility of a single diattenuating component is excluded because of the required condition $Q = 0$ (which means that more than one diattenuating components should add together in such a way as to cancel the resulting diattenuation and polarizance vectors).

If $Q > 0$ and $D = P$, at least one of the r components should exhibit nonzero diattenuation (in which case its diattenuation and polarizance vectors are parallel to $\mathbf{D}$

TABLE 7.1

Number q of Retarders and Number t of Diattenuators as Parallel Components of a Mueller Matrix in Terms of the Values of D, P, r and m_{00} [$Q \equiv \max(D, P)$]

$Q = 0$			
	$r = 2$	$r = 3$	$r = 4$
$m_{00} = 1$	$q = 2,\ t = 0$	$q = 3,\ t = 0$	$q = 4,\ t = 0$
$m_{00} < 1$	$q = 2$ and $t = 0$ or $q = 0$ and $t = 2$	$q = 3$ and $t = 0$ or $q = 1$ and $t = 2$ or $q = 0$ and $t = 3$	$q = 4$ and $t = 0$ or $q = 2$ and $t = 2$ or $q = 1$ and $t = 3$ or $q = 0$ and $t = 4$
$Q > 0,\ D = P$			
	$r = 2$	$r = 3$	$r = 4$
$m_{00} \leq \dfrac{1}{1+D}$	$0 \leq q \leq 1,\ 1 \leq t \leq 2$ $q + t = 2$	$0 \leq q \leq 2,\ 1 \leq t \leq 3$ $q + t = 3$	$0 \leq q \leq 3,\ 1 \leq t \leq 4$ $q + t = 4$
$Q > 0,\ D \neq P$			
	$r = 2$	$r = 3$	$r = 4$
$m_{00} \leq \dfrac{1}{1+Q}$	$q = 0,\ t = 2$	$0 \leq q \leq 1,\ 2 \leq t \leq 3$ $q + t = 3$	$0 \leq q \leq 2,\ 2 \leq t \leq 4$ $q + t = 4$

and **P** respectively), so that the number q of pure retarding components (retarders) is limited by $0 \le q \le r-1$.

If $Q > 0$ and $D \ne P$, at least two of the r components should exhibit nonzero diattenuation, so that the number q of pure retarding components (retarders) is limited by $0 \le q \le r-2$. The largest value of m_{00} compatible with passivity of **M** is given by $\tilde{m}_{00} = 1/(1+Q)$. When **M** is in its passive form $\tilde{\mathbf{M}} = \tilde{m}_{00}\hat{\mathbf{M}}$, the diattenuating components should necessarily exhibit mutually parallel diattenuating vectors (if $Q = D$) or mutually parallel polarizance vectors (if $Q = P$). The smaller m_{00} is (with $m_{00} < \tilde{m}_{00}$) the less strict the passivity constraints are and the broader the family of physically realizable decompositions is, so that when $m_{00} \le 1/2$ all mathematically achievable arbitrary decompositions are passive-realizable.

7.6 Characteristic Decomposition of a Mueller Matrix

Starting from the arbitrary decomposition of a given Mueller matrix **M**, different parallel decompositions can be generated in terms of pure and nonpure components. This can easily be done by taking a given particular form of the arbitrary decomposition and grouping some of the components into a nonpure component as, for instance

$$\begin{gathered}\mathbf{M} = p_0\mathbf{M}_{J0} + p_1\mathbf{M}_{J1} + p_2\mathbf{M}_{J2} + p_3\mathbf{M}_{J3} = p_0\mathbf{M}_{J0} + p_1\mathbf{M}_{J1} + p_2'\mathbf{M}_2' \\ \left[p_2' = 1 - p_0 - p_1 = p_2 + p_3 \qquad \mathbf{M}_2' = \frac{p_2\mathbf{M}_{J2} + p_3\mathbf{M}_{J3}}{p_2 + p_3}\right]\end{gathered} \tag{7.22}$$

in which case $\mathbf{M}_2'$ is a depolarizing Mueller matrix whose covariance matrix $\mathbf{H}_2'$ has $\operatorname{rank}\mathbf{H}_2' = 2$.

In Section 5.14.3, we have analyzed the so-called *characteristic decomposition* (Gil 2007) based on a particular grouping of the components of the spectral decomposition. In particular, by considering the diagonalization of **H**,

$$\mathbf{H} = \mathbf{U}\operatorname{diag}(\lambda_0, \lambda_1, \lambda_2, \lambda_3)\,\mathbf{U}^\dagger \tag{7.23}$$

and by setting

$$\begin{gathered}\begin{pmatrix}\lambda_0 & 0 & 0 & 0\\ 0 & \lambda_1 & 0 & 0\\ 0 & 0 & \lambda_2 & 0\\ 0 & 0 & 0 & \lambda_3\end{pmatrix} = (\lambda_0 - \lambda_1)\begin{pmatrix}1&0&0&0\\0&0&0&0\\0&0&0&0\\0&0&0&0\end{pmatrix} + (\lambda_1 - \lambda_2)\begin{pmatrix}1&0&0&0\\0&1&0&0\\0&0&0&0\\0&0&0&0\end{pmatrix} \\ + (\lambda_2 - \lambda_3)\begin{pmatrix}1&0&0&0\\0&1&0&0\\0&0&1&0\\0&0&0&0\end{pmatrix} + \lambda_3\begin{pmatrix}1&0&0&0\\0&1&0&0\\0&0&1&0\\0&0&0&1\end{pmatrix}\end{gathered} \tag{7.24}$$

we get

$$\begin{aligned}
\mathbf{H} &= \frac{\lambda_0-\lambda_1}{m_{00}}\mathbf{H}_{J0}+2\frac{\lambda_1-\lambda_2}{m_{00}}\mathbf{H}_1+3\frac{\lambda_2-\lambda_3}{m_{00}}\mathbf{H}_2+4\frac{\lambda_3}{m_{00}}\mathbf{H}_{\Delta 0}\\
\mathbf{H}_{J0} &\equiv m_{00}\left[\mathbf{U}\mathrm{diag}(1,0,0,0)\mathbf{U}^{\dagger}\right]= m_{00}\left[\hat{\mathbf{u}}_0\otimes\hat{\mathbf{u}}_0^{\dagger}\right]\\
\mathbf{H}_1 &\equiv m_{00}\left[\frac{1}{2}\mathbf{U}\mathrm{diag}(1,1,0,0)\mathbf{U}^{\dagger}\right]=m_{00}\left[\frac{1}{2}\sum_{i=0}^{1}\hat{\mathbf{u}}_i\otimes\hat{\mathbf{u}}_i^{\dagger}\right]\\
\mathbf{H}_2 &\equiv m_{00}\left[\frac{1}{3}\mathbf{U}\mathrm{diag}(1,1,1,0)\mathbf{U}^{\dagger}\right]=m_{00}\left[\frac{1}{3}\sum_{i=0}^{2}\hat{\mathbf{u}}_i\otimes\hat{\mathbf{u}}_i^{\dagger}\right]\\
\mathbf{H}_{\Delta 0} &\equiv m_{00}\left[\frac{1}{4}\mathbf{I}_4\right]\qquad\left[\mathbf{I}_4\equiv\mathrm{diag}(1,1,1,1)=\sum_{i=0}^{3}\hat{\mathbf{u}}_i\otimes\hat{\mathbf{u}}_i^{\dagger}\right]
\end{aligned}\tag{7.25}$$

where $\hat{\mathbf{u}}_i$ are the eigenvectors of $\mathbf{H}$ (i.e., the column vectors of $\mathbf{U}$), $\mathbf{H}_{J0}$ is a pure component, and $\mathbf{H}_i$ $(i=1,2,3)$ are nonpure components with $\mathrm{rank}\,\mathbf{H}_i=i+1$ and with $i+1$ degenerate nonzero eigenvalues. It is remarkable that $\mathbf{H}_{J0}$ is precisely the parallel component of the spectral decomposition (7.3) associated with the largest coefficient $p_0=\lambda_0/m_{00}$.

The characteristic decomposition has the following expression in terms of the corresponding Mueller components and the indices of polarimetric purity P_i $(i=1,2,3)$ (introduced in Section 6.3 and referred to with the acronym IPP):

$$\mathbf{M}=P_1\,m_{00}\,\hat{\mathbf{M}}_{J0}+(P_2-P_1)m_{00}\,\hat{\mathbf{M}}_1+(P_3-P_2)m_{00}\,\hat{\mathbf{M}}_2+(1-P_3)m_{00}\,\hat{\mathbf{M}}_{\Delta 0}\tag{7.26}$$

so that, as noted in Section 5.14.3, any given Mueller matrix can be interpreted as the parallel composition shown in Figure 7.2:

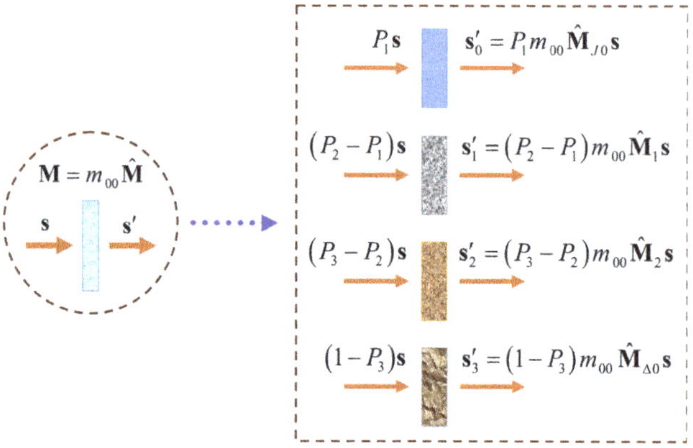

FIGURE 7.2
Characteristic decomposition of a Mueller matrix. The Mueller matrix $\mathbf{M}$ of the system is expressed as a convex linear combination of the Mueller matrices of a pure component, a 2D depolarizer, a 3D depolarizer, and a perfect depolarizer. The coefficients are determined by the indices of polarimetric purity.

- a *characteristic pure component* $\mathbf{M}_{J0} \equiv m_{00}\hat{\mathbf{M}}_{J0}$, determined by the eigenvector $\mathbf{u}_0$ of $\mathbf{H}$ whose associated eigenvalue is the largest one. The relative cross section of $\mathbf{M}_{J0}$ with respect to the complete system is given by the first index of polarimetric purity P_1 of $\mathbf{M}$;
- a *2D depolarizer* $\mathbf{M}_1 \equiv m_{00}\hat{\mathbf{M}}_1$, constituted by an equiprobable mixture of the first two spectral components. The relative cross section of $\mathbf{M}_1$ with respect to the complete system is given by the difference $P_2 - P_1$ between the first two IPP;
- a *3D depolarizer* $\mathbf{M}_2 \equiv m_{00}\hat{\mathbf{M}}_2$, constituted by an equiprobable mixture of the first three spectral components. The relative cross section of $\mathbf{M}_2$ with respect to the complete system is given by the difference $P_3 - P_2$; and
- a *perfect depolarizer* $\mathbf{M}_{\Delta 0} \equiv m_{00}\,\mathrm{diag}(1,0,0,0)$, whose relative cross section with respect to the complete system is given by the difference $1 - P_3$.

The following situations can be distinguished depending on the values of the IPP:

- $P_1 = 1$ (i.e., $\mathrm{rank}\,\mathbf{H} = 1$ and $P_1 = P_2 = P_3 = P_\Delta = 1$). The only non-vanishing component of the characteristic decomposition is the pure one $\mathbf{M}_{J0}$, that is, $\mathbf{M}$ is itself a pure Mueller matrix: $\mathbf{M} = \mathbf{M}_{J0}$.
- $P_1 < P_2 = 1$ (i.e., $\mathrm{rank}\,\mathbf{H} = 2$) then the 3D depolarizer $\mathbf{M}_2$ and the perfect depolarizer $\mathbf{M}_{\Delta 0}$ have no contribution in the characteristic decomposition, so that $\mathbf{M} = P_1\mathbf{M}_{J0} + (1 - P_1)\mathbf{M}_1$. Note also that, in this case, the 2D depolarizer $\mathbf{M}_1$ has necessarily nonzero contribution. Furthermore, when $P_1 = 0$, then the contribution of the characteristic component $\mathbf{M}_{J0}$ vanishes, so that $\mathbf{M} = \mathbf{M}_1$.
- $P_2 < P_3 = 1$ (i.e., $\mathrm{rank}\,\mathbf{H} = 3$), then the perfect depolarizer $\mathbf{M}_{\Delta 0}$ has no contribution in the characteristic decomposition: $\mathbf{M} = P_1\mathbf{M}_{J0} + (P_2 - P_1)\mathbf{M}_1 + (1 - P_2)\mathbf{M}_2$. Note also that, in this case, the 3D depolarizer $\mathbf{M}_2$ has necessarily nonzero contribution while the contributions of the other components may be zero or not depending on the values of P_2 and P_1.
- $P_3 < 1$ (i.e., $\mathrm{rank}\,\mathbf{H} = 4$), then the perfect depolarizer $\mathbf{M}_{\Delta 0}$ has a nonzero contribution in the characteristic decomposition, while other components may also contribute or not, depending on the values of P_2 and P_1.

Notice that the passivity of $\mathbf{M}$ does not ensure the passivity of the components $\mathbf{M}_{J0}$, $\mathbf{M}_1$ and $\mathbf{M}_2$. In general, a passive-realizable characteristic decomposition of a given $\mathbf{M}$ can be achieved through replacing m_{00} by the MIC $\tilde{m}_{00,q}$ of the passive form of the component 'q' with largest diattenuation (when $D \geq P$) or largest polarizance (when $P \geq D$).

7.6.1 Application of the Characteristic Decomposition to an Experimental Example

As an example of the application of the characteristic decomposition, consider the following experimental normalized Mueller matrix:

$$\hat{\mathbf{M}}_s = \begin{pmatrix} 1 & 0.0607 & -0.0060 & 0.0005 \\ 0.0418 & 0.2775 & -0.0230 & 0 \\ -0.0015 & 0.0039 & -0.2756 & 0.0329 \\ -0.0017 & -0.0120 & -0.0199 & -0.2013 \end{pmatrix}$$

corresponding to a reflection measurement of an irradiated pig skin (Boulvert et al. 2005). Its covariance matrix, readily found to be

$$\hat{\mathbf{H}}_s = \begin{pmatrix} 0.3450 & -0.0072+i0.0001 & 0.0006+i0.0034 & -0.1192+i0.0132 \\ -0.0072-i0.0001 & 0.1759 & -0.0186-i0.0032 & -0.0014-i0.0026 \\ 0.0006-i0.0034 & -0.0186+i0.0032 & 0.1854 & 0.0043+i0.0001 \\ -0.1192-i0.0132 & -0.0014+i0.0026 & 0.0043-i0.0001 & 0.2938 \end{pmatrix}$$

has the following eigenvalues: 0.4422, 0.2050, 0.1919 and 0.1608. The characteristic decomposition of $\hat{\mathbf{M}}_s$, determined from the eigenvalues and eigenvectors of $\hat{\mathbf{H}}_s$, can be put into the form

$$\hat{\mathbf{M}}_s = p_{J0}\hat{\mathbf{M}}_{J0} + p_1\hat{\mathbf{M}}_1 + p_2\hat{\mathbf{M}}_2 + p_{\Delta 0}\hat{\mathbf{M}}_{\Delta 0}$$

where the weights, calculated from the indices of purity of $\hat{\mathbf{M}}_s$, are 0.5363, 0.0296, 0.0703 and 0.3637. The characteristic pure component (coinciding with the largest-eigenvalue component of the spectral decomposition) is

$$\hat{\mathbf{M}}_{J0} = \begin{pmatrix} 1 & 0.2092 & -0.0177 & 0.0038 \\ 0.2096 & 0.9988 & -0.0393 & -0.0256 \\ 0.0106 & -0.0332 & -0.9711 & 0.1092 \\ -0.0082 & -0.0308 & -0.1081 & -0.9713 \end{pmatrix}$$

the two depolarizing components are

$$\hat{\mathbf{M}}_1 = \begin{pmatrix} 1 & 0.1145 & 0.0814 & -0.0397 \\ 0.0084 & 0.3819 & -0.4942 & 0.0888 \\ -0.0091 & 0.4586 & -0.5961 & 0.1012 \\ -0.0257 & -0.0376 & 0.0495 & -0.0068 \end{pmatrix} \quad \hat{\mathbf{M}}_2 = \begin{pmatrix} 1 & 0.0866 & -0.0422 & 0.0068 \\ -0.0871 & 0.3282 & -0.0079 & 0.0400 \\ -0.0406 & -0.0026 & -0.3182 & 0.0467 \\ 0.0099 & -0.0397 & 0.0475 & 0.3131 \end{pmatrix}$$

and $\hat{\mathbf{M}}_{\Delta 0} = \mathrm{diag}(1,0,0,0)$ is the normalized perfect depolarizer. The inspection of the weights listed above shows that the second and the third, $p_1 = 0.0296$ and $p_2 = 0.0703$, corresponding to the respective depolarizing components, are significantly smaller than the first and the last one, $p_{J0} = 0.5363$ and $p_{\Delta 0} = 0.3637$, of comparable magnitudes. Consequently, the characteristic decomposition indicates that we are essentially witnessing (strong) isotropic depolarization, that is, the original matrix $\mathbf{M}_s$ can be represented to a good approximation as the weighted sum of the pure characteristic component and the perfect depolarizer $\hat{\mathbf{M}}_{\Delta 0}$,

$$\hat{\mathbf{M}}_s \approx p_{J0}\hat{\mathbf{M}}_{J0} + p_{\Delta 0}\hat{\mathbf{M}}_{\Delta 0}$$

If we assume that retardance, diattenuation and polarizance of the medium are contained in the characteristic component $\hat{\mathbf{M}}_{J0}$ and polar-decompose it in the order $\mathbf{M}_R\hat{\mathbf{M}}_D$, we get for the diattenuation and retardance magnitudes $D = 0.2100$ and $R = 3.032$, whereas their respective unit vectors are $\hat{\mathbf{D}} = (0.9963, -0.0842, 0.0181)^T$ and $\hat{\mathbf{R}} = (0.9997, -0.0184, -0.0143)^T$. We see that the sample exhibits moderate diattenuation while its retardance value, close to π, is approximately that of a half-wave plate. Next, the two direction vectors show that diattenuation and retardance are both almost perfectly oriented along the S_1 axis of the Poincaré sphere, indicating that they are of linear polarimetric nature. The polarimetric behavior of the irradiated pig skin can thus be described as that of an imperfect half-wave plate with moderate diattenuation properties, featuring strong isotropic depolarization.

7.7 Polarimetric Subtraction of Mueller Matrices

The potential applications of the polarimetric subtraction inspired early efforts to address its mathematical formulation in terms of Mueller matrices. Successive approaches to this issue have been presented within the scope of polarization optics (Gil and Correas 2003; Correas et al. 2003; Gil 2007; Gil and San José 2013), including experimental validations (Foldyna et al. 2009; Ossikovski et al. 2014). Furthermore, several procedures for the polarimetric subtraction have also been dealt with in works relative to radar polarimetry. The subtraction of a volume scattering matrix from measured data was proposed by Freeman and Durden (1998); their method was later extended by Yamaguchi et al. (2005) to consider several models for volume scattering. Nevertheless, these works proposed subtraction procedures disregarding possible negative eigenvalues of the remainder covariance matrix. The necessity of consistent subtractions yielding a positive semi-definite remainder covariance matrix was pointed out by van Zyl (1993) and by van Zyl et al. (2011).

7.7.1 Condition for Polarimetric Subtractability

From the arbitrary decomposition approach dealt with in Section 7.3, it is straightforward to deduce that, given two Mueller matrices $\mathbf{M}$ and $\mathbf{M}_m$ with respective associated covariance matrices $\mathbf{H}$ and $\mathbf{H}_m$, the decomposition (Figure 7.3)

$$m_{00}\hat{\mathbf{M}} = k\left(m_{00,m}\hat{\mathbf{M}}_m\right) + (1-k)\left(m_{00,X}\hat{\mathbf{M}}_X\right) \qquad (0 < k < 1) \tag{7.27}$$

where $\mathbf{M}_X \equiv m_{00,X}\hat{\mathbf{M}}_X$ is a Mueller matrix, is physically realizable (leaving aside passivity contraints) if and only if the image subspace of $\mathbf{H}_m$ is contained in the image subspace of $\mathbf{H}$, i.e. (Gil and San José 2013),

$$\operatorname{im}\mathbf{H}_m \subseteq \operatorname{im}\mathbf{H} \tag{7.28}$$

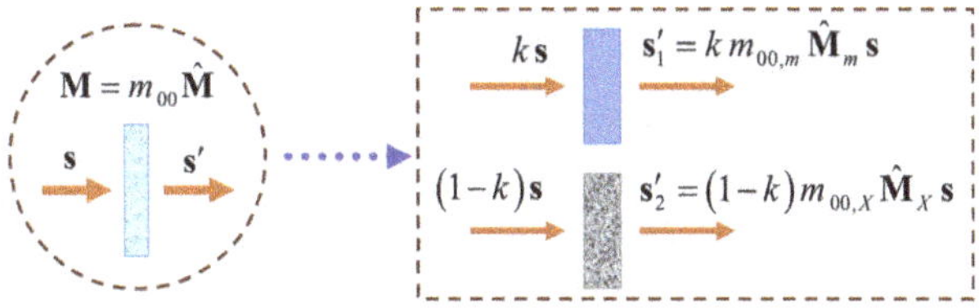

FIGURE 7.3
Provided that the condition $\operatorname{rank}(\mathbf{H}_m + \mathbf{H}) = \operatorname{rank}\mathbf{H}$ is satisfied, the parallel decomposition $m_{00}\hat{\mathbf{M}} = k\left(m_{00,m}\hat{\mathbf{M}}_m\right) + (1-k)\left(m_{00,X}\hat{\mathbf{M}}_X\right)$ is physically realizable.

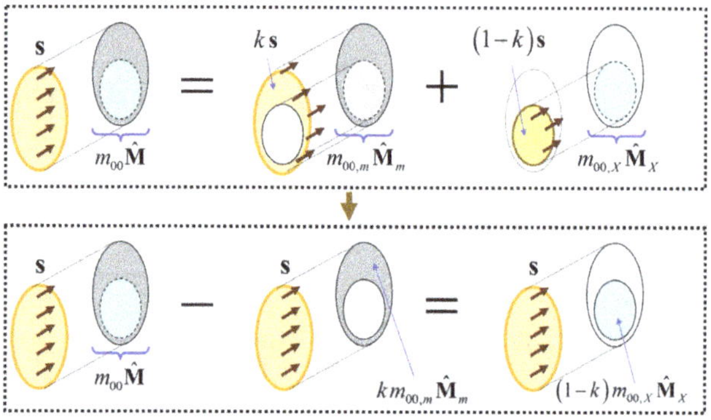

FIGURE 7.4
Polarimetric subtraction of a Mueller matrix $\mathbf{M}_m$ from the whole Mueller matrix $\mathbf{M}$. The procedure provides the test for checking if the subtraction is physically realizable or not and, in the affirmative case, it gives the value of the relative cross section k of $\mathbf{M}_m$, as well as of the Mueller matrix $\mathbf{M}_X$ of the remainder. (Gil and San José 2013).

or, equivalently

$$\operatorname{rank}(\mathbf{H}_m + \mathbf{H}) = \operatorname{rank}\mathbf{H} \tag{7.29}$$

In other words, the eigenvectors of $\mathbf{H}_m$ with nonzero eigenvalues can be written as linear combinations of the eigenvectors of $\mathbf{H}$ with nonzero eigenvalues. Let us denote $\operatorname{rank}\mathbf{H} \equiv r$ and $\operatorname{rank}\mathbf{H}_m \equiv m$, and observe that when $r = 4$, then Eq. (7.29) is satisfied. Furthermore, condition (7.29) implies that the inequality $m \le r$ must necessarily be satisfied.

Provided that the subtractability condition (7.29) holds, the coefficient k of $\mathbf{M}_m$, representing the relative cross section of the medium described by $\mathbf{M}_m$ (with respect to the whole medium described by $\mathbf{M}$), can be calculated through the procedure indicated in the next subsections (Gil and San José 2013). The difference Mueller matrix $\mathbf{M}_X$ is obtained as (see Figure 7.4)

$$\mathbf{M}_X = \frac{m_{00}\hat{\mathbf{M}} - k\,m_{00,m}\hat{\mathbf{M}}_m}{1-k} \tag{7.30}$$

7.7.2 Polarimetric Subtraction of a Pure Component

Consider now the particular, but important, case where the subtrahend is a pure component with associated pure Mueller matrix $\mathbf{M}_{J0}(\mathbf{H}_{J0})$. Then, providing the test of subtractability (7.29) is passed, i.e., $\operatorname{rank}(\mathbf{H}_{J0} + \mathbf{H}) = \operatorname{rank}\mathbf{H} \equiv r$, the coefficient k_0 and the Mueller matrix $\mathbf{M}_X$ of the difference (or remainder) system are calculated from the following expressions (Gil and San José 2013):

$$k_0 = \frac{1}{m_{00,0}\displaystyle\sum_{j=0}^{r-1}\frac{1}{\lambda_j}\left|\left(\mathbf{U}^{\dagger}\mathbf{w}_0\right)_j\right|^2} = \frac{1}{m_{00,0}\left(\hat{\mathbf{w}}_0^{\dagger}\mathbf{H}^{-}\hat{\mathbf{w}}_0\right)} \qquad \mathbf{M}_X = \frac{m_{00}\hat{\mathbf{M}} - k_0\,m_{00,0}\hat{\mathbf{M}}_{J0}}{1-k_0} \tag{7.31}$$

where $\mathbf{U}$ is the unitary matrix that diagonalizes $\mathbf{H}$, $\hat{\mathbf{w}}_0$ is the unit covariance vector of the pure subtrahend $\mathbf{H}_{J0} = m_{00,0}(\hat{\mathbf{w}}_0 \otimes \hat{\mathbf{w}}_0^{\dagger})$, λ_j are the eigenvalues of $\mathbf{H}$, $\mathbf{H}^{-}$ is the pseudoinverse of $\mathbf{H}$ (defined as $\mathbf{H}^{-} = \mathbf{U}\mathbf{D}^{-}\mathbf{U}^{\dagger}$, $\mathbf{D}^{-}$ being the diagonal matrix whose r first diagonal elements are $1/\lambda_0, 1/\lambda_1, ..., 1/\lambda_{r-1}$ and the last $4-r$ elements are zero (Gil and San José 2019)) and $\mathbf{M} \equiv m_{00}\hat{\mathbf{M}}$ is the Mueller matrix of the minuend, whose associated covariance matrix is $\mathbf{H}$.

Observe that the subtraction of $\mathbf{M}_{J0}$ from $\mathbf{M}$ is also physically realizable for any coefficient $k < k_0$. When k_0 is applied, then the subtraction produces a reduction of the rank

$$\operatorname{rank}\mathbf{H}_X = \operatorname{rank}\mathbf{H} - \operatorname{rank}\mathbf{H}_{J0} = r-1 \tag{7.32}$$

while, for values $k < k_0$

$$\mathbf{H}'_X = \frac{1}{1-k}\left(\mathbf{H} - k\mathbf{H}_{J0}\right) = \frac{1}{1-k}\left[\mathbf{H} + (k_0 - k)\mathbf{H}_{J0}\right] - \frac{k_0}{1-k}\mathbf{H}_{J0} \tag{7.33}$$

so that $\operatorname{rank}\mathbf{H}'_X = \operatorname{rank}\mathbf{H} = r$, which means that a relative portion $k_0 - k$ of $\mathbf{M}_{J0}$ remains unsubtracted into the remainder matrix.

The iterative procedure for the arbitrary decomposition shown in Section 7.3 constitutes, by itself, a method for subtracting pure Mueller matrices from a given depolarizing Mueller matrix.

7.7.3 Polarimetric Subtraction of a Pure Component from a Rank 2 Mixture

One of the relatively frequently encountered situations in experimental polarimetry is the need for subtraction of a pure component, known from experiment or simulation, from a rank-two measured depolarizing Mueller matrix. This case occurs, for instance, in measurements whereby the spot of the probing light covers simultaneously two areas, the area of interest that we shall call generically *the pattern* and the surrounding area, or *the substrate*. The problem to solve in such a *finite-spot-size* experiment is therefore to subtract the substrate response (determined from an additional measurement on a pattern-free area) from the pattern–substrate mixture in order to get the pattern response alone. Due to the specific interest of this situation, an explicit derivation of the subtraction procedure is developed in what follows. It is obviously consistent with the general method dealt with in Section 7.7.2.

Let the Mueller matrix of the mixture be $\mathbf{M}$ and that of the substrate $\mathbf{M}_s$. Then $\operatorname{rank}[\mathbf{H}(\mathbf{M})] = 2$ whereas $\operatorname{rank}[\mathbf{H}(\mathbf{M}_s)] = 1$ and one looks for a parameter p such that the pattern matrix $(1-p)\hat{\mathbf{M}}_p = \hat{\mathbf{M}} - p\hat{\mathbf{M}}_s$ be pure (or nondepolarizing),

$$\operatorname{rank}\left[\mathbf{H}\left(\mathbf{M}_p\right)\right] = \operatorname{rank}\left[\mathbf{H}\left(\mathbf{M} - p\mathbf{M}_s\right)\right] = 1 \tag{7.34}$$

The parameter p can be found from the general expression (7.31); however, in this specific case, explicit formulas can be obtained. If we denote by h_{ij} and s_{ij} the elements of $\mathbf{H}(\hat{\mathbf{M}})$ and $\mathbf{H}(\hat{\mathbf{M}}_s)$, then the above relation can be explicitly written,

$$\operatorname{rank}\left[\mathbf{H}\left(\hat{\mathbf{M}} - p\hat{\mathbf{M}}_s\right)\right] = \operatorname{rank}\begin{pmatrix} h_{00} - p s_{00} & h_{10}^* - p s_{10}^* & h_{20}^* - p s_{20}^* & h_{30}^* - p s_{30}^* \\ h_{10} - p s_{10} & h_{11} - p s_{11} & h_{21}^* - p s_{21}^* & h_{31}^* - p s_{31}^* \\ h_{20} - p s_{20} & h_{21} - p s_{21} & h_{22} - p s_{22} & h_{32}^* - p s_{32}^* \\ h_{30} - p s_{30} & h_{31} - p s_{31} & h_{32} - p s_{32} & h_{33} - p s_{33} \end{pmatrix} = 1 \tag{7.35}$$

In order for the above to hold, it is necessary that all minors of order higher than one of the difference matrix vanish. For instance, this condition expressed for the upper principal 2×2 minor reads,

$$\begin{aligned} h_{00}h_{11} - h_{01}h_{10} - p\left(h_{11}s_{00} + h_{00}s_{11} - h_{01}s_{10} - h_{10}s_{01}\right) + p^2\left(s_{00}s_{11} - s_{01}s_{10}\right) = \\ = h_{00}h_{11} - h_{01}h_{10} - p\left(h_{11}s_{00} + h_{00}s_{11} - h_{01}s_{10} - h_{10}s_{01}\right) = 0 \end{aligned} \tag{7.36}$$

where it has been taken into account the fact that $s_{00}s_{11} - s_{01}s_{10} = 0$ since $\mathrm{rank}[\mathbf{H}(\mathbf{M}_s)] = 1$. The parameter p is uniquely determined from the above equation (or from any other analogous equation resulting from zeroing one of the remaining minor determinants) under the two conditions $\mathrm{rank}[\mathbf{H}(\mathbf{M})] = 2$ whereas $\mathrm{rank}[\mathbf{H}(\mathbf{M}_s)] = 1$. After having determined p, the normalized pattern Mueller matrix $\hat{\mathbf{M}}_p$ is finally obtained from $\hat{\mathbf{M}}_p = (\hat{\mathbf{M}} - p\hat{\mathbf{M}}_s)/(1-p)$.

In practical situations, the parameter m_{00} of $\mathbf{M}$ is known either from an absolute polarimetric measurement, or from the specific information on the sample that is available to the experimentalist; similarly, the known nature of the substrate usually provides an estimate for the MIC $m_{00,s}$ of $\mathbf{M}_s$. Thus, when appropriate, the above expression for the calculation of $\hat{\mathbf{M}}_p$ in terms of $\hat{\mathbf{M}}$, $\hat{\mathbf{M}}_s$ and p, can be transformed into

$$\begin{gathered} m_{00,p}\hat{\mathbf{M}}_p = \left(m_{00}\hat{\mathbf{M}} - k\,m_{00,s}\hat{\mathbf{M}}_s\right)\Big/(1-k) \\ k = p\,m_{00}/m_{00,s} \qquad m_{00,p} = \left(m_{00} - k m_{00,s}\right)/(1-k) \end{gathered} \tag{7.37}$$

where the effective coefficient k and the unknown $m_{00,p}$ are readily obtained from p, m_{00} and $m_{00,s}$.

As indicated above, the matrix $\mathbf{M}$ is commonly determined experimentally and, typically, $\mathrm{rank}[\mathbf{H}(\mathbf{M})] > 2$; one should then favor the determination of p from the minor of $\mathbf{H}\left(\hat{\mathbf{M}} - p\hat{\mathbf{M}}_s\right)$ featuring largest elements for better noise resilience. Similarly, one typically has $\mathrm{rank}[\mathbf{H}(\mathbf{M}_s)] > 1$ so, the substrate matrix $\mathbf{M}_s$ must be filtered to a nondepolarizing estimate $\mathbf{M}_{se}$ such that $\mathrm{rank}[\mathbf{H}(\mathbf{M}_{se})] = 1$ strictly (see Chapter 11) before applying the above relations. A practical example for the application of this procedure can be found in Foldyna et al. 2009.

Another practically important variant of the above situation occurs when the substrate matrix $\mathbf{M}_s$ is not known (because its independent measurement is impossible to perform, for instance), but instead some symmetries obeyed by the elements of $\mathbf{M}_s$ are known. Thus, most often, the substrate represents a plane isotropic surface and $\mathbf{M}_s$ is consequently of block-diagonal form with well-known symmetries ($\mathbf{M}_{s00} = \mathbf{M}_{s11}$, $\mathbf{M}_{s22} = \mathbf{M}_{s33}$, $\mathbf{M}_{s01} = \mathbf{M}_{s10}$ and $\mathbf{M}_{s23} = -\mathbf{M}_{s32}$). On the other hand, since $\mathrm{rank}[\mathbf{H}(\mathbf{M}_p)] = \mathrm{rank}[\mathbf{H}(\mathbf{M}_s)] = 1$, the pattern and substrate covariance matrices are projectors and one can consequently write their respective elements in the form $a_i a_j^*$ and $b_i b_j^*$, where the pair of unnormalized covariance vectors $\mathbf{a}$ and $\mathbf{b}$ with respective components a_k and b_k $\mathbf{M}_s$ are to be determined by the procedure. Further, since $(k = 0,1,2,3)$ is assumed block-diagonal with special symmetries, its covariance matrix $\mathbf{H}(\mathbf{M}_s)$ has only four non-vanishing elements and $b_1 = b_2 = 0$. The covariance matrix $\mathbf{H}(\mathbf{M})$ of the pattern–substrate mixture, expressed through the elements of $\mathbf{H}(\mathbf{M}_p)$ and $\mathbf{H}(\mathbf{M}_s)$, is

$$\mathbf{H}(\mathbf{M})=\mathbf{H}(\mathbf{M}_p+\mathbf{M}_s)=\begin{pmatrix} |a_0|^2+|b_0|^2 & a_0 a_1^* & a_0 a_2^* & a_0 a_3^*+b_0 b_3^* \\ a_1 a_0^* & |a_1|^2 & a_1 a_2^* & a_1 a_3^* \\ a_2 a_0^* & a_2 a_1^* & |a_2|^2 & a_2 a_3^* \\ a_3 a_0^*+b_3 b_0^* & a_3 a_1^* & a_3 a_2^* & |a_3|^2+|b_3|^2 \end{pmatrix} \tag{7.38}$$

since the equality $\mathbf{M}=\mathbf{M}_p+\mathbf{M}_s$ holds for unnormalized Mueller matrices (where $\mathbf{M}_p \equiv m_{00}(1-p)\hat{\mathbf{M}}_p$ and $\mathbf{M}_s \equiv m_{00}\, p\hat{\mathbf{M}}_s$, with $\hat{\mathbf{M}}_p$, $\hat{\mathbf{M}}_s$ and p being unknown). The above matrix can be directly identified with the elements h_{ij} of $\mathbf{H}(\mathbf{M})$ determined from the experimentally obtained $\mathbf{M}$. Thus, we have a system of ten equations for the determination of a_k and b_k.

$$\begin{gathered} |a_0|^2+|b_0|^2=h_{00} \qquad |a_1|^2=h_{11} \qquad |a_2|^2=h_{22} \qquad |a_3|^2+|b_3|^2=h_{33} \\ a_1 a_0^*=h_{10} \quad a_2 a_0^*=h_{20} \quad a_3 a_0^*+b_3 b_0^*=h_{30} \quad a_2 a_1^*=h_{21} \quad a_3 a_1^*=h_{31} \quad a_3 a_2^*=h_{32} \end{gathered} \tag{7.39}$$

Since both vectors **a** and **b** are defined within a phase factor only, we can assume without restricting the generality that a_1 and b_0 are real. With this assumption it is easy to find that

$$\begin{gathered} a_0=h_{10}^*/\sqrt{h_{11}}=h_{01}/\sqrt{h_{11}} \qquad a_1=\sqrt{h_{11}} \qquad a_2=h_{21}/\sqrt{h_{11}} \qquad a_3=h_{31}/\sqrt{h_{11}} \\ b_0=\sqrt{\left(h_{00}h_{11}-|h_{10}|^2\right)/h_{11}} \qquad b_3=\left(h_{30}h_{11}-h_{31}h_{10}\right)/\sqrt{h_{11}\left(h_{00}h_{11}-|h_{10}|^2\right)} \end{gathered} \tag{7.40}$$

From a_k and b_k one readily finds $\mathbf{H}(\mathbf{M}_p)$ and $\mathbf{H}(\mathbf{M}_s)$ and eventually, $\hat{\mathbf{M}}_p$ and $\hat{\mathbf{M}}_s$ which solves the problem of determining the two pure polarimetric components from the measurement of their mixture. If the value of the parameter p is sought too, then it is enough to consider the relation $\hat{\mathbf{M}}=(1-p)\hat{\mathbf{M}}_p+p\hat{\mathbf{M}}_s$ in terms of the normalized Mueller matrices. Indeed, since this normalized superposition relation is proportional to the unnormalized one, $\mathbf{M}=\mathbf{M}_p+\mathbf{M}_s$, the parameter p is simply given by the ratio of the first elements of $\mathbf{M}_s$ and $\mathbf{M}$. On the other hand, the trace of the covariance matrix $\mathbf{H}(\mathbf{M})$ being equal to the first element of its associated Mueller matrix $\mathbf{M}$, the ratio in question equals the ratio of $\mathrm{tr}[\mathbf{H}(\mathbf{M}_s)]$ to $\mathrm{tr}[\mathbf{H}(\mathbf{M})]$. With the help of Eq. (7.38) one immediately obtains

$$p=\frac{\mathrm{tr}\left[\mathbf{H}(\mathbf{M}_s)\right]}{\mathrm{tr}\left[\mathbf{H}(\mathbf{M})\right]}=\frac{|\mathbf{b}|^2}{|\mathbf{a}|^2+|\mathbf{b}|^2}=\frac{|b_0|^2+|b_3|^2}{|a_0|^2+|a_1|^2+|a_2|^2+|a_3|^2+|b_0|^2+|b_3|^2} \tag{7.41}$$

where the components a_k and b_k of **a** and **b** are given by Eqs. (7.40)

As previously, the mixture matrix $\mathbf{M}$ is an experimentally obtained one and, consequently, $\mathrm{rank}[\mathbf{H}(\mathbf{M})]>2$ because of the inevitable presence of measurement noise. Since the set of ten equations is overdetermined for the six unknowns a_k and b_k, different subsets of equations will lead to different solutions. To avoid this, one can select the subset of equations with largest coefficients for better noise resilience. For instance, it is easy to see that the choice of real a_1 and b_0 we have made leads to an optimal set when $||a_1|>|a_2|$ and $|b_0|>|b_3|$. Another option is to reduce the rank of $\mathbf{M}$ to two by rank-reduction filtering as explained in Section 11.2. An example of the practical application of this *unmixing* method (using a different, but equivalent, algebraic expression in terms of the arbitrary decomposition of a Mueller matrix) is reported in (Ossikovski et al. 2014).

7.7.4 Polarimetric Subtraction of a Depolarizing Component

In the above sections, we have considered the case where the subtrahend is a pure Mueller matrix, so that the solution is directly given by Eqs. (7.31), provided the subtractability condition (7.29) is satisfied. Let us now consider the general problem of subtracting a depolarizing Mueller matrix $\mathbf{M}_m(\mathbf{H}_m) \equiv m_{00,m}\hat{\mathbf{M}}_m$ (where $m \equiv \operatorname{rank}\mathbf{H}_m$, with $1 < m$), from another depolarizing Mueller matrix $\mathbf{M}(\mathbf{H}) \equiv m_{00}\hat{\mathbf{M}}$ (with $r \equiv \operatorname{rank}\mathbf{H}$), resulting in a difference matrix $\mathbf{M}_X = (\mathbf{M} - k\mathbf{M}_m)/(1-k)$ (with $0 < k < 1$), that satisfies the covariance conditions to be a Mueller matrix.

As shown in Section 7.7.1, the subtraction is physically consistent if and only if $\operatorname{rank}(\mathbf{H}_m + \mathbf{H}) = \operatorname{rank}\mathbf{H}$. Let us tackle the problem of calculating the coefficient k and the remainder Mueller matrix $\mathbf{M}_X$ when such a condition of subtractability is satisfied. As it has been proven, the solution is achieved through the following procedure (Gil and San José 2013):

Consider the diagonalization of $\mathbf{H}$

$$\mathbf{H} = m_{00}\,\mathbf{U}\,\mathbf{D}\,\mathbf{U}^{\dagger} \qquad \mathbf{D} \equiv \operatorname{diag}\left(\hat{\lambda}_0, \hat{\lambda}_1, \hat{\lambda}_2, \hat{\lambda}_3\right) \qquad \left[\hat{\lambda}_i = \lambda_i / m_{00} \qquad m_{00} = \operatorname{tr}\mathbf{H}\right] \tag{7.42}$$

where the unitary matrix $\mathbf{U}$ is taken so that the eigenvalues are ordered as $\lambda_0 \geq \lambda_1 \geq \lambda_2 \geq \lambda_3$. Thus $\operatorname{rank}\mathbf{D} = \operatorname{rank}\mathbf{H} = r$, in such a manner that the last $4-r$ diagonal elements of $\mathbf{D}$ are zero. Now, by recalling the definition of the pseudoinverse of $\mathbf{D}$

$$\mathbf{D}^{-} \equiv \operatorname{diag}\left(1/\hat{\lambda}_0, \ldots, 1/\hat{\lambda}_{r-1}, 0, \ldots, 0\right) \tag{7.43}$$

we define the positive semidefinite Hermitian matrix

$$\mathbf{H}'_m \equiv \sqrt{\mathbf{D}^{-}}\,\mathbf{U}^{\dagger}\,\mathbf{H}_m\,\mathbf{U}\sqrt{\mathbf{D}^{-}} \tag{7.44}$$

whose rank is equal to the rank of the covariance matrix $\mathbf{H}_m = m_{00,m}\hat{\mathbf{H}}_m$ of the subtrahend, $\operatorname{rank}\mathbf{H}'_m = \operatorname{rank}\mathbf{H}_m = m$. $\mathbf{H}'_m$ can be diagonalized as

$$\mathbf{H}'_m = m_{00,m}\,\mathbf{U}'\,\mathbf{D}'_m\,\mathbf{U}'^{\dagger} \tag{7.45}$$

where the columns of the unitary matrix $\mathbf{U}'$ are the eigenvectors of $\mathbf{H}'_m$ and $\mathbf{D}'_m \equiv \operatorname{diag}(\hat{\lambda}'_0, \ldots, \hat{\lambda}'_{m-1}, 0, \ldots 0)$ is the diagonal matrix whose elements $\hat{\lambda}'_i$ (with $\hat{\lambda}'_0 \geq \hat{\lambda}'_1 \ldots \geq \hat{\lambda}'_{m-1} > 0$) are the ordered eigenvalues of $\hat{\mathbf{H}}'_m$. Then the difference covariance matrix is given by (Gil and San José 2013)

$$\begin{aligned}\mathbf{H}_X &= \mathbf{U}\sqrt{\mathbf{D}}\left[m_{00}\mathbf{I}_r - k\mathbf{H}'_m\right]\sqrt{\mathbf{D}}\,\mathbf{U}^{\dagger} = \mathbf{U}\sqrt{\mathbf{D}}\left[m_{00}\,\mathbf{U}'\mathbf{I}_r\,\mathbf{U}'^{\dagger} - k\,m_{00,m}\mathbf{U}'\mathbf{D}'_m\,\mathbf{U}'^{\dagger}\right]\sqrt{\mathbf{D}}\,\mathbf{U}^{\dagger} \\ &= \mathbf{U}\sqrt{\mathbf{D}}\,\mathbf{U}'\mathbf{D}_{r-\xi}\,\mathbf{U}'^{\dagger}\sqrt{\mathbf{D}}\,\mathbf{U}^{\dagger} \qquad \left[\mathbf{D}_{r-\xi} \equiv m_{00}\,\mathbf{I}_r - k\,m_{00,m}\,\mathbf{D}'_m\right]\end{aligned} \tag{7.46}$$

where $\mathbf{I}_r$ is the diagonal matrix whose first r elements are 1 and whose last $4-r$ elements are zero, and the diagonal matrix $\mathbf{D}_{r-\xi}$ has the following structure:

$$\begin{aligned}
&\left(\mathbf{D}_{r-\xi}\right)_i = m_{00}\left(1 - k m_{00,m}\hat{\lambda}' \big/ m_{00i}\right) && (i = 0,\ldots,m-1) \\
&\left(\mathbf{D}_{r-\xi}\right)_i = m_{00} && (i = m,\ldots,r-1) \\
&\left(\mathbf{D}_{r-\xi}\right)_i = 0 && (i \geq r)
\end{aligned} \tag{7.47}$$

Moreover, since $\mathbf{U}$ is unitary (hence $\mathbf{U}$ and $\mathbf{U}^{\dagger}$ are non-singular), and due to the particular forms of $\sqrt{\mathbf{D}}$ and $\mathbf{U}'$,

$$\operatorname{rank}\mathbf{H}_X = \operatorname{rank}\left(\mathbf{D}_{r-\xi}\right) \tag{7.48}$$

Therefore, provided the coefficient k satisfies $k \leq m_{00}/(m_{00,m}\hat{\lambda}'_0)$, $\mathbf{D}_{r-\xi}$ has non-negative elements. Thus, by choosing the following value for the coefficient k:

$$k = m_{00}\Big/\left(m_{00,m}\hat{\lambda}'_0\right) \tag{7.49}$$

the rank of the remainder covariance matrix $\mathbf{H}_X$ has the following value:

$$\operatorname{rank}\mathbf{H}_X = r - \xi \tag{7.50}$$

where ξ is the multiplicity of the largest eigenvalue $\hat{\lambda}'_0$ of $\hat{\mathbf{H}}'_m$.

When all the eigenvalues of $\hat{\mathbf{H}}'_m$ are different, then $\xi = 1$. In fact, ξ is precisely the maximum number of pure components of an achievable arbitrary decomposition of $\mathbf{M}_m$ that are also components (with the same relative proportion of the respective coefficients) of an achievable arbitrary decomposition of $\mathbf{M}$. Consequntly, the algebraic conditions to perform a polarimetric subtraction with $\xi > 1$ are strongly restrictive in comparison to the case where $\xi = 1$ (Gil and San José 2013).

Obviously, when $m = 1$, the above procedure is consistent with those described in Sections 7.7.2 and 7.7.3 for the common case of subtracting a pure component. In fact, when $m = 1$, then $\xi = 1$ and

$$(m = 1) \qquad k = \frac{m_{00}}{m_{00,m}\hat{\lambda}'_0} = \frac{1}{m_{00,m}\left(\hat{\mathbf{w}}_m^{\dagger}\mathbf{H}^{-}\hat{\mathbf{w}}_m\right)} \tag{7.51}$$

where $\mathbf{w}_m = \sqrt{m_{00,m}}\,\hat{\mathbf{w}}_m$ is the covariance vector associated with $\mathbf{H}_m$ and $\mathbf{H}^{-} = \mathbf{U}\mathbf{D}^{-}\mathbf{U}^{\dagger}$. When $m = k$ and the multiplicity of λ'_0 is equal to r, then the minuend and the subtrahend matrices are equal ($\mathbf{M}_m = \mathbf{M}$). Furthermore, it should be noted that the subtraction is also physically realizable for values $k < m_{00}\big/(m_{00,m}\hat{\lambda}'_0)$, in which case a portion of $\mathbf{M}_m$ remains embedded in $\mathbf{M}$ after the subtraction, and consequently $\operatorname{rank}\mathbf{H}_X = \operatorname{rank}\mathbf{H} = r$.

In summary, providing $\operatorname{rank}(\mathbf{H}_m + \mathbf{H}) = \operatorname{rank}\mathbf{H}$, the maximum coefficient for the polarimetric subtraction of a Mueller matrix $\mathbf{M}_m(\mathbf{H}_m) \equiv m_{00,m}\hat{\mathbf{M}}_m$ (in general depolarizing) from a depolarizing Mueller matrix $\mathbf{M}(\mathbf{H}) \equiv m_{00}\,\hat{\mathbf{M}}$ is given by $k = m_{00} / (m_{00,m}\,\hat{\lambda}'_0)$, where λ'_0 is the largest eigenvalue of the covariance matrix $\mathbf{H}'_m$ defined in Eq. (7.44) from $\mathbf{H}_m$ and $\mathbf{H}$.

The difference, or remainder, Mueller matrix is then calculated as

$$\mathbf{M}_X = (\mathbf{M} - k\,\mathbf{M}_m)/(1-k) = \left(m_{00}\,\hat{\mathbf{M}} - k\,m_{00,m}\,\hat{\mathbf{M}}_m\right)/(1-k) \tag{7.52}$$

with $k \le m_{00} / (m_{00,m}\,\hat{\lambda}'_0)$, so that for $k = m_{00} / (m_{00,m}\,\hat{\lambda}'_0)$ the subtraction is complete, i.e., $\operatorname{rank}\mathbf{H}_X = \operatorname{rank}\mathbf{H} - \xi$, where ξ is the multiplicity of the eigenvalue λ'_0.

8

Serial Decompositions of Depolarizing Mueller Matrices

8.1 Introduction

A powerful tool for the analysis of experimental Mueller matrices is provided by their representation through serial combinations of simple components as retarders, diattenuators and canonical depolarizers. In this manner, the sample is characterized polarimetrically by means of the properties of the individual components. The main methods for the serial decomposition of a pure (or nondepolarizing) Mueller matrix, like the polar decomposition and the dual linear retarder transformation, have already been presented in Chapter 4. Although certain serial transformations of depolarizing Mueller matrices have been introduced in Chapters 5 and 6, the actual chapter presents a comprehensive analysis of the serial decompositions of depolarizing Mueller matrices. In accordance with the guidelines followed throughout this book, the purpose of this chapter is to collect, review and compare the main algebraic procedures, while also including certain original results wherever appropriate.

Serial decompositions consist of representing a general Mueller matrix as a product of particular Mueller matrices. The physical meaning of serial decompositions is that the whole system is considered as a cascade of polarization components, so that the incoming electromagnetic beam interacts sequentially with them. This arrangement of the components constitutes the *serial equivalent system* (Figure 8.1).

At a first thought, it is tempting to apply the polar decomposition or the singular value decomposition to a depolarizing Mueller matrix, but it easily turns out that, in general, such decompositions lead to unphysical components (Gil 2007). Nevertheless, a number of decompositions of a general Mueller matrix into specific products of Mueller matrices associated with certain simple components have been developed and are studied in the next sections.

Obviously, the representation of a serial combination of media through the product of their respective Mueller matrices is valid inasmuch as the media exhibit homogeneous polarimetric behavior over the spot size of the probing electromagnetic wave. The limits to the validity of the use of the Mueller matrix product to represent serial compositions, resulting from the inhomogeneous behavior of the corresponding serial components in experimental devices, have been studied by Chironi and Iemmi (2021).

DOI: 10.1201/9780367815578-8

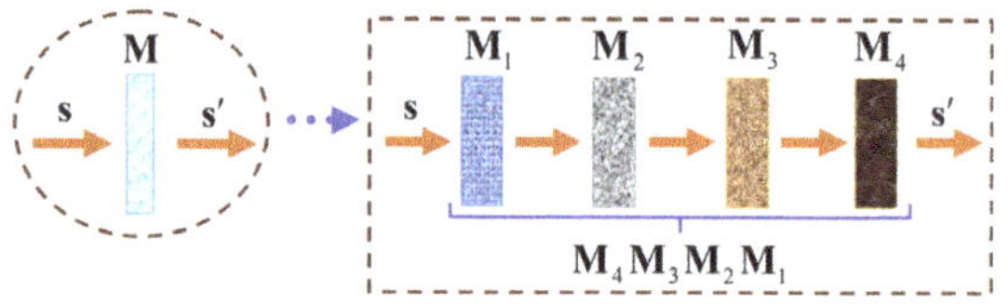

FIGURE 8.1
Serial decomposition of a Mueller matrix **M**. The electromagnetic beam passes sequentially through the components of the equivalent system, so that $\mathbf{M} = \mathbf{M}_4\,\mathbf{M}_3\,\mathbf{M}_2\,\mathbf{M}_1$.

8.2 Generalized Polar Decomposition

Lu and Chipman (1996) developed a useful procedure to decompose a Mueller matrix into a serial combination of a diattenuator, a retarder and an enpolarizing depolarizer without diattenuation. This *generalized polar decomposition* (also called *Lu-Chipman decomposition* or *forward decomposition*) of a Mueller matrix is formulated as

$$\mathbf{M} \equiv m_{00}\begin{pmatrix} 1 & \mathbf{D}^T \\ \mathbf{P} & \mathbf{m} \end{pmatrix} = m_{00}\,\hat{\mathbf{M}}_{\Delta P}\,\mathbf{M}_R\,\hat{\mathbf{M}}_D$$
$$\left[\hat{\mathbf{M}}_{\Delta P} \equiv \begin{pmatrix} 1 & \mathbf{0}^T \\ \mathbf{P}_{\Delta P} & \mathbf{m}_{\Delta P} \end{pmatrix} \quad \mathbf{M}_R \equiv \begin{pmatrix} 1 & \mathbf{0}^T \\ \mathbf{0} & \mathbf{m}_R \end{pmatrix} \quad \hat{\mathbf{M}}_D \equiv \begin{pmatrix} 1 & \mathbf{D}^T \\ \mathbf{D} & \mathbf{m}_D \end{pmatrix}\right] \tag{8.1}$$

where:

- The normalized symmetric matrix $\hat{\mathbf{M}}_D$ represents a normal diattenuator (see Section 4.3) whose diattenuation-polarizance vector $\mathbf{D}$ is equal to that of $\mathbf{M}$ and whose symmetric submatrix $\mathbf{m}_D$ is given by Eq. (4.75): $\mathbf{m}_D \equiv (\sin\kappa_D)\mathbf{I}_3 + (1-\sin\kappa_D)\hat{\mathbf{D}}\otimes\hat{\mathbf{D}}^T$ (where $\mathbf{I}_3 \equiv \mathrm{diag}(1,1,1)$, $\cos\kappa_D \equiv D$ and $\hat{\mathbf{D}} \equiv \mathbf{D}/D$).
- The orthogonal matrix $\mathbf{M}_R$ represents a pure retarder ($\mathbf{m}_R^{-1} = \mathbf{m}_R^T$, $\det\mathbf{m}_R = +1$).
- The normalized matrix $\hat{\mathbf{M}}_{\Delta P}$, with $\mathbf{m}_{\Delta P}^T = \mathbf{m}_{\Delta P}$, represents a depolarizer with nonzero polarizance and zero diattenuation.

Note that the serial combination of the retarder $\mathbf{M}_R$ and the normalized normal diattenuator $\hat{\mathbf{M}}_D$ represents a normalized pure Mueller matrix $\hat{\mathbf{M}}_J \equiv \mathbf{M}_R\,\hat{\mathbf{M}}_D$, so that $\hat{\mathbf{M}}_J$ can be expressed in any of the possible serial decompositions studied in Sections 4.4–4.6, namely, the polar decomposition (in either of the two possible forms), the general serial decomposition, and the symmetric decomposition. In fact, decomposition (8.1) can also be expressed as (Morio and Goudail 2004) (Fig 8.2)

$$\mathbf{M} = m_{00}\,\hat{\mathbf{M}}_{\Delta P}\,\mathbf{M}_R\,\hat{\mathbf{M}}_D = m_{00}\,\hat{\mathbf{M}}_{\Delta P}\,\hat{\mathbf{M}}'_D\,\mathbf{M}_R = m_{00}\,\mathbf{M}_R\,\hat{\mathbf{M}}'_{\Delta P}\,\hat{\mathbf{M}}_D$$
$$\left(\hat{\mathbf{M}}'_D \equiv \mathbf{M}_R\,\hat{\mathbf{M}}_D\,\mathbf{M}_R^T \qquad \hat{\mathbf{M}}'_{\Delta P} \equiv \mathbf{M}_R^T\,\hat{\mathbf{M}}_{\Delta P}\,\mathbf{M}_R\right) \tag{8.2}$$

By applying the decomposition to $\mathbf{M}^T$ (which, as shown in Section 5.8, is a Mueller matrix inasmuch as $\mathbf{M}$ is) we get

$$\mathbf{M}^T \equiv m_{00}\begin{pmatrix} 1 & \mathbf{P}^T \\ \mathbf{D} & \mathbf{m}^T \end{pmatrix} = m_{00}\hat{\mathbf{M}}^T_{\Delta D}\,\mathbf{M}'^T_R\,\hat{\mathbf{M}}_P \tag{8.3}$$

where the normalized depolarizer $\hat{\mathbf{M}}^T_{\Delta D}$ exhibits zero diattenuation and nonzero polarizance, $\mathbf{M}'^T_R$ represents a retarder, and $\hat{\mathbf{M}}_P$ represents a normal diattenuator. By taking the transpose in Eq. (8.3) we deduce that, given a Mueller matrix $\mathbf{M}$, the following alternative *reverse decomposition* is also possible (Ossikovski et al. 2007) (recall that $\hat{\mathbf{M}}_P = \hat{\mathbf{M}}^T_P$):

$$\mathbf{M} = m_{00}\,\hat{\mathbf{M}}_P\,\mathbf{M}'_R\,\hat{\mathbf{M}}_{\Delta D}$$
$$\left[\hat{\mathbf{M}}_{\Delta D} \equiv \begin{pmatrix} 1 & \mathbf{D}^T_{\Delta D} \\ \mathbf{0} & \mathbf{m}_{\Delta D} \end{pmatrix} \quad \mathbf{M}'_R \equiv \begin{pmatrix} 1 & \mathbf{0}^T \\ \mathbf{0} & \mathbf{m}'_R \end{pmatrix} \quad \hat{\mathbf{M}}_P \equiv \begin{pmatrix} 1 & \mathbf{P}^T \\ \mathbf{P} & \mathbf{m}_P \end{pmatrix}\right] \tag{8.4}$$

where the normalized depolarizer $\hat{\mathbf{M}}_{\Delta D}$ exhibits nonzero diattenuation and zero polarizance.

In analogy to the forward decomposition, the reverse decomposition (8.4) can be transformed into the following alternative forms (Ossikovski et al. 2007) (Figure 8.2):

$$\mathbf{M} = m_{00}\,\hat{\mathbf{M}}_P\,\mathbf{M}'_R\,\hat{\mathbf{M}}_{\Delta D} = m_{00}\,\mathbf{M}'_R\,\hat{\mathbf{M}}'_P\,\hat{\mathbf{M}}_{\Delta D} = m_{00}\,\hat{\mathbf{M}}_P\,\hat{\mathbf{M}}'_{\Delta D}\,\mathbf{M}'_R$$
$$\left[\hat{\mathbf{M}}'_P \equiv \mathbf{M}'^T_R\,\hat{\mathbf{M}}_P\,\mathbf{M}'_R \qquad \hat{\mathbf{M}}'_{\Delta D} \equiv \mathbf{M}'_R\hat{\mathbf{M}}_{\Delta D}\,\mathbf{M}'^T_R\right] \tag{8.5}$$

Observe that the normalized normal diattenuator $\hat{\mathbf{M}}_D$ is completely determined by the diattenuation vector $\mathbf{D}$ of $\mathbf{M}$, whereas the calculation of $\hat{\mathbf{M}}_{\Delta P}$ and $\mathbf{M}_R$ requires to distinguish between non-singular and singular Mueller matrices.

It should be stressed that in both forward and reverse formulations, the depolarizer involves depolarizing, enpolarizing or diattenuating, as well as retarding properties in a coupled manner. Lu and Chipman's (1996) detailed procedure for the calculation of $\hat{\mathbf{M}}_{\Delta P}$, $\mathbf{M}_R$ and $\hat{\mathbf{M}}_D$, is described in the following subsections, while the calculation of $\hat{\mathbf{M}}_{\Delta D}$, $\mathbf{M}'_R$ and $\hat{\mathbf{M}}_P$ is achieved by applying the same procedure to $\mathbf{M}^T$.

The algebraic procedure has to be generalized to the case of negative-determinant Mueller matrices ($\det\mathbf{M} < 0$) that typically occur in certain strongly depolarizing systems (e.g., in turbid media at high concentrations of scatterers, such as milk). The method consists in

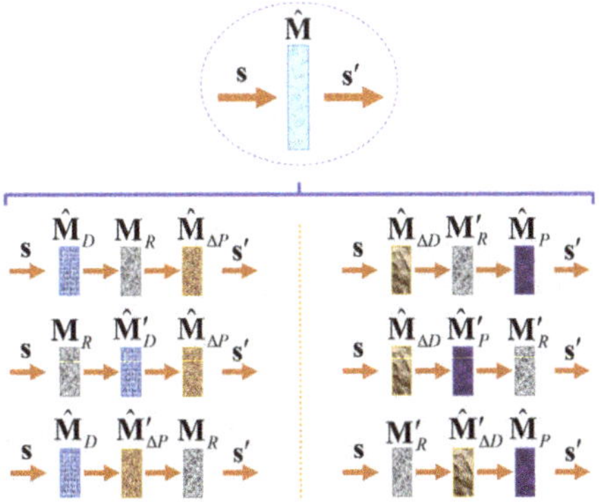

FIGURE 8.2
Forward and reverse decompositions of a depolarizing Mueller matrix $\mathbf{M} \equiv m_{00}\hat{\mathbf{M}}$.

applying the substitution rules, $\mathbf{M}_R \to \mathbf{S}\,\mathbf{M}_R$ and $\hat{\mathbf{M}}_{\Delta P} \to \hat{\mathbf{M}}_{\Delta P}\,\mathbf{S}$, respectively, to the retarder $\mathbf{M}_R$ and to the depolarizer $\hat{\mathbf{M}}_{\Delta P}$, originally provided by the generalized polar decomposition. In these substitutions, $\mathbf{S}$ is the sign-flip matrix given by $\mathbf{S} = \mathrm{diag}(1, \varepsilon_1, \varepsilon_2, -\varepsilon_1\varepsilon_2)$ where $\varepsilon_1 = \pm 1$ and $\varepsilon_2 = \pm 1$ are "selectable" sign factors. The sign-flip matrix $\mathbf{S}$ changes the signs of the determinants of $\mathbf{M}_R$ and $\hat{\mathbf{M}}_{\Delta P}$ and ensures that $\det \mathbf{M}_R = +1$ in order for $\mathbf{M}_R$ to physically represent a retarder (Ossikovski et al. 2008). The above procedure is efficient for diagonal (or for close to diagonal, in practice) depolarizer matrices which are the most common ones among experimentally occurring depolarizers. The procedures handling the most general case of nondiagonal depolarizer matrices are more involved; we refer the reader to the work of Ossikovski and Vizet (2017).

8.2.1 Forward Decomposition of a Non-singular Mueller Matrix

When $\det \mathbf{M} \neq 0$, then $\hat{\mathbf{M}}_D$ is non-singular , consequently $D < 1$, so that the polarizance vector $\mathbf{P}_{\Delta P}$ of $\hat{\mathbf{M}}_{\Delta P}$ is obtained from

$$\mathbf{P}_{\Delta P} = (\mathbf{P} - \mathbf{m}\mathbf{D})/\sin^2\kappa_D \qquad \left[\sin\kappa_D \equiv \sqrt{1-D^2}\right] \tag{8.6}$$

Analogously, the normalized diattenuator $\hat{\mathbf{M}}_P$ of the reverse decomposition (8.4) is completely determined from the polarizance vector $\mathbf{P}$ of $\mathbf{M}$, while the diattenuation vector $\mathbf{D}_{\Delta D}$ of $\hat{\mathbf{M}}_{\Delta D}$ is given by

$$\mathbf{D}_{\Delta D} = (\mathbf{D} - \mathbf{m}^T\mathbf{P})/\sin^2\kappa_P \qquad \left[\sin\kappa_P \equiv \sqrt{1-P^2}\right] \tag{8.7}$$

Further, by considering Eq. (3.62), the inverse $(\hat{\mathbf{M}}_D)^{-1}$ of $\hat{\mathbf{M}}_D$ can be calculated as follows from the diattenuation vector $\mathbf{D}$ of $\mathbf{M}$:

$$(\hat{\mathbf{M}}_D)^{-1} = \mathbf{G}\,\mathbf{M}_D\,\mathbf{G}/\sin^2\kappa_D \qquad \left[\sin\kappa_D \equiv \sqrt{1-D^2}\right] \tag{8.8}$$

(Note that $(\hat{\mathbf{M}}_D)^{-1}$ is not normalized in the sense that $[(\hat{\mathbf{M}}_D)^{-1}]_{00} = 1/\sin^2 k_D \neq 1$, which justifies the notation used in order to avoid confusion with the normalized version of $(\hat{\mathbf{M}}_D)^{-1}$ whose expression is $\mathbf{G}\mathbf{M}_D\mathbf{G}$). By using this result, it is possible to obtain the non-diattenuating matrix $\mathbf{M}_f \equiv \mathbf{M}(\hat{\mathbf{M}}_D)^{-1}$ from

$$\mathbf{M}_f \equiv \mathbf{M}\,(\hat{\mathbf{M}}_D)^{-1} = m_{00}\,\hat{\mathbf{M}}_{\Delta P}\,\mathbf{M}_R = m_{00}\begin{pmatrix} 1 & \mathbf{0}^T \\ \mathbf{P}_{\Delta P} & \mathbf{m}_{\Delta P}\,\mathbf{m}_R \end{pmatrix} \tag{8.9}$$

(The name *generalized polar decomposition* stems from the fact that two matrices $\mathbf{m}_{\Delta P}$ and $\mathbf{m}_R$ are precisely the polar components of the non-singular submatrix $\mathbf{m}_f \equiv \mathbf{m}_{\Delta P}\,\mathbf{m}_R$). Next, we have for $\mathbf{m}_{\Delta P}$

$$\mathbf{m}_{\Delta P}^2 = \mathbf{m}_f\,\mathbf{m}_f^T = \mathbf{m}_{R2}\,\mathrm{diag}\left(q_1^2, q_2^2, q_3^2\right)\mathbf{m}_{R2}^T \tag{8.10}$$

where $\mathbf{m}_{R2}$ is the proper orthogonal matrix ($\mathbf{m}_{R2}^T = \mathbf{m}_{R2}^{-1}$ and $\det \mathbf{m}_{R2} = +1$) whose columns are the eigenvectors of the positive definite symmetric matrix $\mathbf{m}_{\Delta P}^2$, and q_i^2 $(i = 1,2,3)$, with

$0 < q_3^2 \le q_2^2 \le q_1^2$, are the corresponding ordered eigenvalues of $\mathbf{m}_{\Delta P}^2$. Note that q_i are the singular values of $\mathbf{m}_f$. Finally, matrices $\mathbf{m}_{\Delta P}$ and $\mathbf{m}_R$ are obtained from

$$\begin{aligned} \mathbf{m}_{\Delta P} &= \mathbf{m}_{R2}\,\mathrm{diag}\left(q_1, q_2, \varepsilon q_3\right)\mathbf{m}_{R2}^T \\ \mathbf{m}_R &= \mathbf{m}_{\Delta P}^{-1}\,\mathbf{m}_f = \mathbf{m}_{R2}\,\mathrm{diag}\left(1/q_1, 1/q_2, \varepsilon/q_3\right)\mathbf{m}_{R2}^T\,\mathbf{m}_f \end{aligned} \qquad \left[\varepsilon \equiv (\det \mathbf{M})/|\det \mathbf{M}|\right] \tag{8.11}$$

Observe also that $\mathbf{m}_R$ obtained through this calculation procedure is a proper rotation matrix (i.e., $\det \mathbf{m}_R = 1$), and therefore, $\mathbf{M}_R$ is the Mueller matrix of a retarder.

8.2.2 Forward Decomposition of a Singular Mueller Matrix

When $\mathbf{M}$ is singular, one of the following possibilities holds: $\hat{\mathbf{M}}_{\Delta P}$ is singular; $\hat{\mathbf{M}}_D$ is singular, or both $\hat{\mathbf{M}}_{\Delta P}$ and $\hat{\mathbf{M}}_D$ are singular. In what follows, we analyze separately these three alternatives.

1. $\det \hat{\mathbf{M}}_{\Delta P} = 0,\ \det \hat{\mathbf{M}}_D > 0.$

   As in the generic case, $\mathbf{P}_{\Delta P}$ is obtained by means of Eq. (8.6), and the inverse of $\hat{\mathbf{M}}_D$ can be calculated from the diattenuation vector $\mathbf{D}$ of $\mathbf{M}$ by means of Eq. (8.8).
   Then, the matrix $\mathbf{M}_f \equiv m_{00}\,\hat{\mathbf{M}}_{\Delta P}\,\mathbf{M}_R = \mathbf{M}\,(\hat{\mathbf{M}}_D)^{-1}$ is determined from $\mathbf{M}$ and $(\hat{\mathbf{M}}_D)^{-1}$,

$$\mathbf{M}_f = m_{00}\,\hat{\mathbf{M}}_{\Delta P}\,\mathbf{M}_R = m_{00}\begin{pmatrix} 1 & \mathbf{0}^T \\ \mathbf{P}_{\Delta P} & \mathbf{m}_{\Delta P}\,\mathbf{m}_R \end{pmatrix} \tag{8.12}$$

   where the submatrix $\mathbf{m}_f = \mathbf{m}_{\Delta P}\,\mathbf{m}_R$ is necessarily singular and $\mathbf{m}_{\Delta P}^2$ can be calculated from

$$\mathbf{m}_{\Delta P}^2 = \mathbf{m}_f\,\mathbf{m}_f^T \tag{8.13}$$

   Further, since $\mathbf{m}_{\Delta P}^2$ is symmetric and positive semidefinite, it can be diagonalized as

$$\mathbf{m}_{\Delta P}^2 = \mathbf{m}_{R2}\,\mathrm{diag}\left(q_1^2, q_2^2, 0\right)\mathbf{m}_{R2}^T \tag{8.14}$$

   where $\mathbf{m}_{R2}$ is an orthogonal matrix whose columns are the eigenvectors of $\mathbf{m}_{\Delta P}^2$. Moreover, the starting assumption $\det \hat{\mathbf{M}}_{\Delta P} = 0$ entails $\det \mathbf{m}_{\Delta P} = 0$, and therefore, the smallest singular value of $\mathbf{m}_{\Delta P}$ is necessarily zero ($q_3 = 0$), while q_1 and q_2 can be taken so as to satisfy $0 \le q_2^2 \le q_1^2$ without loss of generality.
   Thus, the singular value decomposition of $\mathbf{m}_f$ is given by

$$\mathbf{m}_f = \mathbf{m}_{R2}\,\mathrm{diag}\left(q_1, q_2, 0\right)\mathbf{m}_{R1} \qquad \left[\mathbf{m}_{R2} \equiv \left(\hat{\mathbf{v}}_1\,\hat{\mathbf{v}}_2\,\hat{\mathbf{v}}_3\right) \quad \mathbf{m}_{R1}^T \equiv \left(\hat{\mathbf{w}}_1\,\hat{\mathbf{w}}_2\,\hat{\mathbf{w}}_3\right)\right] \tag{8.15}$$

   where the column vectors $\hat{\mathbf{v}}_i$ of the orthogonal matrix $\mathbf{m}_{R2}$ are the eigenvectors of $\mathbf{m}_f\,\mathbf{m}_f^T = \mathbf{m}_{R2}\,\mathrm{diag}\,(q_1^2, q_2^2, 0)\,\mathbf{m}_{R2}^T$ (or *left-singular vectors* of $\mathbf{m}_f$) and the column vectors $\hat{\mathbf{w}}_i$ of $\mathbf{m}_{R1}^T$ are the eigenvectors of $\mathbf{m}_f^T\,\mathbf{m}_f = \mathbf{m}_{R1}^T\,\mathrm{diag}\,(q_1^2, q_2^2, 0)\,\mathbf{m}_{R1}$ (or *right-singular vectors* of $\mathbf{m}_f$). Depending on the number of zero singular values, the following cases can be distinguished (Lu and Chipman 1996):

a) $0 = q_3 = q_2 = q_1$

Then, simply

$$\hat{\mathbf{M}}_{\Delta P} = \begin{pmatrix} 1 & \mathbf{0}^T \\ \mathbf{P}_{\Delta P} & \mathbf{0} \end{pmatrix} \qquad \mathbf{M}_R = \mathbf{I}_4 \tag{8.16}$$

b) $0 = q_3 = q_2 < q_1^2$

In this case,

$$\mathbf{m}_{\Delta P} = \mathbf{m}_f\, \mathbf{m}_f^T \Big/ \sqrt{\operatorname{tr}\left(\mathbf{m}_f\, \mathbf{m}_f^T\right)} \tag{8.17}$$

while $\mathbf{m}_R$ is not completely determined. By choosing $\mathbf{m}_R$ so as to exhibit minimum retardance, $\mathbf{m}_R$ is determined by the following normalized retardance vector $\hat{\mathbf{R}}$ whose retardance $\Delta \equiv R \equiv |\mathbf{R}|$ is obtained from

$$\cos\Delta = \operatorname{tr}\mathbf{m}_f \Big/ \sqrt{\operatorname{tr}\left(\mathbf{m}_f\, \mathbf{m}_f^T\right)} \qquad \hat{\mathbf{R}} = \left(\hat{\mathbf{v}}_1 \wedge \hat{\mathbf{w}}_1\right) \big/ \left|\hat{\mathbf{v}}_1 \wedge \hat{\mathbf{w}}_1\right| \tag{8.18}$$

c) $0 = q_3 < q_2^2 \le q_1^2$

In this case, by applying the condition for $\mathbf{m}_R$ to have the minimum retardance,

$$\mathbf{m}_{\Delta P} = (q_1 + q_2)\left[\mathbf{m}_f\, \mathbf{m}_f^T + q_1 q_2 \mathbf{I}_3\right]^{-1} \mathbf{m}_f\, \mathbf{m}_f^T \qquad \mathbf{m}_R = \hat{\mathbf{v}}_1\, \hat{\mathbf{w}}_1^T + \hat{\mathbf{v}}_2\, \mathbf{w}_2^T + \frac{\hat{\mathbf{v}}_1 \wedge \hat{\mathbf{v}}_2}{\left|\hat{\mathbf{v}}_1 \wedge \hat{\mathbf{v}}_2\right|} \frac{\left(\hat{\mathbf{w}}_1 \wedge \hat{\mathbf{w}}_2\right)^T}{\left|\hat{\mathbf{w}}_1 \wedge \hat{\mathbf{w}}_2\right|} \tag{8.19}$$

2. $\det \hat{\mathbf{M}}_{\Delta P} \neq 0, \det \hat{\mathbf{M}}_D = 0.$

In this case, $D = 1$ and $P < 1$ so that $\mathbf{m}_D = \hat{\mathbf{D}} \otimes \hat{\mathbf{D}}^T$ and consequently,

$$\mathbf{M} = m_{00} \begin{pmatrix} 1 & \hat{\mathbf{D}}^T \\ \mathbf{P}_{\Delta P} + \mathbf{m}_{\Delta P}\, \mathbf{m}_R\, \hat{\mathbf{D}} & \left(\mathbf{P}_{\Delta P} + \mathbf{m}_{\Delta P}\, \mathbf{m}_R\, \hat{\mathbf{D}}\right) \otimes \hat{\mathbf{D}}^T \end{pmatrix} \tag{8.20}$$

that is, $\mathbf{M}$ is a type-I matrix that has the form of the following depolarizing analyzer:

$$\mathbf{M}_{P\hat{D}} = m_{00} \begin{pmatrix} 1 & \hat{\mathbf{D}}^T \\ \mathbf{P} & \mathbf{P} \otimes \hat{\mathbf{D}}^T \end{pmatrix} = m_{00}\left(\mathbf{s}_P \otimes \mathbf{s}_{\hat{D}}^T\right) \qquad \mathbf{s}_P \equiv \begin{pmatrix} 1 \\ \mathbf{P} \end{pmatrix} \quad \mathbf{s}_{\hat{D}} \equiv \begin{pmatrix} 1 \\ \hat{\mathbf{D}} \end{pmatrix} \tag{8.21}$$

which corresponds to the decomposition

$$\mathbf{M}_{P\hat{D}} = m_{00} \begin{pmatrix} 1 & \mathbf{0}^T \\ \mathbf{P} & \mathbf{0} \end{pmatrix} \begin{pmatrix} 1 & \hat{\mathbf{D}}^T \\ \hat{\mathbf{D}} & \hat{\mathbf{D}} \otimes \hat{\mathbf{D}}^T \end{pmatrix} \tag{8.22}$$

where $\mathbf{M}_R = \mathbf{I}_4$.

3. $\det \hat{\mathbf{M}}_{\Delta P} = 0, \det \hat{\mathbf{M}}_D = 0.$

In this case $D = P = 1$ and therefore, the Mueller matrix **M** corresponds to a perfect polarizer (in general, non-normal). That is, the depolarizer is the identity matrix, so that the Lu-Chipman decomposition becomes the polar decomposition of a perfect polarizer

$$\mathbf{M}_{\hat{P}\hat{D}} = m_{00}\begin{pmatrix} 1 & \mathbf{0}^T \\ \mathbf{0} & \mathbf{m}_R \end{pmatrix}\begin{pmatrix} 1 & \hat{\mathbf{D}}^T \\ \hat{\mathbf{D}} & \hat{\mathbf{D}}\otimes\hat{\mathbf{D}}^T \end{pmatrix} = m_{00}\begin{pmatrix} 1 & \hat{\mathbf{D}}^T \\ \hat{\mathbf{P}} & \hat{\mathbf{P}}\otimes\hat{\mathbf{D}}^T \end{pmatrix} \qquad \left(\hat{\mathbf{P}} \equiv \mathbf{m}_R\,\hat{\mathbf{D}}\right) \tag{8.23}$$

8.2.3 Reverse Decomposition of a Mueller Matrix

The calculation of the components of the reverse decomposition (8.4) of a Mueller matrix **M** can be performed by using the above described forward decomposition procedure, but instead applied to

$$\mathbf{M}^T = m_{00}\,\hat{\mathbf{M}}_{\Delta D}^T\,\mathbf{M}_R^T\,\hat{\mathbf{M}}_P \tag{8.24}$$

(recall that $\hat{\mathbf{M}}_P^T = \hat{\mathbf{M}}_P$) so that, once the components $\hat{\mathbf{M}}_{\Delta D}^T$, $\mathbf{M}_R^T$ and $\hat{\mathbf{M}}_P$ have been calculated, the reverse decomposition is obtained as

$$\mathbf{M} = m_{00}\,\hat{\mathbf{M}}_P\,\mathbf{M}_R\,\hat{\mathbf{M}}_{\Delta D} \tag{8.25}$$

8.3 Symmetric Decomposition

As seen in the previous section, the Lu-Chipman decomposition does not exhibit symmetric ordering of the components and, hence, admits two alternative forms, namely the forward and reverse decompositions. In fact, the left (right) depolarizer, besides retaining certain retarding properties (Gil and San José 2016a), has an asymmetric structure because it exhibits nonzero polarizance (diattenuation) and zero diattenuation (polarizance). Nevertheless, it is possible to formulate alternative serial decompositions where the components are disposed in a symmetric way and the central depolarizer is defined without the necessity of any a priori knowledge of the real order, thus fitting directly into any physical situation.

The *symmetric decomposition* of a depolarizing Mueller matrix **M** is based on the fact that any Mueller matrix can be expressed in the normal form $\mathbf{M} = \mathbf{M}_{J2}\,\mathbf{M}_\Delta\,\mathbf{M}_{J1}$ (Figure 8.3) where $\mathbf{M}_{J1}$ and $\mathbf{M}_{J2}$ are pure Mueller matrices and $\mathbf{M}_\Delta$ has the form of one of the two alternative canonical depolarizing Mueller matrices $\mathbf{M}_{\Delta d}$ and $\mathbf{M}_{\Delta nd}$ introduced and studied in detail in Section 5.7. (Ossikovski 2010a)

$$\mathbf{M}_{\Delta d} \equiv \operatorname{diag}\left(d_0, d_1, d_2, \varepsilon d_3\right) \qquad 0 \le d_3 \le d_2 \le d_1 \le d_0 \qquad \varepsilon \equiv \det\mathbf{M}/\left|\det\mathbf{M}\right| \tag{8.26}$$

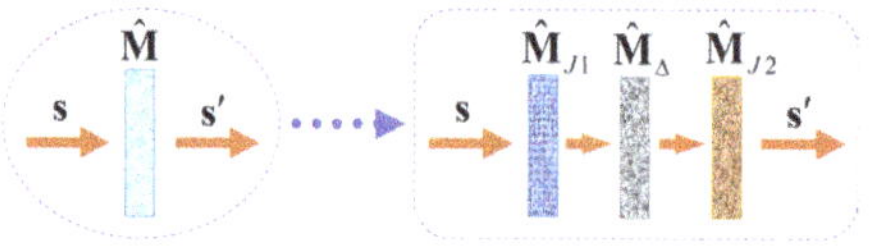

FIGURE 8.3
Symmetric decomposition of a depolarizing Mueller matrix.

$$\mathbf{M}_{\Delta nd} \equiv \begin{pmatrix} 2a_0 & -a_0 & 0 & 0 \\ a_0 & 0 & 0 & 0 \\ 0 & 0 & a_2 & 0 \\ 0 & 0 & 0 & a_2 \end{pmatrix} \qquad 0 \le a_2 \le a_0 \tag{8.27}$$

The parameters (d_0, d_1, d_2, d_3) and (a_0, a_0, a_2, a_2) are the square roots of the (non-negative) eigenvalues $(\rho_0, \rho_1, \rho_2, \rho_3)$ of the N-matrix $\mathbf{N} \equiv \mathbf{G}\mathbf{M}^T\mathbf{G}\mathbf{M}$ associated with $\mathbf{M}$ for type-I and type-II Mueller matrices respectively.

In this section we present the procedure for the calculation of the Mueller matrices $\mathbf{M}_{J1}$ and $\mathbf{M}_{J2}$ of the pure components, as well as that of the canonical depolarizer for both type-I and type-II Mueller matrices (Ossikovski 2009; 2010a). To do so, the first step is determining whether $\mathbf{M}$ is of type-I (i.e., $\mathbf{N}$ is diagonalizable) or of type-II ($\mathbf{N}$ is not diagonalizable).

A matrix $\mathbf{N}$ is said to be diagonalizable if $\mathbf{N}$ is similar to a diagonal matrix, i.e., if there exists an invertible matrix $\mathbf{A}$ such that $\mathbf{A}^{-1}\mathbf{N}\mathbf{A}$ is a diagonal matrix. Despite the fact that the diagonalizability test is dealt with in a large number of linear algebra treatises (see e.g., Horn and Johnson 2012), the inspection of the properties of the N-matrices $\mathbf{N} \equiv \mathbf{G}\mathbf{M}^T\mathbf{G}\mathbf{M}$ and $\mathbf{N}' \equiv \mathbf{G}\mathbf{M}\mathbf{G}\mathbf{M}^T$ provides a simple criterion to discriminate between type-I and type-II Mueller matrices (Gopala Rao et al, 1998b). In accordance with the definitions for type-I and type-II introduced in Section 5.7, if $\mathbf{N} \neq \mathbf{0}$, $\mathbf{N}' \neq \mathbf{0}$, and both $\mathbf{N}$ and $\mathbf{N}'$ have respective totally polarized eigenstates, then $\mathbf{M}$ is of type-II; $\mathbf{M}$ is of type-I in all remaining cases.

The polar decomposition of the two pure components of the normal form of $\mathbf{M}$ leads to the *symmetric decomposition* (Ossikovski 2009)

$$\mathbf{M} = \mathbf{M}_{J2}\mathbf{M}_{\Delta}\mathbf{M}_{J1} = \mathbf{M}_{D2}\mathbf{M}_{R2}\mathbf{M}_{\Delta}\mathbf{M}_{R1}\mathbf{M}_{D1} \tag{8.28}$$

The calculation of the components of the symmetric decomposition requires different procedures for type-I and type-II Mueller matrices. In addition, when $\mathbf{M}$ is of type-I, different cases have to be distinguished depending on whether the matrices $\mathbf{N}$ and $\mathbf{N}'$ are identical to the zero matrix or not (Ossikovski 2009). Let us recall that $\mathbf{M}_{Di} = \mathbf{M}_{Di}^T$, and observe that the equality $\mathbf{M}_{Di}\mathbf{G}\mathbf{M}_{Di} = \mathbf{0}$ holds if and only if $D_i = 1$. Therefore, leaving aside the trivial case $\mathbf{M} = \mathbf{0}$, $\mathbf{N} = \mathbf{0}$ holds if and only if $D_2 = 1$ (since $\mathbf{N}$ is given by a matrix product containing $\mathbf{M}_{D2}\mathbf{G}\mathbf{M}_{D2}$) whereas $\mathbf{N}' = \mathbf{0}$ occurs if and only if $D_1 = 1$ ($\mathbf{N}'$ is given by a matrix product containing $\mathbf{M}_{D1}\mathbf{G}\mathbf{M}_{D1}$).

Another property of Mueller matrices which underlies the symmetric decomposition is that, as demonstrated by Van der Mee (1993), Sridhar and Simon (1994) and Gopala Rao et al. (1998a) the respective eigenspaces of the two matrices $\mathbf{N}$ and $\mathbf{N}'$ contain each a unique eigenvector representing a Stokes vector and, in addition, their common eigenvalue spectrum is real and non-negative. In the following developments we use the eigenvalue-eigenvector equations for matrices $\mathbf{M}^T\mathbf{G}\mathbf{M}\mathbf{G}$ and $\mathbf{M}\mathbf{G}\mathbf{M}^T\mathbf{G}$ which are similar by $\mathbf{G}$ to $\mathbf{N}$ and $\mathbf{N}'$ respectively and therefore, share the same eigenvalue spectrum. Their respective eigenvectors (which are Stokes vectors) are **G** times the eigenvectors of **N** and **N′**, so that both pairs of eigenvectors physically represent polarization states of the same nature (partially or totally polarized).

8.3.1 Symmetric Decomposition of a Type-I Mueller Matrix

As indicated above, one needs to distinguish between the following cases (Ossikovski 2009).

8.3.1.1 $\mathbf{N} \neq \mathbf{0}$ *and* $\mathbf{N}' \neq \mathbf{0}$

When both N-matrices, **N** and **N′**, corresponding to the type-I Mueller matrix **M** are different from the zero matrix, then both $\mathbf{M}_{D1}$ and $\mathbf{M}_{D2}$ are non-singular matrices (i.e., $D_1 < 1$ and $D_2 < 1$). It should be noted that, in general, $\mathbf{D}_i \neq \mathbf{D}$, $\mathbf{D}_i \neq \mathbf{P}$, and $D_i \neq D$, $D_i \neq P$ $(i = 1,2)$ where **D** and **P** are the diattenuation and polarizance vectors of **M** with respective absolute values D and P.

Let us take $\mathbf{M}_{D1}$ and $\mathbf{M}_{D2}$ as having their maximum achievable intensity coefficients (ensuring passivity), that is $\mathbf{M}_{D1} = \hat{\mathbf{M}}_{D1}/(1+D_1)$ and $\mathbf{M}_{D2} = \hat{\mathbf{M}}_{D2}/(1+D_2)$, so that their respective inverse matrices are given by

$$\mathbf{M}_{Di}^{-1} = \frac{1}{1-D_i}\begin{pmatrix} 1 & -\mathbf{D}_i^T \\ -\mathbf{D}_i & \mathbf{m}_{Di} \end{pmatrix} = \frac{1+D_i}{1-D_i}\mathbf{G}\mathbf{M}_{Di}\mathbf{G}\,(i=1,2)$$

$$\left[\mathbf{m}_{Di} \equiv \sin\kappa_i\,\mathbf{I}_3 + (1-\sin\kappa_i)\left(\hat{\mathbf{D}}_i \otimes \hat{\mathbf{D}}_i^T\right) \quad \hat{\mathbf{D}}_i \equiv \mathbf{D}_i/D_i \quad \sin\kappa_i \equiv \sqrt{1-D_i^2} \quad (i=1,2)\right] \tag{8.29}$$

Then, Eq. (8.28) can be rewritten as

$$\mathbf{M}\mathbf{M}_{D1}^{-1} = \left(\frac{1+D_i}{1-D_i}\right)\mathbf{M}\mathbf{G}\mathbf{M}_{D1}\mathbf{G} = \mathbf{M}_{D2}\mathbf{M}_{R2}\mathbf{M}_{\Delta d}\mathbf{M}_{R1} \tag{8.30}$$

or, after appropriate regrouping of factors,

$$\left(\frac{1+D_i}{1-D_i}\right)(\mathbf{M}\mathbf{G})(\mathbf{M}_{D1}\mathbf{G}) = \mathbf{M}_{D2}\mathbf{M}' \quad [\mathbf{M}' \equiv \mathbf{M}_{R2}\mathbf{M}_{\Delta d}\mathbf{M}_{R1}] \tag{8.31}$$

When **M** is a type-I Mueller matrix, then **M′** has both zero diattenuation and polarizance, so that Eq. (8.31) can be rewritten as

$$\frac{1}{1-D_1}\mathbf{M}\mathbf{G}\begin{pmatrix} 1 & -\mathbf{D}_1^T \\ \mathbf{D}_1 & -\mathbf{m}_{D1} \end{pmatrix} = \frac{d_0}{1+D_2}\begin{pmatrix} 1 & \mathbf{D}_2^T\mathbf{m}' \\ \mathbf{D}_2 & \mathbf{m}_{D2}\mathbf{m}' \end{pmatrix} \tag{8.32}$$

By equating the first columns one obtains

$$\frac{1}{1-D_1}\mathbf{M}\mathbf{G}\,\mathbf{s}_{D1} = \frac{d_0}{1+D_2}\mathbf{s}_{D2} \quad \left[\mathbf{s}_{D1} \equiv \begin{pmatrix} 1 \\ \mathbf{D}_1 \end{pmatrix} \quad \mathbf{s}_{D2} \equiv \begin{pmatrix} 1 \\ \mathbf{D}_2 \end{pmatrix}\right] \tag{8.33}$$

By applying the same procedure on $\mathbf{M}^T$ we get

$$\frac{1}{1+D_2}\mathbf{M}^T\mathbf{G}\,\mathbf{s}_{D2}=\frac{d_0}{1-D_1}\mathbf{s}_{D1} \tag{8.34}$$

Therefore, the combination of Eqs. (8.33) and (8.34) results in the following pair of eigenvalue-eigenvector equations, with common eigenvalue d_0^2 (Ossikovski 2009):

$$\left(\mathbf{M}^T\mathbf{GMG}\right)\mathbf{s}_{D1}=d_0^2\,\mathbf{s}_{D1}\qquad\left(\mathbf{MGM}^T\mathbf{G}\right)\mathbf{s}_{D2}=d_0^2\,\mathbf{s}_{D2} \tag{8.35}$$

The diattenuation vectors $\mathbf{D}_1$ and $\mathbf{D}_2$ characterizing $\mathbf{M}_{D1}$ and $\mathbf{M}_{D2}$ are calculated from $\mathbf{M}$ by solving equations (8.35) or, equivalently, by solving one of the two following pairs of sequential equations (Ossikovski 2009):

$$\begin{aligned}&\left(\mathbf{M}^T\mathbf{GMG}\right)\mathbf{s}_{D1}=d_0^2\,\mathbf{s}_{D1}\qquad \mathbf{s}_{D2}=\mathbf{MG}\mathbf{s}_{D1}/\left(\mathbf{MG}\mathbf{s}_{D1}\right)_0\\&\left(\mathbf{MGM}^T\mathbf{G}\right)\mathbf{s}_{D2}=d_0^2\mathbf{s}_{D2}\qquad \mathbf{s}_{D1}=\mathbf{M}^T\mathbf{G}\mathbf{s}_{D2}/\left(\mathbf{M}^T\mathbf{G}\mathbf{s}_{D2}\right)_0\end{aligned} \tag{8.36}$$

Once the normal diattenuators of the equivalent system have been obtained from $\mathbf{D}_1$ and $\mathbf{D}_2$, $\mathbf{M}'$ is given by

$$\mathbf{M}'=\mathbf{M}_{D2}^{-1}\,\mathbf{M}\mathbf{M}_{D1}^{-1} \tag{8.37}$$

or, in explicit form,

$$\mathbf{M}'=c\begin{pmatrix}1-\mathbf{D}^T\mathbf{D}_1-\mathbf{P}^T\mathbf{D}_2+\mathbf{D}_2^T\mathbf{m}\mathbf{D}_1 & \mathbf{0}^T\\ \mathbf{0} & \left(\mathbf{D}_2\otimes\mathbf{D}_1^T\right)-\left(\mathbf{D}_2\otimes\mathbf{D}^T\right)\mathbf{m}_{D1}-\mathbf{m}_{D2}\left(\mathbf{P}\otimes\mathbf{D}_1^T\right)+\mathbf{m}_{D2}\mathbf{m}\mathbf{m}_{D1}\end{pmatrix}$$

$$\left[c\equiv m_{00}\frac{(1+D_1)(1+D_2)}{(1-D_1)(1-D_2)}\qquad m'_{00}=c\left(1-\mathbf{D}^T\mathbf{D}_1-\mathbf{P}^T\mathbf{D}_2+\mathbf{D}_2^T\mathbf{m}\mathbf{D}_1\right)\right] \tag{8.38}$$

and, since the diattenuation vector $\mathbf{D}'$ and the polarizance vector $\mathbf{P}'$ of $\mathbf{M}'$ are both equal to the zero vector $(\mathbf{D}'=\mathbf{P}'=\mathbf{0})$, the following equalities necessary hold:

$$\left(\mathbf{D}_2^T\mathbf{P}-1\right)\mathbf{D}_1+\mathbf{m}_{D1}\left(\mathbf{D}-\mathbf{m}^T\mathbf{D}_2\right)=\mathbf{0}\qquad\left(\mathbf{D}_1^T\mathbf{D}-1\right)\mathbf{D}_2+\mathbf{m}_{D2}\left(\mathbf{P}-\mathbf{m}\mathbf{D}_1\right)=\mathbf{0} \tag{8.39}$$

so that $\mathbf{m}_{D1}(\mathbf{D}-\mathbf{m}^T\mathbf{D}_2)$ and $\mathbf{m}_{D2}(\mathbf{P}-\mathbf{m}\,\mathbf{D}_1)$ are necessarily parallel to $\mathbf{D}_1$ and $\mathbf{D}_2$ respectively. By taking into account the property $\mathbf{m}_{Di}^{-1}\mathbf{D}_i=\mathbf{m}_{Di}\mathbf{D}_i=\mathbf{D}_i$, Eq. (8.39) transforms to

$$\mathbf{D}_1=\left(\mathbf{D}-\mathbf{m}^T\mathbf{D}_2\right)/\left(1-\mathbf{D}_2^T\mathbf{P}\right)\qquad\mathbf{D}_2=\left(\mathbf{P}-\mathbf{m}\mathbf{D}_1\right)/\left(1-\mathbf{D}_1^T\mathbf{D}\right) \tag{8.40}$$

Eqs. (8.40) represent a nonlinear system for $\mathbf{D}_1$ and $\mathbf{D}_2$, equivalent to the eigenvalue-eigenvector relations (8.36).

Next, $\mathbf{M}' = \mathbf{M}_{R2}\mathbf{M}_{\Delta}\mathbf{M}_{R1}$, i.e.,

$$\mathbf{M}' = \begin{pmatrix} 1 & \mathbf{0}^T \\ \mathbf{0} & \mathbf{m}_{R2} \end{pmatrix} d_0 \begin{pmatrix} 1 & \mathbf{0}^T \\ \mathbf{0} & \text{diag}(\hat{d}_1, \hat{d}_2, \varepsilon \hat{d}_3) \end{pmatrix} \begin{pmatrix} 1 & \mathbf{0}^T \\ \mathbf{0} & \mathbf{m}_{R1} \end{pmatrix} = d_0 \begin{pmatrix} 1 & \mathbf{0}^T \\ \mathbf{0} & \mathbf{m}_{R2}\,\text{diag}(\hat{d}_1, \hat{d}_2, \varepsilon \hat{d}_3)\,\mathbf{m}_{R1} \end{pmatrix} \tag{8.41}$$

where $\varepsilon \equiv \det \mathbf{M}/|\det \mathbf{M}|$. By equating m'_{00} in Eqs. (8.38) and (8.41), the parameter $d_0 = m'_{00}$ is obtained,

$$d_0 = m_{00} \frac{(1+D_1)(1+D_2)}{(1-D_1)(1-D_2)} \left(1 - \mathbf{D}^T\mathbf{D}_1 - \mathbf{P}^T\mathbf{D}_2 + \mathbf{D}_2^T\mathbf{m}\mathbf{D}_1\right) \qquad (D_1 < 1 \quad D_2 < 1) \tag{8.42}$$

and finally, the Mueller matrices of the two retarders $\mathbf{M}_{R1}$ and $\mathbf{M}_{R2}$ and that of the canonical type-I depolarizer $\mathbf{M}_{\Delta d} = d_0\,\text{diag}(1, \hat{d}_1, \hat{d}_2, \varepsilon \hat{d}_3)$ can be calculated through the singular value decomposition (SVD) of $\mathbf{M}' = d_0\,\hat{\mathbf{M}}'$.

Obviously, once the equivalent diattenuators and retarders of the symmetric decomposition have been calculated, the pure components $\mathbf{M}_{J1}$ and $\mathbf{M}_{J2}$ can be determined and can then be submitted to any of the possible serial decompositions of pure systems considered in Chapter 4.

As indicated above, the eigenvalue-eigenvector equations (8.35) admit unique solutions, whenever existing. More specifically, since all steps of the proof of the symmetric decomposition are reversible, such decomposition is unique inasmuch as the SVD of $\mathbf{m}'$ is. The singular values are commonly taken to be ordered, which is equivalent to taking $d_0 \geq d_1 \geq d_2 \geq d_3$ for the elements of $\mathbf{M}_{\Delta d}$. Despite the fact that such a choice corresponds quite often to physical reality, a depolarizing medium described by the intrinsic depolarizer $\mathbf{M}_{\Delta d}$ may as well exhibit $d_2 < d_3$ for instance. In such case, the corresponding two diagonal elements of $\mathbf{m}_{\Delta d}$ should be permuted, and a permutation of the corresponding pairs of columns and rows within the orthogonal factors $\mathbf{m}_{R1}$ and $\mathbf{m}_{R2}$ respectively is required as well.

Let us also observe that the SVD is invariant under both (1) a permutation of any two diagonal elements of $\mathbf{m}_{\Delta d}$ combined with that of the two corresponding pairs of columns and rows within the left and right orthogonal factors respectively, and (2) a sign change of the elements of one or more columns and rows from the left and right orthogonal factors respectively. These two invariance operations can be applied jointly to the decomposition $\mathbf{m}' = \mathbf{m}_{R2}\,\mathbf{m}_{\Delta d}\,\mathbf{m}_{R1}$ to obtain the orthogonal factor $\mathbf{m}_{R1}$ having the largest trace, thus achieving uniqueness. Indeed, recalling that the retardance of a retarder $\mathbf{M}_R$ is given by $\cos\Delta = (\text{tr}\,\mathbf{m}_R - 1)/2 = \text{tr}\,\mathbf{M}_R/2 - 1$ (see Eq. (4.22)), the maximal trace corresponds to the minimal retardance Δ_m. It should be likewise kept in mind that condition $\det \mathbf{M}_{R1} = \det \mathbf{M}_{R2} = +1$ must hold while applying the two invariance operations. Even though, naturally, all equivalent forms of the SVD satisfying the condition $\det \mathbf{M}_{R1} = \det \mathbf{M}_{R2} = +1$ are equally physical, i.e., they all result in physically realizable matrices $\mathbf{M}_{R1}$, $\mathbf{M}_{R2}$ and $\mathbf{M}_{\Delta d}$, the advantage of the minimum retardance choice is that it is also consistent with the correct asymptotic behavior of the decomposition. In fact, if the measured $\mathbf{M}$ is such that $\mathbf{M} \approx \mathbf{M}'_{\Delta d}$ where $\mathbf{M}'_{\Delta d}$ is a diagonal depolarizer, then one would naturally expect to obtain $\mathbf{M}_{R1} \approx \mathbf{M}_{R2} \approx \mathbf{I}_4$ and $\mathbf{M}_{\Delta d} \approx \mathbf{M}'_{\Delta d}$ from the symmetric decomposition, and these two sets of approximate equalities are guaranteed by the application of the minimum retardance criterion associated with the SVD procedure (Ossikovski 2009).

Note that the degenerate depolarizer case where $d_1 = d_2$ needs special treatment. Degeneracy typically occurs in transmission measurement configuration at normal incidence in which the depolarizer takes a rotationally invariant form. (The equivalent situation in normal reflection or backward scattering, $d_1 = -d_2$, can be formally reduced to the transmission, $d_1 = d_2$, one.) Under such degeneracy condition, the decomposition of $\mathbf{M}' = \mathbf{M}_{R2}\mathbf{M}_{\Delta}\mathbf{M}_{R1}$ through the SVD procedure becomes non-unique because $\mathbf{M}'$ is invariant upon rotation of the depolarizer $\mathbf{M}_{\Delta}$ by an arbitrary angle θ, i.e., $\mathbf{M}' = \mathbf{M}_{R2}\mathbf{M}_{\Delta}\mathbf{M}_{R1} = \mathbf{M}_{R2}\mathbf{M}_{G}(\theta)\mathbf{M}_{\Delta}\mathbf{M}_{G}^{T}(\theta)\mathbf{M}_{R1}$ where $\mathbf{M}_{G}(\theta)$ is the rotation matrix through θ (see Section 3.3.6). Due to the potential presence of the arbitrary rotation matrix, the two retarder matrices $\mathbf{M}_{R2}$ and $\mathbf{M}_{R1}$ returned by the SVD may exhibit a spurious circular retardance contribution (expressed by θ) that has no real physical meaning. The spurious circular retardance "added" by the SVD is efficiently removed from the two retarders by decomposing them down to linear and circular factors (Vizet and Ossikovski 2018).

Finally, let us consider the case where $\det\mathbf{M} < 0$, so that one of the two orthogonal factors $\mathbf{m}_{Ri}$ resulting from the SVD is such that $\det\mathbf{m}_{Ri} = -1$. The necessary condition $\det\mathbf{m}_{R1} = \det\mathbf{m}_{R2} = 1$ (i.e., $\det\mathbf{M}_{R1} = \det\mathbf{M}_{R2} = 1$) is recovered by setting $\varepsilon = -1$ (Xing 1992) and multiplying $\mathbf{M}_{Ri}$ by $\mathrm{diag}(1,1,1,-1)$.

Application of the symmetric decomposition to an experimental example

As an example, consider the experimental normalized Mueller matrix of a polyacrylamide gel measured in transmission with visible light (Arteaga and Canillas 2009),

$$\hat{\mathbf{M}}_g = \begin{pmatrix} 1 & -0.0312 & 0.0029 & -0.0066 \\ -0.0214 & 0.7678 & -0.0370 & 0.0204 \\ -0.0055 & 0.0230 & 0.1043 & -0.7735 \\ 0.0014 & 0.0390 & 0.7972 & 0.1920 \end{pmatrix}$$

None of the matrices $\mathbf{N}$ or $\mathbf{N}'$ turns to be zero and the two diagonalizing Stokes vectors, solutions to the eigenvalue equations (8.35) are found to be

$$\mathbf{s}_{D1} = (1, -0.0364, -0.0023, -0.0304)^T \qquad \mathbf{s}_{D2} = (1, 0.0071, -0.0280, 0.0105)^T$$

After constructing the normalized normal diattenuators from the two Stokes vectors $\mathbf{s}_{D1}$ and $\mathbf{s}_{D2}$,

$$\hat{\mathbf{M}}_{gD1} = \begin{pmatrix} 1 & -0.0364 & -0.0023 & -0.0304 \\ -0.0364 & 0.9995 & 0 & 0.0006 \\ -0.0023 & 0 & 0.9989 & 0 \\ -0.0304 & 0.0006 & 0 & 0.9993 \end{pmatrix}$$

$$\hat{\mathbf{M}}_{gD2} = \begin{pmatrix} 1 & 0.0071 & -0.0280 & 0.0105 \\ 0.0071 & 0.9996 & -0.0001 & 0 \\ -0.0280 & -0.0001 & 0.9999 & -0.0001 \\ 0.0105 & 0 & -0.0001 & 0.9996 \end{pmatrix}$$

the diattenuation-free matrix $\mathbf{M}'_g$ obtained from Eq. (8.37) is found to be

$$\mathbf{M}'_g = \begin{pmatrix} 1.0009 & 0 & 0 & 0 \\ 0 & 0.7688 & -0.0371 & 0.0201 \\ 0 & 0.0225 & 0.1045 & -0.7749 \\ 0 & 0.0393 & 0.7985 & 0.1924 \end{pmatrix}$$

As expected, this matrix features zero diattenuation and polarizance. Finally, the SVD decomposition of $\mathbf{M}'_g$ provides the canonical diagonal depolarizer,

$$\mathbf{M}_{g\Delta d} = \text{diag}(1.0009, 0.7698, 0.7561, 0.8464)$$

together with the two elliptical retarders,

$$\mathbf{M}_{gR1} = \begin{pmatrix} 1 & 0 & 0 & 0 \\ 0 & 0.9977 & -0.0649 & -0.0201 \\ 0 & 0.0291 & 0.6758 & -0.7365 \\ 0 & 0.0614 & 0.7342 & 0.6761 \end{pmatrix} \qquad \mathbf{M}_{gR2} = \begin{pmatrix} 1 & 0 & 0 & 0 \\ 0 & 0.9989 & -0.0232 & 0.0397 \\ 0 & 0.0406 & 0.8491 & -0.5267 \\ 0 & -0.0215 & 0.5278 & 0.8491 \end{pmatrix}$$

Notice that the *minimum retardance principle* discussed before has been applied to the two retarders and consequently, the elements of the diagonal depolarizer are not in descending order ($1 > d_1 > d_2 < d_3$). Physically, the matrix $\mathbf{M}_g$ depolarizes more strongly linearly polarized states rather than circularly polarized ones. This observation will be confirmed and phenomenologically interpreted in Chapter 9 where the same matrix $\mathbf{M}_g$ will be analyzed in the framework of the differential formalism.

It is also physically informative to calculate the pairs of diattenuation and birefringence vectors from the two diattenuator–retarder matrix pairs. More simply, the diattenuation vectors can be obtained directly from the two Stokes vectors $\mathbf{s}_{D1}$ and $\mathbf{s}_{D2}$; their magnitudes are $D_1 = 0.0475$ and $D_2 = 0.0302$, and they are directed along the respective unit vectors $\hat{\mathbf{D}}_1 = (-0.7669, -0.0480, -0.6399)^T$ and $\hat{\mathbf{D}}_2 = (0.2317, -0.9108, 0.3416)^T$ (in Poincaré sphere coordinates). We see that both diattenuations are extremely weak and do not exhibit any preferential spatial orientation. As for the retardance vectors derived from the two retarder matrices, their magnitudes are $R_1 = 0.8301$ and $R_2 = 0.5575$, whereas their respective directions are given by $\hat{\mathbf{R}}_1 = (-0.9964, 0.0553, -0.0637)^T$ and $\hat{\mathbf{R}}_2 = (-0.9965, -0.0578, -0.0603)^T$. The medium exhibits quite strong retardance values with the two retardance vectors being almost collinear and directed in practice along the S_1 axis of the Poincaré sphere. Consequently, the retardance is essentially a linear one and the two retardance values can be approximately added into a total retardance of $R = 1.3876$. This value, as well as the linear nature of the retardance will be found in perfect agreement with those determined within the framework of the differential formalism presented in Chapter 9. Thus, this experimental example shows the consistency in the information obtained from the various matrix decompositions (symmetric and differential, in the present case).

8.3.1.2 $\mathrm{N} \neq \mathbf{0}$ *and* $\mathrm{N}' = \mathbf{0}$

In this case, $D_1 = 1$ and $D_2 < 1$, so that $\mathbf{M}$ is a type-I matrix that necessarily corresponds to a depolarizing analyzer (Ossikovski 2009) that can be written as the direct product

$\mathbf{M}_{P\hat{D}} = \mathbf{s}_P \otimes \mathbf{s}_{\hat{D}}^T$ of a partially polarized Stokes vector $\mathbf{s}_P$ ($\mathbf{s}_P^T \mathbf{G}\, \mathbf{s}_P > 0$) and a totally polarized Stokes vector $\mathbf{s}_{\hat{D}}$ ($\mathbf{s}_{\hat{D}}^T \mathbf{G}\, \mathbf{s}_{\hat{D}} = 0$). The calculation of $\mathbf{s}_P$ and $\mathbf{s}_{\hat{D}}$ from $\mathbf{M}_{P\hat{D}}$ is straightforward. Furthermore, $\mathbf{M}_{P\hat{D}}$ is a type-I Mueller matrix that can also be written as

$$\mathbf{M}_{P\hat{D}} = \mathbf{M}_P\, \mathbf{M}_{\Delta 0}\, \mathbf{M}_{\hat{D}} \tag{8.43}$$

where $\mathbf{M}_{\hat{D}}$ is a normal perfect polarizer characterized by the polarization vector $\hat{\mathbf{D}}$ of $\mathbf{s}_{\hat{D}}$, $\mathbf{M}_{\Delta 0} = \mathrm{diag}\,(d_0, 0, 0, 0)$ is a perfect depolarizer, and $\mathbf{M}_P$ is a normal diattenuator defined from the polarization vector $\mathbf{P}$ of $\mathbf{s}_P$.

8.3.1.3 $N = 0$ and $N' \neq 0$

In analogy to the previous case, $D_2 = 1$, $\mathbf{M}$ is a type-I matrix that corresponds to a depolarizing polarizer that can be written as the direct product $\mathbf{M}_{\hat{P}D} = \mathbf{s}_{\hat{P}} \otimes \mathbf{s}_D^T$ of a totally polarized Stokes vector $\mathbf{s}_{\hat{P}}$ ($\mathbf{s}_{\hat{P}}^T \mathbf{G}\, \mathbf{s}_{\hat{P}} = 0$) and a partially polarized Stokes vector $\mathbf{s}_D$ ($\mathbf{s}_D^T \mathbf{G}\, \mathbf{s}_D > 0$). Furthermore, $\mathbf{M}_{\hat{P}D}$ can also be written as

$$\mathbf{M}_{\hat{P}D} = \mathbf{M}_{\hat{P}}\, \mathbf{M}_{\Delta 0}\, \mathbf{M}_D \tag{8.44}$$

where $\mathbf{M}_{\hat{P}}$ is a normal perfect polarizer defined from the polarization vector $\hat{\mathbf{P}}$ of $\mathbf{s}_{\hat{P}}$, $\mathbf{M}_{\Delta 0}$ is a perfect depolarizer, and $\mathbf{M}_D$ is a normal diattenuator defined from the polarization vector $\mathbf{D}$ of $\mathbf{s}_D$.

8.3.1.4 $N = N' = 0$

In this case $\mathbf{M}$ is a pure Mueller matrix that corresponds to a perfect polarizer (in general non-normal), which can always be written as the direct product $\mathbf{M} = \mathbf{s}_{\hat{P}} \otimes \mathbf{s}_{\hat{D}}^T$ of two totally polarized Stokes vectors $\mathbf{s}_{\hat{P}}$ and $\mathbf{s}_{\hat{D}}$.

8.3.2 Symmetric Decomposition of a Type-II Mueller Matrix

Let us consider the eigenvalue-eigenvector equations

$$\left(\mathbf{M}\mathbf{G}\mathbf{M}^T\mathbf{G}\right)\mathbf{s}_{Pp} = x_0^2\, \mathbf{s}_{Pp} \qquad \left(\mathbf{M}^T\mathbf{G}\mathbf{M}\mathbf{G}\right)\mathbf{s}_{Dp} = x_0^2\, \mathbf{s}_{Dp} \qquad \left[\mathbf{s}_{Pp} \equiv \begin{pmatrix} 1 \\ \mathbf{P}_p \end{pmatrix} \quad \mathbf{s}_{Dp} \equiv \begin{pmatrix} 1 \\ \mathbf{D}_p \end{pmatrix}\right] \tag{8.45}$$

where the notation for the eigenvectors differs from $\mathbf{s}_{D1}$ and $\mathbf{s}_{D2}$ respectively in order to avoid confusion between $\mathbf{P}_p$ and the polarizance-diattenuation vector $\mathbf{D}_1$ of the first diattenuator $\mathbf{M}_{D1}$ of the equivalent system and between $\mathbf{D}_p$ and $\mathbf{D}_2$. In fact, as indicated below, $P_p = D_p = 1$, whereas $D_1 < 1$ and $D_2 < 1$, so that necessarily $\mathbf{P}_p \neq \mathbf{D}_1$ and $\mathbf{D}_p \neq \mathbf{D}_2$.

As demonstrated by Gopala Rao et al. (1998b), if $\mathbf{N} \neq \mathbf{0}$ and $\mathbf{N}' \neq \mathbf{0}$, and the solutions of Eqs. (8.45) are such that $P_p = D_p = 1$ (i.e., the eigenstates $\mathbf{s}_{Pp}$ and $\mathbf{s}_{Dp}$ correspond to totally polarized states), then $\mathbf{M}$ is a type-II Mueller matrix. Thus, the first step for determining the components of the symmetric decomposition of a type-II Mueller matrix consists in calculating the pure polarization states $\mathbf{s}_{Pp}$ and $\mathbf{s}_{Dp}$ by solving Eqs. (8.45). Then, let us construct the following pure Mueller matrix of a polarizer-analyzer (in general non-normal):

$$\hat{\mathbf{M}}_{\hat{P}p\hat{D}p} \equiv \left(\mathbf{s}_{Pp} \otimes \mathbf{s}_{Dp}^T\right) = \begin{pmatrix} 1 & \hat{\mathbf{D}}_p^T \\ \hat{\mathbf{P}}_p & \hat{\mathbf{P}}_p \otimes \hat{\mathbf{D}}_p^T \end{pmatrix} \tag{8.46}$$

Prior to continuing the calculation of the components of the symmetric decomposition of **M**, it is worth considering the following parallel decomposition of the second canonical type-II depolarizer $\mathbf{M}_{\Delta II}$:

$$\mathbf{M}_{\Delta II} \equiv \begin{pmatrix} a_0+a & -a & 0 & 0 \\ a & a_0-a & 0 & 0 \\ 0 & 0 & a_2 & 0 \\ 0 & 0 & 0 & a_2 \end{pmatrix} = p\mathbf{M}_{\Delta d2} + (1-p)\mathbf{M}_p$$
$$\left[\begin{array}{ll} \mathbf{M}_{\Delta d2} \equiv (a_0+a)\begin{pmatrix} 1 & 0 & 0 & 0 \\ 0 & 1 & 0 & 0 \\ 0 & 0 & a_2/a_0 & 0 \\ 0 & 0 & 0 & a_2/a_0 \end{pmatrix} & \mathbf{M}_p \equiv (a_0+a)\begin{pmatrix} 1 & -1 & 0 & 0 \\ 1 & -1 & 0 & 0 \\ 0 & 0 & 0 & 0 \\ 0 & 0 & 0 & 0 \end{pmatrix} \\ p \equiv a_0/(a_0+a) \qquad (1-p) = a/(a_0+a) & (0 \le a_2 \le a_0 \quad 0 < a) \end{array}\right] \tag{8.47}$$

where $\mathbf{M}_{\Delta d2} = (a_0+a)\,\mathrm{diag}(1,1,P_1,P_1)$, $P_1 \equiv a_2/a_0$ being its first index of polarimetric purity, so that $\mathrm{rank}\,\mathbf{H}_{\Delta d2} = 2$, $\mathbf{H}_{\Delta d2}$ being the covariance matrix associated with $\mathbf{M}_{\Delta d2}$.

Let us now observe that $\mathbf{M}_{\Delta II}$ is a degenerate Mueller matrix with a unique eigenstate $\mathbf{s}_{1+} \equiv (1,1,0,0)^T$ with a doubly degenerate eigenvalue a_0, whereas the remaining doubly degenerate eigenvalue a_2 corresponds to the unphysical eigenvectors $(0,0,1,0)^T$ and $(0,0,0,1)^T$. Furthermore $\hat{\mathbf{M}}_p$ can be expressed as $\hat{\mathbf{M}}_p = \mathbf{s}_{1+} \otimes \mathbf{s}_{1-}^T$, where $\mathbf{s}_{1+} \equiv (1,1,0,0)^T$ and $\mathbf{s}_{1-} \equiv (1,-1,0,0)^T$, so that $\mathbf{s}_{1+}$ is an eigenstate of both $\mathbf{M}_{\Delta II}$ and $\mathbf{M}_p$. It is also straightforward to see that $\mathbf{s}_{1+}$ is likewise the unique eigenstate of the matrix $\mathbf{M}_{\Delta II}\,\mathbf{G}\mathbf{M}_{\Delta II}^T\mathbf{G}$ for the eigenvalue a_0^2. In the case of $\mathbf{M}_{\Delta II}^T$, which can be expressed as

$$\mathbf{M}_{\Delta II}^T = p\mathbf{M}_{\Delta d2} + (1-p)\mathbf{M}_p^T \qquad \left[\mathbf{M}_p^T = (a_0+a)\left(\mathbf{s}_{1-} \otimes \mathbf{s}_{1+}^T\right)\right] \tag{8.48}$$

the unique eigenstate, common to both $\mathbf{M}_{\Delta II}^T$ and $\mathbf{M}_p^T$, is $\mathbf{s}_{1-}$. This same Stokes vector is also the unique eigenstate of the matrix $\mathbf{M}_{\Delta II}^T\,\mathbf{G}\mathbf{M}_{\Delta II}\,\mathbf{G}$ for the eigenvalue a_0^2.

Note that, from Eq. (8.47), the following polarimetric subtraction can properly be performed:

$$\mathbf{M}_{\Delta d2} = \frac{1}{p}\left[\mathbf{M}_{\Delta II} - (1-p)\mathbf{M}_p\right] = \frac{1}{p}\left[\mathbf{M}_{\Delta II}^T - (1-p)\mathbf{M}_p^T\right] \tag{8.49}$$

Further, the symmetric decomposition can be written as follows in terms of the central depolarizer $\mathbf{M}_{\Delta II}$:

$$\begin{aligned} \mathbf{M} &= \mathbf{M}_{DO}\,\mathbf{M}_{RO}\,\mathbf{M}_{\Delta II}\,\mathbf{M}_{RI}\,\mathbf{M}_{DI} \\ &= p\mathbf{M}_{DO}\,\mathbf{M}_{RO}\,\mathbf{M}_{\Delta d2}\,\mathbf{M}_{RI}\,\mathbf{M}_{DI} + (1-p)\mathbf{M}_{DO}\,\mathbf{M}_{RO}\,\mathbf{M}_p\,\mathbf{M}_{RI}\,\mathbf{M}_{DI} \end{aligned} \tag{8.50}$$

where the second addend can be represented as

$$(a_0+a)\mathbf{M}_{DO}\,\mathbf{M}_{RO}\left(\mathbf{s}_{1+}\otimes\mathbf{s}_{1-}^T\right)\mathbf{M}_{RI}\,\mathbf{M}_{DI}=(a_0+a)\left(\mathbf{M}_{DO}\,\mathbf{M}_{RO}\,\mathbf{s}_{1+}\right)\otimes\left(\mathbf{M}_{DI}\,\mathbf{M}_{RI}^T\,\mathbf{s}_{1-}\right)^T \quad (8.51)$$

so that the following equality holds:

$$\mathbf{M}_{DO}\,\mathbf{M}_{RO}\,\mathbf{M}_p\,\mathbf{M}_{RI}\,\mathbf{M}_{DI}=m_{00}\left(\mathbf{s}_{Pp}\otimes\mathbf{s}_{Dp}^T\right)=m_{00}\,\hat{\mathbf{M}}_{\hat{P}p\hat{D}p} \quad (8.52)$$

because the unique eigenstate of the degenerate matrix $\mathbf{MGM}^T\mathbf{G}$ is precisely

$$\mathbf{s}_{Pp}=\hat{\mathbf{M}}_{DO}\,\mathbf{M}_{RO}\,\mathbf{s}_{1+} \quad (8.53)$$

whereas the unique eigenstate of the degenerate matrix $\mathbf{M}^T\mathbf{G}\mathbf{M}\mathbf{G}$ is

$$\mathbf{s}_{Dp}=\hat{\mathbf{M}}_{DI}\,\mathbf{M}_{RI}\,\mathbf{s}_{1-} \quad (8.54)$$

and these two eigenstates are common to $\hat{\mathbf{M}}_{\hat{P}p\hat{D}p}$ and $\hat{\mathbf{M}}_{\hat{P}p\hat{D}p}^T$ respectively, as follows from the previous paragraph on $\mathbf{M}_{\Delta II}$. The square root of the common eigenvalue x_0^2 for those eigenstates is $x_0=a_0$.

Let us now recall Eq. (8.50) and formulate it as

$$\hat{\mathbf{M}}=p\,\hat{\mathbf{M}}_2+(1-p)\hat{\mathbf{M}}_{\hat{P}p\hat{D}p} \qquad \hat{\mathbf{M}}_2\equiv\frac{\mathbf{M}_{DO}\,\mathbf{M}_{RO}\,\mathbf{M}_{\Delta d2}\,\mathbf{M}_{RI}\,\mathbf{M}_{DI}}{\left(\mathbf{M}_{DO}\,\mathbf{M}_{RO}\,\mathbf{M}_{\Delta d2}\,\mathbf{M}_{RI}\,\mathbf{M}_{DI}\right)_{00}} \quad (8.55)$$

the coefficient p being related to a_0 and the positive parameter a by means of $p=a_0/(a_0+a)$.

Obviously, since $\hat{\mathbf{M}}$ and $\hat{\mathbf{M}}_{\hat{P}p\hat{D}p}$ are known matrices, the coefficient p (and hence the parameter a) can be calculated through the polarimetric subtraction procedure shown in Section 7.7:

$$p=1/\left(\hat{\mathbf{w}}_0^{\dagger}\mathbf{H}^{-}\,\hat{\mathbf{w}}_0\right) \quad (8.56)$$

where $\hat{\mathbf{w}}_0$ is the covariance vector of the covariance matrix $\hat{\mathbf{H}}_{\hat{P}p\hat{D}p}\equiv\hat{\mathbf{w}}_0\otimes\hat{\mathbf{w}}_0^{\dagger}$ associated with $\hat{\mathbf{M}}_{\hat{P}p\hat{D}p}$, and $\mathbf{H}^{-}$ is the pseudoinverse of the covariance matrix $\hat{\mathbf{H}}$ associated with $\hat{\mathbf{M}}$.

The matrix $\mathbf{M}_2\equiv m_{00}\,\hat{\mathbf{M}}_2$ is determined from

$$\hat{\mathbf{M}}_2=\frac{1}{p}\left[\hat{\mathbf{M}}-(1-p)\hat{\mathbf{M}}_{\hat{P}p\hat{D}p}\right] \quad (8.57)$$

Once the type-I matrix $\mathbf{M}_2$ has been obtained, the next step is calculating $\mathbf{M}_{DI}$, $\mathbf{M}_{DO}$, $\mathbf{M}_{RI}$, $\mathbf{M}_{RO}$ and $\mathbf{M}_{\Delta d2}$ through the procedure indicated above for type-I matrices. Then, from Eq. (8.52) we get

$$\mathbf{M}=\mathbf{M}_{DO}\,\mathbf{M}_{RO}\,\mathbf{M}_{\Delta II}\,\mathbf{M}_{RI}\,\mathbf{M}_{DI} \quad (8.58)$$

where the central second canonical type-II depolarizer $\mathbf{M}_{\Delta II} \equiv p\mathbf{M}_{\Delta d2} + (1-p)\mathbf{M}_p$ has the form indicated in Eq. (8.47). Note that $\mathbf{M}_{RI}$ and $\mathbf{M}_{RO}$ are both determined only within a horizontal linear retarder of the form

$$\mathbf{M}_{RL0}(\varphi) = \begin{pmatrix} 1 & 0 & 0 & 0 \\ 0 & 1 & 0 & 0 \\ 0 & 0 & \cos\varphi & \sin\varphi \\ 0 & 0 & -\sin\varphi & \cos\varphi \end{pmatrix} \quad (8.59)$$

since $\mathbf{M}_{RL0}(\varphi)$ commutes with $\mathbf{M}_{\Delta II}$.

Let us finally consider the procedure for transforming the components of the decomposition (8.58) into

$$\mathbf{M} = \mathbf{M}_{D2}\,\mathbf{M}_{R2}\,\mathbf{M}_{\Delta nd}\,\mathbf{M}_{R1}\,\mathbf{M}_{D1} \quad (8.60)$$

in terms of the canonical type-II Mueller matrix $\mathbf{M}_{\Delta nd}$. As demonstrated in Section 5.7.2, $\mathbf{M}_{\Delta II} = \mathbf{M}_{Da}^{-1}\,\mathbf{M}_{\Delta nd}\,\mathbf{M}_{Da}$, where the auxiliary matrix $\mathbf{M}_{Da}$ and its inverse $\mathbf{M}_{Da}^{-1}$ are given by Eq. (5.34) in terms of a_0, and a. Therefore, Eq. (8.58) can be written as

$$\mathbf{M} = \mathbf{M}_{DO}\,\mathbf{M}_{RO}\left(\mathbf{M}_{Da}^{-1}\,\mathbf{M}_{\Delta nd}\,\mathbf{M}_{Da}\right)\mathbf{M}_{RI}\,\mathbf{M}_{DI} = \mathbf{M}_{J2}\,\mathbf{M}_{\Delta nd}\,\mathbf{M}_{J1} \quad (8.61)$$

where $\mathbf{M}_{J1} \equiv \mathbf{M}_{Da}\,\mathbf{M}_{RI}\,\mathbf{M}_{DI}$ and $\mathbf{M}_{J2} \equiv \mathbf{M}_{DO}\,\mathbf{M}_{RO}\,\mathbf{M}_{Da}^{-1}$, which can be further decomposed as

$$\mathbf{M}_{J2} = \mathbf{M}_{D2}\,\mathbf{M}_{R2} = \mathbf{M}_{R2}\,\mathbf{M}'_{D2} \qquad \mathbf{M}_{J1} = \mathbf{M}_{R1}\,\mathbf{M}_{D1} = \mathbf{M}'_{D1}\,\mathbf{M}_{R1} \quad (8.62)$$

through the polar decomposition procedure for pure Mueller matrices.

8.3.3 Synthetic View of the Symmetric Decomposition Procedure

Next, we present a synthetic guide to applying the procedure for the symmetric decomposition of a depolarizing Mueller matrix $\mathbf{M}$.

1. Calculate the N-matrices $\mathbf{N} \equiv \mathbf{G}\mathbf{M}^T\mathbf{G}\mathbf{M}$ and $\mathbf{N}' \equiv \mathbf{M}\mathbf{G}\mathbf{M}^T\mathbf{G}$ associated with the given Mueller matrix $\mathbf{M}$.
2. If $\mathbf{N} \neq \mathbf{0}$ and $\mathbf{N}' \neq \mathbf{0}$, solve the eigenvalue-eigenvector equations (8.45),
   2.1. **Type-I**. If $P_p < 1$ and $D_p < 1$ (i.e., if $\mathbf{M}$ is type-I).
      2.1.1. Calculate $\mathbf{M}_{D1}$, $\mathbf{M}_{D2}$ and $\mathbf{M}' = \mathbf{M}_{D2}^{-1}\,\mathbf{M}\mathbf{M}_{D2}^{-1}$
      2.1.2. Calculate d_0 by means of Eq. (8.42), and obtain $\mathbf{M}_{R1}$, $\mathbf{M}_{R2}$ and (d_1, d_2, d_3) through the singular value decomposition of $\mathbf{M}' = \mathbf{M}_{R2}\,\mathbf{M}_{\Delta d}\,\mathbf{M}_{R1}$. If $\mathbf{M}$ is non-singular, calculate $\varepsilon = \det\mathbf{M}/|\det\mathbf{M}|$, so that $\mathbf{M}_{\Delta d} = (d_0, d_1, d_2, \varepsilon d_3)$. By permuting the elements of $\mathbf{M}_{\Delta d}$ and the rows and columns of $\mathbf{M}_{R1}$ and $\mathbf{M}_{R2}$, determine the pair $\mathbf{M}_{R1}$, $\mathbf{M}_{R2}$ satisfying the criterion of minimum retardance.
   2.2. **Type-II**. If $P_p = D_p = 1$ (i.e., if $\mathbf{M}$ is type-II).
      2.2.1. Calculate the eigenvalues a_0^2 and a_2^2 of $\mathbf{N}$

2.2.2. Calculate the eigenstates $\mathbf{s}_{Pp}$ and $\mathbf{s}_{Dp}$ from the eigenvalue-eigenvector Eqs. (8.45), with $x_0 = a_0$, and construct the singular pure Mueller matrix $\hat{\mathbf{M}}_{\hat{P}p\hat{D}p} = \mathbf{s}_{Pp} \otimes \mathbf{s}_{Dp}^T$

2.2.3. Calculate the coefficient p by means of Eq. (8.56) (in accordance with the subtraction procedure given in Section 7.7).

2.2.4. Calculate $\hat{\mathbf{M}}_2 = [\hat{\mathbf{M}} - (1-p)\, \hat{\mathbf{M}}_{\hat{P}p\hat{D}p}]/p$.

2.2.5. Calculate the pure components of the symmetric decomposition of the type-I Mueller matrix $\mathbf{M}_2 \equiv m_{00}\, \hat{\mathbf{M}}_2 = \mathbf{M}_{DO}\, \mathbf{M}_{RO}\, \mathbf{M}_{\Delta d2}\, \mathbf{M}_{RI}\, \mathbf{M}_{DI}$.

2.2.6. Obtain the symmetric decomposition of $\mathbf{M}$ through $\mathbf{M} = \mathbf{M}_{J2}\, \mathbf{M}_{\Delta nd}\, \mathbf{M}_{J1}$, where $\mathbf{M}_{J1} \equiv \mathbf{M}_{Da}\, \mathbf{M}_{RI}\, \mathbf{M}_{DI}$ and $\mathbf{M}_{J2} \equiv \mathbf{M}_{DO}\, \mathbf{M}_{RO}\, \mathbf{M}_{Da}^{-1}$.

3. If $\mathbf{N} \neq \mathbf{0}$ and $\mathbf{N}' = \mathbf{0}$ (i.e., $D_1 \equiv D_p = 1$ and $D_2 \equiv P_p < 1$), then $\mathbf{M}$ is the type-I Mueller matrix $\mathbf{M} = \mathbf{M}_{P\hat{D}} = \mathbf{M}_P\, \mathbf{M}_{\Delta d}\, \mathbf{M}_{\hat{D}}$, where $\mathbf{M}_P$ represents a normal diattenuator, $\mathbf{M}_{\Delta d}$ represents an intrinsic (or diagonal) depolarizer, and $\mathbf{M}_{\hat{D}}$ represents a normal perfect polarizer. Therefore, $\mathbf{M}$ represents a depolarizing analyzer ($P_\Delta(\mathbf{M}) < 1$, $D(\mathbf{M}) = 1$, $P(\mathbf{M}) < 1$).
4. If $\mathbf{N} = \mathbf{0}$ and $\mathbf{N}' \neq \mathbf{0}$ (i.e., $D_1 \equiv D_p < 1$ and $D_2 \equiv P_p = 1$), then $\mathbf{M}$ is the type-I Mueller matrix: $\mathbf{M} = \mathbf{M}_{\hat{P}D} = \mathbf{M}_{\hat{P}}\, \mathbf{M}_{\Delta d}\, \mathbf{M}_D$, where $\mathbf{M}_{\hat{P}}$ represents a normal perfect polarizer, $\mathbf{M}_{\Delta d}$ represents an intrinsic depolarizer, and $\mathbf{M}_D$ represents a normal diattenuator. Therefore, $\mathbf{M}$ represents a depolarizing polarizer ($P_\Delta(\mathbf{M}) < 1$, $D(\mathbf{M}) < 1$, $P(\mathbf{M}) = 1$).
5. If $\mathbf{N} = \mathbf{N}' = \mathbf{0}$, then $\mathbf{M}$ corresponds to a perfect polarizer ($D(\mathbf{M}) = P(\mathbf{M}) = 1 \Rightarrow P_\Delta(\mathbf{M}) = 1$).

8.3.4 Procedure for the Numerical Calculation of the Components of the Normal Form

In addition to the procedure presented in the previous section for the obtainment of the components of the symmetric decomposition of any given Mueller matrix $\mathbf{M}$ (which obviously also allows for the calculation of the components of the normal form $\mathbf{M} = \mathbf{M}_{J2}\, \mathbf{M}_\Delta\, \mathbf{M}_{J1}$), an alternative efficient algorithm for the direct numerical calculation of $\mathbf{M}_\Delta$, $\mathbf{M}_{J1}$ and $\mathbf{M}_{J2}$ has been developed by San José et al. (2020). The algorithm includes two consecutive stages, namely (1) the conversion of the starting $\mathbf{M}$ to a bi-diagonal Mueller matrix $\mathbf{M}_B$ through a non-singular Mueller transformation, and (2) a simple iterative procedure that leads to the calculation of the components for both type-I and type-II Mueller matrices $\mathbf{M}$ without the necessity of a previous classification of $\mathbf{M}$ into such categories.

The efficiency and easy-to-implement features of the numerical algorithm have been demonstrated through its application to various experimental Mueller matrices (San José et al. 2020).

8.4 Passivity Constraints in Serial Decompositions of Depolarizing Mueller Matrices

In general, the media and devices submitted to polarimetric measurements are assumed to be *passive*, in the sense that they do not increase the intensity of electromagnetic beams interacting with them. General passivity constraints have been considered in Section 5.9. Here, we deal with the implications of the passivity conditions on the physical realizability of the Lu-Chipman and of the normal form of depolarizing Mueller matrices. In spite of the undeniable interest and applications of these decompositions, there are certain families of passive Mueller matrices which are not compatible with the simultaneous passivity of all the respective serial components. In other words, the Lu-Chipman decomposition and the normal form are not always physically realizable by means of passive components (Gil 2013b).

8.4.1 Passivity Constraints in the Lu-Chipman Decomposition

To ensure the passivity of the components in the Lu-Chipman decomposition, the global factor m_{00} must be factored and the factors have to be distributed among the three components. We observe that $\mathbf{M}_R$ is passive, with constant intensity coefficient equal to one for any incoming polarization state, whereas the passivity of the diattenuator requires multiplying $\hat{\mathbf{M}}_D$ by a factor $c_D \leq 1/(1+D)$. Moreover, the passivity of $\hat{\mathbf{M}}_{\Delta P}$ requires multiplying it by a factor $c_{\Delta P} \leq 1/(1+P_{\Delta P})$. Consequently, the simultaneous passivity of the components implies the inequality $(1+D)(1+P_{\Delta P})\, m_{00} \leq 1$, and, by using the equality $P_{\Delta P} = |\mathbf{P} - \mathbf{m}\mathbf{D}|/(1-D^2)$, we get

$$|\mathbf{P} - \mathbf{m}\mathbf{D}| \leq (1-D)\left[1/m_{00} - (1+D)\right] \tag{8.63}$$

This condition is not satisfied for certain passive Mueller matrices (Gil 2007; Devlaminck and Terrier 2010). Let us, for example, consider the following Mueller matrix:

$$\mathbf{M}_{DP} \equiv m_{00} \begin{pmatrix} 1 & \mathbf{D}^T \\ \mathbf{P} & \mathbf{P} \otimes \mathbf{D}^T \end{pmatrix} \quad \left(0 < D \leq 1 \quad 0 < P \leq 1 \quad PD < 1\right) \tag{8.64}$$

(note that the case of a perfect polarizer ($D = P = 1$) is excluded). The Lu-Chipman decomposition of $\mathbf{M}_{DP}$ is given by

$$\mathbf{M}_{DP} = m_{00}\, \hat{\mathbf{M}}_{\Delta P}\, \hat{\mathbf{M}}_D \qquad \hat{\mathbf{M}}_{\Delta P} = \begin{pmatrix} 1 & \mathbf{0}^T \\ \mathbf{P} & \mathbf{0} \end{pmatrix} \quad \hat{\mathbf{M}}_D = \begin{pmatrix} 1 & \mathbf{D}^T \\ \mathbf{D} & \mathbf{m}_D \end{pmatrix} \tag{8.65}$$

(The equivalent retarder $\mathbf{M}_R$ has been omitted because, in this case, $\mathbf{M}_R$ is simply equal to the identity matrix.)

Thus, the required passivity for all the components entails the condition $m_{00}(1+D)(1+P) \leq 1$, which is more restrictive than the pair of inequalities $m_{00}(1+D) \leq 1$ and $m_{00}(1+P) \leq 1$. If we denote by Q the largest value of the pair (D, P), any matrix $\mathbf{M}_{DP}$ with $\tilde{m}_{00} = 1/(1+Q)$ satisfies the passivity condition, but its Lu-Chipman decomposition necessarily contains at least one non-passive component (recall that the cases $D = 0$ or $P = 0$ have been excluded).

As an example, consider the Mueller matrix

$$\tilde{\mathbf{M}}_I \equiv \frac{1}{2}\begin{pmatrix} 1 & 1 & 0 & 0 \\ 1/2 & 1/2 & 0 & 0 \\ 0 & 0 & 0 & 0 \\ 0 & 0 & 0 & 0 \end{pmatrix} = \frac{1}{2}\hat{\mathbf{M}}_{\Delta P}\, \hat{\mathbf{M}}_D \qquad \hat{\mathbf{M}}_{\Delta P} = \begin{pmatrix} 1 & 0 & 0 & 0 \\ 1/2 & 0 & 0 & 0 \\ 0 & 0 & 0 & 0 \\ 0 & 0 & 0 & 0 \end{pmatrix} \qquad \hat{\mathbf{M}}_D = \begin{pmatrix} 1 & 1 & 0 & 0 \\ 1 & 1 & 0 & 0 \\ 0 & 0 & 0 & 0 \\ 0 & 0 & 0 & 0 \end{pmatrix} \tag{8.66}$$

whose Lu-Chipman decomposition cannot be physically performed by using passive components only.

8.4.2 Passivity Constraints on the Normal Form Components

Since multiplication by the matrix of a retarder does not affect the transmittance properties, the passivity constraints only depend on the two diattenuators $\mathbf{M}_{Di}$ and on the depolarizer $\mathbf{M}_{\Delta}$ of the symmetric decomposition $\mathbf{M}=\mathbf{M}_{D2}\,\mathbf{M}_{R2}\,\mathbf{M}_{\Delta}\,\mathbf{M}_{R1}\,\mathbf{M}_{D1}=\mathbf{M}_{R2}\left(\mathbf{M}'_{D2}\,\mathbf{M}_{\Delta}\,\mathbf{M}'_{D1}\right)\mathbf{M}_{R1}$, where $D'_1=D_1$ and $D'_2=D_2$.

In case of type-I Mueller matrices, the passivity of the components requires the fulfillment of the condition $m_{00}\le d_0/[(1+D_2)(1+D_1)]$ (Gil 2013b). As with the Lu-Chipman decomposition, passive Mueller matrices of the form $\mathbf{M}_{DP}$ in Eq. (8.64) constitute a family of examples of Mueller matrices whose symmetric decomposition necessarily includes at least one non-passive component. The symmetric decomposition of $\mathbf{M}_{DP}$ is given by

$$\mathbf{M}_{DP}=m_{00}\,\hat{\mathbf{M}}_{DO}\,\hat{\mathbf{M}}_{\Delta}\,\hat{\mathbf{M}}_{DI}$$
$$m_{00}=\frac{1}{(1+Q)}\qquad \hat{\mathbf{M}}_{DO}=\begin{pmatrix}1 & \mathbf{P}^T\\ \mathbf{P} & \mathbf{m}_P\end{pmatrix}\qquad \hat{\mathbf{M}}_{\Delta}=\begin{pmatrix}1 & \mathbf{0}^T\\ \mathbf{0} & \mathbf{0}\end{pmatrix}\qquad \hat{\mathbf{M}}_{DI}=\begin{pmatrix}1 & \mathbf{D}^T\\ \mathbf{D} & \mathbf{m}_D\end{pmatrix}\tag{8.67}$$

where the passivity of all components requires the fulfillment of the condition $m_{00}\le 1/[(1+D)(1+P)]$ (with $P\ne 0$, $D\ne 0$), which is incompatible with the starting value $m_{00}=1/(1+Q)$. The type-I Mueller matrix $\tilde{\mathbf{M}}_{\mathrm{I}}$ in Eq. (8.66) is again a particular example of a passive Mueller matrix such that the passivity of $\mathbf{M}_{D1}$ and $\mathbf{M}_{D2}$ cannot be achieved simultaneously (Gil 2013b).

Furthermore, examples of type-II Mueller matrices whose symmetric decomposition is not physically passive-realizable can also be given as, for instance,

$$\tilde{\mathbf{M}}_{\mathrm{II}}=\tilde{m}_{00}\begin{pmatrix}1 & b+(b^2-1)/2 & 0 & 0\\ b-(b^2-1)/2 & b^2 & 0 & 0\\ 0 & 0 & 0 & 0\\ 0 & 0 & 0 & 0\end{pmatrix}=\tilde{m}_{00}\,\hat{\mathbf{M}}_D\,\hat{\mathbf{M}}_{\Delta nd}\,\hat{\mathbf{M}}_D\qquad\left[\begin{array}{c}\tilde{m}_{00}=\dfrac{2}{3-b^2+2b}\\ 0<b\le 1\end{array}\right]$$
$$\hat{\mathbf{M}}_D=\begin{pmatrix}1 & b & 0 & 0\\ b & 1 & 0 & 0\\ 0 & 0 & \sqrt{1-b^2} & 0\\ 0 & 0 & 0 & \sqrt{1-b^2}\end{pmatrix}\qquad \hat{\mathbf{M}}_{\Delta nd}=\begin{pmatrix}1 & -1/2 & 0 & 0\\ 1/2 & 0 & 0 & 0\\ 0 & 0 & 0 & 0\\ 0 & 0 & 0 & 0\end{pmatrix}\tag{8.68}$$

whose normal form $\tilde{\mathbf{M}}_{\mathrm{II}}=\tilde{m}_{00}\,\hat{\mathbf{M}}_D\,\hat{\mathbf{M}}_{\Delta nd}\,\hat{\mathbf{M}}_D$ is incompatible with the simultaneous passivity of the three serial components.

Thus, we conclude that there exist passive Mueller matrices that do not allow for a normal form in terms of passive components only.

8.5 Invariant-Equivalent Mueller Matrices

In this section, we compare the diattenuation D, the polarizance P and the degree of polarimetric purity P_Δ (also called the depolarization index) of a given Mueller matrix **M** with those of its components resulting from the different ways of its serial decomposition.

As we have seen in the previous section, the diattenuations D_1 and D_2 of the right and left components of normal form are respectively different from the diattenuation D and polarizance P of **M**. In addition, the depolarization index of the central depolarizer is different from that of **M** (Ossikovski 2010b). Moreover, the diattenuator $\mathbf{M}_D$ of the forward Lu-Chipman decomposition has the same diattenuation as **M**, but both the polarizance $P_{\Delta P}$ and the depolarization index of the depolarizer $\mathbf{M}_{\Delta P}$ are different from those of **M**. Conversely, with regard to the reverse decomposition, the diattenuator $\mathbf{M}_P$ has the same polarizance as **M**, whereas both the diattenuation $D_{\Delta D}$ and the depolarization index of the depolarizer $\mathbf{M}_{\Delta D}$ are different from those of **M**. Let us also observe that, although **M** can be expressed as $\mathbf{M} = \mathbf{M}_P\,\mathbf{M}''\,\mathbf{M}_D$ through the simultaneous application of the forward and reverse Lu-Chipman decompositions, the central depolarizer represented by $\mathbf{M}''$ retains some amount of residual diattenuation, polarizance, as well as retardance properties, while its depolarization index is different from that of **M**. Therefore, both the Lu-Chipman and the normal form decompositions do not provide a complete decoupling between the diattenuation, the polarizance and the depolarization of **M**. This is a natural consequence of the fact that diattenuation and polarizance are consubstantial to certain depolarizing properties. Let us recall that even pure diattenuators (also called partial polarizers) induce depolarization for some incident partially polarized states (Simon 1990).

8.5.1 Invariant-Equivalent Transformations

As seen in Section 6.4, dual retarder transformations $\mathbf{M}_{R2}\mathbf{M}\,\mathbf{M}_{R1}$ ($\mathbf{M}_{R1}$ and $\mathbf{M}_{R2}$ being Mueller matrices of retarders, i.e., Mueller orthogonal matrices) leave unchanged the following quantities (called *physical invariants* of **M**) (Gil 2016c):

$$m_{00},\ P,\ D,\ P_P,\ P_S,\ a_1,\ a_2,\ a_3,\ P_1\ ,P_2,\ P_3\ ,P_\Delta,\ \mathbf{P}^T\mathbf{m}\mathbf{D},\ \det\mathbf{M} \tag{8.69}$$

(among which different sets of ten independent parameters can be identified). Moreover, since the action of the retarders does not affect the passivity of **M**, passivity is preserved under dual retarder transformations. Obviously, single retarder, dual rotation and single rotation transformations are particular forms of dual retarder transformations, so that all of them satisfy the invariance properties of the dual retarder transformations.

Given a Mueller matrix **M**, an infinite number of *invariant-equivalent* Mueller matrices can be obtained through particular dual retarder transformations (Gil 2013a, 2016c)

$$\mathbf{M}_E(\mathbf{M}) = \mathbf{M}_{RO}\,\mathbf{M}\mathbf{M}_{RI} = m_{00}\begin{pmatrix} 1 & \mathbf{D}^T\mathbf{m}_{RI} \\ \mathbf{m}_{RO}\mathbf{P} & \mathbf{m}_{RO}\,\mathbf{m}\,\mathbf{m}_{RI} \end{pmatrix} \tag{8.70}$$

The 16 elements of $\mathbf{M}_E(\mathbf{M})$ are, in general, nonzero. Thus, apart from the ten independent invariant quantities, namely m_{00}, P, D, a_1, a_2, a_3, $P_1\ ,P_2$, $\mathbf{P}^T\mathbf{m}\mathbf{D}$ and $\det\mathbf{M}$, each particular invariant-equivalent matrix $\mathbf{M}_E(\mathbf{M})$ also contains the information relative to four parameters given by the respective azimuths and ellipticity angles of the transformed diattenuation and polarizance vectors

$$\mathbf{D}_E(\mathbf{M}) \equiv \mathbf{M}_E^T(1,0,0,0)^T = \mathbf{m}_{RI}\,\mathbf{D} \qquad \mathbf{P}_E(\mathbf{M}) \equiv \mathbf{M}_E(1,0,0,0)^T = \mathbf{m}_{RO}\,\mathbf{P} \tag{8.71}$$

(with $|\mathbf{D}_E| = D$ and $|\mathbf{P}_E| = P$), as well as two additional parameters necessary to recover the 16 elements of $\mathbf{M}$ and whose interpretation will be considered in Section 8.6 in the light of the arrow form decomposition.

Next, we explore some particularly interesting forms of dual retarder transformations.

8.5.2 Reduced Forms of a Mueller Matrix

Let us first observe that the forward Lu-Chipman decomposition (8.1) $\mathbf{M} = m_{00}\,\hat{\mathbf{M}}_{\Delta P}\,\mathbf{M}_R\,\hat{\mathbf{M}}_D$ can be modified through the diagonalization $\mathbf{m}_{\Delta P} = \mathbf{m}_{Rf}\,\mathbf{m}_{L\Delta P}\,\mathbf{m}_{Rf}^T$ of the symmetric matrix $\mathbf{m}_{\Delta P}$, and through the polar decomposition of the resulting pure component located on the right-hand side:

$$\mathbf{M} = m_{00}\left(\mathbf{M}_{Rf}\,\hat{\mathbf{M}}_{L\Delta P}\,\mathbf{M}_{Rf}^T\right)\mathbf{M}_R\,\hat{\mathbf{M}}_D = m_{00}\,\mathbf{M}_{Rf}\,\hat{\mathbf{M}}_{L\Delta P}\left(\mathbf{M}_{Rf}^T\,\mathbf{M}_R\right)\hat{\mathbf{M}}_D = m_{00}\,\mathbf{M}_{Rf}\left(\hat{\mathbf{M}}_{L\Delta P}\hat{\mathbf{M}}'_D\right)\mathbf{M}'_R$$

$$\hat{\mathbf{M}}_{L\Delta P} \equiv \begin{pmatrix} 1 & \mathbf{0}^T \\ \mathbf{P}_{L\Delta P} & \mathbf{m}_{L\Delta P} \end{pmatrix} \quad \mathbf{m}_{L\Delta P} \equiv \begin{pmatrix} l_{\Delta P1} & 0 & 0 \\ 0 & l_{\Delta P2} & 0 \\ 0 & 0 & \varepsilon l_{\Delta P3} \end{pmatrix} \quad 0 \le l_{\Delta P3} \le l_{\Delta P2} \le l_{\Delta P1} \quad \varepsilon \equiv \frac{\det \mathbf{m}_{\Delta P}}{|\det \mathbf{m}_{\Delta P}|}$$

$$\mathbf{P}_{L\Delta P} = \frac{1}{1-D^2}\,\mathbf{m}_{Rf}^T(\mathbf{P} - \mathbf{m}\mathbf{D}) \qquad \mathbf{M}'_R \equiv \mathbf{M}_{Rf}^T\,\mathbf{M}_R \qquad \hat{\mathbf{M}}'_D \equiv \mathbf{M}'_R\,\hat{\mathbf{M}}_D\,\mathbf{M}_R'^T \tag{8.72}$$

where the enpolarizing depolarizer $\hat{\mathbf{M}}_{L\Delta P}$, hereafter called the *forward reduced form* of $\mathbf{M}$ (Gil and San José 2016a), has nonzero elements only along its first column and on its diagonal. Therefore, the Mueller matrix $\mathbf{M}_{\Gamma f}(\mathbf{M}) \equiv m_{00}\,\hat{\mathbf{M}}_{L\Delta P}\,\hat{\mathbf{M}}'_D = \mathbf{M}_{Rf}^T\,\mathbf{M}\mathbf{M}_R'^T$ is invariant-equivalent to $\mathbf{M}$ and has the form

$$\mathbf{M}_{\Gamma f}(\mathbf{M}) = \begin{pmatrix} 1 & \mathbf{D}'^T \\ \mathbf{P}_{L\Delta P} + \mathbf{m}_{L\Delta P}\,\mathbf{D}' & \mathbf{P}_{L\Delta P} \otimes \mathbf{D}'^T + \mathbf{m}_{L\Delta P}\,\mathbf{m}'_D \end{pmatrix} \qquad \begin{pmatrix} \mathbf{D}' \equiv \mathbf{m}'_R\,\mathbf{D} \\ \mathbf{m}'_D \equiv \mathbf{m}'_R\,\mathbf{m}_D\,\mathbf{m}_R'^T \end{pmatrix} \tag{8.73}$$

where $|\mathbf{D}'| = D$ and $|\mathbf{P}_{L\Delta P} + \mathbf{m}_{L\Delta P}\,\mathbf{D}'| = P$.

Analogous results are obtained from the reverse decomposition, which can be expressed as follows:

$$\mathbf{M} = m_{00}\,\hat{\mathbf{M}}_P\,\mathbf{M}_R\left(\mathbf{M}_{Rr}\,\hat{\mathbf{M}}_{L\Delta D}\,\mathbf{M}_{Rr}^T\right) = m_{00}\,\hat{\mathbf{M}}_P\left(\mathbf{M}_R\,\mathbf{M}_{Rr}\right)\hat{\mathbf{M}}_{L\Delta D}\,\mathbf{M}_{Rr}^T = m_{00}\,\mathbf{M}''_R\left(\hat{\mathbf{M}}'_P\,\hat{\mathbf{M}}_{L\Delta D}\right)\mathbf{M}_{Rr}^T$$

$$\hat{\mathbf{M}}_{L\Delta D} \equiv \begin{pmatrix} 1 & \mathbf{D}_{L\Delta D}^T \\ \mathbf{0} & \mathbf{m}_{L\Delta D} \end{pmatrix} \quad \mathbf{m}_{L\Delta D} \equiv \operatorname{diag}(l_{\Delta D1}, l_{\Delta D2}, \varepsilon l_{\Delta D3}) \quad 0 \le l_{\Delta D3} \le l_{\Delta D2} \le l_{\Delta D1} \quad \varepsilon \equiv \frac{\det \mathbf{m}_{\Delta D}}{|\det \mathbf{m}_{\Delta D}|}$$

$$\mathbf{D}_{L\Delta D} = \frac{1}{1-P^2}\,\mathbf{m}_{Rr}\left(\mathbf{D} - \mathbf{m}^T\mathbf{P}\right) \qquad \mathbf{M}''_R \equiv \mathbf{M}_R\,\mathbf{M}_{Rr} \qquad \hat{\mathbf{M}}'_P \equiv \mathbf{M}_R''^T\,\hat{\mathbf{M}}_P\,\mathbf{M}''_R \tag{8.74}$$

in such a manner that the invariant-equivalent matrix $\mathbf{M}_{\Gamma r}(\mathbf{M}) \equiv m_{00}\,\hat{\mathbf{M}}'_P\,\hat{\mathbf{M}}_{L\Delta D}$ is given by the product of a normal diattenuator $\hat{\mathbf{M}}'_P$, whose polarizance is equal to that of $\mathbf{M}$, and a diattenuating depolarizer $\hat{\mathbf{M}}_{L\Delta D}$, hereafter called the *reverse reduced form* of $\mathbf{M}$ (Gil and San José 2016a), whose nonzero elements are located only on its diagonal and along its first row.

Let us now consider the eigenvalue-eigenvector spectra of the forward and reverse reduced forms of $\mathbf{M}$

$$\mathbf{M}_{L\Delta P} \equiv m_{00}\begin{pmatrix} 1 & 0 & 0 & 0 \\ P_{L\Delta P1} & l_{\Delta P1} & 0 & 0 \\ P_{L\Delta P2} & 0 & l_{\Delta P2} & 0 \\ P_{L\Delta P3} & 0 & 0 & \varepsilon l_{\Delta P3} \end{pmatrix} \qquad \mathbf{M}_{L\Delta D} \equiv m_{00}\begin{pmatrix} 1 & D_{L\Delta D1} & D_{L\Delta D2} & D_{L\Delta D3} \\ 0 & l_{\Delta D1} & 0 & 0 \\ 0 & 0 & l_{\Delta D2} & 0 \\ 0 & 0 & 0 & \varepsilon l_{\Delta D3} \end{pmatrix} \tag{8.75}$$

and observe that they are degenerate in the sense that each one has the following respective unique eigenstate (Gil and San José 2016a):

$$\mathbf{s}_{L\Delta P} = \begin{pmatrix} 1 \\ P_{L\Delta P1}/(1-l_{\Delta P1}) \\ P_{L\Delta P2}/(1-l_{\Delta P2}) \\ P_{L\Delta P3}/(1-\varepsilon l_{\Delta P3}) \end{pmatrix} \quad [1 > l_{\Delta P1}] \qquad \mathbf{s}_{L\Delta D} = \begin{pmatrix} 1 \\ 0 \\ 0 \\ 0 \end{pmatrix} \tag{8.76}$$

both with the same eigenvalue m_{00}.

8.5.3 Invariant-Equivalent Transformation Induced by the Symmetric Decomposition

By inspecting the symmetric decomposition of a Mueller matrix $\mathbf{M}$, written as

$$\mathbf{M} = \mathbf{M}_{J2}\,\mathbf{M}_{\Delta}\,\mathbf{M}_{J1} = \mathbf{M}_{R2}\,\mathbf{M}'_{D2}\,\mathbf{M}_{\Delta}\,\mathbf{M}'_{D1}\,\mathbf{M}_{R1} \tag{8.77}$$

we find that the following invariant-equivalent Mueller matrix can be defined:

$$\mathbf{M}_S(\mathbf{M}) \equiv \mathbf{M}'_{D2}\,\mathbf{M}_{\Delta}\,\mathbf{M}'_{D1} = \mathbf{M}^T_{R2}\,\mathbf{M}\,\mathbf{M}^T_{R1} \tag{8.78}$$

8.5.4 Other Invariant-Equivalent Transformations

Among the infinite dual retarder transformations of a given $\mathbf{M}$ producing respective invariant-equivalent Mueller matrices, and in addition to the one induced by the symmetric decomposition considered in the previous subsection, some interesting particular cases that have deserved attention in the literature are listed below.

The *kernel form* of $\mathbf{M}$ is defined through the factorization $\mathbf{M} = \mathbf{M}_{Rb}(\mathbf{G}_b\,\mathbf{D}_k\,\mathbf{G}_a)\mathbf{M}_{Ra}$ where $\mathbf{M}_{Ra}$ and $\mathbf{M}_{Rb}$ are orthogonal Mueller matrices (thus corresponding to respective retarders); $\mathbf{G}_a\mathbf{M}_{Ra}$ and $\mathbf{M}_{Rb}\mathbf{G}_b$ are the orthogonal matrices (in general non-Mueller matrices) that perform the singular value decomposition of $\mathbf{M}$; $\mathbf{G}_a$ and $\mathbf{G}_b$ are two orthogonal non-Mueller matrices, and $\mathbf{D}_k$ is the diagonal matrix (in general, a non-Mueller matrix) whose diagonal entries are the singular values of $\mathbf{M}$. The Mueller matrix $\mathbf{M}_K = \mathbf{M}_{Rb}^T \mathbf{M} \mathbf{M}_{Ra}^T = \mathbf{G}_b \mathbf{D}_k \mathbf{G}_a$ is the so-called the *kernel form* of $\mathbf{M}$, whose peculiar features are described in Gil (2007).

The *tridiagonal form* $\mathbf{M}_T(\mathbf{M})$ has the genuine property that its six elements m_{t02}, m_{t03}, m_{t13}, m_{t20}, m_{t30} and m_{t31} are zero. The dual retarder transformation that converts $\mathbf{M}$ to $\mathbf{M}_T(\mathbf{M})$ was introduced by San José and Gil (2020b) and applied to the characterization of passivity of Mueller matrices.

The *arrow form* $\mathbf{M}_A(\mathbf{M})$ is defined through the singular value decomposition of the 3×3 submatrix $\mathbf{m}$ of $\mathbf{M}$ and has the genuine property that $\mathbf{m}(\mathbf{M}_A)$ takes a diagonal form, while $\mathbf{M}_A(\mathbf{M})$ preserves the depolarizing and enpolarizing properties of $\mathbf{M}$. The peculiar properties of $\mathbf{M}_A(\mathbf{M})$ are dealt with specifically in the next section.

8.6 The Arrow Decomposition of a Mueller Matrix

Due to its particular interest, here we focus on certain specific features of the invariant-equivalent *arrow form* $\mathbf{M}_A(\mathbf{M})$ of a Mueller matrix $\mathbf{M}$, which corresponds to the *arrow decomposition* of $\mathbf{M}$.

As seen in Section 5.12, the definition of the arrow form is based on the singular value decomposition of the submatrix $\mathbf{m}$ of $\mathbf{M}$

$$\mathbf{m} = \mathbf{m}_{RO}\,\mathbf{m}_A\,\mathbf{m}_{RI} \qquad \begin{bmatrix} \mathbf{m}_{Ri}^{-1} = \mathbf{m}_{Ri}^T \quad \det\mathbf{m}_{Ri} = +1 \quad (i = I, O) \\ \mathbf{m}_A \equiv \mathrm{diag}(a_1, a_2, \varepsilon a_3) \quad \varepsilon \equiv \det\mathbf{m}/|\det\mathbf{m}| \end{bmatrix} \tag{8.79}$$

where the non-negative parameters (a_1, a_2, a_3) are the singular values of $\mathbf{m}$.

The arrow matrix $\mathbf{M}_A(\mathbf{M})$ is then defined as

$$\mathbf{M}_A(\mathbf{M}) \equiv \mathbf{M}_{RO}^T\,\mathbf{M}\,\mathbf{M}_{RI}^T = m_{00}\begin{pmatrix} 1 & \mathbf{D}_A^T \\ \mathbf{P}_A & \mathbf{m}_A \end{pmatrix} \qquad \begin{bmatrix} \mathbf{m}_A \equiv \mathrm{diag}(a_1, a_2, \varepsilon a_3) \quad a_1 \geq a_2 \geq a_3 \\ \mathbf{D}_A = \mathbf{m}_{RI}\,\mathbf{D} \qquad \mathbf{P}_A = \mathbf{m}_{RO}^T\,\mathbf{P} \\ \mathbf{M}_{Ri} = \begin{pmatrix} 1 & \mathbf{0}^T \\ \mathbf{0} & \mathbf{m}_{Ri} \end{pmatrix} \quad (i = I, O) \end{bmatrix} \tag{8.80}$$

Thus, the *arrow decomposition* of $\mathbf{M}$ is formulated as $\mathbf{M} = \mathbf{M}_{RO}\,\mathbf{M}_A\,\mathbf{M}_{RI}$ (Figure 5.4)

Recall that, to avoid ambiguity in the definition of $\mathbf{M}_A(\mathbf{M})$, the convention $a_1 \geq a_2 \geq a_3$ (hence, $1 \geq a_1 \geq a_2 \geq a_3 \geq 0$) with the sign of εa_3 equal to the sign of $\det\mathbf{m}$ is taken (thus ensuring that $\mathbf{M}_{RI}$ and $\mathbf{M}_{RO}$ correspond to retarders, i.e., $\det\mathbf{M}_{RI} = \det\mathbf{M}_{RO} = +1$)

From its very definition, the arrow form $\mathbf{M}_A(\mathbf{M})$ has peculiar and residual retardance properties (note that even the diagonal retarders $\mathbf{M}_{Rdi}$ $(i = 1, 2, 3)$ in Eqs. (4.45) and (5.115) contain at least two negative diagonal elements so that their arrow forms coincide with the identity matrix). The arrow form contains, in a simple and condensed manner, complete information on the set of six independent physical invariants $(m_{00}, D, P, P_1, P_2, P_3)$, together with the two orientation angles $(\varphi_{DA}, \chi_{DA})$ of the diattenuation vector $\mathbf{D}_A$ and the two orientation angles $(\varphi_{PA}, \chi_{PA})$ of the polarizance vector $\mathbf{P}_A$ of $\mathbf{M}_A$ respectively. The diattenuation and polarizance vectors of $\mathbf{M}$ are related to those of $\mathbf{M}_A$ through the respective rotations $\mathbf{D}_A = \mathbf{m}_{RI}^T \mathbf{D}_A$ and $\mathbf{P} = \mathbf{m}_{RO} \mathbf{P}_A$.

Therefore, since $\mathbf{M}_A(\mathbf{M})$ is invariant-equivalent to $\mathbf{M}$ and exhibits (at least) six zero elements, it has the peculiar property that it contains all the information on the physical invariants of $\mathbf{M}$ in a particularly simple and condensed manner. The set of ten parameters constituted by the mean intensity coefficient (MIC) m_{00}, together with the three vectors $\mathbf{D}_A$, $\mathbf{P}_A$ and $\mathbf{P}_S \equiv (a_1, a_2, \varepsilon a_3)^T/\sqrt{3}$, determine the ten invariant quantities.

The constitutive vectors $\mathbf{r}$, $\mathbf{q}$, $\boldsymbol{\eta}$, $\boldsymbol{\tau}$, $\bar{\mathbf{R}}_+$ and $\bar{\mathbf{R}}_-$ defined in Section 5.13.4 for general Mueller matrices, vanish for $\mathbf{M}_A(\mathbf{M})$. Moreover,

$$P_S(\mathbf{M}) = P_S(\mathbf{M}_A) = \sqrt{(a_1^2 + a_2^2 + a_3^2)/3} = k(\mathbf{M}_A) \tag{8.81}$$

k being the absolute value of the constitutive vector $\mathbf{P}_S \equiv \mathbf{k}(\mathbf{M}_A) = (a_1, a_2, \varepsilon a_3)^T/\sqrt{3}$, which agrees with the expression for the degree of polarimetric purity in terms of the components of purity (see Section 6.2)

$$P_\Delta^2(\mathbf{M}) = P_\Delta^2(\mathbf{M}_A) = \left(P^2 + D^2 + 3P_S^2\right)/3 \tag{8.82}$$

Furthermore, the anisotropy coefficients of $\mathbf{M}_A(\mathbf{M})$ (see Section 6.7) take the simplified form

$$\alpha_i = \left(D_{Ai} + P_{Ai}\right)/\Sigma_A \qquad \left[\Sigma_A = \sqrt{3\left(1 - P_S^2\right) + 2\mathbf{P}_A^T \mathbf{D}_A}\right] \tag{8.83}$$

where D_{Ai}, P_{Ai} are the respective components of the diattenuation and polarizance vectors $\mathbf{D}_A$ and $\mathbf{P}_A$ of $\mathbf{M}_A(\mathbf{M})$. Therefore, the degree of anisotropy $P_\alpha(\mathbf{M}_A)$ is given by

$$P_\alpha^2(\mathbf{M}_A) = \frac{\left|\mathbf{P}_A + \mathbf{D}_A\right|^2}{3\left(1 - P_\Delta^2\right) + \left|\mathbf{P}_A + \mathbf{D}_A\right|^2} = \frac{D^2 + P^2 + 2\left|\mathbf{D}_A^T \mathbf{P}_A\right|}{3\left(1 - P_S^2\right) + 2\left|\mathbf{D}_A^T \mathbf{P}_A\right|} \tag{8.84}$$

In the special case where $P(\mathbf{M}) = D(\mathbf{M}) = 0$, the arrow form of $\mathbf{M}$ corresponds to an *intrinsic depolarizer* (or diagonal depolarizer) given by the diagonal matrix $\mathbf{M}_A = m_{00}\,\mathrm{diag}(1, a_1, a_2, \varepsilon a_3)$ which coincides with the canonical type-I depolarizer $\mathbf{M}_{\Delta d}$, and which is free from both dichroic and birefringence anisotropies.

In the particular case of $\mathbf{M}$ being a pure Mueller matrix $\mathbf{M}_J$, its four singular values $(p_1^2, p_1^2, p_1 p_2, p_1 p_2)$ are determined by the extremal intensity coefficients $p_1^2 = m_{00}(1 + D)$ and $p_2^2 = m_{00}(1 - D)$. If the decreasing order choice $a_1 \geq a_2 \geq a_3$ is disregarded, the singular value decomposition of the submatrix $\mathbf{m}$ of $\mathbf{M}_J$ is not unique, and the corresponding normalized arrow matrix $\hat{\mathbf{M}}_{JA}(\mathbf{M}_J)$ adopts one of the three following alternative forms:

$$\hat{\mathbf{M}}_{JA1} = \begin{pmatrix} 1 & \cos\kappa & 0 & 0 \\ \cos\kappa & 1 & 0 & 0 \\ 0 & 0 & \sin\kappa & 0 \\ 0 & 0 & 0 & \sin\kappa \end{pmatrix} \quad \hat{\mathbf{M}}_{JA2} = \begin{pmatrix} 1 & 0 & \cos\kappa & 0 \\ 0 & \sin\kappa & 0 & 0 \\ \cos\kappa & 0 & 1 & 0 \\ 0 & 0 & 0 & \sin\kappa \end{pmatrix}$$
$$\hat{\mathbf{M}}_{JA3} = \begin{pmatrix} 1 & 0 & 0 & \cos\kappa \\ 0 & \sin\kappa & 0 & 0 \\ 0 & 0 & \sin\kappa & 0 \\ \cos\kappa & 0 & 0 & 1 \end{pmatrix} \tag{8.85}$$

where $\cos\kappa \equiv D = P$ $(0 \le \kappa \le \pi/2)$ is the diattenuation-polarizance of $\mathbf{M}_J$. Obviously, the forms $\mathbf{M}_{JA2}$ and $\mathbf{M}_{JA3}$ can be reduced to the canonical arrow form $\mathbf{M}_{JA} \equiv \mathbf{M}_{JA1}$ through the appropriate single retarder transformations.

It is worth comparing the previous arrow forms of nondepolarizing Mueller matrices with the matrices $\mathbf{M}_L$, $\mathbf{M}_{L'}$ and $\mathbf{M}_C$ defined in Section 6.7 as nondepolarizing matrices preserving the state of polarization of the canonical Stokes vectors. We observe that, as expected, the arrow form eliminates the parameters Δ_L, $\Delta_{L'}$ and Δ_C, so that each canonical anisotropy is given by a unique parameter related to the respective dichroism.

8.7 Serial-Parallel Decompositions

Serial and parallel decompositions can be combined to obtain a useful framework for both the analysis of measured Mueller matrices and the development of theoretical models.

This section is devoted to the serial-parallel decomposition based on the normal form of $\mathbf{M}$. An additional interesting formulation, not included in this section, is the *structured decomposition*, which makes use of the arrow form of $\mathbf{M}$ and that has the peculiarity that the enpolarizing properties are concentrated in two parallel components whose joint polarimetric behavior is modeled through the so-called *enpolarizing ellipsoid* (Gil and San José 2021b).

By taking advantage of the simple forms of the Mueller matrices $\mathbf{M}_{\Delta d}$ and $\mathbf{M}_{\Delta nd}$ of type-I and type-II canonical depolarizers, the parallel decomposition of the latter provides a framework for the systematic parallel decomposition of an arbitrary Mueller matrix $\mathbf{M}$ and hence, an appropriate basis for a suitable analysis of polarimetric measurements based on equivalent systems constituted by simple components.

To start, notice that each parallel decomposition of the central depolarizer $\mathbf{M}_\Delta$ results in a parallel decomposition of $\mathbf{M}$. Thus, the serial-parallel decompositions studied in this section are based on three steps: (1) the symmetric serial decomposition of $\mathbf{M}$, (2) the parallel decomposition of $\mathbf{M}_\Delta$ and (3) the obtainment of the corresponding parallel decomposition of $\mathbf{M}$ from that of the corresponding canonical forms $\mathbf{M}_\Delta$ ($\mathbf{M}_{\Delta d}$ or $\mathbf{M}_{\Delta nd}$).

Serial parallel decompositions of a given $\mathbf{M}$ are closely related to the synthesis-decomposition procedure discussed in Section 6.8 where the rank of the covariance matrix $\mathbf{H}_\Delta$ associated with $\mathbf{M}_\Delta$ plays a key role as a descriptor of the minimum number of pure components of $\mathbf{M}$. Moreover, a detailed and systematic study of serial parallel decompositions, including analytic general expressions for the cases where $1 \le \operatorname{rank}\mathbf{H}_\Delta \le 3$, as well as a procedure for calculating the components of $\mathbf{M}$ when $\operatorname{rank}\mathbf{H}_\Delta = 4$ (in which case $\mathbf{M}$ is necessarily of type-I), can be found in (Gil et al. 2013). The main results are summarized in the following sections.

8.7.1 Normal Serial-Parallel Decomposition of a Type-I Mueller Matrix

By applying the homogeneous arbitrary decomposition to the central type-I depolarizer of the normal form of $\mathbf{M}$, one obtains (Gil et al. 2013)

$$\mathbf{M} = \mathbf{M}_{J2}\left(\sum_{i=0}^{r-1} p_i \mathbf{M}_i\right)\mathbf{M}_{J1} = \sum_{i=0}^{r-1} p_i \left(\mathbf{M}_{J2}\mathbf{M}_i\mathbf{M}_{J1}\right)$$
$$\left[\sum_{i=0}^{r-1} p_i = 1 \quad r \equiv \operatorname{rank}\mathbf{H}(\mathbf{M}) = \operatorname{rank}\mathbf{H}(\mathbf{M}_{\Delta d})\right] \tag{8.86}$$

where $\mathbf{M}_i$ are the pure parallel components of $\mathbf{M}_{\Delta d}$. Notice that all components have equal mean intensity coefficients (MICs) $(\mathbf{M}_i)_{00} = d_0$. Recall that, as seen in Section 7.3, the MICs of the components can be taken with respective different values, provided the coefficients p_i of the expansion are replaced by $k_i = p_i d_0/(\mathbf{M}_i)_{00}$. Therefore, although the arbitrary decompositions considered below are presented in their homogeneous forms, all of them can be easily transformed to have components with different MICs.

In general, in accordance with the formulation of the arbitrary decomposition in Section 7.3, an infinity of arbitrary parallel decompositions of $\mathbf{M}_{\Delta d}$ constituted by sets of pure retarders is achievable. The convex combination of each set of these pure retarders, denoted as $(\mathbf{M}_{R0}, \mathbf{M}_{R1}, \mathbf{M}_{R2}, \mathbf{M}_{R3})$, ensures the cancellation of the off-diagonal elements. Thus, the arbitrary parallel decompositions of $\mathbf{M}_{\Delta d}$ into sums of pure retarders are not restricted by additional constraints derived from the passivity conditions and, in consequence, the application of the procedure for calculating the different components is simplified. Thus, the serial-parallel decomposition of $\mathbf{M}$ can be formulated in terms of four pure components of the form $d_0\mathbf{M}_{J2}\mathbf{M}_{Ri}\mathbf{M}_{J1}$ $(i = 0,1,2,3)$,

$$\mathbf{M} = \sum_{i=0}^{r-1} k_i\left[\mathbf{M}_{J2}\left(d_0\mathbf{M}_{Ri}\right)\mathbf{M}_{J1}\right] \qquad \left[\sum_{i=0}^{r-1} k_i = 1 \qquad d_0\left(\mathbf{M}_{J2}\mathbf{M}_{Ri}\mathbf{M}_{J1}\right)_{00} = m_{00}\right] \tag{8.87}$$

Note that each pure component $d_0\mathbf{M}_{J2}\mathbf{M}_{Ri}\mathbf{M}_{J1}$ can be further reduced to the polar form or to the dual linear retarder transformation considered in Chapter 4.

A particular form of a serial-parallel decomposition of a type-I Mueller matrix $\mathbf{M}$ in terms of retarders is achieved through the spectral decomposition of $\mathbf{M}_{\Delta d}$ in Eq. (5.32) (see Section 5.7.1). Further, as seen in Section 7.3, certain equivalent systems composed of pure diattenuators are also feasible when $d_0 < 1$; since increasing values of the diattenuations of the components are allowed for decreasing values of d_0, when $d_0 \le 1/2$ then all arbitrary decompositions of $\mathbf{M}_{\Delta d}$ are passive-realizable regardless of the values of the diattenuations of the components. As an example, let us consider the Mueller matrix of an intrinsic (or diagonal) depolarizer decomposed into a parallel combination of four passive perfect polarizers (Gil et al. 2013),

$$\frac{1}{2}\mathbf{M}_{\Delta 0} \equiv \frac{1}{2}\begin{pmatrix} 1 & 0 & 0 & 0 \\ 0 & 0 & 0 & 0 \\ 0 & 0 & 0 & 0 \\ 0 & 0 & 0 & 0 \end{pmatrix}$$

$$= \frac{1}{4}\left\{ \frac{1}{2}\begin{pmatrix} 1 & 1 & 0 & 0 \\ 1 & 1 & 0 & 0 \\ 0 & 0 & 0 & 0 \\ 0 & 0 & 0 & 0 \end{pmatrix} + \frac{1}{2}\begin{pmatrix} 1 & -1 & 0 & 0 \\ 1 & -1 & 0 & 0 \\ 0 & 0 & 0 & 0 \\ 0 & 0 & 0 & 0 \end{pmatrix} + \frac{1}{2}\begin{pmatrix} 1 & 1 & 0 & 0 \\ -1 & -1 & 0 & 0 \\ 0 & 0 & 0 & 0 \\ 0 & 0 & 0 & 0 \end{pmatrix} + \frac{1}{2}\begin{pmatrix} 1 & -1 & 0 & 0 \\ -1 & 1 & 0 & 0 \\ 0 & 0 & 0 & 0 \\ 0 & 0 & 0 & 0 \end{pmatrix} \right\}$$

$$(8.88)$$

where the common factor $1/4$ indicates that the portion of the incoming intensity is equal for all four components, whereas the coefficients $1/2$ are intrinsic to each perfect linear polarizer. Thus, we observe that this equivalent system is consistent with the fact that the total intensity is shared among the four components and that all the components $\mathbf{M}_i$ satisfy the passivity condition $d_0(1+D_{\Delta d,i}) \leq 1$ (where $D_{\Delta d,i}$ is the diattenuation of $\mathbf{M}_{\Delta d,i}$). Nevertheless, it should be stressed that this kind of decomposition into pure polarizers is not passive-realizable when $d_0 > 1/2$.

Thus, in accordance with the arbitrary decomposition formalism, any covariance matrix of a pure system whose only nonzero eigenvalue is associated with an eigenvector contained in $\mathrm{im}\,\mathbf{H}_{\Delta d}$ can be considered as a potential component of the parallel decomposition. If this pure component exhibits nonzero polarizance (recall that, for pure elements, polarizance equals diattenuation), the passivity conditions must be checked prior to validating it as a potential parallel component. Once the first physically realizable parallel component has been chosen, its coefficient p_i (or k_i) in the expansion is determined by the procedure described Section 7.3.

To analyze the possible parallel decompositions, the respective Mueller matrices of the central depolarizer are denoted as $\mathbf{M}_{\Delta r}$ where $r = 1, 2, 3, 4$ indicates the rank of $\mathbf{H}_{\Delta d}$.

When $r = 1$, then $\mathbf{M}_{\Delta d1} = d_0 \mathbf{I}_4$, and obviously the parallel decomposition of $\mathbf{M}_{\Delta d1}$ makes no sense because the starting system is pure. When $r = 2$, it has been demonstrated that any homogeneous arbitrary decomposition of $\mathbf{M}_{\Delta d2}$ can be expressed as the following convex combination of two elements (Gil et al. 2013):

$$\mathbf{M}_{\Delta d2} \equiv \mathrm{diag}\left(d_0, d_0, d_2, d_2\right) = p_1 \mathbf{M}_{J1} + p_2 \mathbf{M}_{J2}$$

$$\mathbf{M}_{J1} = \frac{d_0}{2p_1}\begin{pmatrix} 1+c_\rho c_{2\alpha} & s_\rho s_{2\alpha} c_\mu & 0 & 0 \\ s_\rho s_{2\alpha} c_\mu & 1+c_\rho c_{2\alpha} & 0 & 0 \\ 0 & 0 & c_\rho+c_{2\alpha} & s_\rho s_{2\alpha} s_\mu \\ 0 & 0 & -s_\rho s_{2\alpha} s_\mu & c_\rho+c_{2\alpha} \end{pmatrix}$$

$$\mathbf{M}_{J2} = \frac{d_0}{2p_2}\begin{pmatrix} 1-c_\rho c_{2\alpha} & -s_\rho s_{2\alpha} c_\mu & 0 & 0 \\ -s_\rho s_{2\alpha} c_\mu & 1-c_\rho c_{2\alpha} & 0 & 0 \\ 0 & 0 & c_\rho-c_{2\alpha} & -s_\rho s_{2\alpha} s_\mu \\ 0 & 0 & s_\rho s_{2\alpha} s_\mu & c_\rho-c_{2\alpha} \end{pmatrix}$$

$$\begin{bmatrix} p_1 \equiv \left(1+c_\rho c_{2\alpha}\right)/2 \\ p_2 = 1-p_1 = \left(1-c_\rho c_{2\alpha}\right)/2 \\ s_x \equiv \sin x \quad c_x \equiv \cos x \end{bmatrix}$$

The angles α and μ can take arbitrary values, whereas the value of ρ is determined by $\sin\rho \equiv \sqrt{d_0^2 - d_2^2}/d_0$, $\cos\rho \equiv d_2/d_0$.

As a particular case of Eq. (8.89), $\mathbf{M}_{\Delta d2}$ can be decomposed into two linear retarders whose fast axes are mutually crossed ($\mu = \pi/2$, $\alpha = \pi/4$). Another possible choice where the components are pure diattenuators is obtained by setting $\mu = 0$, $\alpha = \pi/4$. This last decomposition is passive-realizable only when $d_0 \le (1 + d_2^2)/2$.

When $r = 3$, the analytic general expressions for the arbitrary Mueller components of $\mathbf{M}_{\Delta d3}$ and for their respective coefficients are rather complicated and can be found in (Gil et al. 2013). When $r = 4$, a specific algorithm for calculating the arbitrary decomposition of $\mathbf{M}_{\Delta d4}$ can be used (Gil et al. 2013).

An interesting particular example of a parallel decomposition of $\mathbf{M}_{\Delta d4}$ is given by

$$\mathbf{M}_{\Delta d4} = \sum_{i=0}^{3} p_{\Delta d4,i}\,\mathbf{M}_{\Delta d4,i}\quad \begin{cases} \mathbf{M}_{\Delta d4,0} = \dfrac{d_0}{c_0}\begin{pmatrix} t+u\,c_{2\alpha} & r\,s_{2\alpha}c_\mu & 0 & 0 \\ r\,s_{2\alpha}c_\mu & t+u\,c_{2\alpha} & 0 & 0 \\ 0 & 0 & u+t\,c_{2\alpha} & r s_{2\alpha}s_\mu \\ 0 & 0 & -r s_{2\alpha}s_\mu & u+t\,c_{2\alpha} \end{pmatrix} \\[2ex] \mathbf{M}_{\Delta d4,1} = \dfrac{d_0}{c_1}\begin{pmatrix} t-u\,c_{2\alpha} & -r\,s_{2\alpha}c_\mu & 0 & 0 \\ -r\,s_{2\alpha}c_\mu & t-u\,c_{2\alpha} & 0 & 0 \\ 0 & 0 & u-t\,c_{2\alpha} & -r s_{2\alpha}s_\mu \\ 0 & 0 & r s_{2\alpha}s_\mu & u-t\,c_{2\alpha} \end{pmatrix} \\[2ex] \mathbf{M}_{\Delta d4,2} = d_0\,\mathrm{diag}(1,-1,1,-1) \\[1ex] \mathbf{M}_{\Delta d4,3} = d_0\,\mathrm{diag}(1,-1,-1,1) \end{cases}$$

$$\begin{gathered} c_0 \equiv t+u\,c_{2\alpha}\quad c_1 \equiv t\text{-}u\,c_{2\alpha}\quad c_2 \equiv d_0 - d_1 + d_2 - \varepsilon d_3 = 4\lambda_{\Delta d2}\quad c_3 \equiv d_0 - d_1 - d_2 + \varepsilon d_3 = 4\lambda_{\Delta d3} \\ t \equiv d_0 + d_1 = 2(\lambda_{\Delta d0} + \lambda_{\Delta d1})\quad u \equiv d_2 + d_3 = 2(\lambda_{\Delta d0} - \lambda_{\Delta d1})\quad r \equiv \sqrt{t^2 - u^2} = 4\sqrt{\lambda_{\Delta d0}\lambda_{\Delta d1}} \\ p_{\Delta d4,i} = \frac{c_i}{4d_0}\quad (i = 0,1,2,3)\qquad [s_x \equiv \sin x\quad c_x \equiv \cos x] \end{gathered} \tag{8.90}$$

The above example can be readily applied to $\mathbf{M}_{\Delta d3}$ by setting $\lambda_{\Delta d3} = 0$, so that $c_3 = 0$, while the general arbitrary decomposition (8.89) of $\mathbf{M}_{\Delta d2}$ is retrieved by setting $\lambda_{\Delta d2} = \lambda_{\Delta d3} = 0$.

8.7.2 Normal Serial-Parallel Decomposition of a Type-II Mueller Matrix

As we have seen, any type-II Mueller matrix can be expressed as $\mathbf{M} = \mathbf{M}_{J2}\,\mathbf{M}_{\Delta nd}\,\mathbf{M}_{J1}$, where $\mathbf{M}_{\Delta nd}$ is the type-II canonical depolarizer.

An important difference with respect to the previous case of type-I Mueller matrices comes from the fact that, since both the polarizance P and the diattenuation D of $\mathbf{M}_{\Delta nd}$ are not zero ($P = D = 1/2$), at least one parallel component of $\mathbf{M}_{\Delta nd}$ must be a pure diattenuator (in general non-normal). Furthermore, as noted in Section 5.7.2, the cases of rank $\mathbf{H}_{\Delta nd} = 1$ and rank $\mathbf{H}_{\Delta nd} = 4$ are not physically achievable (Ossikovski 2010a). Consequently, the symmetric decomposition of a type-II Mueller matrix $\mathbf{M}$ results in the serial-parallel decomposition (8.86) into two or three pure components, where $\mathbf{M}_i$ are the pure parallel components of $\mathbf{M}_{\Delta nd}$.

As in the case of type-I matrices, each pure component $\mathbf{M}_{J2}\,\mathbf{M}_i\,\mathbf{M}_{J1}$ of the arbitrary decomposition of a type-II matrix can be further reduced to the polar form or to the dual linear retarder transformation considered in Chapter 4.

A particular form of serial-parallel decomposition of a type-II Mueller matrix $\mathbf{M}$ is achieved through the spectral decomposition in Eq. (5.43) of $\mathbf{M}_{\Delta nd}$ (see Section 5.7.2).

As done for type-I Mueller matrices, we consider the two different cases according to the rank r of the covariance matrix $\mathbf{H}(\mathbf{M}_{\Delta nd})$. In what follows, we denote by $\mathbf{M}_{\Delta ndr}$ (where $r = 2, 3$) the respective depolarizer matrices. When $r = 2$, then $0 < a_2 = a_0$ and the general arbitrary parallel decompositions of $\mathbf{M}_{\Delta nd2}$ can be expressed as the following convex combination of two elements (Gil et al. 2013):

$$
\begin{aligned}
\mathbf{M} = \frac{1}{2}(2a_0)&\begin{pmatrix} 1 & -c_\alpha^2 & s_{2\alpha}c_\mu/\sqrt{2} & s_{2\alpha}c_\mu/\sqrt{2} \\ c_\alpha^2 & -c_{2\alpha} & s_{2\alpha}c_\mu/\sqrt{2} & s_{2\alpha}c_\mu/\sqrt{2} \\ s_{2\alpha}c_\mu/\sqrt{2} & -s_{2\alpha}c_\mu/\sqrt{2} & s_\alpha^2 & 0 \\ s_{2\alpha}s_\mu/\sqrt{2} & -s_{2\alpha}s_\mu/\sqrt{2} & 0 & s_\alpha^2 \end{pmatrix} \\
+\frac{1}{2}(2a_0)&\begin{pmatrix} 1 & -s_\alpha^2 & -s_{2\alpha}c_\mu/\sqrt{2} & -s_{2\alpha}c_\mu/\sqrt{2} \\ s_\alpha^2 & c_{2\alpha} & -s_{2\alpha}c_\mu/\sqrt{2} & -s_{2\alpha}c_\mu/\sqrt{2} \\ -s_{2\alpha}c_\mu/\sqrt{2} & s_{2\alpha}c_\mu/\sqrt{2} & c_\alpha^2 & 0 \\ -s_{2\alpha}s_\mu/\sqrt{2} & s_{2\alpha}s_\mu/\sqrt{2} & 0 & c_\alpha^2 \end{pmatrix} \begin{bmatrix} s_x \equiv \sin x \\ c_x \equiv \cos x \end{bmatrix}
\end{aligned}
\tag{8.91}
$$

When $r = 3$, the general analytic expressions for the arbitrary Mueller components of $\mathbf{M}_{\Delta nd3}$ and for their respective coefficients are rather complicated and can be found in (Gil et al. 2013), where some interesting particular cases are also studied.

9

Differential Jones and Mueller Matrices

9.1 Introduction

The Jones and Mueller calculi developed in Chapters 3–6 describe an arbitrary sequence of optical components sequentially crossed by light. Every matrix from the sequence yields the electric field or, more specifically, the state of polarization at the output of the individual component, given the input electric field (in the form of a Jones or a Stokes vector). However, none of the Jones or Mueller matrices considered up to now describes the state of polarization *inside* the component, that is, in between its input and output.

To get a formal description of "what's going in there" one needs, first of all, to operate a conceptual change by considering an optical component not just as a "black box" with an input and an output, but rather as a *continuous* medium modifying the state of polarization of light crossing it at every point along its path-length. Second, one must complement the *discrete* Jones and Mueller formalisms, picturing the propagation of polarized light through the sequence of components as successive matrix multiplications, with differential matrix equations describing the *continuous* evolution of the state of polarization as it travels through the medium. This particular picture leads to the so-called differential Jones and Mueller matrix formalisms that we shall address in this chapter. Note that the differential approach is applicable to media probed by electromagnetic waves of arbitrary wavelength, so the term "light" used here is most generally synonymous of "electromagnetic wave."

The first part of the chapter is devoted to the introduction of the differential Jones and Mueller matrices, expressed in terms of the phenomenological elementary polarization properties of the medium. Next, the differential decomposition of depolarizing Mueller matrices is developed, namely, the mathematical existence, the physical realizability and the multiplicity of the Mueller matrix logarithm are addressed. Finally, after discussing the algebraic structure of the differential Mueller matrix, the model of the homogeneous depolarizing medium, allowing for immediate physical interpretation of polarimetric experimental data in terms of mean values and variances-covariances of the elementary polarization properties, is introduced. The most important theoretical results are illustrated on Jones or Mueller matrices taken from the literature.

DOI: 10.1201/9780367815578-9

9.2 Differential Jones Matrices and Elementary Polarization Properties

9.2.1 Evolution Equation for Continuous Media

Consider a continuous medium whose spatial extension is characterized by the z (linear) coordinate specifying the position of the state of polarization along the path of light ($z = 0$ corresponds to the entrance position or the "input"). In accordance with the fundamental multiplicative property of the Jones (as well as Mueller) calculus, we can express the Jones matrix $\mathbf{T}(z+\Delta z)$ of the medium at the position $z+\Delta z$, Δz being an infinitesimal change in position, as the product of the Jones matrix $\mathbf{T}(z)$ at z and the matrix $\mathbf{T}(z, z+\Delta z)$ of the infinitesimal slab of the medium, of thickness Δz, enclosed between the coordinates z and $z+\Delta z$,

$$\mathbf{T}(z+\Delta z) = \mathbf{T}(z, z+\Delta z)\mathbf{T}(z) \tag{9.1}$$

The above equation can be rewritten as

$$\frac{\mathbf{T}(z+\Delta z)-\mathbf{T}(z)}{\Delta z}\mathbf{T}^{-1}(z) = \frac{\mathbf{T}(z, z+\Delta z)-\mathbf{I}_2}{\Delta z} \tag{9.2}$$

after subtraction of $\mathbf{T}(z)$ from both sides, multiplication by $\mathbf{T}^{-1}(z)$ and division by the slab thickness Δz, provided the Jones matrix $\mathbf{T}(z)$ is non-singular for all z position values (here, $\mathbf{I}_2$ is the 2×2 identity matrix). In the limit $\Delta z \to 0$, the fraction on the left-hand side of Eq. (9.2) will obviously tend to the derivative $d\mathbf{T}(z)/dz$ of the Jones matrix whereas the expression on the right-hand side will, in its turn, tend to a well-defined limit, called *differential Jones matrix*, or N-matrix (Jones 1948) (not to be confused with the matrix $\mathbf{N} \equiv \mathbf{G}\mathbf{M}^T\mathbf{G}\mathbf{M}$ associated with the Mueller matrix $\mathbf{M}$),

$$\mathbf{N}(z) = \underbrace{\lim}_{\Delta z\to 0}\frac{\mathbf{T}(z, z+\Delta z)-\mathbf{I}_2}{\Delta z} = \underbrace{\lim}_{\Delta z\to 0}\frac{\mathbf{T}(z+\Delta z)-\mathbf{T}(z)}{\Delta z}\mathbf{T}^{-1}(z) = \frac{d\mathbf{T}(z)}{dz}\mathbf{T}^{-1}(z) \tag{9.3}$$

Eventually, Eq. (9.3) can be solved for the Jones matrix derivative,

$$\frac{d\mathbf{T}(z)}{dz} = \mathbf{N}(z)\,\mathbf{T}(z) \tag{9.4}$$

effectively providing the evolution equation for the Jones matrix $\mathbf{T}(z)$ along the position z.

The evolution Eq. (9.4) is a first-order linear homogeneous matrix equation. Given the differential Jones matrix $\mathbf{N}(z)$ one can, in principle, integrate Eq. (9.4) to determine the Jones matrix $\mathbf{T}(z)$ at any position z subject to the initial condition $\mathbf{T}(0) = \mathbf{I}_2$.

Eq. (9.4) describes the evolution of not only the Jones matrix $\mathbf{T}(z)$, but also that of the Jones vector $\boldsymbol{\varepsilon}(z)$. Indeed, the substitution of the relation

$$\boldsymbol{\varepsilon}(z) = \mathbf{T}(z)\,\boldsymbol{\varepsilon}(0) \tag{9.5}$$

into Eq. (9.4) results in an analogous evolution equation for $\boldsymbol{\varepsilon}(z)$,

$$\frac{d\boldsymbol{\varepsilon}(z)}{dz} = \mathbf{N}(z)\,\boldsymbol{\varepsilon}(z) \tag{9.6}$$

after an obvious simplification by $\boldsymbol{\varepsilon}(0)$.

Eq. (9.5) provides the state of polarization at the *fixed* position z (that is, at the "output") knowing the Jones matrix of the medium (or of the component) of thickness z and given the state of polarization at $z = 0$ (at the "input"). Consequently, the Jones matrix $\mathbf{T}(z)$ entering Eq. (9.5) appears to be a *global* (or *integral*) descriptor of the polarization properties of the medium (or of the optical component) over the finite path-length (or thickness) z traveled by light. On the contrary, Eq. (9.6), as well as its analogue, Eq. (9.4), describe, respectively, the evolution of the state of polarization and of the Jones matrix at the *current* position z along the light path knowing the differential matrix $\mathbf{N}(z)$ for *all* positions between zero and z. Thus, unlike $\mathbf{T}(z)$, $\mathbf{N}(z)$ is a *local* (or *differential*) descriptor of the polarization properties of the medium (or of the optical component) it refers to.

9.2.2 Definition of the Elementary Polarization Properties

Like $\mathbf{T}(z)$, the Jones differential matrix $\mathbf{N}(z)$ is a 2×2 complex matrix. Every such matrix $\mathbf{A}$ can be expanded on the well-known Pauli basis

$$\boldsymbol{\sigma}_0 = \begin{pmatrix} 1 & 0 \\ 0 & 1 \end{pmatrix} \quad \boldsymbol{\sigma}_1 = \begin{pmatrix} 1 & 0 \\ 0 & -1 \end{pmatrix} \quad \boldsymbol{\sigma}_2 = \begin{pmatrix} 0 & 1 \\ 1 & 0 \end{pmatrix} \quad \boldsymbol{\sigma}_3 = \begin{pmatrix} 0 & -i \\ i & 0 \end{pmatrix} \tag{9.7}$$

resulting in the expansion

$$\mathbf{A} = a_0 \boldsymbol{\sigma}_0 + \sum_{k=1}^{3} a_k \boldsymbol{\sigma}_k \tag{9.8}$$

or, in component form,

$$\mathbf{A} = \begin{pmatrix} a_0 + a_1 & a_2 - ia_3 \\ a_2 + ia_3 & a_0 - a_1 \end{pmatrix} \tag{9.9}$$

The last three of the generally complex expansion coefficients a_i, arranged in the ordered triplet (a_1, a_2, a_3), are usually referred to as the *Pauli vector* $\mathbf{a}$ of the matrix $\mathbf{A}$.

In accordance with Eq. (9.9), the Pauli expansion of the differential Jones matrix $\mathbf{N}$ can be cast in the form

$$\mathbf{N} = \frac{1}{2} \begin{pmatrix} 2\mathrm{A} + \mathrm{L} & \mathrm{L}' - i\mathrm{C} \\ \mathrm{L}' + i\mathrm{C} & 2\mathrm{A} - \mathrm{L} \end{pmatrix} \tag{9.10}$$

in which A is the *isotropic attenuation* of the medium and L, L′ and C are its *complex elementary polarization properties* in "spectroscopic" notations (Schellman and Jensen 1987). Represented explicitly in their algebraic form, the three complex polarization properties are

$$\mathrm{L} = \mathrm{LD} + i\mathrm{LB} \qquad \mathrm{L}' = \mathrm{LD}' + i\mathrm{LB}' \qquad \mathrm{C} = \mathrm{CD} + i\mathrm{CB} \tag{9.11}$$

where "D" stands for *dichroism* and "B" for *birefringence*. Thus, LD and LB are the *linear* dichroism and birefringence with respect to the X and Y laboratory axes; LD′ and LB′ are the analogous quantities with respect to the pair of axes making 45° degrees with X and

Y (usually referred to as ±45° axes) whereas CD and CB are the *circular* dichroism and birefringence. Finally, the isotropic attenuation A can be decomposed into *absorption* AA and *retardation* AR,

$$\mathrm{A} = \mathrm{AA} + i\,\mathrm{AR} \tag{9.12}$$

As unambiguously suggested by their names, all expansion coefficients of $\mathbf{N}$ have a well-defined physical meaning. To see this, express the Jones matrix $\mathbf{T}(z, z+\Delta z)$ of the infinitesimal slab of thickness $\Delta z \to 0$ from Eq. (9.3),

$$\mathbf{T}(z, z+\Delta z) = \mathbf{I}_2 + \mathbf{N}(z)\,\Delta z \tag{9.13}$$

The substitution of $\mathrm{A}\boldsymbol{\sigma}_0$ for $\mathbf{N}$ into Eq. (9.13) yields

$$\mathbf{T}_{\mathrm{A}}(\Delta z) = \begin{pmatrix} 1+\mathrm{A}\Delta z & 0 \\ 0 & 1+\mathrm{A}\Delta z \end{pmatrix} = \begin{pmatrix} \exp(\mathrm{A}\Delta z) & 0 \\ 0 & \exp(\mathrm{A}\Delta z) \end{pmatrix} = \exp(\mathrm{A}\Delta z)\,\mathbf{I}_2 \tag{9.14}$$

where we have used the exponential function series expansion valid in the limit $\Delta z \to 0$ and have omitted, for simplicity, the explicit z-dependence of $\mathbf{T}$ and A. The matrix $\mathbf{T}_{\mathrm{A}}$ simply multiplies the incident Jones vector by the complex scalar $\exp(\mathrm{A}\,\Delta z)$: thus, the amplitude of the electric field is attenuated by the factor $\exp(\mathrm{AA}\,\Delta z)$ (the absorption AA being assumed negative for a *passive* medium) whereas its phase is retarded by $\exp(i\mathrm{AR}\,\Delta z)$, AR being the *isotropic retardation* (or the *global phase shift*). The state of polarization of the electric field depends only on the ratio of the components of the Jones vector and therefore remains unaffected by the action of $\mathbf{T}_{\mathrm{A}}$.

Similarly, the substitution of $\mathrm{L}\,\boldsymbol{\sigma}_1$ for the differential matrix $\mathbf{N}$ into Eq. (9.13) results in

$$\mathbf{T}_{\mathrm{L}}(\Delta z) = \begin{pmatrix} 1+\frac{1}{2}\mathrm{L}\,\Delta z & 0 \\ 0 & 1-\frac{1}{2}\mathrm{L}\,\Delta z \end{pmatrix} = \begin{pmatrix} \exp(\frac{1}{2}\mathrm{L}\,\Delta z) & 0 \\ 0 & \exp(-\frac{1}{2}\mathrm{L}\,\Delta z) \end{pmatrix} \tag{9.15}$$

The eigenvectors of this Jones matrix are obviously

$$\mathbf{e}_x \equiv \begin{pmatrix} 1 \\ 0 \end{pmatrix} \quad \mathbf{e}_y \equiv \begin{pmatrix} 0 \\ 1 \end{pmatrix} \tag{9.16}$$

and correspond to horizontal (H) and vertical (V) polarizations along the laboratory axes X and Y respectively. The ratio of their associated eigenvalues, $\exp(\pm\frac{1}{2}\mathrm{L}\,\Delta z)$, is equal to $\exp(\text{-}\mathrm{L}\,\Delta z) = \exp(\text{-}\mathrm{LD}\,\Delta z)\,\exp(-i\,\mathrm{LB}\,\Delta z)$: thus, the amplitude of the electric field along the Y axis is attenuated by the factor $\exp(-\mathrm{LD}\,\Delta z)$ with respect to that of the X axis, LD being the linear dichroism, while its phase is retarded by $\exp(-i\,\mathrm{LB}\,\Delta z)$, LB being the linear birefringence. Consequently, the matrix $\mathbf{T}_{\mathrm{L}}$ is that of a linear diattenuator-retarder whose eigenaxes are oriented along the X and Y laboratory axes.

In exactly the same way one obtains the following Jones matrix:

$$\mathbf{T}_{\mathrm{L'}}(\Delta z) = \begin{pmatrix} 1 & \frac{1}{2}\mathrm{L'}\Delta z \\ \frac{1}{2}\mathrm{L'}\Delta z & 1 \end{pmatrix} \tag{9.17}$$

after substituting $L'\boldsymbol{\sigma}_2$ for **N** into Eq. (9.13). Its eigenvectors,

$$\mathbf{e}_{+\pi/4} \equiv \frac{1}{\sqrt{2}}\begin{pmatrix}1\\1\end{pmatrix} \qquad \mathbf{e}_{-\pi/4} \equiv \frac{1}{\sqrt{2}}\begin{pmatrix}1\\-1\end{pmatrix} \tag{9.18}$$

correspond to +45° (P) and –45° (Q) polarization states along the ±45° laboratory axes whereas the ratio of its eigenvalues, $1 \pm \frac{1}{2}L'\Delta z = \exp(\pm\frac{1}{2}L'\Delta z)$, equal to $\exp(-L'\Delta z) = \exp(-LD'\Delta z)\exp(-i\,LB'\Delta z)$, expresses the change in amplitude and phase of the electric field along these same axes. The matrix $\mathbf{T}_{L'}$ corresponds to a linear diattenuator-retarder whose eigenaxes are rotated at 45° with respect to the laboratory (X and Y) ones.

Finally, the substitution of $C\boldsymbol{\sigma}_3$ for **N** into Eq. (9.13) yields for the Jones matrix of the infinitesimal slab

$$\mathbf{T}_C(\Delta z) = \begin{pmatrix} 1 & -\frac{1}{2}i\,C\,\Delta z \\ \frac{1}{2}i\,C\,\Delta z & 1 \end{pmatrix} \tag{9.19}$$

whose eigenvectors

$$\mathbf{e}_r \equiv \frac{1}{\sqrt{2}}\begin{pmatrix}1\\i\end{pmatrix} \qquad \mathbf{e}_l \equiv \frac{1}{\sqrt{2}}\begin{pmatrix}1\\-i\end{pmatrix} \tag{9.20}$$

represent right (R) and left (L) circularly polarized states and whose associated eigenvalues are $1 \pm \frac{1}{2}C\,\Delta z = \exp(\pm\frac{1}{2}C\,\Delta z)$. As in the previous two cases, the ratio $\exp(-C\,\Delta z) = \exp(-CD\,\Delta z)\exp(-i\,CB\,\Delta z)$ of the eigenvalues reflects the change in amplitude (through the circular dichroism CD) and in phase (through the circular birefringence CB) between right- (R-) and left- (L-) circularly polarization eigenstates. Therefore, the matrix $\mathbf{T}_C$ represents a circular diattenuator–retarder.

Eq. (9.13) indicates that the dimension of the elementary polarization properties L, L′ and C (as well as that of the isotropic attenuation) is [length^{-1}] that is, they are defined per unit length traveled by light in the medium. This observation is fully consistent with the phenomenological definition of the elementary polarization properties as differences of the propagation constants (that is, of the wavevector lengths) for two orthogonal polarizations,

$$P = i\frac{2\pi}{\lambda}\Delta\tilde{n} = i\frac{2\pi}{\lambda}\left(\tilde{n}_p - \tilde{n}_q\right) = \frac{2\pi}{\lambda}\left(k_p - k_q\right) + i\frac{2\pi}{\lambda}\left(n_p - n_q\right) \tag{9.21}$$

Here, P = L, L′ or C is the complex elementary polarization property; $\tilde{n}_p$ ($\tilde{n}_q$) is the anisotropic complex refractive index "seen" by the state of polarization p (q), orthogonal to q (p); n and k are the respective real refractive index and extinction coefficient ($\tilde{n} = n - ik$ is the complex refractive index) and λ is the wavelength of the probing light. Written explicitly, the series of expressions (9.21) reads

$$\begin{aligned}
LD &= \frac{2\pi}{\lambda}\left(k_H - k_V\right) = \frac{2\pi}{\lambda}\Delta k_{HV} & LB &= \frac{2\pi}{\lambda}\left(n_H - n_V\right) = \frac{2\pi}{\lambda}\Delta n_{HV} \\
LD' &= \frac{2\pi}{\lambda}\left(k_P - k_Q\right) = \frac{2\pi}{\lambda}\Delta k_{PQ} & LB' &= \frac{2\pi}{\lambda}\left(n_P - n_Q\right) = \frac{2\pi}{\lambda}\Delta n_{PQ} \\
CD &= \frac{2\pi}{\lambda}\left(k_R - k_L\right) = \frac{2\pi}{\lambda}\Delta k_{RL} & CB &= \frac{2\pi}{\lambda}\left(n_R - n_L\right) = \frac{2\pi}{\lambda}\Delta n_{RL}
\end{aligned} \tag{9.22}$$

We readily see from Eqs. (9.22) that the dichroic (D) properties of the medium are due to the anisotropy Δk of its extinction coefficient resulting in different absorptions for two orthogonal polarization states, whereas the birefringent (B) properties stem from the anisotropy Δn of the real refractive index entailing different phase shifts for two orthogonal polarization states. The phenomenological definition (9.21) can be also given an equivalent, "operational," formulation in terms of directly measurable quantities (Fresnel coefficients) particularly suitable for reflection measurement configurations (Arteaga et al. 2014). Notice that, in accordance with Eqs. (9.22), the three complex elementary properties, L, L′ and C, can be geometrically associated with the respective axes of the Poincaré sphere, HV, PQ and RL (also denoted as S_1, S_2 and S_3), that correspond physically to XY, ±45° and circular states of polarization).

It should be noted that the six elementary polarization properties, defined phenomenologically by Eqs. (9.22), can be grouped into three pairs in a way alternative to that of Eqs. (9.11), namely (Schellman and Jensen 1987; Arteaga and Canillas 2010)

$$\mathrm{L} = \mathrm{LB} - i\mathrm{LD}\,\mathrm{L}' = \mathrm{LB}' - i\mathrm{LD}'\,\mathrm{C} = \mathrm{CB} - i\mathrm{CD} \tag{9.23}$$

Similarly, the isotropic attenuation A can be defined as

$$\mathrm{A} = \mathrm{AR} - i\mathrm{AA} \tag{9.24}$$

instead of as in Eq. (9.12). The above definitions of L, L′ and C (and A) differ from those of Eq. (9.11) (and of Eq. (9.12)) by the factor i; consequently, the N-matrix from Eq. (9.10), as well as Eq. (9.21), must be divided by the imaginary unit if referring to the definitions from Eqs. (9.23) and (9.24).

Returning to Eqs. (9.15), (9.17) and (9.19), it is easy to see that "stacking" two (or more) infinitesimal slabs containing one elementary property each amounts to having a unique infinitesimal slab simultaneously exhibiting the properties involved. For instance, it follows immediately from Eq. (9.13) that

$$\mathbf{T}_\mathrm{L}(\Delta z)\,\mathbf{T}_\mathrm{C}(\Delta z) = \mathbf{T}_\mathrm{C}(\Delta z)\,\mathbf{T}_\mathrm{L}(\Delta z) = \mathbf{I}_2 + (\mathrm{L}\boldsymbol{\sigma}_1 + \mathrm{C}\boldsymbol{\sigma}_3)\Delta z = \mathbf{T}_{\mathrm{L,C}}(\Delta z) = \mathbf{T}_{\mathrm{C,L}}(\Delta z) \tag{9.25}$$

to the first order in the slab thickness Δz. In other words, unlike finite-thickness Jones matrices, the Jones matrices of infinitesimal slabs commute and are equivalent to a single (infinitesimal) slab whose N-matrix is the sum of the N-matrices of the original slabs.

The additivity of N-matrices (or differential Jones matrices) at the infinitesimal level allows one to obtain the explicit forms of some important infinitesimal Jones matrices. For instance, let us construct the Jones matrix of an (infinitesimal) slab "cut out" of a continuous medium exhibiting dichroic (D) properties only. Putting AR = LB = LB′ = CB = 0 into Eq. (9.10) and substituting into Eq. (9.13) provides

$$\mathbf{T}_\mathrm{D}(\Delta z) = \begin{pmatrix} 1 + \left(\mathrm{AA} + \frac{1}{2}\mathrm{LD}\right)\Delta z & \frac{1}{2}\left(\mathrm{LD}' - i\mathrm{CD}\right)\Delta z \\ \frac{1}{2}\left(\mathrm{LD}' + i\mathrm{CD}\right)\Delta z & 1 + \left(\mathrm{AA} - \frac{1}{2}\mathrm{LD}\right)\Delta z \end{pmatrix} \tag{9.26}$$

The matrix $\mathbf{T}_\mathrm{D}(\Delta z)$ features real diagonal elements and complex conjugate off-diagonal ones. More specifically, Eq. (9.26) represents the most general form of a 2×2 (infinitesimal)

Hermitian matrix parameterized by four real parameters (AA, LD, LD′ and CD). Physically, this matrix corresponds to a general (or elliptical) diattenuator (see Chapter 4). Consequently, the specification of the D elementary properties fully defines the infinitesimal elliptical diattenuator. Alternatively, the substitution of AA = LD = LD′ = CD = 0 into Eqs. (9.10) and (9.13) yields for the infinitesimal Jones matrix of a purely birefringent (B), non-absorbing medium,

$$\mathbf{T}_{\mathrm{B}}(\Delta z) = \begin{pmatrix} 1 + i\left(\mathrm{AR} + \frac{1}{2}\mathrm{LB}\right)\Delta z & \frac{1}{2}\left(i\mathrm{LB}' + \mathrm{CB}\right)\Delta z \\ \frac{1}{2}\left(i\mathrm{LB}' - \mathrm{CB}\right)\Delta z & 1 + i\left(\mathrm{AR} - \frac{1}{2}\mathrm{LB}\right)\Delta z \end{pmatrix} \tag{9.27}$$

To the first order in Δz, the matrix $\mathbf{T}_{\mathrm{B}}(\Delta z)$ features orthogonal columns with unit length each. This matrix is the most general form of a 2×2 (infinitesimal) unitary matrix, characterized by four real parameters (AR, LB, LB′ and CB) and, according to Chapter 4, represents physically a (infinitesimal) general (or elliptical) retarder. Thus, an infinitesimal slab of continuous medium exhibiting only B properties behaves as an elliptical retarder. Note that when $\Delta z \to 0$ both $\mathbf{T}_{\mathrm{D}}(\Delta z)$ and $\mathbf{T}_{\mathrm{B}}(\Delta z)$ matrices tend to the identity, in conformity with the initial condition $\mathbf{T}(0) = \mathbf{I}_2$.

In summary, the differential Jones matrix **N** fully describes the local polarization properties of a continuous medium. Its expansion on the Pauli basis allows for its physically meaningful parameterization in terms of a (complex) isotropic attenuation and six elementary polarization properties (LD, LB, LD′, LB′, CD and CB), grouped in three complex pairs (L, L′ and C).

9.2.3 Elementary Polarization Properties and Jones Matrices of Homogeneous Media

If the continuous medium is *homogeneous* that is, if its differential Jones matrix **N** is z-independent, the integration of the evolution equation (9.4) over a finite path-length z is straightforward. The formal solution is given by

$$\mathbf{T}(z) = \exp\left(\int_0^z \mathbf{N}\,dz\right) = \exp(\mathbf{N}z) \tag{9.28}$$

subject to the initial condition $\mathbf{T}(0) = \mathbf{I}_2$. Physically, the elementary polarization properties of a homogeneous medium remain constant along the light path and their cumulated amounts grow linearly with the path-length z.

The use of the Pauli-algebra analogue of the well-known Euler formula,

$$\exp\left(i\sum_{k=1}^{3} a_k \boldsymbol{\sigma}_k\right) = \cos a\,\boldsymbol{\sigma}_0 + i\sin a\left(\sum_{k=1}^{3} a_k \boldsymbol{\sigma}_k\right) \tag{9.29}$$

where $a = \sqrt{a_1^2 + a_2^2 + a_3^2}$ is the magnitude of vector **a**, allows one to exponentiate the general matrix **N** from Eq. (9.10) after its substitution into Eq. (9.28). The resulting Jones matrix of a homogeneous medium of finite thickness z, exhibiting all three pairs L, L′ and C of elementary polarization properties (as well as complex isotropic attenuation A) is (Jones 1948; Huard 1997)

$$\mathbf{T}(z)=\exp(\mathbf{N}z)=\exp(\mathrm{A}z)\begin{pmatrix}\cosh\frac{\mathrm{T}z}{2}+\frac{\mathrm{L}}{\mathrm{T}}\sinh\frac{\mathrm{T}z}{2} & \frac{\mathrm{L}'-i\mathrm{C}}{\mathrm{T}}\sinh\frac{\mathrm{T}z}{2}\\ \frac{\mathrm{L}'+i\mathrm{C}}{\mathrm{T}}\sinh\frac{\mathrm{T}z}{2} & \cosh\frac{\mathrm{T}z}{2}-\frac{\mathrm{L}}{\mathrm{T}}\sinh\frac{\mathrm{T}z}{2}\end{pmatrix} \tag{9.30}$$

where $\mathrm{T}=\sqrt{\mathrm{L}^2+\mathrm{L}'^2+\mathrm{C}^2}$.

If, instead of the definitions (9.11) and (9.12) for L, L′, C and A, those of Eqs. (9.23) and (9.24) are used, then Eq. (9.30) takes the alternative form

$$\mathbf{T}(z)=\exp(i\mathrm{A}z)\begin{pmatrix}\cos\frac{\mathrm{T}z}{2}+\frac{i\mathrm{L}}{\mathrm{T}}\sin\frac{\mathrm{T}z}{2} & \frac{\mathrm{C}-i\mathrm{L}'}{\mathrm{T}}\sin\frac{\mathrm{T}z}{2}\\ \frac{\mathrm{C}+i\mathrm{L}'}{\mathrm{T}}\sin\frac{\mathrm{T}z}{2} & \cos\frac{\mathrm{T}z}{2}-\frac{i\mathrm{L}}{\mathrm{T}}\sin\frac{\mathrm{T}z}{2}\end{pmatrix} \tag{9.31}$$

Eq. (9.30) represents the most general physical parameterization of the Jones matrix of a homogeneous medium. Proceeding as in Section 9.2.2, it is a straightforward exercise to study important special cases of the general expression. Thus, if the medium of thickness z exhibits the L elementary property only, its Jones matrix will be

$$\mathbf{T}_{\mathrm{L}}(z)=\begin{pmatrix}\cosh\frac{\mathrm{L}z}{2}+\sinh\frac{\mathrm{L}z}{2} & 0\\ 0 & \cosh\frac{\mathrm{L}z}{2}-\sinh\frac{\mathrm{L}z}{2}\end{pmatrix}=\begin{pmatrix}\exp\left(\frac{1}{2}\mathrm{L}z\right) & 0\\ 0 & \exp\left(-\frac{1}{2}\mathrm{L}z\right)\end{pmatrix} \tag{9.32}$$

to be compared to that of the infinitesimal slab reported in Eq. (9.15). The cases "L′ only" and "C only" are dealt with in the same way. In all three cases, the eigenvectors remain unchanged with respect to those of Section 9.2.2, see Eqs. (9.16), (9.18) and (9.20), whereas the associated eigenvalues are obtained by simply replacing the infinitesimal thickness Δz by the finite one, z.

If the medium exhibits D-properties only (and AR = 0 also), its Jones matrix

$$\mathbf{T}_{\mathrm{D}}(z)=\exp(\mathrm{AA}\,z)\begin{pmatrix}\cosh\frac{\mathrm{D}z}{2}+\frac{\mathrm{LD}}{\mathrm{D}}\sinh\frac{\mathrm{D}z}{2} & \frac{\mathrm{LD}'-i\mathrm{CD}}{\mathrm{D}}\sinh\frac{\mathrm{D}z}{2}\\ \frac{\mathrm{LD}'+i\mathrm{CD}}{\mathrm{D}}\sinh\frac{\mathrm{D}z}{2} & \cosh\frac{\mathrm{D}z}{2}-\frac{\mathrm{LD}}{\mathrm{D}}\sinh\frac{\mathrm{D}z}{2}\end{pmatrix} \tag{9.33}$$

is Hermitian and represents physically an elliptical diattenuator. The quantity $\mathrm{D}=\sqrt{\mathrm{LD}^2+\mathrm{LD}'^2+\mathrm{CD}^2}$ represents the total (or net) dichroism. Conversely, if only B-properties are present (and AA = 0 also, i.e., the medium is non-absorbing) the resulting Jones matrix

$$\mathbf{T}_{\mathrm{B}}(z)=\exp(i\mathrm{AR}\,z)\begin{pmatrix}\cos\frac{\mathrm{B}z}{2}+\frac{i\mathrm{LB}}{\mathrm{B}}\sin\frac{\mathrm{B}z}{2} & \frac{\mathrm{CB}+i\mathrm{LB}'}{\mathrm{B}}\sin\frac{\mathrm{B}z}{2}\\ \frac{-\mathrm{CB}+i\mathrm{LB}'}{\mathrm{B}}\sin\frac{\mathrm{B}z}{2} & \cos\frac{\mathrm{B}z}{2}-\frac{i\mathrm{LB}}{\mathrm{B}}\sin\frac{\mathrm{B}z}{2}\end{pmatrix} \tag{9.34}$$

is unitary and describes an elliptical retarder, $B = \sqrt{LB^2 + LB'^2 + CB^2}$ being the total (or net) birefringence.

9.2.4 Extraction of the Elementary Polarization Properties from the Jones Matrix

In practice, one typically measures (or models) the Jones matrix of a continuous medium and, consequently, one wants to know what the elementary polarization properties of the medium are. For a homogeneous medium, Eq. (9.28) is readily solved for the differential matrix **N**,

$$\mathbf{N} = \frac{1}{z} \ln \mathbf{T}(z) \tag{9.35}$$

provided the Jones matrix $\mathbf{T}(z)$ is non-singular. This relation can be extended as is to the *inhomogeneous* case also; in such case the differential matrix **N** will contain the elementary polarization properties of an *equivalent homogeneous* medium having the same Jones matrix **T**. After setting the path-length z to unit ($z = 1$) in Eq. (9.35) we can write in general

$$\mathbf{N} = \ln \mathbf{T} \tag{9.36}$$

for any non-singular **T**. Notice that now the N-matrix contains the elementary properties cumulated over the finite path-length from zero to $z = 1$, that is, the dimension of the properties is not [length^{-1}] like in Eq. (9.35) but they are rather dimensionless (the B-properties are measured in radians or sometimes, in degrees).

The most common mathematical procedure for evaluating the matrix logarithm entering Eq. (9.36) is based on the diagonalization of **T**,

$$\mathbf{T} = \mathbf{W}\,\Lambda\,\mathbf{W}^{-1} \tag{9.37}$$

where

$$\Lambda = \operatorname{diag}(\lambda_1, \lambda_2) \tag{9.38}$$

is a diagonal matrix containing the eigenvalues λ_1 and λ_2 of **T** and **W** is a non-singular matrix formed by stacking in columns the eigenvectors of **T**. If, prior to the diagonalization, the Jones matrix **T** is normalized to unit determinant – that is, if it is divided by $\sqrt{\det \mathbf{T}}$ (assuming the common convention of taking the *representative* Jones matrix **T** satisfying $\det \mathbf{T} \in \mathbb{R}$; see Section 3.2.1), the two eigenvalues λ_1 and λ_2 will then be necessarily of the form $\lambda_1 = \lambda \exp(i\varphi)$ and $\lambda_2 = \lambda^{-1} \exp(-i\varphi)$ with λ real and positive and ϕ real. Without going into further detail, the differential matrix **N** is then given by

$$\mathbf{N} = \mathbf{W} \ln\Lambda\, \mathbf{W}^{-1} + p\,\mathbf{W}\,\Delta\Lambda\,\mathbf{W}^{-1} = \mathbf{N}_0 + p\,\Delta\mathbf{N} \tag{9.39}$$

where p is an integer ($p = 0, \pm 1, \pm 2 \ldots$);

$$\ln \Lambda = \operatorname{diag}(\ln \lambda_1, \ln \lambda_2) = \operatorname{diag}(\ln \lambda + i\varphi, -\ln \lambda - i\varphi) = (\ln \lambda + i\varphi) \operatorname{diag}(1, -1) \tag{9.40}$$

is the so-called principal logarithm of $\mathbf{\Lambda}$ and

$$\Delta\mathbf{\Lambda} = \pi i \,\mathrm{diag}(1,-1) \tag{9.41}$$

Finally, identification of the matrix **N** thus obtained with its explicit definition (9.10) immediately yields the three pairs of elementary polarization properties L, L′ and C. Note that, if the Jones matrix **T** is non-diagonalizable, its matrix logarithm can be still evaluated (by using a modified procedure), but this particular case is of academic interest only and will not be treated here. It should also be noted that an alternative procedure not using the matrix logarithm, but rather based on the direct identification of the Jones matrix **T** with its parameterized form (9.30), can also be used (Arteaga and Canillas 2010).

Eq. (9.39) shows that, because of the matrix logarithm being a multivalued function, the matrix **N** is not unique, but rather depends on the choice of the integer p, called *order*. The multiplicity of solutions for **N** for a given **T** is physically due to the periodicity of the birefringent (B) properties of the medium. Possibly the simplest example is that of the waveplate or linear retarder (see Chapter 4): its Jones matrix defines the retardance value only within $2\pi p$ where p is the waveplate order. To determine the value of p, a measurement at a different wavelength has to be performed. In the general case where both birefringent (B) and dichroic (D) properties are present simultaneously in the medium, the values of all elementary properties depend on the order p in an intricate manner. In the special case where only D properties are present in the medium, see Eq. (9.33), all B properties are zero at zero order ($p = 0$) and change their values to $2\pi\, p$ at orders p different from zero.

To illustrate the above procedure for finding the elementary polarization properties, consider the following Jones matrix:

$$\mathbf{T} = \tfrac{1}{2}\begin{pmatrix} 11 & 5 \\ 7 & 9 \end{pmatrix} \tag{9.42}$$

taken from R. C. Jones' seventh paper (Jones 1948). This matrix is non-singular and can therefore represent the polarimetric response of some continuous homogeneous medium (provided it is further normalized by a suitable factor in order to satisfy the passivity condition; see Section 3.3.4). The application of Eq. (9.39) to **T** yields

$$\mathbf{N}_0 = \begin{pmatrix} 0.1155 & 0.5776 \\ 0.8087 & -0.1155 \end{pmatrix} \tag{9.43}$$

and

$$\Delta\mathbf{N} = \pi \begin{pmatrix} 0.1667i & 0.8333i \\ 1.1667i & -0.1667i \end{pmatrix} \tag{9.44}$$

(Notice that $\mathbf{N}_0$ turns out to be real while $\Delta\mathbf{N}$ is imaginary, which is clearly a special case since both are generally complex.) The identification of $\mathbf{N}_0$ with Eq. (9.10) produces the following list of elementary polarization properties at zero order ($p = 0$):

$$\begin{array}{lll} \mathrm{LD} = 0.2310 & \mathrm{LD}' = 1.3863 & \mathrm{CD} = 0 \\ \mathrm{LB} = 0 & \mathrm{LB}' = 0 & \mathrm{CB} = -0.2310 \end{array} \tag{9.45}$$

whereas the first-order properties ($p = 1$), found from $\mathbf{N}_0 + \Delta\mathbf{N}$, are

$$\begin{array}{lll} \text{LD} = 0.2310 & \text{LD}' = 1.3863 & \text{CD} = 1.0472 \\ \text{LB} = 1.0472 & \text{LB}' = 6.2832 & \text{CB} = -0.2310 \end{array} \tag{9.46}$$

Thus, at zero order, the medium exhibits general linear dichroism (both XY and ±45°) as well as circular birefringence (or optical activity), the remaining three of the properties being identically zero. At first order, all properties are nonzero; the general linear dichroism and the circular birefringence values remain unchanged whereas the value of the ±45° birefringence "jumps" from zero to 2π. Unlike the ±45° birefringence, both the linear birefringence and circular dichroism "jump" from zero values to $1.0472 = \pi/3$. The medium exhibits a complex anisotropic behavior and, to determine the value of the order p, an additional polarimetric measurement at a different wavelength is necessary.

9.3 Differential Mueller Matrices

9.3.1 Differential Mueller Matrix of a Nondepolarizing Medium

We have seen in the previous chapters that all relations from the Jones calculus can be reformulated equivalently in the framework of the more general Mueller formalism. In particular, the evolution Eqs. (9.4) and (9.6) take the equivalent forms (Azzam 1978),

$$\frac{d\mathbf{M}_J(z)}{dz} = \mathbf{m}(z)\mathbf{M}_J(z) \tag{9.47}$$

and

$$\frac{d\mathbf{s}(z)}{dz} = \mathbf{m}(z)\,\mathbf{s}(z) \tag{9.48}$$

for the pure (or nondepolarizing) Mueller matrix $\mathbf{M}_J$ and the Stokes vector $\mathbf{s}$ at the position z. Here, $\mathbf{m}$ is the differential Mueller matrix that is generally z-dependent, like $\mathbf{M}_J$ and $\mathbf{s}$; see Figure 9.1. (Note that, although the notation $\mathbf{m}$ has also been used to refer to the 3×3

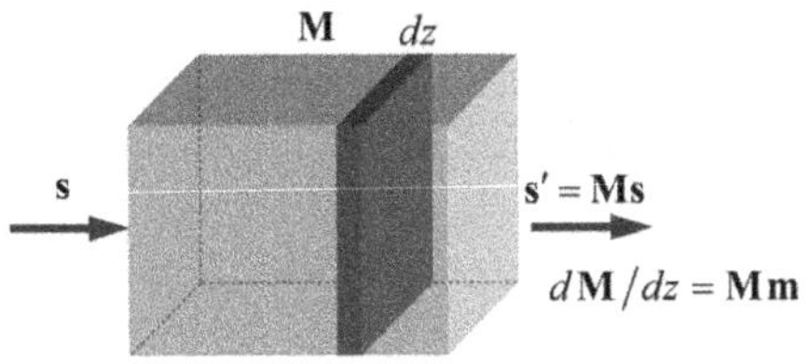

FIGURE 9.1
The polarimetric action of a differential slab of a continuous medium is characterized by the differential Mueller matrix **m**.

submatrix of a Mueller matrix **M**, we shall use it to refer to the differential Mueller matrix throughout this chapter, both meanings being easily distinguishable from the context.)

Since the differential Jones matrix **N** is described by eight real parameters (the six elementary properties and the isotropic absorption and retardation), **m** must depend correspondingly on the first seven of them (the isotropic retardation, or global phase shift, is not a measurable quantity within the Mueller formalism; see Chapter 3).

Consequently, **m** having 16 real elements in general, it must obey certain symmetries. To identify them, recall that every nondepolarizing Mueller matrix *with unit determinant* satisfies the relationship (Xing 1992)

$$\mathbf{G}\,\mathbf{M}_J^T\,\mathbf{G} = \mathbf{M}_J^{-1} \tag{9.49}$$

where $\mathbf{G} = \mathrm{diag}(1,-1,-1,-1)$ is the Minkowski metric matrix (see Eq. (3.62)). If we define the G-transpose of **M** by the expression $\mathbf{M}^G = \mathbf{G}\mathbf{M}^T\mathbf{G}$, then Eq. (9.49) expresses the fact that a unit-determinant pure (or nondepolarizing) $\mathbf{M}_J$ is G-orthogonal, $\mathbf{M}_J^G = \mathbf{M}_J^{-1}$. The use of this fundamental relation in conjunction with the definition of the differential Mueller matrix,

$$\mathbf{m} = \frac{d\mathbf{M}_J}{dz}\mathbf{M}_J^{-1} \tag{9.50}$$

allows one to show readily that **m** obeys the relation

$$\mathbf{G}\mathbf{m}^T\mathbf{G} = \mathbf{m}^G = -\mathbf{m} \tag{9.51}$$

that is, **m** is a G-antisymmetric matrix and is, consequently, parameterized by only six parameters, namely the six elementary polarization properties. Notice that the seventh parameter, the isotropic absorption, did not appear explicitly since we considered a unit-determinant $\mathbf{M}_J$; if instead one substitutes $\alpha\mathbf{M}_J$ into Eq. (9.49) where α is the z-dependent transmission of the medium, one readily finds that one should then add, to the G-antisymmetric differential matrix **m**, the matrix $\beta\mathbf{I}_4$ where $\beta = \alpha^{-1}\, d\alpha/dz$ ($\mathbf{I}_4$ is the 4×4 identity matrix here). However, the isotropic absorption does not interact with the elementary polarization properties and is rarely measured in polarimetric experiments; we shall therefore set it to zero for simplicity and not discuss it anymore.

To establish the explicit form of the differential Mueller matrix **m**, use can be made of the definition of the Mueller matrix $\mathbf{M}_J$ of a nondepolarizing medium in terms of its Jones matrix **T** (see Chapter 3),

$$\mathbf{M}_J = \mathcal{L}\left(\mathbf{T}\otimes\mathbf{T}^*\right)\mathcal{L}^{-1} \tag{9.52}$$

where "⊗" stands for the Kronecker product and the transformation matrix $\mathcal{L}$ is given by

$$\mathcal{L} = \frac{1}{\sqrt{2}}\begin{pmatrix} 1 & 0 & 0 & 1 \\ 1 & 0 & 0 & -1 \\ 0 & 1 & 1 & 0 \\ 0 & i & -i & 0 \end{pmatrix} \tag{9.53}$$

Substituting $\mathbf{M}_J$ from Eq. (9.52) into Eq. (9.50) and taking advantage of the distributive property of the Kronecker product, $(\mathbf{A}\otimes\mathbf{B})(\mathbf{C}\otimes\mathbf{D}) = \mathbf{AC}\otimes\mathbf{BD}$, results in the following expression for differential Mueller matrix **m**:

$$\mathbf{m} = \mathcal{L}\left(\mathbf{N}\otimes\mathbf{I}_2 + \mathbf{I}_2\otimes\mathbf{N}^*\right)\mathcal{L}^{-1} \tag{9.54}$$

in terms of the differential Jones matrix **N** (here, $\mathbf{I}_2$ is the 2×2 identity matrix) (Barakat 1996a). The substitution of the parameterized **N** from Eq. (9.10) (with A = 0) into Eq. (9.54) yields for **m**,

$$\mathbf{m} = \begin{pmatrix} 0 & \mathrm{Re\,L} & \mathrm{Re\,L'} & \mathrm{Re\,C} \\ \mathrm{Re\,L} & 0 & \mathrm{Im\,C} & -\mathrm{Im\,L'} \\ \mathrm{Re\,L'} & -\mathrm{Im\,C} & 0 & \mathrm{Im\,L} \\ \mathrm{Re\,C} & \mathrm{Im\,L'} & -\mathrm{Im\,L} & 0 \end{pmatrix} = \begin{pmatrix} 0 & \mathrm{LD} & \mathrm{LD'} & \mathrm{CD} \\ \mathrm{LD} & 0 & \mathrm{CB} & -\mathrm{LB'} \\ \mathrm{LD'} & -\mathrm{CB} & 0 & \mathrm{LB} \\ \mathrm{CD} & \mathrm{LB'} & -\mathrm{LB} & 0 \end{pmatrix} \tag{9.55}$$

after taking into account the definitions (9.11) of the complex polarization properties L, L′ and C.

As expected, **m** is indeed G-antisymmetric, i.e., it satisfies Eq. (9.51). The differential matrix **m** can be further decomposed into symmetric and antisymmetric parts,

$$\mathbf{m} = \begin{pmatrix} 0 & \mathrm{LD} & \mathrm{LD'} & \mathrm{CD} \\ \mathrm{LD} & 0 & 0 & 0 \\ \mathrm{LD'} & 0 & 0 & 0 \\ \mathrm{CD} & 0 & 0 & 0 \end{pmatrix} + \begin{pmatrix} 0 & 0 & 0 & 0 \\ 0 & 0 & \mathrm{CB} & -\mathrm{LB'} \\ 0 & -\mathrm{CB} & 0 & \mathrm{LB} \\ 0 & \mathrm{LB'} & -\mathrm{LB} & 0 \end{pmatrix} = \mathbf{m}_\mathrm{D} + \mathbf{m}_\mathrm{B} \tag{9.56}$$

separating the dichroic (D) from the birefringent (B) properties.

Eq. (9.55) represents the most general differential Mueller matrix of a continuous nondepolarizing medium exhibiting all six elementary polarization properties (Azzam 1978). Like **N**, **m** is also z-dependent, in general.

9.3.2 The Mueller Matrix of a Homogeneous Nondepolarizing Medium

When the nondepolarizing continuous medium is *homogeneous*, that is, its elementary polarization properties and therefore, its differential Mueller matrix do not depend on the position z along the light path, Eq. (9.47) is readily integrated exactly as in the Jones matrix case (see Section 9.2.3),

$$\mathbf{M}_J(z) = \exp\left(\int_0^z \mathbf{m}\,dz\right) = \exp(\mathbf{m}\,z) \tag{9.57}$$

We see that the cumulative polarization properties of the homogeneous medium grow linearly with the distance z traveled by the electromagnetic wave into the medium.

Although realizable in closed form, the exponentiation of the general differential matrix **m** from Eq. (9.55) results in a very cumbersome expression for the associated Mueller matrix $\mathbf{M}_J$ (Schellman and Jensen 1987) and will not be reproduced here. However, it is very easy to see that the exponentiation of the *dichroic* (D) and *birefringence* (B) parts of $\mathbf{m}$ produces the Mueller matrices of an elliptical diattenuator and an elliptical retarder respectively. Indeed, by substituting consecutively $\mathbf{m}_\mathrm{D}$ and $\mathbf{m}_\mathrm{B}$ into Eq. (9.57) and taking the transpose, one readily obtains (at $z = 1$ for simplicity)

$$\mathbf{M}_\mathrm{D}^T = \exp\left(\mathbf{m}_\mathrm{D}^T\right) = \exp\left(\mathbf{m}_\mathrm{D}\right) = \mathbf{M}_\mathrm{D} \qquad \mathbf{M}_\mathrm{B}^T = \exp\left(\mathbf{m}_\mathrm{B}^T\right) = \exp\left(-\mathbf{m}_\mathrm{B}\right) = \mathbf{M}_\mathrm{B}^{-1} \tag{9.58}$$

where we have used the fact that the two differential matrices $\mathbf{m}_\mathrm{D}$ and $\mathbf{m}_\mathrm{B}$ are symmetric and antisymmetric respectively. Thus, the associated Mueller matrices $\mathbf{M}_\mathrm{D}$ and $\mathbf{M}_\mathrm{B}$ turn out to be symmetric and orthogonal respectively that is, they represent a general (elliptical) diattenuator and a general (elliptical) retarder (see Chapter 4). The explicit expressions of $\mathbf{M}_\mathrm{D}$ and $\mathbf{M}_\mathrm{B}$ in terms of the elementary polarization properties are, however, quite involved (Azzam 1978) and will not be reported.

It is also of potential interest to establish when the two differential matrices $\mathbf{m}_\mathrm{D}$ and $\mathbf{m}_\mathrm{B}$ commute. By using their explicit forms from Eq. (9.56), it is a straightforward exercise to check that their commutator

$$\left[\mathbf{m}_\mathrm{D}, \mathbf{m}_\mathrm{B}\right] = \mathbf{m}_\mathrm{D}\,\mathbf{m}_\mathrm{B} - \mathbf{m}_\mathrm{B}\,\mathbf{m}_\mathrm{D} \tag{9.59}$$

vanishes if and only if

$$\mathrm{LD} : \mathrm{LB} = \mathrm{LD}' : \mathrm{LB}' = \mathrm{CD} : \mathrm{CB} \tag{9.60}$$

that is, when the dichroic (D) and birefringent (B) vectors $\vec{\mathrm{D}}$ and $\vec{\mathrm{B}}$ defined respectively as

$$\vec{\mathrm{D}} = \left(\mathrm{LD}, \mathrm{LD}', \mathrm{CD}\right)^T \quad \text{and} \quad \vec{\mathrm{B}} = \left(\mathrm{LB}, \mathrm{LB}', \mathrm{CB}\right)^T \tag{9.61}$$

in Poincaré polarization space are parallel, $\vec{\mathrm{D}} \parallel \vec{\mathrm{B}}$. (Note that the above "arrow" notation for dichroic and birefringence vectors has been introduced in order to avoid any confusion with other vectorial quantities used, like the diattenuation vector $\mathbf{D}$.)

In virtue of the factorization property of the exponential of two commuting matrices, we then obtain

$$\mathbf{M}_J = \exp\left(\mathbf{m}_\mathrm{D} + \mathbf{m}_\mathrm{B}\right) = \exp\left(\mathbf{m}_\mathrm{D}\right)\exp\left(\mathbf{m}_\mathrm{B}\right) = \mathbf{M}_\mathrm{D}\,\mathbf{M}_\mathrm{B} = \exp\left(\mathbf{m}_\mathrm{B}\right)\exp\left(\mathbf{m}_\mathrm{D}\right) = \mathbf{M}_\mathrm{B}\,\mathbf{M}_\mathrm{D} \tag{9.62}$$

that is, Mueller matrix $\mathbf{M}_J$ factors down to the product of a general diattenuator $\mathbf{M}_\mathrm{D}$ and a general retarder $\mathbf{M}_\mathrm{B}$ that commute. This is a special case of the polar decomposition of a nondepolarizing Mueller matrix (Gil and Bernabéu 1987) whereby the two polar factors $\mathbf{M}_\mathrm{D}$ and $\mathbf{M}_\mathrm{B}$ do not commute in general (see Section 4.4). In particular, when the medium exhibits a single nonzero elementary polarization property, say $\mathrm{L} = \mathrm{LD} + i\,\mathrm{LB}$, then the condition (9.60) is trivially satisfied and one gets by inserting Eq. (9.55) with $\mathrm{LD}' = \mathrm{LB}' = \mathrm{CD} = \mathrm{CB} = 0$ into Eq. (9.62) that

$$\mathbf{M}_{\mathrm{L}} = \begin{pmatrix} \cosh \mathrm{LD} & \sinh \mathrm{LD} & 0 & 0 \\ \sinh \mathrm{LD} & \cosh \mathrm{LD} & 0 & 0 \\ 0 & 0 & \cos \mathrm{LB} & \sin \mathrm{LB} \\ 0 & 0 & -\sin \mathrm{LB} & \cos \mathrm{LB} \end{pmatrix}$$
$$= \begin{pmatrix} \cosh \mathrm{LD} & \sinh \mathrm{LD} & 0 & 0 \\ \sinh \mathrm{LD} & \cosh \mathrm{LD} & 0 & 0 \\ 0 & 0 & 1 & 0 \\ 0 & 0 & 0 & 1 \end{pmatrix} \begin{pmatrix} 1 & 0 & 0 & 0 \\ 0 & 1 & 0 & 0 \\ 0 & 0 & \cos \mathrm{LB} & \sin \mathrm{LB} \\ 0 & 0 & -\sin \mathrm{LB} & \cos \mathrm{LB} \end{pmatrix} = \mathbf{M}_{\mathrm{D}}\mathbf{M}_{\mathrm{B}} = \mathbf{M}_{\mathrm{B}}\mathbf{M}_{\mathrm{D}} \tag{9.63}$$

i.e., the polar factors of a Mueller matrix exhibiting a single property only commute. The matrix $\mathbf{M}_{\mathrm{L}}$ is easily identifiable as being proportional to the well-known 'psi-delta' matrix describing a plane isotropic surface in reflection ellipsometry (Azzam and Bashara 1987).

9.3.3 Extraction of the Elementary Polarization Properties from a Nondepolarizing Mueller Matrix

For a *homogeneous and nondepolarizing* medium whose Mueller matrix $\mathbf{M}_J$ is known (either from measurement or model) the elementary polarization properties, contained in the differential matrix $\mathbf{m}$, are simply obtained by taking the matrix logarithm of $\mathbf{M}_J$,

$$\mathbf{m} = \ln \mathbf{M}_J \tag{9.64}$$

as follows directly from Eq. (9.57). (As before, we have put $z = 1$ assuming "integrated," dimensionless values of the properties.) Note that the expression (9.64) is likewise applicable to Mueller matrices of *inhomogeneous* (but still nondepolarizing) media in which case the elementary polarization properties are to be interpreted as being those of an *equivalent homogeneous* medium (with an equivalent, z-independent, differential matrix $\mathbf{m}$) having the same Mueller matrix as the original inhomogeneous medium.

Like in the Jones matrix case (see Section 9.2.4) the most common mathematical procedure for evaluating the matrix logarithm is based on the diagonalization of $\mathbf{M}_J$,

$$\mathbf{M}_J = \mathbf{V}\mathbf{D}\mathbf{V}^{-1} \tag{9.65}$$

where $\mathbf{V}$ is a non-singular, generally complex-valued, matrix containing the eigenvectors of $\mathbf{M}_J$ whereas the diagonal matrix $\mathbf{D}$ contains the eigenvalues μ_i of $\mathbf{M}_J$ (the context prevents confusion between the matrix $\mathbf{D}$ and the eponymous diattenuation vector). With the help of Eq. (9.52) defining the nondepolarizing $\mathbf{M}_J$ from its associated Jones matrix $\mathbf{T}$ and the properties of the Kronecker product, it is easy to establish the following relations between the eigenvalues μ_i of $\mathbf{M}_J$ and those, λ_1 and λ_2, of $\mathbf{T}$:

$$\mu_1 = |\lambda_1|^2 \quad \mu_2 = |\lambda_2|^2 \quad \mu_3 = \lambda_1\lambda_2^* \quad \mu_4 = \lambda_1^*\lambda_2 \tag{9.66}$$

We see that, in general, two of the eigenvalues of a nondepolarizing Mueller matrix are real and non-negative whereas the other two form a complex conjugate pair, $\mu_{3,4} = \mu\exp(\pm i\varphi)$

with $\mu > 0$. In the special case where the eigenvalues of **T** are both real, those of $\mathbf{M}_J$ are likewise real, the last two then forming a doubly degenerate eigenvalue.

The differential matrix **m** is obtained in a way similar to that of Eq. (9.39),

$$\mathbf{m} = \mathbf{V}\left(\ln\mathbf{D}\right)\mathbf{V}^{-1} + p\,\mathbf{V}\left(\Delta\mathbf{D}\right)\mathbf{V}^{-1} = \mathbf{m}_0 + p\,\Delta\mathbf{m} \tag{9.67}$$

where

$$\ln\mathbf{D} = \mathrm{diag}\left(\ln\mu_1, \ln\mu_2, \ln\mu_3, \ln\mu_4\right) = \mathrm{diag}\left(\ln\mu_1, \ln\mu_2, \ln\mu + i\varphi, \ln\mu - i\varphi\right) \tag{9.68}$$

is the principal logarithm of **D**, and

$$\Delta\mathbf{D} = 2\pi i\,\mathrm{diag}\left(0, 0, 1, -1\right) \tag{9.69}$$

p being the order of the differential matrix (p = 0, ±1, ±2,…) (Devlaminck and Ossikovski 2014).

Unlike in the Jones matrix case, no normalization of $\mathbf{M}_J$ to a unit determinant is necessary prior to applying the procedure. (Note that the two matrices $\mathbf{m}_0$ and $\Delta\mathbf{m}$ are real as expected despite **W** and ln**D** being generally complex.)

The above expressions show that there are multiple differential matrices **m**, parameterized by p, corresponding to a given nondepolarizing $\mathbf{M}_J$ because of the periodicity of the birefringent (B) properties contained in **m**. The special case of all eigenvalues of $\mathbf{M}_J$ being real requires particular considerations for handling the multiplicity (i. e. for $p \neq 0$) that we shall not discuss here. We shall only notice that, in this case, the associated Jones matrix **T** can be obtained from $\mathbf{M}_J$ by the well-known relations (Chipman 2009; Arteaga and Canillas 2010) and the elementary polarization properties can then be extracted from the matrix **N**, instead of from **m**, following the procedure given in Section 9.2.4.

As an example, consider the following normalized Mueller matrix:

$$\mathbf{M} = \begin{pmatrix} 1 & -0.0291 & -0.0111 & 0.0295 \\ -0.0287 & 0.9993 & -0.0190 & 0.0049 \\ -0.0092 & 0.0188 & 0.9957 & 0.0802 \\ 0.0305 & -0.0073 & -0.0804 & 0.9963 \end{pmatrix} \tag{9.70}$$

taken from (Arteaga and Canillas 2010). It is readily verified that **M** is nondepolarizing (as well as non-singular), either by checking the validity of Eq. (9.49) (after normalizing **M** so as to satisfy det **M** = 1) or by determining the rank of the coherency matrix associated with **M** (see Section 5.3).

Consequently, **M** potentially represents the polarimetric response of some continuous medium and its differential matrix can be obtained from Eqs. (9.65–9.69). One finds

$$\mathbf{m}_0 = \begin{pmatrix} 0 & -0.0289 & -0.0102 & 0.0300 \\ -0.0289 & 0 & -0.0189 & 0.0061 \\ -0.0102 & 0.0189 & 0 & 0.0805 \\ 0.0300 & -0.0061 & -0.0805 & 0 \end{pmatrix} \tag{9.71}$$

and

$$\Delta\mathbf{m} = 2\pi\begin{pmatrix} 0 & 0.0761 & -0.1354 & 0.2256 \\ 0.0761 & 0 & -0.3401 & 0.0161 \\ -0.1354 & 0.3401 & 0 & 0.9793 \\ 0.2256 & -0.0161 & -0.9793 & 0 \end{pmatrix} \tag{9.72}$$

As expected, both $\mathbf{m}_0$ and $\Delta\mathbf{m}$ are G-antisymmetric, i.e., they satisfy Eq. (9.51). The zero-order ($p = 0$) elementary polarization properties are read directly from the entries of $\mathbf{m}_0$, in correspondence with Eq. (9.55),

$$\begin{array}{lll} \mathrm{LD} = -0.0289 & \mathrm{LD}' = -0.0102 & \mathrm{CD} = 0.0300 \\ \mathrm{LB} = 0.0805 & \mathrm{LB}' = -0.0061 & \mathrm{CB} = -0.0189 \end{array} \tag{9.73}$$

The first-order ($p = 1$) properties are found from $\mathbf{m}_0 + \Delta\mathbf{m}$,

$$\begin{array}{lll} \mathrm{LD} = 0.4493 & \mathrm{LD}' = -0.8606 & \mathrm{CD} = 1.4473 \\ \mathrm{LB} = 6.2337 & \mathrm{LB}' = -0.1070 & \mathrm{CB} = -2.1560 \end{array} \tag{9.74}$$

The medium exhibits a complex polarimetric behavior featuring all six elementary polarization properties in small, but not negligible, amounts (for $p = 0$). The values of the zero-order properties are, generally, smaller than those of the first (and higher) order(s); in the absence of any additional measurements making possible the determination the order p of the sample under study, it is typically the zero-order properties that are reported from the experiment. It should be also kept to mind that, if the medium is known to be nondepolarizing, then its experimental Mueller matrix should be "filtered" from experimental noise and possible measurement artifacts (see Chapter 11) prior to applying the above procedure.

9.4 The Differential Decomposition of Mueller Matrices

9.4.1 Differential Mueller Matrix of a Depolarizing Medium

When the continuous medium exhibits *depolarization*, its Mueller matrix (assumed nonsingular and normalized to unit determinant) does not satisfy Eq. (9.49); consequently, unlike in the nondepolarizing case, its differential matrix $\mathbf{m}$, defined by Eq. (9.50), is not G-antisymmetric. However, any matrix $\mathbf{m}$ can be decomposed into G-antisymmetric and G-symmetric parts $\mathbf{m}_a$ and $\mathbf{m}_s$,

$$\mathbf{m} = \mathbf{m}_a + \mathbf{m}_s \tag{9.75}$$

where

$$\mathbf{m}_a = \frac{1}{2}\left(\mathbf{m} - \mathbf{G}\mathbf{m}^T\mathbf{G}\right)\mathbf{m}_s = \frac{1}{2}\left(\mathbf{m} + \mathbf{G}\mathbf{m}^T\mathbf{G}\right) \tag{9.76}$$

On the other hand, under some quite general conditions that will be given later, one can define the matrix logarithm $\mathbf{L}$ of the depolarizing Mueller matrix $\mathbf{M}$ of the medium,

$$\mathbf{L} = \ln \mathbf{M} \tag{9.77}$$

and decompose $\mathbf{L}$, like $\mathbf{m}$ just before, into G-antisymmetric and G-symmetric parts $\mathbf{L}_m$ and $\mathbf{L}_u$,

$$\mathbf{L}_m = \frac{1}{2}\left(\mathbf{L} - \mathbf{G}\mathbf{L}^T\mathbf{G}\right) \qquad \mathbf{L}_u = \frac{1}{2}\left(\mathbf{L} + \mathbf{G}\mathbf{L}^T\mathbf{G}\right) \tag{9.78}$$

Now, if the differential matrix $\mathbf{m}$ is z-independent, it follows from its definition (9.50) that $\mathbf{m}$ can be identified directly, like in the nondepolarizing case, with the Mueller matrix logarithm $\mathbf{L}$,

$$\mathbf{m} = \mathbf{L} \equiv \ln \mathbf{M} \tag{9.79}$$

or, if written in part-wise manner,

$$\mathbf{m}_a = \mathbf{L}_m \qquad \mathbf{m}_s = \mathbf{L}_u \tag{9.80}$$

where we have assumed, as before, a medium of unit thickness or path-length ($z = 1$). If $\mathbf{m}$ depends on z, the simple identification between $\mathbf{m}$ and $\mathbf{L}$ does not hold in general; nevertheless, the Mueller matrix logarithm $\mathbf{L}$ can still be evaluated and be given a physical meaning. Thus, the G-antisymmetric part $\mathbf{L}_m$ can be interpreted as the *polarization part* of $\mathbf{L}$ whose entries are the *mean values* of the elementary polarization properties of the depolarizing medium in accordance with Eq. (9.55), whereas the G-symmetric part $\mathbf{L}_u$ is the *depolarization part* of $\mathbf{L}$ containing the *uncertainties* of the elementary properties. Indeed, when $\mathbf{L}_u = \mathbf{0}$ the Mueller matrix $\mathbf{M}$ is nondepolarizing; the G-antisymmetric polarization part $\mathbf{L}_m$ then coincides with $\mathbf{L}$ and contains the elementary polarization properties of the medium (be it homogeneous or not), as discussed in the previous section.

Consequently, any (non-singular) nondepolarizing Mueller matrix is representable in the form $\mathbf{M}_J = \exp(\mathbf{L}_m)$ and can be physically associated with a continuous medium whose nonzero elementary polarization properties ($\mathbf{L}_m \neq \mathbf{0}$) feature zero uncertainties ($\mathbf{L}_u = \mathbf{0}$). Conversely, if $\mathbf{L}_m = \mathbf{0}$ then $\mathbf{L}$ will coincide with the G-symmetric depolarization part $\mathbf{L}_u$. The exponential of a G-symmetric matrix being itself G-symmetric, the matrix $\mathbf{M}_\Delta = \exp(\mathbf{L}_u)$ thus corresponds physically to the Mueller matrix of a *generalized depolarizer*, i.e., to a Mueller matrix with opposite diattenuation and polarizance vectors whose 3×3 submatrix is symmetric (recall that, by definition, the generalized depolarizer is characterized by a G-symmetric Mueller matrix). This allows us to picture physically a certain class of generalized depolarizers, namely those with positive eigenvalues, as continuous media whose elementary polarization properties exhibit zero mean values ($\mathbf{L}_m = \mathbf{0}$) but nonzero uncertainties ($\mathbf{L}_u \neq \mathbf{0}$). (As we shall see in the next section, not all generalized depolarizers allow for a matrix logarithm and are physically representable in this way.)

Eqs. (9.77) and (9.78), accompanied by the above physical interpretation, represent the differential decomposition of a Mueller matrix $\mathbf{M}$ (Ossikovski 2011; Ortega-Quijano and Arce-Diego 2011). This decomposition extends the scope of the fundamental relations (9.47), (9.50) and (9.64) from pure (nondepolarizing) Mueller matrices $\mathbf{M}_J$ to depolarizing

Mueller matrices **M**. The generally depolarizing **M** may physically describe not only a continuous medium whose elementary polarization (and depolarization) properties are z-independent, but also a continuous medium whose properties vary along the path-length or even a cascade of discontinuous media or structures (Ortega-Quijano et al. 2012). In particular, if the elementary properties of the depolarizing **M** are z-independent, the differential decomposition coincides with Eqs. (9.79) and (9.80), that is, it provides the differential matrix **m** of the medium.

If **M** is *nondepolarizing*, the differential decomposition likewise reduces to Eqs. (9.79) and (9.80), with $\mathbf{m} = \mathbf{m}_a$ and $\mathbf{m}_s = \mathbf{0}$, even for z-dependent properties (in which case **m** thus obtained is to be interpreted as an *equivalent* z-independent differential matrix; see Section 9.3.3).

The physical meaning of the differential decomposition becomes clear if Eq. (9.77) is inverted and rewritten in the following equivalent form:

$$\mathbf{M} = \exp(\mathbf{L}) = \underbrace{\lim}_{n\to\infty}\left(\mathbf{I}_4 + \frac{1}{n}\mathbf{L}\right)^n \tag{9.81}$$

On the other hand, it is straightforward to derive from the fundamental relation (9.47) that the Mueller matrix of an infinitesimal slab of thickness Δz is given by

$$\mathbf{M}(z, z+\Delta z) = \mathbf{I}_4 + \mathbf{m}(z)\Delta z \tag{9.82}$$

(compare Eq. (9.82) with Eq. (9.13) from the Jones matrix formalism).

In the light of Eq. (9.82), Eq. (9.81) allows for an immediate physical interpretation: it represents the Mueller matrix **M** of the medium, assumed of unit thickness (z = 1), as a series concatenation of an infinite number $n \to \infty$ of identical infinitesimal slabs, of thickness $1/n$ each, whose differential Mueller matrices, contributing to the elementary properties of the medium, equal the Mueller matrix logarithm **L**.

Since $\mathbf{L} = \mathbf{L}_m + \mathbf{L}_u$ in general, i.e., **L** contains both polarizing ($\mathbf{L}_m$) and depolarizing ($\mathbf{L}_u$) parts, the slabs are generally depolarizing and, the product of depolarizing Mueller matrices being (another) depolarizing Mueller matrix, **M** is consequently also depolarizing. Thus, according to Eq. (9.81), the differential decomposition appears as a *continuous* serial (or product) decomposition with infinitesimal depolarizing Mueller matrix factors. When the Mueller matrix is *nondepolarizing*, the infinitesimal factors are also nondepolarizing ($\mathbf{L}_u = \mathbf{0}$) and vice versa.

In summary, the differential decomposition of **M** regards **M** as resulting from the polarimetric response of a *uniformly depolarizing* continuous medium, polarimetrically equivalent to the real medium, whose differential matrix equals the matrix logarithm **L** of **M**. Here, by uniformly depolarizing medium we mean a medium whose polarization and depolarization parts, $\mathbf{L}_m$ and $\mathbf{L}_u$, both cumulate linearly with the position z, i.e., whose Mueller matrix is of the form $\mathbf{M}(z) = \exp(\mathbf{L}_m z + \mathbf{L}_u z)$. This general interpretation of the differential decomposition extends its range of applicability even to Mueller matrices measured in reflection configuration, under the condition of the mathematical existence and the physical realizability of the matrix logarithm (discussed in the following sections).

It should also be noted that the last expression in Eq. (9.81) can be solved for **L**

$$\mathbf{L} = \underbrace{\lim}_{n\to\infty} n\left(\sqrt[n]{\mathbf{M}} - \mathbf{I}_4\right) \tag{9.83}$$

resulting in an alternative procedure for evaluating **L**, called Mueller matrix roots decomposition (Chipman 2009; Noble and Chipman 2012; Noble et al. 2012). It can be shown that, under quite general conditions on **M**, the matrix **L** thus determined coincides with the Mueller matrix logarithm (Ossikovski 2012b).

9.4.2 Existence and Multiplicity of the Mueller Matrix Logarithm

It is clear from the preceding section that the crucial step in performing the differential decomposition of the Mueller matrix **M** is the evaluation of its logarithm **L** from Eq. (9.79). Most generally, the requirement for the logarithm of **M** to be real (and to be, therefore, attributable a physical meaning), imposes a set of conditions on the eigenvalue spectrum of **M**. Thus, the Mueller matrix being real, its characteristic equation has real coefficients and consequently, **M** has either four real eigenvalues or two real eigenvalues and another two forming a complex conjugate pair. (**M** cannot have two complex conjugate pairs of eigenvalues since its largest-modulus eigenvalue is necessarily real and non-negative on physical grounds.)

Next, if we interpret, somewhat artificially, a double negative eigenvalue as being a special case of a complex conjugate pair whose argument is equal to π, then it can be shown that for the logarithm **L** to be real it is necessary that the real eigenvalues of **M** be all positive (Higham 2008). This is what we shall call the *existence condition* for the Mueller matrix logarithm. A closer look at Eqs. (9.66) indicates that a non-singular *nondepolarizing* Mueller matrix always satisfies the above existence condition. This observation is in perfect consistency with the established facts that any non-singular Jones matrix allows for a (complex-valued) logarithm and that any nondepolarizing Mueller matrix is equivalent to a Jones matrix within a common phase factor.

For a non-singular *depolarizing* **M**, Eqs. (9.66) are no more applicable since a depolarizing Mueller matrix does not have a single Jones matrix counterpart; the above existence condition is therefore not automatically fulfilled and must be checked. Nevertheless, whenever existing, the logarithm can be still obtained through the diagonalization of **M**, as explained in Section 9.3.3 treating the nondepolarizing case. In particular, if we write $\mathbf{L}_0$ and $\Delta\mathbf{L}$ instead of $\mathbf{m}_0$ and $\Delta\mathbf{m}$ in Eqs. (9.67–9.69), these remain applicable for a depolarizing Mueller matrix **M** exhibiting two real (μ_1 and μ_2) and two complex conjugate (μ_3 and μ_4) eigenvalues, even though the eigenvalue relations (9.66) do not hold.

The matrix logarithm yields, in general, multiple solutions, parameterized by the order parameter p. The difference with the nondepolarizing case is that the matrix $\Delta\mathbf{L}$ *must be G-antisymmetric* for the multiple solutions to be physically meaningful; this condition is satisfied whenever the polarizing and depolarizing parts $\mathbf{L}_m$ (of zero order) and $\mathbf{L}_u$ of **L** commute. However, if the matrix $\Delta\mathbf{L}$ does not exhibit G-antisymmetry then the matrix logarithm is single-valued ($p = 0$) and the elementary polarization properties (as well as their uncertainties), determined from the principal matrix logarithm, are unique (Devlaminck and Ossikovski 2014). When **M** has four real distinct eigenvalues the first equality in Eq. (9.68) applies and the solution is likewise unique ($p = 0$). The case of a double real eigenvalue needs special treatment and will be addressed shortly in the following paragraph, as well as in Section 9.4.5 in larger detail.

These considerations are best illustrated on several examples. Consider the normalized Mueller matrix of a polyacrylamide gel measured in transmission (Arteaga and Canillas 2009),

$$\mathbf{M}_g = \begin{pmatrix} 1 & -0.0312 & 0.0029 & -0.0066 \\ -0.0214 & 0.7678 & -0.0370 & 0.0204 \\ -0.0055 & 0.0230 & 0.1043 & -0.7735 \\ 0.0014 & 0.0390 & 0.7972 & 0.1920 \end{pmatrix} \tag{9.84}$$

This Mueller matrix is depolarizing as can be seen, for instance, by evaluating its depolarization index P_Δ (see Section 5.10.3) equal to 0.79<1. Next, $\mathbf{M}_g$ possesses two real and positive eigenvalues accompanied by a complex conjugate pair (1.0029; 0.7665; 0.1474 ± i 0.7847) and, according to the existence condition formulated previously, it admits a real matrix logarithm that can be evaluated.

The application of Eqs. (9.67–9.69), supplemented by Eqs. (9.78), provides the polarization and depolarization parts of the principal matrix logarithm $\mathbf{L}_g$ of $\mathbf{M}_g$,

$$\begin{gathered} \mathbf{L}_{gm} = \begin{pmatrix} 0 & -0.0300 & 0.0016 & 0.0006 \\ -0.0300 & 0 & -0.0596 & -0.0153 \\ 0.0016 & 0.0596 & 0 & -1.3860 \\ 0.0006 & 0.0153 & 1.3860 & 0 \end{pmatrix} \\ \mathbf{L}_{gu} = \begin{pmatrix} 0 & -0.0055 & 0.0053 & -0.0049 \\ 0.0055 & -0.2628 & 0.0001 & 0.0035 \\ -0.0053 & 0.0001 & -0.3009 & 0.0214 \\ 0.0049 & 0.0035 & 0.0214 & -0.1481 \end{pmatrix} \end{gathered} \tag{9.85}$$

as well as the matrix "period" $\mathbf{\Delta L}_g$,

$$\mathbf{\Delta L}_g = 2\pi \begin{pmatrix} 0 & 0.0001 & 0.0051 & -0.0012 \\ -0.0002 & -0.0004 & -0.0431 & -0.0075 \\ -0.0022 & 0.0425 & -0.0550 & -0.9852 \\ 0.0051 & 0.0148 & 1.0161 & 0.0554 \end{pmatrix} \tag{9.86}$$

Notice that we have artificially set the first element of $\mathbf{L}_{gu}$ to zero by subtracting its value from the diagonal elements of $\mathbf{L}_{gu}$; as follows from Eq. (9.79), doing so simply amounts to renormalizing $\mathbf{M}_g$ without affecting the polarization and depolarization properties. (We have also omitted for simplicity the zero subscript denoting the principal matrix logarithm.)

The inspection of the matrix $\mathbf{\Delta L}_g$ shows that, although being relatively close to, it is not G-antisymmetric within the experimental uncertainty (of the order of several 0.001 typically). Consequently, the only physically meaningful matrix logarithm is the principal one and the mean values of the elementary polarization properties of the medium can be read directly from the entries of the matrix $\mathbf{L}_{gm}$, in correspondence with Eq. (9.55),

$$\begin{array}{lll} \mathrm{LD} = -0.0300 & \mathrm{LD}' = 0.0016 & \mathrm{CD} = 0.0006 \\ \mathrm{LB} = -1.3860 & \mathrm{LB}' = 0.0153 & \mathrm{CB} = -0.0596 \end{array} \tag{9.87}$$

The medium exhibits very strong linear birefringence (LB), accompanied by a much lower amount of optical activity (CB), as well as by very weak linear dichroism (LD) and ±45°

birefringence (LB′); the values of the rest of the elementary properties are comparable to or below the typical measurement uncertainty. As for the uncertainty matrix $\mathbf{L}_{gu}$, we notice that it is very close to a diagonal matrix; however, any physical interpretation has to be postponed so far to Section 9.5 addressing the statistical interpretation of the differential decomposition.

Consider now the frequently encountered normalized Mueller matrix

$$\mathbf{S}=\begin{pmatrix} 1 & a & 0 & 0 \\ a & 1 & 0 & 0 \\ 0 & 0 & b & c \\ 0 & 0 & -c & b \end{pmatrix} \tag{9.88}$$

where $a^2+b^2+c^2<1$, physically representing, for instance, an optically thick depolarizing substrate (in oblique transmission) or a depolarizing (e.g., optically rough) surface (in oblique reflection). The matrix **S** is the depolarizing analogue of the matrix $\mathbf{M}_\mathrm{L}$ from Section 9.3.2; see Eq. (9.63).

The inequality $a^2+b^2+c^2<1$ entails depolarization; if replaced by equality, it transforms **S** into the well-known nondepolarizing psi-delta matrix from ellipsometry. The eigenvalues of **S** are $1\pm a$ (two real and positive) and $b\pm ic$ (a complex conjugate pair) allowing for the direct application of Eqs. (9.67–9.69). The final result is

$$\mathbf{L}_{\mathrm{S}\,m}=\begin{pmatrix} 0 & \ln\sqrt{(1+a)/(1-a)} & 0 & 0 \\ \ln\sqrt{(1+a)/(1-a)} & 0 & 0 & 0 \\ 0 & 0 & 0 & \arg(b+ic) \\ 0 & 0 & -\arg(b+ic) & 0 \end{pmatrix}$$
$$\mathbf{L}_{\mathrm{S}\,u}=\begin{pmatrix} 0 & 0 & 0 & 0 \\ 0 & 0 & 0 & 0 \\ 0 & 0 & \ln\sqrt{(b^2+c^2)/(1-a^2)} & 0 \\ 0 & 0 & 0 & \ln\sqrt{(b^2+c^2)/(1-a^2)} \end{pmatrix} \tag{9.89}$$

for the G-antisymmetric (or polarization) and G-symmetric (or depolarization) parts of the matrix logarithm $\mathbf{L}_\mathrm{S}$ of **S** and

$$\Delta\mathbf{L}_\mathrm{S}=2\pi\begin{pmatrix} 0 & 0 & 0 & 0 \\ 0 & 0 & 0 & 0 \\ 0 & 0 & 0 & 1 \\ 0 & 0 & -1 & 0 \end{pmatrix} \tag{9.90}$$

for the matrix "period" $\Delta\mathbf{L}_\mathrm{S}$ (which turns out to be clearly G-antisymmetric). As before, we have "centered" $\mathbf{L}_{\mathrm{S}\,u}$ to zero first element by the subtraction of $\ln\sqrt{1-a^2}$ from its main diagonal. The matrix $\mathbf{L}_{\mathrm{S}\,u}$ is strictly diagonal and depends on the single parameter $\boldsymbol{\Delta}^{\ln}=\ln\sqrt{(b^2+c^2)/(1-a^2)}$.

The parameter $\Delta^{\ln}$ expresses the amount of depolarization and can be called "logarithmic depolarization"; its value is negative when $a^2+b^2+c^2<1$ (since then $(b^2+c^2)/(1-a^2)<1$ and so, $\ln[(b^2+c^2)/(1-a^2)]<0$) and becomes zero if $a^2+b^2+c^2=1$: then, the depolarization part $\mathbf{L}_{\mathrm{S}u}$ of $\mathbf{L}_{\mathrm{S}}$ vanishes identically for a nondepolarizing Mueller matrix **S** and its matrix logarithm $\mathbf{L}_{\mathrm{S}}=\mathbf{L}_{\mathrm{S}m}$ (containing the elementary polarization properties) is strictly G-antisymmetric.

The only two non-vanishing elementary polarization properties are

$$\mathrm{LD}=\ln\sqrt{(1+a)/(1-a)}\,\mathrm{LB}=\arg(b+ic)+2\pi p \tag{9.91}$$

that is, a medium described by the matrix **S** exhibits the L complex property only. The matrix $\mathbf{\Delta L}_{\mathrm{S}}$ being G-antisymmetric, the matrix logarithm is consequently multivalued and given by Eq. (9.67); the LB values are thus periodic by the order parameter p with a period of 2π. This is a special case of a more general situation: the multiplicity of the elementary polarization properties physically stems from the periodicity of the birefringent (B) property. Even if the zero-order B property vanishes, the periodicity, and consequently the multiplicity of solutions for the elementary properties, persist. Thus, if $c=0$ and $b>0$, Eq. (9.91) yields $\mathrm{LB}=2\pi p$. However, $b<0$ if (while $c=0$) then $\mathrm{LB}=\pi+2\pi p$. Note that the two cases where $c=0$ correspond to a double eigenvalue (equal to b) of **S**: they can be considered special cases of taking the logarithm of a Mueller matrix featuring a double eigenvalue.

Sometimes it is desirable "to convert" the dichroic (D) and birefringent (B) properties provided by the differential decomposition into their "integral" counterparts, diattenuation (D) and retardance (R), yielded by the various serial (product) decompositions (see Chapter 8). The relationships between respective magnitudes are (Kumar et al. 2012)

$$D=\tanh\mathrm{D}=\tanh\sqrt{\mathrm{LD}^2+\mathrm{LD}'^2+\mathrm{CD}^2} \tag{9.92}$$

and

$$R=\mathrm{B}=\sqrt{\mathrm{LB}^2+\mathrm{LB}'^2+\mathrm{CB}^2} \tag{9.93}$$

assuming a unit path-length. (Notice the notation D (italic) for diattenuation, not to be confused with dichroism D!) As for the directions of the corresponding vectors, they coincide; that is, the diattenuation vector **D** is directed along the dichroism unit vector given by $\vec{\mathrm{D}}_u=(\mathrm{LD/D},\mathrm{LD'/D},\mathrm{CD/D})^T$ whereas the retardance vector **R** is collinear with the birefringence one, $\vec{\mathrm{B}}_u=(\mathrm{LB/B},\mathrm{LB'/B},\mathrm{CB/B})^T$. The application of the above conversion relations to the elementary polarization properties of the matrix **S** results in

$$D=\tanh\mathrm{LD}=\tanh\left[\ln\sqrt{(1+a)/(1-a)}\right]=a \tag{9.94}$$

and

$$R=\mathrm{LB}=\arg(b+ic)+2\pi p \tag{9.95}$$

with the directions of the respective vectors both being given by the unit vector $(1,0,0)^T$ that is, both being directed along the HV axis of the Poincaré sphere, as expected for the L property, the only one elementary polarization property present in **S**.

Consider now the matrix logarithms of the two canonical forms of a general depolarizer (Ossikovski 2010a). The normalized diagonal canonical form (see Section 5.7),

$$\mathbf{M}_{\Delta d} = \mathrm{diag}(1, a, b, c) \tag{9.96}$$

has a single-valued G-symmetric matrix logarithm

$$\mathbf{L}_{\Delta d} = \mathbf{L}_{\Delta du} = \mathrm{diag}(0, \ln a, \ln b, \ln c) \tag{9.97}$$

provided all three diagonal elements a, b and c of $\mathbf{M}_{\Delta d}$ are positive. Since a, b and c do not exceed the unit for $\mathbf{M}_{\Delta d}$ not to overpolarize, the diagonal entries of $\mathbf{L}_{\Delta du}$ are negative or zero. As we shall see in Section 9.5, this is a general property of the diagonal elements of any $\mathbf{L}_u$ matrix, centered to zero first element. The G-antisymmetric part $\mathbf{L}_{\Delta dm}$ of $\mathbf{L}_{\Delta d}$ is identically zero and consequently, all elementary polarization properties assume zero mean values as stated before.

The non-diagonal depolarizer canonical form (see Section 5.7),

$$\mathbf{M}_{\Delta nd} = \begin{pmatrix} 2 & -1 & 0 & 0 \\ 1 & 0 & 0 & 0 \\ 0 & 0 & a & 0 \\ 0 & 0 & 0 & a \end{pmatrix} \tag{9.98}$$

where $0 \le a \le 1$ cannot be diagonalized down to Eq. (9.65) and Eqs. (9.67–9.69) cannot be used. Nevertheless, its matrix logarithm can be evaluated by using the classic series expansion

$$\ln \mathbf{M} = \ln(\mathbf{I}_4 + \mathbf{M} - \mathbf{I}_4) = (\mathbf{M} - \mathbf{I}_4) - \frac{1}{2}(\mathbf{M} - \mathbf{I}_4)^2 + \frac{1}{3}(\mathbf{M} - \mathbf{I}_4)^3 - \ldots \tag{9.99}$$

which converges for $\mathbf{M}_{\Delta nd}$ provided $0 \le a \le 1$. The closed-form result is

$$\mathbf{L}_{\Delta nd} = \mathbf{L}_{\Delta ndu} = \begin{pmatrix} 0 & -1 & 0 & 0 \\ 1 & -2 & 0 & 0 \\ 0 & 0 & \ln a - 1 & 0 \\ 0 & 0 & 0 & \ln a - 1 \end{pmatrix} \tag{9.100}$$

after subtraction of the first element (equal to one) from the main diagonal (for "centering" $\mathbf{L}_{\Delta nd}$). The matrix logarithm does not exist for $a = 0$. As expected, $\mathbf{L}_{\Delta nd}$ is G-symmetric (the logarithm of a G-symmetric matrix is G-symmetric) and non-diagonal, like $\mathbf{M}_{\Delta nd}$; its diagonal entries are non-positive, in accordance with the general property mentioned previously. The G-antisymmetric (polarization) part of $\mathbf{L}_{\Delta nd}$ vanishes identically and therefore all elementary polarization properties assume zero mean values, like in the case of the diagonal canonical depolarizer.

9.4.3 Local Physical Realizability

As we have seen before, the differential decomposition allows one to interpret a generally depolarizing medium with a Mueller matrix $\mathbf{M}$ as being a stack of an infinite number of infinitesimal slabs. For this interpretation to have a physical meaning, every slab must represent a physically realizable Mueller matrix in the Cloude sense (see Section 5.6). We shall call this physical realizability *local*, in opposition to the *global* realizability defined by $\mathbf{M}$ itself.

It is clear that local realizability entails a global one, since the realizability of each infinitesimal slab warrants the realizability of the entire stack, i.e., that of $\mathbf{M}$ (because the product of realizable Mueller matrices is a realizable Mueller matrix). The converse is, however, not true: from the physical realizability of $\mathbf{M}$ does not follow the realizability of its differential decomposition.

The Mueller matrix of an infinitesimal slab is given by Eq. (9.82) in terms of the differential matrix $\mathbf{m}$ or by the bracketed expression in Eq. (9.81) in terms of the matrix logarithm $\mathbf{L}$ (recall that $\mathbf{m}$ and $\mathbf{L}$ coincide only if $\mathbf{m}$ is z-independent). It can be shown that the physical realizability of the infinitesimal slab depends only on the G-symmetric part $\mathbf{m}_a$ (or $\mathbf{L}_u$) of $\mathbf{m}$ (or of $\mathbf{L}$) (Ossikovski and Devlaminck 2014). In other words, the differential decomposition of a *nondepolarizing* Mueller matrix $\mathbf{M}$ is always physically realizable which is in full agreement with the practice of using the matrix logarithm for the extraction of the elementary polarization properties (see Section 9.3.3).

For a *depolarizing* $\mathbf{M}$ to be *locally* physically realizable, i.e., for its differential decomposition to have a physical meaning, it is necessary and sufficient that the lower 3×3 principal minor $\mathbf{C}_3$ of the coherency matrix $\mathbf{C}$ of $\mathbf{m}$ (or of $\mathbf{L}$) be positive semi-definite (Ossikovski and Devlaminck 2014). In explicit form, the elements of $\mathbf{C}_3$ expressed through those of $\mathbf{m}$ (or of $\mathbf{L}$) read (Van der Mee 1993)

$$\mathbf{C}_3 = \frac{1}{4}\begin{pmatrix} \mathbf{m}_{11}+\mathbf{m}_{22}-\mathbf{m}_{33}-\mathbf{m}_{44} & \mathbf{m}_{23}+\mathbf{m}_{32}+i\mathbf{m}_{14}-i\mathbf{m}_{41} & \mathbf{m}_{24}+\mathbf{m}_{42}-i\mathbf{m}_{13}+i\mathbf{m}_{31} \\ \mathbf{m}_{23}+\mathbf{m}_{32}-i\mathbf{m}_{14}+i\mathbf{m}_{41} & \mathbf{m}_{11}-\mathbf{m}_{22}+\mathbf{m}_{33}-\mathbf{m}_{44} & \mathbf{m}_{34}+\mathbf{m}_{43}+i\mathbf{m}_{12}-i\mathbf{m}_{21} \\ \mathbf{m}_{24}+\mathbf{m}_{42}+i\mathbf{m}_{13}-i\mathbf{m}_{31} & \mathbf{m}_{34}+\mathbf{m}_{43}-i\mathbf{m}_{12}+i\mathbf{m}_{21} & \mathbf{m}_{11}-\mathbf{m}_{22}-\mathbf{m}_{33}+\mathbf{m}_{44} \end{pmatrix} \tag{9.101}$$

If the *reduced coherency matrix* $\mathbf{C}_3$ is evaluated rather from the G-symmetric matrix $\mathbf{m}_s$ (or $\mathbf{L}_u$) instead of from $\mathbf{m}$ (or $\mathbf{L}$) and if, furthermore, the first element of $\mathbf{m}_s$ (or of $\mathbf{L}_u$) is set to zero by the subtraction of its initial value from the main diagonal, then Eqs. (9.101) slightly simplify to

$$\mathbf{C}_3(\mathbf{m}_s) = \frac{1}{2}\begin{pmatrix} \tfrac{1}{2}(\mathbf{m}_{s22}-\mathbf{m}_{s33}-\mathbf{m}_{s44}) & \mathbf{m}_{s23}+i\mathbf{m}_{s14} & \mathbf{m}_{s24}-i\mathbf{m}_{s13} \\ \mathbf{m}_{s23}-i\mathbf{m}_{s14} & \tfrac{1}{2}(-\mathbf{m}_{s22}+\mathbf{m}_{s33}-\mathbf{m}_{s44}) & \mathbf{m}_{s34}+i\mathbf{m}_{s12} \\ \mathbf{m}_{s24}+i\mathbf{m}_{s13} & \mathbf{m}_{s34}-i\mathbf{m}_{s12} & \tfrac{1}{2}(-\mathbf{m}_{s22}-\mathbf{m}_{s33}+\mathbf{m}_{s44}) \end{pmatrix} \tag{9.102}$$

Conceptually, the reduced coherency matrix $\mathbf{C}_3$ can be considered a "differential" analogue of Cloude's coherency matrix $\mathbf{C}$ of $\mathbf{M}$ (Cloude 1989). Note that Eqs. (9.101) and Eqs. (9.102) imply that the reduced coherency matrix can be calculated indifferently from the depolarization part $\mathbf{m}_s$ of the differential matrix of directly from $\mathbf{m}$ itself, that is $\mathbf{C}_3(\mathbf{m}_s) = \mathbf{C}_3(\mathbf{m})$.

Typically, positive semi-definiteness is checked by computing the eigenvalues of $\mathbf{C}_3$ (all three of them must be non-negative). However, a much-easier-to-verify *necessary condition* for local physical realizability follows from Eq. (9.102). Indeed, positive semi-definiteness of $\mathbf{C}_3$ requires non-negativity of its three diagonal elements, which, according to Eq. (9.102), implies that the triplet $(-\mathbf{m}_{s22}, -\mathbf{m}_{s33}, -\mathbf{m}_{s44})$ must satisfy the triangle inequality. Consequently, this triplet must geometrically correspond to three segments with non-negative lengths forming a triangle which means that all three diagonal entries of the matrix $\mathbf{m}_s$ (or of $\mathbf{L}_u$) must be negative or zero. Of course, this condition is necessary but not sufficient, i.e., if one of the three diagonal elements of $\mathbf{m}_s$ (or of $\mathbf{L}_u$) is positive, then the differential decomposition is not physically realizable but the converse is not true.

As an example, consider checking the local physical realizability of the experimental matrix $\mathbf{M}_g$ from the previous section. A glance at the matrix $\mathbf{L}_{gu}$ reported in Eq. (9.85) shows that the necessary realizability condition is fulfilled: all (but the first one) diagonal entries of $\mathbf{L}_{gu}$ are negative (the first one is zero by construction). Since this is a necessary condition only, we have to proceed further. Applying Eq. (9.102) to $\mathbf{L}_{gu}$ yields the following reduced coherency matrix:

$$\mathbf{C}_3\left(\mathbf{L}_{gu}\right)=\begin{pmatrix} 0.0466 & 0.0001-i0.0025 & 0.0018-i0.0027 \\ 0.0001+i0.0025 & 0.0275 & 0.0107-i0.0028 \\ 0.0018+i0.0027 & 0.0107+i0.0028 & 0.1039 \end{pmatrix} \tag{9.103}$$

This matrix, being Hermitian by construction, has three real eigenvalues, 0.0257, 0.0466 and 0.1057, which turn out to be positive: the experimental Mueller matrix $\mathbf{M}_g$ is *locally* physically realizable (or equivalently, the differential decomposition of $\mathbf{M}_g$ is physically realizable).

Similarly, for the Mueller matrix $\mathbf{S}$ from the previous section, one obtains the following particularly simple reduced coherency matrix:

$$\mathbf{C}_3\left(\mathbf{L}_{\mathbf{S}u}\right)=\frac{1}{2}\operatorname{diag}\left(-\Delta_{\ln},0,0\right) \tag{9.104}$$

where $\Delta_{\ln}=\ln\sqrt{(b^2+c^2)/(1-a^2)}$ is the logarithmic depolarization. In virtue of the inequality $a^2+b^2+c^2<1$ obeyed by the coefficients of the matrix $\mathbf{S}$, the only nonzero element of $\mathbf{C}_3$ is positive and $\mathbf{C}_3$ is (trivially) positive semi-definite: consequently, the matrix $\mathbf{S}$ is locally physically realizable.

Consider now the local physical realizability of the diagonal canonical depolarizer $\mathbf{M}_{\Delta d}=\operatorname{diag}(1,a,b,c)$ with all its three coefficients a, b and c being positive to allow for differential decomposition. Its Cloude coherency matrix $\mathbf{C}$ is

$$\mathbf{C}\left(\mathbf{M}_{\Delta d}\right)=\frac{1}{4}\operatorname{diag}\left(1+a+b+c,\ 1+a-b-c,\ 1-a+b-c,\ 1-a-b+c\right) \tag{9.105}$$

(see Section 5.7.1) whereas its reduced coherency matrix $\mathbf{C}_3$ from Eq. (9.102) is

$$\mathbf{C}_3\left(\mathbf{L}_{\Delta d\,u}\right)=\frac{1}{4}\operatorname{diag}\left(\ln\frac{a}{bc},\ \ln\frac{b}{ac},\ \ln\frac{c}{ab}\right) \tag{9.106}$$

The positive semi-definiteness of **C**, warranting *global* physical realizability of $\mathbf{M}_{\Delta d}$, simply reduces to the non-negativity of its diagonal elements,

$$1+a+b+c \ge 0 \qquad 1+a \ge b+c \qquad 1+b \ge a+c \qquad 1+c \ge a+b \tag{9.107}$$

whereas that of $\mathbf{C}_3$, implying the *local* physical realizability of $\mathbf{M}_{\Delta d}$, reduces in a similar manner to

$$a \ge bc \qquad b \ge ac \qquad c \ge ab \tag{9.108}$$

Since local physical realizability entails global one, as shown in the beginning of the section on physical grounds, Eqs. (9.107) must follow from Eqs. (9.108). Indeed, the first of Eqs. (9.108) can be rewritten as $1+a \ge 1+bc = (1-b)(1-c)+b+c \ge b+c$ since b and c do not exceed the unit from which the second of Eqs. (9.107) follows. The validity of the last two of Eqs. (9.107) follows from the last pair of Eq. (9.108) in a similar manner while the first of Eqs. (9.107) is fulfilled trivially. The converse is clearly not true, i.e., Eqs. (9.107) do not imply Eqs. (9.108).

Notice also that, although being physically obvious for $\mathbf{M}_{\Delta d}$ not to overpolarize, the condition that all three coefficients a, b and c must not exceed the unit may be also formally deduced from the necessary condition for the positive semi-definiteness of $\mathbf{C}_3$ mentioned before, namely, all diagonal elements of the G-symmetric part $\mathbf{L}_{\Delta du}$ of $\mathbf{L}_{\Delta d}$ must be non-positive. A look at $\mathbf{L}_{\Delta du}$ from Eq. (9.97) immediately shows that this necessary condition is satisfied only if $a,\ b,\ c \le 1$.

The situation is similar with the non-diagonal canonical depolarizer $\mathbf{M}_{\Delta nd}$, given by Eq. (9.98), with the coefficient a such that $0 < a \le 1$ (the value $a = 0$ is excluded to ensure the existence of the matrix logarithm). Its coherency matrix and reduced coherency matrix are readily found to be, respectively,

$$\mathbf{C}(\mathbf{M}_{\Delta nd}) = \frac{1}{2}\begin{pmatrix} 1+a & 0 & 0 & 0 \\ 0 & 1-a & 0 & 0 \\ 0 & 0 & 1 & -i \\ 0 & 0 & i & 1 \end{pmatrix} \tag{9.109}$$

and

$$\mathbf{C}_3(\mathbf{L}_{\Delta ndu}) = \frac{1}{2}\begin{pmatrix} \ln(1/a) & 0 & 0 \\ 0 & 1 & -i \\ 0 & i & 1 \end{pmatrix} \tag{9.110}$$

Clearly, the condition $0 < a \le 1$ warrants both local and global physical realizability since ensuring positive semi-definiteness of both $\mathbf{C}_3$ and **C** (note that the eigenvalues of the lower 2×2 minor present in both coherency matrices are zero and one and so, alone they are insufficient to confirm or to infirm physical realizability). Notice that the matrix **C** from Eq. (9.109) is positive semi-definite over the wider range $-1 < a \le 1$ rather than just over $0 < a \le 1$; again, local physical realizability (holding over $0 < a \le 1$) entails global one but not the opposite.

9.4.4 Algebraic Structure of the Differential Mueller Matrix Formalism

The differential representation of a depolarizing Mueller matrix **M** splits its differential matrix **m** (and its matrix logarithm **L**) into polarization and depolarization parts, $\mathbf{m}_a$ and $\mathbf{m}_s$ ($\mathbf{L}_m$ and $\mathbf{L}_u$) of opposite G-symmetries, having distinct physical meanings (mean values and uncertainties of the elementary polarization properties). If we consider **M** as being known only within a positive factor (which corresponds physically to disregarding the isotropic absorption; see Section 9.3.1) then the G-symmetric part $\mathbf{m}_s$ (and $\mathbf{L}_u$) is defined within a term of the form $\alpha\mathbf{I}$ (**I** is the identity matrix) as we have seen previously, and we may assume without loss of generality that the first element of $\mathbf{m}_s$ (and of $\mathbf{L}_u$) is identically zero.

Consequently, in accordance with the second of Eqs. (9.76), this G-symmetric part is entirely parameterized by nine elements (or "degrees of freedom"). Conversely, the first of Eqs. (9.76) shows that the G-antisymmetric part is fully defined by six parameters (physically corresponding to the six elementary polarization properties).

As a consequence, the G-antisymmetric and G-symmetric parts of **m** (and of **L**) can be expanded on two bases (Chipman 2000; Noble et al. 2012; Devlaminck 2013) consisting of two respective matrix sets, $\{\mathbf{P}_k, \mathbf{Q}_k\}$ and $\{\mathbf{D}_k, \mathbf{R}_k, \mathbf{\Delta}_k\}$, the first containing six and the second, nine matrices ($k = 1, 2, 3$). These sets are

$$\mathbf{P}_1 = \begin{pmatrix} 0 & 1 & 0 & 0 \\ 1 & 0 & 0 & 0 \\ 0 & 0 & 0 & 0 \\ 0 & 0 & 0 & 0 \end{pmatrix} \quad \mathbf{P}_2 = \begin{pmatrix} 0 & 0 & 1 & 0 \\ 0 & 0 & 0 & 0 \\ 1 & 0 & 0 & 0 \\ 0 & 0 & 0 & 0 \end{pmatrix} \quad \mathbf{P}_3 = \begin{pmatrix} 0 & 0 & 0 & 1 \\ 0 & 0 & 0 & 0 \\ 0 & 0 & 0 & 0 \\ 1 & 0 & 0 & 0 \end{pmatrix}$$
$$\mathbf{Q}_1 = \begin{pmatrix} 0 & 0 & 0 & 0 \\ 0 & 0 & 0 & 0 \\ 0 & 0 & 0 & 1 \\ 0 & 0 & -1 & 0 \end{pmatrix} \quad \mathbf{Q}_2 = \begin{pmatrix} 0 & 0 & 0 & 0 \\ 0 & 0 & 0 & -1 \\ 0 & 0 & 0 & 0 \\ 0 & 1 & 0 & 0 \end{pmatrix} \quad \mathbf{Q}_3 = \begin{pmatrix} 0 & 0 & 0 & 0 \\ 0 & 0 & 1 & 0 \\ 0 & -1 & 0 & 0 \\ 0 & 0 & 0 & 0 \end{pmatrix} \tag{9.111}$$

for the G-antisymmetric (polarization) part and

$$\mathbf{D}_1 = \begin{pmatrix} 0 & -1 & 0 & 0 \\ 1 & 0 & 0 & 0 \\ 0 & 0 & 0 & 0 \\ 0 & 0 & 0 & 0 \end{pmatrix} \quad \mathbf{D}_2 = \begin{pmatrix} 0 & 0 & -1 & 0 \\ 0 & 0 & 0 & 0 \\ 1 & 0 & 0 & 0 \\ 0 & 0 & 0 & 0 \end{pmatrix} \quad \mathbf{D}_3 = \begin{pmatrix} 0 & 0 & 0 & -1 \\ 0 & 0 & 0 & 0 \\ 0 & 0 & 0 & 0 \\ 1 & 0 & 0 & 0 \end{pmatrix}$$
$$\mathbf{R}_1 = \begin{pmatrix} 0 & 0 & 0 & 0 \\ 0 & 0 & 0 & 0 \\ 0 & 0 & 0 & 1 \\ 0 & 0 & 1 & 0 \end{pmatrix} \quad \mathbf{R}_2 = \begin{pmatrix} 0 & 0 & 0 & 0 \\ 0 & 0 & 0 & 1 \\ 0 & 0 & 0 & 0 \\ 0 & 1 & 0 & 0 \end{pmatrix} \quad \mathbf{R}_3 = \begin{pmatrix} 0 & 0 & 0 & 0 \\ 0 & 0 & 1 & 0 \\ 0 & 1 & 0 & 0 \\ 0 & 0 & 0 & 0 \end{pmatrix} \tag{9.112}$$
$$\mathbf{\Delta}_1 = \begin{pmatrix} 0 & 0 & 0 & 0 \\ 0 & 0 & 0 & 0 \\ 0 & 0 & -1 & 0 \\ 0 & 0 & 0 & -1 \end{pmatrix} \quad \mathbf{\Delta}_2 = \begin{pmatrix} 0 & 0 & 0 & 0 \\ 0 & -1 & 0 & 0 \\ 0 & 0 & 0 & 0 \\ 0 & 0 & 0 & -1 \end{pmatrix} \quad \mathbf{\Delta}_3 = \begin{pmatrix} 0 & 0 & 0 & 0 \\ 0 & -1 & 0 & 0 \\ 0 & 0 & -1 & 0 \\ 0 & 0 & 0 & 0 \end{pmatrix}$$

for the G-symmetric (depolarization) part. We shall refer to the matrix sets from Eqs. (9.111) and (9.112) as *polarization* and *depolarization basis* respectively.

For instance, the general differential matrix of a nondepolarizing medium, given by Eq. (9.56), is strictly G-antisymmetric and can be expanded as

$$\begin{aligned}\mathbf{m} &= \mathrm{LD}\,\mathbf{P}_1 + \mathrm{LD}'\,\mathbf{P}_2 + \mathrm{CD}\,\mathbf{P}_3 + \mathrm{LB}\,\mathbf{Q}_1 + \mathrm{LB}'\,\mathbf{Q}_2 + \mathrm{CB}\,\mathbf{Q}_3 \\ &= (\mathrm{Re\,L})\mathbf{P}_1 + (\mathrm{Re\,L}')\mathbf{P}_2 + (\mathrm{Re\,C})\mathbf{P}_3 + (\mathrm{Im\,L})\mathbf{Q}_1 + (\mathrm{Im\,L}')\mathbf{Q}_2 + (\mathrm{Im\,C})\mathbf{Q}_3 \\ &= \sum_{k=1}^{3}(\mathrm{Re\,P}_k)\mathbf{P}_k + \sum_{k=1}^{3}(\mathrm{Im\,P}_k)\mathbf{Q}_k\end{aligned} \tag{9.113}$$

the expansion coefficients simply coinciding with the elementary polarization properties. As seen from the expansion, the set of $\mathbf{P}_k$ matrices describes the dichroic (D) properties whereas the $\mathbf{Q}_k$ matrices parameterize the birefringent (B) properties. The D- and B-properties of $\mathbf{m}$ can be equivalently represented as real and imaginary parts of the complex elementary properties $\mathrm{P}_1 = \mathrm{L}$, $\mathrm{P}_2 = \mathrm{L}'$ and $\mathrm{P}_3 = \mathrm{C}$ in accordance with their definition (9.11). The same expansion holds for the G-antisymmetric part $\mathbf{m}_a$ of $\mathbf{m}$ of a depolarizing medium, provided the *mean values* of the elementary properties are used as expansion coefficients (i.e., the mean values of the properties of an equivalent uniformly depolarizing continuous medium).

Notice that the polarization basis $\{\mathbf{P}_k, \mathbf{Q}_k\}$ is the Mueller formalism analogue of the Pauli basis set used for expanding the differential Jones matrix in Section 9.2.2. It can be formally obtained from the latter by using Eq. (9.54) transforming differential Jones matrices into their associated Mueller counterparts.

Similarly, any G-symmetric part of a differential matrix or of a matrix logarithm can be expanded on the basis given by Eq. (9.112). For instance, the matrix logarithms $\mathbf{L}_{\Delta d}$ and $\mathbf{L}_{\Delta nd}$ of the two canonical depolarizers, respectively given by Eqs. (9.97) and (9.100), can be written compactly as

$$\mathbf{L}_{\Delta d} = \ln\sqrt{\frac{a}{bc}}\,\boldsymbol{\Delta}_1 + \ln\sqrt{\frac{b}{ac}}\,\boldsymbol{\Delta}_2 + \ln\sqrt{\frac{c}{ab}}\,\boldsymbol{\Delta}_3 \tag{9.114}$$

and

$$\mathbf{L}_{\Delta nd} = \ln(1/a)\boldsymbol{\Delta}_1 + \boldsymbol{\Delta}_2 + \boldsymbol{\Delta}_3 + \mathbf{D}_1 \tag{9.115}$$

Likewise, the polarizing and depolarizing parts of the matrix logarithm of the matrix $\mathbf{S}$ from Eqs. (9.89) expand on the respective basis sets as

$$\mathbf{L}_{\mathbf{S}\,m} = \ln\sqrt{\frac{1+a}{1-a}}\,\mathbf{P}_1 + \arg(b+ic)\mathbf{Q}_1 \tag{9.116}$$

and

$$\mathbf{L}_{\mathbf{S}\,u} = \ln\sqrt{\frac{1-a^2}{b^2+c^2}}\,\boldsymbol{\Delta}_1 \tag{9.117}$$

Notice that all expansion coefficients in Eqs. (9.114, 9.115 and 9.117) associated with the diagonal depolarization matrices $\mathbf{\Delta}_k$ are non-negative, as follows from the necessary condition for local physical realizability derived in the previous section. The physical meaning of these expansion coefficients will be clarified in the Section 9.5.3.

Now, if we denote by

$$(\mathbf{A},\mathbf{B}) = \mathbf{AB} + \mathbf{BA} \tag{9.118}$$

the *anticommutator* of the two matrices **A** and **B**, then the basis matrices of the G-symmetric (depolarization) part can be readily expressed through those of the G-antisymmetric (polarization) part,

$$\mathbf{D}_k = (\mathbf{P}_i, \mathbf{Q}_j) = -(\mathbf{Q}_i, \mathbf{P}_j) \tag{9.119}$$

$$\mathbf{R}_k = (\mathbf{P}_i, \mathbf{P}_j) = (\mathbf{Q}_i, \mathbf{Q}_j) \tag{9.120}$$

$$\mathbf{\Delta}_k = \mathbf{Q}_k^2 = \mathbf{P}_k^2 - \mathbf{I} \tag{9.121}$$

where the indices i, j and k correspond to cyclic permutations of 1, 2 and 3. It follows from the above relations that the square of any linear combination of matrices from the basis set $\{\mathbf{P}_k, \mathbf{Q}_k\}$ can be expanded on the basis set $\{\mathbf{D}_k, \mathbf{R}_k, \mathbf{\Delta}_k\}$. Consequently, these relations between basis sets simply express the fact that any squared G-antisymmetric matrix is a G-symmetric matrix.

Similarly, if we write for the *commutator* of **A** and **B**

$$[\mathbf{A},\mathbf{B}] = \mathbf{AB} - \mathbf{BA} \tag{9.122}$$

then the following commutation relations between subsets of matrices from each one of the two basis sets can be readily established:

$$[\mathbf{P}_k, \mathbf{Q}_k] = 0 \tag{9.123}$$

$$[\mathbf{D}_k, \mathbf{R}_k] = 0 \tag{9.124}$$

Furthermore, the matrix $\mathbf{\Delta}_k$ commutes with any of the matrices $\mathbf{P}_k$, $\mathbf{Q}_k$, $\mathbf{D}_k$ and $\mathbf{R}_k$. Since, according to Eq. (9.113), the $\mathbf{P}_k$ and $\mathbf{Q}_k$ matrices correspond respectively to the dichroic (D) and birefringent (B) components of the k-th elementary polarization property (k = 1: L; k = 2: L′; k = 3: C) associated with the corresponding axis of the Poincaré sphere (k = 1: HV; k = 2: PQ; k = 3: RL) then, in virtue of the above commutation properties, the depolarization matrices $\mathbf{D}_k$, $\mathbf{R}_k$ and Δ_k must be also related to the same axis. We shall confirm the correctness of this observation in Section 9.5.3.

9.4.5 Relation of the Differential Decomposition to the Product Decompositions of Mueller Matrices

We have seen that the differential decomposition of a depolarizing Mueller matrix **M** is a special kind of a *continuous product decomposition*, in contrast to the classic discrete product (serial) decompositions from Chapter 8. Being of similar algebraic nature, the differential and product decompositions must be consequently, formally related to each other. Indeed, it is relatively easy to show that the various product decompositions can be reduced to the so-called Lorentz decomposition (Ossikovski 2010a),

$$\mathbf{M} = \mathbf{U}\mathbf{H} \tag{9.125}$$

where the two matrix factors $\mathbf{U}$ and $\mathbf{H}$, being respectively G-orthogonal ($\mathbf{U}^G = \mathbf{U}^{-1}$) and G-symmetric ($\mathbf{H}^G = \mathbf{H}$), physically represent a nondepolarizing Mueller matrix and a generalized depolarizer. The Lorentz decomposition is clearly a "depolarizing analogue" of the well-known polar decomposition of a nondepolarizing Mueller matrix (Gil and Bernabéu 1987) into symmetric and orthogonal factors (physically representing a general diattenuator and an elliptical retarder).

Assuming implicit z-dependence for all matrices entering Eq. (9.125), the calculation of the differential matrix $\mathbf{m}$ of the Lorentz-decomposed depolarizing Mueller matrix $\mathbf{M}$ is straightforward,

$$\mathbf{m} = \frac{d\mathbf{M}}{dz}\mathbf{M}^{-1} = \frac{d\mathbf{U}}{dz}\mathbf{U}^{-1} + \mathbf{U}\frac{d\mathbf{H}}{dz}\mathbf{H}^{-1}\mathbf{U}^{-1} = \mathbf{m}_\mathrm{U} + \mathbf{U}\,\mathbf{m}_\mathrm{H}\mathbf{U}^{-1} \tag{9.126}$$

where $\mathbf{m}_\mathrm{U}$ and $\mathbf{m}_\mathrm{H}$ are the differential matrices of the nondepolarizing and depolarizing factors $\mathbf{U}$ and $\mathbf{H}$ respectively. Since $\mathbf{U}$ is G-orthogonal, its differential matrix $\mathbf{m}_\mathrm{U}$ is G-antisymmetric (see Section 9.3.1) whereas $\mathbf{m}_\mathrm{H}$ does not have definite G-symmetry in general. However, it can be easily shown that when $\mathbf{H}$ commutes with its derivative then its differential matrix $\mathbf{m}_\mathrm{H}$ is G-symmetric. If furthermore $\mathbf{U}$ commutes with both $\mathbf{H}$ and its derivative that is, if $\mathbf{U}$ commutes with $\mathbf{m}_\mathrm{H}$, then Eq. (9.126) reduces to

$$\mathbf{m} = \mathbf{m}_\mathrm{U} + \mathbf{m}_\mathrm{H} \tag{9.127}$$

Since $\mathbf{m}_\mathrm{U}$ and $\mathbf{m}_\mathrm{H}$ are respectively G-antisymmetric and G-symmetric, they are to be identified with the G-antisymmetric (polarization) and G-symmetric (depolarization) parts $\mathbf{m}_a$ and $\mathbf{m}_s$ of the differential matrix $\mathbf{m}$ in this special case.

Eq. (9.126) thus expresses the differential matrix of $\mathbf{M}$ through those of its two polar factors. It is mainly of theoretical rather than of practical significance since the z-dependence of the matrix factors is usually unknown. In practice, one deals with the differential decomposition in terms of the Mueller matrix logarithm rather than with the differential Mueller matrix itself. Let $\mathbf{L}_\mathrm{U}$ and $\mathbf{L}_\mathrm{H}$ be the matrix logarithms of the polar factors $\mathbf{U}$ and $\mathbf{H}$ respectively. The application of the well-known Baker-Campbell-Hausdorff (BCH) formula makes it possible to relate the matrix logarithm $\mathbf{L}$ of $\mathbf{M}$ to those of its polar factors $\mathbf{U}$ and $\mathbf{H}$,

$$\mathbf{L} = \mathbf{L}_\mathrm{U} + \mathbf{L}_\mathrm{H} + \frac{1}{2}[\mathbf{L}_\mathrm{U}, \mathbf{L}_\mathrm{H}] + \frac{1}{12}[\mathbf{L}_\mathrm{U}, [\mathbf{L}_\mathrm{U}, \mathbf{L}_\mathrm{H}]] + \frac{1}{12}[[\mathbf{L}_\mathrm{U}, \mathbf{L}_\mathrm{H}], \mathbf{L}_\mathrm{H}] + \ldots \tag{9.128}$$

Because of the specific G-symmetries of the polar factors $\mathbf{U}$ (G-orthogonal) and $\mathbf{H}$ (G-symmetric), their logarithms $\mathbf{L}_\mathrm{U}$ and $\mathbf{L}_\mathrm{H}$ are respectively G-antisymmetric and G-symmetric. It can be easily seen that their nested commutators, entering Eq. (9.128), are also of definite G-symmetry. The separation of G-symmetric and G-symmetric parts in Eq. (9.128) allows one to express the $\mathbf{L}_m$ (polarization) and $\mathbf{L}_u$ (depolarization) parts of the Mueller matrix logarithm $\mathbf{L}$ through the logarithms of the two polar factors,

$$\mathbf{L}_m = \mathbf{L}_\mathrm{U} + \frac{1}{12}\left[\left[\mathbf{L}_\mathrm{U},\ \mathbf{L}_\mathrm{H}\right],\ \mathbf{L}_\mathrm{H}\right] + \ldots \qquad \mathbf{L}_u = \mathbf{L}_\mathrm{H} + \frac{1}{2}\left[\mathbf{L}_\mathrm{U},\ \mathbf{L}_\mathrm{H}\right] + \frac{1}{12}\left[\mathbf{L}_\mathrm{U}\ ,\left[\mathbf{L}_\mathrm{U},\ \mathbf{L}_\mathrm{H}\right]\right] + \ldots \tag{9.129}$$

In the important special case where the polar factors $\mathbf{U}$ and $\mathbf{H}$ commute, so will their logarithms $\mathbf{L}_\mathrm{U}$ and $\mathbf{L}_\mathrm{H}$, and Eqs. (9.129) simplify considerably to

$$\mathbf{L}_m = \mathbf{L}_\mathrm{U} \qquad \mathbf{L}_u = \mathbf{L}_\mathrm{H} \tag{9.130}$$

that is, the polarization and depolarization parts of the Mueller matrix logarithm are simply given by the logarithms of the two polar factors, $\mathbf{L} = \mathbf{L}_m + \mathbf{L}_u = \mathbf{L}_\mathrm{U} + \mathbf{L}_\mathrm{H}$.

As an illustration of the above theoretical developments, consider the Mueller matrix $\mathbf{S}$ from Section 9.4.2. Its Lorentz decomposition factors are readily shown to be

$$\mathbf{U}_\mathrm{S} = \begin{pmatrix} 1 & a & 0 & 0 \\ a & 1 & 0 & 0 \\ 0 & 0 & b/\Delta & c/\Delta \\ 0 & 0 & -c/\Delta & b/\Delta \end{pmatrix} \tag{9.131}$$

and

$$\mathbf{H}_\mathrm{S} = \mathrm{diag}(1, 1, \Delta, \Delta) \tag{9.132}$$

where $\Delta = \sqrt{(b^2 + c^2)/(1 - a^2)}$ is the depolarization. Since the two polar factors obviously commute, in virtue of Eq. (9.130) their respective logarithms $\mathbf{L}_\mathrm{U}$ and $\mathbf{L}_\mathrm{H}$ provide the polarization and depolarization parts of the matrix logarithm of $\mathbf{S}$, that is, $\mathbf{L}_\mathrm{S} = \mathbf{L}_\mathrm{U} + \mathbf{L}_\mathrm{H}$,

$$\mathbf{L}_{\mathrm{S}\,m} = \mathbf{L}_\mathrm{U} = \begin{pmatrix} 0 & \ln\sqrt{(1+a)/(1-a)} & 0 & 0 \\ \ln\sqrt{(1+a)/(1-a)} & 0 & 0 & 0 \\ 0 & 0 & 0 & \arg(b/\Delta + ic/\Delta) \\ 0 & 0 & -\arg(b/\Delta + ic/\Delta) & 0 \end{pmatrix}$$

$$\mathbf{L}_{\mathrm{S}\,u} = \mathbf{L}_\mathrm{H} = \mathrm{diag}\left(0,\ 0,\ \ln\sqrt{\frac{b^2+c^2}{1-a^2}},\ \ln\sqrt{\frac{b^2+c^2}{1-a^2}}\right) = \mathrm{diag}(0, 0, \Delta_{\ln}, \Delta_{\ln}) \tag{9.133}$$

where $\Delta_{\ln} = \ln\sqrt{(b^2 + c^2)/(1 - a^2)} = \ln \Delta$ is the logarithmic depolarization. Notice that, because of the inequality $0 < \Delta \le 1$ one necessarily has $\Delta_{\ln} \le 0$. Since $\arg[(b/\Delta) + i(c/\Delta)] = \arg(b + ic)$ for $\Delta > 0$, Eqs. (9.133) coincide with Eqs. (9.89) from Section 9.4.2, as expected.

Note that Eq. (9.127) also holds for the present example since, whatever the z-dependencies of the polar factors, $\mathbf{H}$ being diagonal, it necessarily commutes with its derivative, as well as $\mathbf{U}$ commutes with both $\mathbf{H}$ and its derivative. However, in the absence of any knowledge on the z-dependence of the polar factors only their matrix logarithms can be determined, i.e., they can only be treated in terms of equivalent uniformly depolarizing media.

The polarization properties of $\mathbf{S}$ derived from the two decompositions, the Lorentz and the differential one, coincide. Indeed, the diattenuation D and retardance R obtained from the nondepolarizing Mueller matrix factor $\mathbf{U_S}$ are respectively $D = a$ and $R = \arg(b + ic)$ and, in virtue of the relations (9.94) and (9.95), they are in perfect correspondence with the values of the LD and LB properties extracted from the polarization part $\mathbf{L}_{\mathbf{S}m}$ of $\mathbf{L_S}$.

The depolarizations provided by the two decompositions can be easily put in correspondence as well. The *depolarization power* (Lu and Chipman 1996) of a diagonal depolarizer $\mathbf{M}_{\Delta d}$ given by Eq. (9.96) is entirely determined from its three coefficients a, b and c whereas that of its matrix logarithm $\mathbf{L}_{\Delta d}$ depends on $\ln a$, $\ln b$ and $\ln c$; see, Eq. (9.97). It is therefore sufficient to replace the coefficients a, b and c by $\exp(l_a)$, $\exp(l_b)$ and $\exp(l_c)$ where l_a, l_b and l_c are the respective diagonal elements of $\mathbf{L}_{\Delta d}$ in the various depolarization index expressions (Ossikovski 2010b) to achieve the correspondence sought (Kumar et al. 2012). For instance, the depolarization index P_Δ of $\mathbf{H_S}$ is

$$P_\Delta(\mathbf{H_S}) = \sqrt{\frac{1^2 + \Delta^2 + \Delta^2}{3(1)^2}} = \sqrt{\frac{1 + 2\Delta^2}{3}} \tag{9.134}$$

and obviously coincides with that derived from its matrix logarithm $\mathbf{L_H}$ with the help of the above correspondence,

$$P_\Delta(\mathbf{L_H}) = \sqrt{\frac{\exp^2(0) + \exp^2(\Delta_{\ln}) + \exp^2(\Delta_{\ln})}{3\exp^2(0)}} = \sqrt{\frac{1^2 + \Delta^2 + \Delta^2}{3}} = P_\Delta(\mathbf{H_S}) \tag{9.135}$$

since $\exp(\Delta_{\ln}) = \Delta$. Note however, that strict numerical coincidence between depolarization index values, respectively calculated from the product and from the differential decompositions, holds only for diagonal matrices (since the matrix logarithm of a diagonal matrix is the matrix of the logarithms of its diagonal elements).

Eventually, the intimate relationship between differential and polar decompositions can be used for obtaining the multiple solutions $\mathbf{L}_0 + p\Delta\mathbf{L}$ of the Mueller matrix logarithm in the special case where the Mueller matrix $\mathbf{M}$ has a double real eigenvalue, mentioned in Section 9.4.2.

Recall that the condition for the existence of physically meaningful multiple solutions is the G-antisymmetry of the matrix "period" $\Delta\mathbf{L}$. Since $\Delta\mathbf{L}$ is G-antisymmetric when the polarization and depolarization parts $\mathbf{L}_m$ (of zero order) and $\mathbf{L}_u$ of the Mueller matrix logarithm commute, Eqs. (9.130) can be applied to obtain the matrices $\mathbf{L}_m$ and $\mathbf{L}_u$ of the differential decomposition from $\mathbf{L_U}$ and $\mathbf{L_H}$, the matrix logarithms of the polar factors $\mathbf{U}$ and $\mathbf{H}$ of $\mathbf{M}$. As the multiplicity results from the nondepolarizing polar factor $\mathbf{U}$ only, its multivalued logarithm $\mathbf{L_U}$ can be handled at the Jones matrix level (see Section 9.2.4), converted back to Mueller space by using Eq. (9.52) and identified with $\mathbf{L}_m$, thus successfully solving the problem of the logarithm of $\mathbf{M}$ with a double real eigenvalue. The inconvenience of this approach is the necessity for performing the polar decomposition of $\mathbf{M}$ as an intermediate step. Finally, notice that the example of the matrix $\mathbf{S}$ treated above is a trivial case of this method.

9.5 The Differential Mueller Matrix of a Homogeneous Depolarizing Medium

9.5.1 The Differential Mueller Matrix of a Fluctuating Homogeneous Medium

In Section 9.4.1 we have interpreted the G-antisymmetric (polarization) and G-symmetric (depolarization) parts, $\mathbf{m}_a$ and $\mathbf{m}_s$, of the differential matrix $\mathbf{m}$ associated with the depolarizing Mueller matrix $\mathbf{M}$, as respectively containing the mean values and the uncertainties of the elementary polarization properties of the medium described by $\mathbf{M}$. Whereas by mean values of the properties one means the properties of an equivalent nondepolarizing homogeneous medium, it is so far unclear what the exact physical meaning of their uncertainties is.

The G-symmetric part $\mathbf{m}_s$ of $\mathbf{m}$ accounts for the presence of *depolarization* in the polarimetric experiment. Most generally, the phenomenon of depolarization occurs because of spatial or temporal *averaging* of the polarimetric response of the medium by the measurement equipment. Thus in order to interpret the depolarization part $\mathbf{m}_s$ of the differential Mueller matrix, we need to build a microscopic model of the depolarization phenomenon based on averaging the elementary polarization properties of the medium. For this purpose, assume that the complex elementary polarization property P_k ($\mathrm{P}_1 = \mathrm{L}$, $\mathrm{P}_2 = \mathrm{L}'$ or $\mathrm{P}_3 = \mathrm{C}$) of the initially nondepolarizing medium is not constant, but rather fluctuates around its mean value P_{k0} by the amount $\Delta\mathrm{P}_k$ (the fluctuation) (Devlaminck 2013; 2015),

$$\mathrm{P}_k = \mathrm{P}_{k\,0} + \Delta\mathrm{P}_k \tag{9.136}$$

If we denote by the brackets $\langle\ldots\rangle$ the averaging process (of any, spatial or temporal, nature), we further assume that the fluctuation $\Delta\mathrm{P}_k$ has zero mean value, i.e., $\langle\Delta\mathrm{P}_k\rangle = 0$, as well as that both the mean value P_{k0} and the fluctuation $\Delta\mathrm{P}_k$ of the elementary property P_k are z-independent that is, the medium is *homogeneous along the light path*. As a result, the differential matrix of a fluctuating nondepolarizing medium takes the form

$$\mathbf{m} = \mathbf{m}_0 + \Delta\mathbf{m} \tag{9.137}$$

where $\langle\Delta\mathbf{m}\rangle = \mathbf{0}$ (and consequently, $\langle\mathbf{m}\rangle = \mathbf{m}_0$) since, in virtue of Eq. (9.55), $\mathbf{m}$ has the real and imaginary parts of the elementary polarization properties P_i for its elements. (Note that the meaning attributed to the notations $\mathbf{m}_0$ and $\Delta\mathbf{m}$ in this section is totally different from that given to them in Section 9.3.3 discussing the multiple solutions of the matrix logarithm.)

Now, consider the Mueller matrix $\mathbf{M}(z)$ of a slab of fluctuating medium of small thickness z ($z << 1$) whose differential matrix is given by Eq. (9.137). In virtue of Eq. (9.57), $\mathbf{M}(z)$ can be written as

$$\mathbf{M}(z) = \exp\left[(\mathbf{m}_0 + \Delta\mathbf{m})z\right] \tag{9.138}$$

where both the *mean* values $\mathbf{m}_0$ and the *fluctuations* $\Delta\mathbf{m}$ of the elementary properties have been assumed z-independent as mentioned before. Expanding $\mathbf{M}(z)$ in series while retaining terms up to second order in z and averaging the result yields

$$\langle\mathbf{M}(z)\rangle = \mathbf{I}_4 + \mathbf{m}_0 z + \frac{1}{2}\left(\mathbf{m}_0^2 + \langle\Delta\mathbf{m}^2\rangle\right)z^2 \tag{9.139}$$

Inserting the averaged Mueller matrix into Eq. (9.50) provides for the differential matrix $\mathbf{m}(z)$ of the averaged fluctuating medium

$$\mathbf{m}(z) = \frac{d\langle \mathbf{M}(z)\rangle}{dz}\langle \mathbf{M}(z)\rangle^{-1} = \mathbf{m}_0 + \langle \Delta\mathbf{m}^2\rangle z \tag{9.140}$$

The two terms of $\mathbf{m}(z)$ from Eq. (9.140) are G-antisymmetric and G-symmetric respectively and thus, are to be identified with the G-antisymmetric (polarization) and G-symmetric (depolarization) parts $\mathbf{m}_a$ and $\mathbf{m}_s$ of the differential matrix (Ossikovski and Arteaga 2014),

$$\mathbf{m}_a = \mathbf{m}_0 = \langle \mathbf{m}\rangle \qquad \mathbf{m}_s = \langle \Delta\mathbf{m}^2\rangle z \tag{9.141}$$

We see that the polarization part of the differential matrix indeed contains the mean values $\mathbf{m}_0 = \langle \mathbf{m}\rangle$ of the elementary polarization properties, whereas the depolarization part is proportional to their variance matrix $\langle \Delta\mathbf{m}^2\rangle$ (containing respectively their variances and covariances, i.e., the averages of the squared moduli and of the conjugate cross-products of their fluctuations).

More generally, Eq. (9.141) also confirms the physical origin of the depolarization phenomenon as being due to the fluctuations of the medium whose polarimetric response is averaged by the measurement process: the absence of fluctuations ($\Delta\mathbf{m} = \mathbf{0}$) entails $\langle \Delta\mathbf{m}^2\rangle = \mathbf{0}$, that is, the absence of depolarization. However, unlike the polarization part of $\mathbf{m}$ being z-independent, the depolarization grows with the path-length z even though the depolarizing fluctuating medium is homogeneous, i.e., neither the mean values $\mathbf{m}_0$ of the properties nor their fluctuations $\Delta\mathbf{m}$ depend on z. It is important to stress that this difference in behavior results from the use of the specific microscopic model of a homogeneous fluctuating medium, see Eq. (9.138); other assumptions will lead to other z-dependencies (Devlaminck 2013; 2015). Most generally, the scope of a given depolarization model may be limited to a certain class of media only and its validity can be established only experimentally.

Another consequence of the difference in z-dependence of the polarization and depolarization parts of the differential matrix $\mathbf{m}$ is the more complex relation of $\mathbf{m}$ to the Mueller matrix logarithm $\mathbf{L}$. Indeed, we can identify $\mathbf{L}$ with $\mathbf{m}$ only if $\mathbf{m}$ is z-independent (see Section 9.4.1) which is not the case here. More specifically, the expression for the Mueller matrix $\mathbf{M}$ of the depolarizing medium provided by the integration (from $z = 0$ to $z = 1$) of the evolution equation (9.47) with the z-dependent differential matrix $\mathbf{m} = \mathbf{m}_a + \mathbf{m}_s$ being given by Eq. (9.141),

$$\mathbf{M} = \exp\left(\mathbf{m}_0 + \frac{1}{2}\langle \Delta\mathbf{m}^2\rangle\right) \tag{9.142}$$

is, in general, strictly valid only for commuting $\mathbf{m}_0$ and $\langle \Delta\mathbf{m}^2\rangle$. Consequently, the relations following from Eq. (9.142) and from $\ln\mathbf{M} = \mathbf{L}$,

$$\mathbf{m}_0 = \langle \mathbf{m}\rangle = \mathbf{L}_m \qquad \langle \Delta\mathbf{m}^2\rangle = 2\mathbf{L}_u \tag{9.143}$$

are only approximate, in the general case. We shall come back to this point in Section 9.5.4 and shall see, in particular, that the practical range of the validity of Eqs. (9.143) is quite large despite their approximate nature.

Eqs. (9.143) give a clear physical meaning to the Mueller matrix logarithm **L** within the framework of the fluctuating homogeneous medium model: the polarization part $\mathbf{L}_m$ of **L** contains the mean values of the elementary polarization properties whereas the depolarization one, $\mathbf{L}_u$, is proportional to the variance matrix of the properties. By using Eq. (9.55) for **m**, one obtains the following explicit expression for the variance matrix $\langle \mathbf{\Delta m}^2 \rangle$:

$$\langle \mathbf{\Delta m}^2 \rangle = \begin{pmatrix} \langle \Delta D^2 \rangle & \mathrm{Im}\langle \Delta L' \Delta C^* \rangle & \mathrm{Im}\langle \Delta C\, \Delta L^* \rangle & \mathrm{Im}\langle \Delta L\, \Delta L'^* \rangle \\ -\mathrm{Im}\langle \Delta L' \Delta C^* \rangle & \langle |\Delta L|^2 \rangle - \langle \Delta B^2 \rangle & \mathrm{Re}\langle \Delta L\, \Delta L'^* \rangle & \mathrm{Re}\langle \Delta C\, \Delta L^* \rangle \\ -\mathrm{Im}\langle \Delta C\, \Delta L^* \rangle & \mathrm{Re}\langle \Delta L\, \Delta L'^* \rangle & \langle |\Delta L'|^2 \rangle - \langle \Delta B^2 \rangle & \mathrm{Re}\langle \Delta L' \Delta C^* \rangle \\ -\mathrm{Im}\langle \Delta L\, \Delta L'^* \rangle & \mathrm{Re}\langle \Delta C\, \Delta L^* \rangle & \mathrm{Re}\langle \Delta L' \Delta C^* \rangle & \langle |\Delta C|^2 \rangle - \langle \Delta B^2 \rangle \end{pmatrix} \quad (9.144)$$

in which $\Delta D^2 = \Delta LD^2 + \Delta LD'^2 + \Delta CD^2$ and $\Delta B^2 = \Delta LB^2 + \Delta LB'^2 + \Delta CB^2$. Notice that, being the averaged square of a G-antisymmetric matrix, the variance matrix is G-symmetric (the averaging process has no incidence on the G-symmetry). With the usual convention for zero first element achieved by the subtraction of $\langle \Delta D^2 \rangle$ from the main diagonal, the centered variance matrix $\langle \mathbf{\Delta m}_c^2 \rangle$ takes the form

$$\langle \mathbf{\Delta m}_c^2 \rangle = \begin{pmatrix} 0 & \mathrm{Im}\langle \Delta L' \Delta C^* \rangle & \mathrm{Im}\langle \Delta C\, \Delta L^* \rangle & \mathrm{Im}\langle \Delta L\, \Delta L'^* \rangle \\ -\mathrm{Im}\langle \Delta L' \Delta C^* \rangle & -\langle |\Delta L'|^2 \rangle - \langle \Delta C^2 \rangle & \mathrm{Re}\langle \Delta L\, \Delta L'^* \rangle & \mathrm{Re}\langle \Delta C\, \Delta L^* \rangle \\ -\mathrm{Im}\langle \Delta C\, \Delta L^* \rangle & \mathrm{Re}\langle \Delta L\, \Delta L'^* \rangle & -\langle |\Delta C|^2 \rangle - \langle |\Delta L|^2 \rangle & \mathrm{Re}\langle \Delta L' \Delta C^* \rangle \\ -\mathrm{Im}\langle \Delta L\, \Delta L'^* \rangle & \mathrm{Re}\langle \Delta C\, \Delta L^* \rangle & \mathrm{Re}\langle \Delta L' \Delta C^* \rangle & -\langle |\Delta L|^2 \rangle - \langle |\Delta L'|^2 \rangle \end{pmatrix} \quad (9.145)$$

This matrix contains the variances, $\langle |\Delta P_i|^2 \rangle = \langle \Delta PD_i^2 \rangle + \langle \Delta PB_i^2 \rangle$, as well as the real and imaginary parts of the covariances, $\langle \Delta P_j\, \Delta P_k^* \rangle = \langle \Delta PD_j\, \Delta PD_k^* \rangle + \langle \Delta PB_j\, \Delta PB_k^* \rangle + i\left(\langle \Delta PB_j\, \Delta PB_k^* \rangle - \langle \Delta PD_j\, \Delta PB_k^* \rangle\right)$, of the three complex elementary properties P_i. This content becomes even clearer upon the calculation of the reduced coherency matrix $\mathbf{C}_3$ of $\langle \mathbf{\Delta m}_c^2 \rangle$ with the help of Eqs. (9.102),

$$\mathbf{C}_3\left(\langle \mathbf{\Delta m}_c^2 \rangle\right) = \frac{1}{2}\begin{pmatrix} \langle |\Delta L|^2 \rangle & \langle \Delta L\, \Delta L'^* \rangle & \langle \Delta L\, \Delta C^* \rangle \\ \langle \Delta L' \Delta L^* \rangle & \langle |\Delta L'|^2 \rangle & \langle \Delta L' \Delta C^* \rangle \\ \langle \Delta C\, \Delta L^* \rangle & \langle \Delta C\, \Delta L'^* \rangle & \langle |\Delta C|^2 \rangle \end{pmatrix} = \frac{1}{2}\langle \Delta\vec{P} \otimes \Delta\vec{P}^{\dagger} \rangle \quad (9.146)$$

where we have introduced the (complex) vector of the elementary polarization properties $\vec{P}$,

$$\vec{P} \equiv (L, L', C)^T = \vec{D} + i\vec{B} \quad (9.147)$$

that we shall call *(complex) polarization vector*. The same expression for the coherency matrix $\mathbf{C}_3$ is obtained if $\mathbf{C}_3$ is evaluated from Eqs. (9.101) with $\langle \mathbf{\Delta m}^2 \rangle$ being given by

Eq. (9.145) that is, $\mathbf{C}_3(\langle\boldsymbol{\Delta}\mathbf{m}^2\rangle) = \mathbf{C}_3(\langle\boldsymbol{\Delta}\mathbf{m}_c^2\rangle)$. Note that the explicit forms of the matrices $\langle\boldsymbol{\Delta}\mathbf{m}^2\rangle$ and $\mathbf{C}_3(\langle\boldsymbol{\Delta}\mathbf{m}^2\rangle)$ do not depend on the specific definition adopted for the complex elementary properties L, L′ and C, i.e., both definitions (9.11) and (9.23) lead to the same expressions. Eventually, notice that the diagonal elements of the centered variance matrix $\langle\boldsymbol{\Delta}\mathbf{m}_c^2\rangle$ reported in Eq. (9.145) are always non-positive, in full agreement with a general property of the G-symmetric part of the differential matrix we have discussed before (see Section 9.4.3).

The matrix in Eq. (9.146) has the mathematical form of a *covariance matrix* $\langle\Delta\vec{P}\otimes\Delta\vec{P}^{+}\rangle$ of three complex random variables. On the other hand, the variance matrix $\langle\boldsymbol{\Delta}\mathbf{m}^2\rangle$ of the elementary properties and the depolarization part of the $\mathbf{m}_s$ of the differential matrix $\mathbf{m}$ are proportional in virtue of the second of Eqs. (9.141). Since, furthermore, $\mathbf{C}_3(\mathbf{m}_s) = \mathbf{C}_3(\mathbf{m})$, as mentioned in Section 9.4.3, then $\mathbf{C}_3(\langle\boldsymbol{\Delta}\mathbf{m}^2\rangle)$ and $\mathbf{C}_3(\mathbf{m})$ are also proportional. Consequently, Eq. (9.146) implies that the reduced coherency matrix $\mathbf{C}_3(\mathbf{m})$ of the differential matrix $\mathbf{m}$ of the medium is, in turn, proportional to the covariance matrix $\langle\Delta\vec{P}\otimes\Delta\vec{P}^{+}\rangle$ of the three complex elementary polarization properties L, L′ and C, considered random complex variables or, equivalently, of the complex polarization vector $\vec{P}$ (considered a vector of random variables).

A very important consequence of the matrix proportionality established above is the local physical realizability of the fluctuating homogeneous medium model. We have seen in Section 9.4.3 that a given Mueller matrix $\mathbf{M}$ is locally physically realizable if and only if the reduced coherency matrix $\mathbf{C}_3(\mathbf{m})$ of its differential matrix is positive semi-definite. Within the framework of the fluctuating medium model, $\mathbf{C}_3(\mathbf{m})$ is proportional to the covariance matrix of the three complex elementary polarization properties. However, every covariance matrix is positive semi-definite by construction and consequently, the fluctuating medium model is always locally physically realizable.

9.5.2 Statistical and Geometrical Interpretation of the Differential Mueller Matrix

Eq. (9.146) reveals the statistical meaning of the depolarizing fluctuating medium model: the reduced coherency matrix $\mathbf{C}_3(\mathbf{m})$ of the differential matrix $\mathbf{m}$ of the fluctuating medium and the covariance matrix $\langle\Delta\vec{P}\otimes\Delta\vec{P}^{+}\rangle$ of the complex polarization vector $\vec{P}$ (containing the elementary polarization properties considered random complex variables), are proportional to one another.

Thus the knowledge of the differential matrix $\mathbf{m}$ allows one to interpret the elementary polarization properties of the depolarizing medium in statistical terms. For instance, suppose that a given property P_i is "sharp," i.e., its fluctuation is zero, $\Delta P_i = 0$. Then $\langle|\Delta P_i|^2\rangle = 0$, and the corresponding row and column of the reduced coherency matrix $\mathbf{C}_3(\langle\boldsymbol{\Delta}\mathbf{m}_c^2\rangle)$ given by Eq. (9.146) (and proportional to $\mathbf{C}_3(\mathbf{m})$) will vanish identically. In practice, it is the converse statement that is of interest: if the *i*-th row and column of the reduced coherency matrix are zero, then the property P_i is not fluctuating.

The statistical interpretation can be further developed by defining the (complex) correlation $c(P_j, P_k)$ between two properties P_j and P_k,

$$c(\mathrm{P}_j,\mathrm{P}_k)=\frac{\langle\Delta\mathrm{P}_j\,\Delta\mathrm{P}_k^*\rangle}{\sqrt{\langle|\Delta\mathrm{P}_j|^2\rangle\langle|\Delta\mathrm{P}_k|^2\rangle}}=\frac{\mathrm{Re}\langle\Delta\mathrm{P}_j\,\Delta\mathrm{P}_k^*\rangle}{\sqrt{\langle|\Delta\mathrm{P}_j|^2\rangle\langle|\Delta\mathrm{P}_k|^2\rangle}}+i\frac{\mathrm{Im}\langle\Delta\mathrm{P}_j\,\Delta\mathrm{P}_k^*\rangle}{\sqrt{\langle|\Delta\mathrm{P}_j|^2\rangle\langle|\Delta\mathrm{P}_k|^2\rangle}} \quad (9.148)$$

If the correlation is unimodular, i.e., if $|c(\mathrm{P}_j,\mathrm{P}_k)|=1$, then the properties P_j and P_k (considered random variables) are *totally correlated*, that is, a proportionality relationship between their respective *m*-th statistical realizations $P_j^{(m)}$ and $P_k^{(m)}$ exists, $P_j^{(m)}=c\,P_k^{(m)}$. The generally complex proportionality coefficient *c* is real if furthermore $c(\mathrm{P}_j,\mathrm{P}_k)=\pm1$ and imaginary if $c(\mathrm{P}_j,\mathrm{P}_k)=\pm i$. If, instead of being unimodular, the correlation vanishes, $c(\mathrm{P}_j,\mathrm{P}_k)=0$, then the properties P_i and P_j are *totally uncorrelated.*

Since the centered variance matrix $\langle\mathbf{\Delta m}_c^2\rangle$ is real, it must be expressible in terms of real parameters instead of the three complex properties. If we define the three two-component real random vectors of the elementary polarization properties as

$$\vec{\mathrm{L}}=(\mathrm{LD},\mathrm{LB})^T \qquad \vec{\mathrm{L}}'=(\mathrm{LD}',\mathrm{LB}')^T \qquad \vec{\mathrm{C}}=(\mathrm{CD},\mathrm{CB})^T \quad (9.149)$$

then $\langle\mathbf{\Delta m}_c^2\rangle$ takes the following "geometrical" form:

$$\langle\mathbf{\Delta m}_c^2\rangle=\begin{pmatrix} 0 & -\langle(\Delta\vec{\mathrm{L}}'\times\Delta\vec{\mathrm{C}})_n\rangle & -\langle(\Delta\vec{\mathrm{C}}\times\Delta\vec{\mathrm{L}})_n\rangle & -\langle(\Delta\vec{\mathrm{L}}\times\Delta\vec{\mathrm{L}}')_n\rangle \\ \langle(\Delta\vec{\mathrm{L}}'\times\Delta\vec{\mathrm{C}})_n\rangle & -\langle|\Delta\vec{\mathrm{L}}'|^2\rangle-\langle|\Delta\vec{\mathrm{C}}|^2\rangle & \langle\Delta\vec{\mathrm{L}}.\Delta\vec{\mathrm{L}}'\rangle & \langle\Delta\vec{\mathrm{C}}.\Delta\vec{\mathrm{L}}\rangle \\ \langle(\Delta\vec{\mathrm{C}}\times\Delta\vec{\mathrm{L}})_n\rangle & \langle\Delta\vec{\mathrm{L}}.\Delta\vec{\mathrm{L}}'\rangle & -\langle|\Delta\vec{\mathrm{C}}|^2\rangle-\langle|\Delta\vec{\mathrm{L}}|^2\rangle & \langle\Delta\vec{\mathrm{L}}'.\Delta\vec{\mathrm{C}}\rangle \\ \langle(\Delta\vec{\mathrm{L}}\times\Delta\vec{\mathrm{L}}')_n\rangle & \langle\Delta\vec{\mathrm{C}}.\Delta\vec{\mathrm{L}}\rangle & \langle\Delta\vec{\mathrm{C}}.\Delta\vec{\mathrm{L}}\rangle & -\langle|\Delta\vec{\mathrm{L}}|^2\rangle-\langle|\Delta\vec{\mathrm{L}}'|^2\rangle \end{pmatrix} \quad (9.150)$$

Here, the notation $\vec{\mathrm{A}}.\vec{\mathrm{B}}$ means scalar ("dot") product, whereas $(\vec{\mathrm{A}}\times\vec{\mathrm{B}})_n$ means the projection of the vector ("cross") product $\vec{\mathrm{A}}\times\vec{\mathrm{B}}$ onto the normal $\vec{n}$ of the oriented plane defined by $\vec{\mathrm{A}}$ and $\vec{\mathrm{B}}$; since $\vec{\mathrm{A}}\times\vec{\mathrm{B}}$ is parallel to $\vec{n}$ by the very definition of the vector product, then $(\vec{\mathrm{A}}\times\vec{\mathrm{B}})_n$ is simply the oriented length of $\vec{\mathrm{A}}\times\vec{\mathrm{B}}$. Further, the first row and column of the matrix $\langle\mathbf{\Delta m}_c^2\rangle$ can alternatively be expressed in terms of the dichroic and birefringent vectors $\vec{\mathrm{D}}$ and $\vec{\mathrm{B}}$,

$$\langle\mathbf{\Delta m}_c^2\rangle=\begin{pmatrix} 0 & -\langle(\Delta\vec{\mathrm{D}}\times\Delta\vec{\mathrm{B}})_1\rangle & -\langle(\Delta\vec{\mathrm{D}}\times\Delta\vec{\mathrm{B}})_2\rangle & -\langle(\Delta\vec{\mathrm{D}}\times\Delta\vec{\mathrm{B}})_3\rangle \\ \langle(\Delta\vec{\mathrm{D}}\times\Delta\vec{\mathrm{B}})_1\rangle & -\langle|\Delta\vec{\mathrm{L}}'|^2\rangle-\langle|\Delta\vec{\mathrm{C}}|^2\rangle & \langle\Delta\vec{\mathrm{L}}.\Delta\vec{\mathrm{L}}'\rangle & \langle\Delta\vec{\mathrm{C}}.\Delta\vec{\mathrm{L}}\rangle \\ \langle(\Delta\vec{\mathrm{D}}\times\Delta\vec{\mathrm{B}})_2\rangle & \langle\Delta\vec{\mathrm{L}}.\Delta\vec{\mathrm{L}}'\rangle & -\langle|\Delta\vec{\mathrm{C}}|^2\rangle-\langle|\Delta\vec{\mathrm{L}}|^2\rangle & \langle\Delta\vec{\mathrm{L}}'.\Delta\vec{\mathrm{C}}\rangle \\ \langle(\Delta\vec{\mathrm{D}}\times\Delta\vec{\mathrm{B}})_3\rangle & \langle\Delta\vec{\mathrm{C}}.\Delta\vec{\mathrm{L}}\rangle & \langle\Delta\vec{\mathrm{L}}'.\Delta\vec{\mathrm{C}}\rangle & -\langle|\Delta\vec{\mathrm{L}}|^2\rangle-\langle|\Delta\vec{\mathrm{L}}'|^2\rangle \end{pmatrix} \quad (9.151)$$

where the subscripts denote projections on the respective axes of the Poincaré sphere (1: HV; 2: PQ; and 3: RL). We shall call the vector $\Delta\vec{\mathrm{P}}_i$ the *fluctuation vector* of the property

P_i. Like Eq. (9.146) expressing the reduced covariance matrix, Eqs. (9.150) and (9.151) can be used to interpret existing correlations between elementary properties in terms of their respective fluctuation vectors. Thus, if

$$\frac{\langle \Delta\vec{P}_j.\Delta\vec{P}_k \rangle}{\sqrt{\langle |\Delta\vec{P}_j|^2 \rangle \langle |\Delta\vec{P}_k|^2 \rangle}} = \pm 1 \quad \textit{and} \quad \left(\langle \Delta\vec{P}_j \times \Delta\vec{P}_k \rangle \right)_n = \vec{0} \tag{9.152}$$

then the two properties P_j and P_k are totally correlated and furthermore, their fluctuation vectors are parallel, i.e., $\Delta\vec{P}_j \parallel \Delta\vec{P}_k$. Next, if

$$\frac{\left\langle \left(\Delta\vec{P}_j \times \Delta\vec{P}_k \right)_n \right\rangle}{\sqrt{\langle |\Delta\vec{P}_j|^2 \rangle \langle |\Delta\vec{P}_k|^2 \rangle}} = \pm 1 \quad \text{and} \quad \langle \Delta\vec{P}_j.\Delta\vec{P}_k \rangle = 0 \tag{9.153}$$

then P_j and P_k are again totally correlated and their fluctuation vectors are orthogonal, $\Delta\vec{P}_j \perp \Delta\vec{P}_k$. (Note that it does not follow from the last two of Eqs. (9.152) and (9.153) *alone* that $\Delta\vec{P}_j \parallel \Delta\vec{P}_k$ or that $\Delta\vec{P}_j \perp \Delta\vec{P}_k$, since the fluctuation vectors $\Delta\vec{P}_j$ and $\Delta\vec{P}_k$ are random.) Conversely, the relations $\Delta\vec{P}_j \parallel \Delta\vec{P}_k$ or $\Delta\vec{P}_j \perp \Delta\vec{P}_k$ entail Eqs. (9.152) and (9.153) respectively. These are examples of two geometrically interpretable limiting cases of total correlation of a pair of properties. In the general case of pairwise total correlation, the two fluctuation vectors $\Delta\vec{P}_j$ and $\Delta\vec{P}_k$ satisfy an analogue of the well-known Lagrange identity,

$$\frac{\langle \Delta\vec{P}_j.\Delta\vec{P}_k \rangle^2}{\langle |\Delta\vec{P}_j|^2 \rangle \langle |\Delta\vec{P}_k|^2 \rangle} + \frac{\left\langle \left(\Delta\vec{P}_j \times \Delta\vec{P}_k \right)_n \right\rangle^2}{\langle |\Delta\vec{P}_j|^2 \rangle \langle |\Delta\vec{P}_k|^2 \rangle} = 1 \tag{9.154}$$

and geometrically subtend a constant (i.e., statistical-realization-invariant) angle α. (Notice that the above relation is equivalent to $|c(P_j, P_k)| = 1$ in the statistical picture.) The squared cosine and sine of α are given respectively by the two left-hand-side terms of Eq. (9.154). The two special cases given by Eqs. (9.152) and (9.153) correspond to $\alpha = 0°$ (or 180°) and $\alpha = 90°$ (or 270°) respectively. When P_j and P_k are only partially correlated the angle α fluctuates, like the lengths of $\Delta\vec{P}_j$ and $\Delta\vec{P}_k$, and the equality (9.154) becomes an inequality (≤ 1).

Eventually, if $\Delta\vec{L} \parallel \Delta\vec{L}' \parallel \Delta\vec{C}$ then $\Delta\vec{D} \parallel \Delta\vec{B}$ also and, according to Eq. (9.151), the first row and column of $\langle \mathbf{\Delta m}_c^2 \rangle$ vanish. For the converse to be true (i.e., $\Delta\vec{D} \parallel \Delta\vec{B}$ from vanishing first row and column of $\langle \mathbf{\Delta m}_c^2 \rangle$), it is further necessary that

$$\frac{\langle \Delta\vec{L}.\Delta\vec{L}' \rangle}{\sqrt{\langle |\Delta\vec{L}|^2 \rangle \langle |\Delta\vec{L}'|^2 \rangle}} = \frac{\langle \Delta\vec{L}'.\Delta\vec{C} \rangle}{\sqrt{\langle |\Delta\vec{L}'|^2 \rangle \langle |\Delta\vec{C}|^2 \rangle}} = \frac{\langle \Delta\vec{C}.\Delta\vec{L} \rangle}{\sqrt{\langle |\Delta\vec{C}|^2 \rangle \langle |\Delta\vec{L}|^2 \rangle}} = \pm 1 \tag{9.155}$$

Consequently, the simultaneous vanishing of the first row and column of the variance matrix *alone* is not sufficient. Note that what we called geometrical picture is, strictly speaking, statistical as well, the random complex variables being replaced by random real vectors.

To illustrate the above, apply the fluctuating medium depolarization model to the experimental Mueller matrix $\mathbf{M}_g$ given by Eq. (9.84) and decomposed differentially in Section 9.4.2. According to Eqs. (9.143), the mean values of its elementary polarization properties are given by the G-antisymmetric part (or mean value part) $\mathbf{L}_{gm}$ of its matrix logarithm $\mathbf{L}_g$, whereas the variance matrix of the properties equals twice the G-symmetric part (or uncertainty part) $\mathbf{L}_{gu}$ of $\mathbf{L}_g$,

$$\mathbf{m}_{g0} = \mathbf{L}_{gm} = \begin{pmatrix} 0 & -0.0300 & 0.0016 & 0.0006 \\ -0.0300 & 0 & -0.0596 & -0.0153 \\ 0.0016 & 0.0596 & 0 & -1.3860 \\ 0.0006 & 0.0153 & 1.3860 & 0 \end{pmatrix}$$

$$\left\langle \Delta\mathbf{m}_g^2 \right\rangle = 2\mathbf{L}_{gu} = \begin{pmatrix} 0 & -0.0111 & 0.0106 & -0.0098 \\ 0.0111 & -0.5257 & 0.0002 & 0.0071 \\ -0.0106 & 0.0002 & -0.6018 & 0.0429 \\ 0.0098 & 0.0071 & 0.0429 & -0.2963 \end{pmatrix} \tag{9.156}$$

where two matrices $\mathbf{L}_{gm}$ and $\mathbf{L}_{gu}$ are taken from Eqs. (9.85). In virtue of Eq. (9.146) and the second of Eq. (9.156), quadrupling the elements of the reduced coherency matrix $\mathbf{C}_3(\mathbf{L}_{gu})$ reported in Eq. (9.103) provides the covariance matrix $\langle \Delta\vec{\mathrm{P}}_g \otimes \Delta\vec{\mathrm{P}}_g^{\dagger} \rangle$ of the elementary polarization properties,

$$\left\langle \Delta\vec{\mathrm{P}}_g \otimes \Delta\vec{\mathrm{P}}_g^{\dagger} \right\rangle = 2\mathbf{C}_3\left(\left\langle \Delta\mathbf{m}_g^2 \right\rangle\right) = 4\mathbf{C}_3\left(\mathbf{L}_{gu}\right) =$$
$$= \begin{pmatrix} 0.1862 & 0.0002 - i0.0098 & 0.0071 - i0.0106 \\ 0.0002 + i0.0098 & 0.1100 & 0.0429 - i0.0111 \\ 0.0071 + i0.0106 & 0.0429 + i0.0111 & 0.4156 \end{pmatrix} \tag{9.157}$$

The mean values of the properties, being simply equal to the respective entries of the first matrix from Eqs. (9.156), were already reported in Eq. (9.87). The second matrix from Eqs. (9.156) contains the full information about the depolarization in the form of the (real) variance matrix $\langle \Delta\mathbf{m}_g^2 \rangle$. The same amount of information is contained in the (complex) covariance matrix (9.157) comprising the variances and the covariances of the elementary properties. The latter can be represented in a table, in a way similar to that of the mean values of the properties,

$$\begin{aligned} &\left\langle |\Delta \mathrm{L}|^2 \right\rangle = 0.1862 \quad \left\langle |\Delta \mathrm{L}' \Delta \mathrm{C}^*| \right\rangle = 0.0429 - i0.0111 \\ &\left\langle |\Delta \mathrm{L}'|^2 \right\rangle = 0.1100 \quad \left\langle |\Delta \mathrm{C} \Delta \mathrm{L}^*| \right\rangle = 0.0071 + i0.0106 \\ &\left\langle |\Delta \mathrm{C}|^2 \right\rangle = 0.4156 \quad \left\langle |\Delta \mathrm{L} \Delta \mathrm{L}'^*| \right\rangle = 0.0002 - i0.0098 \end{aligned} \tag{9.158}$$

If the geometrical rather than the statistical picture is preferred, one readily derives from the identification of the variance matrix given by the second of Eqs. (9.156), with Eqs. (9.150) and (9.151), that

$$\begin{array}{lll}\left\langle|\Delta\vec{L}|^2\right\rangle=0.1862 & \langle\Delta\vec{L}'.\Delta\vec{C}\rangle=0.0429 & \left\langle(\Delta\vec{L}'\times\Delta\vec{C})_n\right\rangle=\left\langle(\Delta\vec{D}\times\Delta\vec{B})_1\right\rangle=0.0111\\ \left\langle|\Delta\vec{L}'|^2\right\rangle=0.1100 & \langle\Delta\vec{C}.\Delta\vec{L}\rangle=0.0071 & \left\langle(\Delta\vec{C}\times\Delta\vec{L})_n\right\rangle=\left\langle(\Delta\vec{D}\times\Delta\vec{B})_2\right\rangle=-0.0106\\ \left\langle|\Delta\vec{C}|^2\right\rangle=0.4156 & \langle\Delta\vec{L}.\Delta\vec{L}'\rangle=0.0002 & \left\langle(\Delta\vec{L}\times\Delta\vec{L}')_n\right\rangle=\left\langle(\Delta\vec{D}\times\Delta\vec{B})_3\right\rangle=0.0098\end{array} \tag{9.159}$$

The first columns of Eqs. (9.158) and (9.159) show that all three pairs of elementary properties are fluctuating, the C property exhibiting the highest variance or equivalently, the largest mean squared length of its fluctuation vector. Physically, the higher the variances, the larger the depolarization is. Since the variances are contained in the diagonal elements of the matrix $\langle\mathbf{\Delta m}_c^2\rangle$ as shown by Eq. (9.150) and $\langle\mathbf{\Delta m}_c^2\rangle$ is proportional to the uncertainty part $\mathbf{L}_u$ of the Mueller matrix logarithm in virtue of the second of Eq. (9.143), this justifies the use of the exponentiated diagonal elements of $\mathbf{L}_u$ into the evaluation of the various depolarization indices (see Section 9.4.5).

The correlations between pairs of properties, evaluated from Eq. (9.148) and Eqs. (9.158), are

$$c(\mathrm{L}',\mathrm{C})=0.2004-i0.0518 \qquad c(\mathrm{C},\mathrm{L})=0.0254+i0.0381 \qquad c(\mathrm{L},\mathrm{L}')=0.0017-i0.0658 \tag{9.160}$$

whereas their respective absolute values are

$$|c(\mathrm{L}',\mathrm{C})|=0.2070 \qquad |c(\mathrm{C},\mathrm{L})|=0.0458 \qquad |c(\mathrm{L},\mathrm{L}')|=0.0686 \tag{9.161}$$

We see that the first pair, {L′, C}, is only weakly correlated whereas the last two pairs, {C, L} and {L, L′}, are almost uncorrelated. The same holds for the pairs formed by the fluctuation vectors, $\Delta\vec{L}$, $\Delta\vec{L}'$ and $\Delta\vec{C}$, of the elementary properties, i.e., there are only weakly pronounced angular relationships between them (that is, they do not subtend definite angles). This practical lack of geometrical relationship between $\Delta\vec{L}$, $\Delta\vec{L}'$ and $\Delta\vec{C}$ automatically entails almost no geometrical relationship between the dichroic and birefringent fluctuation vectors, $\Delta\vec{D}$ and $\Delta\vec{B}$, as well. (Notice that, unlike the mean squared lengths of the fluctuation vectors of the three pairs of properties, those of the dichroic and birefringent properties, $\left\langle|\Delta\vec{D}|^2\right\rangle$ and $\left\langle|\Delta\vec{B}|^2\right\rangle$ are not measurable. Indeed, they are contained in the "uncentered" variance matrix $\langle\mathbf{\Delta m}^2\rangle$ given by Eq. (9.144) (written there as $\langle\Delta\mathrm{D}^2\rangle$ and $\langle\Delta\mathrm{B}^2\rangle$); however, the matrix $\langle\mathbf{\Delta m}^2\rangle$ is known only within an additive matrix proportional to the identity, $\beta\,\mathbf{I}$, since being itself proportional to the uncertainty part $\mathbf{L}_u$ of the matrix logarithm, see the second of Eqs. (9.143).)

9.5.3 Interpretation of Canonical, General and Rotationally Invariant Depolarizers

The two canonical forms of general depolarizers are readily interpretable within the framework of the fluctuating medium model. The centered variance matrix $\langle\mathbf{\Delta m}_{\Delta dc}^2\rangle$ of the diagonal canonical depolarizer $\mathbf{M}_{\Delta d}=\mathrm{diag}(1,a,b,c)$ is readily obtained from the second of Eqs. (9.143) and the matrix logarithm $\mathbf{L}_{\Delta d}$ of $\mathbf{M}_{\Delta d}$ reported in Eq. (9.114),

$$\langle \Delta \mathbf{m}^2_{\Delta dc} \rangle = 2\mathbf{L}_{\Delta d} = \ln\frac{a}{bc}\,\Delta_1 + \ln\frac{b}{ac}\,\Delta_2 + \ln\frac{c}{ab}\,\Delta_3 \tag{9.162}$$

in the form of an algebraic expansion on the depolarization basis (see Section 9.4.4). Identification of Eq. (9.162) with Eq. (9.145) shows that $\langle \Delta \mathbf{m}^2_{\Delta dc} \rangle$ can also be represented as

$$\langle \Delta \mathbf{m}^2_{\Delta dc} \rangle = \langle |\Delta \mathrm{L}|^2 \rangle \Delta_1 + \langle |\Delta \mathrm{L}'|^2 \rangle \Delta_2 + \langle |\Delta \mathrm{C}|^2 \rangle \Delta_3 \tag{9.163}$$

and consequently

$$\langle |\Delta \mathrm{L}|^2 \rangle = \ln\frac{a}{bc} \qquad \langle |\Delta \mathrm{L}'|^2 \rangle = \ln\frac{b}{ac} \qquad \langle |\Delta \mathrm{C}|^2 \rangle = \ln\frac{c}{ab} \tag{9.164}$$

The only three parameters of the diagonal canonical depolarizer are the variances of the three complex elementary properties. The equality of any two of the three coefficients of the depolarizer, say $a = b$, entails the equality of their respective variances, $\langle |\Delta \mathrm{L}|^2 \rangle = \langle |\Delta \mathrm{L}'|^2 \rangle$, as follows directly from Eqs. (9.164). If moreover $a = b = c$, then $\langle |\Delta \mathrm{L}|^2 \rangle = \langle |\Delta \mathrm{L}'|^2 \rangle = \langle |\Delta \mathrm{C}|^2 \rangle$: the *isotropic depolarizer* has the variances of all its three complex properties equal. In the limit $a = b = c = \varepsilon \to 0$ the isotropic depolarizer tends to the *perfect depolarizer* whose output is a totally unpolarized state whatever the polarization state at its input. One obtains $\langle |\Delta \mathrm{L}|^2 \rangle = \langle |\Delta \mathrm{L}'|^2 \rangle = \langle |\Delta \mathrm{C}|^2 \rangle = -\ln \varepsilon \to +\infty$: the perfect depolarizer features infinite variances of the elementary polarization properties, revealing the practical impossibility to be constructed from a continuous fluctuating medium. However, being physically realizable, the perfect depolarizer can be produced by a suitable parallel combination of optical components.

From Eqs. (9.146) and (9.106) one gets for the covariance matrix of the canonical diagonal depolarizer,

$$\langle \Delta \vec{\mathrm{P}}_{\Delta d} \otimes \Delta \vec{\mathrm{P}}^{\dagger}_{\Delta d} \rangle = \mathrm{diag}\left(\langle |\Delta \mathrm{L}|^2 \rangle, \langle |\Delta \mathrm{L}'|^2 \rangle, \langle |\Delta \mathrm{C}|^2 \rangle\right) = 4\mathbf{C}_3\left(\mathbf{L}_{\Delta du}\right) = \mathrm{diag}\left(\ln\frac{a}{bc}, \ln\frac{b}{ac}, \ln\frac{c}{ab}\right) \tag{9.165}$$

The covariance matrix being likewise diagonal, all covariances and correlations of pairs of complex elementary properties are identically zero, i.e., the three complex properties are totally uncorrelated. The same holds for the fluctuation vectors of the properties, as well as for the dichroic and birefringent fluctuation vectors, as follows from the diagonal variance matrix $\langle \Delta \mathbf{m}^2_{\Delta dc} \rangle$ reported in Eq. (9.163).

A practically important special case of a diagonal canonical depolarizer is the G-symmetric polar factor $\mathbf{H}_{\mathrm{S}}$, reported in Eq. (9.132), of the matrix $\mathbf{S}$ introduced in Section 9.4.2. A depolarizer of this particular form often occurs in oblique transmission and reflection polarimetry on optically thick (with respect to the wavelength of the probing light) samples such as multilayer stacks deposited on thick substrates. The algebraic expansion of the matrix logarithm of this depolarizer (i.e., its variance matrix) is given by Eq. (9.117) whereas, in virtue of Eq. (9.104), the covariance matrix of its elementary polarization properties is

$$\langle \Delta \vec{\mathrm{P}}_{\mathrm{H}} \otimes \Delta \vec{\mathrm{P}}^{\dagger}_{\mathrm{H}} \rangle = \mathrm{diag}\left(\langle |\Delta \mathrm{L}|^2 \rangle, 0, 0\right) = 4\mathbf{C}_3\left(\mathbf{L}_{\mathrm{S}u}\right) = \mathrm{diag}\left(-2\Delta_{\mathrm{ln}}, 0, 0\right) \tag{9.166}$$

where $-2\Delta_{\ln} = \ln[(1-a^2)/(b^2+c^2)]$. The only non-vanishing depolarization term is the variance of the L property, in perfect agreement with the fact the matrix **S** exhibits the L property only (see Section 9.4.2). When $a^2+b^2+c^2=1$ then $\langle|\Delta L|^2\rangle = 0$: the variance and covariance matrices given by Eqs. (9.117) and (9.166) vanish identically, the L property is "sharp" and the matrix **S** is nondepolarizing (**S** is a "psi-delta" matrix).

Similarly, from Eq. (9.115) one gets for the centered variance matrix of the non-diagonal canonical depolarizer the expression

$$\langle \Delta\mathbf{m}^2_{\Delta ndc}\rangle = 2\mathbf{L}_{\Delta nd} = \left[\ln\left(1/a^2\right)\right]\boldsymbol{\Delta}_1 + 2\boldsymbol{\Delta}_2 + 2\boldsymbol{\Delta}_3 + 2\mathbf{D}_1 \tag{9.167}$$

that can be identified as a special case of Eq. (9.145),

$$\langle \Delta\mathbf{m}^2_{\Delta ndc}\rangle = \langle|\Delta L|^2\rangle\boldsymbol{\Delta}_1 + \langle|\Delta L'|^2\rangle\boldsymbol{\Delta}_2 + \langle|\Delta C|^2\rangle\boldsymbol{\Delta}_3 - \mathrm{Im}\langle\Delta L'\Delta C^*\rangle\mathbf{D}_1 \tag{9.168}$$

resulting in the identifications

$$\langle|\Delta L|^2\rangle = \ln\left(1/a^2\right) \quad \langle|\Delta L'|^2\rangle = \langle|\Delta C|^2\rangle = -\mathrm{Im}\langle\Delta L'\Delta C^*\rangle = 2 \tag{9.169}$$

All three variances of the complex elementary properties are generally nonzero; those of L′ and C are equal and independent of the parameter a, whereas the variance of L vanishes if $a = 1$. Furthermore, L′ and C are totally correlated; indeed, according to Eqs. (9.148) and (9.169) their correlation is

$$c(L',C) = \frac{\langle\Delta L'\Delta C^*\rangle}{\sqrt{\langle|\Delta L'|^2\rangle\langle|\Delta C|^2\rangle}} = i\frac{\mathrm{Im}\langle\Delta L'\Delta C^*\rangle}{\sqrt{\langle|\Delta L'|^2\rangle\langle|\Delta C|^2\rangle}} = -i \tag{9.170}$$

and obviously $|c(L',C)| = 1$. Since $-\mathrm{Im}\langle\Delta L'\Delta C^*\rangle = \langle(\Delta\vec{L}'\times\Delta\vec{C})_n\rangle$ and $\mathrm{Re}\langle\Delta L'\Delta C^*\rangle = \langle\Delta\vec{L}'\,\Delta\vec{C}^*\rangle = 0$, L′ and C also satisfy Eqs. (9.153) from the geometrical picture,

$$\frac{\langle(\Delta\vec{L}'\times\Delta\vec{C})_n\rangle}{\sqrt{\langle|\Delta\vec{L}'|^2\rangle\langle|\Delta\vec{C}|^2\rangle}} = +1 \quad \text{and} \quad \langle\Delta\vec{L}'.\Delta\vec{C}\rangle = 0 \tag{9.171}$$

that is, their fluctuation vectors are orthogonal, $\Delta\vec{L}' \perp \Delta\vec{C}$. The equality of the mean squared lengths and the orthogonality of the two fluctuation vectors $\Delta\vec{L}'$ and $\Delta\vec{C}$ are characteristic properties of the non-diagonal canonical depolarizer. Note that the identifications from Eq. (9.169) and the correlation from Eq. (9.170) can be as well deduced directly from the covariance matrix of the elementary properties,

$$\langle\Delta\vec{P}_{\Delta nd}\otimes\Delta\vec{P}^{\dagger}_{\Delta nd}\rangle = \begin{pmatrix}\langle|L|^2\rangle & 0 & 0\\ 0 & \langle|L'|^2\rangle & \langle L'C^*\rangle\\ 0 & \langle CL'^*\rangle & \langle|C|^2\rangle\end{pmatrix} = 4\mathbf{C}_3\left(\mathbf{L}_{\Delta ndu}\right) = \begin{pmatrix}\ln\left(1/a^2\right) & 0 & 0\\ 0 & 2 & -2i\\ 0 & 2i & 2\end{pmatrix} \tag{9.172}$$

which is easily obtained from Eq. (9.110).

The two variance matrices (9.163) and (9.168) of canonical depolarizers are special cases of the variance matrix of the general depolarizer expressed by Eq. (9.145) and representable in the algebraic form

$$\begin{aligned}\left\langle \mathbf{\Delta m}_{\Delta c}^2 \right\rangle &= \left\langle |\Delta L|^2 \right\rangle \mathbf{\Delta}_1 + \left\langle |\Delta L'|^2 \right\rangle \mathbf{\Delta}_2 + \left\langle |\Delta C|^2 \right\rangle \mathbf{\Delta}_3 + \\ &+ \mathrm{Re}\left\langle \Delta L' \Delta C^* \right\rangle \mathbf{R}_1 + \mathrm{Re}\left\langle \Delta C\, \Delta L^* \right\rangle \mathbf{R}_2 + \mathrm{Re}\left\langle \Delta L\, \Delta L'^* \right\rangle \mathbf{R}_3 - \\ &- \mathrm{Im}\left\langle \Delta L' \Delta C^* \right\rangle \mathbf{D}_1 - \mathrm{Im}\left\langle \Delta C\, \Delta L^* \right\rangle \mathbf{D}_2 - \mathrm{Im}\left\langle \Delta L\, \Delta L'^* \right\rangle \mathbf{D}_3 \\ &= \sum_{k=1}^{3} \left\langle |\Delta P_k|^2 \right\rangle \mathbf{\Delta}_k + \sum_{k=1}^{3} \mathrm{Re}\left\langle \Delta P_i\, \Delta P_j^* \right\rangle \mathbf{R}_k - \sum_{k=1}^{3} \mathrm{Im}\left\langle \Delta P_i\, \Delta P_j^* \right\rangle \mathbf{D}_k\end{aligned} \tag{9.173}$$

where the subscripts i, j and k are cyclic permutations of 1, 2 and 3. This representation reveals the physical meaning within the fluctuating medium model of the three groups of matrices, $\mathbf{\Delta}_k$, $\mathbf{R}_k$ and $\mathbf{D}_k$, forming the depolarization basis (9.112): they expand the general variance matrix, with the variances and the covariances (real and imaginary parts) of the complex elementary properties as expansion coefficients. Furthermore, the variance matrix (9.173) having been obtained as the averaged square of the fluctuations of the differential matrix $\mathbf{m}$ given by Eq. (9.113), the depolarization basis set $\{\mathbf{D}_i, \mathbf{R}_i, \mathbf{\Delta}_i\}$ is obtained by squaring linear combinations of the polarization basis set $\{\mathbf{P}_i, \mathbf{Q}_i\}$ as noted in Section 9.4.4.

Further, since the matrix $\mathbf{\Delta}_k$ of the depolarization basis is related to the property P_k belonging to the k-axis of the Poincaré sphere (k = 1: HV; k = 2: PQ; k = 3: RL), then so are the matrices $\mathbf{R}_k$ and $\mathbf{D}_k$ of the rest of the basis. This observation enables a straightforward interpretation of the commutation relations (9.124) from Section 9.4.4: the expansion matrices of the real and imaginary parts of the covariances of the properties belonging to the same k-axis commute in the same way as the expansion matrices of the real and imaginary parts of the k-th property commute (see Eq. (9.123)).

To summarize, polarization, $\mathbf{P}_k$ and $\mathbf{Q}_k$, and depolarization, $\mathbf{D}_k$ and $\mathbf{R}_k$, basis matrices belonging to the same k-axis commute among themselves since associated with real and imaginary parts of complex quantities (mean values and covariances of the properties), whereas the depolarization basis matrix $\mathbf{\Delta}_k$ commutes with all the rest since related to the respective variance which is a real quantity.

In a large number of practical cases, the general depolarizer with variance matrix (9.173) is rotationally invariant, that is, its polarimetric response in transmission does not depend on the specific choice of the axes in its cross-section plane (the plane normal to the light propagation direction). If one denotes by $\mathbf{R}_\varphi$ the rotation matrix

$$\mathbf{R}_\varphi = \begin{pmatrix} 1 & 0 & 0 & 0 \\ 0 & \cos 2\varphi & -\sin 2\varphi & 0 \\ 0 & \sin 2\varphi & \cos 2\varphi & 0 \\ 0 & 0 & 0 & 1 \end{pmatrix} \tag{9.174}$$

rotating the coordinate axes in the cross-section plane by the angle ϕ, then the rotation of the Mueller matrix $\mathbf{M} = \exp(\mathbf{L})$ is equivalent to the rotation of its matrix logarithm $\mathbf{L}$ since $\mathbf{R}_\varphi \mathbf{M} \mathbf{R}_\varphi^{-1} = \mathbf{R}_\varphi \exp(\mathbf{L}) \mathbf{R}_\varphi^{-1} = \exp(\mathbf{R}_\varphi \mathbf{L} \mathbf{R}_\varphi^{-1})$. The matrix logarithm $\mathbf{L}$ being proportional to the variance matrix $\langle \mathbf{\Delta m}_\Delta^2 \rangle$ in virtue of Eq. (9.143), the condition for rotational invariance of the polarimetric response of the general depolarizer takes the form

$$\mathbf{R}_{\varphi}\left\langle \Delta\mathbf{m}_{\Delta}^{2}\right\rangle \mathbf{R}_{\varphi}^{-1} = \left\langle \Delta\mathbf{m}_{\Delta}^{2}\right\rangle \tag{9.175}$$

or equivalently

$$\left[\mathbf{R}_{\varphi}, \left\langle \Delta\mathbf{m}_{\Delta}^{2}\right\rangle\right] = \mathbf{0} \tag{9.176}$$

The variance matrix of the rotationally invariant depolarizer satisfying Eq. (9.175) or (9.176) is of the form

$$\left\langle \Delta\mathbf{m}_{\Delta rot}^{2}\right\rangle = \langle|\Delta \mathrm{L}|^{2}\rangle \boldsymbol{\Delta}_{1} + \langle|\Delta \mathrm{L}'|^{2}\rangle \boldsymbol{\Delta}_{2} + \langle|\Delta \mathrm{C}|^{2}\rangle \boldsymbol{\Delta}_{3} - \mathrm{Im}\left\langle \Delta \mathrm{L}\, \Delta \mathrm{L}'^{*}\right\rangle \mathbf{D}_{3} \tag{9.177}$$

that is, the variances of the two complex linear properties L and L′ coincide, $\langle|\Delta \mathrm{L}|^{2}\rangle = \langle|\Delta \mathrm{L}'|^{2}\rangle$, and the only non-vanishing covariance is the imaginary part of that of L and L′, $\mathrm{Im}\langle \Delta \mathrm{L}\, \Delta \mathrm{L}'^{*}\rangle$. If furthermore, $|\mathrm{Im}\langle \Delta \mathrm{L}\, \Delta \mathrm{L}'^{*}\rangle| = \langle|\Delta \mathrm{L}|^{2}\rangle = \langle|\Delta \mathrm{L}'|^{2}\rangle$, then the two linear properties are totally correlated and the variance matrix (9.177) is reducible to the non-diagonal canonical form (9.168) by a permutation of its second and third rows and columns (which physically amounts to the exchange of its L and C properties by a similarity transformation by a retarder). On the other hand, if $\mathrm{Im}\left\langle \Delta \mathrm{L}\, \Delta \mathrm{L}'^{*}\right\rangle = 0$ then the variance matrix takes the degenerate diagonal canonical form,

$$\left\langle \Delta\mathbf{m}_{\Delta rot}^{2}\right\rangle = \mathrm{diag}\left(0, -\langle|\Delta \mathrm{L}|^{2}\rangle - \langle|\Delta \mathrm{C}|^{2}\rangle, -\langle|\Delta \mathrm{L}|^{2}\rangle - \langle|\Delta \mathrm{C}|^{2}\rangle, -2\langle|\Delta \mathrm{L}|^{2}\rangle\right) \tag{9.178}$$

common to a large class of turbid scattering media such as, for instance, milk solutions in water at not too high concentrations (Ossikovski et al. 2008). Finally, if furthermore $\langle|\Delta \mathrm{L}|^{2}\rangle = \langle|\Delta \mathrm{L}'|^{2}\rangle = \langle|\Delta \mathrm{C}|^{2}\rangle$, then the rotationally invariant depolarizer becomes isotropic. This example shows that both diagonal and non-diagonal depolarizers can be "produced" from a rotationally invariant continuous fluctuating medium.

9.5.4 Relation between the Differential Mueller Matrix and the Mueller Matrix Logarithm

In Section 9.5.1 we mentioned that Eqs. (9.143) relating the matrix of the mean values, $\mathbf{m}_0 = \langle \mathbf{m} \rangle$, and the variance matrix, $\langle \Delta\mathbf{m}^2 \rangle$, to the G-antisymmetric and G-symmetric parts of the Mueller matrix logarithm hold strictly only if $\mathbf{m}_0$ and $\langle \Delta\mathbf{m}^2 \rangle$ commute. Indeed, Eqs. (9.143) result from the evolution of the integral of the *z*-dependent differential matrix of the fluctuating medium; if $\mathbf{m}_0$ and $\langle \Delta\mathbf{m}^2 \rangle$ do not commute, then the integral is not simply given by the closed-form expression (9.142) but rather appears in the form of an infinite series of nested commutators of $\mathbf{m}_0$ and $\langle \Delta\mathbf{m}^2 \rangle$, called Magnus expansion (Ossikovski and De Martino 2015). Without going into further detail, keeping the first terms in the Magnus expansion amounts to replacing the generally approximate Eqs. (9.143) by the following recursive relations for $\mathbf{m}_0$ and $\langle \Delta\mathbf{m}^2 \rangle$:

$$\mathbf{m}_0 = \mathbf{L}_m - \frac{1}{240}\left[\langle \Delta\mathbf{m}^2 \rangle, \left[\langle \Delta\mathbf{m}^2 \rangle, \mathbf{m}_0\right]\right] \langle \Delta\mathbf{m}^2 \rangle = 2\mathbf{L}_u - \frac{1}{6}\left[\langle \Delta\mathbf{m}^2 \rangle, \mathbf{m}_0\right] \tag{9.179}$$

Eqs. (9.179) can then be solved iteratively, by using the approximate Eqs. (9.143) as starting guesses for $\mathbf{m}_0$ and $\langle \Delta\mathbf{m}^2 \rangle$. It is clear that when $\mathbf{m}_0$ and $\langle \Delta\mathbf{m}^2 \rangle$commute, Eqs. (9.143) and (9.179) produce identical solutions.

The commutators in the right-hand sides of Eqs. (9.179) appear as correction terms to the approximate solution given by Eq. (9.143). In practice, these correction terms are usually small with respect to $\mathbf{L}_m$ and $2\mathbf{L}_u$ (in the sense of any matrix norm) and can be neglected in the large majority of cases of "not-too-depolarizing" Mueller matrices. By a "not-too-depolarizing" Mueller matrix we mean a matrix whose depolarization index P_Δ is higher than 0.7 as a rule of a thumb. Consider, for instance, the Mueller matrix $\mathbf{M}_g$ of the polyacrylamide gel reported in Eq. (9.84). Its depolarization index is $P_\Delta(\mathbf{M}_g) = 0.79$ as noted before, and the matrices $\mathbf{m}_{g0}$ and $\langle \Delta\mathbf{m}_g^2 \rangle$containing the mean values and variances-covariances of the elementary properties, obtained from the approximate Eqs. (9.143), are reported in Eqs. (9.156). If Eqs. (9.179) are used instead of Eqs. (9.143), then the two matrices $\mathbf{m}_{g0}$ and $\langle \Delta\mathbf{m}_g^2 \rangle$ become

$$\mathbf{m}_{g0} = \begin{pmatrix} 0 & -0.0300 & 0.0017 & 0.0006 \\ -0.0300 & 0 & -0.0596 & -0.0153 \\ 0.0017 & 0.0596 & 0 & -1.3856 \\ 0.0006 & 0.0153 & 1.3856 & 0 \end{pmatrix}$$

$$\left\langle \Delta\mathbf{m}_g^2 \right\rangle = \begin{pmatrix} 0 & -0.0085 & 0.0120 & -0.0071 \\ 0.0085 & -0.5255 & -0.0006 & 0.0066 \\ -0.0120 & -0.0006 & -0.5912 & -0.0228 \\ 0.0071 & 0.0066 & -0.0228 & -0.3067 \end{pmatrix} \tag{9.180}$$

Entry-wise comparison of Eqs. (9.180) and Eqs. (9.156) shows that the differences in the corresponding elements of $\mathbf{m}_{g0}$ are well beyond the typical experimental error (about 0.005) whereas those of $\langle \Delta\mathbf{m}_g^2 \rangle$, although generally larger than the former, can be likewise neglected since being still of the order of 0.01. Similar conclusions can be drawn about the "corrected" covariance matrix $\langle \Delta\vec{P}_g \otimes \Delta\vec{P}_g^\dagger \rangle$ of the properties,

$$\left\langle \Delta\vec{P}_g \otimes \Delta\vec{P}_g^\dagger \right\rangle = 2\mathbf{C}_3\left(\left\langle \Delta\mathbf{m}_g^2 \right\rangle\right) = \begin{pmatrix} 0.1862 & -0.0006 - i0.0071 & 0.0066 - i0.0120 \\ -0.0006 + i0.0071 & 0.1205 & -0.0228 - i0.0085 \\ 0.0066 + i0.0120 & -0.0228 + i0.0085 & 0.4050 \end{pmatrix} \tag{9.181}$$

derived from $\langle \Delta\mathbf{m}_g^2 \rangle$ from Eq. (9.180), when compared to the "uncorrected" one reported in Eq. (9.157). Strictly speaking, the modification of the elements of the covariance matrix $\langle \Delta\vec{P}_g \otimes \Delta\vec{P}_g^\dagger \rangle$ may impact the local physical realizability of $\mathbf{M}_g$ by affecting the positive semi-definiteness of $\langle \Delta\vec{P}_g \otimes \Delta\vec{P}_g^\dagger \rangle$. However, this potential effect is again negligible in practice. Thus, the eigenvalues of $\langle \Delta\vec{P}_g \otimes \Delta\vec{P}_g^\dagger \rangle$ given by Eq. (9.157) are 0.1029, 0.1863 and 0.4227, whereas those of $\langle \Delta\vec{P}_g \otimes \Delta\vec{P}_g^\dagger \rangle$ from Eq. (9.181) are 0.1174, 0.1864 and 0.4079; the positive semi-definiteness of both the "uncorrected" and "corrected" covariance matrices is verified. Finally, note that if the "corrected" covariance matrix turns out not to be positive semidefinite while the "uncorrected" one is, would be simply an indication that the fluctuating medium model is inapplicable to the medium in question.

9.5.5 Spatial Evolution of Depolarization within the Differential Mueller Matrix Formalism

In Section 9.5.1 we expressed the general depolarizing Mueller matrix as an exponential of its associated differential matrix; see Eq. (9.142). Written for a medium of thickness z, the expression reads

$$\mathbf{M}(z) = \exp\left(\mathbf{m}_0 z + \frac{1}{2}\langle \boldsymbol{\Delta}\mathbf{m}^2 \rangle z^2\right) \tag{9.182}$$

We mentioned that Eq. (9.182) is strictly valid only if the mean values $\mathbf{m}_0 = \langle \mathbf{m} \rangle$ of the elementary polarization properties and the variance matrix $\langle \boldsymbol{\Delta}\mathbf{m}^2 \rangle$ of their fluctuations $\boldsymbol{\Delta}\mathbf{m}$ commute. In practice, Eq. (9.182) holds whenever the second term in the exponent is small enough so that the higher-order exponent terms in z, arising from the matrix exponentiation, could be neglected. This condition is satisfied if the thickness z is much smaller than the inverse of the root mean square (r.m.s.) amplitude a of the fluctuations of depolarizing medium, quantified by $a = \sqrt{\| \langle \boldsymbol{\Delta}\mathbf{m}^2 \rangle \|}$ where $\| \ldots \|$ denotes any matrix norm, i.e., if $z \ll 1/a$. (Recall that the dimension of the differential Mueller matrix is inverse length as follows from Eq. (9.47).). Put in physical terms, the smaller the fluctuations of the medium the larger the thickness range over which Eq. (9.182) is valid and vice versa.

Eq. (9.182) shows that, whereas the polarization properties cumulate linearly with the thickness z (exactly like in the nondepolarizing case), the depolarization ones, formally originating from the non-zero variance matrix $\langle \boldsymbol{\Delta}\mathbf{m}^2 \rangle$, evolve quadratically with z. In the special case of the canonical depolarizer discussed in Section 9.5.3, the mean values of all polarization properties are zero, $\mathbf{m}_0 = \langle \mathbf{m} \rangle = \mathbf{0}$, and their variance matrix $\langle \boldsymbol{\Delta}\mathbf{m}^2 \rangle$ is given by Eq. (9.178) from Section 9.5.3. Eq. (9.182) then takes the special form

$$\mathbf{M}_{\Delta rot}(z) = \exp\left\{\mathrm{diag}\left[0, -\alpha_{L'}(z) - \alpha_C(z), -\alpha_L(z) - \alpha_C(z), -\alpha_L(z) - \alpha_{L'}(z)\right]\right\} \tag{9.183}$$

where $\alpha_P(z)$, the z-dependent logarithmic depolarization of the property P (P = L, L′ or C), is given by

$$\alpha_P(z) = \frac{1}{2}\langle |\Delta P|^2 \rangle z^2 \tag{9.184}$$

in which $\langle |\Delta P|^2 \rangle$ is the variance of P. (We have assumed that the Mueller matrix has been normalized so as to make vanish the first entry of the diagonal exponent in Eq. (9.183)). By taking the matrix logarithm of $\mathbf{M}_{\Delta rot}(z)$, $\mathbf{L}_u = \ln[\mathbf{M}_{\Delta rot}(z)]$, and plotting the logarithmic depolarizations against z for various thickness values, one readily gets the variances $\langle |\Delta L|^2 \rangle$, $\langle |\Delta L'|^2 \rangle$ and $\langle |\Delta C|^2 \rangle$ of three elementary polarization properties. These can be further correlated with the concentration of scatterers. The validity of Eq. (9.184) has thus been checked experimentally for different scattering media: on an ensemble of scattering microspheres embedded in a solid host matrix (Agarwal et al. 2015), as well as on stacks of stretched plastic sheets (Yoo et al. 2017)). In the latter case the mean values of the elementary polarization properties are generally non-zero($\mathbf{m}_0 \neq \mathbf{0}$) because of the stretch-induced birefringence in the sheets, and the more general Eq. (9.182), rather than just Eq. (9.183), is validated.

Eq. (9.182) was derived by explicitly assuming that the fluctuations $\Delta\mathbf{m}$ of the elementary properties are z-independent (see Section 9.5.1). In practice, Eq. (9.182) remains valid even if the fluctuations vary with the position along the path-length provided the characteristic length ζ of their variation, called *correlation length*, is much larger than the thickness z of the medium, i.e., $\zeta \gg z$. (From a formal viewpoint, strict z-independence of $\Delta\mathbf{m}$ implies infinite correlation length, $\zeta \to \infty$, so that the condition $\zeta \gg z$ is fulfilled.) It has been shown both theoretically (Devlaminck 2015) and experimentally (Charbois and Devlaminck 2016) that, in the other limiting case, $\zeta \ll z$, physically corresponding to large thicknesses, the evolution of the depolarization is not quadratic like that given by Eq. (9.182) but is rather linear with z. Indeed, at sufficiently large z values such that $z \gg \zeta$ one can set $\zeta \to 0$ and approximate asymptotically the now z-dependent variance matrix $\langle \Delta\mathbf{m}^2(z)\rangle$ by a delta-function, i.e., $\langle \Delta\mathbf{m}^2(z)\rangle \to \langle \Delta\mathbf{m}_0^2\rangle \delta(z)$. The asymptotic solution of Eq. (9.47) is then

$$\mathbf{M}(z \gg \zeta) \to \mathbf{M}(z)_{\zeta\to 0} \to \exp\left(\mathbf{m}_0 z + \langle \Delta\mathbf{m}^2\rangle_0\, z\right) \tag{9.185}$$

We emphasize the asymptotic, rather than exact, nature of Eq. (9.185). Formally, an extra condition for the validity of Eq. (9.185) (besides $z \gg \zeta$) is $\zeta \ll a$ where $a = 1/\langle \Delta\mathbf{m}^2\rangle_0$ is the amplitude of the fluctuations; however, the delta-function limit of the variance implies $\zeta \to 0$ so that this condition is fulfilled. Eventually, in the intermediate case, $\zeta \sim z$, a smooth transition between quadratic and linear regimes takes place with the increase of z; a general closed-form expression is unavailable, except for the special case of the canonical depolarizer discussed below (Charbois and Devlaminck 2016). Note that, experimentally, one is usually in the small thickness regime described by Eq. (9.182) since scattering or absorption severely reduce the transmitted light intensity with the increase of z.

Eq. (9.182) correctly describes the onset of depolarization with thickness in solids where the brackets $\langle ... \rangle$ denote spatial averaging. Experimentally, the spatial averaging takes place over the finite spot size or the finite angular aperture of the illuminating or collecting arm of the measurement equipment (Ossikovski and Hingerl 2016). In liquids however, time averaging, besides spatial one, is likewise active because of the Brownian motion of the scatterers; even if spatial averaging is absent, depolarization will still occur due to the dynamic (non-stationary) nature of the liquid (Charbois and Devlaminck 2016). Formally, time averaging of the fluctuations $\Delta\mathbf{m}$ of the elementary polarization properties of the medium can be replaced by spatial one in virtue of the ergodic hypothesis; however, the evolution of $\Delta\mathbf{m}$ with z now has to be described by a stochastic equation accounting for the Brownian process. Assuming that the correlation length of the elementary property P is ζ_P and solving the related stochastic equation, it is possible to obtain a closed-form expression for the logarithmic depolarizations of the canonical depolarizer (Charbois and Devlaminck 2016),

$$\alpha_P(z) = \frac{1}{2}\langle |\Delta P|^2\rangle \zeta_P^2 \left[2\frac{z}{\zeta_P} + \alpha - 3 + 2(2-\alpha)\exp\left(-\frac{z}{\zeta_P}\right) + (\alpha - 1)\exp\left(-2\frac{z}{\zeta_P}\right)\right] \tag{9.186}$$

In Eq. (9.186), α $(0 \leq \alpha \leq 1)$ is a coefficient characterizing the ratio of spatial to time averaging. The first two terms of the Taylor series expansion of Eq. (9.186) at small thicknesses ($z \ll \zeta_P$) read

$$\alpha_p(z \ll \zeta_P) = \frac{1}{2}\langle|\Delta P|^2\rangle \zeta_P^2 \left[\alpha\left(\frac{z}{\zeta_P}\right)^2 + \left(\frac{2}{3} - \alpha\right)\left(\frac{z}{\zeta_P}\right)^3 + \dots\right] \tag{9.187}$$

If $\alpha \neq 0$, i.e., if spatial averaging is active along with time averaging, then the onset of depolarization is a linear combination of quadratic and cubic terms in z. In the limiting case $\alpha = 0$ only time averaging is active; at small thicknesses, the onset of polarization is cubic rather than quadratic,

$$\alpha_P(z \ll \zeta_P)_{\alpha=0} = \frac{1}{3}\frac{\langle|\Delta P|^2\rangle}{\zeta_P} z^3 \tag{9.188}$$

Eqs. (9.187) and (9.188) thus show that, at small thicknesses, the quadratic and cubic terms are characteristic of spatial and time averaging respectively. As mentioned, spatial averaging is the only one active in solids whereas, in liquids, it may or may not accompany the always-present time averaging. Notice that time averaging can never be "switched off" in a liquid: even if the cubic term vanishes for the specific value of $\alpha = 2/3$, the expansion of Eq. (9.186) to the fourth order in z will bring out a non-vanishing quartic term. (In a solid, one effectively "switches off" spatial averaging through the use of point-like collimated illumination and detection light beams, resulting in a nondepolarizing polarimetric response of the medium.) However, it is easy to see that the logarithmic depolarizations tend asymptotically to quadratic behavior at large correlation lengths provided spatial averaging is present ($\alpha \neq 0$). Indeed, if we formally put $\zeta_P \to \infty$ into Eq. (9.186) or (9.187) we get

$$\alpha_p(z)_{\zeta_P \to \infty} \to \alpha\frac{1}{2}\langle|\Delta P|^2\rangle z^2 \tag{9.189}$$

in agreement with Eq. (9.184), since the limit $\zeta_P \to \infty$ physically means z-independent fluctuations. If, furthermore, $\alpha = 1$, i.e., if spatial and time averaging both generate identical variances, then the expression in Eq. (9.189) formally coincides with that in Eq. (9.184) for the solid case. Therefore, the depolarization evolution in liquids tends to that in solids in the limit of z-independent fluctuations ($\zeta_P \to \infty$) and indistinguishability between spatial and time averaging ($\alpha = 1$).

Eventually, in the large thickness limit, $z \gg \zeta_P$, one gets from Eq. (9.186)

$$\alpha_P(z \gg \zeta_P) = \frac{1}{2}\langle|\Delta P|^2\rangle \zeta_P^2 \left(2\frac{z}{\zeta_P} + \alpha - 3\right) \tag{9.190}$$

i.e., the evolution with z is linear whatever the value of α. Note that the measurement of the slope, $\langle|\Delta P|^2\rangle\zeta_P$, and of the z-axis intercept, $\zeta_P(3 - \alpha)/2$, allows one to determine experimentally the variance $\langle|\Delta P|^2\rangle$ and the correlation length ζ_P of the property P, if the spatial-to-time averaging ratio α is known.

If instead one formally sets $\zeta_P \to 0$ (which, of course, entails $z \gg \zeta_P$) and further assumes that $\langle|\Delta P|^2\rangle$ varies in such a way as the product $\langle|\Delta P|^2\rangle\zeta_P = \langle|\Delta P|^2\rangle_0$ does not vanish and remains finite, Eq. (9.186) becomes

$$\alpha_P(z)_{\zeta_P \to 0} \to \langle|\Delta P|^2\rangle_0 \, z \tag{9.191}$$

i.e., one recovers the delta-function asymptotic approximation discussed before. Indeed, Eq. (9.191) is obviously a special case of the second term of the exponent in Eq. (9.185). Therefore, the linear behavior of depolarization at large thicknesses ($z \gg \zeta_P$) is common to all scattering media (liquids and solids).

For certain liquids, there may exist non-negligible coupling between the B- and D-components of the property P (e.g., between LB and LD) due to the fact that both properties may have a common physical origin (e.g., the scattering particles may exhibit interrelated refractive index and extinction coefficient anisotropies). The coupling coefficient between components then acts as spatial frequency that generates oscillations in the depolarization evolution curves; the resulting expressions (Charbois and Devlaminck 2016; Gill et al. 2021) are however, very cumbersome and will not be reported here.

The validity of Eq. (9.186) has been verified experimentally for several liquids (diluted milk and liquor (Charbois and Devlaminck 2016)), as well as a water suspension of polystyrene spheres (Gill et al. 2021) at various concentrations of scatterers. The experimental determination of ζ_P and of $\left(\langle|\Delta P|^2\rangle\zeta_P\right)^{-1}$ appears to be of significant practical interest since these two parameters have been shown to be respectively proportional to the mean free path and to the momentum relaxation length of the photons propagating through the medium (Gill et al. 2021). Note that, like in solids, exploring the linear depolarization regime in liquids at large thicknesses ($z \gg \zeta_P$) is quite challenging experimentally since requiring a powerful light source in order to overcome excessive scattering or absorption.

Like all matrix decompositions, the differential decomposition is potentially capable of providing a better physical insight into the complex light-matter interaction phenomena without the need of any structural information about the optical system under study. In this sense, it is believed to be a precious algebraic tool in the polarimetric characterization and in the physical understanding of complex optical structures and media such as anisotropic materials, nanostructures, organic and biological samples, etc.

10

Geometric Representation of Mueller Matrices

10.1 Introduction

Due to the power of graphic and visual codes for the interpretation and analysis of polarimetric measurements, the study of geometric representations of Mueller matrices constitutes a way to geometrize the polarization algebra, as well as to get a better insight into the physical behavior of the measured sample. Therefore, geometric and visual representations can provide useful tools for the study and analysis of experimental Mueller matrices, with potential applications to several fields of significant practical interest such as, for example, imaging of biological tissues, remote sensing, scattering by turbid media, etc.

Experimental polarimetry allows for the determination of the 16 elements of the Mueller matrix of the sample under study through the analysis of at least four polarization states resulting from the transformation by the sample of four independent, totally polarized, incident states. Therefore, a natural geometric view of a Mueller matrix is given by the corresponding Poincaré sphere mapping, that is, by the specific way a given Mueller matrix maps the totally polarized states lying on the surface of the sphere.

An early contribution to this kind of geometric approach is due to Williams (1986) who considered the mapping by pure Mueller matrices as well as by certain depolarizers. Le Roy-Brehonnet et al. (1997) and Le Roy-Brehonnet and Le Jeune (1997) analyzed the Poincaré sphere mapping by a number of nondepolarizing and depolarizing systems. The case of diagonal depolarizers was studied by Brosseau (1998), who observed the ellipsoidal shape of the transformed surface. As demonstrated by Lu and Chipman (1998) such an ellipsoidal shape is also produced by any depolarizing Mueller matrix.

Certain alternative kinds of mapping, like the so-called *DoP surfaces* and *DoP maps*, based on the representation of the output degree of polarization and the output intensity (both represented in terms of the input Poincaré sphere coordinates), were proposed by DeBoo et al. (2004). Two alternative complementary mappings of a given Mueller matrix, namely the *P*-object and the *I*-object, were defined as solid objects by Ferreira et al. (2006) on the basis of the respective transformation of the degree of polarization and of the intensity. Further, Tudor and Manea (2011) studied the mapping by nondepolarizing devices of partially polarized incident states with a fixed degree of polarization, and found that the mapping results in an ellipsoid. Later, by taking advantage of the symmetric serial decomposition of a depolarizing Mueller matrix **M**, Ossikovski et al. (2013) developed a procedure for the full geometric representation of **M** by means of two or three complementary ellipsoids.

DOI: 10.1201/9780367815578-10

The present chapter summarizes the main results on the geometric characterization of Mueller matrices. Due to the variety and certain complexity of the results, as well as to provide guidance to the reader, it is worth listing in advance the subjects of the sections to follow:

- Definition of the *I*-image and *P*-image of **M** that stem respectively from the intensity and from the degree of polarization of the emerging states and constitute complementary geometric views of the action of **M**.
- Definition of the *characteristic forward and reverse ellipsoids* of **M**, which provide a complete geometric representation of **M**.
- Analysis of the *P*-images (ellipsoids) of some special depolarizing Mueller matrices, providing better understanding on how the characteristic ellipsoids are generated.
- Definition of the canonical ellipsoid of **M** and obtainment, from the symmetric decomposition of **M**, of a particularly significant geometric representation of **M** through a set of three ellipsoids, in such manner that the topological differences between the characteristic ellipsoids of type-I and type-II arise naturally in a meaningful way.
- Analysis of the topological properties of the characteristic ellipsoids of **M** in relation to the minimum number of parallel components of **M**, determined by the achievable values of the rank of the covariance matrix **H** associated with **M**.
- Geometric representation of nondepolarizing Mueller matrices through a pair of vectors and study of the associated ellipsoids.
- Analysis of some interesting experimental results based on the characteristic ellipsoids of the corresponding Mueller matrices.

10.2 *P*-Image and *I*-Image of a Mueller Matrix

As seen in Section 1.5.3, the Stokes parameters (s_0, s_1, s_2, s_3) satisfy the characteristic inequalities $s_0 > 0$ and $s_0^2 \geq s_1^2 + s_2^2 + s_3^2$, and therefore, define a 4D cone whose s_0 axis is the generatrix. The cut of this cone with the plane $s_0 = 1$ results in the three-coordinate representation provided by the Poincaré sphere (Figure 10.1). As discussed in Section 1.7, the Poincaré sphere constitutes an abstract mathematical representation (whose reference axes $S_1 S_2 S_3$ are different from those of the laboratory reference frame) defined as a solid sphere of unit radius where all polarization states have unit intensities. Let us recall that points on the surface represent totally polarized states (pure states), whereas points inside the sphere represent partially polarized states (mixed states), and the origin represents the unpolarized state. Normalized Stokes vectors being parameterizable as

$$\hat{\mathbf{s}} \equiv \frac{\mathbf{s}}{s_0} \equiv \begin{pmatrix} 1 \\ \hat{s}_1 \\ \hat{s}_2 \\ \hat{s}_3 \end{pmatrix} \equiv \begin{pmatrix} 1 \\ \mathcal{P}\mathbf{u} \end{pmatrix} \equiv \begin{pmatrix} 1 \\ \mathbf{p} \end{pmatrix} \qquad \mathbf{u} \equiv \begin{pmatrix} u_1 \\ u_2 \\ u_3 \end{pmatrix} \equiv \begin{pmatrix} \cos 2\chi \cos 2\varphi \\ \cos 2\chi \sin 2\varphi \\ \sin 2\chi \end{pmatrix} \tag{10.1}$$

where φ (with $0 \leq \varphi < \pi$) and χ (with $-\pi/4 \leq \chi \leq \pi/4$) are the azimuth and ellipticity angle, respectively, of the average polarization ellipse, while $\mathcal{P}$ is the degree of polarization (see

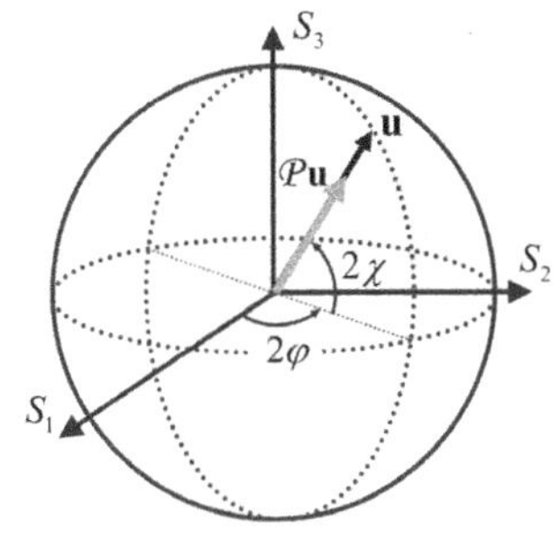

FIGURE 10.1
Poincaré sphere representation of normalized Stokes vectors: **u** is the Poincaré vector of the 2D polarization state, ϕ and χ are the azimuth and ellipticity angle of the average polarization ellipse, and $\mathcal{P}$ is the degree of polarization.

Section 1.5.3), the Poincaré sphere is the solid sphere constituted by the points defined by the *polarization vector* $\mathbf{p} \equiv (\hat{s}_1, \hat{s}_2, \hat{s}_3) = (\mathcal{P}u_1, \mathcal{P}u_2, \mathcal{P}u_3)$.

Normalized totally polarized states are denoted as

$$\hat{\mathbf{s}}_p \equiv \frac{\mathbf{s}_p}{s_0} \equiv \begin{pmatrix} 1 \\ \hat{s}_{p1} \\ \hat{s}_{p2} \\ \hat{s}_{p3} \end{pmatrix} \equiv \begin{pmatrix} 1 \\ \mathbf{u} \end{pmatrix} \equiv \begin{pmatrix} 1 \\ \mathbf{p}_p \end{pmatrix} \qquad \mathbf{u} \equiv \begin{pmatrix} u_1 \\ u_2 \\ u_3 \end{pmatrix} \equiv \begin{pmatrix} \cos 2\chi \cos 2\varphi \\ \cos 2\chi \sin 2\varphi \\ \sin 2\chi \end{pmatrix} \tag{10.2}$$

Given a Mueller matrix **M**, incident Stokes vectors $\hat{\mathbf{s}}$ (covering the entire solid Poincaré sphere) are transformed into emerging Stokes vectors $\mathbf{s}'$ through the linear map $\mathbf{s}' = \mathbf{M}\hat{\mathbf{s}}$. The *P*-object (or *polarization object*) is defined as the solid object given by the end-points of vectors $\mathbf{p}_O$ whose Poincaré vector $\mathbf{u}'$ is that of $\mathbf{s}'$ and whose absolute value is the degree of polarization $\mathcal{P}'$ of $\mathbf{s}'$ (Ferreira et al. 2006; Figure 10.2a)

$$\mathbf{p}_O \equiv (s'_1, s'_2, s'_3)^T / s'_0 \qquad \left[\mathbf{s}' \equiv (s'_0, s'_1, s'_2, s'_3)^T = \mathbf{M}\hat{\mathbf{s}}\right] \tag{10.3}$$

Transformation (10.3) is independent of the mean intensity coefficient m_{00} of **M**, so that the normalized Mueller matrix $\hat{\mathbf{M}} \equiv \mathbf{M}/m_{00}$ is an appropriate representative of **M** for this kind of mapping.

Unlike the transformations by Stokes matrices, which include arbitrary (odd and even) permutations of the canonical basis $\hat{\mathbf{u}}_1 \equiv (1,0,0)^T$, $\hat{\mathbf{u}}_2 \equiv (0,1,0)^T$, $\hat{\mathbf{u}}_3 \equiv (0,0,1)^T$, the transformation (10.3), where **M** represents a physically realizable Mueller matrix, excludes improper rotations (odd permutations of the canonical basis), so that it is called *completely positive mapping* (Sudha et al. 2008; Simon et al. 2010a; Gamel and James 2011). Each fixed value of the degree of polarization $\mathcal{P}$ of the incident states defines a *P-sphere* of radius $\mathcal{P}$, which is transformed into a *P-surface* by **M**. The *P*-surface of **M** corresponding to $\mathcal{P} = 1$ will be called the *P*-image of **M** and is constituted by the points

$$\mathbf{p}_I \equiv \frac{1}{s'_0}(s'_1, s'_2, s'_3)^T \qquad \left[\mathbf{s}' \equiv (s'_0, s'_1, s'_2, s'_3)^T = \mathbf{M}\hat{\mathbf{s}}_p\right] \tag{10.4}$$

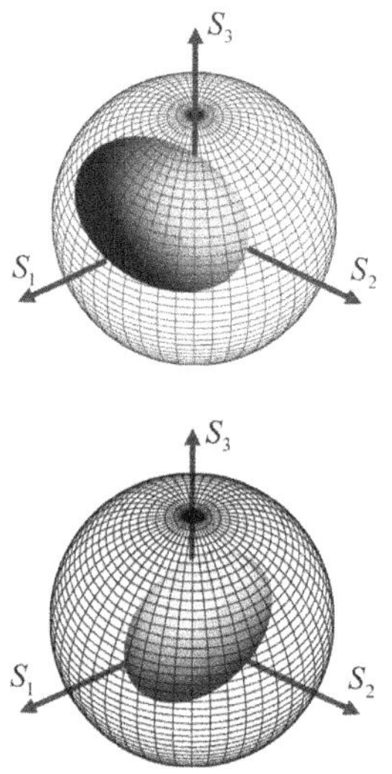

FIGURE 10.2
(a) Polarization object (*P*-object) is a solid object constituted by the end points of vectors $\mathbf{P}_O$. (b) Intensity object (*I*-object) is a solid object constituted by the end points of vectors whose direction is that of $\mathbf{P}_O$ and whose absolute value is the corresponding intensity s_0'.

Obviously, *P*-images do not protrude the unit sphere; otherwise, the system would be *overpolarizing* since violating the so-called Stokes criterion (Van der Mee 1993). Let us recall that, as seen in Section 5.6, Mueller matrices must satisfy the more restrictive *Cloude's criterion* (or *covariance criterion*) (Cloude 1986).

At this point it should also be recalled that any given Mueller matrix **M** fully determines the Mueller matrix $\mathbf{M}^r = \mathbf{X}\mathbf{M}^T\mathbf{X}$ (with $\mathbf{X} \equiv \mathrm{diag}(1,1,-1,1)$, see Section 5.8), representing the interaction whereby the directions of the incident and emerging light beams are interchanged. Thus, we will distinguish between the *forward Mueller matrix* **M** (*forward interaction*) and its corresponding *reverse Mueller matrix* $\mathbf{M}^r$ (*reverse interaction*). Note also that the shapes of the *P*-object and the *P*-image of $\mathbf{M}^r$ coincide with those of $\mathbf{M}^T$, respectively. Obviously, the notions of *forward* and *reverse* are relative and are established via a mere convention. What is important to keep in mind is that the geometric objects associated with **M** and $\mathbf{M}^T$ are different in general. Therefore, both the *P*-image and the *I*-image (defined below) of **M** are different from those of $\mathbf{M}^T$ so that, in general, a complete geometric characterization of **M** requires considering the mappings by both **M** and $\mathbf{M}^T$. Consequently, wherever appropriate, the geometric mappings will be called *forward* (as for instance the *forward P*-image) or *reverse* when referring to the action of **M** and $\mathbf{M}^T$, respectively. In particular, as shown in the next section, the forward and reverse *P*-images of **M** are given by the respective so-called forward and reverse characteristic ellipsoids of **M**.

The *P*-images associated with pure Mueller matrices are given by points on the unit sphere, while Mueller matrices producing depolarization in forward interaction have associated *P*-images in the form of respective ellipsoids (see Section 10.3) with some or all points lying inside the unit sphere. Media exhibiting maximal polarizance, $P = 1$, are either perfect polarizers or depolarizing polarizers, and have associated *P*-images given by respective single points on the unit sphere, determined by the end point of the unit polarizance vector **P** of **M**.

A complementary geometric view of the action of a Mueller matrix **M** is provided by its associated *I*-object, which is a solid object defined by the end points of vectors given by (Ferreira et al. 2006; Figure 10.2b)

$$\mathbf{i}_O \equiv s'_0 \frac{(s'_1, s'_2, s'_3)^T}{\sqrt{s_1'^2 + s_2'^2 + s_2'^2}} \qquad \left[\mathbf{s}' \equiv (s'_0, s'_1, s'_2, s'_3)^T = \mathbf{M}\hat{\mathbf{s}}\right] \tag{10.5}$$

Note that, unlike in the definition of *P*-object, *P*-surfaces and *P*-Image, **M** cannot be replaced by $\hat{\mathbf{M}}$ in the above definition of the *I*-object.

The *I*-image of **M** is defined as the surface limiting the *I*-object, which corresponds to the mapping $\mathbf{i}_I$ generated by totally polarized incident states $\hat{\mathbf{s}}_p$, i.e., the *I*-image of **M** is constituted by the points

$$\mathbf{i}_I \equiv s'_0 \frac{(s'_1, s'_2, s'_3)^T}{\sqrt{s_1'^2 + s_2'^2 + s_2'^2}} \qquad \left[\mathbf{s}' \equiv (s'_0, s'_1, s'_2, s'_3)^T = \mathbf{M}\hat{\mathbf{s}}_p\right] \tag{10.6}$$

The passivity constraints affecting **M** impose that the *I*-image does not protrude the unit sphere. The *I*-image of a transparent medium (characterized by $m_{00} = 1$) is a unit sphere; the *I*-images of media without diattenuation are spherical, and the *I*-images of media with diattenuation are non-spherical surfaces.

10.3 Characteristic Ellipsoids of a Mueller Matrix

In this section, we take advantage of the Lu-Chipman and reverse decompositions of a Mueller matrix **M** (Lu and Chipman 1998; Ossikovski et al. 2007) to present a procedure for the geometric representation of Mueller matrices based on the identification and the analysis of the ellipsoids that can be defined from **M**.

The higher the polarizance, the more inhomogeneous the topological density of the image points in both *P*-image and *I*-image. The density increases according to the proximity of the image point to the intersection of the image surface with the direction of the polarizance vector **P**. Nevertheless, in what follows we shall focus on the shape and location of the *P*-images in the Poincaré space $S_1S_2S_3$ and not in the topological density, nor in the detailed relative positions of the image points.

Let us now consider the forward version $\mathbf{M} = m_{00}\,\hat{\mathbf{M}}_{\Delta P}\mathbf{M}_R\hat{\mathbf{M}}_D$ of the Lu-Chipman decomposition (see Section 8.2) and observe that, leaving aside the case of $\hat{\mathbf{M}}_D$ being singular, the *entrance pure Mueller matrix* $\mathbf{M}_{JI} \equiv \mathbf{M}_R\mathbf{M}_D$ maps the Poincaré sphere onto itself (regardless of the fact that the topological density of the image points on the spherical *P*-image is not necessarily homogeneous). As shown by Lu and Chipman (1998), the normalized Mueller matrix $\hat{\mathbf{M}}_{\Delta P}$ can be written as

$$\hat{\mathbf{M}}_{\Delta P} \equiv \begin{pmatrix} 1 & \mathbf{0}^T \\ \mathbf{P}_{\Delta P} & \mathbf{I}_3 \end{pmatrix}\begin{pmatrix} 1 & \mathbf{0}^T \\ \mathbf{0} & \mathbf{m}_{\Delta P} \end{pmatrix} \tag{10.7}$$

so that, for any incident totally polarized Stokes vector $\hat{\mathbf{s}}_p$, the Stokes vector $\mathbf{s}'$ exiting $\hat{\mathbf{M}}_{\Delta P}$ is given by

$$\hat{\mathbf{M}}_{\Delta P}\,\hat{\mathbf{s}}_p = \begin{pmatrix} 1 \\ \mathbf{P}_{\Delta P} + \mathbf{m}_{\Delta P}\,\mathbf{u} \end{pmatrix} = \begin{pmatrix} 0 \\ \mathbf{P}_{\Delta P} \end{pmatrix} + \begin{pmatrix} 1 \\ \mathbf{m}_{\Delta P}\,\mathbf{u} \end{pmatrix} \qquad \left[\hat{\mathbf{s}}_p \equiv \begin{pmatrix} 1 \\ \mathbf{u} \end{pmatrix} \right] \tag{10.8}$$

Note that the right matrix factor in Eq. (10.7) (representing a nonenpolarizing depolarizer) maps the Poincaré sphere onto an ellipsoid whose semiaxes are given by the absolute values $(l_{\Delta P1}, l_{\Delta P2}, l_{\Delta P3})$ of the eigenvalues of the symmetric matrix $\mathbf{m}_{\Delta P} = \mathbf{m}_{Rf}\,\mathbf{m}_{L\Delta P}\,\mathbf{m}_{Rf}^T$ (see Section 8.5.2), while the directions of the principal axes are given by the eigenvectors of $\mathbf{m}_{\Delta P}$. The left matrix factor in Eq. (10.7) induces geometrically a global displacement by $\mathbf{P}_{\Delta P}$ of the so-called *forward ellipsoid* $E_{\Delta P}$ of $\mathbf{M}$ (Figure 10.3).

Once the P-image of $\hat{\mathbf{M}}_{\Delta P}$ has been analyzed and has been identified as the rotated and displaced ellipsoid $E_{\Delta P}$, the P-image of $\mathbf{M}$ follows from the decomposition

$$\hat{\mathbf{M}} = \hat{\mathbf{M}}_{\Delta P}\,\hat{\mathbf{M}}_{JD} \qquad \left[\hat{\mathbf{M}}_{JD} \equiv \mathbf{M}_R\,\hat{\mathbf{M}}_D \right] \tag{10.9}$$

Provided $D < 1$ (the case where $D = 1$, i.e., $0 = \det \hat{\mathbf{M}}_{JD} = \det \hat{\mathbf{M}}_D$, will be considered below), the P-image of the normalized non-singular pure Mueller matrix $\hat{\mathbf{M}}_{JD}$ is the unit sphere itself and therefore, the forward ellipsoid $E_{\Delta P}$ of $\mathbf{M}$ coincides with that of $\hat{\mathbf{M}}_{\Delta P}$.

The geometric information fully defining $E_{\Delta P}$ consists of nine independent parameters, namely: its three semiaxes $(l_{\Delta P1}, l_{\Delta P2}, l_{\Delta P3})$; the three orientation angles $(\varphi_{\Delta P}, \chi_{\Delta P}, \Delta_{\Delta P})$ of the symmetry axes of $E_{\Delta P}$ with respect to the Poincaré reference frame $S_1S_2S_3$, and the three coordinates of the geometric center of $E_{\Delta P}$ (with respect to $S_1S_2S_3$), given by the components $(P_{\Delta P1}, P_{\Delta P2}, P_{\Delta P3})$ of the polarizance vector $\mathbf{P}_{\Delta P}$.

The above analysis has been performed assuming that $D < 1$. For the sake of completeness, let us now consider the case where $D = 1$ and recall that then $\mathbf{M}$ reduces to a depolarizing analyzer of the form

$$\mathbf{M}_{P\hat{D}} = m_{00} \begin{pmatrix} 1 & \hat{\mathbf{D}}^T \\ \mathbf{P} & \mathbf{P} \otimes \hat{\mathbf{D}}^T \end{pmatrix} \qquad \left[D \equiv |\hat{\mathbf{D}}| = 1 \right] \tag{10.10}$$

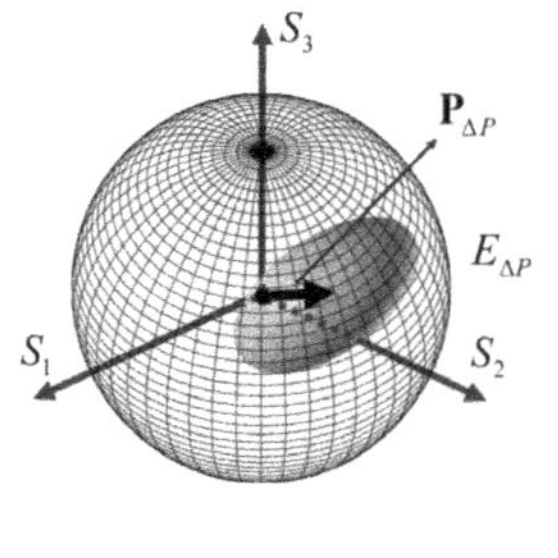

FIGURE 10.3
Forward ellipsoid $E_{\Delta P}$ associated with $\mathbf{M}$. Its semiaxes are given by the absolute values $(l_{\Delta P1}, l_{\Delta P2}, l_{\Delta P3})$ of the eigenvalues of the symmetric matrix $\mathbf{m}_{\Delta P}$; its orientation is given by a rotation $\mathbf{m}_{Rf}$ that tilts the semiaxes with respect to the Poincaré reference axes $S_1S_2S_3$, while the displacement of the origin of $E_{\Delta P}$ with respect to the origin of the Poincaré sphere is determined by the polarizance vector $\mathbf{P}_{\Delta P}$ of $\mathbf{M}_{\Delta P}$.

so that the ellipsoid $E_{\Delta P}$ degenerates into a single point located at the end point of the polarizance vector **P** (lying inside the unit sphere; see Table 10.4). Analogously, when $P = 1$, **M** is a depolarizing polarizer,

$$\hat{\mathbf{M}}_{\hat{P}D} = \begin{pmatrix} 1 & \mathbf{D}^T \\ \hat{\mathbf{P}} & \hat{\mathbf{P}} \otimes \mathbf{D}^T \end{pmatrix} \qquad \left[P \equiv |\hat{\mathbf{P}}| = 1 \right] \tag{10.11}$$

in which case $E_{\Delta P}$ degenerates into a single point located on the surface of the unit sphere at the end point of the polarizance vector $\hat{\mathbf{P}}$ (regardless of the value of D).

Obviously, the nine geometric parameters that, in general, determine the shape and location of $E_{\Delta P}$ are not sufficient to recover the 15 elements of the normalized Mueller matrix $\hat{\mathbf{M}}$. Nevertheless, complete information on $\hat{\mathbf{M}}$ can be achieved by additionally considering the P-image of $\hat{\mathbf{M}}^T$ (which, as seen in Section 5.8, is closely related to the reverse Mueller matrix $\hat{\mathbf{M}}^r$). By considering the reverse decomposition of a normalized Mueller matrix $\hat{\mathbf{M}}$ whose polarizance P is less than one (the case where $P = 1$ has already been considered),

$$\hat{\mathbf{M}} = \hat{\mathbf{M}}_P \mathbf{M}_R \hat{\mathbf{M}}_{\Delta D} \qquad \left[\hat{\mathbf{M}}_{\Delta D} \equiv \begin{pmatrix} 1 & \mathbf{D}_{\Delta D}^T \\ \mathbf{0} & \mathbf{m}_{\Delta D} \end{pmatrix} \right] \tag{10.12}$$

$\hat{\mathbf{M}}^T$ can be expressed as

$$\hat{\mathbf{M}}^T = \hat{\mathbf{M}}_{\Delta D}^T \hat{\mathbf{M}}_{JP} \qquad \left[\hat{\mathbf{M}}_{JP} \equiv \mathbf{M}_R^T \hat{\mathbf{M}}_P \right] \tag{10.13}$$

Since the P-image of the non-singular pure Mueller matrix $\hat{\mathbf{M}}_{JP}$ is the unit sphere itself, the P-image of $\hat{\mathbf{M}}^T$ coincides with that of $\hat{\mathbf{M}}_{\Delta D}^T$, which, in analogy to the analysis of the P-image of **M**, results in the *reverse ellipsoid* $E_{\Delta D}$, dual to $E_{\Delta P}$ and likewise located inside the unit sphere. Its semiaxes are given by the absolute values $(l_{\Delta D1}, l_{\Delta D2}, l_{\Delta D3})$ of the real eigenvalues of $\mathbf{m}_{\Delta D}$; its symmetry axes are misaligned with respect to the Poincaré reference frame $S_1S_2S_3$ by the effect of the rotation matrix $\mathbf{m}_{Rr}$ that performs the diagonalization $\mathbf{m}_{\Delta D} = \mathbf{m}_{Rr} \mathbf{m}_{L\Delta P} \mathbf{m}_{Rr}^T$, and its displacement from the origin is determined by the vector $\mathbf{D}_{\Delta D}$.

Since $\hat{\mathbf{M}}$ depends in general on up to 15 independent parameters, there exist certain constraining relations between the nine geometric parameters of $E_{\Delta P}$ and those of $E_{\Delta D}$ that reduce the number of independent geometric quantities to 15. These constraints will appear explicitly when analyzing the P-images of **M** and $\mathbf{M}^T$ by means of the symmetric decomposition instead of the Lu-Chipman decomposition.

In summary, the pair of *characteristic ellipsoids* $E_{\Delta P}$ and $E_{\Delta D}$ provides a complete geometric representation of a given normalized depolarizing Mueller matrix $\hat{\mathbf{M}}$ (Figure 10.4).

10.4 Ellipsoids Associated with Some Special Mueller Matrices

To achieve better understanding of the geometric properties of the mapping ellipsoid (or P-image) corresponding to a given Mueller matrix, some particularly interesting cases are studied in the following subsections.

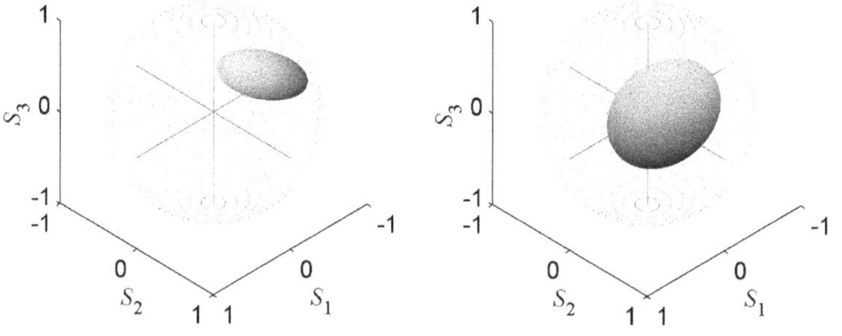

FIGURE 10.4
(a) *Forward ellipsoid* $E_{\Delta P}$ defined as the *P*-image of **M**; and (b) *reverse ellipsoid* $E_{\Delta D}$ defined as the *P*-image of $\mathbf{M}^T$ (Ossikovski et al. 2013).

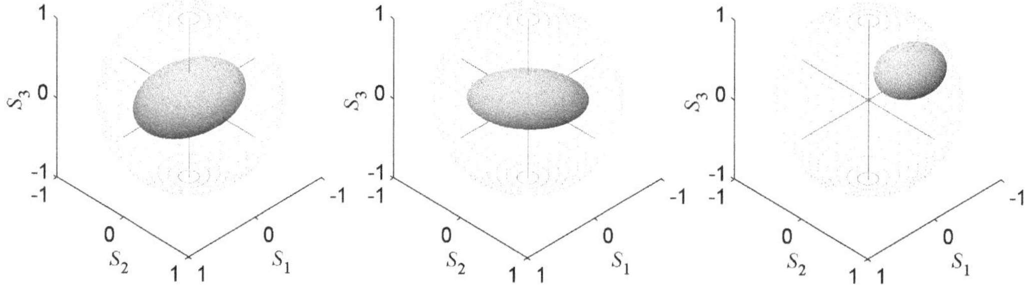

FIGURE 10.5
Examples of ellipsoids $E_{\Delta d}$ (left, canonical type-I depolarizer), $E_{R\Delta d}$ (middle, canonical type-I depolarizer followed by a retarder) and $E_{P\Delta d}$ (right, canonical type-I depolarizer followed by a normal diattenuator) (Ossikovski et al. 2013).

10.4.1 Intrinsic Depolarizer

The Mueller matrix of an intrinsic (or diagonal) depolarizer has the generic form $\mathbf{M}_{\Delta I} = d_0 \operatorname{diag}(1, \hat{d}_1, \hat{d}_2, \hat{d}_3)$, where $0 \le d_0 \le 1$ and $\hat{d}_i$ $(i = 1,2,3)$ are of arbitrary sign (either positive or negative), not necessarily ordered, and satisfy the covariance conditions (see Eq. (5.27))

$$1 \ge -\hat{d}_1 - \hat{d}_2 - \hat{d}_3 \quad 1 \ge -\hat{d}_1 + \hat{d}_2 + \hat{d}_3 \quad 1 \ge \hat{d}_1 - \hat{d}_2 + \hat{d}_3 \quad 1 \ge \hat{d}_1 + \hat{d}_2 - \hat{d}_3 \tag{10.14}$$

(Recall that passivity is ensured by the additional constraint $d_0 \le 1$). A particular form of intrinsic depolarizer is the canonical type-I depolarizer $\mathbf{M}_{\Delta d} \equiv d_0 \operatorname{diag}(1, \hat{d}_1, \hat{d}_2, \varepsilon\hat{d}_3)$, which is defined with the specific choice $\hat{d}_1 \ge \hat{d}_2 \ge \hat{d}_3 \ge 0$ and $\varepsilon \equiv \det\mathbf{M}/|\det\mathbf{M}|$ (see Section 5.7.1).

Obviously, since $\mathbf{M}_{\Delta I}$ is symmetric, $\mathbf{M}_{\Delta I}$ and $\mathbf{M}_{\Delta I}^T$ have a common characteristic ellipsoid, denoted by $E_{\Delta I}$, with semiaxes $\left(|\hat{d}_1|, |\hat{d}_2|, |\hat{d}_3|\right)$, which is both aligned and centered with respect to Poincaré's reference frame (Figure 10.5). The characteristic ellipsoid of $\mathbf{M}_{\Delta d}$ is denoted by $E_{\Delta d}$.

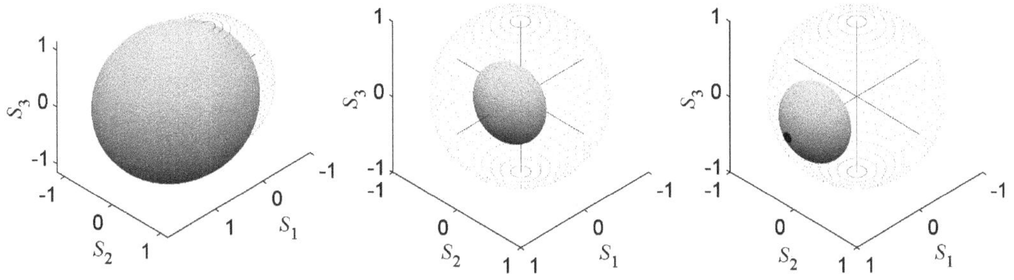

FIGURE 10.6
Poincaré sphere mapping by $\mathbf{A}_1$ (left), $\mathbf{M}_{\Delta dII}\mathbf{A}_1$ (center) and $\mathbf{M}_{\Delta nd} = \mathbf{A}_2\mathbf{M}_{\Delta dII}\mathbf{A}_1$ (right). The single contact point of $E_{\Delta nd}$ with the unit sphere is indicated with a black dot (Ossikovski et al. 2013).

10.4.2 Canonical Type-I Depolarizer Followed by a Retarder

In this case, the Mueller matrix can be written as $\mathbf{M}_{R\Delta d} \equiv \mathbf{M}_R\,\mathbf{M}_{\Delta d}$ and its P-image results in an ellipsoid $E_{R\Delta d}$ with semiaxes $(|\hat{d}_1|,|\hat{d}_2|,|\hat{d}_3|)$, so that its shape is identical to that of $E_{\Delta d}$, and is still centered, but is further rotated about the axis determined by the direction of the retardance vector $\mathbf{R}$ of $\mathbf{M}_R$, by an angle equal to the retardance $\Delta = \arccos[(\mathrm{tr}\,\mathbf{M}_R/2)-1]$ (Figure 10.6).

10.4.3 Canonical Type-I Depolarizer Followed by a Normal Diattenuator

The corresponding normalized Mueller matrix is given by $\hat{\mathbf{M}}_{P\Delta d} \equiv \hat{\mathbf{M}}_P\,\hat{\mathbf{M}}_{\Delta d}$ ($\hat{\mathbf{M}}_P$ representing a normal diattenuator), so that the resulting P-image is an ellipsoid, $E_{P\Delta d}$, with the following characteristics:

(1) The action of $\mathbf{M}_P$ causes a displacement of the center of $E_{P\Delta d}$ from the origin (note that the image point of the origin does not coincide, in general, with the end point of vector $\mathbf{P}$).
(2) Due to the distorting effect of $\mathbf{M}_P$, the semiaxes of $E_{P\Delta d}$ are different from those of $E_{\Delta d}$.

10.4.4 Type-II Canonical Depolarizer

To inspect the mapping produced by the type-II canonical depolarizer $\mathbf{M}_{\Delta nd}$ (see Section 5.7), let us consider its normalized version $\hat{\mathbf{M}}_{\Delta nd} \equiv \mathbf{M}_{\Delta nd}/2a_0$, and observe that it can be factored in the following manner:

$$\hat{\mathbf{M}}_{\Delta nd} = \mathbf{A}_2\hat{\mathbf{M}}_{\Delta dII}\mathbf{A}_1$$

$$\mathbf{A}_2 \equiv \begin{pmatrix} 1 & 0 & 0 & 0 \\ 1/2 & 1 & 0 & 0 \\ 0 & 0 & 1 & 0 \\ 0 & 0 & 0 & 1 \end{pmatrix} \quad \hat{\mathbf{M}}_{\Delta dII} \equiv \begin{pmatrix} 1 & 0 & 0 & 0 \\ 0 & 1/4 & 0 & 0 \\ 0 & 0 & k & 0 \\ 0 & 0 & 0 & k \end{pmatrix} \quad \mathbf{A}_1 \equiv \begin{pmatrix} 1 & -1/2 & 0 & 0 \\ 0 & 1 & 0 & 0 \\ 0 & 0 & 1 & 0 \\ 0 & 0 & 0 & 1 \end{pmatrix} \quad \left[k \equiv \frac{a_2}{2a_0}\right] \tag{10.15}$$

Despite the fact that $\mathbf{A}_1$ and $\mathbf{A}_2$ are not Mueller matrices, they are useful for the analysis of the P-image of $\mathbf{M}_{\Delta nd}$. Indeed, since the mapping by $\mathbf{M}_{\Delta dII}$ alone would produce a centered and aligned ellipsoid, it has to be considered in combination with the pre- and post- mapping by $\mathbf{A}_1$ and $\mathbf{A}_2$, respectively. As shown in Figure 10.6, $\mathbf{A}_1$ produces an anisotropic expansion combined with displacement along the S_1 axis. The resulting surface is an ellipsoid that is subsequently distorted by $\mathbf{M}_{\Delta dII}$, so that the ellipsoid originating from the combined mapping by $\mathbf{M}_{\Delta dII}\,\mathbf{A}_1$ remains inside the Poincaré sphere (Figure 10.6). Finally, $\mathbf{A}_2$ produces a displacement along the S_1 axis in such a manner that the P-image of $\mathbf{M}_{\Delta nd}$, as a whole, is given by the ellipsoid of revolution $E_{\Delta nd}$ defined by the equation

$$\frac{\left(u_1'-2/3\right)^2}{\left(1/3\right)^2}+\frac{u_2'^2}{\left(a_2/\sqrt{3}a_0\right)^2}+\frac{u_3'^2}{\left(a_2/\sqrt{3}a_0\right)^2}=1 \qquad \left[u_i'\equiv\left(\mathbf{M}\hat{\mathbf{s}}_p\right)_i/\left(\mathbf{M}\hat{\mathbf{s}}_p\right)_0 \quad (i=1,2,3)\right] \tag{10.16}$$

Thus, $E_{\Delta nd}$ has its symmetry axis along S_1, with semiaxes $(1/3, a_2/\sqrt{3}a_0, a_2/\sqrt{3}a_0)$ and the center of $E_{\Delta nd}$ has the coordinates $(2/3, 0, 0)$.

The reverse ellipsoid $E_{\Delta ndr}$ of $\mathbf{M}_{\Delta nd}$ (defined as the P-image of $\mathbf{M}^T_{\Delta nd}$) likewise has its symmetry axis along S_1, with the same semiaxes $(1/3, a_2/\sqrt{3}a_0, a_2/\sqrt{3}a_0)$ as $E_{\Delta nd}$, but the coordinates of its center are $(-2/3, 0, 0)$.

$E_{\Delta nd}$ touches the unit sphere at the single point $(1,0,0)$, whereas $E_{\Delta ndr}$ touches the unit sphere at the single point $(-1,0,0)$. As evidenced in previous works (Ossikovski et al. 2009; Ossikovski and Gil 2017), the property of the P-image of touching the unit sphere at a single point is a peculiarity of type-II Mueller matrices (leaving aside the case of singular type-I Mueller matrices, which will be analyzed in a later section).

10.4.5 Type-II Canonical Depolarizer Followed by a Retarder

Consider the Mueller matrix $\mathbf{M}_{R\Delta nd} \equiv \mathbf{M}_R\,\mathbf{M}_{\Delta nd}$ of a type-II canonical depolarizer followed by a retarder represented by $\mathbf{M}_R$. The P-image of $\mathbf{M}_{R\Delta nd}$ results in an ellipsoid $E_{R\Delta nd}$ whose shape is identical to that of $E_{\Delta nd}$, but rotated by an angle equal to the retardance $\Delta = \arccos[(\mathrm{tr}\mathbf{M}_R/2)-1]$ about the direction defined by the retardance vector $\mathbf{R}$ of $\mathbf{M}_R$ (Figure 10.7). The center of the ellipsoid $E_{R\Delta nd}$ goes to a new position, but its distance, equal to $2/3$, from the origin of Poincare's reference frame $S_1S_2S_3$ remains unchanged. Moreover, since $\mathbf{M}_R$ does not modify the degree of polarization of the incoming states (and, consequently, preserves the shape of the ellipsoid), $E_{R\Delta nd}$ possesses one, and only one, point touching the unit sphere (the contact points with the unit sphere are indicated in figures by black dots).

10.4.6 Type-II Canonical Depolarizer Followed by a Normal Diattenuator

The P-image of the normalized Mueller matrix $\hat{\mathbf{M}}_{P\Delta nd} \equiv \hat{\mathbf{M}}_P\,\hat{\mathbf{M}}_{\Delta nd}$ of a type-II canonical depolarizer $\hat{\mathbf{M}}_{\Delta nd}$ followed by a normal diattenuator $\hat{\mathbf{M}}_P$ is an ellipsoid $E_{P\Delta nd}$ generated by the transformation by $\hat{\mathbf{M}}_P$ of the characteristic ellipsoid $E_{\Delta nd}$ of $\hat{\mathbf{M}}_{\Delta nd}$, so that $E_{P\Delta nd}$ has the following features:

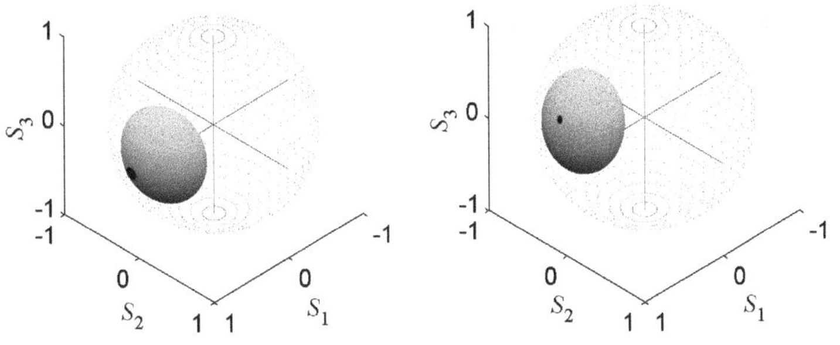

FIGURE 10.7
The type-II canonical ellipsoid $E_{\Delta nd}$ (left) is rotated by $\mathbf{M}_R$ resulting in the ellipsoid $E_{R\Delta nd}$ (right) (Ossikovski et al. 2013).

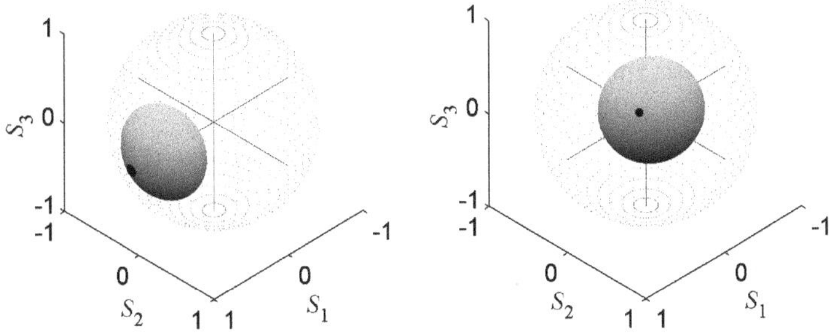

FIGURE 10.8
The type-II canonical ellipsoid $E_{\Delta nd}$ (left) is displaced and distorted by the normal diattenuator $\hat{\mathbf{M}}_p$ resulting in the ellipsoid $E_{P\Delta nd}$ (right). Both $E_{\Delta nd}$ and $E_{P\Delta nd}$ have the genuine property of having a single point that touches the unit sphere (Ossikovski et al. 2013).

(1) $\hat{\mathbf{M}}_P$ produces a global displacement of $E_{\Delta nd}$;
(2) due to the distorting effect of $\hat{\mathbf{M}}_P$, the ellipsoid $E_{\Delta nd}$ further undergoes an anisotropic contraction; and
(3) since $\hat{\mathbf{M}}_P$ does not reduce the degree of polarization of any totally polarized incident state and $E_{\Delta nd}$ has a single point touching the unit sphere, then $E_{P\Delta nd}$ also contains a single point touching the unit sphere (Figure 10.8).

10.5 Characteristic Ellipsoids of a Depolarizing Mueller Matrix

Let us consider the normal form decomposition $\mathbf{M} = \mathbf{M}_{J2}\mathbf{M}_{\Delta}\mathbf{M}_{J1}$ and recall that the mapping by the *entrance pure* Mueller matrix $\mathbf{M}_{J1} \equiv \mathbf{M}_{R1}\mathbf{M}_{D1}$ maps the Poincaré sphere onto itself (except for the limiting case of $\mathbf{M}_{J1}$ being singular, where the transformed object is a single point on the sphere).

As with the procedure for the calculation of the Mueller matrix components of the symmetric decomposition of **M**, the analysis of the *characteristic ellipsoids* derived from the symmetric decomposition depends on whether **M** is of type-I or of type-II, on whether **M** is singular or not, as well as on the properties of the N-matrices $\mathbf{N} \equiv \mathbf{G}\mathbf{M}^T\mathbf{G}\mathbf{M}$ and $\mathbf{N}' \equiv \mathbf{M}\mathbf{G}\mathbf{M}^T\mathbf{G}$.

10.5.1 Characteristic Ellipsoids of a Type-I Mueller Matrix

Let us first consider the most common case in practice of a non-singular type-I Mueller matrix **M**. The possible cases with respect to the invertibility of the two diattenuators from the symmetric decomposition of **M** are analyzed next.

10.5.1.1 $N \neq 0$ and $N' \neq 0$ ($D_1 < 1$, $D_2 < 1$)

Since the entrance Mueller matrix $\mathbf{M}_{J1}$ is non-singular, it maps the Poincaré sphere onto itself and, thus, the mapping by **M** can be analyzed through the sequential effects of the canonical depolarizer $\mathbf{M}_{\Delta d}$, the retarder $\mathbf{M}_{R2}$ and the diattenuator $\mathbf{M}_{D2}$. As seen in Sections 10.4.1 and 10.4.2, $\mathbf{M}_{\Delta d}$ maps the Poincaré sphere onto a centered, aligned ellipsoid $E_{\Delta d}$ with semiaxes $(|\hat{d}_1|,|\hat{d}_2|,|\hat{d}_3|)$ directed along the reference axes S_1,S_2,S_3. This *type-I canonical ellipsoid* $E_{\Delta d}$ is first rotated by an angle equal to the retardance $\Delta_2 = \arccos[(\operatorname{tr}\mathbf{M}_{R2}/2)-1]$ about the axis defined by the retardance vector $\mathbf{R}_2$ of $\mathbf{M}_{R2}$, and is then displaced and distorted by the effect of the diattenuator $\mathbf{M}_{D2}$, resulting in an eccentric transformed ellipsoid E_{I2} (Figure 10.9). Obviously, since the *P*-image of **M** does not depend on the particular serial decomposition of **M** considered, E_{I2} is precisely the *forward ellipsoid* $E_{\Delta P}$ of the type-I Mueller matrix **M**.

As seen in Section 10.3, the geometric parameters of the type-I forward ellipsoid $E_{\Delta P}$ are given by the nine quantities constituted by: the three semiaxes $(l_{\Delta P1},l_{\Delta P2},l_{\Delta P3})$; the three orientation angles $(\varphi_{\Delta P},\chi_{\Delta P},\Delta_{\Delta P})$ of $E_{\Delta P}$ with respect to the Poincaré reference frame $S_1S_2S_3$, and the three coordinates of the center of $E_{\Delta P}$, given by the components of the polarizance vector $\mathbf{P}_{\Delta P}$.

Observe that, once the canonical ellipsoid $E_{\Delta d}$ has been determined, the lengths of the semiaxes of $E_{\Delta P}$ can be obtained from the nine parameters $(\hat{d}_1,\hat{d}_2,\hat{d}_3)$, $\mathbf{P}_{\Delta P}$ and $(\varphi_{\Delta P},\chi_{\Delta P},\Delta_{\Delta P})$ and, therefore, the information about the three semiaxes of $E_{\Delta P}$ is redundant. It should also be noted that the image of the center (0,0,0) of the Poincaré sphere, given by the end point

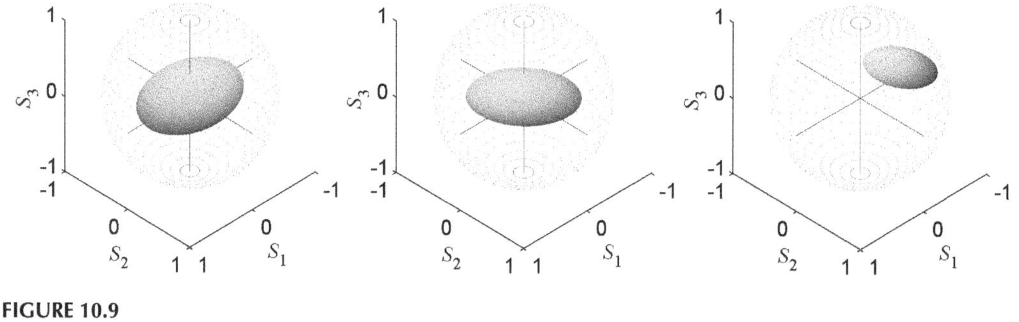

FIGURE 10.9
Examples of mapping by $\mathbf{M}_{\Delta d}$ (type-I canonical ellipsoid $E_{\Delta d}$, left), $\mathbf{M}_{R2}\mathbf{M}_{\Delta d}$ (middle) and $\mathbf{M}_{D2}\mathbf{M}_{R2}\mathbf{M}_{\Delta d}$ (ellipsoid $E_{\Delta P}$, right) corresponding to a given type-I Mueller matrix **M** (Ossikovski et al. 2013).

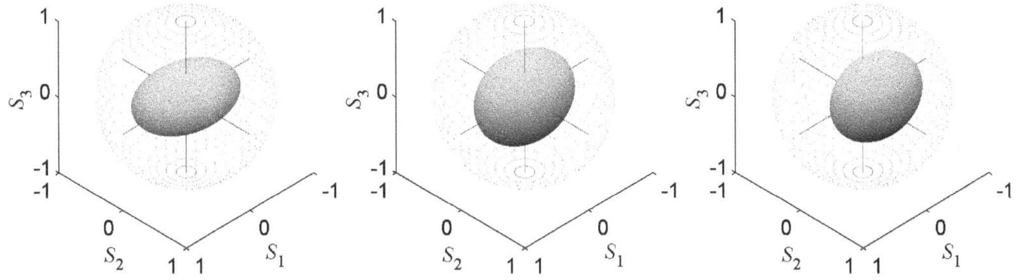

FIGURE 10.10
Examples of mapping by $\mathbf{M}_{\Delta d}$ (type-I canonical ellipsoid $E_{\Delta d}$, left), $\mathbf{M}_{R1}^T \mathbf{M}_{\Delta d}$ (middle) and $\mathbf{M}_{D1} \mathbf{M}_{R1}^T \mathbf{M}_{\Delta d}$ (ellipsoid $E_{\Delta D}$, right) corresponding to a given type-I Mueller matrix $\mathbf{M}$ (Ossikovski et al. 2013).

TABLE 10.1
Geometric Parameters of a Non-singular Type-I Mueller Matrix **M**

Mueller matrix	Ellipsoid	Number of parameters	Parameters	Geometric meaning
$\mathbf{M}_{\Delta d}$	$E_{\Delta d}$	3	$\hat{d}_1, \hat{d}_2, \hat{d}_3$	*Lengths of the semiaxes of* $E_{\Delta d}$
$\mathbf{M}$	$E_{\Delta P}$	3	$(P_{\Delta P1}, P_{\Delta P2}, P_{\Delta P3})$	Coordinates of the center of $E_{\Delta P}$, determined by the polarizance vector $\mathbf{P}_{\Delta P}$ of $\mathbf{M}_{\Delta P}$
		3	$(\varphi_{\Delta P}, \chi_{\Delta P}, \Delta_{\Delta P})$	Orientation angles of the symmetry axes of $E_{\Delta P}$ determined by the retardance vector $\mathbf{R}_f$ of $\mathbf{M}_{Rf}$
$\mathbf{M}^T$	$E_{\Delta D}$	3	$(D_{\Delta D1}, D_{\Delta D2}, D_{\Delta D3})$	Coordinates of the center of $E_{\Delta D}$, determined by the polarizance vector $\mathbf{D}_{\Delta D}$ of $\mathbf{M}_{\Delta D}^T$
		3	$(\varphi_{\Delta D}, \chi_{\Delta D}, \Delta_{\Delta D})$	Orientation angles of the symmetry axes of $E_{\Delta D}$, determined by the retardance vector $\mathbf{R}_r$ of $\mathbf{M}_{Rr}$

of the polarizance vector $\mathbf{D}_2$ of the diattenuator $\mathbf{M}_{D2}$, does not coincide, in general, with the center of $E_{\Delta P}$ (i.e., $\mathbf{P}_{\Delta P} \neq \mathbf{D}_2$).

In Section 10.3, we have mentioned that the nine geometric parameters of the forward ellipsoid $E_{\Delta P}$ are not sufficient to determine the 15 quantities that, in general, characterize a normalized type-I Mueller matrix. The missing geometric information can be completed through the *reverse ellipsoid* $E_{\Delta D}$ generated by $\mathbf{M}^T$ whose symmetric decomposition is

$$\mathbf{M}^T = \mathbf{M}_{D1} \mathbf{M}_{R1}^T \mathbf{M}_{\Delta d} \mathbf{M}_{R2}^T \mathbf{M}_{D2} \tag{10.17}$$

The *P*-image of the non-singular pure Mueller matrix $\mathbf{M}_{J2}^T \equiv \mathbf{M}_{R2}^T \mathbf{M}_{D2}$ reproduces the unit sphere; the action of $\mathbf{M}_{\Delta d}$ gives the canonical ellipsoid $E_{\Delta d}$ which is first rotated by the effect of the retarder $\mathbf{M}_{R1}^T$ and then displaced and distorted by the effect of the diattenuator $\mathbf{M}_{D1}$, resulting in a transformed ellipsoid E_{J1} which is precisely the *reverse ellipsoid* $E_{\Delta D}$ of $\mathbf{M}$ (Figure 10.10).

We stress that, in general, the coordinates of the displaced center of $E_{\Delta D}$ (which are given by the components of $\mathbf{P}_{\Delta D}$), together with the three angles that determine its spatial orientation $(\varphi_{\Delta D}, \chi_{\Delta D}, \Delta_{\Delta D})$, are independent from the geometric parameters of the type-I canonical ellipsoid $E_{\Delta d}$ and those of the forward ellipsoid $E_{\Delta P}$.

Although the above analysis is based on the specific forms $\mathbf{M}_{J1} = \mathbf{M}_{R1} \mathbf{M}_{D1}$ and $\mathbf{M}_{J2} = \mathbf{M}_{D2} \mathbf{M}_{R2}$ for the polar decomposition of the pure components, a similar analysis can be performed by considering the alternative forms of the polar decompositions $\mathbf{M}_{J1} = \mathbf{M}_{P1}\mathbf{M}_{R1}$ and $\mathbf{M}_{J2} = \mathbf{M}_{R2}\mathbf{M}_{P2}$ thus resulting in four possible combinations. Obviously, the specific choice does not affect the shape and location of the characteristic ellipsoids $E_{\Delta P}$ and $E_{\Delta D}$.

In summary, the geometric parameterization of a non-singular type-I normalized Mueller matrix $\hat{\mathbf{M}}$ by means of the characteristic parameters of the three associated *characteristic ellipsoids* $E_{\Delta d}$, $E_{\Delta P}$ and $E_{\Delta D}$ (generated by $\mathbf{M}_{\Delta d}$, $\mathbf{M}$ and $\mathbf{M}^T$ respectively) is given by the 15 quantities listed in Table 10.1.

When $\mathbf{M}$ is singular and $\mathbf{N} \neq \mathbf{0}$ and $\mathbf{N}' \neq \mathbf{0}$ (in which case $\mathbf{M}$ is termed a *singular depolarizer*), then at least one of the semiaxes is zero, so that the type-I canonical ellipsoid $E_{\Delta d}$ degenerates into an ellipse (one null semiaxis), into a line segment (two null semiaxes) or into a single point (the three semiaxes are zero). The Mueller matrix of the degenerate case of three null semiaxes can be written as

$$\mathbf{M} = \mathbf{M}_{D2}\,\mathbf{M}_{\Delta 0}\,\mathbf{M}_{D1} = m_{00}\begin{pmatrix} 1 & \mathbf{D}_1^T \\ \mathbf{D}_2 & \mathbf{D}_2 \otimes \mathbf{D}_1^T \end{pmatrix} = m_{00}\begin{pmatrix} 1 \\ \mathbf{D}_2 \end{pmatrix} \otimes \left(1, \mathbf{D}_1^T\right) \qquad [D_1 < 1 \quad D_2 < 1] \tag{10.18}$$

where $\mathbf{M}_{\Delta 0} \equiv \mathrm{diag}(1,0,0,0)$ is the perfect depolarizer.

It should be noted that, despite the fact that in general $D_1 \neq D$ and $D_2 \neq P$, in the case of singular Mueller matrices it holds that $D_1 = 1$ if and only if $D = 1$, while $D_2 = 1$ if and only if $P = 1$. The degenerate ellipsoids $E_{\Delta d}$ are represented in Table 10.4 ($D < 1, D_1 < 1; P < 1, D_2 < 1$).

10.5.1.2 $\mathbf{N} \neq \mathbf{0}$ *and* $\mathbf{N}' = \mathbf{0}$ *($D_1 = 1$, $D_2 < 1$)*

As seen in Section 8.3.1.2, in this case the system is a depolarizing analyzer whose symmetric decomposition reduces to the type-I Mueller matrix

$$\mathbf{M} = \mathbf{M}_{D2}\,\mathbf{M}_{\Delta 0}\,\mathbf{M}_{D1} = m_{00}\begin{pmatrix} 1 & \hat{\mathbf{D}}_1^T \\ \mathbf{D}_2 & \mathbf{D}_2 \otimes \hat{\mathbf{D}}_1^T \end{pmatrix} = m_{00}\begin{pmatrix} 1 \\ \mathbf{D}_2 \end{pmatrix} \otimes \left(1, \hat{\mathbf{D}}_1^T\right) \qquad [D_1 = 1 \quad D_2 < 1] \tag{10.19}$$

The canonical ellipsoid $E_{\Delta d}$ degenerates into the single point $(0,0,0)$, The forward ellipsoid $E_{\Delta P}$ degenerates into the single point defined by the end point of $\mathbf{D}_2$ (inside the unit sphere) and the reverse ellipsoid $E_{\Delta D}$ degenerates into the single point defined by the end point of $\mathbf{D}_1$ (on the surface of the unit sphere); see Table 10.4 ($D = D_1 = 1; P < 1, D_2 < 1$).

10.5.1.3 $N = 0$ *and* $N' \neq 0$ *($D_1 < 1$, $D_2 = 1$)*

In this case, $\mathbf{M}_{D2}$ is a perfect polarizer and $\mathbf{M}_{D1}$ is a partial polarizer (which can be chosen to be normal). In analogy to the previous case, $\mathbf{M}$ is a type-I Mueller matrix that can be decomposed as

$$\mathbf{M} = \mathbf{M}_{D2}\,\mathbf{M}_{\Delta 0}\,\mathbf{M}_{D1} = m_{00}\begin{pmatrix} 1 & \mathbf{D}_1^T \\ \hat{\mathbf{D}}_2 & \hat{\mathbf{D}}_2 \otimes \mathbf{D}_1^T \end{pmatrix} = m_{00}\begin{pmatrix} 1 \\ \hat{\mathbf{D}}_2 \end{pmatrix} \otimes \left(1, \mathbf{D}_1^T\right) \qquad (D_1 < 1 \quad D_2 = 1) \tag{10.20}$$

so that $\mathbf{M}$ is a depolarizing polarizer. The canonical ellipsoid $E_{\Delta d}$ degenerates into the single point $(0,0,0)$, the forward ellipsoid $E_{\Delta P}$ degenerates into the single point defined by the end point of $\mathbf{D}_2$ (on the surface of the unit sphere) and the reverse $E_{\Delta D}$ degenerates into a single point defined by the end point of $\mathbf{D}_1$ (inside the unit sphere); see Table 10.4 ($D < 1$, $D_1 < 1$; $P = D_2 = 1$).

10.5.1.4 $D_1 = D_2 = 1$

As seen in Section 8.3.1.4, in this case the system is equivalent to a perfect polarizer, in general non-normal, and corresponds to a singular pure Mueller matrix $\mathbf{M}_J$ whose forward and reverse ellipsoids degenerate into respective single points on the surface of the Poincaré sphere; see Table 10.4 ($D = D_1 = 1$; $P = D_2 = 1$).

10.5.2 Characteristic Ellipsoids of a Type-II Mueller Matrix

In accordance with the analysis performed in Section 8.3.2, the diattenuators $\mathbf{M}_{D1}$ and $\mathbf{M}_{D2}$ of the symmetric decomposition are not singular (i.e., $D_1 < 1$ and $D_2 < 1$). Therefore, since the P-image of the entrance non-singular pure Mueller matrix $\mathbf{M}_{J1}$ is the unit sphere itself, we analyze the sequential effects of the type-II canonical depolarizer $\mathbf{M}_{\Delta nd}$, the retarder $\mathbf{M}_{R2}$ and the diattenuator $\mathbf{M}_{D2}$. As already mentioned, $\mathbf{M}_{\Delta nd}$ maps the Poincaré sphere onto a displaced ellipsoid of revolution $E_{\Delta nd}$ with semiaxes $(1/3, a_2/\sqrt{3}a_0, a_2/\sqrt{3}a_0)$ aligned with respect to the reference axes $S_1S_2S_3$ and touching the Poincaré sphere at the single point $(1,0,0)$. The retarder $\mathbf{M}_{R2}$ rotates the type-II canonical ellipsoid $E_{\Delta nd}$ about the retardance vector $\mathbf{R}_2$ of $\mathbf{M}_{R2}$ by an angle equal to the retardance $\Delta_2 = \arccos[(\mathrm{tr}\mathbf{M}_{R2}/2)-1]$, and finally the diattenuator $\mathbf{M}_{D2}$ produces a displacement together with a distortion, resulting in a transformed ellipsoid E_{II2} that still contains a single contact point with the unit sphere. Obviously, E_{II2} is just the forward ellipsoid $E_{\Delta P}$ of the type-II Mueller matrix $\mathbf{M}$ (Figure 10.11).

The geometric parameters of $E_{\Delta nd}$ and $E_{\Delta P}$ are given by the seven quantities constituted by the ratio between the two different semiaxes of $E_{\Delta nd}$, the three coordinates of the displaced origin of $E_{\Delta P}$ (which are given by the components of the polarizance vector $\mathbf{P}_{\Delta P}$ of $\mathbf{M}_{\Delta P}$ studied in Section 10.3), and the three angles $(\varphi_{\Delta P}, \chi_{\Delta P}, \Delta_{\Delta P})$ determining its spatial orientation. Note that the lengths of the semiaxes of $E_{\Delta P}$ can be obtained from these seven parameters and thus, providing them is redundant. Moreover, observe that the image of the center of the Poincaré sphere, given by $\mathbf{D}_2$, does not coincide with the center of $E_{\Delta P}$, given by $\mathbf{P}_{\Delta P}$.

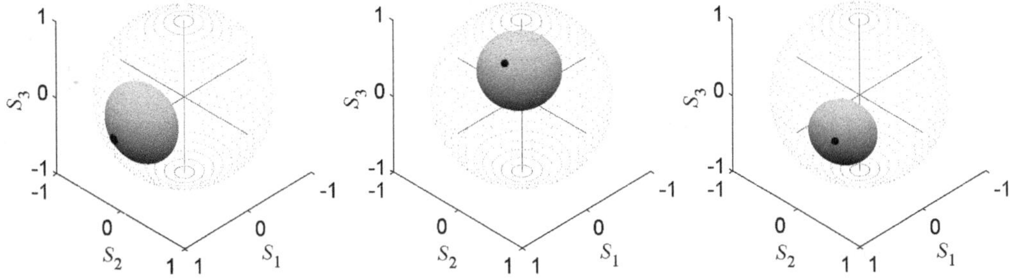

FIGURE 10.11
Examples of ellipsoids $E_{\Delta nd}$ (left), $E_{\Delta P}$ (middle) and $E_{\Delta D}$ (right) of a given type-II Mueller matrix $\mathbf{M}$ (Ossikovski et al. 2013).

TABLE 10.2
Geometric Parameterization of a Non-singular Type-II Mueller Matrix **M**

Mueller matrix	Ellipsoid	Number of parameters	Parameters	Geometric meaning
$\mathbf{M}_{\Delta nd}$	$E_{\Delta nd}$	1	$a_2/\sqrt{3}a_0$	*Length of the minor semiaxis of the ellipsoid of revolution $E_{\Delta nd}$ (the length of the major semiaxis is 1/3)*
$\mathbf{M}$	$E_{\Delta P}$	3	$(P_{\Delta P1}, P_{\Delta P2}, P_{\Delta P3})$	Coordinates of the center of $E_{\Delta P}$, determined by the polarizance vector $\mathbf{P}_{\Delta P}$ of $\mathbf{M}_{\Delta P}$
		3	$(\varphi_{\Delta P}, \chi_{\Delta P}, \Delta_{\Delta P})$	Orientation angles of the symmetry axes of $E_{\Delta P}$, determined by the retardance vector $\mathbf{R}_f$ of $\mathbf{M}_{Rf}$
$\mathbf{M}^T$	$E_{\Delta D}$	3	$(D_{\Delta D1}, D_{\Delta D2}, D_{\Delta D3})$	Coordinates of the center of $E_{\Delta D}$, determined by the polarizance vector $\mathbf{D}_{\Delta D}$ of $\mathbf{M}_{\Delta D}^T$
		3	$(\varphi_{\Delta D}, \chi_{\Delta D}, \Delta_{\Delta D})$	Orientation angles of the symmetry axes of $E_{\Delta D}$, determined by the retardance vector $\mathbf{R}_r$ of $\mathbf{M}_{Rr}$.

Completing the above seven parameters to the 13 quantities that characterize a general normalized type-II Mueller matrix can be achieved by additionally considering the mapping by the transposed Mueller matrix

$$\mathbf{M}^T = \mathbf{M}_{D1}\mathbf{M}_{R1}^T\mathbf{M}_{\Delta nd}^T\mathbf{M}_{R2}^T\mathbf{M}_{D2} \tag{10.21}$$

The mapping by the non-singular pure Mueller matrix $\mathbf{M}_{J2}^T \equiv \mathbf{M}_{R2}^T\mathbf{M}_{D2}$ reproduces the unit sphere; the eccentric canonical ellipsoid $E_{\Delta nd}$ generated by $\mathbf{M}_{\Delta nd}^T$ is then first rotated by the effect of the retarder $\mathbf{M}_{R1}^T$, and then displaced and distorted by the effect of the diattenuator $\mathbf{M}_{D1}^T$, resulting in a transformed ellipsoid $E_{\Delta D}$ which, like $E_{\Delta P}$, contains a single point touching the unit sphere (Figure 10.11).

Observe also that, in general, the three coordinates of the displaced origin of $E_{\Delta D}$ (which are given by the components of the polarizance vector $\mathbf{P}_{\Delta D}$ of the enpolarizing depolarizer $\mathbf{M}_{\Delta D}^T$ studied in Section 10.3), and the three angles that determine its spatial orientation $(\varphi_{\Delta D}, \chi_{\Delta D}, \Delta_{\Delta D})$ are independent from the geometric parameters of $E_{\Delta nd}$ and $E_{\Delta P}$.

If **M** is singular (i.e., it represents a type-II singular depolarizer), then necessarily $a_2 = 0$, so that the canonical ellipsoid degenerates into a line segment (see Table 10.5).

In summary, the geometric parameterization of a non-singular type-II normalized Mueller matrix $\hat{\mathbf{M}}$ by means of the characteristic parameters of the three associated *characteristic ellipsoids* $E_{\Delta nd}$, $E_{\Delta P}$ and $E_{\Delta D}$ (generated by $\mathbf{M}_{\Delta nd}$, $\mathbf{M}$ and $\mathbf{M}^T$ respectively) is given in general by the 13 quantities listed in Table 10.2.

10.6 Topological Properties of the Characteristic Ellipsoids

In accordance with the analysis performed in Section 6.8, the depolarization properties of the canonical depolarizer $\mathbf{M}_\Delta$ can be parameterized in terms of its corresponding indices of polarimetric purity (IPP). Despite of the fact that the IPP of $\mathbf{M}_\Delta$ are, in general, different from those of $\mathbf{M}$, recall that $\operatorname{rank}\mathbf{H}(\mathbf{M}_\Delta) = \operatorname{rank}\mathbf{H}(\mathbf{M})$ and thus, the number of nonzero IPP of $\mathbf{M}_\Delta$ equals the number of nonzero IPP of $\mathbf{M}$ (that is, the minimum number of possible parallel components of $\mathbf{M}_\Delta$ is equal to that of $\mathbf{M}$; see Chapter 7). Therefore, the canonical ellipsoid E_Δ (denoted, when appropriate, as $E_{\Delta d}$ or $E_{\Delta nd}$ for type-I and type-II Mueller matrices respectively), besides its attractive visual simplicity, further provides a meaningful geometric representation of the depolarizing properties of $\mathbf{M}$.

To get a deeper understanding of the information provided by the canonical ellipsoid E_Δ, we next consider the number c of the possible contact points of E_Δ with the unit sphere and its relation to $\operatorname{rank}\mathbf{H}(\mathbf{M})$. The contents of the next subsections are closely related to those of Section 5.15, where this subject is considered in the light of the corresponding Jones generators.

10.6.1 Polarizers and Analyzers

As a first step, let us consider separately the special cases where the polarizance or the diattenuation are equal to the unit.

10.6.1.1 Depolarizing Polarizers

When $P = 1$ and $D < 1$, $\mathbf{M}$ has the form of a depolarizing polarizer $\mathbf{M}_{\hat{P}D} = \mathbf{M}_{\hat{P}}\mathbf{M}_{\Delta 0}\mathbf{M}_D$ where $\mathbf{M}_{\hat{P}}$ corresponds to a normal perfect polarizer ($P = 1$), $\mathbf{M}_{\Delta 0} \equiv \operatorname{diag}(d_0, 0, 0, 0)$ is a perfect depolarizer and $\mathbf{M}_D$ is a normal diattenuator (with $D < 1$). Recall that, as seen in Section 8.3, and in accordance with the convention used throughout this book, this kind of Mueller matrix $\mathbf{M}$ is necessarily of type-I.

The type-I canonical ellipsoid E_Δ degenerates into the single point $(0,0,0)$, corresponding to the origin of the Poincaré reference frame. Ellipsoid $E_{\Delta P}$ degenerates into a single point on the surface of the unit sphere, whereas ellipsoid $E_{\Delta D}$ degenerates into a single point inside the unit sphere (see Table 10.4).

10.6.1.2 Depolarizing Analyzers

When $P < 1$ and $D = 1$, $\mathbf{M}$ is a type-I matrix having the form of a depolarizing analyzer $\mathbf{M}_{P\hat{D}} = \mathbf{M}_P\mathbf{M}_{ID}\mathbf{M}_{\hat{D}}$, where $\mathbf{M}_P$ corresponds to a normal diattenuator (with $P < 1$), $\mathbf{M}_{\Delta 0} \equiv \operatorname{diag}(d_0, 0, 0, 0)$ is a perfect depolarizer, and $\mathbf{M}_{\hat{D}}$ is a normal pure polarizer ($D = 1$).

Like with depolarizing polarizers, the canonical ellipsoid E_Δ of the depolarizing analyzer degenerates into the single point $(0,0,0)$. Ellipsoid $E_{\Delta D}$ degenerates into a single point on the surface of the unit sphere, whereas the ellipsoid $E_{\Delta P}$ degenerates into a single point inside the unit sphere (see Table 10.4).

10.6.1.3 Perfect Polarizers

When $P = D = 1$, $\mathbf{M}$ is a perfect polarizer, in general nonnormal, that depends on only four parameters, and $\hat{\mathbf{M}}$ is t otally determined by its characteristic ellipsoids $E_{\Delta P}$ and $E_{\Delta D}$, which degenerate into respective single points on the surface of the unit sphere (see Table 10.4).

10.6.2 rank H = 1

In this case, $d_0 = d_1 = d_2 = d_3$. $\mathbf{M}$ is necessarily pure (and, hence, type-I). Thus, provided that $P < 1$ and $D < 1$, $\mathbf{M}$ maps the Poincaré sphere onto itself. The canonical ellipsoid, as well as the characteristic ellipsoids, coincide with the unit sphere (case 1: $c = \infty$, in Section 5.15.1).

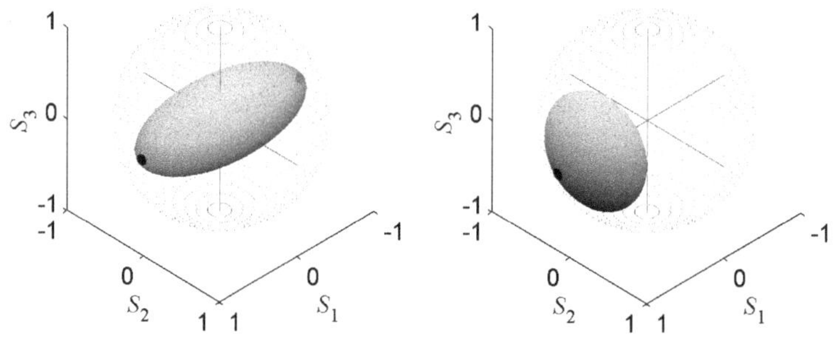

FIGURE 10.12
(a) Ellipsoid $E_{\Delta d}$ for a type-I canonical Mueller matrix $\mathbf{M}_{\Delta d}$ with rank $\mathbf{H}(\mathbf{M}_{\Delta d}) = 2$. $E_{\Delta d}$ touches the unit sphere at two antipodal points. (b) Ellipsoid $E_{\Delta nd}$ for a type-II canonical Mueller matrix $\mathbf{M}_{\Delta nd}$ with rank $\mathbf{H}(\mathbf{M}_{\Delta nd}) = 2$. $E_{\Delta nd}$ touches the unit sphere at a single point (Ossikovski et al. 2013).

The intersection of the direction of the polarizance vector $\mathbf{P}$ with the unit sphere determines the point of highest topological density of transformed points, and the absolute value P of $\mathbf{P}$ gives a measure of the asymmetry of the density distribution of the transformed points (recall that, for pure Mueller matrices, polarizance equals diattenuation: $P = D$).

10.6.3 rank H = 2

10.6.3.1 Type-I

In this case, $d_0 = d_1$, $d_2 = d_3$ (recall the convention taken for the type-I canonical depolarizer $\mathbf{M}_{\Delta d} = \mathrm{diag}(d_0, d_1, d_2, \varepsilon d_3)$ with $d_0 \geq d_1 \geq d_2 \geq d_3 \geq 0$ and $\varepsilon = \det \mathbf{M}_{\Delta d} / |\det \mathbf{M}_{\Delta d}|$), so that, provided that $P < 1$ and $D < 1$, the canonical ellipsoid $E_{\Delta d}$ touches the unit sphere at two antipodal points along the S_1 Poincaré axis (Figure 10.12a). Furthermore, the characteristic ellipsoids $E_{\Delta P}$ and $E_{\Delta D}$ touch the sphere at two points (case 2: $c = 2$, in Section 5.15.1).

10.6.3.2 Type-II

In this case, $a_2 = a_0$, so that $E_{\Delta nd}$ has semiaxes $(1/3, 1/\sqrt{3}, 1/\sqrt{3})$ and touches the unit sphere at the single point (1,0,0) (Figure 10.12b). Moreover, $E_{\Delta P}$ and $E_{\Delta D}$ touch the sphere at respective single points (case 3: $c = 1$, in Section 5.15.1).

10.6.4 rank H = 3

10.6.4.1 Type-I

In this case, $d_0 - d_1 = d_2 - d_3 \equiv q$, so that, provided that $P < 1$ and $D < 1$, the lengths of the three semiaxes $\hat{d}_1, \hat{d}_2, \hat{d}_3$ of $E_{\Delta d}$ are smaller than one and hence, $E_{\Delta d}$ does not touch the unit sphere at any point (Figure 10.13a). Analogously, ellipsoids $E_{\Delta P}$ and $E_{\Delta D}$ do not touch the unit sphere at any point (case 4: $c = 0$, in Section 5.15.1).

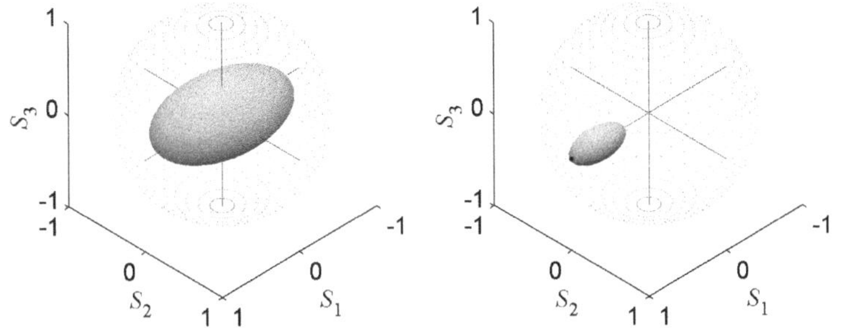

FIGURE 10.13
(a) Ellipsoid $E_{\Delta d}$ for a type-I canonical Mueller matrix $\mathbf{M}_{\Delta d}$ with rank $\mathbf{H}(\mathbf{M}_{\Delta d}) = 3$. $E_{\Delta d}$ is strictly inside the unit sphere. (b) Ellipsoid $E_{\Delta nd}$ for a type-II canonical Mueller matrix $\mathbf{M}_{\Delta nd}$ with rank $\mathbf{H}(\mathbf{M}_{\Delta nd}) = 3$. $E_{\Delta nd}$ touches the unit sphere at a single point (Ossikovski et al. 2013).

10.6.4.2 Type-II

In this case, $a_2 \neq a_0$, so that $E_{\Delta nd}$ has semiaxes $(1/3, a_2/\sqrt{3}a_0, a_2/\sqrt{3}a_0)$ and touches the unit sphere at the single point $(1,0,0)$ (Figure 10.13b). Ellipsoids $E_{\Delta P}$ and $E_{\Delta D}$ touch the sphere at respective single points (case 3: $c = 1$, in Section 5.15.1).

10.6.5 rank $\mathbf{H} = 4$

In this case the system is necessarily of type-I. Provided that $P < 1$ and $D < 1$, all parameters d_i $(i = 0,1,2,3)$ are different, so that the length of the three semiaxes of $E_{\Delta d}$ are smaller than one and hence, $E_{\Delta d}$ does not touch the unit sphere at any point. Furthermore, ellipsoids $E_{\Delta P}$, and $E_{\Delta D}$ do not touch the unit sphere at any point either (case 4: $c = 0$, in Section 5.15.1).

In summary, leaving aside the special cases of polarizers and analyzers, if the canonical ellipsoid E_{Δ} of $\mathbf{M}$ touches the Poincaré sphere at two or more points or does not touch it at all, then $\mathbf{M}$ is type-I, whereas when E_{Δ} touches the sphere at a single point, then $\mathbf{M}$ is type-II. The same holds for ellipsoids $E_{\Delta P}$ and $E_{\Delta D}$, which, provided that $P < 1$ and $D < 1$, correspond to a type-II Mueller matrix when they have a single contact point with the surface of the unit sphere, and correspond to a type-I Mueller matrix in the remaining cases. Table 10.3 summarizes the case analysis performed above.

10.6.6 Poincaré Sphere Mapping by Singular Mueller Matrices

As seen in previous sections, singular Mueller matrices exhibit peculiar features that can be studied by means of their normal forms $\mathbf{M} = \mathbf{M}_{J2}\mathbf{M}_{\Delta}\mathbf{M}_{J1}$ (see Section 5.7).

In spite of the practical difficulties in determining if a measured Mueller matrix can be considered as being singular or not because of the limited precision of the experimental data, Mueller matrices with $\det \mathbf{M} = 0$ play an important role in Mueller algebra and exhibit peculiar features that can be studied by means of their normal forms $\mathbf{M} = \mathbf{M}_{J2}\mathbf{M}_{\Delta}\mathbf{M}_{J1}$ (see Section 5.7). Given a Mueller matrix with $\det \mathbf{M} = 0$ and $m_{00} > 0$ (the trivial limiting case $\mathbf{M} = \mathbf{0}$ is disregarded), one of the following polarimetric situations takes place (D, D_1 and D_2 denote the diattenuations of $\mathbf{M}$, $\mathbf{M}_{J1}$ and $\mathbf{M}_{J2}$ respectively):

TABLE 10.3
Summary of the Topological Properties of the Canonical Ellipsoid E_Δ of a Mueller Matrix in Terms of the Rank of its Associated Covariance Matrix **H**(**M**)

	Type	Parameters of E_Δ	Number of free parameters	Topological properties of E_Δ
$P = 1$ and $D < 1$	Type-I Depolarizing polarizer	$\hat{d}_k = 0$ $(k = 1,2,3)$	0	$E_{\Delta d}$ degenerates into the single point $(0,0,0)$. $E_{\Delta P}$ degenerates into a single point on the surface of the unit sphere. $E_{\Delta D}$ degenerates into a single point inside the unit sphere.
$P < 1$ and $D = 1$	Type-I Depolarizing-analyzer	$\hat{d}_k = 0$ $(k = 1,2,3)$	0	$E_{\Delta d}$ degenerates into the single point $(0, 0, 0)$. $E_{\Delta P}$ degenerates into a single point inside the unit sphere. $E_{\Delta D}$ degenerates into a single point on the surface of the unit sphere.
$P = 1$ and $D = 1$	Type-I Pure polarizer	$\hat{d}_i = 1$ $(i = 0,1,2,3)$	0	$E_{\Delta d}$ is the unit sphere. $E_{\Delta P}$ and $E_{\Delta D}$ degenerate into respective single points on the surface of the unit sphere.
rank**H** $= 1$ $P < 1, D < 1$	Type-I Non-depolarizing	$\hat{d}_i = 1$ $(i = 0,1,2,3)$	0	$E_{\Delta d}$, $E_{\Delta P}$ and $E_{\Delta D}$ coincide with the unit sphere.
rank**H** $= 2$ $P < 1, D < 1$	Type-I depolarizing	$\hat{d}_0 = \hat{d}_1 = 1$ $\hat{d}_2 = \hat{d}_3 < 1$	1	$E_{\Delta d}$ touches the unit sphere at two antipodal points. $E_{\Delta P}$ and $E_{\Delta D}$ touch the sphere at respective pairs of points.
	Type-II depolarizing	$a_2 = a_0$	0	$E_{\Delta nd}$ touches the unit sphere at the single point $(1,0,0)$. $E_{\Delta P}$ and $E_{\Delta D}$ touch the sphere at respective single points.
rank**H** $= 3$ $P < 1, D < 1$	Type-I depolarizing	$\hat{d}_0 - \hat{d}_1 =$ $= \hat{d}_2 - \hat{d}_3$	2	$E_{\Delta d}$ $E_{\Delta P}$ and $E_{\Delta D}$ do not touch the unit sphere at any point.
	Type-II depolarizing	$a_2 \neq a_0$	1	$E_{\Delta nd}$ touches the unit sphere at the single point $(1,0,0)$. $E_{\Delta P}$ and $E_{\Delta D}$ touch the sphere at respective single points.
rank**H** $= 4$ $P < 1$, $D < 1$	Type-I depolarizing	$\hat{d}_i \neq \hat{d}_j$	3	$E_{\Delta d}$ $E_{\Delta P}$ and $E_{\Delta D}$ do not touch the unit sphere at any point.

(1) $\mathbf{M}_{J1}$ is singular while $\mathbf{M}_{J2}$ is non-singular, which occurs if and only if $D = D_1 = 1$ and $D_2 = P < 1$, in which case **M** is a depolarizing analyzer, hence type-I, of the form $\mathbf{M} = \hat{\mathbf{M}}_{D2}\mathbf{M}_{\Delta d}\hat{\mathbf{M}}_{\hat{D}1}$ with $\mathbf{M}_{\Delta d} = \mathrm{diag}(d_0, 0, 0, 0)$.

(2) $\mathbf{M}_{J2}$ is singular while $\mathbf{M}_{J1}$ is non-singular, which occurs if and only if $D = D_1 < 1$ and $D_2 = P = 1$, in which case **M** is a depolarizing polarizer, hence type-I, of the form $\mathbf{M} = \hat{\mathbf{M}}_{\hat{D}2}\mathbf{M}_{\Delta d}\hat{\mathbf{M}}_{D1}$ with $\mathbf{M}_{\Delta d} = \mathrm{diag}(d_0, 0, 0, 0)$.

(3) Both $\mathbf{M}_{J1}$ and $\mathbf{M}_{J2}$ are singular, which occurs if and only if $D_1 = D = D_2 = P = 1$, so that $\mathbf{M}$ is pure and corresponds to a perfect polarizer, in general non-normal.
(4) Both $\mathbf{M}_{J1}$ and $\mathbf{M}_{J2}$ are non-singular, while $\mathbf{M}_\Delta = \mathbf{M}_{\Delta d}$ with $0 = d_3 \le d_2 \le d_1 < d_0$ (that is, the volume of the canonical ellipsoid is zero, so that the volume coefficient, $V = \varepsilon \hat{d}_1 \hat{d}_2 \hat{d}_3$, defined in Section 6.4.1.1 equals zero. Therefore, $\mathbf{M}$ is a type-I singular depolarizer.
(5) Both $\mathbf{M}_{J1}$ and $\mathbf{M}_{J2}$ are non-singular, while $\mathbf{M}_\Delta = \mathbf{M}_{\Delta nd}$ with $V = 0$ (i.e., $a_2 = 0$) so that $\mathbf{M}$ is a type-II singular depolarizer.

A complete geometric representation and classification of singular Mueller matrices through their Poincaré sphere mapping is described in Tables 10.4 and 10.5

10.7 Five-Vector Representation

The arrow decomposition $\mathbf{M} = m_{00}\, \mathbf{M}_{RO}\, \hat{\mathbf{M}}_A\, \mathbf{M}_{RI}$ of a pure or depolarizing Mueller matrix $\mathbf{M}$ where

$$\mathbf{M}_A = \mathbf{M}_{RO}^T\, \mathbf{M}\mathbf{M}_{RI}^T = m_{00}\begin{pmatrix} 1 & \mathbf{D}_A^T \\ \mathbf{P}_A & \mathrm{diag}(a_1, a_2, \varepsilon a_3) \end{pmatrix} \quad \begin{bmatrix} \mathbf{D}_A = \mathbf{m}_{RI}\,\mathbf{D} & a_1 \ge a_2 \ge a_3 \\ \mathbf{P}_A = \mathbf{m}_{RO}^T\,\mathbf{P} & \varepsilon \equiv \det \mathbf{m}/|\det \mathbf{m}| \end{bmatrix} \tag{10.22}$$

provides a straightforward representation of $\mathbf{M}$ in terms of the scalar quantity m_{00} (mean intensity coefficient of both $\mathbf{M}$ and $\hat{\mathbf{M}}_A$) and five three-dimensional real vectors whose absolute values are less or equal to one, namely the *Poincaré retardance vectors* $\bar{\mathbf{R}}_I, \bar{\mathbf{R}}_O$ (determining $\mathbf{M}_{RI}$ and $\mathbf{M}_{RO}$ respectively; see Section 4.2.2.1, Eq. (4.26)); the *diattenuation* and *polarizance vectors* $\mathbf{D}$ and $\mathbf{P}$ (or alternatively, $\mathbf{D}_A = \mathbf{m}_{RI}^T\mathbf{D}$ and $\mathbf{P}_A = \mathbf{m}_{RO}^T\mathbf{P}$) and the *singular vector* $\mathbf{P}_S \equiv (a_1, a_2, \varepsilon a_3)^T/\sqrt{3}$.

Recall that, as seen in Section 4.2.2.1, $\bar{\mathbf{R}}_I = (\Delta_I/\pi)\mathbf{u}_{RI}$ and $\bar{\mathbf{R}}_O = (\Delta_O/\pi)\mathbf{u}_{RO}$, where $\mathbf{u}_{RI}$ and $\mathbf{u}_{RO}$ are the unit Poincaré vectors of the fast eigenstates of $\mathbf{M}_{RI}$ and $\mathbf{M}_{RO}$ respectively, while Δ_I and Δ_O are the corresponding retardances. Thus, given a measured Mueller matrix $\mathbf{M}$, its associated vectors $\bar{\mathbf{R}}_I$, $\bar{\mathbf{R}}_O$, $\mathbf{D}$, $\mathbf{P}$ and $\mathbf{P}_S$ provide a complete and meaningful representation of $\hat{\mathbf{M}}$ (Figure 10.14a). Observe that, while the above five-vector representation can be performed directly for any given $\mathbf{M}$, the arbitrary choice of a set of five vectors $\bar{\mathbf{R}}_I$, $\bar{\mathbf{R}}_O$, $\mathbf{D}$, $\mathbf{P}$, $\mathbf{P}_S$ does not always correspond to a Mueller matrix (because of the covariance conditions).

The statistical parameterization of $\mathbf{M}_A$ (see Section 5.13.2) provides an alternative way to represent $\mathbf{M}$ by means of the scalar parameter m_{00} together with the five *structural vectors* $\bar{\mathbf{R}}_I$, $\bar{\mathbf{R}}_O$, $\mathbf{D}_\Gamma$, $\mathbf{P}_\Gamma$, $\boldsymbol{\sigma}$ (Figure 10.14b), where

$$\begin{gathered} \mathbf{D}_\Gamma \equiv (\boldsymbol{\mu}+\boldsymbol{\nu})/2 \qquad \mathbf{P}_\Gamma \equiv (\boldsymbol{\mu}-\boldsymbol{\nu})/2 \qquad \boldsymbol{\sigma} \equiv \left(\sigma_1^2, \sigma_2^2, \sigma_3^2\right)^T/\sigma_0^2 \\ \left[m_{00} = \sigma_0^2 + \sigma_1^2 + \sigma_2^2 + \sigma_3^2 = \sigma_0^2(1+|\boldsymbol{\sigma}|) \qquad \sigma_0^2 = \mathrm{tr}\,\mathbf{M}_A \ge \sigma_1^2 \ge \sigma_2^2 \ge \sigma_3^2\right] \end{gathered} \tag{10.23}$$

It is remarkable that for any arbitrary set of structural vectors satisfying $|\mathbf{D}|_\Gamma + |\mathbf{P}|_\Gamma \le 1$ (see Section 5.13.3) the corresponding matrix $\mathbf{M}$ satisfies the covariance conditions to be a

TABLE 10.4
Classification and Geometric Properties of Polarizers and Analyzers

	Polarizance and diattenuation	Mueller matrix	Forward ellipsoid $E_{\Delta P}$	Canonical ellipsoid E_{Δ}	Reverse ellipsoid $E_{\Delta P}$
Depolarizing polarizer	$P=1$ $D<1$	$\mathbf{M}_{\hat{P}D}$ $m_{00}\begin{pmatrix} 1 & \mathbf{D}^T \\ \hat{\mathbf{P}} & \hat{\mathbf{P}}\otimes\mathbf{D}^T \end{pmatrix}$	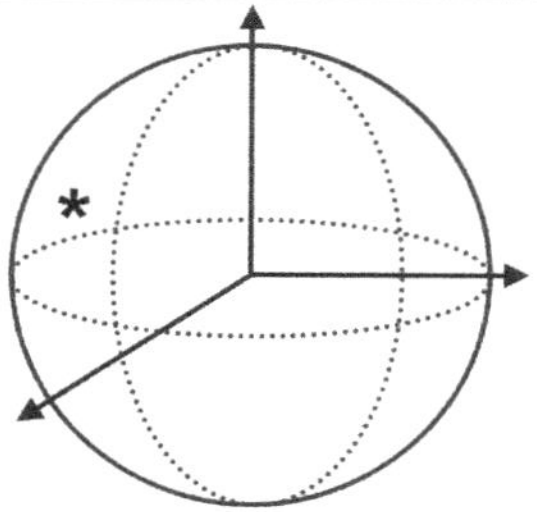	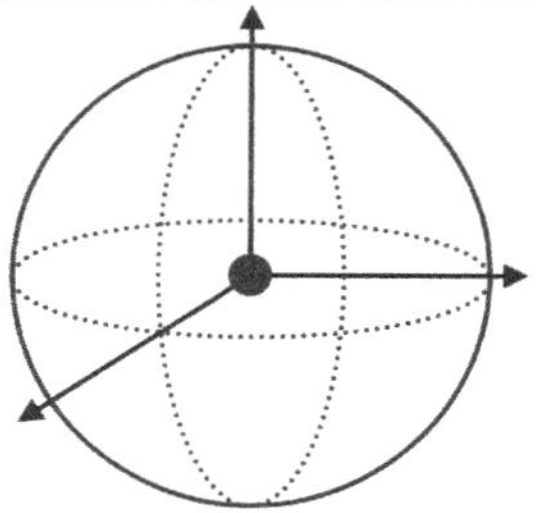	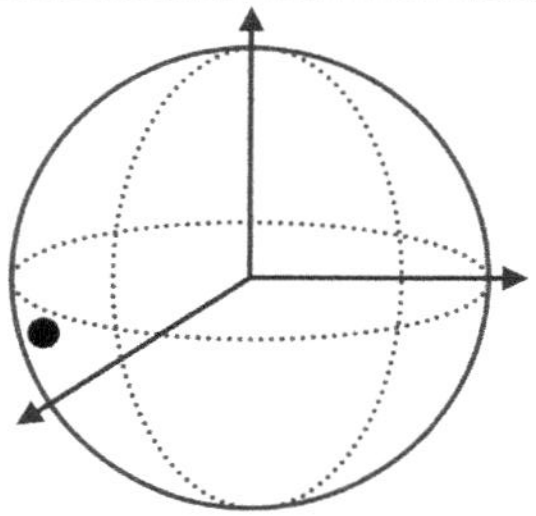
Depolarizing analyzer	$P<1$ $D=1$	$\mathbf{M}_{P\hat{D}}$ $m_{00}\begin{pmatrix} 1 & \hat{\mathbf{D}}^T \\ \mathbf{P} & \mathbf{P}\otimes\hat{\mathbf{D}}^T \end{pmatrix}$	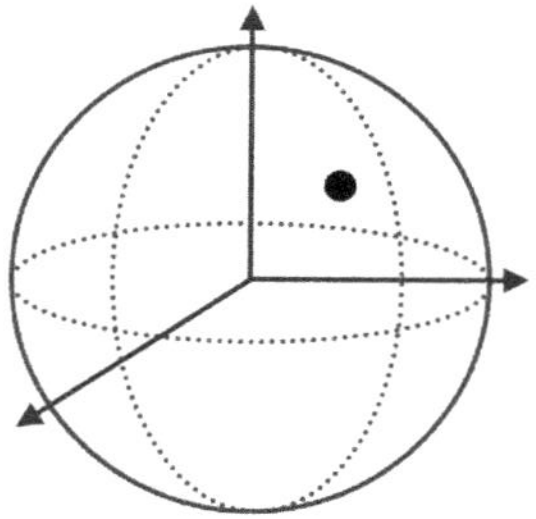	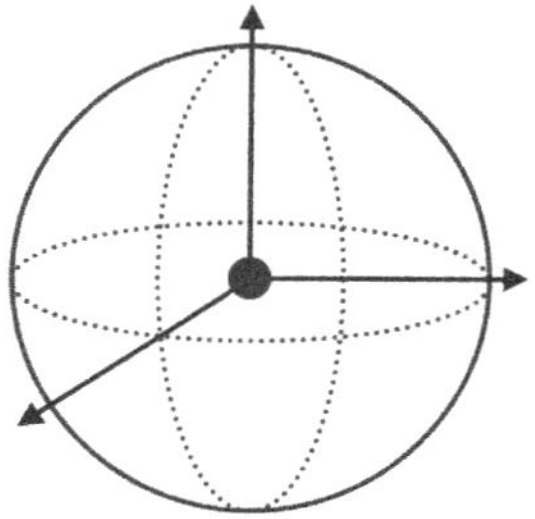	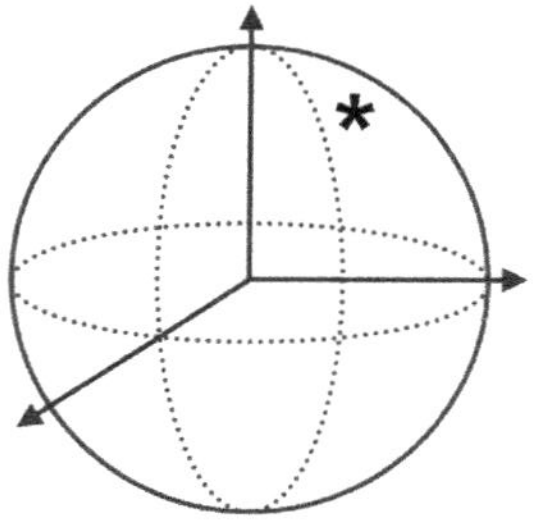
Perfect polarizer	$P=1$ $D=1$	$\mathbf{M}_{\hat{P}\hat{D}}$ $m_{00}\begin{pmatrix} 1 & \hat{\mathbf{D}}^T \\ \hat{\mathbf{P}} & \hat{\mathbf{P}}\otimes\hat{\mathbf{D}}^T \end{pmatrix}$	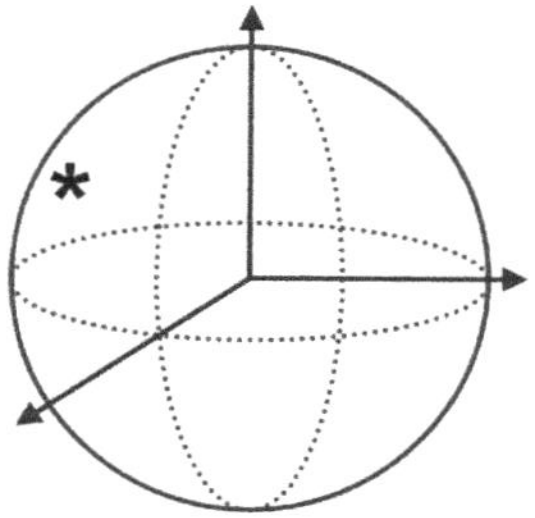	Undetermined	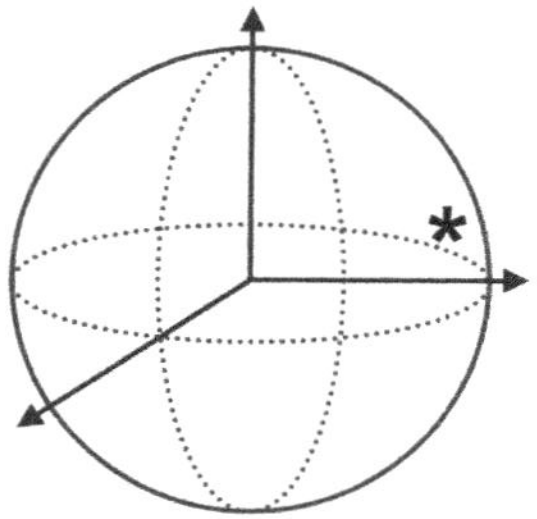

Source: Gil et al. 2016.
Note: Points located on the surface of the unit sphere are represented by asterisks (*), while points inside the unit sphere are represented by dots (.).

TABLE 10.5
Classification and Geometric Properties of Singular Depolarizers

	Polarizance and diattenuation	Canonical depolarizer	Forward ellipsoid $E_{\Delta P}$	Canonical ellipsoid E_{Δ}	Reverse ellipsoid $E_{\Delta P}$
Type-I Singular depolarizer	$P<1$ $D<1 \ \det \mathbf{M}_{\Delta d}=0$	$\mathbf{M}_{\Delta d}$ $\begin{pmatrix} d_0 & 0 & 0 & 0 \\ 0 & d_1 & 0 & 0 \\ 0 & 0 & d_2 & 0 \\ 0 & 0 & 0 & 0 \end{pmatrix}$ $d_2 \le d_1 < d_0$	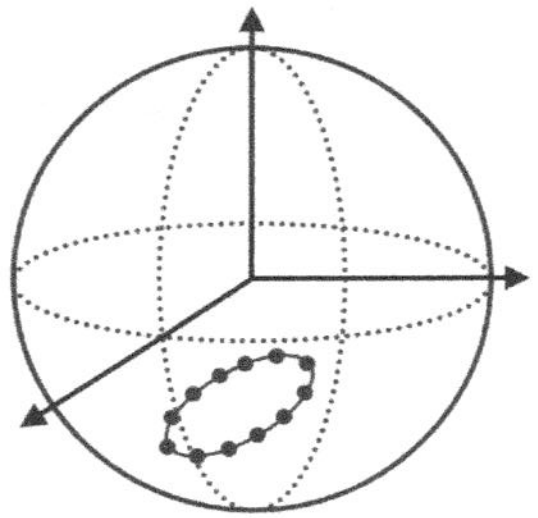	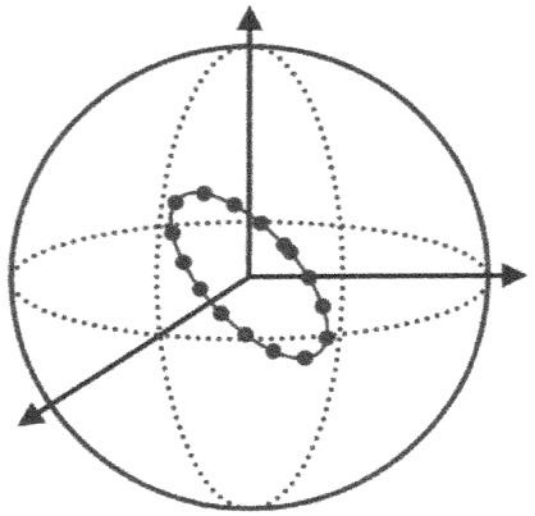	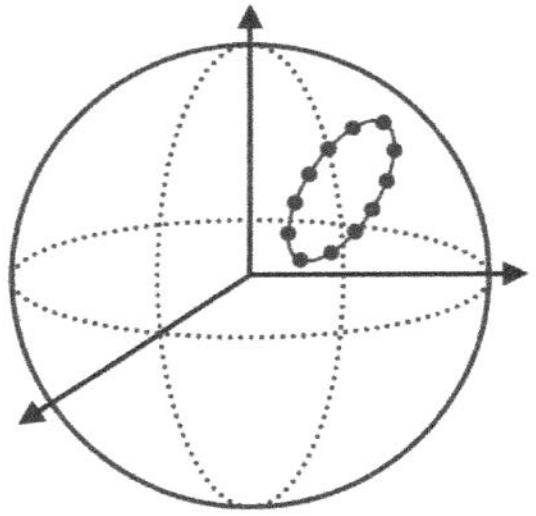
		$\mathbf{M}_{\Delta d}$ $\begin{pmatrix} d_0 & 0 & 0 & 0 \\ 0 & d_0 & 0 & 0 \\ 0 & 0 & d_2 & 0 \\ 0 & 0 & 0 & 0 \end{pmatrix}$ $d_2 < d_0$	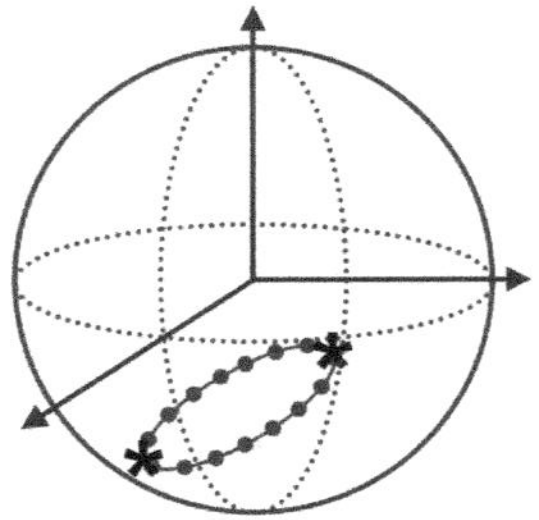	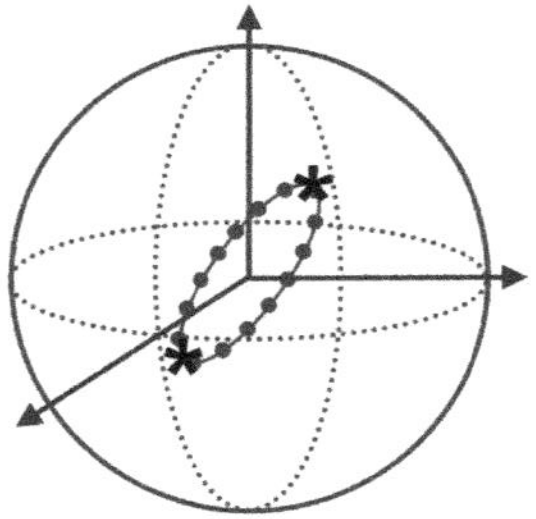	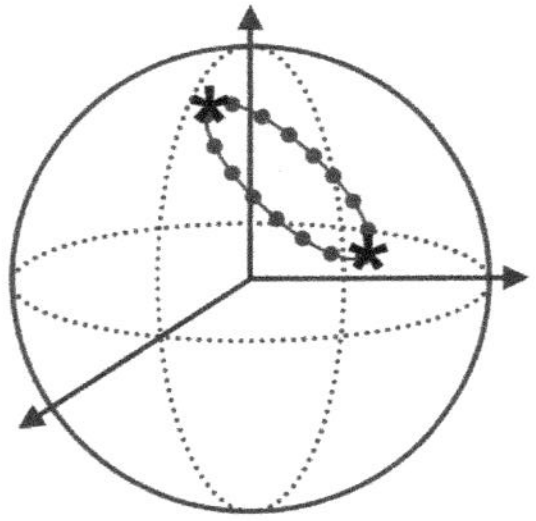
		$\mathbf{M}_{\Delta d}$ $\begin{pmatrix} d_0 & 0 & 0 & 0 \\ 0 & d_1 & 0 & 0 \\ 0 & 0 & 0 & 0 \\ 0 & 0 & 0 & 0 \end{pmatrix}$ $d_1 < d_0$	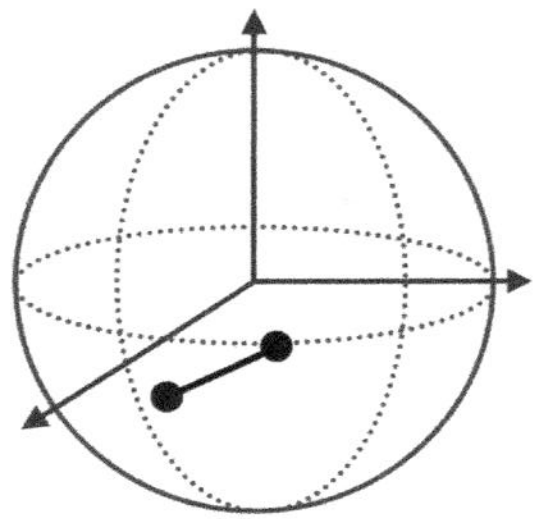	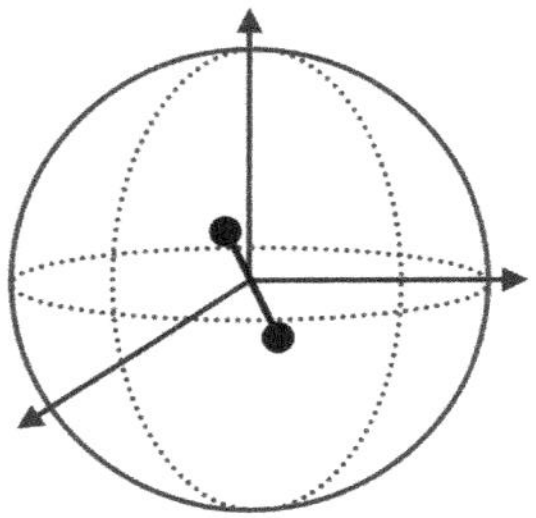	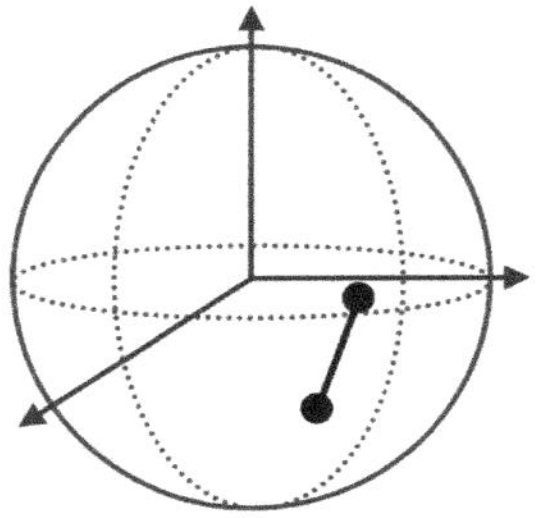

TABLE 10.5 (Continued)
Classification and Geometric Properties of Singular Depolarizers

	Polarizance and diattenuation	Canonical depolarizer	Forward ellipsoid $E_{\Delta P}$	Canonical ellipsoid E_{Δ}	Reverse ellipsoid $E_{\Delta P}$
		$\mathbf{M}_{\Delta d}$ $d_0\begin{pmatrix}1&0&0&0\\0&1&0&0\\0&0&0&0\\0&0&0&0\end{pmatrix}$	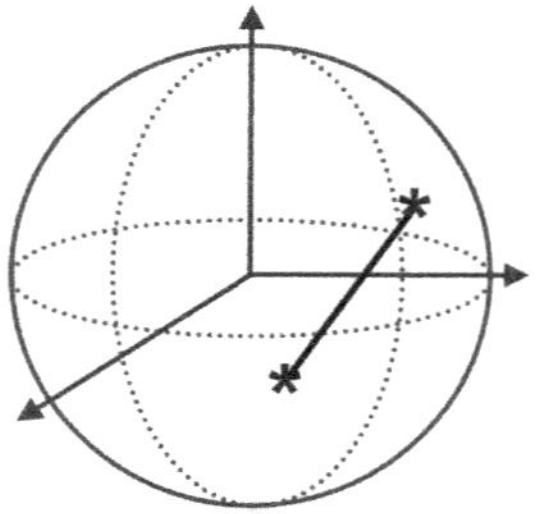	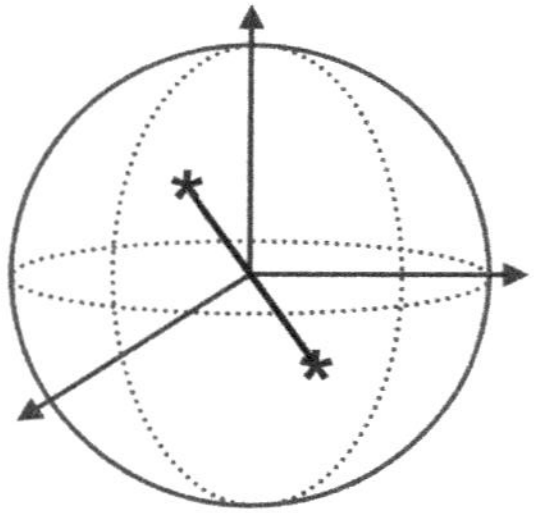	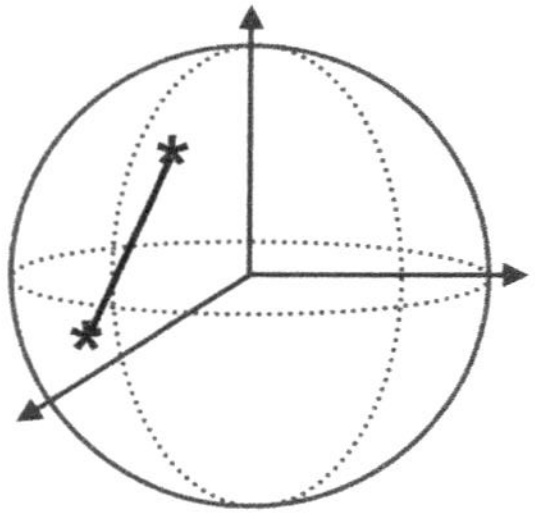
		$\mathbf{M}_{\Delta 0}$ $\begin{pmatrix}d_0&0&0&0\\0&0&0&0\\0&0&0&0\\0&0&0&0\end{pmatrix}$	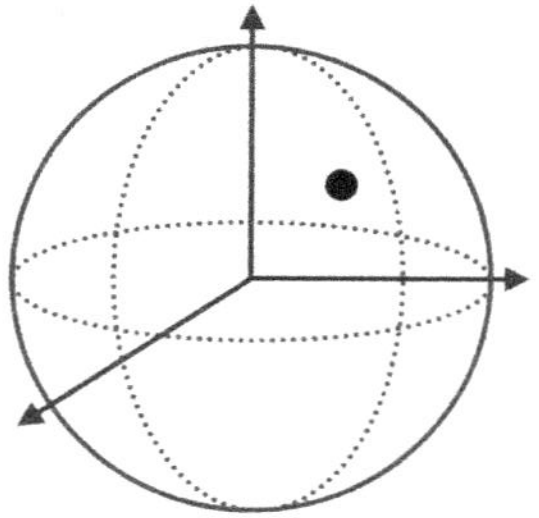	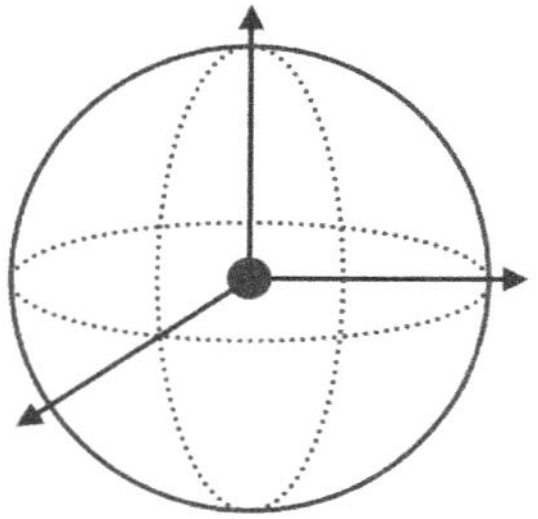	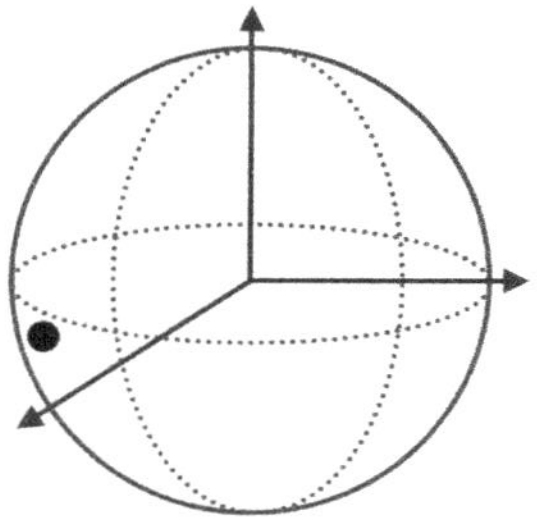
Type-II Singular depolarizer	$P<1$ $D<1$ $\det \mathbf{M}_{\Delta nd}=0$	$\mathbf{M}_{\Delta nd}$ $\frac{m_{00}}{2}\begin{pmatrix}2&-1&0&0\\-1&0&0&0\\0&0&0&0\\0&0&0&0\end{pmatrix}$	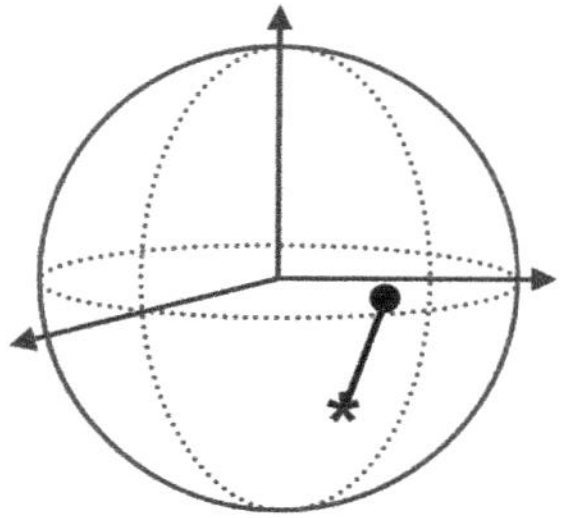	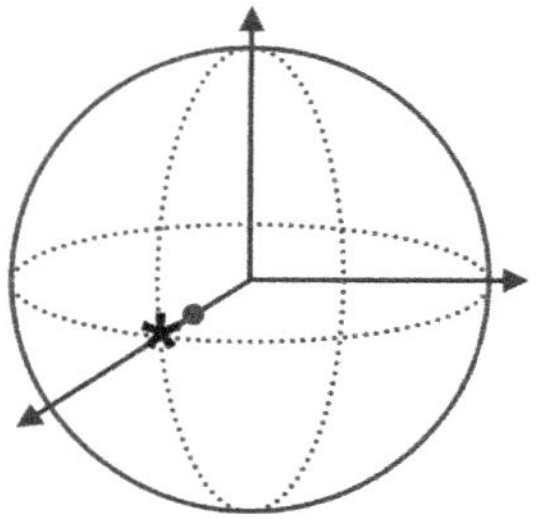	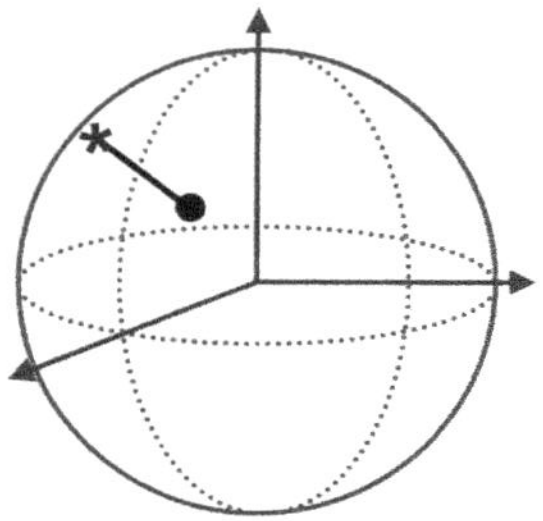

Source: Gil et al. 2016.

Note: Points located on the surface of the unit sphere are represented by asterisks (*), while points inside the unit sphere are represented by dots (.).

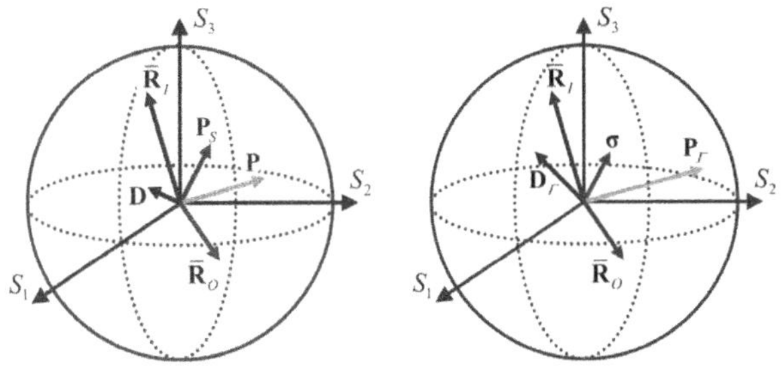

FIGURE 10.14
(a) Five-vector representation of a depolarizing Mueller matrix **M**: entrance and exit Poincaré retardance vectors $\bar{\mathbf{R}}_I$, $\bar{\mathbf{R}}_O$; diattenuation, **D**; polarizance, **P** and singular vector $\mathbf{P}_S$. (b) Representation of **M** through the five structural vectors $\bar{\mathbf{R}}_I$, $\bar{\mathbf{R}}_O$, $\mathbf{D}_\Gamma$, $\mathbf{P}_\Gamma$, $\boldsymbol{\sigma}$.

Mueller matrix, and therefore a given 4×4 real matrix **M** is a Mueller matrix if and only if its associated vectors $\mathbf{D}_\Gamma$ and $\mathbf{P}_\Gamma$ satisfy $|\mathbf{D}|_\Gamma + |\mathbf{P}|_\Gamma \leq 1$. Regarding the remaining structural vectors $\bar{\mathbf{R}}_I, \bar{\mathbf{R}}_O, \boldsymbol{\sigma}$, whose absolute values are less or equal to one, they are not limited by any additional restriction beyond the fact that, from the very definition of $\boldsymbol{\sigma}$, $\sigma_0^2 \geq \sigma_1^2 \geq \sigma_2^2 \geq \sigma_3^2$, together with the passivity constraint $|\boldsymbol{\sigma}| \leq 1/[\sigma_0^2(1+Q)] - 1$, with $Q \equiv \max(P, D)$.

10.8 Geometric View of Depolarization, Diattenuation and Polarizance

Given a normalized Mueller matrix $\hat{\mathbf{M}}$, depolarization, diattenuation, polarizance and retardance properties are related to certain geometric features as described below.

10.8.1 Depolarization

From the point of view of the five-vector approach, the depolarization, measured through the degree of polarimetric purity (or depolarization index) P_Δ, is independent of $\bar{\mathbf{R}}_I$ and $\bar{\mathbf{R}}_O$, and is given by the following weighted quadratic average of the absolute vales of **D**, **P** and $\mathbf{P}_S$:

$$P_\Delta = \sqrt{D^2/3 + P^2/3 + P_S^2} \tag{10.24}$$

In the case of a pure Mueller matrix, $P_\Delta = 1$ and $\mathbf{P}_A = \mathbf{D}_A = (D, 0, 0)^T$.

If $P < 1$, then $P_\Delta < 1$ is satisfied when the forward ellipsoid $E_{\Delta P}$ and the canonical ellipsoid E_Δ do not touch the unit sphere.

If $D < 1$, then $P_\Delta < 1$ is satisfied when the reverse ellipsoid $E_{\Delta D}$ and the canonical ellipsoid E_Δ do not touch the unit sphere.

If $P = 1$, then the forward ellipsoid $E_{\Delta P}$ degenerates into a single point touching the unit sphere, and $P_\Delta < 1$ is satisfied when the reverse ellipsoid $E_{\Delta D}$ (single point) and the canonical ellipsoid E_Δ do not touch the unit sphere.

If $D = 1$, then the reverse ellipsoid $E_{\Delta D}$ degenerates into a single point touching the unit sphere, and $P_\Delta < 1$ is satisfied when the forward ellipsoid $E_{\Delta P}$ (single point) and the canonical ellipsoid E_Δ do not touch the unit sphere.

10.8.2 Polarizance and Diattenuation

If $P > 0$, the center of the forward ellipsoid $E_{\Delta P}$ is displaced with respect to the origin $(0,0,0)$ of the Poincaré reference frame. Analogously, when $D > 0$, the center of the reverse ellipsoid $E_{\Delta D}$ is displaced with respect to the origin $(0,0,0)$. Moreover, by considering the five-vector approach, polarizance P and diattenuation D of $\mathbf{M}$ are respectively given by $P = |\mathbf{P}|$ and $D = |\mathbf{D}|$.

10.9 Geometric Representation of Nondepolarizing Mueller Matrices

10.9.1 Two-Vector Representation

In accordance with the analysis performed in Section 4.4, the polar decomposition $\hat{\mathbf{M}}_J = \mathbf{M}_R \hat{\mathbf{M}}_D = \hat{\mathbf{M}}_P \mathbf{M}_R$ of a normalized pure Mueller matrix $\hat{\mathbf{M}}_J$ leads to the equality $\mathbf{P} = \mathbf{m}_R \mathbf{D}$, and therefore $\hat{\mathbf{M}}_J$ is completely characterized by the diattenuation vector $\mathbf{D}$ and the Poincaré retardance vector $\bar{\mathbf{R}}$, or alternatively by $\mathbf{P}$ and $\bar{\mathbf{R}}$ (Gil and San José 2016b). In the special case where $D = 0$, then $\mathbf{P} = \mathbf{D} = \mathbf{0}$ and consequently, $\mathbf{M}_J = m_{00}\mathbf{M}_R$, so that $\bar{\mathbf{R}}$ determines completely $\hat{\mathbf{M}}_J$, which corresponds to a pure retarder with mean intensity coefficient m_{00}.

Given $\bar{\mathbf{R}}$ and $\mathbf{D}$ (with $0 < D$), the direction of $\mathbf{P}$ is obtained by means of the rotation of $\mathbf{D}$ about the direction defined by $\bar{\mathbf{R}}$ by the angle (retardance) $\Delta = 2\arccos\{[\mathrm{tr}(\mathbf{M}_J\mathbf{M}_D^{-1})]/2 - 1\}$. If $D = 1$, then $\mathbf{M}_R$ is not completely determined, but can be chosen as having the minimum achievable retardance Δ_m, in which case $\bar{\mathbf{R}}$ is orthogonal to $\mathbf{D}$ and Δ_m is given by $\cos\Delta_m = \mathbf{P}^T\mathbf{D}$.

When $\bar{\mathbf{R}}$ and $\mathbf{D}$ are parallel, then necessarily $\mathbf{P} = \mathbf{D}$ and the system is a diattenuating retarder ($\mathbf{M}_J$ is a normal matrix). If, in addition, $D = 1$, then $\mathbf{M}_J$ has the form of a normal perfect polarizer ($\bar{\mathbf{R}} = \mathbf{0}$ and $\mathbf{M}_R = \mathbf{I}_4$).

In summary, the two-vector approach constitutes a meaningful geometric representation of $\hat{\mathbf{M}}_J$ by means of the pair of vectors $\mathbf{D}$ and $\bar{\mathbf{R}}$ (Gil and San José 2016b). Some examples of the two-vector representation of $\mathbf{M}_J$ are shown in Figure 10.15.

10.9.2 Ellipsoid of a Nondepolarizing Mueller Matrix

The geometric representation of depolarizing Mueller matrices through the characteristic ellipsoids developed in the previous sections is based only on considerations on the shape of the corresponding P-images, which are obtained by the mapping of totally polarized incident states. Any given pure Mueller matrix $\mathbf{M}_J$ is necessarily of type-I, with $\mathbf{M}_{\Delta d} = d_0\mathbf{I}_4$ so that, provided $\det\mathbf{M}_J > 0$, all ellipsoids $E_{\Delta P}$, $E_{\Delta D}$ and $E_{\Delta d}$, coincide with the unit sphere itself and, hence, do not provide information about the six characteristic parameters of $\hat{\mathbf{M}}_J$. Thus, the geometric parameterization of pure Mueller matrices (or of slightly depolarizing Mueller matrices in practice) requires a particular approach. For this purpose, Tudor and Manea (2011) considered the P-surfaces obtained through the mapping by $\mathbf{M}_J$ of P-spheres defined by incident states of polarization with a fixed degree of polarization $\mathcal{P}$ (with $0 < \mathcal{P} < 1$).

The polar decomposition $\mathbf{M}_J = \mathbf{M}_P\mathbf{M}_R = \mathbf{M}_R\mathbf{M}_D$ (with $\mathbf{M}_D = \mathbf{M}_R^T\mathbf{M}_P\mathbf{M}_R$) shows that any P-surface of $\mathbf{M}_J$ can be decomposed into two sequential steps given by the mapping by the

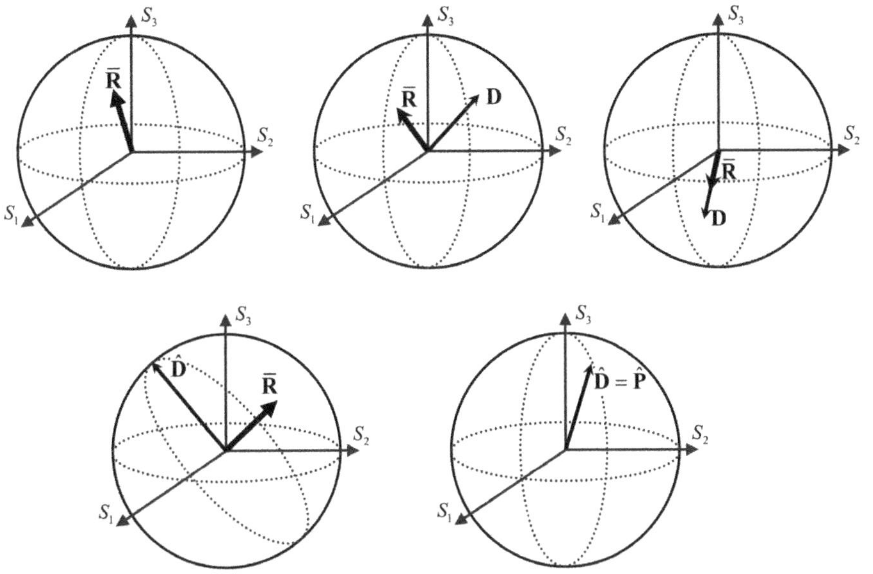

FIGURE 10.15
Two-vector representation of a pure Mueller matrix: (a) retarder; (b) non-normal diattenuator; (c) diattenuating retarder; (d) non-normal polarizer; (e) normal polarizer (Gil and San José 2016b).

retarder $\mathbf{M}_R$ followed by the mapping by the normal diattenuator $\mathbf{M}_P$ (or conversely, the mapping by $\mathbf{M}_D$ followed by the mapping by $\mathbf{M}_R$).

By using the *operatorial Pauli algebraic formulation*, Tudor and Manea (2011) found that the P-surface of $\mathbf{M}_D$ is an ellipsoid of revolution whose symmetry axis is just $\hat{\mathbf{D}} \equiv \mathbf{D}/D$. Furthermore, E_D is oblate with respect to $\hat{\mathbf{D}}$. This result is also obtained below by using the conventional Stokes-Mueller algebra.

Consider the incident states with Stokes vectors

$$\hat{\mathbf{s}} \equiv \begin{pmatrix} 1 \\ \mathcal{P}\mathbf{u} \end{pmatrix} \qquad |\mathbf{u}| = 1 \qquad 0 < \mathcal{P} < 1 \tag{10.25}$$

that describe the P-sphere corresponding to the fixed degree of polarization $\mathcal{P}$. By considering the general serial decomposition in Eq. (4.130), the action of $\mathbf{M}_J$ on $\hat{\mathbf{s}}$ can be formulated as

$$\begin{gathered} \hat{\mathbf{M}}_J\,\hat{\mathbf{s}} = \mathbf{M}_{R2}\,\hat{\mathbf{M}}_{DL0}\,\mathbf{M}_{R1}\,\hat{\mathbf{s}} = \begin{pmatrix} 1 & \mathbf{0}^T \\ \mathbf{0} & \mathbf{m}_{R2} \end{pmatrix} \begin{pmatrix} 1 & \mathbf{D}_0^T \\ \mathbf{D}_0 & \mathbf{m}_{D_0} \end{pmatrix} \begin{pmatrix} 1 & \mathbf{0}^T \\ \mathbf{0} & \mathbf{m}_{R1} \end{pmatrix} \begin{pmatrix} 1 \\ \mathcal{P}\mathbf{u} \end{pmatrix} \\ \left[\mathbf{D}_0 \equiv (D,0,0)^T \qquad \mathbf{m}_{D_0} \equiv \mathrm{diag}\left(1, \sin\kappa_D, \sin\kappa_D\right) \qquad \sin\kappa_D \equiv \sqrt{1-D^2} \right] \end{gathered} \tag{10.26}$$

where the action of retarder $\mathbf{M}_{R1}$ maintains unchanged the input P-sphere. Therefore, the P-surface of $\mathbf{M}_J$, being equal to that of $\mathbf{M}_{R2}\hat{\mathbf{M}}_{DL0}$, can be obtained through the successive actions of the horizontal normal linear diattenuator $\hat{\mathbf{M}}_{DL0}$ and the retarder $\mathbf{M}_{R2}$.

Let us first observe that when $D=1$ (i.e., when $\mathbf{M}_J$ is singular) the P-surfaces of $\mathbf{M}_J$ and $\mathbf{M}_J^T$ degenerate into single points, respectively given by the end points of vectors $\mathbf{m}_{R2}\mathbf{D}_0$ and $\mathbf{m}_{R1}^T\mathbf{D}_0$.

When $D<1$ (i.e., $\mathbf{M}_J$ is non-singular), the transformation of $\hat{\mathbf{s}}$ by $\hat{\mathbf{M}}_{DL0}$ can be expressed as

$$\hat{\mathbf{M}}_{DL0}\,\hat{\mathbf{s}} = \begin{pmatrix} 1 & \mathbf{D}_0^T \\ \mathbf{D}_0 & \mathbf{m}_{D_0} \end{pmatrix}\begin{pmatrix} 1 \\ \mathcal{P}\mathbf{u} \end{pmatrix} = \hat{\mathbf{M}}_{\Delta L0}\begin{pmatrix} 1 \\ \mathbf{u} \end{pmatrix}$$

$$\hat{\mathbf{M}}_{\Delta L0} \equiv \begin{pmatrix} 1 & \mathcal{P}\mathbf{D}_0^T \\ \mathbf{D}_0 & \mathcal{P}\mathbf{m}_{D_0} \end{pmatrix} = \mathbf{A}_{D_0}\hat{\mathbf{M}}_{\Delta\mathcal{P}}\mathbf{A}_{\mathcal{P}_0}$$

$$\mathbf{A}_{D_0} \equiv \begin{pmatrix} 1 & 0 & 0 & 0 \\ D & 1 & 0 & 0 \\ 0 & 0 & 1 & 0 \\ 0 & 0 & 0 & 1 \end{pmatrix} \quad \hat{\mathbf{M}}_{\Delta\mathcal{P}} \equiv \mathcal{P}\left(1-D^2\right)\begin{pmatrix} 1/\mathcal{P}\left(1-D^2\right) & 0 & 0 & 0 \\ 0 & 1 & 0 & 0 \\ 0 & 0 & 1 & 0 \\ 0 & 0 & 0 & 1 \end{pmatrix} \quad \mathbf{A}_{\mathcal{P}_0} \equiv \begin{pmatrix} 1 & \mathcal{P}D & 0 & 0 \\ 0 & 1 & 0 & 0 \\ 0 & 0 & 1 & 0 \\ 0 & 0 & 0 & 1 \end{pmatrix} \tag{10.27}$$

and by inserting on the right of $\hat{\mathbf{M}}_{DL0}$ the identity matrix in the form

$$\mathbf{I}_4 = \hat{\mathbf{M}}_{\mathcal{P}L0}^{-1}\hat{\mathbf{M}}_{\mathcal{P}L0} \qquad \hat{\mathbf{M}}_{\mathcal{P}L0} \equiv \begin{pmatrix} 1 & \mathcal{P}D & 0 & 0 \\ \mathcal{P}D & 1 & 0 & 0 \\ 0 & 0 & \sqrt{1-\mathcal{P}^2D^2} & 0 \\ 0 & 0 & 0 & \sqrt{1-\mathcal{P}^2D^2} \end{pmatrix} \tag{10.28}$$

we obtain

$$\hat{\mathbf{M}}_{DL0}\,\hat{\mathbf{s}} = \mathbf{A}_{D_0}\hat{\mathbf{M}}_{\Delta\mathcal{P}}\mathbf{A}_{\mathcal{P}_0}\hat{\mathbf{M}}_{\mathcal{P}L0}^{-1}\hat{\mathbf{M}}_{\mathcal{P}L0}\begin{pmatrix} 1 \\ \mathbf{u} \end{pmatrix} \tag{10.29}$$

The pure non-singular diattenuator $\hat{\mathbf{M}}_{\mathcal{P}L0}$ preserves the unit sphere constituted by the end points of the unit vectors $\mathbf{u}$, and therefore, the P-surface of $\mathbf{M}_{DL0}$ is equal to the P-image of the matrix $\mathbf{A}_{D_0}\hat{\mathbf{M}}_{\Delta\mathcal{P}}\mathbf{A}_{\mathcal{P}_0}\hat{\mathbf{M}}_{\mathcal{P}L0}^{-1}$, which has the form

$$\mathbf{A}_{D_0}\hat{\mathbf{M}}_{\Delta P}\mathbf{A}_{P_0}\hat{\mathbf{M}}_{PL0}^{-1} = \begin{pmatrix} 1 & 0 & 0 & 0 \\ q & \mathcal{P}g^2 & 0 & 0 \\ 0 & 0 & \mathcal{P}g & 0 \\ 0 & 0 & 0 & \mathcal{P}g \end{pmatrix} = \begin{pmatrix} 1 & 0 & 0 & 0 \\ h & 1 & 0 & 0 \\ 0 & 0 & 1 & 0 \\ 0 & 0 & 0 & 1 \end{pmatrix}\begin{pmatrix} 1 & 0 & 0 & 0 \\ 0 & \mathcal{P}g^2 & 0 & 0 \\ 0 & 0 & \mathcal{P}g & 0 \\ 0 & 0 & 0 & \mathcal{P}g \end{pmatrix} \tag{10.30}$$

$$\left[q \equiv \frac{D\left(1-\mathcal{P}^2\right)}{1-\mathcal{P}^2D^2} \qquad g \equiv \frac{1-D^2}{1-\mathcal{P}^2D^2}\right]$$

The P-surface (for a fixed value $0<\mathcal{P}<1$) is an ellipsoid of revolution E_D whose semiaxes (a,b,b) are given by

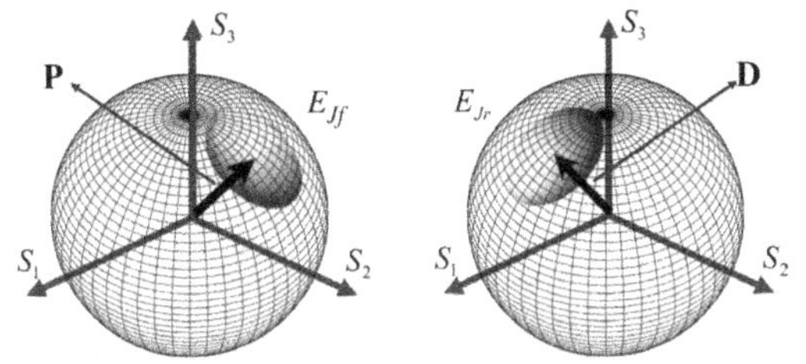

FIGURE 10.16
The axis of rotation of the *forward P-ellipsoid* E_{Jf} of $\mathbf{M}_J$ lies along the direction of the polarizance vector $\mathbf{P}$; E_{Jf} is oblate with respect to this direction. The *reverse P-ellipsoid* E_{Jr} has the same shape as E_{Jf}, but its symmetry axis lies along the direction of the diattenuation vector $\mathbf{D}$.

$$a = \mathcal{P}\left(1-D^2\right)/\left(1-\mathcal{P}^2D^2\right) \qquad b = \sqrt{\mathcal{P}a} \tag{10.31}$$

and whose geometric center is displaced at the distance $q \equiv D\,(1-\mathcal{P}^2)/(1-\mathcal{P}^2D^2)$ along the positive branch of the axis S_1 from the origin $(0,0,0)$ of the Poincaré reference frame $S_1S_2S_3$. Moreover, the eccentricity e of the ellipsoid E_D is $e = D\sqrt{(1-\mathcal{P}^2)/(1-D^2)}$. It is remarkable that a, b, q and e depend only on the diattenuation-polarizance D of $\mathbf{M}_J$ and on the fixed value chosen for $\mathcal{P}$.

The P-surface of $\mathbf{M}_J$ is obtained by applying the rotation given by $\mathbf{M}_{R2}$ to E_D and to the vector $(q,0,0)$ that runs from the origin $(0,0,0)$ to the center of E_D. The mapping results in a new ellipsoid E_{Jf}, called the *forward P-ellipsoid* of $\mathbf{M}_J$, whose shape is identical to that of E_D and whose center is at the same distance q from the origin. However, the axis of revolution of E_{Jf} (which again passes through the origin) is given by the direction of the polarizance vector $\mathbf{P}$ of $\mathbf{M}_J$ (Figure 10.16) which, in turn, is determined by the corresponding azimuth φ_P and ellipticity angle χ_P. Any fixed value $\mathcal{P}$ of the degree of polarization of the input P-sphere (with $0 < \mathcal{P} < 1$) as, for instance, $\mathcal{P} = 1/2$ or $\mathcal{P} = 1/\sqrt{2}$, can be chosen to determine a representative P-sphere. The larger the value of $\mathcal{P}$, the larger the volume of E_{Jf} and the smaller the shift q. As expected, in the limiting case where $\mathcal{P} = 1$, E_{Jf} coincides with the unit sphere and $q = 0$.

Since, in general, $\hat{\mathbf{M}}_J$ depends on up to six independent parameters, the set of three parameters D, φ_P and χ_P is not sufficient to achieve a complete geometric characterization of $\hat{\mathbf{M}}_J$, so that the *reverse P-ellipsoid* E_{Jr} of $\mathbf{M}_J$, defined as the forward P-ellipsoid of $\hat{\mathbf{M}}_J^T$, can additionally be considered. The shape of E_{Jr} is identical to those of E_{Jf} and E_D, with the same distance of the center from the origin, but its symmetry axis is given by the direction of the diattenuation vector $\mathbf{D}$ of $\hat{\mathbf{M}}_J$ (i.e., the polarizance vector of $\hat{\mathbf{M}}_J^T$) which in turn is determined by the corresponding azimuth φ_D and ellipticity angle χ_D. Nevertheless, the set of five parameters D, φ_P, χ_P, φ_D and χ_D is still insufficient for a complete geometric characterization of $\hat{\mathbf{M}}_J$, and it is necessary to consider at least one additional parameter. As shown in the previous subsection, the two-vector representation provides complete information on $\hat{\mathbf{M}}_J$, which can be considered as composed of the forward ellipsoid E_{Jf} (characterized by the diattenuation D and the angles φ_P, χ_P that determine the direction of the axis of symmetry of E_{Jf}), together with the three-dimensional vector $\bar{\mathbf{R}}$. In other words, the diattenuation vector $\mathbf{D}$ of the two-vector representation of $\hat{\mathbf{M}}_J$ can be straightforwardly geometrized by means of the ellipsoid E_{Jf}, providing a visual characterization of $\hat{\mathbf{M}}_J$.

10.10 Experimental Examples

To illustrate the applications of the geometric representation of Mueller matrices by means of the characteristic ellipsoids, namely the canonical ellipsoid together with the forward and reverse ellipsoids, this section is devoted to the analysis of several experimental depolarizing Mueller matrices taken from the literature and representing the polarimetric properties of different kinds of material samples.

Figure 10.17 shows the characteristic ellipsoids of a sample consisting of a rusty steel specimen whose type-I normalized Mueller matrix

$$\hat{\mathbf{M}} = \begin{pmatrix} 1.0000 & 0.0170 & 0.0203 & 0.0061 \\ -0.0199 & 0.6348 & 0.0092 & 0.0031 \\ -0.0076 & 0.0140 & -0.6112 & 0.0038 \\ -0.0057 & -0.0002 & -0.0058 & -0.4220 \end{pmatrix} \quad (10.32)$$

was measured by Le Roy-Bréhonnet and Le Jeune (1997).

Only small diattenuations and retardances are observed and the sample roughly behaves as an intrinsic (or diagonal) depolarizer. The characteristic ellipsoids $E_{\Delta P}$ and $E_{\Delta D}$ do not exhibit any significant displacements or distortions, so that they remain quite similar to $E_{\Delta d}$.

Figure 10.18 shows the characteristic ellipsoids of the Mueller matrix shown below, corresponding to an organic sample composed of a suspension of 2.0 μm polystyrene spheres in aqueous solution of glucose measured in backscattering geometry by Manhas et al. (2006).

$$\hat{\mathbf{M}} = \begin{pmatrix} 1.000 & -0.115 & -0.066 & 0.023 \\ -0.111 & 0.759 & -0.061 & -0.001 \\ -0.018 & 0.151 & -0.435 & -0.139 \\ -0.046 & 0.006 & 0.128 & -0.334 \end{pmatrix} \quad (10.33)$$

Both $E_{\Delta P}$ and $E_{\Delta D}$ exhibit misalignments and distortions with respect to $E_{\Delta d}$, but remain roughly centered.

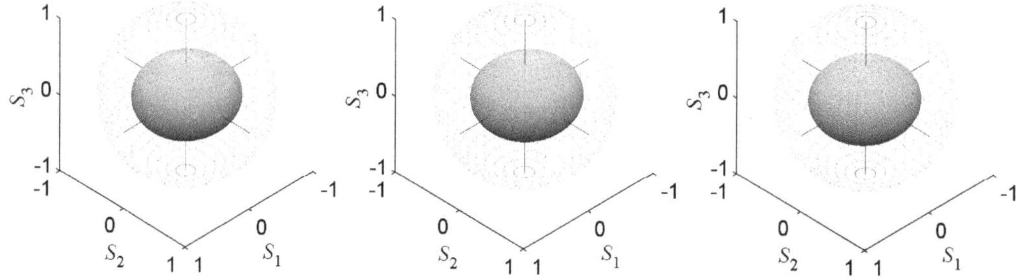

FIGURE 10.17
Characteristic ellipsoids $E_{\Delta d}$ (left), $E_{\Delta P}$ (middle) and $E_{\Delta D}$ (right) of an inorganic sample of rusty steel (Ossikovski et al. 2013).

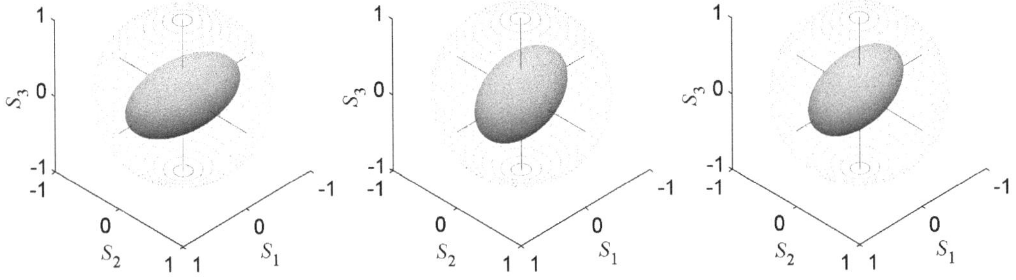

FIGURE 10.18
Characteristic ellipsoids $E_{\Delta d}$ (left), $E_{\Delta P}$ (middle) and $E_{\Delta D}$ (right) of an organic sample composed of a suspension of 2.0 μm polystyrene spheres in aqueous solution of glucose (Ossikovski et al. 2013).

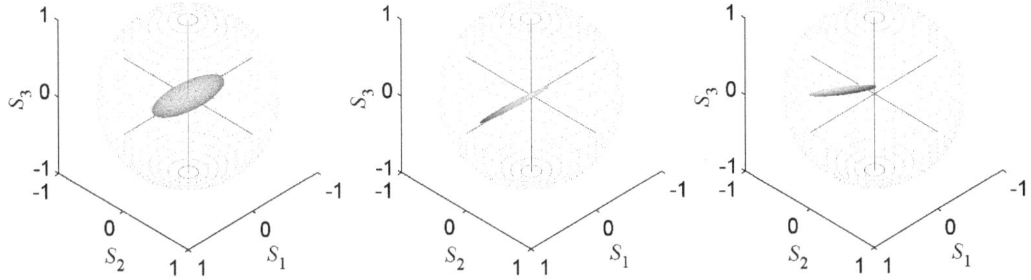

FIGURE 10.19
Characteristic ellipsoids $E_{\Delta d}$ (left), $E_{\Delta P}$ (middle) and $E_{\Delta D}$ (right) of a biological sample of cancerous tissue (Ossikovski et al. 2013).

Figure 10.19 shows the geometric representation of the experimental Mueller matrix

$$\mathbf{M}_C^{\text{corr}} = \begin{pmatrix} 0.91 & 0.48 & -0.22 & 0.13 \\ 0.51 & 0.52 & -0.11 & 0.16 \\ 0.05 & -0.01 & -0.10 & -0.07 \\ 0.06 & 0.02 & -0.08 & -0.08 \end{pmatrix} \tag{10.34}$$

corresponding to a biological sample of cancerous tissue measured by Pereda-Cubián et al. (2005). Very strong polarization anisotropies are observed, so that the characteristic ellipsoids $E_{\Delta P}$ and $E_{\Delta D}$ present important eccentricities, as well as distortions and misalignments with respect to the type-I canonical ellipsoid $E_{\Delta d}$.

These examples show how the polarizing and depolarizing properties of a Mueller matrix can be visually perceived with the help of simple geometrical representations.

11

Filtering of Measured Mueller Matrices

11.1 Introduction

Mueller polarimetry has a very wide scope of applications because it provides, through nondestructive technologies, experimental measures of Mueller matrices associated with many kinds of material samples and for very different measurement configurations (angles of incidence and observation, spectral profile of the light probe, spot size, measurement time, etc.). Unavoidably, such processes have a limited precision and, therefore, filtering techniques that lead to physically realizable and interpretable Mueller matrices are necessary and deserve particular consideration.

Furthermore, the light emerging from the sample under analysis exhibits total purity (degree of polarization equal to one) for measurement times smaller than the polarization time of the emerging light (Voipio, T. et al. 2010; Shevchenko et al. 2017), which in turn depends on the polarization time of the incident light on one hand, and the inhomogeneity, spectral response and other statistical aspects of the nature of the sample on the other. In many experimental conditions, the polarization time of light exiting the sample is much shorter than the measurement time, leading to effective depolarization properties that are encoded in the corresponding measured depolarizing Mueller matrices (i.e., with $P_{\Delta} < 1$). Consequently, it is necessary to have at hand appropriate criteria for filtering the measured Mueller matrices that allow for the reduction of errors and, therefore, obtaining results best fitted to physically realizable situations.

Without going into detail of the analysis of the specific sources of errors and of the particular approaches for data filtering, which constitute a wide specific corpus within the field of Polarimetry (Cloude 1989; Goldstein and Chipman 1990; Tyo 2000; Guyot et al. 2003; Aiello et al. 2006; Zallat et al. 2006; Boulvert et al. 2009; Anna et al. 2011; Ossikovski 2012a; Faisan et al. 2013; Ossikovski and Arteaga 2019b), this chapter is devoted to a brief discussion of the main procedures for filtering measured Mueller matrices.

The fundamental aspects of filtering are considered in Section 11.2. In general, when the measured Mueller matrix $\mathbf{M}_e$ exhibits weak depolarizing properties, the determination of the nondepolarizing Mueller matrix $\mathbf{M}_{est}$, that is, the closest estimate, in a given sense, to the experimental $\mathbf{M}_e$, becomes a particularly interesting and relevant matter in polarimetry. The concept of *integral decomposition* procedure for the determination of the pure Mueller matrix estimate of a sample, considered as nondepolarizing or weakly depolarizing but whose measured Mueller matrix is affected by errors, is studied in Section 11.3.

Regarding Mueller polarimetry and the subsequent filtering, it should be recalled that pure Mueller matrices depend on up to seven independent parameters, so that when

DOI: 10.1201/9780367815578-11

complete Mueller polarimetry is applied to nondepolarizing samples, within experimental error, the information encoded in the 16 parameters of the measured **M** involves certain redundancies (see Chapter 3). Therefore, in such cases the nondepolarizing Mueller matrix can be measured even through incomplete polarimetry (Arteaga and Ossikovski 2019; Ossikovski and Arteaga 2019a). Incomplete Mueller polarimetry is dealt with in Section 11.4 as a preliminary step to the description of some rank-reduction filtering procedures in Sections 11.5 and 11.6 that can be considered and implemented as an alternative to those discussed in Sections 11.2 and 11.3.

11.2 Covariance Filtering of Measured Mueller Matrices

A well-known criterion for filtering experimental Mueller matrices is directly derived from the covariance criterion; negative eigenvalues of the covariance matrix **H** associated with a measured Mueller matrix **M** are necessarily due to experimental errors or noise. Typically, the magnitudes of the normalized negative eigenvalues $\hat{\hat{\lambda}}_i$ are close to zero, so that these can be replaced by zero (hence forcing the fulfillment of the covariance criterion), and the filtered Mueller matrix $\mathbf{M}_f$ can be obtained from the modified covariance matrix $\mathbf{H}_f$ (*covariance filtering*). Furthermore, when the experimentalists, guided by their preliminary knowledge of the nature of the sample under measurement, expect that rank**H** has a particular value r (i.e., the sample should behave as a parallel combination of r pure components), the last $4-r$ normalized eigenvalues of $\mathbf{H} = m_{00}\,\mathbf{U}\,\mathrm{diag}(\hat{\lambda}_0, \hat{\lambda}_1, \hat{\lambda}_2, \hat{\lambda}_3)\,\mathbf{U}^\dagger$ will have magnitudes close to zero. Then, these can be replaced by zeros and a filtered $\mathbf{M}_f$ can be obtained from the recalculated $\mathbf{H}_f = m_{00}\,\mathbf{U}\,\mathrm{diag}(\hat{\lambda}_{f0}, \hat{\lambda}_{f1}, \hat{\lambda}_{f2}, \hat{\lambda}_{f3})\,\mathbf{U}^\dagger$ with up to three "filtered" eigenvalues equal to zero (*rank criterion*).

Consider now the specific (but common in practice) case where the sample is assumed nondepolarizing. The first step for filtering the measured Mueller matrix **M** consists of replacing negative eigenvalues of **H** by respective zeros; if the resulting $\mathbf{H}_f$ has only one significant normalized eigenvalue $\hat{\lambda}_{f0} \approx 1$ (and hence $\hat{\lambda}_{f1} \approx \hat{\lambda}_{f2} \approx \hat{\lambda}_{f3} \approx 0$), the filtering procedure is completed by recalculating the nondepolarizing estimate $\mathbf{H}_e$ as $\mathbf{H}_e = m_{00}\,\mathbf{U}\,\mathrm{diag}(1,0,0,0)\,\mathbf{U}^\dagger$. Nevertheless, it can occur that, in spite of the conviction that the sample is nondepolarizing, the value of $\hat{\lambda}_{f1}$ is not negligible. In that case, the following question arises: since the arbitrary decomposition provides infinite possible sets of parallel components of **M**, which is the pure Mueller matrix $\mathbf{M}_J$ with the largest coefficient $k_{\max}$ in the arbitrary decomposition of **M**? The response to this question comes from Eq. (7.8b), which allows for expressing the coefficient as

$$k_0 = 1\Big/\left[m_{00,0}\left(\hat{\mathbf{w}}_0^\dagger \mathbf{H}^- \hat{\mathbf{w}}_0\right)\right] \tag{11.1}$$

so that it is achieved when $\hat{\mathbf{w}}_0$ matches the eigenvector $\hat{\mathbf{u}}_0$ of **H** with largest eigenvalue, i.e., $k_{\max} = \hat{\lambda}_0$.

At this point, it is appropriate to consider the characteristic decomposition (see Section 7.6), and observe that it provides a privileged view of the structure of polarimetric purity, akin to the decomposition of a 2D polarization state into a pure and a fully random component. Due to the fact that **H** is a 4×4 covariance matrix, two additional peculiar mixed states, $\mathbf{M}_1$ and $\mathbf{M}_2$ appear combined with the pure characteristic component $\mathbf{M}_{J0}$ and the perfect depolarizer $\mathbf{M}_{\Delta 0}$ (see Section 7.6). Thus, both spectral and characteristic decompositions lead to the common nondepolarizing estimate $\mathbf{M}_{J0}$ (termed the *Cloude nondepolarizing estimate* and associated to the above mentioned eigenstate $\hat{\mathbf{u}}_0$). Nevertheless, as seen in Section 7.3, the coefficient $P_1 = \hat{\lambda}_0 - \hat{\lambda}_1$ of $\mathbf{M}_{J0}$ in the characteristic decomposition is just the first IPP of **M** and, in general, is different from its coefficient $\hat{\lambda}_0$ in the spectral decomposition. The origin of the difference between these coefficients comes from the fact that a portion of the pure component $\mathbf{M}_{J0}$ is included into the 2D, 3D and 4D depolarizers of the characteristic decomposition. In other words, if one is interested in eliminating the 4D, 3D and 2D polarimetric noise, the coefficient P_1 provided by the characteristic decomposition must be taken instead of $\hat{\lambda}_0$.

In summary, an objective procedure for the covariance filtering of the measured Mueller matrix is provided by the characteristic decomposition, which expresses the scaled structure of polarimetric purity as a combination of the pure component $\mathbf{M}_{J0}$, with coefficient P_1; a 2D depolarizer, with coefficient $P_2 - P_1$; a 3D depolarizer, with coefficient $P_3 - P_2$; and a perfect depolarizer, with coefficient $1 - P_3$.

The above *rank* and *characteristic* criteria correspond to physically realizable decompositions, provided the passivity criterion is additionally satisfied by the components; otherwise, the filtering procedure could still be useful, but lacks a rigorous physical support in the sense that the representative Mueller matrix may not correspond to the real structure of the sample under measurement. Therefore, a particular passive realization of the arbitrary decomposition (among the infinite possible ones) can be chosen instead of the spectral decomposition, while a parallel decomposition akin to the characteristic one can be formulated in terms of mixtures of the components of the said particular decomposition.

As an illustration of the above, consider the following normalized matrix

$$\hat{\mathbf{M}}_p = \begin{pmatrix} 1 & -0.0229 & 0.0027 & 0.0058 \\ -0.0186 & 0.9956 & -0.0361 & 0.0318 \\ -0.0129 & 0.0392 & 0.2207 & -0.9656 \\ 0.0014 & 0.0280 & 0.9706 & 0.2231 \end{pmatrix}$$

obtained from a transmission measurement of a chiral turbid phantom (Ghosh et al. 2008). Its covariance matrix

$$\hat{\mathbf{H}}_p = \begin{pmatrix} 0.4885 & -0.0083 + i0.0094 & 0.0066 - i0.0073 & 0.1109 - i0.4840 \\ -0.0083 - i0.0094 & 0.0022 & -0.0006 - i0.0013 & -0.0130 + i0.0067 \\ 0.0066 + i0.0073 & -0.0006 + i0.0013 & 0 & 0.0097 - i0.0065 \\ 0.1109 + i0.4840 & -0.0130 - i0.0067 & 0.0097 + i0.0065 & 0.5093 \end{pmatrix}$$

has for eigenvalues 0.9962, 0.0040, 0.0008 and –0.0010. Because of unavoidable experimental noise, one of the eigenvalues of $\hat{\mathbf{H}}_p$ is slightly negative. Consequently, $\hat{\mathbf{H}}_p$ is not positive-semidefinite and, according to Cloude's realizability criterion (*covariance criterion*), the experimental Mueller matrix $\hat{\mathbf{M}}_p$ is not physically realizable and needs to be filtered prior to applying any decomposition on it. The filtering procedure consists in keeping the positive eigenvalue terms only in the spectral decomposition of $\hat{\mathbf{M}}_p$,

$$\hat{\mathbf{M}}_{pf} = p_0\hat{\mathbf{M}}_{J0} + p_1\hat{\mathbf{M}}_{J1} + p_2\hat{\mathbf{M}}_{J2} = \begin{pmatrix} 1 & -0.0222 & 0.0023 & 0.0059 \\ -0.0193 & 0.9937 & -0.0358 & 0.0314 \\ -0.0127 & 0.0396 & 0.2204 & -0.9651 \\ 0.0017 & 0.0281 & 0.9689 & 0.2229 \end{pmatrix}$$

where the weights p_k are given by the remaining positive eigenvalues of $\hat{\mathbf{H}}_p$ and the pure components are obtained from the corresponding eigenvectors of $\hat{\mathbf{H}}_p$. Notice that the filtered matrix $\hat{\mathbf{M}}_{pf}$ features only very slight differences with respect to the original one, $\hat{\mathbf{M}}_p$, which points at the low noise level of the experiment.

A closer inspection of the list of eigenvalues of the covariance matrix shows that the second but largest one (0.0040) is already by more than two orders of magnitude smaller than the largest one (0.9962). This observation suggests that $\hat{\mathbf{M}}_p$ corresponds in practice to a pure (nondepolarizing) optical system and, to a very good approximation, can be filtered down to the Cloude estimate $\hat{\mathbf{M}}_{J0}$,

$$\hat{\mathbf{M}}_p \approx \hat{\mathbf{M}}_{J0} = \begin{pmatrix} 1 & -0.0211 & 0.0009 & 0.0091 \\ -0.0208 & 0.9988 & -0.0362 & 0.0320 \\ -0.0095 & 0.0395 & 0.2226 & -0.9739 \\ 0.0024 & 0.0281 & 0.9740 & 0.2238 \end{pmatrix}$$

in a fashion similar to that discussed in the example from Section 7.3. Applying the polar decomposition (taken in the order $\mathbf{M}_R\hat{\mathbf{M}}_D$) to the pure $\hat{\mathbf{M}}_{J0}$ provides the diattenuation and retardance values, $D = 0.0230$ and $R = 1.3462$, as well as their respective directions, $\hat{\mathbf{D}} = (-0.9179, 0.0411, 0.3946)^T$ and $\hat{\mathbf{R}} = (-0.9992, -0.0020, -0.0388)^T$. We see that the medium features very weak diattenuation but significant retardance, both being approximately directed along the S_1 axis of the Poincaré sphere, thus indicating the presence of essentially linear birefringence. However, the S_3-axis component of the retardance vector $\mathbf{R}$ is non-zero, showing that some (weak) circular birefringence is likewise present, in agreement with the phantom being chiral. Note that, since the measured sample physically represents a continuous medium, the differential formalism developed in Chapter 9 can be likewise applied to it with success.

This example shows that one can generally distinguish between two types of filtering. Thus, the physical realizability filtering reconstructs a noisy or biased experimental Mueller matrix $\mathbf{M}$ featuring negative covariance matrix eigenvalues to a physically realizable one, $\mathbf{M}_f$, having only positive covariance matrix eigenvalues. The second type of filtering, that can be termed *nondepolarizing estimation*, or most generally, *rank-reduction filtering*, replaces the original $\mathbf{M}$ by the largest-eigenvalue pure component $\mathbf{M}_{J0}$ of its spectral decomposition or, more generally, by a truncated sum of its spectral decomposition whose covariance-matrix rank is necessarily lower than that of the original $\mathbf{M}$ (see the example from Section 7.3). In addition to the Cloude estimate, other possible rank-reduction criteria are described in further sections, like the *constant-amplitude integral decomposition* (Section 11.3), the *virtual experiment estimate* (Section 11.5) and the *instrument-dependent method* (Section 11.6).

Eventually, note that not only the spectral decomposition, but any parallel decomposition, as well as different procedures, such as the virtual experiment method (Ossikovski 2012a), can be likewise used for both types of filtering. Whatever their mathematical formulation, all these approaches share the common feature of being fundamentally based on the positive semi-definiteness of the covariance matrix **H** of **M**.

11.3 Integral Decomposition of a Mueller Matrix

The interpretation of experimental depolarizing Mueller matrices in terms of elementary polarization properties and their uncertainties within the framework of the so-called differential formalism (see Chapter 9) has been the subject of many theoretical efforts pioneered by Jones (1948) and Azzam (1978). However, as shown in Chapter 9, not all depolarizing Mueller matrices allow for differential decomposition (Ossikovski 2011; Devlaminck and Ossikovski 2014), either because of mathematical incompatibility, or due to lack of physical realizability.

Nevertheless, as follows from the statistical definition of the Mueller matrix of a random medium (see Section 5.15), any normalized depolarizing Mueller matrix can be interpreted in terms of a set of six polarization properties and their fluctuations (or uncertainties). Among other contents, the present section summarizes the works of Ossikovski, Arteaga and Kuntman on the Mueller matrix integral decomposition approach (Ossikovski and Arteaga 2015; Ossikovski et al. 2019).

11.3.1 General Integral Decomposition of a Mueller Matrix

Consider the Jones matrix generator **T** of the medium (see Section 5.15) and its fluctuation $\Delta\mathbf{T}$ about its mean value $\mathbf{T}_0$, i.e., $\mathbf{T} = \mathbf{T}_0 + \Delta\mathbf{T}$ with $\langle\Delta\mathbf{T}\rangle = 0$ where the brackets $\langle\ldots\rangle$ denote spatial or temporal averaging. Then the Mueller matrix **M** admits the so-called *integral decomposition* (Brosseau and Barakat 1991; Ossikovski and Arteaga 2015)

$$\mathbf{M} = \mathbf{M}_0 + \Delta\mathbf{M} \tag{11.2}$$

where $\mathbf{M}_0$ represents the mean nondepolarizing part of **M** while $\Delta\mathbf{M}$ arises because of the presence of depolarization. If no loss of spatial or temporal coherence occurs during the measurement of **M**, the resulting experimental Mueller matrix is expected to be $\mathbf{M}_0$ (within experimental error).

Note that the integral decomposition may also be expressed in the general form of a parallel decomposition, $\mathbf{M} = q\mathbf{M}_0 + (1-q)\Delta\mathbf{M},\ 0 < q \leq 1$. However, in what follows, the condensed form (11.2) is used because of the physical meaning of the residual $\Delta\mathbf{M}$ (Ossikovski and Arteaga 2015; Ossikovski et al. 2019), so that the parallel decomposition coefficients are considered as being included in the respective matrices.

Unlike the original Mueller matrix **M** (in general, depolarizing), $\mathbf{M}_0$ is nondepolarizing and, therefore, can always be submitted to the differential decomposition from which the elementary polarization properties can be extracted (Devlaminck and Ossikovski 2014). To get a physical interpretation of the components $\mathbf{M}_0$ and $\Delta\mathbf{M}$ in terms of basic polarization properties, the representation used by Van de Hulst (1957) can be redefined as follows in terms of the three pairs of elementary polarization properties:

$$\mathrm{L} = \mathrm{LB} - i\mathrm{LD} \quad \mathrm{L}' = \mathrm{LB}' - i\mathrm{LD}' \quad \mathrm{C} = \mathrm{CB} - i\mathrm{CD} \tag{11.3}$$

where the linear-0° (L), linear-45° (L′), and circular (C) elementary anisotropies are composed of the respective birefringence (B) and dichroism (D) components (Ossikovski and Arteaga 2014).

Any non-singular Jones matrix **T** can be expressed as

$$\mathbf{T} = \frac{a}{2}\begin{pmatrix} 2\cos(\mathrm{T}/2) - i\mathrm{L}\,\mathrm{sinc}(\mathrm{T}/2) & (\mathrm{C} - i\mathrm{L}')\mathrm{sinc}(\mathrm{T}/2) \\ -(\mathrm{C}+i\mathrm{L}')\mathrm{sinc}(\mathrm{T}/2) & 2\cos(\mathrm{T}/2) + i\mathrm{L}\,\mathrm{sinc}(\mathrm{T}/2) \end{pmatrix} \quad \left[\mathrm{sinc}(\mathrm{T}/2) = \frac{\sin(\mathrm{T}/2)}{\mathrm{T}/2}\right] \tag{11.4}$$

where $a = \sqrt{\det \mathbf{T}}$ is directly related to the mean transmissivity or reflectivity (Barakat 1987a) and $\mathrm{T} = \sqrt{\mathrm{L}^2 + \mathrm{L}'^2 + \mathrm{C}^2}$ is the overall anisotropy amplitude (Arteaga and Canillas 2010). Note that if **T** is singular then it physically represents an ideal (generally, elliptical) perfect polarizer whose parameters (namely transmission axis orientation) can be determined by the methods from Section 4.3.2.

A set of integral polarization properties PI_k of **T** can be defined as (Ossikovski and Arteaga 2015; Ossikovski et al. 2019)

$$\mathrm{PI}_k = \mathrm{P}_k\,\mathrm{sinc}(\mathrm{T}/2) \quad k = 1,2,3 \tag{11.5}$$

where PI_k represent the integral linear-0°, linear-45° and circular anisotropies, $\mathrm{PI}_1 \equiv \mathrm{LI}$, $\mathrm{PI}_2 \equiv \mathrm{LI}'$, $\mathrm{PI}_3 \equiv \mathrm{CI}$, and $\mathrm{P}_1 \equiv \mathrm{L}$, $\mathrm{P}_2 \equiv \mathrm{L}'$, $\mathrm{P}_3 \equiv \mathrm{C}$. Then **T** can be re-parameterized in the following manner:

$$\mathbf{T} = \frac{a}{2}\begin{pmatrix} 2\mathrm{AI} - i\mathrm{LI} & -i(\mathrm{LI}' + i\mathrm{CI}) \\ -i(\mathrm{LI}' - i\mathrm{CI}) & 2\mathrm{AI} + i\mathrm{LI} \end{pmatrix} \quad \left[\mathrm{AI} \equiv \cos\frac{\mathrm{T}}{2}\right] \tag{11.6}$$

where $\mathrm{AI} \equiv \cos(\mathrm{T}/2)$ is the integral isotropic absorption.

Note that the application of the normalization $\bar{\mathbf{T}} = \mathbf{T}/\sqrt{\det \mathbf{T}}$ enables the direct identification of the elementary (differential) or integral polarization properties from Eq. (11.4).

As follows from the very definition of the integral polarization properties through Eq. (11.5), for small values of any of the two sets PI_k or P_k, the integral and elementary (differential) values of the properties are close to each other $(\mathrm{PI}_k \approx \mathrm{P}_k)$ and, in particular, $\mathrm{LI} = \mathrm{LI}' = \mathrm{CI} = 0$ whenever $\mathrm{L} = \mathrm{L}' = \mathrm{C} = 0$ and vice versa.

The definitions of AI and PI_k allow for expressing $\bar{\mathbf{T}}$ as the following expansion:

$$\bar{\mathbf{T}} = \mathrm{AI}\,\boldsymbol{\sigma}_0 - \frac{i}{2}\sum_{k=1}^{3}\mathrm{PI}_k\,\boldsymbol{\sigma}_k \tag{11.7}$$

where $\boldsymbol{\sigma}_0$ is the 2×2 identity matrix, while $\boldsymbol{\sigma}_i$ $(i = 1,2,3)$ are the Pauli matrices, labeled as usual in polarization optics (recall that in quantum physics subscripts 2, 3 are commonly interchanged with respect to the ones below),

$$\boldsymbol{\sigma}_1=\begin{pmatrix}1&0\\0&-1\end{pmatrix}\quad\boldsymbol{\sigma}_2=\begin{pmatrix}0&1\\1&0\end{pmatrix}\quad\boldsymbol{\sigma}_3=\begin{pmatrix}0&-i\\i&0\end{pmatrix}\tag{11.8}$$

Then, in virtue of linearity, when considering a "fluctuating" Jones matrix $\bar{\mathbf{T}}=\bar{\mathbf{T}}_0+\Delta\bar{\mathbf{T}}$, its components $\bar{\mathbf{T}}_0$ and $\Delta\bar{\mathbf{T}}$ can be expanded as

$$\bar{\mathbf{T}}_0=\mathrm{AI}_0\,\boldsymbol{\sigma}_0-\frac{i}{2}\sum_{k=1}^{3}\mathrm{PI}_{0k}\,\boldsymbol{\sigma}_k\qquad\Delta\bar{\mathbf{T}}=\Delta\mathrm{AI}\,\boldsymbol{\sigma}_0-\frac{i}{2}\sum_{k=1}^{3}\Delta\mathrm{PI}_k\,\boldsymbol{\sigma}_k\tag{11.9}$$

where PI_{0k} and $\Delta\mathrm{PI}_k$ are respectively the mean values and the fluctuations of the integral polarization properties PI_k (note that AI_0 and $\Delta\mathrm{AI}_0$ represent the analogous quantities for the integral isotropic absorption). Recall that, as seen in Section 3.4.3, the coefficients of the expansions in Eq. (11.9) are precisely the components of the respective coherency vectors $\bar{\mathbf{c}}_0$ and $\Delta\bar{\mathbf{c}}$ associated with the Jones matrices $\bar{\mathbf{T}}_0$ and $\Delta\bar{\mathbf{T}}$.

Therefore, the pure and depolarizing parts $\bar{\mathbf{M}}_0$ and $\Delta\bar{\mathbf{M}}$ of the integral decomposition (11.2), are generated respectively by $\bar{\mathbf{T}}_0$ and $\Delta\bar{\mathbf{T}}$ through the statistical definition of $\mathbf{M}$ (see Eq. (5.6)) (recall that, as seen in Section 3.3.5.3, $\det\mathbf{M}_0=\left|\det\mathbf{T}_0\right|^4$)

$$\begin{gathered}\mathbf{M}=\sqrt[4]{\det\mathbf{M}_0}\,\bar{\mathbf{M}}\qquad\bar{\mathbf{M}}=\mathcal{L}\langle\bar{\mathbf{T}}\otimes\bar{\mathbf{T}}^*\rangle\mathcal{L}^{\dagger}=\bar{\mathbf{M}}_0+\Delta\bar{\mathbf{M}}\\\left[\bar{\mathbf{M}}_0=\mathcal{L}\langle\bar{\mathbf{T}}_0\otimes\bar{\mathbf{T}}_0^*\rangle\mathcal{L}^{\dagger}\qquad\Delta\bar{\mathbf{M}}=\mathcal{L}\langle\Delta\bar{\mathbf{T}}\otimes\Delta\bar{\mathbf{T}}^*\rangle\mathcal{L}^{\dagger}\right]\end{gathered}\tag{11.10}$$

The integral decomposition can also be expressed as $\bar{\mathbf{C}}(\mathbf{M})=\bar{\mathbf{C}}_0+\Delta\bar{\mathbf{C}}$, in terms of the corresponding coherency matrices (see Section 5.3), with $\bar{\mathbf{C}}_0=\mathbf{C}(\bar{\mathbf{M}}_0)$ and $\Delta\bar{\mathbf{C}}=\mathbf{C}(\Delta\bar{\mathbf{M}})$. Thus, in virtue of Eq. (11.9), the lower 3×3 submatrix $\Delta\bar{\mathbf{C}}_3$ of $\Delta\bar{\mathbf{C}}$ is precisely the covariance matrix of the fluctuations $\Delta\mathrm{PI}_k$ of the integral polarization properties PI_k,

$$\Delta\bar{\mathbf{C}}_3=\frac{1}{4}\begin{pmatrix}\langle|\Delta\mathrm{LI}|^2\rangle&\langle\Delta\mathrm{LI}\;\Delta\mathrm{LI}'^*\rangle&\langle\Delta\mathrm{LI}\;\Delta\mathrm{CI}^*\rangle\\\langle\Delta\mathrm{LI}^*\;\Delta\mathrm{LI}'\rangle&\langle|\Delta\mathrm{LI}'|^2\rangle&\langle\Delta\mathrm{LI}'\;\Delta\mathrm{CI}^*\rangle\\\langle\Delta\mathrm{LI}^*\;\Delta\mathrm{CI}\rangle&\langle\Delta\mathrm{LI}'^*\Delta\mathrm{CI}\rangle&\langle|\Delta\mathrm{CI}|^2\rangle\end{pmatrix}\tag{11.11}$$

The variances and covariances of the integral polarization properties can therefore be written explicitly as follows in terms of the elements $\Delta\bar{m}_{ij}$ of $\Delta\bar{\mathbf{M}}$:

$$\begin{aligned}&\langle|\Delta\mathrm{LI}|^2\rangle=\Delta\bar{m}_{00}+\Delta\bar{m}_{11}-\Delta\bar{m}_{22}-\Delta\bar{m}_{33}&&\langle\Delta\mathrm{LI}\;\Delta\mathrm{LI}'^*\rangle=2(\Delta\bar{m}_{12}+i\Delta\bar{m}_{03})\\&\langle|\Delta\mathrm{LI}'|^2\rangle=\Delta\bar{m}_{00}-\Delta\bar{m}_{11}+\Delta\bar{m}_{22}-\Delta\bar{m}_{33}&&\langle\Delta\mathrm{LI}\;\Delta\mathrm{CI}^*\rangle=2(\Delta\bar{m}_{13}-i\Delta\bar{m}_{02})\\&\langle|\Delta\mathrm{CI}|^2\rangle=\Delta\bar{m}_{00}-\Delta\bar{m}_{11}-\Delta\bar{m}_{22}+\Delta\bar{m}_{33}&&\langle\Delta\mathrm{LI}'\;\Delta\mathrm{CI}^*\rangle=2(\Delta\bar{m}_{23}+i\Delta\bar{m}_{01})\end{aligned}\tag{11.12}$$

When a given property PI_k is deterministic $(\Delta\mathrm{PI}_k=0)$, then its corresponding row and column in $\Delta\mathbf{C}_3$ both become zero.

Note that the mean values PI_{0k} of the integral properties can either be directly identified from the Jones matrix $\bar{\mathbf{T}}_0$ associated with the nondepolarizing component $\bar{\mathbf{M}}_0$ of $\bar{\mathbf{M}}$ in Eq. (11.9), or alternatively, PI_{0k} can be evaluated from the elements $\bar{c}_{0,ij}$ $(i,j=0,1,2,3)$ of the coherency matrix $\bar{\mathbf{C}}_0$ of $\bar{\mathbf{M}}_0$ from

$$\mathrm{PI}_{0k}=2i\bar{c}_{0,k0}\Big/\sqrt{\bar{c}_{0,00}^2-\bar{c}_{0,10}^2-\bar{c}_{0,20}^2-\bar{c}_{0,30}^2}\qquad(k=1,2,3)\tag{11.13}$$

Consequently, both the set of mean values and the covariance matrix of the integral polarization properties that completely characterize an arbitrary depolarizing Mueller matrix **M** can be obtained from the integral decomposition of **M** (11.2). As seen in Section 11.3, Eq. (11.2) can be satisfied in an infinite number of ways; for the integral decomposition to be physically meaningful, the nondepolarizing part $\mathbf{M}_0$ of **M** must represent a certain nondepolarizing estimate $\mathbf{M}_J$ of **M**. For instance, if the IPP of **M** take values close enough to 1, then the coefficients $P_2 - P_1$, $P_3 - P_2$ and $1 - P_3$ of the depolarizing components of the characteristic decomposition of **M** are close enough to zero and, in accordance with the analysis performed in Section 11.2, $\mathbf{M}_0$ can be identified with the pure $\mathbf{M}_{J0}$ component of the characteristic decomposition, whose relative weight q with respect to the whole Mueller matrix **M** is given by the first index of polarimetric purity P_1. Note that the relative weight may be alternatively identified with the largest normalized eigenvalue $\hat{\lambda}_0$ of **C**, but, because of the structure of the polarimetric noise (see Section 11.2) a portion $\hat{\lambda}_0 - P_1 = \hat{\lambda}_1$ should be considered as integrated in the depolarizing part.

From the point of view of the integral decomposition, different physical situations can be considered, among which the two limiting cases are: (1) one of the eigenvalues (λ_0) of **C** is large enough compared with the other three, so that, as discussed in the previous paragraph, the first spectral component, or characteristic pure component, $\mathbf{M}_{J0}$, is a natural candidate for the pure estimate of **M**, and (2) the four eigenvalues of **C** have similar values, in which case the weight of the depolarizing component of the integral decomposition takes necessarily a large value, close to the maximum possible one (i.e., $1-q \approx 1-P_3 \approx 1$, that is, the three IPP are close to zero and **M** is close to that of a perfect depolarizer $\mathbf{M}_{\Delta 0}$) and, therefore, the identification of a pure estimate of **M** has no physical sense.

Intermediate physical situations can be found depending on the relative values of the IPP, and it is the experimentalist who can make best use of the a priori information available on the sample, together with the measured mean intensity coefficient m_{00}, as well as the analysis of the passivity and the physical realizability of the possible parallel decompositions of **M**. Recall that the spectral decomposition of certain Mueller matrices exhibiting significant depolarization is not passive-realizable (i.e., not all the spectral components are passive), in which case the characteristic component $\mathbf{M}_{J0}$ is not a realistic estimate because it does not correspond to a physically realizable decomposition.

In general, as seen in Section 7.7.2, when the nature of the pure component is known (or suspected) the method of polarimetric subtraction can be used (Gil and San José 2013), and a positive constant q such that $(\mathbf{M} - q\mathbf{M}_J)/(1-q)$ be a physically realizable Mueller matrix can be determined, so that the integral decomposition (11.2) is secured by putting $\mathbf{M}_0 = \mathbf{M}_J$ and $\Delta\mathbf{M} = (\mathbf{M} - q\mathbf{M}_J)/(1-q)$.

Regarding the application of the integral decomposition to experimental examples, the following important points should be addressed. First, as indicated above, in order to enable the direct identification of the elementary (differential) or integral polarization properties from the unit-determinant Jones matrix $\bar{\mathbf{T}}$ in Eq. (11.4) and to make possible the extraction of these properties (in terms of means, variances and covariances) from the integral decomposition, **M** should be normalized as $\bar{\mathbf{M}} = \mathbf{M}/\sqrt[4]{\det \mathbf{M}_0}$, effectively ensuring $\det \bar{\mathbf{M}}_0 = 1$ after normalization.

Second, it is very important to realize that, besides fully defining the integral properties PI_k of **M**, the integral decomposition (11.2) can also be used to determine the mean values of the elementary (differential) properties of **M** through the nondepolarizing estimate $\mathbf{M}_0$. Indeed, by using either the matrix logarithm (Ossikovski 2011) (see Chapter 9) or the analytic inversion (Arteaga and Canillas 2010) methods, parameters L_0, L'_0 and C_0 can be obtained from $\mathbf{M}_0$. However, these values will be generally different from those

TABLE 11.1
Comparison between the Integral and Differential Decompositions and Properties of Depolarizing Mueller Matrices

Decomposition	Mathematical existence	Physical realizability	Uniqueness	Integral properties	Differential properties
integral	always	always (for a realizable **M**)	non-unique	mean values; variances & covariances	mean values only
differential	not always (depends on **M** spectrum)	not always (depends on ln**M**)	unique (given by ln**M**)	mean values only	mean values; variances & covariances

Source: Ossikovski and Arteaga 2015.

directly determined from the matrix logarithm of **M** (see Chapter 9), whenever mathematically existing. Conversely, it is only the differential decomposition that can provide the variances and covariances of the elementary properties L, L′ and C (see Chapter 9), and therefore fully define them (Ossikovski and Arteaga 2014). Thus, whereas both differential and integral decompositions fully describe polarimetric properties in statistical terms, the integral decomposition can further provide estimates for the mean values of the elementary (differential) properties albeit not for their variances and covariances. Table 11.1 summarizes the general features of the integral and differential decompositions.

Third, as follows from the formulation of the integral decomposition in terms of coherency matrices, the rank of the coherency matrix $\Delta\mathbf{C}$ of the depolarizing residual $\Delta\mathbf{M}$ is $\text{rank}\,\Delta\mathbf{C} = r-1$, r being the rank of the initial coherency matrix $\mathbf{C}$. When $\text{rank}\,\mathbf{C} = 4$ (maximum), then $\text{rank}\,\Delta\mathbf{C} = 3$, which means that $\Delta\mathbf{C}$ has exactly one linearly dependent row (or column). This is physically meaningful since the four complex Pauli components of $\Delta\bar{\mathbf{T}}$ expressed on the right of Eqs. (11.19) that is, the fluctuations ΔLI, ΔLI′, ΔCI and ΔAI of the three integral (complex) anisotropy properties and of the integral isotropic absorption, depend only on the fluctuations (and mean values) of the three pairs of elementary (differential) properties L, L′ and C. In particular, the fluctuation ΔAI of the integral isotropic absorption is entirely expressible through ΔL, ΔL′ and ΔC. The rank deficiency of $\Delta\mathbf{C}$ justifies the use of its lower 3×3 submatrix $\Delta\mathbf{C}_3$ only, instead of considering the entire 4×4 matrix $\Delta\mathbf{C}$.

Concerning the information on the mutual coherence between integral properties, it is directly given by the 3×3 submatrix $\bar{\mathbf{C}}_3$ of the coherency matrix $\bar{\mathbf{C}}$ of $\bar{\mathbf{M}}$,

$$\bar{\mathbf{C}}_3 = \frac{1}{4}\begin{pmatrix} \langle |\text{LI}|^2\rangle & \langle \text{LI}\ \text{LI}'^*\rangle & \langle \text{LI}\ \text{CI}^*\rangle \\ \langle \text{LI}^*\ \text{LI}'\rangle & \langle |\text{LI}'|^2\rangle & \langle \text{LI}'\ \text{CI}^*\rangle \\ \langle \text{LI}^*\ \text{CI}\rangle & \langle \text{LI}'^*\text{CI}\rangle & \langle |\text{CI}|^2\rangle \end{pmatrix} \tag{11.14}$$

The expressions for the matrix elements are formally analogous to those from Eqs. (11.11) but without the delta symbols (Δ). If a property is identically zero, then its row and column vanish in $\bar{\mathbf{C}}_3$. Observe that the expression (11.11) for $\Delta\bar{\mathbf{C}}_3$ could have been used as an alternative starting point in deriving relations $\bar{\mathbf{C}}(\mathbf{M}) = \bar{\mathbf{C}}_0 + \Delta\bar{\mathbf{C}}$ and Eq. (11.11) by setting $\text{PI}_k = \text{PI}_{0k} + \Delta\text{PI}_k$ into it and performing the averaging.

In summary, regardless of whether or not an arbitrary experimental depolarizing Mueller matrix **M** admits differential decomposition, the integral decomposition of **M** can always be interpreted physically in terms of the six scalar integral polarization properties encoded in PI_k together with their (three) variances and (six) covariances, provided the decomposition is secured by determining a nondepolarizing estimate $\mathbf{M}_0$ of $\mathbf{M}$ through one of the existing approaches. Note that the set of 16 independent parameters contained in a general $\mathbf{M}$ is completed by adding the MIC m_{00} to the above mentioned 15 quantities.

11.3.2 Anisotropic Integral Decomposition of a Mueller Matrix

The present section is devoted to the application of the integral decomposition to the particular, but important in practice, case of a weakly anisotropic depolarizing medium, leading to a computationally efficient type of integral decomposition that allows for the obtainment of a nondepolarizing estimate, as well as for the determination of the elementary polarization properties in terms of mean values and variances-covariances of their fluctuations. Let us first recall the expression of a Jones matrix $\mathbf{T}$ in terms of its corresponding coherency vector $\mathbf{c} = (c_0, c_1, c_2, c_3)^T$ (see Section 3.4.3)

$$\mathbf{T} = \begin{pmatrix} c_0 + c_1 & c_2 - ic_3 \\ c_2 - ic_3 & c_0 - c_1 \end{pmatrix} \tag{11.15}$$

and assume that all components c_k $(k = 1,\ 2,\ 3)$, except c_0, fluctuate so that $c_k = c_{mk} + \Delta c_k$, where $c_{mk} = \langle c_k \rangle$ are the mean values of c_k, while Δc_k are their fluctuations or uncertainties, which satisfy $\langle \Delta c_k \rangle = 0$. Then $\mathbf{T}$ becomes the Jones generator (Ossikovski and Gil 2017) of a depolarizing Mueller matrix $\mathbf{M}$ whose coherency matrix $\mathbf{C}$ is given by $\mathbf{C} = \langle \mathbf{c} \otimes \mathbf{c}^{+} \rangle = \mathbf{C}_m + \Delta\mathbf{C}$, where $\mathbf{C}_m = \mathbf{c}_m \otimes \mathbf{c}_m^{+}$ and $\Delta\mathbf{C} = \langle \Delta\mathbf{c} \otimes \Delta\mathbf{c}^{+} \rangle$ are the mean and the fluctuating matrix components of $\mathbf{C}$ respectively. The coherency vector $\mathbf{c}$ of the Jones generator $\mathbf{T}$ can accordingly be decomposed as

$$\mathbf{c} = \mathbf{c}_m + \Delta\mathbf{c} = (c_0, c_{m1}, c_{m2}, c_{m3})^T + (0, \Delta c_1, \Delta c_2, \Delta c_3)^T \tag{11.16}$$

into its mean, $\mathbf{c}_m$, and fluctuating, $\Delta\mathbf{c}$, vector components, so that the fluctuating part $\Delta\mathbf{C}$ of $\mathbf{C}$ takes the form

$$\Delta\mathbf{C} = \begin{pmatrix} 0 & 0 & 0 & 0 \\ 0 & \langle |\Delta c_1|^2 \rangle & \langle \Delta c_1 \Delta c_2^* \rangle & \langle \Delta c_1 \Delta c_3^* \rangle \\ 0 & \langle \Delta c_2 \Delta c_1^* \rangle & \langle |\Delta c_2|^2 \rangle & \langle \Delta c_2 \Delta c_3^* \rangle \\ 0 & \langle \Delta c_3 \Delta c_1^* \rangle & \langle \Delta c_3 \Delta c_2^* \rangle & \langle |\Delta c_3|^2 \rangle \end{pmatrix} \tag{11.17}$$

and the corresponding *anisotropic integral decomposition*, in terms of Mueller matrices, is given by

$$\mathbf{M} = \mathbf{M}_m + \Delta\mathbf{M} \quad \left[\mathbf{M}_m \equiv \mathbf{M}(\mathbf{C}_m) \quad \Delta\mathbf{M} \equiv \mathbf{M}(\Delta\mathbf{C})\right] \tag{11.18}$$

The explicit expression for the mean coherency vector $\mathbf{c}_m$ is then obtained as

$$\mathbf{c}_m = \frac{1}{2\sqrt{t}}\begin{pmatrix} m_{00}+m_{11}+m_{22}+m_{33} \\ m_{01}+m_{10}+i(m_{23}-m_{32}) \\ m_{02}+m_{20}-i(m_{13}-m_{31}) \\ m_{03}+m_{30}+i(m_{12}-m_{21}) \end{pmatrix} = \frac{1}{2\sqrt{t}}\begin{pmatrix} t \\ \alpha_A \\ \beta_A \\ \gamma_A \end{pmatrix} \tag{11.19}$$

where m_{ij} are the elements of $\mathbf{M}_0 = \mathcal{L}\langle \mathrm{T}_0 \otimes \mathrm{T}_0^* \rangle \mathcal{L}^\dagger$, the positive parameter t stands for trace of $\mathbf{M}_0$, whereas α_A, β_A and γ_A are the complex counterparts of the anisotropy coefficients of $\mathbf{M}$ (Arteaga et al. 2011) defined in Section 6.7.

Moreover, the variances and covariances of the fluctuations Δc_k are given by

$$\begin{aligned}
\langle|\Delta c_1|^2\rangle &= \frac{m_{00}+m_{11}-m_{22}-m_{33}}{4} - |c_{m1}|^2 & \langle \Delta c_1 \Delta c_2^* \rangle &= \frac{m_{12}+m_{21}+i(m_{03}-m_{30})}{4} - c_{m1}c_{m2}^* \\
\langle|\Delta c_2|^2\rangle &= \frac{m_{00}-m_{11}+m_{22}-m_{33}}{4} - |c_{m2}|^2 & \langle \Delta c_1 \Delta c_3^* \rangle &= \frac{m_{13}+m_{31}-i(m_{02}-m_{20})}{4} - c_{m1}c_{m3}^* \\
\langle|\Delta c_3|^2\rangle &= \frac{m_{00}-m_{11}-m_{22}+m_{33}}{4} - |c_{m3}|^2 & \langle \Delta c_2 \Delta c_3^* \rangle &= \frac{m_{23}+m_{32}+i(m_{01}-m_{10})}{4} - c_{m2}c_{m3}^*
\end{aligned} \tag{11.20}$$

and eventually, the depolarizing residual $\Delta\mathbf{M}$ of $\mathbf{M}$ can be obtained from $\Delta\mathbf{M} = \mathbf{M} - \mathbf{M}_m$.

The *anisotropic integral decomposition* $\mathbf{M} = \mathbf{M}_m + \Delta\mathbf{M}$ is uniquely defined by Eqs. (11.19) and (11.20), so that its physical meaning is univocally determined from the assumption that the three anisotropy components c_k $(k = 1,2,3)$ of the Jones generator $\mathbf{T}$ in Eq. (11.15) fluctuate, while c_0 (the isotropic one) remains constant.

A deeper physical insight into the anisotropic integral decomposition is achieved by considering the parameterization (11.4) of $\mathbf{T}$ in terms of its three elementary polarization properties P_k, and assuming that P_k are fluctuating about their respective mean values P_{mk}, i.e., $\mathrm{P}_k = \mathrm{P}_{mk} + \Delta\mathrm{P}_k$ where $\mathrm{P}_1 = \mathrm{L}$, $\mathrm{P}_2 = \mathrm{L}'$ and $\mathrm{P}_3 = \mathrm{C}$. If all mean values P_{mk} and fluctuations $\Delta\mathrm{P}_k$ are small, i.e. $|P_{mk}| \ll 1$ and $|\Delta P_k| \ll 1$, then $T \ll 1$ and $|\Delta T| \ll 1$ (recall that the overall anisotropy amplitude parameter T has been defined as $\mathrm{T} = \sqrt{\mathrm{L}^2 + \mathrm{L}'^2 + \mathrm{C}^2}$), and the expression for the fluctuating Jones generator from Eq. (11.4) takes the simple form

$$\mathbf{T}_{wad} = a\begin{pmatrix} 1 - i\mathrm{L}/2 & -\mathrm{C}/2 - i\mathrm{L}'/2 \\ \mathrm{C}/2 - i\mathrm{L}'/2 & 1 + i\mathrm{L}/2 \end{pmatrix} \tag{11.21}$$

Observe $|P_{mk}| \ll 1$ is physically equivalent to weak depolarization. Therefore, $\mathbf{T}_{wad}$ represents the Jones generator of a weakly anisotropic and weakly depolarizing medium.

By comparing $\mathbf{T}_{wad}$ in Eq. (11.21) with the expression (11.15) of a Jones generator $\mathbf{T}$ in terms of the components of the associated coherency vector $\mathbf{c}$, it follows that the components of $\mathbf{c}_{wad}$ associated with $\mathbf{T}_{wad}$ are precisely $c_0 = a$ and $c_k = -aP_k/2$ $(k = 1,2,3)$, so that, from Eqs. (11.19) and (11.20), the vector $\mathbf{p}_m$, with components P_{mk}, and the variances-covariances $\langle \Delta\mathrm{P}_k \Delta\mathrm{P}_l^* \rangle$ of their fluctuations, are given by

$$\begin{gathered}
\mathbf{p}_m \equiv (\mathrm{L}_m, \mathrm{L}'_m, \mathrm{C}_m)^T = (2i/t)(\alpha_A, \beta_A, \gamma_A)^T \\
\langle \Delta\mathrm{P}_k \Delta\mathrm{P}_l^* \rangle = (16/t)\langle \Delta c_k \Delta c_l^* \rangle \quad [k,l = 1,2,3]
\end{gathered} \tag{11.22}$$

This result shows that the anisotropic integral decomposition fully characterizes a weakly anisotropic and depolarizing medium in terms of the mean values and the variances-covariances of the fluctuations of its elementary polarization properties. Nevertheless, even if both weak anisotropy and weak depolarization conditions are relaxed, Eqs. (11.22) remain still valid (after rescaling) if the amplitude T of the elementary properties P_k does not fluctuate and P_k are replaced by their integral counterparts PI_k defined in Eq. (11.5). Therefore, for this case of *constant-amplitude integral decomposition*, Eqs. (11.22) should be replaced by

$$\begin{array}{l}\mathbf{pI}_m \equiv (\mathrm{LI}_m, \mathrm{LI}'_m, \mathrm{CI}_m)^T = (2i/t')(\alpha_A, \beta_A, \gamma_A)^T \\ \langle \Delta \mathrm{PI}_k \Delta \mathrm{PI}_l^* \rangle = (16t/t'^2)\langle \Delta c_k \Delta c_l^* \rangle \ [k,l = 1,2,3]\end{array} \qquad \left[t' \equiv \sqrt{t^2 - \alpha_A^2 - \beta_A^2 - \gamma_A^2}\right] \quad (11.23)$$

Conversely, in the weak anisotropy and depolarization limit, conditions $T \ll 1$ and $|\Delta T| \ll 1$ allow for the recovery of Eq. (11.22) provided PI_k, t' and t'^2/t are replaced by P_k, t and t respectively.

In summary, as shown throughout this section, if **M** is known to be weakly anisotropic and depolarizing, the anisotropic integral decomposition can be used in practice either for obtaining a nondepolarizing estimate $\mathbf{M}_m$ of **M** or for evaluating the elementary polarization properties of **M** (in terms of their mean values and variances-covariances). Moreover, in the general case where no assumptions on **M** are made, the decomposition evaluates its integral polarization properties.

Illustrative experimental examples of the application of both general and anisotropic integral decompositions, together with the corresponding physical analysis can be found in Ossikovski and Arteaga 2014; 2015; Ossikovski et al. 2019.

11.4 Completing an Experimental Nondepolarizing Mueller Matrix Whose Column or Row Is Missing

A very special case of filtering an experimental Mueller matrix is completing a partially measured matrix of a medium or system, assumed to be pure, to a full 16-element nondepolarizing estimate.

As seen in Chapter 5, if no loss of spatial or temporal coherence takes place during the measurement, the resulting experimental Mueller matrix $\mathbf{M}_J$ is, within experimental error, *nondepolarizing* (or *pure*) in the sense that it transforms any totally polarized incident Stokes vector into a totally polarized emerging one. Therefore, there exist $16 - 7 = 9$ equations that link the 16 elements of $\mathbf{M}_J$. Conversely, when depolarization is not negligible (i.e., if spatial or spectral coherence is either partially or totally lost during the experiment), then all 16 elements of $\mathbf{M}_J$ are independent.

The fact that a nondepolarizing Mueller matrix $\mathbf{M}_J$ depends on only seven parameters suggests that it is susceptible to be measured through *partial* (or *incomplete*) polarimetry. Indeed, conventional partial Mueller polarimeter designs provide experimentally either nine (when both column and row are missing) or 12 (when a column or a row is missing) elements of $\mathbf{M}_J$ (Hauge 1980; Chipman 1995). In the first case it has been shown that the completion problem for a partial nondepolarizing Mueller matrix $\mathbf{M}_J$ features exactly two distinct solutions (Ossikovski and Arteaga 2019a), whereas, in the second one, the solution is unique (Arteaga and Ossikovski 2019). In both cases the completion of the partial 9- or 12-element $\mathbf{M}_J$ to a full 16-element one can be based on the covariance filtering (see Section 11.2).

An alternative approach to the recovery of the full $\mathbf{M}_J$ from a partial experimental counterpart would consist in exploiting the nine relations existing between its elements (see Section 3.3.1), which can be derived from the characteristic property $\mathbf{G}\mathbf{M}_J^T\mathbf{G}\mathbf{M}_J = \sqrt{\det \mathbf{M}_J}\,\mathbf{I}_4$ (see Eq. (3.37)).

The analysis below is focused specifically on the 12-element partial nondepolarizing Mueller matrix case, which has the genuine property that it is the only partial measurement approach for which the uniqueness of the solution is ensured. Furthermore, the significant experimental importance of the 12-element partial Mueller polarimetry (Jellison and Modine 1997) comes also from the fact that it is commonly used to characterize anisotropic nondepolarizing samples by performing the so-called generalized ellipsometry (Azzam and Bashara 1974).

Even though complete Mueller polarimetry is relatively widespread, 12-element partial polarimeters still retain notable scientific and commercial interest because of their simpler optical layout usually offering wider spectral range, increased sensitivity and easier adaptability to imaging operation. Consequently, the development of computationally efficient closed-form algebraic procedures for the recovery of the full $\mathbf{M}_J$ from its 12-element experimental counterpart (having a column or a row missing), without resorting to the evaluation of its associated covariance matrix $\mathbf{H}_J$ or of its equivalent Jones matrix $\mathbf{T}$, is of definite practical interest.

In what follows it will be assumed that $\mathbf{M}_J$ has its last (i.e., fourth) column missing (i.e., undetermined experimentally). If instead the second or the third column of $\mathbf{M}_J$ is missing, then it can be permuted with the fourth one, the recovery procedure applied and the final result permuted back (note that the first column or row of $\mathbf{M}_J$ is always measured by partial polarimeters). Alternatively, if a row instead of a column is missing then the procedure can be applied to the transpose $\mathbf{M}_J^T$ of $\mathbf{M}_J$ and the final result be re-transposed .

Under the indicated assumption, the 12-element partial $\mathbf{M}_J$ (m_{ik} $i = 0,1,2,3,\ k = 0,1,2,$ being the 12 known elements) can be partitioned column-wise as $\mathbf{M} = (\mathbf{m}_0, \mathbf{m}_1, \mathbf{m}_2, \mathbf{v})$ where $\mathbf{m}_k = (m_{0k}, m_{1k}, m_{2k}, m_{3k})^T$, $k = 1,2,3,$ is its k-th column and $\mathbf{v} = (v_0, v_1, v_2, v_3)^T$ is its unknown last column. The condition $\mathbf{G}\mathbf{M}_J^T\mathbf{G}\mathbf{M}_J = \sqrt{\det \mathbf{M}_J}\,\mathbf{I}_4$ readily leads to the following set of three equations for the unknown column $\mathbf{v}$, $\mathbf{m}_1^T\mathbf{G}\mathbf{v} = \mathbf{m}_2^T\mathbf{G}\mathbf{v} = \mathbf{m}_3^T\mathbf{G}\mathbf{v} = 0$, which can be grouped into the single matrix equation (Ossikovski and Arteaga 2019b)

$$\mathbf{A}\mathbf{v} = \begin{pmatrix} m_{00} & -m_{10} & -m_{20} & -m_{30} \\ m_{01} & -m_{11} & -m_{21} & -m_{31} \\ m_{02} & -m_{12} & -m_{22} & -m_{32} \end{pmatrix}\begin{pmatrix} v_0 \\ v_1 \\ v_2 \\ v_3 \end{pmatrix} = \mathbf{0} \qquad \left[\mathbf{A} \equiv \begin{pmatrix} \mathbf{m}_1^T\mathbf{G} \\ \mathbf{m}_2^T\mathbf{G} \\ \mathbf{m}_3^T\mathbf{G} \end{pmatrix} \quad \mathbf{G} \equiv diag(1,-1,-1,-1)\right] \tag{11.24}$$

It is well known from linear algebra that the solution of the above equation is given by $\mathbf{v} = v_0(A_0, A_1, A_2, A_3)^T$, where v_0 plays the role of a proportionality coefficient, while A_m ($m = 0,1,2,3$) are the adjuncts (i.e., the signed minors) obtained by striking out the m-th column of $\mathbf{A}$, that, after the application of certain determinant identities, take the forms (Ossikovski and Arteaga, 2019b)

$$A_0 = \begin{vmatrix} m_{10} & m_{11} & m_{12} \\ m_{20} & m_{21} & m_{22} \\ m_{30} & m_{31} & m_{32} \end{vmatrix} \quad A_1 = \begin{vmatrix} m_{00} & m_{20} & m_{30} \\ m_{20} & m_{21} & m_{22} \\ m_{30} & m_{31} & m_{32} \end{vmatrix} \quad A_2 = -\begin{vmatrix} m_{00} & m_{01} & m_{02} \\ m_{10} & m_{11} & m_{12} \\ m_{30} & m_{31} & m_{32} \end{vmatrix} \quad A_3 = \begin{vmatrix} m_{00} & m_{01} & m_{02} \\ m_{10} & m_{11} & m_{12} \\ m_{20} & m_{21} & m_{22} \end{vmatrix} \tag{11.25}$$

The absolute value $|v_0|$ of the real coefficient v_0 can be determined from the equalities $\mathbf{v}^T\mathbf{G}\mathbf{v} = -\mathbf{m}_0^T\mathbf{G}\mathbf{m}_0 = \mathbf{m}_1^T\mathbf{G}\mathbf{m}_1 = \mathbf{m}_2^T\mathbf{G}\mathbf{m}_2$ obtained by equating the diagonal elements of both sides of $\mathbf{G}\mathbf{M}_J^T\,\mathbf{G}\mathbf{M}_J = \sqrt{\det \mathbf{M}_J}\,\mathbf{I}_4$, so that the following three equivalent expressions for $|v_0|$ are obtained:

$$|v_0| = \sqrt{\frac{\mathbf{m}_0^T\mathbf{G}\mathbf{m}_0}{A_3^2 + A_2^2 + A_1^2 - A_0^2}} = \sqrt{\frac{-\mathbf{m}_1^T\mathbf{G}\mathbf{m}_1}{A_3^2 + A_2^2 + A_1^2 - A_0^2}} = \sqrt{\frac{-\mathbf{m}_2^T\mathbf{G}\mathbf{m}_2}{A_3^2 + A_2^2 + A_1^2 - A_0^2}} \tag{11.26}$$

On experimental data, $|v_0|$ can be calculated through the average of these three theoretically equal products for better noise resilience, i.e., instead of (11.26), rather use the following expression (Ossikovski and Arteaga, 2019b):

$$|v_0| = \sqrt{\frac{\mathbf{m}_0^T\mathbf{G}\mathbf{m}_0 - \mathbf{m}_1^T\mathbf{G}\mathbf{m}_1 - \mathbf{m}_2^T\mathbf{G}\mathbf{m}_2}{3\left(A_3^2 + A_2^2 + A_1^2 - A_0^2\right)}} \tag{11.27}$$

Apparently, two solutions $\pm|v_0|(A_0, A_1, A_2, A_3)^T$ are obtained for the unknown last column $\mathbf{v}$ of $\mathbf{M}_J$. Nevertheless, only the solution $\mathbf{v} = |v_0|(A_0, A_1, A_2, A_3)^T$ is consistent with the characteristic properties of $\mathbf{M}_J$ and therefore, the procedure for the determination of the missing last column $\mathbf{v}$ of $\mathbf{M}_J$ is unique. Experimental application and validation of the above procedure have been dealt with by Ossikovski and Arteaga (2019b).

11.5 Retrieval of a Nondepolarizing Estimate from an Experimental Mueller Matrix Through Virtual Experiment

The so-called *virtual experiment nondepolarizing estimate* (VEE) (Ossikovski 2012a) provides a criterion for rank-reduction filtering, which is alternative to other criteria like the Cloude or the anisotropic integral decomposition nondepolarizing estimates of a measured Mueller matrix $\mathbf{M}_e$.

It is well known that the commonly used Cloude estimate $\mathbf{M}_C$ constitutes the closest to $\mathbf{M}_e$ nondepolarizing Mueller matrix in the least squares sense, i.e., it minimizes the sum-of-squares estimator $\delta_M^2 = \sum_{i,j}(\mathbf{M}_{eij} - \mathbf{M}_{Cij})^2 \to min$ (Anderson and Barakat 1994), where the sum runs over all matrix elements. However, from an experimental point of view, the elements of $\mathbf{M}_e$ are indirectly calculated from the detected intensities $I_{ij} = \mathbf{s}_i^T\mathbf{M}_e\,\mathbf{s}_j$ in which $\mathbf{s}_j$ and $\mathbf{s}_i$ are the input and output Stokes vectors generated respectively by the polarization state generator and analyzer (PSG and PSA) of the polarimeter. This fact suggests that, rather than employing δ_M^2, the estimator $\delta_I^2 = \sum_{i,j}(I_{ij} - \mathbf{s}_i^T\,\mathbf{M}_J\,\mathbf{s}_j)^2 \to min$ can be used for the determination of the VEE $\mathbf{M}_J$ of $\mathbf{M}_e$. The nondepolarizing character of $\mathbf{M}_J$ allows for expressing it in the form $\mathbf{M}_J \equiv \mathcal{L}(\mathbf{T}\otimes\mathbf{T}^*)\mathcal{L}^{-1}$ as a function of its associated Jones matrix $\mathbf{T}$ (see Section 3.3.1).

Analogously, the Stokes vectors $\mathbf{s}_j$ and $\mathbf{s}_i$ entering the expression $\delta_I^2 = \sum_{i,j}(I_{ij} - \mathbf{s}_i^T \mathbf{M}_J \mathbf{s}_j)^2$ are related to their Jones vector counterparts $\boldsymbol{\varepsilon}_j$ and $\boldsymbol{\varepsilon}_i$ through $\mathbf{s}_{i,j} = \mathcal{L}(\boldsymbol{\varepsilon}_{i,j} \otimes \boldsymbol{\varepsilon}_{i,j}^*)$, so that, with the help of some algebra, the estimator δ_I^2 can be reformulated as (Ossikovski 2012a)

$$\delta_I^2 = \sum_{i,j}\left(I_{ij} - 2\left|\boldsymbol{\varepsilon}_i^\dagger \mathbf{T} \boldsymbol{\varepsilon}_j\right|^2\right)^2 = min \tag{11.28}$$

where the unknown Jones matrix $\mathbf{T}$ appears explicitly.

Unlike the minimization of δ_M^2 through covariance filtering procedure, based on the eigenvalue-eigenvector structure of the covariance (or coherency) matrix associated with $\mathbf{M}_e$, leading to the Cloude nondepolarizing estimate $\mathbf{M}_C$, the minimization of δ_I^2 has necessarily to be performed trough a numerical procedure, where the Jones matrix $\mathbf{T}_C$ derived from the Cloude estimate $\mathbf{M}_C$ can be used as the input matrix to the calculation algorithm. Once the Jones matrix $\mathbf{T}$ is determined, the Mueller matrix estimate $\mathbf{M}_J$ is obtained from $\mathbf{M}_J \equiv \mathcal{L}(\mathbf{T} \otimes \mathbf{T}^*)\mathcal{L}^{-1}$.

At first sight, it seems that the values of the elements of $\mathbf{T}$ determined from the above described VEE procedure strongly depend on the choice of the input and output polarization states $\mathbf{s}_j$ and $\mathbf{s}_i$ entering the expression of δ_I^2. This suspicion is true in general, however it has been shown (and numerically checked) that when the polarization states are chosen to be uniformly distributed on the Poincaré sphere, that is to say, their Poincaré vectors are located at the vertices of inscribed regular polyhedra (Platonic solids), then the elements of $\mathbf{T}$ depend neither on the number of states N (N = 4, 6, 8, 12 or 20) nor on the relative orientation of the inscribed polyhedron. This geometrical invariance property enables one to perform the VEE procedure by using only the four following input and output polarization states corresponding to the regular tetrahedron case (Ossikovski 2012a):

$$\mathbf{s}_{i,j} = \begin{pmatrix} 1 & 1 & 1 & 1 \\ a & -a & a & -a \\ a & -a & -a & a \\ a & a & -a & -a \end{pmatrix} \qquad \left[a = \tfrac{1}{\sqrt{3}}\right] \tag{11.29}$$

The Jones vectors $\boldsymbol{\varepsilon}_{i,j}$ are obtained from $\mathbf{s}_{i,j}$ by direct inversion of $\mathbf{s}_{i,j} = \sqrt{2}\mathcal{L}(\boldsymbol{\varepsilon}_{i,j} \otimes \boldsymbol{\varepsilon}_{i,j}^*)$ (see Section 1.5.3). It should be noted that the regular tetrahedron polarization states are actually those that optimize the measurement configuration of "real life" polarimetric experiments in terms of minimization of the experimental error (Tyo et al. 2010). This fact, together with the invariance of δ_I^2 with respect to the spatial orientation and number of vertices of the inscribed regular polyhedra formed by the input and output polarization states point at the robustness and the uniqueness of solution of the virtual experiment approach.

The practical application of the VEE procedure to measured $\mathbf{M}_e$, together with its physical analysis and comparison to the Cloude estimate can be found in (Ossikovski 2012a).

In general, both $\mathbf{M}_C$ and $\mathbf{M}_J$ estimates naturally lead to close results because they minimize respective least squares estimators δ_M^2 and δ_I^2. It is remarkable that both estimators do not depend on any assumption on the spatial location of the depolarization phenomenon that generates a multitude of product decompositions and correspondingly derived nondepolarizing estimates (Ossikovski 2012a).

As discussed by Ossikovski (2012a), the differences in quality of performance of the different estimates depend on the intrinsic physical properties of the material sample, as well as on the characteristics of the polarimetric measurement (noise, spectral and spatial resolution, angular aperture of the probing beam, etc.). The main potential difference between the $\mathbf{M}_C$ and $\mathbf{M}_I$ estimates stems from the different ways of minimization of their respective estimators δ_M^2 and δ_I^2. Since δ_I^2 is minimized numerically, $\mathbf{M}_I$ can be sought in a special form that is suggested by the studied physical problem (e.g., in a block-diagonal or a diagonal form): an approach commonly referred to as *adaptive* polarimetry (Tyo et al. 2010) or *incomplete* polarimetry (Savenkov et al. 2011). Conversely, the minimization of δ_M^2 is based on an algebraic procedure and consequently all elements of the Cloude estimate $\mathbf{M}_C$ take generally nonzero values (of course, the computation of $\mathbf{M}_C$ can be likewise performed numerically for a special form matrix, but then the advantage of using an algebraic procedure would be lost).

A peculiarity of the VEE method is that it has the advantage of being applicable to partial 12-element Mueller matrices with missing (unknown) last column (or row) elements, provided by generalized ellipsometry experiments (see Section 11.4).

An example of application of the VEE method to experimental Mueller polarimetry together with the comparison of the use of δ_M^2 and δ_I^2 is discussed in (Ossikovski 2012a).

11.6 Instrument-Dependent Method for Obtaining a Nondepolarizing Estimate from an Experimental Mueller Matrix

The *instrument-dependent method* (hereafter IDM) for determining a nondepolarizing estimate was introduced by Ossikovski and Arteaga (2019b) as a variant of the general VEE approach by minimizing the least squares distance between the light intensities virtually generated by the measured Mueller matrix and by its nondepolarizing estimate, but taking into account the exact phenomenological description of the polarimetric instrument (complete polarimeter or generalized ellipsometer) used to determine experimentally the Mueller matrix. The IDM can be applied to both complete or partial (12-element) experimental Mueller matrices (see Section 11.4).

In what follows, conventional instrument designs, both partial and complete, will be considered for the recovery of the nondepolarizing estimate of the Mueller matrix of the sample and the Jones matrix associated with it.

In general, Mueller matrix polarimeters measure the Mueller matrix $\mathbf{M}$ of the sample through the $M{\times}N$ positive matrix $\mathbf{I}$ of light intensities, which is given by $\mathbf{I} = \mathbf{AMW}$ where $\mathbf{W}$ and $\mathbf{A}$ are the respective $4{\times}N$ and $M{\times}4$ matrices of the polarization state generator (PSG) and the polarization state analyzer (PSA) of the instrument; N and M being, respectively, the number of polarization states generated by the PSG and analyzed by the PSA (Chipman 1995; Goldstein 2011).

The instrument matrices $\mathbf{W}$ and $\mathbf{A}$ have the respective forms $\mathbf{W} = (\mathbf{s}_i^{(1)}, \mathbf{s}_i^{(2)}, \ldots, \mathbf{s}_i^{(N)})$ and $\mathbf{A} = (\mathbf{s}_o^{(1)}, \mathbf{s}_o^{(2)} \ldots \mathbf{s}_o^{(M)})^T$ where $\mathbf{s}_i^{(k)}$ and $\mathbf{s}_o^{(k)}$ are the input and output Stokes vectors describing the k-th polarization state generated by the PSG and analyzed by the PSA, respectively. This measurement procedure based on $M{\times}N$ discrete polarization states can be readily extended to the commonly used continuous periodic modulation of the polarization states in the PSG and PSA (e.g., in a rotating-retarder or phase-modulated instrument), in which case the instrument matrices are time-dependent, $\mathbf{W}(t) = [\mathbf{s}_i(t)]$ and $\mathbf{A}(t) = [\mathbf{s}_o(t)]^T$. Nevertheless, the discrete forms can still be used provided the continuous periodic signals are sampled in accordance with the Nyquist criterion (Ossikovski and Arteaga 2019b).

The complete measurement of **M** requires that (1) $N \geq 4$ and $M \geq 4$, and (2) the input and output Stokes vectors $\mathbf{s}_i^{(k)}$ and $\mathbf{s}_o^{(k)}$ entering the instrument matrices **W** and **A** must be *complete*, i.e., must not contain any identically vanishing components. The instrument is then termed a *complete* (or *absolute*) Mueller polarimeter.

When $N < 4$ or $M < 4$, or if $\mathbf{s}_i^{(k)}$or $\mathbf{s}_o^{(k)}$ are incomplete, the Mueller polarimeter is *partial*. An important special case of a partial polarimeter is considered in Section 11.4, where $\mathbf{s}_i^{(k)}$ ($\mathbf{s}_o^{(k)}$) contain exactly one vanishing component whereas $\mathbf{s}_o^{(k)}$ ($\mathbf{s}_i^{(k)}$) are complete, so that the instrument measures 12 out of the 16 elements of the Mueller matrix with a column (a row) missing.

The VEE method determines the nondepolarizing estimate $\mathbf{M}_J$ of the experimentally determined Mueller matrix $\mathbf{M}_e$ as the one that would produce light intensities that are as close as possible to those potentially generated by $\mathbf{M}_e$. If least squares estimation is supposed, then one must minimize the normalized residual δ^2 given by (Ossikovski and Arteaga, 2019b)

$$\delta^2 = \frac{1}{MN}\left\|\mathbf{A}\mathbf{M}_e\mathbf{W} - \mathbf{A}\mathbf{M}_J\mathbf{W}\right\|_2^2 = \frac{1}{MN}\left\|\mathbf{A}\left(\mathbf{M}_e - \mathbf{M}_J\right)\mathbf{W}\right\|_2^2 \tag{11.30}$$

where $\| \; \|_2$ stands for the Frobenius matrix norm. The minimization parameters are the four complex elements $t_{ij}, i, j = 1,2$ of the Jones matrix **T** associated with $\mathbf{M}_J \equiv \mathcal{L}\left(\mathbf{T}\otimes\mathbf{T}^*\right)\mathcal{L}^{-1}$

Since the value of the residual defined by Eq. (11.30) is independent of the way the PSG and PSA regular tetrahedra are oriented in the Poincaré sphere polarization space, Eq. (11.29) can be used to model any 16-state Mueller polarimeter that is optimal, i.e., whose four PSG and four PSA polarization states are both uniformly distributed in polarization space, regardless of their specific orientation.

Complete Mueller polarimeters based on continuous time modulation, e.g. the dual rotating compensator (DRC) (Gil 1983; Goldstein 2011) or the four photoelastic modulator (4PEM) (Arteaga et al. 2012) polarimeters, have time-dependent PSG and PSA matrices **W**(t) and **A**(t). When the arrangements in the PSG and in the PSA are mirror images of each other, then $\mathbf{W}(t) = \mathbf{A}^T(t)$ and $\mathbf{s}_i(t) = \mathbf{s}_o(t)$. Then, for the DRC polarimeter the following relation holds (Ossikovski and Arteaga, 2019b):

$$\mathbf{s}_i(t) = \mathbf{s}_o(t) = \left(1, \cos^2\left(R_C/2\right) + \sin^2\left(R_C/2\right)\cos 4\omega_C t, \sin^2\left(R_C/2\right)\sin 4\omega_C t, \sin R_C \sin 2\omega_C t\right)^T \tag{11.31}$$

where R_C and ω_C are the retardance and the angular speed of the PSG (PSA) retarder respectively (Hauge 1980).

For the 4PEM polarimeter (Ossikovski and Arteaga, 2019b),

$$\mathbf{s}_i(t) = \mathbf{s}_o(t) = \left(1, \cos R_{m1}(t), \sin R_{m1}(t)\sin R_{m2}(t), \sin R_{m1}(t)\cos R_{m2}(t)\right)^T \tag{11.32}$$

where $R_{mk}(t) = A_{mk}\sin\omega_{mk}t$ is the sinusoidally varying retardance of the k-th PSG (PSA) photoelastic modulator with amplitude A_{mk} and cyclic frequency ω_{mk} (k = 1, 2) (Arteaga et al. 2012). All input and output Stokes vectors from Eqs. (11.31) and (11.32) feature non-vanishing components, so that both DRC and 4PEM polarimeters are complete.

An important class of partial Mueller polarimeters is constituted by the so-called *generalized ellipsometers*. The explicit forms of matrices **W** and **A** are presented below for two common arrangements, namely the rotating polarizer-compensator ellipsometer (RPCE) and the extended photoelastic modulator ellipsometer (PME) (the latter representing a rotating polarizer-photoelastic modulator ellipsometer design (abbreviated RPPME) (Hauge 1980)).

The PSG of both RPCE and PME instruments is based on a simple rotating polarizer design generating a time-dependent input Stokes vector given by $\mathbf{s}_i(t) = (1, \cos 2\omega_P t, \sin 2\omega_P t, 0)^T$ where ω_P is the angular speed of the polarizer. The PSA of RPCE contains a rotating compensator (a waveplate retarder) followed by a fixed analyzer that analyzes the output Stokes vector $\mathbf{s}_o(t)$ given by Eq. (11.31). Since $\mathbf{s}_o(t)$ is complete whereas $\mathbf{s}_i(t)$ has its fourth component identically equal to zero, the RPCE measures only the first three columns of the Mueller matrix of the sample, i.e., only 12 out of its 16 elements.

The PSA of RPPME consists of a photoelastic modulator (an electrically driven variable retarder) and an analyzer whose transmission axis makes a fixed azimuth of 45° with the modulator fast axis (Hauge 1980). The analyzed output Stokes vector $\mathbf{s}_o(t)$ depends on the azimuth θ_m of the modulator with respect to the incidence plane (the plane defined by the incident and the emerging light beams) and is given by either $\mathbf{s}_{oI}(t) = (1, 0, \pm\cos R_m(t), \sin R_m(t))^T$ if $\theta_m = 0°, 90°$ (called *configuration I*), or by $\mathbf{s}_{oII}(t) = (1, \pm\cos R_m(t), 0, \sin R_m(t))^T$ if $\theta_m = \pm 45°$ (*configuration II*), where $\cos R_m(t) = A_m \sin(\omega_m t)$ is the sinusoidally varying modulator retardance with amplitude A_m and cyclic frequency ω_m.

Due to the fact that the respective second and third components of $\mathbf{s}_{oI}(t)$ and $\mathbf{s}_{oII}(t)$ vanish identically, the achievement of the complete analysis of the output Stokes vector requires performing two consecutive series of measurements in configurations *I* and *II*. Therefore, for the RPPM instrument, the normalized estimator δ^2 takes the form (Ossikovski and Arteaga, 2019b)

$$\delta^2 = \frac{1}{2MN}\left(\left\|\mathbf{A}_I \mathbf{M}_e \mathbf{W} - \mathbf{A}_I \mathbf{M}_J \mathbf{W}\right\|_2^2 + \left\|\mathbf{A}_{II} \mathbf{M} \mathbf{W} - \mathbf{A}_{II} \mathbf{M}_J \mathbf{W}\right\|_2^2\right) \tag{11.33}$$

where $\mathbf{A}_I$ and $\mathbf{A}_{II}$ are the discrete (sampled) counterparts of the time varying PSA matrices $\mathbf{A}_I(t)$ and $\mathbf{A}_{II}(t)$ for configurations *I* and *II*, obtained from $\mathbf{s}_{oI}(t)$ and $\mathbf{s}_{oII}(t)$. Note that, as with the RPCE, the RPPME is a generalized ellipsometer measuring only the first three columns of the Mueller matrix (both RPCE and RPPME have in common a rotating-polarizer-based incomplete PSG). Consequently, the fourth column of the experimental Mueller matrix $\mathbf{M}_e$ is immaterial when $\mathbf{M}_e$ has been partially determined in a generalized ellipsometry experiment.

As indicated above, the Jones matrix $\mathbf{T}$ whose elements are treated as fitting parameters in the process of numerical minimization of the estimator δ^2 can be of special form, accounting for the available *a priori* information on the structure of the sample or on the experimental configuration.

An interesting example corresponds to samples exhibiting mirror symmetry with respect to the incidence plane (in either reflection or in transmission measurement configuration), in which case $\mathbf{T}$ is a diagonal matrix (Van de Hulst 1957). Therefore, the number of real fitting parameters reduces from seven to only $2 \times 2 - 1 = 3$.

Another example is given by the relatively common backscattering (or backreflection) configuration of the instrument, in which the incident and the outgoing light beams follow the same path albeit in opposite directions. It can be easily shown that in this case the Jones matrix of the sample is antisymmetric, $\mathbf{T}^T = -\mathbf{T}$ (Van de Hulst 1957; Arteaga et al. 2014), and the number of real fitting parameters is reduced to $3 \times 2 - 1 = 5$. More generally, this symmetry property is satisfied if a 180°-rotation of the sample about its normal brings it in its initial position (Van de Hulst 1957).

To apply the IDM approach, it is necessary to sample the periodically varying input and output Stokes vectors $\mathbf{s}_i(t)$ and $\mathbf{s}_o(t)$ in order to build the discrete counterparts **W** and **A** of the time-dependent PSG and PSA matrices $\mathbf{W}(t) = (\mathbf{s}_i(t))$ and $\mathbf{A}(t) = (\mathbf{s}_o(t))^T$. If $\omega_{\max}$ is the highest (rotation or modulation) cyclic frequency present in $\mathbf{s}_i(t)$ or in $\mathbf{s}_o(t)$, then it is enough to sample uniformly its period $2\pi/\omega_{\max}$ in only $N = 3$ number of points defining the corresponding time instants $t_k = (k-1)2\pi/(N\omega_{\max})$ $(k = 1,\ 2,\ \ldots, N)$.

For the rotating polarizer PSG and the rotating compensator PSG (or PSA) $\omega_{\max} = 2\omega_P$ and $\omega_{\max} = 4\omega_C$, so that these values can be used in modeling the RPCE and the RPPME instruments, as well as the DRC complete polarimeter. For the phase-modulated PSA in the RPPME instrument $\omega_{\max} = \omega_m$, while for the 4PEM complete polarimeter $\omega_{\max} = \omega_{m1} = \omega_{m2}$. The retardance of the compensator(s) can be set to $R_C = \pi/4$; the modulation amplitude A_{m} of the PSA of the RPPME instrument, as well as those of the PSG and the PSA of the 4PEM complete polarimeter, assumed equal, $A_{m1} = A_{m2}$, can all be set to 2.405 rad (Ossikovski and Arteaga, 2019b).

By using the above values, the matrices **W** and **A** of all four instruments (RPCE, RPPME, and DRC and 4PEM polarimeters) can be computed. In accordance with the fundamental measurement relation $\mathbf{I} = \mathbf{AMW}$, matrices **W** and **A** generate, in the case of $N = 3$ sampling points, a total number of $N \times N = 3 \times 3 = 9$ virtual intensities, which is the minimum number making possible the determination of the seven-parameter Jones matrix **T** from either the complete (16-element), or the partial (12-element), experimental Mueller matrix $\mathbf{M}_e$.

Examples of calculation of nondepolarizing estimates through IDM from experimental Mueller matrices and their comparisons with those obtained by means of other methods can be found in (Ossikovski and Arteaga, 2019b).

Let us finally observe that, to address the more general case of the so called *adaptive polarimetry* based on nonconventional designs generating and analyzing predefined polarization states (Aiello et al. 2006) or using *channels* (Alenin and Tyo 2015) containing responses of linear combinations of Mueller matrix elements, it is necessary to modify and extend the approach presented above.

References

Agarwal, N., et al. 2015. Spatial evolution of depolarization in homogeneous turbid media within the differential Mueller matrix formalism. *Optics Letters* **40**, 5634–5637

Aiello, A., and Woerdman, J. P. 2005. Physical bounds to the entropy-depolarization relation in random light scattering. *Physical Review Letters* **94**, 090406.

Aiello, A., et al. 2006. Maximum-likelihood estimation of Mueller matrices. *Optics Letters* **31**(6), 817–819.

Al Lakki et al. 2021. Complete coherence of random, nonstationary electromagnetic fields. *Optics Letters* **46**(7), 1756–1759.

Alenin, A. S., and Tyo, J. S. 2015. Structured decomposition design of partial Mueller matrix polarimeters. *Journal of the Optical Society of America A* **32**, 1302–1312.

Allen, L., and Padgett, M. J. 2002. Response to Question #79. Does a plane wave carry spin angular momentum? *American Journal of Physics* **70**, 567.

Allen, L., et al. 1992. Orbital angular momentum of light and the transformation of Laguerre-Gaussian laser modes. *Physical Review A* **45**, 8185.

Allen, L., et al. 2003. *Optical angular momentum* (Bristol: Institute of Physics).

Alonso, M. A., and Wolf, E. 2008. The cross-spectral density matrix of a planar, electromagnetic stochastic source as a correlation matrix. *Optics Communications* **281**, 2393–2396.

Anderson, D. G. M., and Barakat, R. 1994. Necessary and sufficient conditions for a Mueller matrix to be derivable from a Jones matrix. *Journal of the Optical Society of America A* **11**, 2305–2319.

Anna, G., et al. 2011. Optimal discrimination of multiple regions with an active polarimetric imager. *Optics Express* **19**, 25367–25378.

Arfken, G. 1970. *Mathematical methods for physicists*. New York: Academic Press.

Arnal, P. M. 1990. *Modelo matricial para el estudio de fenómenos de polarización de la luz*. PhD Dissertation. University of Zaragoza.

Arteaga, O., and Canillas, A. 2009. Pseudopolar decomposition of the Jones and Mueller-Jones exponential polarization matrices. *Journal of the Optical Society of America A* **26**, 783–793.

Arteaga, O., and Canillas, A. 2010. Analytic inversion of the Mueller-Jones polarization matrices for homogeneous media. *Optics Letters* **35**, 559–561.

Arteaga, O., and Nichols, S. 2018. Soleillet's formalism of coherence and partial polarization in 2D and 3D: application to fluorescence polarimetry. *Journal of the Optical Society of America A* **35**, 1254–1260.

Arteaga, O., and Ossikovski, R. 2019. Complete Mueller matrix from a partial polarimetry experiment: the 12-element case. *Journal of the Optical Society of America A* **36**, 416–427.

Arteaga, O., et al. 2011. Anisotropy coefficients of a Mueller matrix. *Journal of the Optical Society of America A* **28**, 548–553.

Arteaga, O., et al. 2012. Mueller matrix polarimetry with four photoelastic modulators: theory and calibration. *Applied Optics* **51**, 6805–6817.

Arteaga, O., et al. 2014. Elementary polarization properties in the backscattering configuration. *Optics Letters* **39**, 6050–6053.

Auñón, J. M., and Nieto-Vesperinas, M. 2013. On two definitions of the three-dimensional degree of polarization in the near field of statistically homogeneous partially coherent sources. *Optics Letters* **38**, 58–60.

Azzam, R. M. A. 1978. Propagation of partially polarized light through anisotropic media with or without depolarization: A differential 4x4 matrix calculus. *Journal of the Optical Society of America* **68**, 1756–1767.

Azzam, R. M. A. 2000. Poincaré sphere representation of the fixed-polarizer rotating-retarder optical system. *Journal of the Optical Society of America A* **17**, 2105–2107.

Azzam, R. M. A. 2011. Three-dimensional polarization states of monochromatic light fields. *Journal of the Optical Society of America A* **28**, 2279–2283.

Azzam, R. M. A., and Bashara, N. M. 1974. Application of generalized ellipsometry to anisotropic crystals. *Journal of the Optical Society of America* **64**, 128–133.
Azzam, R. M. A., and Bashara, N. M. 1987. *Ellipsometry and polarized light*. Amsterdam: North-Holland.
Barakat, R. 1963. Theory of the coherency matrix for light of arbitrary spectral bandwidth. *Journal of the Optical Society of America* **53**, 317–323.
Barakat, R. 1977. Degree of polarization and the principal idempotents of the coherency matrix. *Optics Communications* **23**, 147–150.
Barakat, R. 1983. *n*-Fold polarization measures and associated thermodynamic entropy of *N* partially coherent pencils of radiation. *Optica Acta* **30**, 1171–1182.
Barakat, R. 1985. The statistical properties of partially polarized light. *Optica Acta* **32**, 295–312.
Barakat, R. 1987a. Conditions for the physical realizability of polarization matrices characterizing passive systems. *Journal of Modern Optics* **34**, 1535–1544.
Barakat, R. 1987b. Statistics of the Stokes parameters. *Journal of the Optical Society of America A* **4**, 1256–1263.
Barakat, R. 1996a. Exponential versions of the Jones and Mueller-Jones polarization matrices. *Journal of the Optical Society of America A* **13**, 158–163.
Barakat, R. 1996b. Polarization entropy transfer and relative polarization entropy. *Optics Communications* **123**, 443–448.
Barakat, R., and Brosseau, C. 1992. Von Neumann entropy of N interacting pencils of radiation. *Journal of the Optical Society of America A* **10**, 529–532.
Barakat, R., and Brosseau, C. 1996. Polarization entropy transfer and relative polarization entropy. *Optics Communications* **123**, 443–448.
Barnett, S. M. 2014. Optical Dirac equation. *New Journal of Physics* **16**, 093008.
Barnett, S. M., and Allen, L. 1994. Orbital angular momentum and nonparaxial light beams. *Optics Communications* **110**, 670–678.
Barnett, S. M., Cameron R. P., and Yao A. M. 2012. Duplex symmetry and its relation to the conservation of optical helicity. *Physical Review A* **86** 013845.
Barnett, S. M., et al. 2016. On the natures of the spin and orbital parts of optical angular momentum. *Journal of Optics*, **18** 064004.
Bekshaev, A. Y., and Soskin, M. S. 2007. Transverse energy flows in vectorial fields of paraxial beams with singularities. *Optics Communications*, **271**, 332–348.
Bendat, J. S., and Piersol, A. G. 2010. *Random data: Analysis and measurement procedures*. New York: Wiley.
Bendel, R. B., and Mickey, M. R. 1978. Population correlation matrices for sampling experiments. *Communications in Statististics Simulation and Computation* **7**, 163–182.
Bennett, J. M. 1995. Polarizers. In Bass M., et al., *Handbook of Optics*. New York: McGraw-Hill, vol. 2, chap. 3.
Berestetskii, V. B., Lifshitz, E. M., and Pitaevskii, L. P. 1982. *Quantum electrodynamics* 2nd edn. New York: Pergamon Press.
Berry, M. V. 2009. Optical Currents. *Journal of Optics A: Pure and Applied Optics*, **11**, 094001.
Berry, M. V., and Dennis, M. R. 2001. Polarization singularities in isotropic random vector waves. *Proceedings of the Royal Society A* **457**, 141–155
Berry, M. V., and Dennis, M. R. 2004. Black polarization sandwiches are square roots of zero. *Journal of Optics A: Pure and Applied Optics* **6**, S24.
Beth, R. A. 1936. Mechanical detection and measurement of the angular momentum of light. *Physical Review* **50**, 115–125.
Bhandari, R. 2008. Transpose symmetry of the Jones matrix and topological phases. *Optics Letters* **33**, 854–856.
Bird, G. R., and Shurcliff, W. A. 1959. Pile-of-plates polarizers for the infrared: improvement in analysis and design. *Journal of the Optical Society of America* **49**, 235–237.
Bialynicki-Birula, I., and Bialynicka-Birula, Z. 2011. Canonical separation of angular momentum of light into its orbital and spin parts. *Journal of Optics* **13**, 064014.
Bliokh, K. Y., et al. 2013. Dual electromagnetism: helicity, spin, momentum and angular momentum. *New Journal of Physics* **15**, 033026.

Bliokh, K. Y., et al. 2014. Conservation of the spin and orbital angular momenta in electromagnetism. *New Journal of Physics* **16**, 093037.

Bliokh, K. Y., et al. 2017. Optical Momentum, Spin, and Angular Momentum in Dispersive Media. *Physical Review Letters* **119**, 073901.

Blomstedt, K. 2013. *Electromagnetic coherence theory, universality results, and effective degree of coherence.* PhD dissertation, Aalto University.

Blomstedt, K., et al. 2007. Effective degree of coherence: general theory and application to electromagnetic fields. *Journal of Optics A* **9**, 907–919.

Blomstedt, K., et al. 2015. Effective degree of coherence: a second look. *Journal of the Optical Society of America A* **32**, 718–732.

Boerner, W. M. 2006. *Basic concepts in radar polarimetry.* [Online] Available at http://envisat.esa.int/polsarpro/Manuals/LN_Basic_Concepts.pdf.

Bohley, C., and Scharf, T. 2004. Polarization of light reflected by cholesteric blue phases. *Journal of Optics A: Pure and Applied Optics* **6**, S77–S80.

Bohren, C. F., and Huffman, D. R. 1983. *Absorption and scattering of light by small particles.* New York: Wiley.

Bolshakov, Y., et al. 1996. Polar decompositions in finite dimensional indefinite scalar product spaces: special cases and applications. In Gohberg, I., Lancaster, P., and Shivakumar, P. N. *Recent developments in operator theory and its applications.* Basel: Birkhäuser, 61–94.

Bolshakov, Y., et al. 1997. Errata for: Polar decompositions in finite dimensional indefinite scalar product spaces: special cases and applications. *Integral Equations and Operator Theory* **27**, 497–501.

Born, M., and Wolf, E. 2005. *Principles of optics.* Cambridge: Cambridge University Press.

Boulvert, F., et al. 2005. Analysis of the depolarizing properties of irradiated pig skin. *Journal of Optics A: Pure and Applied Optics* **7**, 21–28.

Boulvert, F., et al. 2009. Decomposition algorithm of an experimental Mueller matrix. *Optics Communications* **282**, 692–704.

Boya, L. J., and Dixit, K. 2008. Geometry of density matrix states. *Physical Review A* **78**, 042108 1–6.

Brosseau, C. 1998. *Fundamentals of polarized light: A statistical optics approach.* New York: Wiley.

Brosseau, C., and Barakat, R. 1991. Jones and Mueller polarization matrices for random media. *Optics Communications* **84**, 127–132.

Brosseau, C., and Bicout, D. 1994. Entropy production in multiple scattering of light by a spatially random medium. *Physical Review E* **50**, 4997–5005.

Brosseau, C., and Dogariu, A. 2006. Symmetry properties and polarization descriptors for an arbitrary electromagnetic wavefield. *Progress in Optics* **49**, 315–380.

Cameron R. P., et al. 2012. Optical helicity, optical spin and related quantities in electromagnetic theory. *New Journal of Physics* **14**, 053050.

Carozzi, et al. 2000. Parameters characterizing electromagnetic wave polarization. *Physical Review E* **61**, 2024–2028.

Charbois, J. M., and Devlaminck, V. 2016. Stochastic model for the differential Mueller matrix of stationary and nonstationary turbid media. *Journal of the Optical Society of America A* **33**, 2414–2424.

Chen, Y., et al. 2020 Polarimetric dimension and nonregularity of tightly focused light beams. *Physical review A* **101**, 053825.

Chipman, R. A. 1987. *Polarization aberrations.* PhD Dissertation. University of Arizona.

Chipman, R. A. 1995. Polarimetry. In *Handbook of optics* 2nd Edition, vol. 2, chap. 22. New York: McGraw-Hill.

Chipman, R. A. 2000. The structure of the Mueller calculus. In David B. Chenault et al. (ed.) *Polarization analysis, measurement and remote sensing III.* SPIE Proceedings 4133. Bellingham: SPIE, 1–9.

Chipman, R. A. 2005. Depolarization index and the average degree of polarization. *Applied Optics* **44**, 2490–2495.

Chipman, R. A. 2009. Mueller matrices. In *Handbook of optics.* 3rd edn. New York: McGraw Hill, vol. 1, chap. 14.

Chironi, E., and Iemmi, C. 2021. Validity of the product rule and its impact on the accuracy of a Mueller matrix polarimeter. *Applied Optics* **60**, 2736–2744.

Clarke, D., and Grainger, J. F. 1974. *Polarized light and optical measurement*. Oxford: Pergamon Press.

Cloude, S. R. 1986. Group theory and polarisation algebra. *Optik* **75**, 26–36.

Cloude, S. R. 1989. Conditions for the physical realizability of matrix operators in polarimetry. In R. Chipman (ed.) *Polarization considerations for optical systems II*. SPIE Proceedings 1166. Bellingham: SPIE, 177–185.

Cloude, S. R. 2009. *Polarisation: Applications in remote sensing*, Oxford: Oxford University Press.

Cloude, S. R. 2013. Depolarization synthesis: understanding the optics of Mueller matrix depolarization. *Journal of the Optical Society of America A* **30**, 691–700.

Cloude, S. R., and Pottier, E. 1995. Concept of polarization entropy in optical scattering. *Optical Engineering* **34**, 1599–1610.

Cloude, S. R. and Pottier, E. 1996. A review of target decomposition theorems in radar polarimetry. *IEEE Transactions on Geoscience and Remote Sensing* **34**, 498–518.

Collett, E. 1971. Mueller-Stokes matrix formulation of Fresnel's equations, *American Journal of Physics* **39**, 517–528.

Compain, E., et al. 1999. General and self-consistent method for the calibration of polarization modulators, polarimeters, and Mueller-matrix ellipsometers. *Applied Optics* **38**, 3490–3502.

Correas, J. M., P. Melero, and J. J. Gil. 2003. Decomposition of Mueller matrices into pure optical media. *Monografías Seminario Matematico García Galdeano* **27**, 23–240.

Crimin, F., et al. 2019. Optical helicity and chirality: Conservation and sources. *Applied Sciences* **9**, 828.

DeBoo, B., et al. 2004. Degree of polarization surfaces and maps for analysis of depolarization. *Optics Express* **12**, 4941–4957.

De Martino, A., et al. 2004. General methods for optimized design and calibration of Mueller polarimeters. *Thin Solid Films* **455**, 112–119.

Dennis, M. R. 2004. Geometric interpretation of the three-dimensional coherence matrix for nonparaxial polarization. *Journal of Optics A: Pure and Applied Optics* **6**, S26–S31.

Dennis, M. R. 2007. A three-dimensional degree of polarization based on Rayleigh scattering. *Journal of the Optical Society of America A* **24**, 2065–2069.

Devlaminck, V. 2013. Physical model of differential Mueller matrix for depolarizing uniform media. *Journal of the Optical Society of America A* **30**, 2196–2204.

Devlaminck, V. 2015. Depolarizing differential Mueller matrix of homogeneous media under Gaussian fluctuation hypothesis. *Journal of the Optical Society of America A* **32**, 1736–1743.

Devlaminck, V., and Ossikovski, R. 2014. Uniqueness of the differential Mueller matrix of uniform homogeneous media. *Optics Letters* **39**, 3149–3152.

Devlaminck, V., and Terrier, P. 2010. Non-singular Mueller matrices characterizing passive systems. *Optik* **121**, 1994–1997.

Dlugnikov, L. N. 1984. Changes in polarization of a light beam with arbitrary autocoherence propagating in a birefringent crystal: Integrated Mueller matrix representation. *Optica Acta* **31**, 803–811.

Ellis, J., and Dogariu A. 2004a. Differentiation of globally unpolarized complex random fields. *Journal of the Optical Society of America A* **21**, 988–993.

Ellis, J., and Dogariu A. 2004b. Complex degree of mutual polarization. *Optics Letters* **29**, 536–538.

Ellis, J., and Dogariu, A. 2005a. On the degree of polarization of random electromagnetic fields. *Optics Communications* **253**, 257–265.

Ellis, J., and Dogariu A. 2005b. Optical polarimetry of random fields. *Physical Review Letters* **95**, 203905. doi: https://doi.org/10.1103/PhysRevLett.95.203905.

Ellis, J., et al. 2004. Correlation matrix of a completely polarized, statistically stationary electromagnetic field. *Optics Letters* **29**, 1536–1538.

Ellis, J., et al. 2005. Degree of polarization of statistically stationary electromagnetic fields. *Optics Communications* **248**, 333–337.

Espinosa-Luna, R., and Bernabéu, E. 2007. On the Q(M) depolarization metric. *Optics Communications* **277**, 256–258.

Espinosa-Luna, R., et al. 2008. Q(M) and the depolarization index scalar metrics. *Applied Optics* **47** 1575–1580

Espinosa-Luna, R., et al. 2010. Dealing depolarization of light in Mueller matrices with scalar metrics. *Optik* **121**, 1058–1068.

Faisan, S., et al. 2013. Estimation of Mueller matrices using non-local means filtering. *Optics Express* **21**, 4424–4438.

Falkoff, D. L., and Macdonald, J. E. 1951. On the Stokes parameters for polarized radiation. *Journal of the Optical Society of America* **41**, 861–862.

Fano, U. 1953. A Stokes-parameter technique for the treatment of polarization in quantum mechanics. *Physical Review* **93**, 121–123.

Fano, U. 1957. Description of states in quantum mechanics by density matrix and operator techniques. *Reviews of Modern Physics* **29**, 74–93.

Ferreira, C., et al. 2006. Geometric modeling of polarimetric transformations. *Monografías Seminario Matematico García Galdeano* **33**, 115–119.

Fernandez-Corbaton I., Zambrana-Puyalto X., and Molina-Terriza G. 2012. Helicity and angular momentum: A symmetry-based framework for the study of light-matter interactions. *Physical Review A* **86**, 042103.

Foldyna, M., et al. 2009. Retrieval of a nondepolarizing component of experimentally determined depolarizing Mueller matrices. *Optics Express* **17**, 12794–12806.

Freeman, A., and Durden, S. L. 1998. A three-component model for polarimetric SAR data. IEEE Transactions on Geoscience Remote Sensing **36**, 963–973.

Friberg, A. T., and Dändliker R. (eds.) 2008. *Advances in information optics and photonics.* Bellingham: SPIE.

Friberg A. T. and Setälä, T. 2016. Electromagnetic theory of optical coherence. *Journal of the Optical Society of America A* **33**, 2431–2442.

Friberg, A. T., and Wolf, E. 1995. Relationships between the complex degrees of coherence in the space–time and in the space–frequency domains. *Optics Letters* **20**, 623–625.

Gabor, D. 1946. Theory of communication. *Journal of Institution of Electrical Engineers* **93**, 429–457.

Gabriel, C. J., and Nedoluha, A. 1971. Transmittance and reflectance of systems of thin and thick layers. *Optica Acta* **18**, 415–423.

Gamel, O., and James, D. F. V. 2011. Causality and the complete positivity of classical polarization maps. *Optics Letters* **36**, 2821–2823.

Gamel, O., and James, D. F. V. 2012. Measures of quantum state purity and classical degree of polarization. *Physical Review A* **86**, 033830.

Gamel, O., and James, D. F. V. 2014. Majorization and measures of classical polarization in three dimensions. *Journal of the Optical Society of America A* **31**, 1620–1626.

Gamo, H. 1964. Matrix treatment of partial coherence. In *Progress in Optics* **3**, 187–332.

Ghosh, N. and Vitkin, I. A. 2011. Tissue polarimetry: concepts, challenges, applications and outlook,. *Journal of Biomedical Optics* **16**, 110801.

Ghosh N., et al. 2008. Mueller matrix decomposition for extraction of individual polarization parameters from complex turbid media exhibiting multiple scattering, optical activity, and linear birefringence. *Journal of Biomedical Optics* **13**, 044036.

Gil, J. J. 1983. *Determination of polarization parameters in matricial representation: Theoretical contribution and development of an automatic measurement device.* PhD Dissertation, University of Zaragoza. Available at http://zaguan.unizar.es/record/10680/files/TESIS-2013–057.pdf

Gil, J. J. 2000a. Characteristic properties of Mueller matrices. *Journal of the Optical Society of America A* **17**, 328–334.

Gil, J. J. 2000b. Mueller matrices. In Moreno, F. and González, F. (eds.) *Light scattering from microstructures.* Berlin: Springer.

Gil, J. J. 2007. Polarimetric characterization of light and media. *European Physical Journal: Applied Physics* **40**, 1–47.

Gil, J. J. 2011. Components of purity of a Mueller matrix. *Journal of the Optical Society of America A* **28**, 1578–1585.

Gil, J. J. 2013a. Mathematical tools for the analysis and exploitation of polarimetric measurements. In *Polarization science and remote sensing VI*. SPIE Proceedings 8873. Bellingham: SPIE, 887307.
Gil, J. J. 2013b. Transmittance constraints in serial decompositions of Mueller matrices: The arrow form of a Mueller matrix. *Journal of the Optical Society of America A* **30**, 701–707.
Gil, J. J. 2014a. Interpretation of the coherency matrix for three-dimensional polarization states. *Physical Review A* **90**, 043858.
Gil, J. J. 2014b. Review on Mueller matrix algebra for the analysis of polarimetric measurements. *Journal of Applied Remote Sensing* **8**, 081599.
Gil, J. J. 2014c. From a nondepolarizing Mueller matrix to a depolarizing Mueller matrix. *Journal of the Optical Society of America A* **31**, 2736–2743.
Gil, J. J. 2015. Intrinsic Stokes parameters for 2D and 3D polarization states. *Journal of the European Optical Society: Rapid Publications* **10**, 15054.
Gil, J. J. 2016a. Structure of polarimetric purity of a Mueller matrix and sources of depolarization. *Optics Communications* **368**, 165–173.
Gil, J. J. 2016b. Components of purity of a three-dimensional polarization state. *Journal of the Optical Society of America A* **33**, 40–43.
Gil, J. J. 2016c. Invariant quantities of a Mueller matrix under rotation and retarder transformations. *Journal of the Optical Society of America A* **33**, 52–58.
Gil, J. J. 2016d. Degrees of mutual coherence of a 3D polarization state. *Journal of Modern Optics* **63**, 1055–1058.
Gil, J. J. 2018. Parametrization of 3x3 unitary matrices based on polarization algebra. *European Physical Journal Plus* **133**, 206.
Gil, J. J. 2020. Sources of asymmetry and the concept of nonregularity of *n*-dimensional density matrices. *Symmetry* **12**, 1002.
Gil, J. J. 2021. Geometric interpretation and general classification of three-dimensional polarization states through the intrinsic Stokes parameters. *Photonics* **8**, 315.
Gil, J. J. 2022. Polarimetric reversibility. To be published
Gil, J. J., and Bernabéu, E. 1982. Diseño de rotores, compensadores y moduladores de retardo a partir de retardadores comerciales. *Optica Pura y Aplicada* **15** 39–43.
Gil, J. J., and Bernabéu, E. 1985. A depolarization criterion in Mueller matrices. Optica Acta **32**, 259–261.
Gil, J. J., and Bernabéu, E. 1986. Depolarization and polarization indices of an optical system. *Optica Acta* **33**, 185–189.
Gil, J. J., and Bernabéu, E. 1987. Obtainment of the polarizing and retardation parameters of a nondepolarizing optical system from its Mueller matrix. *Optik* **76**, 67–71.
Gil, J. J., and Correas, J. 2003. Polarimetric subtraction for obtaining the Mueller matrices of components which appear combined in a whole material sample under measurement. ICO Topical Meeting on Polarization Optics (ICOPO) Polvijärvi, July. Available at www.pepegil.es/Polarimetric-subtraction-ICOPO-2003-.pdf.
Gil, J. J., and San José, I. 2010. 3D polarimetric purity. *Optics Communications* **283**, 4430–4434.
Gil, J. J., and San José, I. 2013. Polarimetric subtraction of Mueller matrices. *Journal of the Optical Society of America A* **30**, 1078–1088.
Gil, J. J., and San José, I. 2014. Explicit algebraic characterization of Mueller matrices. *Optics Letters* **39**, 4041–4044.
Gil, J. J., and San José, I. 2016a. Reduced form of a Mueller matrix. *Journal of Modern Optics* **63**, 1579–1583.
Gil, J. J., and San José, I. 2016b. Two-vector representation of a nondepolarizing Mueller matrix. *Optics Communications* **374**, 133–141.
Gil, J. J., and San José, I. 2016c. Invariant quantities of a nondepolarizing Mueller matrix. *Journal of the Optical Society of America A* **33**, 1307–1312.
Gil, J. J., and San José, I., 2019. Arbitrary decomposition of a Mueller matrix. *OpticsLetters* **44**, 5715–5718.

Gil, J. J., and San José, I., 2021. Universal synthesizer of Mueller matrices based on the symmetry properties of the enpolarizing ellipsoid. *Symmetry* **13**, 983.
Gil, J. J., and San José, I., 2022. Information structure and general characterization of Mueller matrices. *Journal of the Optical Society of America A* **39**, 314–321.
Gil, J. J., et al. 2004. Generalized polarization algebra. *Monografías del Seminario Matemático García de Galdeano* **31**, 161–167.
Gil, J. J., et al. 2013. Serial-parallel decompositions of Mueller matrices. *Journal of the Optical Society of America A* **30**, 32–50.
Gil, J. J., et al. 2016. Singular Mueller matrices. *Journal of the Optical Society of America A* **33**, 600–609.
Gil, J. J., et al. 2017. Structure of polarimetric purity of three-dimensional polarization states. *Physical Review A* **95**, 053856.
Gil, J. J., et al. 2018a. Polarimetric purity and the concept of degree of polarization. *Physical Review A* **97**, 023838.
Gil, J. J., et al. 2018b Nonregularity of three-dimensional polarization states. *Optics Letters* **43**, 4611–4614.
Gil, J. J., et al. 2019a. Intensity and spin anisotropy of three-dimensional polarization states. *Optics Letters* **44**, 3578–3581.
Gil, J. J., et al. 2019b. Sets of orthogonal three-dimensional polarization states and their physical interpretation. *Physical Review A* **100**, 033824.
Gil, J. J., et al. 2021. Effect of polarimetric nonregularity on the spin of three-dimensional polarization states. *New Journal of Physics* **23**, 063059.
Gil, J. J., et al. 2022a. Spin of random stationary light. To be published.
Gil, J. J., et al. 2022b. Physical significance of the determinant of a Mueller matrix. Photonics, Accepted.
Gill, M., et al. 2021d. Equivalence of light transport and depolarization, arXiv:2104.06767v1.
Girgel, S. S. 1991. Structure of the Mueller matrices of depolarized optical systems. *Sovietic Physics Crystallography* **36**, 890–891.
Givens C. R., and Kostinski, A. B. 1993. A simple necessary and sufficient condition on physically realizable Mueller matrices. *Journal of Modern Optics* **40**, 471–481.
Glauber, R. J. 1963. Quantum theory of coherence. *Journal of the Optical Society of America* **130**, 2529–2539.
Goldberg, et al. 2021. Quantum concepts in optical polarization. *Advances in Optics and Photonics* **13**, 1–73
Goldstein, D. H. 1992. Mueller matrix dual-rotating retarder polarimeter. *Appl. Optics* **31**, 6676–6683.
Goldstein, D. H. 2011. *Polarized light*. 3rd edn. Boca Raton: CRC Press.
Goldstein, D. H., and Chipman, R. A. 1990. Error analysis of a Mueller matrix polarimeter, *Journal of the Optical Society of America A* **7**, 693–700.
Goodman, J. W. 2015. *Statistical optics*. 2nd edn. New York: Wiley.
Gopala Rao, A.V., et al. 1998a. On the algebraic characterization of a Mueller matrix in polarization optics. I. Identifying a Mueller matrix from its *N* matrix," *Journal of Modern Optics* **45**, 955–987.
Gopala Rao, A.V., et al. 1998b. On the algebraic characterization of a Mueller matrix in polarization optics. II. Necessary and sufficient conditions for Jones derived Mueller matrices," *Journal of Modern Optics* **45**, 989–999.
Gori, F. 1998. Matrix treatment for partially polarized, partially coherent beams. *Optics Letters* **23**, 241–243.
Gori F., et al. 1998. Orbital angular momentum of light: a simple view. *European Journal of Physics* **19**, 439–444.
Gosh, N., and Vitkin, A. I. 2011. Tissue polarimetry: concepts, challenges, applications, and outlook. *Journal of Biomedical Optics* **16**, 110801.
Guyot S., et al. 2003. Optical properties filtering using Mueller's formalism: Application to laser imaging. *Optik* **114**, 289–297.
Hannay, J. H. 1998. The Majorana representation of polarization, and the Berry phase of light. *Journal of Modern Optics* **45**, 1001–1008.

Hassinen, T. 2013. *Studies on Coherence and Purity of Electromagnetic Fields*. PhD Thesis. University of Eastern Finland. Available at http://epublications.uef.fi/pub/urn_isbn_978-952-61-1307-4/urn_isbn_978-952-61-1307-4.pdf.

Hassinen, T., et al. 2009. Cross-spectral purity of electromagnetic fields. *Optics Letters* **34**, 3866–3868.

Hassinen, T., et al. 2011a. Hanbury Brown–Twiss effect with electromagnetic waves. *Optics Express* **19**, 15188–15195.

Hassinen, T., et al. 2011b. Cross-spectral purity of the Stokes parameters. *Applied Physics B* **10**, 305–308.

Hassinen, T., et al. 2013. Purity of partial polarization in the frequency and time domains. *Optics Letters* **38**, 1221–1223.

Hauge, P. S. 1980. Recent developments in instrumentation in ellipsometry. *Surface Science* **96**, 108–140.

Higham, N. J. 2008. *Functions of matrices: Theory and computation*. Philadelphia: SIAM.

Hingerl, K. and Ossikovski, R. 2016. General approach for modeling partial coherence in spectroscopic Mueller matrix polarimetry. *Optics Letters* **41**, 219–222.

Hioe, F. T. 2006. Isotropy and polarization in a statistically stationary electromagnetic field in three-dimensions. *Journal of Modern Optics* **53**, 1715–1725.

Horn, R. A., and Johnson, C. R. (eds.). 2012. *Matrix analysis*. Cambridge: Cambridge University Press.

Hornreich, R. M., and Shtrikman, S. 1983. Theory of light scattering in cholesteric blue phases. *Physical Review A* **28**, 1791–1807.

Huard, S. 1997. *Polarization of light*. Chichester: Wiley

Hurwitz, H., and Jones, R. C. 1941. A new calculus for the treatment of optical systems, II. Proof of three general equivalence theorems. *Journal of the Optical Society of America* **31**, 493–499.

James, D. F. V. 1994. Change of polarization of light beams on propagation in free space. *Journal of the Optical Society of America A* **11**, 1641–1643.

Jellison, G. E., and Modine, F. A. 1997. Two-modulator generalized ellipsometry: experiment and calibration. *Applied Optics* **36**, 8184–8189.

Jin, I-Q., and Xu, F. 2013. *Polarimetric scattering and SAR information retrieval*. Singapore: Wiley & Sons.

Jones, R. C. 1941. A new calculus for the treatment of optical systems, I. Description and discussion of the calculus. *Journal of the Optical Society of America* **31**, 488–493.

Jones, R. C. 1947. A new calculus for the treatment of optical systems, V. A more general formulation, and description of another calculus. *Journal of the Optical Society of America* **37**, 107–110.

Jones, R. C. 1948. A new calculus for the treatment of optical systems, VII. Properties of the N-matrices. *Journal of the Optical Society of America* **38**, 671–685.

Kim, K. et al. 1987. Relationship between Jones and Mueller matrices for random media. *Journal of the Optical Society of America A* **4**, 433–437.

Kliger, D. S. et al. 1990. *Polarized light in optics and spectroscopy*. Boston: Academic Press, Ch. 5 Sec. 6.

Korotkova, O. 2017. *Random light beams. Theory and applications*. Boca Raton: CRC Press.

Korotkova, O., and Wolf, E. 2005. Generalized Stokes parameters of random electromagnetic beams" *Optics Letters* **30**, 198–200.

Kostinski A. B., and Givens, R. C. 1992. On the gain of a passive linear depolarizing system," *Journal of Modern Optics* **39**, 1947–1952.

Kostinski, A. B., et al. 1988. Optimal reception of partially polarized waves. *Journal of the Optical Society of America A* **5**, 58–64.

Kostinski, A. B., et al. 1993. Constraints on Mueller matrices of polarization optics. *Applied Optics* **32**, 1646–1651.

Kumar, S., et al. 2012. Comparative study of differential matrix and extended polar decomposition formalisms for polarimetric characterization of complex tissue-like turbid media. *Journal of Biomedical Optics* **17**, 105006.

Kuntman, E., et al. 2017a. Vector and matrix states for Mueller matrices of nondepolarizing optical media. *Journal of the Optical Society of America A* **34**, 39–45.

Kuntman, E., et al. 2017b. Formalism of optical coherence and polarization based on material media states. *Physical Review A* **95**, 063819.

Kuntman, E., et al. 2019. Quaternion algebra for Stokes–Mueller formalism. *Journal of the Optical Society of America A* **36**, 492–497.
Lahiri, M. 2009. Polarization properties of stochastic light beams in the space–time and space–frequency domains. *Optics Letters* **34**, 2936–2938.
Lahiri, M. 2013. Concept of purity in the theory of optical polarization. *Optics Letters* **38**, 866–868.
Lahiri, M., and Wolf, E. 2009. Cross-spectral density matrices of polarized light beams. *Optics Letters* **34**, 557–559.
Lahiri, M., and Wolf, E. 2010. Does a light beam of very narrow bandwidth always behave as a monochromatic beam? *Physics Letters A* **374**, 997–1000.
Lamekin, P. I. 2000 Polar forms of mueller matrices of nondepolarizing optical systems. *Optics and Spectroscopy* **88**, 814–820.
Leppänen, L-P., et al. 2016. Connection of electromagnetic degrees of coherence in space–time and space–frequency domains. *Optics Letters* **41**, 1821–1824.
Le Roy-Bréhonnet, F., and Le Jeune, B. 1997. Utilization of Mueller matrix formalism to obtain optical targets depolarization and polarization properties. *Progress in Quantum Electronics* **21**, 109–151.
Le Roy-Bréhonnet, F., et al. 1997. Analysis of depolarizing optical targets by Mueller matrix formalism. *Pure Applied Optics* **6**, 385–404.
López-Martínez, C., and Pottier, E. 2021. *Basic principles of SAR polarimetry*. In Hajnsek I., and Desnos YL. (eds.) *Polarimetric synthetic aperture radar* . Cham: Springer, 1–58.
Loudon, R. 1983. *The quantum theory of light*. Oxford: Clarendon Press.
Lu, S.-Y., and Chipman, R. A. 1994. Homogeneous and inhomogeneous Jones matrices. *Journal of the Optical Society of America A* **11**, 766–773.
Lu, S.-Y., and Chipman, R. A. 1996. Interpretation of Mueller matrices based on polar decomposition. *Journal of the Optical Society of America A* **13**, 1106–1113.
Lu, S.-Y., and Chipman, R. A. 1998. Mueller matrices and the degree of polarization. *Optics Communications* **146**, 11–14.
Luis, A. 2002. Degree of polarization in quantum optics. *Physical Review A* **66**, 013806.
Luis, A. 2005a. Degree of polarization for three-dimensional fields as a distance between correlation matrices. *Optics Communications* **253**, 10–14.
Luis, A. 2005b. Polarization distribution and degree of polarization for three-dimensional quantum light fields. *Physical Review A* **71**, 063815–8.
Luis, A. 2005c. Quantum polarization for three-dimensional fields via Stokes operators. *Physical Review A* **71**, 023810–7.
Luis, A. 2005d. Properties of spatial-angular Stokes parameters. *Optics Communications* **251**, 243–253.
Luis, A. 2007. Degree of coherence for vectorial electromagnetic fields as the distance between correlation matrices. *Journal of the Optical Society of America A* **24**, 1063–1068.
Luis, A. 2010. Coherence and visibility of vectorial light *Journal of the Optical Society of America A* **27**, 1764–1769.
Luis, A. 2016. Polarization in quantum optics. *Progress in Optics* **61**, 283–331.
Luis, A., and Rodil, A. 2014. Polarization versus photon spin. *Optics Express* **22**, 1569–1575.
Lüneburg, E., and Cloude, S. R. 1997. Radar versus optical polarimetry. In Mott, H. and Boerner, W.-M. (eds.) *Wideband interferometric sensing and imaging polarimetry*. SPIE Proceedings **3120**. Bellingham: SPIE, 361–372.
Macías–Romero, C., et al. 2011. Spatial and temporal variations in vector fields. *Optics Express* **19**, 25066–25076.
Mandel, L. 1961. Concept of cross-spectral purity in coherence theory. *Journal of the Optical Society of America* **51**, 1342–1350.
Mandel, L. 1963. Intensity fluctuations of partially polarized light. *Proceedings of the Physical Society* (London) **81**, 1104–1114.
Mandel, L., and Wolf, E. 1976. Spectral coherence and the concept of cross-spectral purity. *Journal of the Optical Society of America* **66**, 529–535.
Mandel, L., and Wolf, E. 1981. Complete coherence in the space-frequency domain. *Optics Communications*, **36**, 247–249.

Mandel L., and Wolf E. 1995. *Optical coherence and quantum optics*. Cambridge: Cambridge University Press.
Manhas, S., et al. 2006. Mueller matrix approach for determination of optical rotation in chiral turbid media in backscattering geometry. *Optics Express* **14**, 190–202.
Marathay, A. S. 1965. Operator formalism in the theory of partial polarization. *Journal of the Optical Society of America* **55**, 969–980.
Martínez-Herrero, R., et al. 2009. *Characterization of partially polarized light fields*. Berlin: Springer.
Meemon, P., et al. 2008. Determination of the coherency matrix of a broadband stochastic electromagnetic light beam. *Journal of Modern Optics* **55**, 2765–2776.
Mehta, C.L., Wolf, E., and Balachandran, A.P. 1966. Some theorems on the unimodular complex degree of optical coherence. *Journal of Mathematical Physics* **7**, 133–138.
Migliaccio, M., et al. 2015a. Three-dimensional polarimetry for wave chaos description. *2015 IEEE Metrology for Aerospace (MetroAeroSpace)*, Benevento, Italy, pp. 235–239. doi: https://doi.org/10.1109/MetroAeroSpace.2015.7180660.
Migliaccio, M., et al. 2015b. An alternative method for evaluating a spatial uniform and isotropic electromagnetic field within a reverberating chamber. *2015 IEEE International Workshop on Measurements & Networking (M&N)*, pp. 1–5. doi: https://doi.org/10.1109/IWMN.2015.7322994.
Migliaccio, M., et al. 2016. The polarization purity of the electromagnetic field in a reverberating chamber. *IEEE Transactions on Electromagnetic compatibility* **58**, 694–700.
Migliaccio, M., et al. 2019. Polarimetric decomposition of the complex electromagnetic field for EMC aerospace testing. *2019 ESA Workshop on Aerospace EMC (Aerospace EMC)*, Budapest, Hungary, pp. 1–6. doi: https://doi.org/10.23919/AeroEMC.2019.8788949.
Migliaccio, M., et al. 2020. Components of purity to describe the polarimetric state of a 3-D field within the reverberating chamber. *IEEE transactions on electromagnetic compatibility* **62**, 2661–2668.
Miranda-Medina, M., et al. 2017. Experimental validation of the partial coherence model in spectroscopic ellipsometry and Mueller matrix polarimetry. *Applied Surface Science* **421**, 656–662.
Morio J., and Goudail, F. 2004. Influence of the order of diattenuator, retarder and polarizer in polar decomposition of Mueller matrices. *Optics Letters* **29**, 2234–2236.
Nagirner, D. I. 1993. Constraints on matrices transforming Stokes vectors. *Astronomy and Astrophysics* **275**, 318–324.
Neugebauer, M., et al. 2015. Measuring the transverse spin density of light. *Physical Review Letters* **114**, 063901.
Noble, H. D., and Chipman, R. A. 2012. Mueller matrix roots algorithm and computational considerations. *Optics Express* **20**, 17–31.
Noble, H. D., et al. 2012. Mueller matrix roots depolarization parameters. *Applied Optics* **51**, 735–744.
Norrman, A., et al. 2017. Dimensionality of random light fields. *Journal of the European Optical Society: Rapid Publications* **13**, 1–5
Norrman, A., et al. 2019. Polarimetric nonregularity of evanescent waves. *Optics Letters* **44**, 215–218.
O'Neill, E. L. 1963. *Introduction to statistical optics*. Reading: Addison-Wesley, chap. 9.
Opatrný, T., and Perina, J. 1993. Non-image-forming polarization optical devices and Lorentz transformations - an analogy. *Physics Letters A* **181**, 199–202.
Ortega-Quijano, N., and Arce-Diego, J. L. 2011. Depolarizing differential Mueller matrices. *Optics Letters* **36**, 2429–2431.
Ortega-Quijano, N., et al. 2012. Experimental validation of Mueller matrix differential decomposition. *Optics Express* **20**, 1151–1163.
Ossikovski, R. 2008. Interpretation of nondepolarizing Mueller matrices based on singular-value decomposition. *Journal of the Optical Society of America A* **25**, 473–482.
Ossikovski, R. 2009. Analysis of depolarizing Mueller matrices through a symmetric decomposition. *Journal of the Optical Society of America A* **26**, 1109–1118.
Ossikovski, R. 2010a. Canonical forms of depolarizing Mueller matrices. *Journal of the Optical Society of America A* **27**, 123–130.
Ossikovski, R. 2010b. Alternative depolarization criteria for Mueller matrices. *Journal of the Optical Society of America A* **27**, 808–814.

Ossikovski, R. 2011. Differential matrix formalism for depolarizing anisotropic media. *Optics Letters* **36**, 2330–2332.
Ossikovski, R. 2012a. Retrieval of a nondepolarizing estimate from an experimental Mueller matrix through virtual experiment. *Optics Letters* **37**, pp.578–580.
Ossikovski, R. 2012b. Differential and product Mueller matrix decompositions: A formal comparison. *Optics Letters* **37**, 220–222.
Ossikovski, R., and Arteaga, O. 2014. Statistical meaning of the differential Mueller matrix of depolarizing homogeneous media. Optics Letters **39**, 4470–4473.
Ossikovski, R., and Arteaga, O. 2015. Integral decomposition and polarization properties of depolarizing Mueller matrices. *Optics Letters* **40**, 954–957.
Ossikovski, R., and Arteaga, O. 2019a. Complete Mueller matrix from a partial polarimetry experiment: The nine-element case. *Journal of the Optical Society of America* **36**, 403–415.
Ossikovski, R., and Arteaga, O. 2019b. Instrument-dependent method for obtaining a nondepolarizing estimate from an experimental Mueller matrix. *Optical Engineering* **58**, 082409.
Ossikovski, R., and De Martino, A. 2015. Differential Mueller matrix of a depolarizing homogeneous medium and its relation to the Mueller matrix logarithm. *Journal of the Optical Society of America A* **32**, 343–348.
Ossikovski, R., and Devlaminck, V. 2014. General criterion for the physical realizability of the differential Mueller matrix. *Optics Letters* **39**, 1216–1219.
Ossikovski, R., and Gil, J. J. 2017. Basic properties and classification of Mueller matrices derived from their statistical definition. *Journal of the Optical Society of America* **34**, 1727–1737.
Ossikovski, R., and Hingerl, K. 2016. General formalism for partial spatial coherence in reflection Mueller matrix polarimetry. *Optics Letters* **41**, 4044–4047.
Ossikovski, R., and Vizet, J. 2017. Polar decompositions of negative-determinant Mueller matrices featuring non-diagonal depolarizers. *Applied Optics* **56**, 8446–8451.
Ossikovski, R., and Vizet, J. 2019. Eigenvalue-based depolarization metric spaces for Mueller matrices. *Journal of the Optical Society of America* **36**, 1173–1186.
Ossikovski, R., et al. 2007. Forward and reverse product decompositions of depolarizing Mueller matrices. *Optics Letters* **32**, 689–691.
Ossikovski, R., et al. 2008. Product decompositions of depolarizing Mueller matrices with negative determinants. *Optics Communications* **281**, 2406–2410.
Ossikovski, R., et al. 2009a. Experimental evidence for naturally occurring nondiagonal depolarizers. *Optics Letters* **34**, 2426–2428.
Ossikovski, R., et al. 2009b. Experimental implementation and properties of Stokes nondiagonalizable depolarizing Mueller matrices. *Optics Letters* **34**, 974–976.
Ossikovski, R., et al. 2013. Poincaré sphere mapping by Mueller matrices. *Journal of the Optical Society of America A* **30**, 2291–2305.
Ossikovski, R., et al. 2014. Application of the arbitrary decomposition to finite spot size Mueller matrix measurements. *Applied Optics* **53**, 6030–6036.
Ossikovski, R., et al. 2019. Anisotropic integral decomposition of depolarizing Mueller matrices. *OSA Continuum*, **2**, 1900–1907.
Parke, N. G. 1948. *Matrix Optics*. PhD Thesis, MIT.
Partanen, H., et al. 2018. Young's interference experiment with electromagnetic narrowband light. *Journal of the Optical Society of America A* **35**, 1379–1384.
Partanen, H., et al. 2019. Spectral measurement of coherence Stokes parameters of broadband random light beams. *Photonics Research* **7**, 669–677.
Pereda Cubián, D., et al. 2005. Characterization of depolarizing optical media by means of entropy factor: application to biological tissues. *Applied Optics* **44**, 358–364.
Perrin, F. 1942. Polarization of light scattering by isotropic opalescent media. *Journal of Chemical Physics* **10**, 415–427.
Petrov, N. I. 2008, Vector and Tensor Polarizations of Light Beams. *Laser Physics* **18**, 522–525.
Petruccelli, J. C., et al. 2010. Two methods for modeling the propagation of the coherence and polarization properties of nonparaxial fields. *Optics Communications* **282**, 4457–4466.

Picardi, M. F., et al. 2018. Angular momenta, helicity, and other properties of dielectric-fiber and metallic-wire modes. *Optica* **5**, 1016–1026.

Pires, H. D. L., and Monken, C. H. 2008. On the statistics of the entropy-depolarization relation in random light scattering. *Optics Express* **16**, 21059–21068.

Poincaré, H. 1892. *Théorie Mathématique de la Lumière*, vol. 2. Paris: G. Carre.

Porras, M. A. 2013. Propagation-induced changes in the instantaneous polarization state, phase, and carrier–envelope phase of few-cycle pulsed beams. *Journal of the Optical Society of America B* **30**, 1652–1659.

Poynting, J. H. 1909 The wave-motion of a revolving shaft, and a suggestion as to the angular momentum in a beam of circularly polarised light. *Proceedings of the. Royal Society A* **82** 560–567.

Puentes, G., et al. 2005. Experimental observation of universality in depolarized light scattering. *Optics Letters* **30**, 3216–3218.

Pye, D. 2001. *Polarised light in science and nature*. London: IOP Publishing.

Qian, X. F., and Eberly, J. H. 2011. Entanglement and classical polarization states. *Optics Letters* **36**, 4110–4112.

Réfrégier, P. 2005. Mutual information-based degrees of coherence of partially polarized light with Gaussian fluctuations *Optics Letters* **30**, 3117–3119.

Réfrégier, P., and Goudail, F. 2005. Invariant degrees of coherence of partially polarized light. *Optics Express* **13**, 6051–6060.

Réfrégier, P., and Roueff, A. 2006. Coherence polarization filtering and relation with intrinsic degrees of coherence. *Optics Letters* **31**, 1175–7.

Réfrégier, P., and Roueff, A. 2007. Intrinsic coherence: A new concept in polarization and coherence theory. *Optics and Photonics News* **18**, 30–35.

Réfrégier, P., and Roueff, A. 2008. Intrinsic degrees of coherence of electromagnetic fields. In Friberg, A. T. and Dändliker R. (eds.) *Advances in Information Optics and Photonics*. Bellingham: SPIE.

Réfrégier, P., et al. 2012. Maximal polarization order of random optical beams: reversible and irreversible polarization variations. *Optics Letters* **37**, 3750–3752.

Réfrégier, P., et al. 2014. Analysis of coherence properties of partially polarized light in 3D and application to disordered media. *Optics Letters* **39**, 2362–2365.

Robson, B. A. 1974. *The theory of polarization phenomena*. Oxford: Clarendon.

Roman, P. 1959. Generalized Stokes parameters for waves with arbitrary form. *Nuovo Cimento* **13**, 974–982.

Roychowdhury, H., and Wolf, E. 2005. Young's interference experiment with light of any state of coherence and polarization. *Optics Communications* **248**, 327.

Saastamoinnen, T., and Tervo, J. 2004. Geometric approach to the degree of polarization for arbitrary fields. *Journal of Modern Optics* **51**, 2039–2045.

Saastamoinen, T., et al. 2005. Electromagnetic coherence theory of laser resonator modes. *Journal of the Optical Society of America A* **22**, 103–108.

Saastamoinen T., et al. 2020. Probing the electromagnetic degree of coherence of light beams with nanoscatterers. *ACS Photonics* **7**, 1030–1035.

Salazar-Ariza, K., and Torres, R. 2018. Trajectories on the Poincaré sphere of polarization states of a beam passing through a rotating linear retarder. *Journal of the Optical Society of America A* **35**, 65–72.

Samson, J. C. 1973. Description of the polarization states of vector processes: Applications to ULF magnetic fields. *Geophysical Journal of the Royal Astronomical Society* **34**, 403–419.

Sánchez-Soto, L. L., et al. 2006. Degrees of polarization for a quantum field. *Journal of Physics: Conference series* **36**, 177–182.

San José, I and Gil, J. J. 2011. Invariant indices of polarimetric purity: Generalized indices of purity for nxn covariance matrices. *Optics Communications* **284**, 38–47.

San José, I., and Gil, J. J. 2020a. Coherency vector formalism for polarimetric transformations. *Optics Communications* **475**, 126230.

San José, I., and Gil, J. J. 2020b. Characterization of passivity in Mueller matrices. *Journal of the Optical Society of America A* **37**, 199–208.

San José, I., and Gil, J. J. 2020c. Retarding parallel components of a Mueller matrix. *Optics Communications* **459**, 124892.

San José, I., et al. 2020. Algorithm for the numerical calculation of the serial components of the normal form of depolarizing Mueller matrices. *Applied Optics* **59**, 2291–2297.

Sarabandy, K. 1992. Derivation of phase statistics from the Mueller matrix. *Radio Science* **27**, 553–560.

Savenkov, S. N., et al. 2005. Conditions for polarization elements to be dichroic and birefringent. *Journal of the Optical Society of America A* **22**, 1447–1452.

Savenkov, S. N., et al. 2006. Generalized matrix equivalence theorem for polarization theory. *Physical Review E* **74**, 056607.

Savenkov, S. N., et al. 2011. Incomplete active polarimetry: Measurement of the block-diagonal scattering matrix. *Journal of Quantitative Spectroscopy and Radiative Transfer* **112**, 1796–1802.

Schellman, J., and Jensen, H. P. 1987. Optical spectroscopy of oriented molecules. *Chemical Reviews* **87**, 1359–1399.

Schönhofer, A., and Kuball, H.-G. 1987. Symmetry properties of the Mueller matrix. *Chemical Physics* **115**, 159–167.

Sekera, Z. 1966. Scattering matrices and reciprocity relationships for various representations of the state of polarization. *Journal of the Optical Society of America* **56**, 1732–1740.

Setälä, T., et al. 2002a. Degree of polarization for optical near fields. *Physical Review E* **66**, 016615.

Setälä, T., et al. 2002b. Degree of polarization in near fields of thermal sources: Effects of surface waves. *Physical Review Letters* **88**, 123902.

Setälä, T., et al. 2004a. Theorems on complete electromagnetic coherence in the space–time domain. *Optics Communications* **238**, 229–236.

Setälä, T., et al. 2004b. Complete electromagnetic coherence in the space–frequency domain. *Optics Letters* **29**, 328–330.

Setälä, T., et al. 2004c. Intensity fluctuations and degree of polarization in three-dimensional thermal light fields. *Optics Letters* **29**, 2587–2589.

Setälä, T., et al. 2006. Contrasts of Stokes parameters in Young's interference experiment and electromagnetic degree of coherence. *Optics Letters* **31**, 2669–2671.

Setälä, T., et al. 2008. Polarization time and length for random optical beams. *Physical Review A* **78**, 033817

Setälä, T., et al. 2009a. Differences between partial polarizations in the space–time and space–frequency domains. *Optics Letters* **34**, 2924–2926.

Setälä, T., et al. 2009b. Degree of polarization in 3D optical fields generated from a partially polarized plane wave. *Optics Letters* **34**, 3394–3396.

Setälä, T., et al. 2010a. Partial polarization in the space-time and space-frequency domains. In Vainos, N. et al. (eds.) International Commission for Optics topical meeting on emerging trends and novel materials in photonics. Melville: American Institute of Physics.

Setälä T., et al. 2010b. *Partial polarization of optical beams: Temporal and spectral descriptions.* In Javidi B. and Fournel T. (eds.) *Information Optics and Photonics.* New York: Springer.

Setälä, T., et al. 2021. Coherence stokes parameters in the description of electromagnetic coherence. *Photonics* **8**, 85.

Sheppard, C. J. R. 2011. Partial polarization in three dimensions. *Journal of the Optical Society of America A* **28**, 2655–2659.

Sheppard, C. J. R. 2012. Geometric representation for partial polarization in three dimensions. *Optics Letters* **37**, 2772–2774.

Sheppard, C. J. R. 2014. Jones and Stokes parameters for polarization in three dimensions. *Physical Review A* **90**, 023809.

Sheppard, C. J. R. 2016a. Parameterization of the Mueller matrix. *Journal of the Optical Society of America A* **33**, 2323–2332.

Sheppard, C. J. R. 2016b. Geometry of the Mueller matrix spectral decomposition", *Journal of the Optical Society of America A* **33**, 1331–1340.

Sheppard, C. J. R., et al. 2016a. Three-dimensional polarization algebra. *Journal of the Optical Society of America A* **33**, 1938–1947.

Sheppard, C. J. R., et al. 2016b. Expressions for parallel decomposition of the Mueller matrix. *Journal of the Optical Society of America A* **33**, 741–751.

Sheppard, C. J. R., et al. 2017. Expressions for parallel decomposition of the Mueller matrix: erratum. *Journal of the Optical Society of America A* **34**, 813–813.

Sheppard, C. J. R., et al. 2019. Eigenvalues of the coherency matrix for exact backscattering *Journal of the Optical Society of America A* **36**, 1540–1550.

Sheppard, C. J. R., et al. 2020a. Eigenvectors of polarization coherency matrices. *Journal of the Optical Society of America A* **37**, 1143–1154.

Sheppard, C. J. R., et al. 2020b. Polarization in reflectance imaging. *Journal of the Optical Society of America A* **37**, 491–500.

Shevchenko, A., et al. 2009. Characterization of polarization fluctuations in random electromagnetic beams. *New Journal of Physics* **11**, 073004.

Shevchenko, A., et al. 2012. On experimental characterization of polarization fluctuation dynamics in random optical beams. *Applied Optics* **51**, C44–C47

Shevchenko A., et al. 2017 Polarization time of unpolarized light *Optica* **4** 64–70.

Shore, B. W. 1990. *The theory of coherent atomic excitation*. New York: Wiley.

Shurcliff, W A. 1962. *Polarized light*. Cambridge, MA: Harvard University Press.

Simmons, J. W., and Guttmann, M. J. 1970. *States, waves and photons*. Reading, MA: Addison-Wesley.

Simon, B. N., et al. 2010a. A complete characterization of pre-Mueller and Mueller matrices in polarization optics. *Journal of the Optical Society of America A* **27**, 188–199.

Simon, B. N., et al. 2010b. Nonquantum entanglement resolves a basic issue in Polarization Optics. *Physical Review Letters* **104**, 023901.

Simon, R. 1982. The connection between Mueller and Jones matrices of polarization optics. *Optics Communications* **42**, 293–297.

Simon, R. 1987. Mueller matrices and depolarization criteria. *Journal of Modern* Optics **34**, 569–575.

Simon, R. 1990. Nondepolarizing systems and degree of polarization. *Optics Communications* **77**, 349–354.

Snik, F., et al. 2014. An overview of polarimetric sensing techniques and technology with applications to different research fields. In Chenault, D. B., and Goldstein, D. H. (eds.) *Polarization: Measurement, analysis, and remote sensing XI*. SPIE Proceedings **9099**, 90990B. doi: https://doi.org/10.1117/12.2053245.

Soleillet, M. P. 1929. Sur les paramètres caractérisant la polarisation partielle de la lumière dans les phenomenes de fluorescence. *Annales de physique* **12**, 23–59.

Sorrentino, A., et al. 2017a. Microwave measurement technique for evaluating the polarization of the electromagnetic field within a reverberating chamber. *2017 International Applied Computational Electromagnetics Society Symposium – Italy (ACES)*, Florence, 2017, pp. 1–2. doi: https://doi.org/10.23919/ROPACES.2017.7916365.

Sorrentino, A., et al. 2017b. An alternative polarimetric representation of the electromagnetic field in a reverberating environment. *2017 International Symposium on Electromagnetic Compatibility – EMC EUROPE*, Angers, 2017, pp. 1–5. doi: https://doi.org/10.1109/EMCEurope.2017.8094747.

Sridhar, R., and Simon, R. 1994. Normal form for Mueller matrices in polarization optics. *Journal of Modern Optics* **41**, 1903–1915.

Stewart, A. M. 2005a. Angular momentum of light. *Journal of Modern Optics* **52**, 1145–1154.

Stewart, A. M. 2005b. Angular momentum of the electromagnetic field – the plane wave paradox resolved. *European Journal of Physics* **26**, 635–641.

Stokes, G. G. 1852. On the composition and resolution o streams of polarized light from different sources. *Transactions of the Cambridge Philosophical Society* **9**, 399–416.

Sudha, and Gopala Rao, A. V. 2001. Polarization elements: A group-theoretical study. *Journal of the Optical Society of America A* **18**, 3130–3134.

Sudha, A. V., et al. 2008. Positive-operator-valued-measure view of the ensemble approach to polarization optics. *Journal of the Optical Society of America A* **25**, 874–880.

Sun, M., et al. 2014. Characterizing the microstructures of biological tissues using Mueller matrix and transformed polarization parameters. *Biomed Optics Express* **5**, 4223–4234.

Swindell, W. 1975. *Polarized light*. Dowden: Hutchinson & Ross.
Tariq, A., et al. 2017. Physically realizable space for the purity-depolarization plane for polarized light scattering media. *Physical Review Letters* **119**, 033202.
Takenaka, H. 1973. A unified formalism for polarization optics by using Group Theory" *Nouvelle Revue d'Optique* **4**, 37–41.
Tervo, J., et al. 2003. Degree of coherence for electromagnetic fields. *Optics Express* **11**, 1137–1143.
Tervo, J., et al. 2004. Theory of partially coherent electromagnetic fields in the space–frequency domain. *Journal of the Optical Society of America A* **21**, 2205–2214.
Tervo, J., et al. 2009. Two-point Stokes parameters: interpretation and properties. *Optics Letters* **34**, 3074–3076.
Tervo, J., et al. 2012. Phase correlations and optical coherence. *Optics letters* **37**, 151–153.
Tudor, T. 2003a. Generalized observables in polarization optics. *Journal of Physics A* **36**, 9577–9590.
Tudor, T. 2003b. Dirac-algebraic approach to the theory of device operators in polarization optics. *Journal of the Optical Society of America A* **20**, 728–732.
Tudor, T. 2008. Interaction of light with the polarization devices: a vectorial Pauli algebraic approach. *Journal of Physics A: Mathematical and Theoretical* **41**, 415303.
Tudor, T., and Manea, V. 2011. Ellipsoid of the polarization degree: a vectorial, pure operatorial Pauli algebraic approach. *Journal of the Optical Society of America B* **28**, 596–601.
Tyo, J. S. 2000. Noise equalization in Stokes parameter images obtained by use of variable-retardance polarimeters. *Optics Letters* **25**, 1198–1200.
Tyo, J. S., et al. 2010. Design and optimization of partial Mueller matrix polarimeters. *Applied Optics* **49**, 2326–2333.
Ulaby, F. T. et al. 1992. Statistical properties of the Mueller matrix of distributed targets. *IEE Proceedings* **139**, 136–146
Vahimaa, P. and Tervo, J. 2004. Unified measures for optical fields: degree of polarization and effective degree of coherence. *Journal of Optics A* **6**, S41–S44.
Van de Hulst, H.C. 1957. *Light scattering from small particles*. New York: Wiley.
Van der Mee, C. V. M. 1993. An eigenvalue criterion for matrices transforming Stokes parameters. *Journal of Mathematical Physics* **34**, 5072–5088.
Van der Mee, C. V. M., and Hovenier, J. W. 1992. Structure of matrices transforming Stokes parameters *Journal of Mathematical Physics* **33**, 3574–3584.
Van Eeckhout, A., et al. 2021a. Unraveling physical information of depolarizers. *Optics Express* **29**, 38811–38823.
Van Eeckhout, A., et al. 2021b. Polarimetric imaging microscopy for advanced inspection of vegetal tissues. *Scientific Re*ports **11**, 3913.
Van Enk, S. J., and Nienhuis, G. 1992. Eigenfunction description of laser beams and orbital angular momentum of light. *Optics Communications* **94** 147–158.
Van Enk, S. J., and Nienhuis G. 1994a. Spin and orbital angular momentum of photons. *Europhysics Letters* **25**, 497–501.
Van Enk, S. J., and Nienhuis, G. 1994b. Commutation rules and eigenvalues of spin and orbital angular momentum of radiation fields. *Journal of Modern Optics* **41**, 963–77.
Van Zyl, J. J., 1993. Application of Cloude's target decomposition theorem to polarimetric imaging radar. *Radar polarimetry*. SPIE Proceedings 1748. Bellingham: SPIE, 184–212.
Van Zyl, J. J., et al. 2011. Model-based decomposition of polarimetric SAR covariance matrices constrained for nonnegative eigenvalues. *IEEE Transactions on Geoscience Remote Sensing* **49**, 3452–3459.
Verdet, E. 1869. *Leçons d'Optique Physique*. ed. Levistal, A. vol. 2. Paris: Imprimerie Impériale.
Vizet, J., and Ossikovski, R. 2018. Symmetric decomposition of experimental Mueller matrices in the degenerate case. *Applied Optics* **57**, 1159–1167.
Voipio, T., et al. 2010. Polarization dynamics and polarization time of random three-dimensional electromagnetic fields. *Physical Review A* **82**, 063807.
Von Neumann, J. 1996. *Mathematical foundations of quantum mechanics* (originally *Mathematische Grundlagen der Quantenmechanik*, Berlin 1932). Princeton: Princeton University Press.

Whitney, C. 1971. Pauli-algebraic operators in polarization optics. *Journal of the Optical Society of America* **61**, 1207–1213.

Wiener, N. 1930. Generalized harmonic analysis. *Acta Mathematica.* **55**, 182–195.

Williams, M. W. 1986. Depolarization and cross polarization in ellipsometry of rough surfaces. *Applied Optics* **25**, 3616–3622.

Wolf, E. 1954. Optics in terms of observable quantities. *Nuovo Cimento* **12** 884–888.

Wolf, E. 1959. Coherence properties of partially polarized electromagnetic radiation. *Nuovo Cimento* **13**, 1165–1181.

Wolf, E. 1960. Correlation between photons in partially polarized light beams. *Proceedings of the Physical Society* (London) **76**, 424–426.

Wolf, E. 1981. New spectral representation of random sources and of the partially coherent fields that they generate. *Optics Communications* **38**, 3–6.

Wolf, E. 1982. New theory of partial coherence in the space-frequency domain. Part 1. Spectra and cross-spectra of steady-state sources. *Journal of the Optical Society of America* **72**, 343–351.

Wolf, E. 2003. Unified theory of coherence and polarization of random electromagnetic beams. *Physics Letters A* **312**, 263–267.

Wolf, E. 2007. *Introduction to the theory of coherence and polarization of light.* Cambridge: Cambridge University Press.

Xing, Z-F. 1992. On the deterministic and nondeterministic Mueller matrix. *Journal of Modern Optics* **39**, 461–484.

Yakovlev, D. D., and Yakovlev, D. A. 2019. Scattering patterns of orthogonally polarized light components for statistically rotationally invariant mosaic birefringent layers. *Optics and Spectroscopy*, **126**, 245–256.

Yamaguchi, Y., et al. 2005. Four-component scattering model for polarimetric SAR image decomposition. *IEEE Transactions on Geoscience Remote Sensing* **43**, 1699–1706.

Yoo, S. H., et al. 2017. Experimental study of thickness dependence of polarization and depolarization properties of anisotropic turbid media using Mueller matrix polarimetry and differential decomposition, *Applied Surface Science* **421**, 870–877.

Yurchenko, V. B. 2002 *American Journal of Physics.* **70**, 568.

Zallat, J., et al. 2006. Optimal configurations for imaging polarimeters: impact of image noise and systematic errors. *Journal of Optics A: Pure and Applied Optics* **8**, 807–814.

Index

Note: Page locators in **bold** and *italics* represent tables and figures on the corresponding pages.

For Product Safety Concerns and Information please contact our EU
representative GPSR@taylorandfrancis.com
Taylor & Francis Verlag GmbH, Kaufingerstraße 24, 80331 München, Germany

www.ingramcontent.com/pod-product-compliance
Lightning Source LLC
Chambersburg PA
CBHW080117220326
41598CB00032B/4870
9781032215112